Experimental Methods in the Physical Sciences

Volume 48

Neutron Scattering - Magnetic and Quantum Phenomena

Experimental Methods in the Physical Sciences

Thomas Lucatorto, John T. Yates,
and Kenneth Baldwin
Editors in Chief

Experimental Methods in the Physical Sciences
Volume 48

Neutron Scattering - Magnetic and Quantum Phenomena

Edited by

Felix Fernandez-Alonso
ISIS Facility
Rutherford Appleton Laboratory
Chilton, Didcot, Oxfordshire
United Kingdom

and

Department of Physics and Astronomy
University College London
London
United Kingdom

David L. Price
Synchrotron Radiation and Neutron Research Group, CEMHTI
Centre National de la Recherche Scientifique
Orléans
France

AMSTERDAM • BOSTON • HEIDELBERG • LONDON
NEW YORK • OXFORD • PARIS • SAN DIEGO
SAN FRANCISCO • SINGAPORE • SYDNEY • TOKYO

ELSEVIER   Academic Press is an imprint of Elsevier

Academic Press is an imprint of Elsevier
125 London Wall, London EC2Y 5AS, UK
525 B Street, Suite 1800, San Diego, CA 92101-4495, USA
225 Wyman Street, Waltham, MA 02451, USA
The Boulevard, Langford Lane, Kidlington, Oxford OX5 1GB, UK

Copyright © 2015 Elsevier Inc. All rights reserved.

No part of this publication may be reproduced or transmitted in any form or by any means, electronic or mechanical, including photocopying, recording, or any information storage and retrieval system, without permission in writing from the publisher. Details on how to seek permission, further information about the Publisher's permissions policies and our arrangements with organizations such as the Copyright Clearance Center and the Copyright Licensing Agency, can be found at our website: www.elsevier.com/permissions.

This book and the individual contributions contained in it are protected under copyright by the Publisher (other than as may be noted herein).

Notices
Knowledge and best practice in this field are constantly changing. As new research and experience broaden our understanding, changes in research methods, professional practices, or medical treatment may become necessary.

Practitioners and researchers must always rely on their own experience and knowledge in evaluating and using any information, methods, compounds, or experiments described herein. In using such information or methods they should be mindful of their own safety and the safety of others, including parties for whom they have a professional responsibility.

To the fullest extent of the law, neither the Publisher nor the authors, contributors, or editors, assume any liability for any injury and/or damage to persons or property as a matter of products liability, negligence or otherwise, or from any use or operation of any methods, products, instructions, or ideas contained in the material herein.

ISBN: 978-0-12-802049-4
ISSN: 1079-4042

For information on all Academic Press publications
visit our website at http://store.elsevier.com/

Contents

List of Contributors	xi
Volumes in Series	xiii
Preface	xvii
Eulogy	xxi
Symbols	xxiii

1. Neutron Optics and Spin Labeling Methods
Janos Major, Bela Farago, Ferenc Mezei

1.1	Introduction	1
1.2	Particle Properties and Interactions of Slow Neutrons	2
1.3	Neutron States and Wave Functions	7
	1.3.1 Wave versus Geometrical Optics in Neutron Scattering Experiments	10
	1.3.2 Summary of High Precision Rules for Neutron Beam Propagation	12
1.4	The Principles of Spin Labeling	13
	1.4.1 Practical Spin Labeling	18
	1.4.2 Choices of Neutron Parameters for Spin Labeling	22
1.5	Neutron Spin-Echo Spectroscopy	24
	1.5.1 NSE Spectroscopy for Nuclear Scattering	26
	1.5.2 NSE Spectroscopy in Magnetism	30
1.6	Neutron Spin-Echo for Elastic Scattering at Small Angles	33
	1.6.1 Neutron Beam Polarizers and Analyzers	34
	1.6.2 Transport of Polarized Neutron Beams and Spin-Injection Devices	35
	1.6.3 Precession Region and Magnetic Shielding	36
	1.6.4 Experimental Results	38
	References	40

2. Quantum Phase Transitions
Sara Haravifard, Zahra Yamani, Bruce D. Gaulin

2.1	Introduction	44
	2.1.1 Classical Phase Transitions	45
	2.1.2 Continuous Phase Transitions and Critical Behavior	46
	2.1.3 Quantum Critical Scaling	48

		2.1.4	Quantum Critical Point	49
		2.1.5	Quantum Critical Region	50
	2.2	Experimental Techniques		51
		2.2.1	General Principles of Neutron Scattering	51
		2.2.2	Neutron Scattering Cross Sections	52
		2.2.3	Correlation and Scattering Functions	54
		2.2.4	Magnetic Cross Section	55
		2.2.5	Instruments	56
	2.3	Extreme Environmental Conditions		59
		2.3.1	Cryogenics	59
		2.3.2	High Magnetic Field	64
		2.3.3	High Pressure	68
	2.4	Quantum Phase Transitions in Spin Dimer Systems		71
		2.4.1	Spin Dimer Systems	71
		2.4.2	$TlCuCl_3$	73
		2.4.3	Field-Induced QPT in $TlCuCl_3$	73
		2.4.4	Pressure-Induced QPT in $TlCuCl_3$	79
	2.5	Quantum Phase Transitions in $J_{eff} = 1/2$ Pyrochlore Magnets		87
		2.5.1	XY Pyrochlore Magnets	87
		2.5.2	$Er_2Ti_2O_7$	88
		2.5.3	Spin Excitations in $Er_2Ti_2O_7$	90
		2.5.4	$Yb_2Ti_2O_7$	93
		2.5.5	Spin Excitations in $Yb_2Ti_2O_7$	95
	2.6	Quantum Phase Transitions in Heavy Fermions		99
		2.6.1	Heavy Fermions	99
		2.6.2	Cerium-Based Heavy-Fermion System	101
		2.6.3	$CeCu_{6-x}Au_x$	102
		2.6.4	$CeT(In_{1-x}M_x)_5$	107
		2.6.5	$CeRhIn_5$	109
		2.6.6	$CeCoIn_5$	111
		2.6.7	$CeIrIn_5$	112
	2.7	Quantum Phase Transitions in Itinerant Magnets		112
		2.7.1	Weak Itinerant Ferromagnets	112
		2.7.2	MnSi	115
		2.7.3	URu_2Si_2	120
	2.8	Quantum Phase Transitions in Transverse Field Ising Systems		126
		2.8.1	Transverse Field Quantum Ising Model	126
		2.8.2	$CoNb_2O_6$	130
	2.9	Closing Remarks		138
	Acknowledgments			138
	References			139

3. High-Temperature Superconductors

Yu Song, Pengcheng Dai

	3.1	Introduction	145
	3.2	Hole-Doped Cuprate Superconductors	147
		3.2.1 Magnetic Order and Spin Waves in the Parent Compounds	147
		3.2.2 Evolution of Magnetic Excitations upon Doping	151

3.2.3 Neutron Scattering versus RIXS 156
3.2.4 The "Resonance" Mode in the Superconducting State 158
3.3 Iron-Based Superconductors 160
3.3.1 Introduction, Compounds, and Phase Diagrams 160
3.3.2 Antiferromagnetically Ordered Metallic Parent Compounds 162
3.3.3 Evolution of Magnetic Order and Magnetic Excitations with Doping 168
3.3.4 Persistence of High-Energy Magnetic Excitations 176
3.3.5 The Resonance Mode in Iron-Based Superconductors 179
3.3.6 Magnetism in $A_xFe_{2-y}Se_2$ (A = Alkali Metal or Tl) Compounds 182
3.3.7 Polarization Dependence of Low-Energy Magnetic Excitations 186
3.4 Summary and Outlook 190
Acknowledgments 193
References 193

4. Magnetic Structures

V. Ovidiu Garlea, Bryan C. Chakoumakos

4.1 Introduction 204
4.2 The Beginnings of Magnetic Structure Determination Using Neutron Diffraction 206
4.3 Fundamentals of Magnetic Scattering 211
4.3.1 Scattering Amplitude from Magnetic Order 211
4.3.2 The Magnetic Propagation Vector Formalism 215
4.3.3 Symmetry Considerations in Magnetic Structures 222
4.4 Practical Aspects in Determination of Magnetic Structures 231
4.4.1 Steps in the Determination of Magnetic Structures 231
4.4.2 Limitations of Neutron Scattering for Complex Magnetic Structures 234
4.4.3 Standard Description for Magnetic Structures 235
4.5 Hierarchy of Magnetic Structures (Important Examples) 235
4.5.1 Three-Dimensional Networks 237
4.5.2 Layered Structures 250
4.5.3 Quasi-One-Dimensional Lattices 266
4.6 Disordered Magnetic Structures 274
4.6.1 Diffuse Scattering from Disordered Alloys 275
4.6.2 Diffuse Scattering from Systems with Reduced Dimensionality 276
4.6.3 Diffuse Magnetic Scattering in Frustrated Magnets 278
4.7 Concluding Remarks 281
Acknowledgments 281
References 282

5. Multiferroics
William D. Ratcliff II, Jeffrey W. Lynn

5.1	Introduction	291
5.2	Symmetry Considerations for Ferroelectrics	293
5.3	Type-I Proper Multiferroics	294
	5.3.1 $HoMnO_3$	294
	5.3.2 $BiFeO_3$	297
	5.3.3 $(Sr-Ba)MnO_3$	300
5.4	Type-II Improper Multiferroics	301
	5.4.1 Spin-Spiral Systems: $TbMnO_3$, $MnWO_4$, and $RbFe(MoO_4)_2$	302
	5.4.2 Exchange-Striction Systems: Ca_3CoMnO_6 and YMn_2O_5	307
5.5	Domains	315
	5.5.1 $BiFeO_3$	315
	5.5.2 $HoMnO_3$	315
5.6	Thin Films and Multilayers	316
	5.6.1 $BiFeO_3$	316
	5.6.2 $TbMnO_3$	319
5.7	Spin Dynamics	322
	5.7.1 $HoMnO_3$	322
	5.7.2 $BiFeO_3$	323
	5.7.3 $(Sr-Ba)MnO_3$	323
	5.7.4 Electromagnons	325
	5.7.5 $MnWO_4$	328
	5.7.6 YMn_2O_5	329
	5.7.7 $RbFe(MoO_4)_2$	330
5.8	Future Directions	332
	Acknowledgments	333
	References	333

6. Neutron Scattering in Nanomagnetism
Boris P. Toperverg, Hartmut Zabel

6.1	Introduction	340
	6.1.1 Topics in Nanomagnetism	340
	6.1.2 Magnetic Neutron Scattering	341
6.2	Theoretical Background	344
	6.2.1 Basic Interactions and Scattering Amplitudes	344
	6.2.2 Scattering Cross Section of Polarized Neutrons	347
	6.2.3 Grazing Incidence Kinematics	352
	6.2.4 Specular Polarized Neutron Reflectivity for 1D Potential	359
	6.2.5 Off-Specular Neutron Scattering	363

6.3	Instrumental Considerations	366
	6.3.1 Design of Neutron Reflectometers	366
	6.3.2 Neutron Optics	370
	6.3.3 Detection and Acquisition of Data	373
	6.3.4 Modeling and Fitting of Data	373
6.4	Case Studies: Static Experiments	374
	6.4.1 Thin Films	374
	6.4.2 Multilayers	396
	6.4.3 Stripes, Islands, and Nanoparticles	403
6.5	Time Dependent Polarized Neutron Scattering	416
6.6	Summary, Conclusion, and Outlook	421
	Acknowledgments	422
	References	423

7. Nuclear Magnetism and Neutrons
Michael Steiner, Konrad Siemensmeyer

7.1	Introduction	436
7.2	Experimental Background	439
	7.2.1 Nuclear Moments—The Neutron Cross Sections	439
	7.2.2 ULT Experimental Methods and Neutron Techniques	441
	7.2.3 Nuclear Polarization Measurement from Neutron Scattering and Transmission	445
	7.2.4 Neutron Diffraction Cryostat for ULT Applications	451
	7.2.5 Sample Requirements	453
	7.2.6 Spontaneous Nuclear Magnetic Order	453
7.3	Experimental Results from Neutron Diffraction on Cu and Ag	461
	7.3.1 Nuclear Magnetic Ordering of Cu and Ag in Zero and Finite Magnetic Field	461
	7.3.2 Structures—Cu and Ag	470
	7.3.3 The Phase Diagrams in a Magnetic Field	472
7.4	Neutron Diffraction Investigations on Solid ^{3}He	474
7.5	Applications of Polarized Nuclei in Neutron Diffraction	482
	7.5.1 Neutron Polarization from Polarized ^{3}He Gas Targets	482
	7.5.2 Contrast Variation by Polarized Nuclei in Neutron Scattering	484
7.6	Summary	486
	References	486

Index	489

List of Contributors

Bryan C. Chakoumakos, Quantum Condensed Matter Division, Oak Ridge National Laboratory, Oak Ridge, Tennessee, USA

Pengcheng Dai, Department of Physics and Astronomy, Rice University, Houston, Texas, USA

Bela Farago, Institut Laue-Langevin, Grenoble, France

V. Ovidiu Garlea, Quantum Condensed Matter Division, Oak Ridge National Laboratory, Oak Ridge, Tennessee, USA

Bruce D. Gaulin, Department of Physics and Astronomy, McMaster University, Hamilton, Ontario, Canada; Canadian Institute for Advanced Research, Toronto, Ontario, Canada; Brockhouse Institute for Materials Research, McMaster University, Hamilton, Ontario, Canada

Sara Haravifard, Department of Physics, Duke University, Durham, North Carolina, USA; The James Franck Institute and Department of Physics, The University of Chicago, Chicago, Illinois, USA; Advanced Photon Source, Argonne National Laboratory, Argonne, Illinois, USA

Jeffrey W. Lynn, NIST Center for Neutron Research, National Institute of Standards and Technology, Gaithersburg, Maryland, USA

Janos Major, Max-Planck-Institut für Metallforschung, Stuttgart, Germany

Ferenc Mezei, European Spallation Source ERIC, Lund, Sweden; HAS Wigner Research Center, Budapest, Hungary

William D. Ratcliff II, NIST Center for Neutron Research, National Institute of Standards and Technology, Gaithersburg, Maryland, USA

Konrad Siemensmeyer, Helmholtz Zentrum Berlin, Berlin, Germany

Yu Song, Department of Physics and Astronomy, Rice University, Houston, Texas, USA

Michael Steiner, Helmholtz Zentrum Berlin, Berlin, Germany

Boris P. Toperverg, Institute for Experimental Condensed Matter Physics, Ruhr-University Bochum, Bochum, Germany; Petersburg Nuclear Physics Institute, St Petersburg, Russia

Zahra Yamani, Canadian Neutron Beam Centre, Chalk River Laboratories, Chalk River, Ontario, Canada

Hartmut Zabel, Institute for Experimental Condensed Matter Physics, Ruhr-University Bochum, Bochum, Germany; Johannes Gutenberg-Universität Mainz, Mainz, Germany

Volumes in Series

Experimental Methods in the Physical Sciences (Formerly Methods of Experimental Physics)

Volume 1. Classical Methods
Edited by Immanuel Estermann

Volume 2. Electronic Methods, Second Edition (in two parts)
Edited by E. Bleuler and R. O. Haxby

Volume 3. Molecular Physics, Second Edition (in two parts)
Edited by Dudley Williams

Volume 4. Atomic and Electron Physics - Part A: Atomic Sources and Detectors; Part B: Free Atoms
Edited by Vernon W. Hughes and Howard L. Schultz

Volume 5. Nuclear Physics (in two parts)
Edited by Luke C. L. Yuan and Chien-Shiung Wu

Volume 6. Solid State Physics - Part A: Preparation, Structure, Mechanical and Thermal Properties; Part B: Electrical, Magnetic and Optical Properties
Edited by K. Lark-Horovitz and Vivian A. Johnson

Volume 7. Atomic and Electron Physics - Atomic Interactions (in two parts)
Edited by Benjamin Bederson and Wade L. Fite

Volume 8. Problems and Solutions for Students
Edited by L. Marton and W. F. Hornyak

Volume 9. Plasma Physics (in two parts)
Edited by Hans R. Griem and Ralph H. Lovberg

Volume 10. Physical Principles of Far-Infrared Radiation
Edited by L. C. Robinson

Volume 11. Solid State Physics
Edited by R. V. Coleman

Volume 12. Astrophysics - Part A: Optical and Infrared Astronomy
Edited by N. Carleton
Part B: Radio Telescopes; Part C: Radio Observations
Edited by M. L. Meeks

Volume 13. Spectroscopy (in two parts)
Edited by Dudley Williams

Volume 14. Vacuum Physics and Technology
Edited by G. L. Weissler and R. W. Carlson

Volume 15. Quantum Electronics (in two parts)
Edited by C. L. Tang

Volume 16. Polymers - Part A: Molecular Structure and Dynamics; Part B: Crystal Structure and Morphology; Part C: Physical Properties
Edited by R. A. Fava

Volume 17. Accelerators in Atomic Physics
Edited by P. Richard

Volume 18. Fluid Dynamics (in two parts)
Edited by R. J. Emrich

Volume 19. Ultrasonics
Edited by Peter D. Edmonds

Volume 20. Biophysics
Edited by Gerald Ehrenstein and Harold Lecar

Volume 21. Solid State Physics: Nuclear Methods
Edited by J. N. Mundy, S. J. Rothman, M. J. Fluss, and L. C. Smedskjaer

Volume 22. Solid State Physics: Surfaces
Edited by Robert L. Park and Max G. Lagally

Volume 23. Neutron Scattering (in three parts)
Edited by K. Skold and D. L. Price

Volume 24. Geophysics - Part A: Laboratory Measurements; Part B: Field Measurements
Edited by C. G. Sammis and T. L. Henyey

Volume 25. Geometrical and Instrumental Optics
Edited by Daniel Malacara

Volume 26. Physical Optics and Light Measurements
Edited by Daniel Malacara

Volume 27. Scanning Tunneling Microscopy
Edited by Joseph Stroscio and William Kaiser

Volume 28. Statistical Methods for Physical Science
Edited by John L. Stanford and Stephen B. Vardaman

Volume 29. Atomic, Molecular, and Optical Physics - Part A: Charged Particles; Part B: Atoms and Molecules; Part C: Electromagnetic Radiation
Edited by F. B. Dunning and Randall G. Hulet

Volume 30. Laser Ablation and Desorption
Edited by John C. Miller and Richard F. Haglund, Jr.

Volume 31. Vacuum Ultraviolet Spectroscopy I
Edited by J. A. R. Samson and D. L. Ederer

Volume 32. Vacuum Ultraviolet Spectroscopy II
Edited by J. A. R. Samson and D. L. Ederer

Volume 33. Cumulative Author Index and Tables of Contents, Volumes 1-32

Volume 34. Cumulative Subject Index

Volume 35. Methods in the Physics of Porous Media
Edited by Po-zen Wong

Volume 36. Magnetic Imaging and its Applications to Materials
Edited by Marc De Graef and Yimei Zhu

Volume 37. Characterization of Amorphous and Crystalline Rough Surface: Principles and Applications
Edited by Yi Ping Zhao, Gwo-Ching Wang, and Toh-Ming Lu

Volume 38. Advances in Surface Science
Edited by Hari Singh Nalwa

Volume 39. Modern Acoustical Techniques for the Measurement of Mechanical Properties
Edited by Moises Levy, Henry E. Bass, and Richard Stern

Volume 40. Cavity-Enhanced Spectroscopies
Edited by Roger D. van Zee and J. Patrick Looney

Volume 41. Optical Radiometry
Edited by A. C. Parr, R. U. Datla, and J. L. Gardner

Volume 42. Radiometric Temperature Measurements. I. Fundamentals
Edited by Z. M. Zhang, B. K. Tsai, and G. Machin

Volume 43. Radiometric Temperature Measurements. II. Applications
Edited by Z. M. Zhang, B. K. Tsai, and G. Machin

Volume 44. Neutron Scattering — Fundamentals
Edited by Felix Fernandez-Alonso and David L. Price

Volume 45. Single-Photon Generation and Detection
Edited by Alan Migdall, Sergey Polyakov, Jingyun Fan, and Joshua Bienfang

Volume 46. Spectrophotometry: Accurate Measurement of Optical Properties of Materials
Edited by Thomas A. Germer, Joanne C. Zwinkels, and Benjamin K. Tsai

Volume 47. Optical Radiometry for Ocean Climate Measurements
Edited by Giuseppe Zibordi, Craig J. Donlon, and Albert C. Parr

Volume 48. Neutron Scattering - Magnetic and Quantum Phenomena
Edited by Felix Fernandez-Alonso and David L. Price

Preface

Just over 80 years ago, a brief letter from James Chadwick to *Nature* [1,2] presented conclusive experimental evidence unveiling the existence of a neutral particle (nearly) isobaric with the proton. The discovery of the henceforth-to-be-known-as "neutron" had profound consequences for both scientific research and the destiny of humankind, as it led to the unleashing of the might of nuclear power in less than a decade [3].

The first use of these "neutral protons" to probe the microscopic underpinnings of the material world around us also dates back to those early years, with pioneering neutron-diffraction experiments at Oak Ridge National Laboratory (USA) in the mid-1940s, and the subsequent development of neutron spectroscopy at Chalk River (Canada) in the 1950s. Since then, neutron-scattering techniques have matured into a robust and increasingly versatile toolkit for physicists, chemists, biologists, materials scientists, engineers, or technologists. At the turn of the last century, the 1994 Nobel Prize in Physics awarded to C.G. Shull and B.N. Brockhouse recognized their ground-breaking efforts toward the development and consolidation of neutron science as a discipline in its own right [4]. This milestone also served to define neutron scattering as the technique *par excellence* to investigate *where atoms are* (structure) and *what atoms do* (dynamics), a popular motto across generations of neutron-scattering practitioners.

Sustained and continued developments in experimental methods over the past few decades have greatly increased the sensitivity and range of applications of neutron scattering. While early measurements probed distances on the order of interatomic spacings (fractions of a nanometer) and characteristic times associated with lattice vibrations (picoseconds), contemporary neutron-scattering experiments can cover length scales from less than 0.01 to 1000s of nanometers, and time scales from the attosecond to the microsecond. These advances have been made possible via a significant expansion of the range of neutron energies available to the experimenter, from micro-electron-volts (particularly at cold sources in research reactors) to hundreds of electron-volts (at pulsed spallation sources), as well as by unabated progress in the implementation of a variety of novel and ingenious ideas such as position- and polarization-sensitive detection or back-scattering and spin-labeling methods. As a result, neutron science has grown beyond traditional research areas, from the conventional determination of crystal structures and lattice dynamics of half-a-century ago (not to forget their magnetic analogs), to high-resolution structural studies of disordered thin films, liquid interfaces, biological structures, macromolecular and supramolecular architectures and devices, or the unraveling of the dynamics and energy-level structure of complex molecular solids, nanostructured materials and surfaces, or magnetic clusters and novel superconductors. Along with these scientific and technical developments, the community of neutron scientists has also expanded and diversified beyond recognition. Whereas the early stages of neutron scattering had its roots in condensed-matter physics and crystallography, present-day users of central neutron-scattering facilities include chemists, biologists, ceramicists, and metallurgists, to name a few, as well as physicists with an increasingly diverse range of transdisciplinary interests, from the foundations of quantum mechanics to soft matter, food science, biology, geology, or archeometry.

This book series seeks to cover in some detail the production and use of neutrons across the aforementioned disciplines, with a particular emphasis on technical and scientific developments over the past two decades. As such, it necessarily builds upon an earlier and very successful three-volume set edited by K. Sköld and D.L. Price, published in the 1980s by *Academic Press* as part of *Methods of Experimental Physics* (currently *Experimental Methods in the Physical Sciences*). Furthermore, with the third-generation spallation sources recently constructed in the US and Japan, or in the advanced construction or planning stage in China and Europe, there has been an increasing interest in time-of-flight and broadband neutron-scattering techniques. Correspondingly, the improved performance of cold moderators at both reactors and spallation sources has extended long-wavelength capabilities to such an extent that a sharp distinction between fission- and accelerator-driven neutron sources may no longer be of relevance to the future of the discipline.

On a more practical front, the chapters that follow are meant to enable you to identify aspects of your work in which neutron-scattering techniques might contribute, conceive the important experiments to be done, assess what is required to carry them out, write a successful proposal to a user facility, and perform these experiments under the guidance and support of the appropriate facility-based scientist. The presentation is aimed at professionals at all levels, from early career researchers to mature scientists who may be insufficiently aware or up-to-date with the breadth of opportunities provided by neutron techniques in their area of specialty. In this spirit, it does not aim to present a systematic and detailed development of the underlying theory, which may be found in superbly written texts such as those of Lovesey [5] or Squires [6]. Likewise, it is not a detailed hands-on manual of experimental methods, which in our opinion is best obtained directly from experienced practitioners or, alternatively, by attending practical training courses at the neutron facilities. As an intermediate (and highly advisable) step, we also note the existence of neutron-focused thematic schools, particularly those at Grenoble [7] and Oxford [8], both of which have been running on a regular basis since the 1990s. With these primary objectives in mind, each chapter focuses on well-defined areas of neutron science and has been written by a leading practitioner or practitioners of the application of neutron methods in that particular field.

In the previous volume, Neutron Scattering — Fundamentals [9], we gave a self-contained survey of the theoretical concepts and formalism of the technique and established the notation used throughout the series. Subsequent chapters reviewed neutron production and instrumentation, respectively, areas which have profited enormously from recent developments in accelerator physics, materials research and engineering, or computing, to name a few. The remaining chapters treated several basic applications of neutron scattering including the structure of complex materials, large-scale structures, and dynamics of atoms and molecules. The appendix went back to some requisite fundamentals linked to neutron–matter interactions, along with a detailed compilation of neutron scattering lengths and cross sections across the periodic table.

The present volume is dedicated to the applications of neutron scattering techniques to magnetic and quantum phenomena. The first chapter deals with neutron optics and spin-labeling methods and also gives a broad introduction to the interaction of neutrons with electronic spins in condensed matter. The following chapters discuss recent developments in the use of neutron scattering to investigate quantum phase transitions, high-temperature superconductors, magnetic structures, multiferroics, nanomagnetism, and nuclear magnetism. A third volume will cover applications in biology, chemistry, and materials science.

In closing this preface, we wish to thank all authors for taking time out of their busy schedules to be part of this venture, Drs T. Lucatorto, A.C. Parr, and K. Baldwin for inviting us to undertake this work, and the staff of *Academic Press* for their encouragement, diligence, and forbearance along the way.

Felix Fernandez-Alonso
David L. Price

REFERENCES

[1] J. Chadwick, Nature 129 (1932) 312.
[2] URL: www.nobelprize.org/nobel_prizes/physics/laureates/1935/ (last accessed on 12.08.15).
[3] A. MacKay, The Making of the Atomic Age, Oxford University Press, Oxford, 1984.
[4] URL: www.nobelprize.org/nobel_prizes/physics/laureates/1994/ (last accessed on 12.08.15).
[5] S.W. Lovesey, Theory of Neutron Scattering from Condensed Matter, vols. I and II, Oxford University Press, Oxford, 1986.
[6] G.L. Squires, Introduction to the Theory of Thermal Neutron Scattering, third ed., Cambridge University Press, Cambridge, 2012.
[7] URL: hercules-school.eu/ (last accessed on 12.08.15).
[8] URL: www.oxfordneutronschool.org/ (last accessed on 12.08.15).
[9] F. Fernandez-Alonso, D.L. Price (Eds.), Neutron Scattering — Fundamentals, Experimental Methods in the Physical Sciences, vol. 44, Academic Press, Amsterdam, 2013.

Eulogy

Janos Major

During the preparation of this book, we were saddened by the news of the premature death of Janos Major, a prominent member of the neutron scattering community and coauthor of the first chapter of this volume.

Janos spent the majority of his career in the Max Planck Institute in Stuttgart, working with Directors Alfred Seeger and then Helmut Dosch. Janos organized essentially all the neutron works in the Dosch department with a focus on the original development of the SERGIS concept together with Gian Felcher and Roger Pynn. This instrument approach exploits neutron spin echo for encoding the momentum transfer of neutrons.

Janos was a wonderful person and an exquisite scientist. Those who worked with him enjoyed his spirited experimental and technical skills, his dedication to science, and his steadfast loyalty to his colleagues and students.

Helmut Dosch
Felix Fernandez-Alonso
David L. Price

Symbols

Note: Other symbols may be used that are unique to the chapter where they occur.

a	Scattering length in center-of-mass frame
b	Bound scattering length
A	Atomic mass (or nucleon) number
$\overline{b}$	Coherent scattering length
b^+	Scattering length for $I + \frac{1}{2}$ state
b^-	Scattering length for $I - \frac{1}{2}$ state
b_i	Incoherent scattering length
b_N	Spin-dependent scattering length
c	Speed of light in vacuum $= 299792458$ m s^{-1} [1]
$\mathbf{D}$	Dynamical matrix
$\mathbf{D}_\perp(\mathbf{Q})$	Magnetic interaction operator
d	Mass density
$\mathbf{d}$	Equilibrium position of atom in unit cell
$d\sigma/d\Omega$	Differential scattering cross section
$d^2\sigma/d\Omega\, dE_f$	Double differential scattering cross section
E	Neutron energy transfer ($E_i - E_f$)
E_n	Neutron energy
E_i, E_f	Initial, final (scattered) neutron energy
e	Elementary charge $= 1.602176565 \times 10^{-19}$ C
$\mathbf{e}^k(\mathbf{q})$	Polarization vector of normal mode k [$\mathbf{e}_d^k(\mathbf{q})$ for a non-Bravais crystal]
$F(\tau)$	Unit cell structure factor
f	Force constant
$f(\mathbf{Q})$	Form factor
$G(\mathbf{r},t)$	Space–time (van Hove) correlation function [$G_d(\mathbf{r},t) + G_s(\mathbf{r},t)$]
$G_d(\mathbf{r},t)$	"Distinct" space–time correlation function
$G_s(\mathbf{r},t)$	"Self" space–time correlation function
$g(\mathbf{r})$	Pair distribution function
g_e	Electron g factor $= -2.00231930436153$
g_n	Neutron g factor $= -3.82608545$
h	Planck constant $= 6.62606957 \times 10^{-34}$ J s
$\hbar$	Planck constant over 2π ($h/2\pi$) $= 1.054571726 \times 10^{-34}$ J s
$\mathbf{I}$	Angular momentum operator for nucleus
$I(\mathbf{Q},t)$	Intermediate scattering function
$I_s(\mathbf{Q},t)$	Self intermediate scattering function
k_B	Boltzmann constant $= 1.3806488 \times 10^{-23}$ J K^{-1}
$\mathbf{k}_i, \mathbf{k}_f$	Initial, final (scattered) neutron wave vector
$\mathbf{k}_n$	Neutron wave vector ($\mathbf{k}_n = 2\pi/\lambda_n$)

Symbols

L_i, L_f	Initial (primary), final (secondary) flight path
l	Unit cell position vector (lowercase "L")
M	Atom mass
$\mathbf{M}$	Magnetization operator
m_e	Electron mass = $9.10938291 \times 10^{-31}$ kg
m_n	Neutron mass = $1.674927351 \times 10^{-27}$ kg
m_p	Proton mass = $1.672621777 \times 10^{-27}$ kg
N	Number of atoms in sample
N_A	Avogadro constant = $6.02214129 \times 10^{23}$ mol^{-1}
N_c	Number of unit cells in crystal
$\mathbf{R}_j$	Position vector for particle j
$\mathbf{Q}$	Scattering vector ($\mathbf{k}_i - \mathbf{k}_f$)
$\mathbf{q}$	Reduced wave vector ($\mathbf{Q} - \tau$)
r_0	Classical electron radius ($e^2/4\pi\varepsilon_0 m_e c^2$) = $2.8179403267 \times 10^{-15}$ m
$\mathbf{S}$	Atomic spin operator
$S(\mathbf{Q})$	Static structure factor $I(\mathbf{Q},0)$
$S(\mathbf{Q},E)$	Dynamic structure factor or scattering function
$S_c(\mathbf{Q},E)$	Coherent dynamic structure factor or scattering function
$S_i(\mathbf{Q},E)$	Incoherent dynamic structure factor or scattering function
$\mathbf{s}$	Electron spin operator
T	Temperature
u	Unified atomic mass unit = $1.660538921 \times 10^{-27}$ kg
$\mathbf{u}^l$	Vibrational amplitude ($\mathbf{u}_d^l$ for non-Bravais crystals)
V	Volume of sample
v_i, v_f	Initial, final (scattered) neutron velocity
v_n	Neutron velocity
v_0	Unit cell volume
$2W$	Exponential argument in the Debye–Waller factor
Z	Atomic (or proton) number
$Z(E)$	Phonon density of states
Γ	Spectral linewidth
γ	Neutron magnetic moment to nuclear magneton ratio (μ_n/μ_N) = -1.91304272
γ_n	Neutron gyromagnetic ratio = 1.83247179×10^8 s^{-1} T^{-1}
Θ	Debye temperature
θ	Bragg angle
λ_i, λ_f	Initial, final (scattered) neutron wavelength
μ	Atomic magnetic moment
μ_N	Nuclear magneton ($e\hbar/2m_p$) = $5.05078353 \times 10^{-27}$ J T^{-1}
μ_B	Bohr magneton ($e\hbar/2m_e$) = $927.400968 \times 10^{-26}$ J T^{-1}
μ_n	Neutron magnetic moment = $-0.96623647 \times 10^{-26}$ J T^{-1}
ν	Frequency ($\omega/2\pi$)
$\rho(\mathbf{r},t)$	Particle density operator
ρ_0	Average number density
Σ	Macroscopic cross section associated with a given cross section σ (see below)
σ_a	Absorption cross section
σ_t	Bound total cross section ($\sigma_s + \sigma_a$)
σ_c	Bound coherent scattering cross section
σ_i	Bound incoherent scattering cross section
σ_s	Bound scattering cross section ($\sigma_c + \sigma_i$)
$\frac{1}{2}\boldsymbol{\sigma}$	Neutron spin (Pauli) operator

τ	Reciprocal lattice vector $\{2\pi[(h/a),(k/b),(l/c)]\}$
Φ	Neutron flux (typically defined as neutrons crossing per unit area per unit time)
ϕ	Scattering angle ($=2\theta$)
χ	Susceptibility
$\chi(\mathbf{Q},E)$	Generalized susceptibility
Ω	Solid angle
ω	Radian frequency associated with neutron energy transfer $E = \hbar\omega$
$\omega_k(\mathbf{q})$	Frequency of normal mode k

REFERENCE

[1] All numerical values have been taken from the CODATA Recommended Values of the Fundamental Physical Constants, as detailed in P.J. Mohr, B.N. Taylor, D.B. Newell, Rev. Mod. Phys. 84 (2010) 1527−1605. Updated (2010) values can be found at physics.nist.gov/cuu/Constants/ (Last accessed on 12.08.15).

Chapter 1

Neutron Optics and Spin Labeling Methods

Janos Major,[1] Bela Farago[2] and Ferenc Mezei[3,4,*]
[1]*Max-Planck-Institut für Metallforschung, Stuttgart, Germany;* [2]*Institut Laue-Langevin, Grenoble, France;* [3]*European Spallation Source ERIC, Lund, Sweden;* [4]*HAS Wigner Research Center, Budapest, Hungary*
*Corresponding author: E-mail: ferenc.mezei@esss.se

Chapter Outline

1.1	Introduction	1	1.5	Neutron Spin-Echo Spectroscopy	24
1.2	Particle Properties and Interactions of Slow Neutrons	2		1.5.1 NSE Spectroscopy for Nuclear Scattering	26
1.3	Neutron States and Wave Functions	7		1.5.2 NSE Spectroscopy in Magnetism	30
	1.3.1 Wave versus Geometrical Optics in Neutron Scattering Experiments	10	1.6	Neutron Spin-Echo for Elastic Scattering at Small Angles	33
	1.3.2 Summary of High Precision Rules for Neutron Beam Propagation	12		1.6.1 Neutron Beam Polarizers and Analyzers	34
1.4	The Principles of Spin Labeling	13		1.6.2 Transport of Polarized Neutron Beams and Spin-Injection Devices	35
	1.4.1 Practical Spin Labeling	18		1.6.3 Precession Region and Magnetic Shielding	36
	1.4.2 Choices of Neutron Parameters for Spin Labeling	22		1.6.4 Experimental Results	38
				1.6.4.1 Spin-Echo Small-Angle Scattering	38
				1.6.4.2 Spin-Echo Reflectometry	38
			References		40

1.1 INTRODUCTION

Neutrons are magnetic particles, they possess a magnetic moment, coupled to its spin $s = 1/2$. The value of this magnetic moment is $\mu = 1.913$ Bohr magneton and its direction is opposite to that of the spin. This property makes neutron radiation particularly well suited for the study of magnetism in condensed matter.

Indeed, the specific interaction of the magnetic moment of the neutrons with the microscopic magnetic fields created by magnetic atoms offers unique opportunities to probe magnetism on the microscale by neutron scattering, often with a sensitivity not equaled by any other microscopic probe. In addition, manipulating and observing the direction of the neutron magnetic moment in spin-polarized neutron beams is a very powerful tool to single out in neutron scattering experiments what is related to the magnetic behavior of the sample in the scattering signal, which is a mixture of contributions of different origins.

Going a step further, there also is another side to the story. The magnetic moment of each neutron can also be used to keep track of other relevant parameters of a neutron propagating in a beam, notably the value and direction of its velocity. In doing this, the neutron magnetic moment is used as a measuring device attached individually to each neutron, which can deliver information on the neutrons individually. Such methods are called "spin labeling" and they can be advantageously used to observe fine changes in the neutron parameters in a scattering process. For this reason, they can offer valuable opportunities for exploring matter by any neutron scattering process independently of whether it is related to the magnetic properties of the sample or not at all.

Thus when we talk about neutron scattering and magnetism, on the one hand, we need to consider exploration of magnetism in a broad variety of materials, and on the other hand, the opportunities the magnetism of the neutron offers to study nonmagnetic and/or magnetic phenomena in another broad variety of materials. While most of the rest of this volume is devoted to the first of these two large subject cases, this chapter focuses on the overview of the second one.

1.2 PARTICLE PROPERTIES AND INTERACTIONS OF SLOW NEUTRONS

For the study of condensed matter by particle radiation, the key parameters are energy and momentum of the particles, in other words their frequency and their wave number. Namely, as commonly pointed out, radiation can most efficiently probe the sample in the space—time domain that is comparable to the frequency and wavelength of the radiation. Thus the subnanometer wavelength of X-rays used by Laue and Bragg allowed them to directly observe crystalline structures on comparable length scale for the first time. Neutron radiation of similar wavelengths in addition has frequencies close to those of atomic scale vibrations in crystalline matter, which allowed Brockhouse to directly observe these atomic scale motions in unprecedented details by neutron scattering. These frequencies are quite low compared to the energy scale of nuclear physical phenomena, so in all what follows; we will only consider slow neutrons in the far nonrelativistic limit.

The very simple and plausible statement about frequency and wavelength of particle radiation already needs some conceptual refinement when quantum mechanics is a visible part of the story—as it is here by the de Broglie relation between velocity and wavelength $v = 2\pi\hbar/m\lambda$ and Planck's relation between energy and frequency $E = \hbar\omega$ (where $\hbar$ is Planck's constant, m the particle mass, λ and $f = 2\pi\omega$ are the radiation wavelength and frequency, respectively). Quantum mechanics is of course the key background of everything we experience around us, but in many cases this does not manifest itself explicitly and we can work with classical mechanics. This is very fortunate; we would never master many aspects of our technology if we would, for example, need to use the Schrödinger equation of a steam engine for understanding and predicting its behavior.

The first important reminder of this kind is that in quantum mechanics there is no absolute frequency or wavelength for a radiation; it depends on the reference frame we compare with. So, just by the existence of gravity, the frequency of a neutron wave will be different by the amount corresponding to the gravitational potential hg (where h is the height and g the value of the local gravitational acceleration), which clearly depends on where h is measured from. This relativity in quantum mechanics has nothing to do with Einstein's relativity. For the very fundamental reason of quantum mechanical relativity of the particle frequency and wavelength (or the more frequently used wave number—often inaccurately called momentum, also in what follows—$k = 2\pi/\lambda$), in comparing energies and wave-numbers of radiation to phenomena on classically defined length and time scales, such as the lattice spacing in a perfect crystal, we must only consider frequency or wave number differences (commonly called "transfers") between two states of the particle radiation. To be more explicit, we know exactly the value of the momentum transfer in a Bragg scattering process, but additional, eventually arbitrary conditions need to be defined for translating these values into absolute numbers for the particle parameters in a reference frame. It is practically more significant to remember this in connection with radiation frequencies, since the zero point of the potential energy cannot be uniquely defined. For all practical purposes, one can consider the sum of all potential energies—including that of a magnetic moment in a magnetic field—to be set equal to zero in the "empty space at infinity," at least in principle on paper.

In order to define the properties of the neutron for the purpose of slow neutron radiation experiments, we need to define its Hamiltonian in the low-energy terrestrial environment of our experiments. It is a curiosity of the history of physics that although the neutron was discovered in 1932, all theoretical efforts remained inconclusive for the magnetic term in the Hamiltonian related to its magnetic moment—until experiments in 1951 decided between the two main theoretically proposed candidates, with Nobel laureates involved in the unsettled theoretical debate for two decades. This actually indicates the fundamental fact that the discovery of the neutron and its magnetic moment fundamentally defied our understanding of electromagnetism. The simple reason for this is that thinking about neutron scattering experiments raised for the first time the issue of overlapping electromagnetic objects.

Although Maxwell's theory does not rule out the overlap in space between two electromagnetic objects, there were no practical cases to study such situations and the exclusion of overlap was assumed in all considerations. For example, considering the magnetic fields inside magnetic matter it was assumed that a probe needs to be considered in a hole inside the magnetic media. It was established that for ellipsoidal holes demagnetization factors can be defined and perpendicularly to a flat disc-shaped hole the magnetic field inside the hole $\mathbf{H_h}$ will be the corresponding component of magnetic induction $\mathbf{B}$ of the magnetic medium, while along the axis of a pin-shaped hole it will be equal to the corresponding component of the $\mathbf{H}$ field of the medium—assuming that the holes do not disturb the magnetization in the medium. Felix Bloch in 1937 tried to address the issue by assuming that the neutron could create different shapes of holes for itself inside magnetic matter [1]. The quantum mechanical reality, on the other hand, is that the neutron wave function simply overlaps with the sample and its magnetization, without causing significant deformation due to the weakness of its interaction, its absence of electric charge, and the small neutron density in beams. Such a situation just happened to emerge for the first time in physics history with the magnetism of the neutron combined with its capability of traversing all kinds of matter, including magnets.

The result of these 20 years of search concluded by experiments published nearly simultaneously in early 1951 by Hughes and Burgy [2] and independently by Shull et al. [3] was the following:

1. The Hamiltonian for the interaction of neutrons with magnetic fields and substances is $H = -\mathbf{\mu} \cdot \mathbf{B}$, where $\mathbf{\mu}$ is the magnetic moment vector of the neutron and $\mathbf{B}$ the magnetic induction vector ($\mathbf{B}$ field) inside or outside magnetic media. This in particular definitively implies that the neutrons do not see or directly interact with the magnetic field $\mathbf{H}$ and with the magnetic moments or magnetic moment density $\mathbf{M}$ inside magnetic materials. The relation $\mathbf{B} = \mathbf{H} + 4\pi\mathbf{M}$ also indicates that the magnetic field $\mathbf{H}$ are identical with $\mathbf{B}$ outside magnetized media that is characterized by the magnetic moment density $\mathbf{M} = 0$.
2. The nature of the neutron magnetic moment is that of an Amperian current loop and incompatible with a dipole represented by a pair of charges.

It is a surprise with respect to electromagnetic theory that Amperian current loops and magnetic dipoles with fictitious magnetic charges at a small distance from each other are not experimentally identical. But they indeed behave very differently if the neutron trajectory crosses volumes with nonzero current density $\mathbf{j}$. By Maxwell equations, $\mathbf{B}$ is determined by the current density distribution $\mathbf{j}$ via $curl\ \mathbf{B} = 4\pi\mathbf{j}/c$ and $div\ \mathbf{B} = 0$ (in Gaussian units where c is the velocity of light). The magnetic moment density corresponding

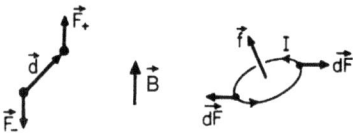

FIGURE 1 The difference between forces acting on a magnetic moment in the dipole (on the left) and Amperian current loop model on the right.

to a given distribution of current densities is determined by the equation $curl\,\mathbf{M} = \mathbf{j}/c$ together with the boundary condition that $\mathbf{M} = 0$ everywhere outside the magnetic media.

The difference between predictions for neutron trajectories by the two microscopic models of magnetic moments is illustrated in Figure 1 [4]. The forces on the neutron traversing a magnetic field for the dipole and current loop models respectively are:

$$\mathbf{F_d} = \mathbf{F}^+ + \mathbf{F}^- = (\mu \cdot grad)\mathbf{B} = grad(\mu \cdot \mathbf{B}) - \mu \times curl\,\mathbf{B} \quad (1)$$

$$\mathbf{F_c} = \oint d\mathbf{F} = \frac{I}{c}\oint d\mathbf{l} \times \mathbf{B} = \frac{I}{c}\int grad(\mathbf{B} \cdot d\mathbf{f}) - \frac{I}{c}\int (div\,\mathbf{B})d\mathbf{f} = grad(\mu \cdot \mathbf{B}) \quad (2)$$

where the definitions of the different forces **F** are shown in Figure 1, together with the line and surface integration over dl and df. The magnetic moments in the two cases are ed and $I\mathbf{f}/c$, respectively, with e standing for the magnetic dipole charge and I for the current in the Amperian loop. Equations (1) and (2) give measurably different results if $curl\,\mathbf{B} \propto \mathbf{j}$ is not zero. The more, Eqn (2) defines a force corresponding to the potential energy $-\mu \cdot \mathbf{B}$. In contrast, the force in Eqn (1) cannot be derived from a potential energy: different trajectories between the same points lead to different changes of the particle energy (i.e., candidate for a *perpetuum mobile*). Fortunately, the experiments by Hughes and Burgy [2] have shown that the neutron acceleration on the surface of a magnetized ferromagnetic layer corresponds to Eqn (2) and contradicts Eqn (1). By going to the end of the arguments based on this observation, one can inversely show that the magnetic moments in the magnetic material interacting with the field created by the neutron magnetic moment can also only be of the nature of Amperian current loops and not that of dipoles.

In sum, the key result of this old controversy is that for the interaction of the neutron magnetic moment with fields created by macroscopic currents and/or microscopic ones on the atomic scale inside magnetic matter can be described by the Zeeman energy type Hamiltonian of

$$H_m = -\mu \cdot \mathbf{B}(\mathbf{r},t) \quad (3)$$

One important mathematical consequence of this equation is that when we calculate the neutron scattering cross section related to this part of the interaction, its Fourier transform in the space variable **r** as a function of the conjugated momentum transfer vector variable **Q** becomes $-\mu \cdot \mathbf{M}_\perp(\mathbf{Q})$, where the subscript $\perp$ indicates the component perpendicular to **Q**. If in Eqn (3) we would have the **H** field instead of **B**, in the cross section formulae we would in contrast end up with $-\mu \cdot \mathbf{M}_\parallel(\mathbf{Q})$, i.e., with the component of **M(Q)** parallel to **Q**. Shull et al. [3] used this difference with respect to the direction of the momentum transfer vector to conclude in favor of the Hamiltonian in Eqn (3).

With this established, the full Hamiltonian for slow neutron propagation through space and matter becomes:

$$H(\mathbf{r},t) = \frac{1}{2}mv^2 - \mu \cdot \mathbf{B}(\mathbf{r},t) + mgh + \frac{2\pi\hbar^2}{m}\sum_i b_i(1 + c_i \mathbf{I}_i \cdot \sigma)\delta(\mathbf{r}-\mathbf{r}_i(t)) \quad (4)$$

Here the summation goes over the Fermi pseudopotentials of all the nuclei i of different elements and isotopes that constitute the materials the neutrons traverse and/or scatter on. The interaction parameters ("scattering lengths") b_i and c_i are tabulated for the most common isotopes, together with the values of the nuclear spins $\mathbf{I}_i$. Finally, σ is the Pauli spin operator for the neutron and $\mathbf{r}_i(t)$ the position of the nucleus i at time t. For neutrons with large wavelength compared to the atomic distances (or, more precisely, for neutron momentum transfers much smaller than the inverse of atomic distances), the summation over the nuclei in Eqn (4) will result in an effective volume average nuclear potential $V_n(\mathbf{r}, t)$, that is determined by the local average of the scattering length density over all nuclei within a volume element. If on average the nuclei are polarized, i.e., their spin has a preferred orientation, the average of the terms including c_i will lead to a nonzero spin-dependent component of the potential that acts on the neutron spin σ as an effective magnetic field. This is called the nuclear pseudomagnetism and has been experimentally well established by the work of Abragam and collaborators [5].

The values of the four terms in Hamiltonian (Eqn (4)) for representative experimental conditions in neutron scattering are in the range of 5 meV for the kinetic energy (at $v = 978$ m s^{-1}), 60 neV for the Zeeman energy (at $B = 0.995$ T), 100 neV for the gravitational potential energy (at $h = 0.98$ m), and for most materials less than 250 neV for the volume average nuclear potential, 250 neV being the actual value for natural Ni metal. These potentials can be converted into the critical neutron velocities below which the neutrons cannot pass the potential energy barrier and totally reflect. This leads for example to critical velocities of 3.40, 4.42, and 6.92 m s^{-1} for the three potential energy terms in Eqn (4) for, respectively, $B = 1$ T, $h = 1$ m, and the isotopic composition and density of natural Ni metal.

It is important to note that the small volume average potential $V_n(\mathbf{r}, t)$ for the nuclear interaction term in Hamiltonian (Eqn (4)) does not imply that it is always a small perturbation. In fact this term is the slow neutron limit of the nuclear strong interaction which can become overwhelming. One overarching reason for this is that the parameters b_i and c_i generally are not real numbers and can have a significant imaginary part, formally turning e.g., an extended plane wave into a decaying wave. This corresponds to absorption of the neutrons by nuclear reactions with the nuclei i. A high density of absorbing nuclei in the sum turns part of the space into obstacles to neutron propagation. The other reason is that by their point-like shape each δ function potential term in Eqn (4) adds a spherical wave centered on its position to the solution of the Schrödinger equation. Constructive interference between the large number of spherical waves in the summation over all nuclei i can represent a much larger potential at certain wave number transfers than the volume average. The probability of particle scattering processes is determined by the absolute square of the Fourier transform $V_n(\mathbf{Q})$ of the potential $V_n(\mathbf{r})$. This will thus scale with the square of the number of particles within a coherence volume element in the sample within which constructive interference of scattering from individual nuclei i takes place (e.g., a small monocrystalline grain in a polycrystalline sample). In contrast, the volume average potential considered above scales proportionally to the number of particles involved. This consideration also applies for the Zeeman term in the Hamiltonian (Eqn (4)): the function $\mathbf{B}(\mathbf{r}, t)$ can show periodic variation inside coherence volumes for constructive interference in a magnetic sample and lead to very strong perturbations in the solution of the Schrödinger equation even if the volume average of the magnetic potential remains in the weak range discussed above. It is worth noting at this stage that in most practical samples the coherence/correlated volumes of the microscopic structure is rather small (well below 1 mm), while in perfect crystals or in certain directions in low dimensional structures it can reach several centimeters or more.

Solving the Schrödinger equation exactly for the Hamiltonian (Eqn (4)) for any practical situation of a neutron experiment is a hopeless exercise, so what we are concerned with is to find good approximations together with their precise conditions of validity. As we will discuss in the next section, these established approximations are very accurate in most of neutron scattering work and in addition much simpler than, for example, for light and X-ray radiations.

1.3 NEUTRON STATES AND WAVE FUNCTIONS

The energies involved in the various terms of Eqn (4) under common neutron scattering experimental conditions as reviewed above fully justify of taking all the potential energy terms as perturbations and work in terms of eigenstates of free particles in vacuum, in absence of magnetic fields and gravity. The complete set of eigenstates of the unperturbed Hamiltonian will then be given by wave functions $|\varphi\rangle = |\mathbf{k}, \chi\rangle$, where the wave number vectors $\mathbf{k}$ of plane waves can take any value in a continuum (and is related to the neutron velocity $\mathbf{v}$ by the relation $\hbar\mathbf{k} = m\mathbf{v}$), and the spin variable χ can take the two eigenstates

$|\uparrow\rangle$ and $|\downarrow\rangle$ with respect to any defined coordinate system. All general neutron states can be reproduced as a superposition of this continuum of eigenstates:

$$\Phi = \int \left(a_\uparrow(\mathbf{k}) e^{i(\mathbf{kr}-\omega t)} |\uparrow\rangle + a_\downarrow(\mathbf{k}) e^{i(\mathbf{kr}-\omega t)} |\downarrow\rangle \right) d\mathbf{k} = \int |\chi(\mathbf{k})\rangle a(\mathbf{k}) e^{i(\mathbf{kr}-\omega t)} d\mathbf{k} \neq$$

$$\neq |\chi\rangle \int a(\mathbf{k}) e^{i(\mathbf{kr}-\omega t)} d\mathbf{k}$$

(5)

where the various amplitude distribution functions a are complex scalars. It is of fundamental importance that the single particle state Φ cannot be assumed to be factorized into spatial and spin variables, as illustrated on the rightmost side following the crucial $\neq$ sign in Eqn (5). Instead, in general, it has a spin variable part that itself is a function of wave number $\mathbf{k}$, and the unfortunately widespread imposition of factorization in the literature is at variance with the basic principles of quantum mechanics. In reality, the exact quantum mechanical state of a single neutron can also be unpolarized (i.e., zero expectation values for all three spin components) as a result of superposition (which nowadays frequently is called—without any change of meaning—"entanglement") over various wave numbers. Such a superposition, usually arbitrarily factorized into a spatial and spin component, is often called a "wave packet" and received a lot of attention in the literature. We will show in what follows an efficient optical scheme for the approximate solution of the Schrödinger equation for Hamiltonian (Eqn (4)) for experimental slow neutron beam work which is in a sense the opposite of the wave packet concept.

Wave packets are often considered to start with a minimum size compatible with the uncertainty principle and due to the dispersive nature of the $\omega(\mathbf{k})$ function they spread in width with propagating in time. However, this spreading just corresponds to the spreading out in space of a group of classical, point-like particles due to their differences in classical velocity. For a numerical example, a minimal size wave packet of 0.1 mm rms width and average velocity of 1000 m s^{-1} will spread by the uncertainty principle over 100-m flight path by only about 30%. In this wave packet, the spread of neutron velocity is less than 1 ppm, so the packet can be considered as monochromatic by any practical standards and measures in neutron beam work (outside perfect crystal interferometry, for which we have excluded the classical picture by definition). So for the purists, the point-like classical particles with at the same time perfectly defined position and velocity can be represented by such very monochromatic wave packets, in order to stay in comfortable terms with the Heisenberg uncertainty principle. In reality, the significance of this principle for nonrelativistic particle beam experiments, is a simple mathematical consequence of Fourier transformation, which enters due to the representation of spatial wave functions as a superposition of the plane wave eigenfunctions of the free particle Hamiltonian. (For this reason the "violation" of the Heisenberg inequality for the so-called "compressed" particle states is also of no significance: it is just elementary mathematics with the Fourier transform of other than Gaussian velocity distribution functions.)

Another aspect of simultaneously measuring parameters that correspond to noncommutative operators in quantum mechanics is the determination of the components of the particle spin **S**. Here, any given quantum mechanical spin wave function $|\chi\rangle$ has perfectly well defined expectation values for the spin components operator $\langle\chi|S_\alpha|\chi\rangle$, $\alpha = x, y, z$ and these components form a vector with absolute value $|\mathbf{S}|$. In particle beam experiments, the result corresponds to probabilities of detection inside well-defined parameter pixels and the particle detection events are uncorrelated and reproducible repetitions. Therefore the statistical precision of the results (probabilities) increases as $\sqrt{N}$ with the number N of events detected, and there is no principal limit on the precision of determining any probabilities, i.e., expectation values. Therefore, while a single particle detection event cannot correspond to perfectly defined value of the three components of a spin operator, by collecting sufficient repetitions, the expectation values of these components can be simultaneously determined with unlimited precision.

Thus we can conclude that instead of speculating about quantum mechanical wave packets, in all experimental situations where particle trajectories are subject to slowly varying environment for their propagation, every **k** component in the particle wave function in the general representations of Eqn (5) (which also includes wave packets) can be perfectly considered as a point-like classical particle with for all practical (i.e., observable) purposes infinitely well-defined classical trajectory $\mathbf{r}(t)$, hence similarly well-defined velocity and wave number (related as $\mathbf{v}(t)=\mathbf{k}(t)/\hbar m$) and spin $\mathbf{S}(t)$ at any instant along the trajectory. Furthermore, the solution of the quantum mechanical equation of motion reduces to representing the particle by a distribution of classical point-like particles following the classical trajectories. The full beam thus becomes a global, combined distribution of such point-like particle distributions representing putative quantum mechanical wave packets in the beam. It needs to be stressed that, while this classical approach is fully well defined and unambiguous, nature provides no hint or recipe for constructing wave packets or their distributions for real life slow particle beam experiments, for example, whether a measured distribution of particle velocities corresponds to a single coherent wave packet following Eqn (5), or it is an incoherent ensemble of some number of more monochromatic wave packets with different average velocities.

Looking at the part of Eqn (5) just before the $\neq$ sign, one can observe that for each value of **k** the neutron spin wave function $|\chi(\mathbf{k})\rangle$ is a simple superposition of $|\uparrow\rangle$ and $|\downarrow\rangle$ spin eigenstates. In the above sense, this means that the expectation values of the three spin components $S_\alpha = \langle\chi(\mathbf{k})|\sigma_\alpha|\chi(\mathbf{k})\rangle$ for $\alpha = x, y, z$ form a classical spin direction vector $\mathbf{S}(\mathbf{k})$ of unit length (full polarization). Due to the Zeeman term in the Hamiltonian, this spin direction vector evolves according to the classical Larmor precessions, following the equation:

$$\frac{dS}{dt} = \gamma_L [S \times B] \quad (6)$$

where for neutrons the Larmor constant $\gamma_L = 29.164$ kHz mT^{-1}. Furthermore, the wave packet will advance in space with the group velocity i.e., in view of the de Broglie equation with the classical particle velocity: $d\omega/dk = v$.

1.3.1 Wave versus Geometrical Optics in Neutron Scattering Experiments

The above description of beam propagation in terms of classical particle trajectories has been developed into high art over centuries of development of optical devices of highest performance, together with full understanding of the limits of conditions where wave optical effects can be neglected. We conclude our general considerations on the propagation of neutron radiation in this light. As well known from classical optics, wave optical effects are due to interference between possible optical paths, which will only be observed (i.e., do not destructively average to zero) if the optical path differences between the various possible geometrical trajectories a particle can take do not exceed considerably its wavelength. Following Huygens' principle, if the interference between the various possible trajectories cancels to zero, we will be left with the geometrical optical trajectory. In a typical scattering experimental setup we work with beams of diameter d collimated over a distance $l \gg d$ between diaphragms. The path differences between beam trajectories within such an envelope will amount to the order of $\Delta = \sqrt{l^2 + d^2} - l \approx d^2/2l$. Deviations from geometrical optics represented by the classical point-like particle trajectories are expected if these path differences are comparable or less than the wavelength of the radiation. In typical neutron scattering experiments $d \sim 1$ cm and $l \sim 10$ m define the orders of magnitude, so we end up with $\Delta \approx 500$ nm, i.e., far above the radiation wavelength of interest 0.05–2 nm. In comparison, for synchrotron X-rays $d \sim 20$ µm and l is similar as above, and we arrive at $\Delta \approx 0.02$ nm, which is rather small compared to the wavelength λ in the range of 0.1 nm = 1 Å. Situations similar to X-rays are also common in experiments with visible light, where λ is much larger, ~ 500 nm.

Thus we can conclude that in typical neutron scattering work, geometrical optics provides a very good approximation for describing the propagation of the radiation across the instruments in sizable cross section beams. This reduces the high precision description of the neutron spin behavior along the particle propagation to a particularly simple and reliable process: follow by the classical Larmor equation of motion the point-like magnetic moment, which experiences at any moment of time a well-defined local $B = B(r(t), t)$ field along a fully well-defined point-like classical particle trajectory $r(t)$.

In contrast in the case of electromagnetic radiations, in particular synchrotron X-rays, wave optics is commonly required for describing not only the scattering samples but inseparably including the scattering instruments themselves. Of course, wave optics needs to be used for describing the neutron scattering processes in condensed matter samples, where the significant optical path differences are due to the Å scale microscopic arrangements of the positions of scattering atoms. The facility of limiting the wave optical approach to the microscopic scattering process while the radiation propagation through the scattering instruments can be treated independently and by the much simpler geometrical optical means is a very important simplification in data interpretation, which helps to make neutron scattering the most quantitatively understood probe in particle scattering study of matter. Geometrical optics is also valid for neutron propagation within the samples before and after the scattering events (in particular in the case of multiple scattering). These events are actually limited in space to the particular coherence volume in the sample, where the scattering takes place. This coherence volume can be as small as a single atom for incoherent scattering, correspond to a perfect monocrystalline grain in a powder or mosaic single crystal, or extend to much of the whole sample volume for a perfect single crystal or in a homogenous layer structure.

Similar considerations also apply to modulating particle radiation in time. As long as the characteristic times of modulation are much longer than the periods corresponding to the frequency range of the radiation, wave propagation effects remain insignificant.

Interference between the various **k** components in the wave packet is the basic principle of the wave packet picture; it is what makes the packet a packet in space. Nevertheless, such interference in real beams has never been unambiguously observed in experiments with Fermion particles [6], since the putative coherence (phase relation) between the amplitudes at two different wave numbers $a(\mathbf{k})$ and $a(\mathbf{k}')$ within the amplitude distribution becomes random over different wave packages corresponding to different individual particles. In an ensemble of particles this is the expression of the random phase statistics: $\langle a^*(\mathbf{k})a(\mathbf{k}')\rangle = 0$ if $\mathbf{k} \neq \mathbf{k}'$. In Boson radiation, this phase coherence can be induced by stimulated emission (lasers) or modulation by external intervention by a macroscopic device, but it is also absent in spontaneous emission. This leaves us with the common, time honored wisdom that any particle can only interfere with itself (except for the cases of stimulated Boson emission and external modulation mentioned above), i.e., on average in an ensemble of particles only diagonal interference terms $a^*(\mathbf{k})a(\mathbf{k})$ can be observed, all the rest averages to zero.

This observation also has a maybe surprising logical consequence: the inherent coherence length of radiation such as neutron beams is infinity. Indeed, if the observable interference only can occur for infinite plane waves with a well-defined wave number **k** interfering with themselves, infinite coherence length is a basic part of the picture. There is no experimental evidence for the opposite, although many publications talk of coherence length in an erroneous context. Any observed interference pattern is an ensemble average over a large number of particles with a wave number distribution in the classical sense. So the interference pattern is a classical average over all the particles observed in the experiment, and it only proves that the inherent coherence of each particle is larger than what appears in the results due to the distribution of the parameters describing the states of each particle in the ensemble. All experimental studies accomplished by now confirmed that such a lower observed limit for inherent particle coherence length was simply the inverse of the width of the proper cut across the classical wave number distribution $f(\mathbf{k})$ of the more or less monochromatic ensemble of detected particles (fully independently of how the monochromatic selection of the particles came about, e.g., by using a monochromator at the generation of the beam or an analyzer before detection, or both, or something else). Particle beams invariably show as much coherence as the fully classical mechanical beam monochromaticity allows us to observe at all, with no sign of any underlying limitation of quantum mechanical or particle property character.

1.3.2 Summary of High Precision Rules for Neutron Beam Propagation

To summarize all the observations and considerations above, the complete set of basic rules governing beam propagation in neutron scattering work is recapitulated below, including all possible interactions of slow neutrons, in particular all possible magnetic effects. Beam propagation between scattering and absorption processes in samples and beam shaping devices and other materials present (diaphragms, collimator, choppers, beam windows, guides, air, etc.) can roughly be called beam optics, which in the case of neutrons are classically deterministic. Probabilistic scattering and absorption by nuclear reactions are the other part of the story. These rules are approximations with more than satisfactorily precision and specific to neutron beam work by having assumed typical geometrical configurations of neutron scattering experiments. Expressions like "infinitely well defined" are used in the sense of comparison with the precision that can at all be achieved in such experiments. These rules do not necessarily hold for scattering work with other radiations, such as light or synchrotron X-ray.

1. Between probabilistic scattering and absorption events inside matter, the neutrons propagate as point-like classical particles with infinitely well-defined trajectories $\mathbf{r}(t)$, each carrying a classical magnetic moment with perfectly well-defined direction at any instance of time. (This is no contradiction to the uncertainty principle: (1) classical distribution of particle parameters also provides measurement uncertainties masking smaller effects and (2) for a large number N of detected particles the principal uncertainty limit is divided by $\sqrt{N}$.)

Neutron Optics and Spin Labeling Methods Chapter | 1   13

2. The magnetic moment direction vector follows the classical Larmor precession motion governed by the Zeeman energy $-\mu \cdot \mathbf{B}(\mathbf{r}(t), t)$, where the magnetic induction field **B** shows time dependence as seen by the point-like neutron along its infinitely well-defined classical trajectory across the magnetic fields both inside and outside materials. This energy represents a conservative potential if the **B** field is in itself time independent $\mathbf{B} = \mathbf{B}(\mathbf{r})$, and the sum of the kinetic and all potential energies remains a constant over the classical trajectory of the point-like neutron.
3. The neutrons also follow classical mechanical trajectories (as determined including the volume average nuclear potential V_n) between quantum probabilistic scattering processes inside matter. The probabilistic beam attenuation due to scattering and absorption by nuclear reactions needs to be factored into the effective description of the classical trajectories $\mathbf{r}(t)$.
4. Neutrons do not interact with magnetic moments in any other way than via the Zeeman interaction with the **B** field, that can equally be well created by macroscopic currents and microscopic ones related to magnetism in matters via the relation $curl\, \mathbf{M} = \mathbf{j}/c$.
5. The neutron scattering processes are of wave mechanical nature, in contrast to the classical point-like particle propagation (1–3) between probabilistic scattering events. They need to be determined using adequate quantum mechanical approaches for both the Zeeman term in the Hamiltonian if the **B** field shows short range variations and the term with the sum of the nuclear interaction potentials of individual nuclei.
6. Scattering events that can be handled by approximate theories such as first Born approximation involve correlated/coherence volumes of the scattering matter, which is most often point-like small ($\ll 1$ mm) on the scale of neutron scattering sample volumes. Exceptions are large perfect crystals (including neutron interferometers made from such crystals), for which the Schrödinger equation must be essentially solved exactly without much approximation.
7. In contrast to the finite, structurally correlated volumes that can coherently contribute to the neutron scattering processes inside materials, the neutron radiation by itself has no inherent limit of coherence lengths. Observations of apparently limited coherence are due to classical averaging of the results over the classical (velocity) distribution of the effectively detected particles. In neutron beams interference only involving a single well-defined wave number **k** component of the initial particle state could be observed by now, classically averaged over the classical distribution of particle states **k**.

1.4 THE PRINCIPLES OF SPIN LABELING

In neutron scattering experiments, the initial velocity vector $\mathbf{v}_i$ of the incoming neutron is compared to final velocity $\mathbf{v}_f$ of the scattered neutron, in order to determine the scattering probability as a function of the neutron energy transfer and momentum transfer (or scattering vector):

$$E = \hbar\omega = \frac{1}{2}m\mathbf{v}_i^2 - \frac{1}{2}m\mathbf{v}_f^2 \tag{7}$$

$$\mathbf{Q} = \frac{1}{\hbar}m\mathbf{v}_i - \frac{1}{\hbar}m\mathbf{v}_f \tag{8}$$

Conventionally this is achieved by creating a monochromatic and collimated incoming beam with a well-defined distribution of the initial velocities $\mathbf{v}_i$, which practically means eliminating from the incoming beam all neutrons that are not close enough to the desired central value $\bar{\mathbf{v}}_i$. The value of the final velocity is then analyzed by an independent process of measuring $\mathbf{v}_f$. Achieving very high resolutions by this conventional approach might already become impractical due to the prohibitively high beam intensity loss in selecting the highly monochromatic incoming beam and analyzing the scattered beam parameters with the same kind of elimination method. In spin labeling, the spin of each individual neutron in the incoming neutron beam will be used to measure the velocity and label the neutron by the result of this measurement for the purpose of comparing it to the final velocity of the same neutron after scattering. (If the neutron did not scatter, it will still carry this initial state information while continuing its trajectory without change of velocity.)

It is worth mentioning from the outset that the practical reason for resorting to spin labeling is to refine resolution by the high sensitivity of the method and/or the gain in intensity due the decoupling of resolution and beam monochromaticity (e.g., in practice 0.001% quasi-elastic energy resolution can be achieved with 10% monochromatic beam). Nevertheless, for defining all the relevant scattering parameters (e.g., the momentum transfer Q in a high-resolution quasi-elastic experiment), the incoming beam usually needs to be monochromatized to a moderate degree and collimated, and the same can apply to the scattered beam too. The incoming beam preparation and scattered beam analysis/detection naturally involve spin polarization and polarization analysis, respectively.

The simplest example is to consider a neutron on a well-defined straight trajectory across a uniform magnetic field $\mathbf{B}$, assuming that the free fall in the earth's gravitational field only negligibly changes the velocity (which is the most frequent case). If we align the neutron spin perpendicular to the magnetic field at a position x_0 along the trajectory, the Larmor precession of the neutron spin will act as a clock attached to the neutron, and the time elapsed by the flight along the trajectory will be measured by the precession angle $\varphi = 2\pi\gamma_L Bt$, where $B = |\mathbf{B}|$. (Note that this expression is numerically correct if the angle is measured in radians and the frequency in Hertz units in the definition of γ_L, as given above.) After advancing a certain distance l along the trajectory, the neutron flight time from position x_0 becomes a label of the velocity of the neutron by the relation $v = l/t$. From a strict point of view, this labeling cannot distinguish between precession angles φ and $\varphi + 2\pi n$, where n is an integer number. We will see below a simple experimental trick how to remove this ambiguity.

The point x_0 and the direction with respect of which the precession angle φ is measured is defined by a neutron spin flipper device, which can turn the neutron spin direction away from the "natural" direction of being parallel to the magnet field, as originally produced by the neutron spin polarizer device, e.g., a magnetized neutron mirror. To envisage this, we need to be familiar with two particular extreme cases of the solution of Eqn (6), beyond the canonical Larmor precession on a conic surface around the direction of a constant field **B** with the angle between **B** and **S** remaining constant: namely the cases of slowly and rapidly varying direction of the magnetic field $\mathbf{B}(t) = \mathbf{B}(\mathbf{r}(t))$ as seen by the neutron advancing along the trajectory $\mathbf{r}(t)$.

1. Adiabatic limit: If the change of the direction of field $\mathbf{B}(t)$ is slow compared to the angular velocity of the Larmor precession $\omega_L(t) = 2\pi\gamma_L\mathbf{B}(t)$, the angle between $\mathbf{B}(t)$ and $\mathbf{S}(t)$ remains a constant.
2. Majorana limit: If the change of the direction of field $\mathbf{B}(t)$ is fast compared to the angular velocity of the Larmor precession $\omega_L(t)$, the direction of **S** remains unchanged during the time of the rapid jump of the direction of $\mathbf{B}(t)$.

The simple rectangular coil device of usually one layer of Al winding (quite transparent to neutrons) makes use of these limits for rotating the neutron spin in a desired direction during the passage of the neutron across the coil [7]. In the case shown in Figure 2, the direct current (DC) through the coil creates a field of the same strength as the external constant field, i.e., inside the coil the resulting magnetic field makes an angle of 45° with the external field direction. When the neutrons enter through the windings inside the coil, 45° jump of the field direction to the resulting field direction inside the coils happens gradually through the thickness of the winding, but fast enough for being in the Majorana limit, i.e., **S** remains in its original direction parallel to the external field. Inside the coil the neutron spin will thus precess around the

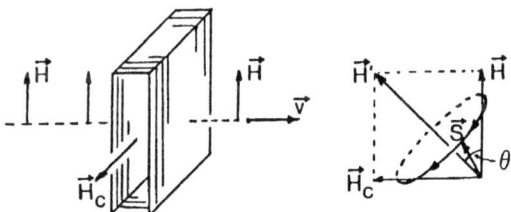

FIGURE 2 Neutron spin flipper coil operated by DC current. The figure shows the details for the case of 90° spin flip, the same device can be used with different parameter settings for 180° or other spin direction rotations.

resulting constant field at 45° and it will exit the coil without substantially changing direction inside the winding. If the parameters are set up such that the neutron spin has the time inside the coil for a half precession, the direction of S on exit from the coil will be parallel to the coil axis and perpendicular to the external field. Thus the exit surface of the coil defines the position x_0, where the Larmor precession labeling starts and the axis of the coil is the direction from which the angle φ is measured. The precession field has to satisfy the adiabatic condition (1) above: if its direction changes along the trajectory, it has to change adiabatically slowly. In practice, the thickness of such a flipper coil can be 0.5—1 cm and the field inside the coil a few Gauss.

In order to determine the change of velocity of Larmor-labeled neutrons, we need to reverse the angle of total precession at a well-defined point and let them precess over a magnetic field with the same strength over an identical length (more precisely, with same product of field and length). The most common way to reverse the angle of precession is to apply a 180° spin flip by the same way shown in Figure 2, just changing the field and orientation of the coil with respect of the external field so that the resulting field inside the coil is perpendicular to the external one and inside the coil the neutron makes a 180° precession around the coil axis. Such a rotation will leave the spin component parallel to the field inside the coil unchanged, and reverse the two others. This corresponds to a mirror reflection of a precessing neutron spin direction with respect of the plane of the coil, i.e., transforming a precession angle φ measured with respect to the direction of coil axis to $-\varphi$. Denoting the precession angle in the second half of the trajectory after the 180° flip by φ', the total precession angle after the two precession sections will be $\varphi_T = \varphi - \varphi'$ and measures the difference of the two flight times, i.e., the change of velocity for each neutron—if and only if the two precession sections in the trajectory before and after the 180° spin reversal are equal with high precision. In practice, often this precision needs to be as good as 1—10 ppm, and exact calibration is crucial.

The sensitivity of the Larmor precession determination of velocity is very high: e.g., for typical cold neutrons with 1000 m s^{-1} velocity (for 3.956 Å wavelength) traversing 2 m of 50 mT field, $\varphi \sim 1,100,000°$, while in the determination of the angular direction of neutron beam polarization 1°—2° precision can readily be achieved. Concerning precision, the physical calibration of the precession system would be a formidable task with this absolute accuracy. But in neutron scattering, we are in fact interested in velocity changes in a process, i.e., we have to only determine the directly observable $\varphi_T = \varphi' - \varphi$ angle, which is the direct measure of this change. Of course, we need a system with impressive stability to keep the directly not observable angles φ and φ' stable with <1 ppm precision relative to each other for weeks (e.g., stray magnetic field gradient must stay stable within 0.001 Gauss m^{-1}).

The more fundamental aspect concerning precision is, however, that the individual labeling of neutrons allows us to observe minute changes in the velocity of a neutron beam which in other techniques are fully masked by the scatter in the initial velocities of the neutrons in the beam. There is no chance to determine the average velocity of a 10% monochromatic beam with 1 ppm precision, but by individual labeling it is common to measure 1 ppm change of the velocity of individual neutrons by directly observing the expectation value φ_T, i.e., of the velocity change itself over all neutrons in the beam. The spin precessions in a beam with neutrons of considerably different velocities start with a common phase at x_0 after the initial 90° flip, but rapidly get out of phase with distance from the starting point, since the neutrons with different velocities spend different times for covering the same distance. On the other hand—without change of neutron velocity—φ and φ' for all neutrons become equal again at one point of the trajectory; namely where the second Larmor precession section becomes equivalent to the first one. At this point, all neutrons possess 0 total precession angle, i.e., all spins point in the same direction, and we observe an echo in the precessing beam polarization, as illustrated in Figure 3. Any velocity change of neutrons results into a value of $\varphi_T \neq 0$, leading to a reduced contribution to the echo signal (proportionally to $\cos \varphi_T$ instead of 1). The evolution of this echo signal as a function of the sensitivity of the velocity labeling as scanning parameter is a directly observed quantity in neutron spin-echo (NSE) spectroscopy [7,8], as discussed in quantitative detail in Section 1.5 below.

In order to study the inelastic scattering by this way, we need to consider a neutron trajectory that corresponds to the changed direction of the neutron at the sample, which should be at the position of the 180° spin flip. One logical reason to apply spin labeling is to allow us to gain intensity by using a much less monochromatic incoming beam than the energy resolution we want to achieve, thus the analysis of the scattered neutron spectrum always goes by the observation of an echo signal. A given energy transfer E of the neutron in the scattering event (cf. Eqn (7)) will more generally provide an echo signal at a point in the trajectory if and only if the individual total precession angles of

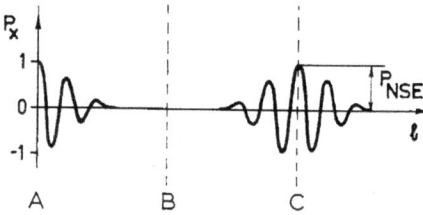

FIGURE 3 The principle of spin echo in a polychromatic neutron beam. The incoming beam polarization is flipped by 90° at point A into the x direction perpendicular to the magnetic field extending from A to C, and starts Larmor precession around the field. At point B, the angle of the Larmor precessions is reversed by a 180° flipper and at point C, the initial precessing polarization is restored by the echo effect with 0 total resulting precession angle (equal amount of forward and backward precessions) for all neutrons independently of velocity, all neutron spins point again in the same x direction, as they started at point A.

neutrons with different incoming velocities become to be in phase, i.e., independent of the incoming neutron velocity:

$$\frac{\partial \varphi_T(v_i, E)}{\partial v_i} = \frac{\partial}{\partial v_i} \gamma_L \left(\frac{(\int Bdl)_2}{\sqrt{v_i^2 - 2E/m}} - \frac{(\int Bdl)_1}{v_i} \right) = 0, \qquad (9)$$

where the subscripts 1 and 2 refer to the field integrals along the first and second precession sections in the trajectory. Performing the differentiation, the solution of Eqn (9) reads

$$\frac{(\int Bdl)_2}{\sqrt{v_i^2 - 2E/m}^3} = \frac{(\int Bdl)_1}{v_i^3} \qquad (10)$$

This NSE condition for the energy variable can be practically well fulfilled by properly choosing the ratio of the magnetic field integrals 1 and 2 for the given average velocity of the incoming beam. The example of determination of the lifetime of the roton excitation in superfluid ^{4}He (whose energy varies little over a substantial Q range) illustrates the process [9]. Here $E = 0.74$ meV, the incoming neutron velocity was chosen to provide initial neutron energy $\frac{1}{2}mv_i^2 = 4.2$ meV on average with 8% precision. The accuracy of the NSE measurement achieved for the line width of the roton excitation was <1 μeV, i.e., less than 0.02% of the incoming energy and 0.13% of the energy change—by far out of reach of other neutron spectroscopy approaches for the study of the elementary excitations.

1.4.1 Practical Spin Labeling

In the previous section, we have considered as example the general case of Larmor labeling for the inelastic/quasi-eleastic scattering along well-defined neutron trajectories. In a practical experiment of this type we need to deal with a large number of neighboring trajectories both in terms of beam cross section (which can be as much as 3×3 cm at the sample and 20×20 cm at a position sensitive neutron detector) and angular divergence of the beam. The neutron path lengths will be different over the ensemble of these trajectories, which will lead to a blurring of the Larmor precession angles φ and φ'. Furthermore, the magnetic field will have some inhomogeneities across the beam cross section, which will lead to the same effect. The largest differences in path lengths would come from neutrons scattered at different positions in the sample, which, e.g., results in some 3-cm path difference between the inside and outside edge of a 3-cm wide sample at scattering angles around 60°. This amounts to about 1% of a thinkable flight path from sample to detector. The high-precision potential of Larmor labeling for velocity measurement can only be taken advantage of by eliminating the effect of such path differences in and around the sample. The simplest and most common way is to keep the magnetic field on the sample small, just enough to ensure adiabatic conditions, which practically means less than about 1 Gauss. In such a field, 3-cm path difference for 1000 m s^{-1} fast neutrons amounts to less than ±15° difference in Larmor precession angle, i.e., only a small impact on the amplitude on the oscillatory echo signal with 360° period (cf. Figure 3). To achieve the trademark high-resolution range, in contrast, the average magnetic field along the few meter Larmor precession sections need to reach >10 mT, i.e., the field needs to display a strong gradient along the trajectory, which also implies some field inhomogeneity across the beam cross section perpendicular to the propagation direction. By careful magnetic design and the use of a set of fine correction coils across the beam, solenoid-type magnets can be tuned to provide in the range of 10 ppm homogeneous field integral $(\int Bdl)_2 - (\int Bdl)_1$ over all possible trajectories within the above-mentioned substantial beam cross sections and corresponding beam divergences. Common examples are the so-called Fresnel corrections coils corresponding to a spiral in a plane, defined in polar coordinates (r, ϕ) by the equation $r = \sqrt{a\phi}$, where a is a constant (Figure 4) [10,11].

A different approach to perform the spin labeling and control the inhomogeneities along the trajectories is to use time-dependent spin modulation and monitor the neutron velocities by the time delay of the arrival of the initial modulation pattern, instead of the amount of Larmor precessions along the trajectory. This can be achieved by using a rather high field DC coil (similar in geometry to the flipper coil in Figure 2) in order to produce the DC operating field **B** for a standard NMR type radio frequency (RF) neutron spin flipper placed inside this high field DC coil. The RF flipper frequency in the range of a

FIGURE 4 Examples of in-beam coils aimed at improving the homogeneity of solenoidal Larmor precession coils in neutron spin-echo spectroscopy. The left-hand side is a 70-mm diameter Fresnel spiral coil first tested and in use since 1979 [10]. On the right-hand side the higher performance "Pythagoras Fresnel" version (two perpendicular structures one after the other) that can handle 20-cm beam diameter at the detector and can help to tune the precession field integrals approaching 10 ppm homogeneity over that large detected beam diameter and corresponding beam divergence [11].

few hundred kilo Hertz corresponds to the Larmor frequency $\gamma_L B$, and the amplitude of the RF signal will be tuned to 90° spin flip. It is a natural feature of such RF flippers that the direction of the flipped spin will be perpendicular to the operating DC field **B**, with a time-dependent initial direction $\varphi(t)$ rotating around **B** with the RF frequency ω_F of the flipper: $\varphi(t) = \varphi_0 + \omega_F t$. If we keep the magnetic field zero outside the RF flippers, this direction of the neutron spin will arrive unchanged to a similar flipper at the end of a timing base of length l_1 of the first spin labeling just before the sample with a time delay corresponding to its velocity $t_i = l_1/v_i$. This second 90° flipper operating synchronously to the first one at the same frequency ω_F will turn the neutron spin into different directions between the original before the first flipper and its opposite, depending of the time delay t_i it arrived with. Thus, as above, the spin direction after this flipper is a label for neutron velocity, but now as a function of time instead of distance along the trajectory. A similar pair of RF flippers can be used to analyze the neutron velocity changes by the echo, with the difference that now the beam propagation happens without Larmor precessions in zero fields and the label is to be read with respect to a clock time provided by the RF flipper frequency. This labeling approach is called resonance neutron spin echo (RNSE) or zero field neutron spin echo (ZFNSE) [12], in contrast to the "classical" NSE with time-independent labeling, as described above.

Beyond flipper coils (DC or RF, both of which are to be tuned to and used for a more or less monochromatic neutron beam) there also are other ways to set the neutron spin into a given direction for labeling purposes. The so called "current sheet" devices will be discussed in more detail in Section 1.6.2. They have the particularity that they can operate with practically "white" (non-monochromatic) beams [37]. For a single beam trajectory, the two approaches (i.e., NSE with spin labeling as a function of space along the trajectory in a DC field configuration and RNSE with spin labeling as a function of time in zero field beam propagation configuration) give by principle identical results. Indeed, the sensitivity on individual particle velocity in beam experiment involving spin precessions depends on the difference between the frequency of the flipper devices used and the average Larmor frequency $\langle \omega_L \rangle$ over the trajectory between flippers. If $\omega_F = \langle \omega_L \rangle$, the Larmor precessions remain synchronous to the RF field in the flippers over the trajectory, i.e., their velocity-dependent time-of-flight has no impact on the final spin direction. This is the basis of many ion beam experiments, e.g., for the fundamental Ramsey split coil flippers. For spin labeling of velocity, the sensitivity is proportional to the difference $|\langle \omega_L \rangle - \omega_F|$. In NSE, the frequency of the DC flippers is 0, and the average Larmor frequency is the scanning parameter, in RNSE, it is the other way round. If we define for a spin labeling unit (which consists of two flippers operating at a common frequency ω_F and a beam propagation section with magnetic field B in between) the effective precession field as $B_{\text{eff}} = |B - \omega_F/\gamma_L|$, the two approaches—or they mix—can be considered in identical terms. In principle, both B and ω_F can be non-zero [13], whose combination can, e.g., be used for extending the dynamic range of the scanning parameter to values very close to 0 (i.e., no sensitivity to neutron velocity, which cannot be practically achieved otherwise). For practical completeness it is to be noted that (1) a flipper can be common to two subsequent spin labeling sections and (2) ω_F can be different of the RF frequency driving a complex flipper: the common bootstrap technique of flippers with two RF coils one after the other can double the effective flipper frequency compared to that of the RF power used [14].

Although the principles are identical, from practical/experimental point of view there are subtle differences between the two spin labeling approaches considered by now. The most fundamental one is the geometrical definition of labeling inhomogeneities in an ensemble of trajectories with finite beam cross section and divergence. In the RNSE technique, the base for time-of-flight velocity encoding is defined by the surfaces of the flipper devices, represented by the operating field coils. For such a piece operating under considerable energy load 0.1 mm effective geometrical precision is already an engineering feat, which implies about 100 ppm blurring of the velocity labels between different trajectories for practical section lengths. On the other hand, by careful µ-metal shielding, the homogeneity of the "zero" magnetic field can be realized with high accuracy as measured by the amount of spurious Larmor precessions in the intended zero-field domains, whose shape is very well defined by the flipper device surfaces. In NSE, the magnetic precession field regions have no sharp, geometrically materialized boundaries, instead the homogeneity is finally achieved by electromagnetic fine-tuning, as discussed above, to about 10 ppm precision, equivalent to 0.01 mm path length differences.

A key feature in spin labeling work is the need to make the neutron carry the label through the scattering process and other interactions with the sample. This is automatically fulfilled for nonmagnetic samples and scattering processes in the absence of strong magnetic fields on the sample, since there is no change of spin direction (although the spin polarization signal is reduced to 1/3 and turned in the opposite direction for nuclear spin incoherent scattering). For magnetic scattering processes and samples in high magnetic fields, the spin direction undergoes specific changes in the scattering process and/or on passage through the sample environment magnet. This calls for special flipper configurations around the sample for "saving" the labeled information (in some cases at the price of 50% loss of the spin label signal/background ratio), which will be discussed in later sections below together with typical experimental examples. Among others, NSE experiments have been successfully conducted with high magnetic fields up to 7 T on the sample [15].

Sample environment equipment can also detrimentally interfere with the information carried by the neutron spin (e.g., heating coils in furnaces, superconducting sample holders or other close structural parts at low temperatures, etc.), but such effects can be avoided by proper design.

Ferromagnetic samples represent a special case, if they are not magnetized to saturation by a sizable magnetic field: ferromagnets (and ferrimagnets) containing a random distribution of magnetic domains depolarize neutron beams, i.e., all or much of the spin label information gets lost. At the price of some additional intensity loss, one can convert the spin label information to intensity modulation by adding a neutron spin analyzer in front of the sample, which converts the spin label for the neutron velocity into beam intensity modulation label as a rapidly oscillating function of velocity [16]. For the analysis, the beam needs to be repolarized after interaction with the sample. An experimental example will be discussed below with more details. A similar scheme of transcribing the spin label for velocity into intensity modulation label in relation with RNSE has been established under the acronym MIEZE [17]. In this case a more complex, focusing intensity labeling scheme is used which provides a time-dependent echo at a focus point after the sample, which can be directly observed by the oscillation of the count rate as a function of time. Without spin labeling after the scattering, in this method, the neutron path length differences between different trajectories (up to the point of absorption in the detector) set the limits of resolution, and they can be in the range of 0.5 mm.

1.4.2 Choices of Neutron Parameters for Spin Labeling

Up to this point, we have considered the most common case of spin labeling: encoding the absolute value of neutron velocity around well-defined neutron trajectories for the purposes of spectroscopy. Two main generalizations have been established by now: using spin labeling to simultaneously determine changes in the absolute value of the velocity of the neutrons in scattering processes over a wide angular range (from 0 to some 145°) and to track changes of the direction of the neutron velocity. Both can be achieved by choosing the shape of effective precession field B_{eff} regions of the labeling units. A rather straightforward solution for high-resolution spectroscopy in a wide range of scattering angles is the use of circular precession field configuration. This can be achieved in NSE by concentric ring-shaped magnets with several meters of diameter around the sample area as the spin precession domain for the scattered neutrons [18]. A similarly straightforward solution for labeling a component of the neutron velocity (instead of its absolute value) is offered by parallelogram-shaped effective precession fields. Figure 5 illustrates the effective fields we need to implement for these alternatives, viewed from above for horizontal scattering plane.

Here we assume that the effective precession field B_{eff} is negligible outside the considered shapes. The time for the neutrons to cross two parallel sides of the parallelogram shape (independently from direction and position) is $t = l/v_\perp = l/(\mathbf{vn})$, where $v_\perp$ is the component of the neutron velocity perpendicular to the parallelogram sides it crosses, and $\mathbf{n}$ is the unit vector perpendicular to these parallelogram sides.

Neutron Optics and Spin Labeling Methods Chapter | 1 23

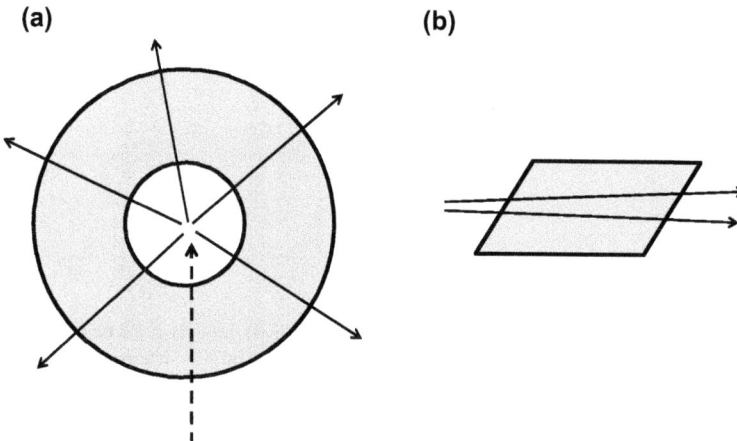

FIGURE 5 Shapes of effective spin labeling Larmor precession fields for different applications discussed in the text. (a) Cylindrical symmetry centered on the sample position with an axis perpendicular in the median scattering plane for continuous wide angle detector coverage from 0 to ±145° for neutron energy labeling [18]. (b) Parallelogram with the neutron beam traversing opposite "tilted" faces compared to the median beam direction for combined neutron energy and momentum labeling [21].

There are two successful established cases for using spin labeling with parallelogram-shaped fields, also called "tilted" or "inclined" fields, where the angle between the front face and the average direction of the neutron beam is called "tilt angle." If one assumes, as customary for "elastic" neutron scattering experiments (such as SANS or reflectometry) that the inelastic scattering processes have negligible cross sections, the change of $v_\perp$ in the scattering process will be the measure of the change of the neutron direction at constant velocity, i.e., of the neutron momentum transfer **Q** [19]. Analogously to the appearance of the neutron spin-echo signal for the case of no neutron energy transfer, as discussed above, we will observe an echo corresponding to the case of no momentum transfer (i.e., straight beam propagation) if the tilt angles of the labeling and analyzing "tilted" field regions, respectively before and after the scattering sample, are of the same value χ. The scanning parameter that determines the sensitivity of the setup to the change of neutron propagation angle in the scattering process is proportional to $\tan \chi$, as it will be discussed below in connection with experimental examples. A most recent development in the design of tilted fields for spin-echo applications was the introduction and testing of superconducting Wollaston prism-type magnetic DC coils, which can achieve as high as 85° tilt angles for NSE configurations [20]. Physically tilting of the RF flipper units in NRSE setups is adequate up to about 70°.

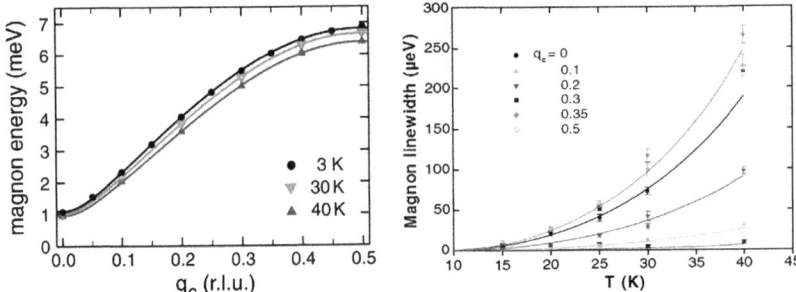

FIGURE 6 Magnon energies and linewidths (inverse lifetimes) throughout the Brillouin zone as a function of temperature in the prototypical antiferromagnet MnF$_2$ as determined by resonance neutron spin-echo spectroscopy installed on triple axis spectrometer for achieving unprecedented sensitivity in the determination of lifetimes and energy shifts of dispersive ($dE/dq \neq 0$) elementary excitation [22].

The final, most general labeling method to be considered here is to use the above tilted field approach for inelastic scattering, i.e., without the elastic scattering assumption we have just discussed. The tilting of the fields makes the spin label depend by the relation $t = l/(\mathbf{v}\mathbf{n})$ on a combination of the absolute value and the direction of $\mathbf{v}$. It has been shown that by the proper choice of the three parameters, the tilt angles of the two field shapes before and after the sample and the ratio of the effective field integrals $\int B_{\text{eff}} dl$ of the two field regions before and after the sample, an echo signal will be created for scattering on an elementary excitation around a given point in the $(\mathbf{Q}, \omega)$ space with a given local slope of the dispersion relation $\partial E/\partial Q_\alpha$ in a given direction α within the scattering plane (where Q_α is the component of $\mathbf{Q}$ in the direction of α) [21]. Adding spin labeling option to a triple axis spectrometer used to select the $(\mathbf{Q}, \omega)$ point in the reciprocal space, the high resolution capability of NSE can be used to determine elementary excitation lifetimes and shifts with otherwise unachievable resolution [22] (Figure 6).

1.5 NEUTRON SPIN-ECHO SPECTROSCOPY

The Larmor precession of neutron spins have found a revolutionary application as invented by F. Mezei [7] in the so-called NSE method. Subtleties of the method are described at various places [8,23], here we summarize only the basic idea. Let us consider a polarized neutron beam that traverses a magnetic field region between point R_1 and R_2, with field direction perpendicular to its polarization. According to the description we gave above, the neutron spin will start to precess. The angle of precession for a neutron with velocity v will be given by:

$$\varphi_1 = \gamma_L \int_{R_2}^{R_1} |\mathbf{B}(\mathbf{r})| \frac{dl}{v} \quad (11)$$

as $\gamma_L = 2916.4$ Hz Gauss^{-1}. For example, neutrons with a wavelength of 4 Å in 10 Gauss field will make a full turn in 3.5-cm distance. A finite width in the velocity distribution will quickly lead to the dephasing of the precession angles of the neutrons in the beam. However, if the neutron beam will traverse now another magnetic field region (R_3, R_4) with opposite field direction, the total angle of precession will read:

$$\varphi_T = \gamma_L \int_{R_2}^{R_1} |\mathbf{B}(\mathbf{r})| \frac{\mathrm{d}l}{v} - \gamma_L \int_{R_3}^{R_4} |\mathbf{B}(\mathbf{r})| \frac{\mathrm{d}l}{v} = \gamma_L \left(\int_{R_2}^{R_1} |\mathbf{B}(\mathbf{r})| \frac{\mathrm{d}l}{v} - \int_{R_3}^{R_4} |\mathbf{B}(\mathbf{r})| \frac{\mathrm{d}l}{v} \right) \tag{12}$$

If the two field integrals are equal, the total precession angle will be zero, independently of the velocity, thus the wavelength, of the neutrons! Placing a polarization analyzer behind the second precession section, it will transmit the neutrons with a probability of $\cos(\varphi_T)$.

This idea can have several applications. For example, placing a sample between the two regions of precession, the scattering on the sample can modify the neutron velocities and the total precession angle might not be zero for all neutrons. If the probability that a neutron scattered with an energy exchange of ω at the scattering vector of Q is $S(Q, \omega)$, then in first order in ω the beam polarization in the echo signal is:

$$P_{\mathrm{NSE}}(Q, \tau_{\mathrm{NSE}}) = P_0 \frac{\int \mathrm{d}\omega S(Q, \omega) \cos(\omega \tau_{\mathrm{NSE}})}{\int \mathrm{d}\omega S(Q, \omega)} \langle \cos(\varphi_T) \rangle_{\mathrm{Traj}} \tag{13}$$

$$P_{\mathrm{NSE}}(Q, \tau_{\mathrm{NSE}}) = P_0 \frac{I(Q, \tau_{\mathrm{NSE}})}{I(Q, 0)}, \tag{14}$$

where

$$\tau_{\mathrm{NSE}} = \frac{\hbar \gamma_L \int |\mathbf{B}(\mathbf{r})| \mathrm{d}l}{m v^3}, \tag{15}$$

$I(Q, t)$ is the intermediate scattering function, which measures the decay of correlation inside the sample with time. τ_{NSE} can be varied by varying the precession field strength and/or choosing different neutron velocities. Typically it is in the range from picosecond to close to microsecond, or the equivalent hundreds of microelectron volt to nanoelectron volt energy exchange. The great advantage of this method is to be able to measure with close to 10^{-5} precision the energy exchange without the need to monochromatize the incoming beam accordingly.

Here we considered only the ω dependence of $S(Q, \omega)$. We can also shape the magnetic field region such as to introduce a dependence of φ_T on Q. In this case, P_{NSE} will be a transform of the scattered intensity also in Q. We need to keep in mind; however, that both dependencies might be present. For example one always uses finite beam size and sample size. The path length difference between parallel and 0.5° inclined trajectory dl/l is already 3.8×10^{-5}, thus too big compared to the precision we are aiming for. Even worse, if the magnetic fields are generated by the most symmetrical cylindrical geometry [10] even in the most optimized case [24], the distribution of the magnetic field integral for different trajectories is not better than 10^{-3}! Luckily, as was established by Mezei [10], in-beam correction elements can improve the situation by about a factor 100, leading to the targeted 10^{-5} precision.

What will be the effect of the remaining inhomogeneity? If for all trajectories φ_T is not strictly zero then in Eqn (13)

$$\langle \cos(\varphi_T) \rangle_{\text{Traj}} < 1.0 \qquad (16)$$

Thus even for a strictly elastic scatterer $P_{\text{NSE}}(Q, \tau_{\text{NSE}})$ will decay as a function of Fourier time τ_{NSE}. The quasi-elastic scattering on the real sample will just make $P_{\text{NSE}}(Q, \tau_{\text{NSE}})$ decay further. We just have to divide the curve measured on the sample by the one measured on an elastic standard scatterer, and we are done with the resolution correction. In fact this emerges from the fact that we are measuring the Fourier transform of $S(Q, \omega)$. In the ω space the resolution correction is a deconvolution, which in the Fourier transformed, time-dependent space becomes a simple division.

1.5.1 NSE Spectroscopy for Nuclear Scattering

Let us consider a few simple cases, for example, identical, non interacting objects largely separated in space compared to their size, like colloidal particles in a liquid media.

By definition

$$I(Q,t) = \left\langle \sum_{i,j} b_i b_j e^{-i\mathbf{Q}\cdot(\mathbf{R}_j(t)-\mathbf{R}_i(0))} \right\rangle \qquad (17)$$

where b_i and b_j are the scattering lengths of the atoms i and j, respectively at position $\mathbf{R}_j(t)$ at time t and $\mathbf{R}_i(0)$ at time zero. Let us decompose these vectors to ones which point to the center of the mass and a vector which points into the object: $\mathbf{R}_j(t) = \mathbf{R}_{\text{CM}}(t) + \mathbf{r}_j(t)$. If the center of the mass motion is independent from the internal motions of the objects, then we have

$$\begin{aligned} I(Q,t) &= \left\langle e^{-Q\cdot(\mathbf{R}_{\text{CM}}(t)-\mathbf{R}_{\text{CM}}(0))} \right\rangle \left\langle \sum_{i,j} b_i b_j e^{-i\mathbf{Q}\cdot(\mathbf{r}_j(t)-\mathbf{r}_i(0))} \right\rangle \\ &= e^{-Q^2 Dt} \left\langle \sum_{i,j} b_i b_j e^{-i\mathbf{Q}\cdot(\mathbf{r}_j(t)-\mathbf{r}_i(0))} \right\rangle \end{aligned} \qquad (18)$$

Here the first factor will describe the center of the mass diffusion and the second one the internal motions.

When the object is really a rigid body, simple Lorentzians will be measured, and by fitting $e^{-Q^2 Dt}$ the diffusion constant D can be extracted and it will be a constant as a function of Q (Figure 7).

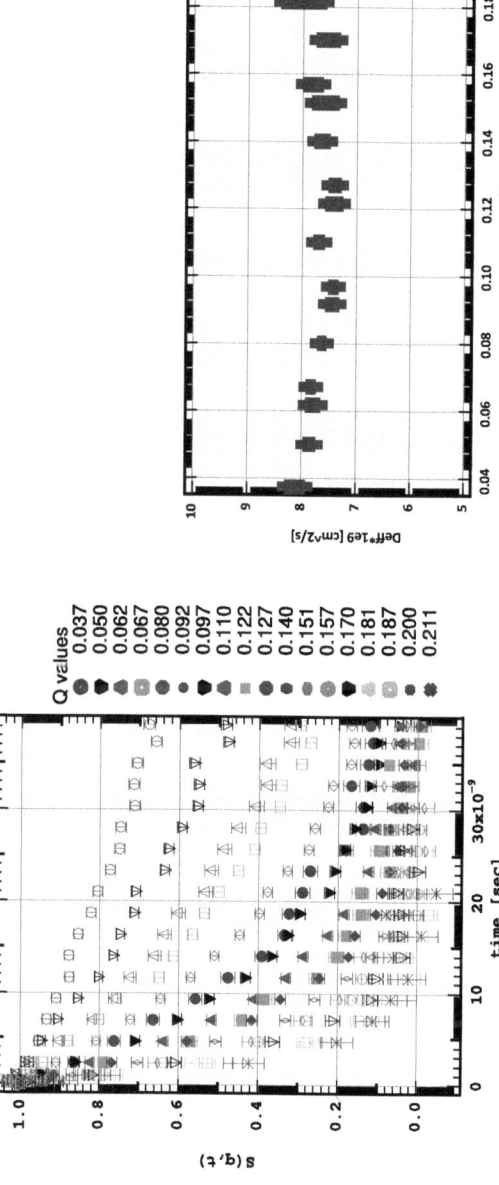

FIGURE 7 Left: $I(Q, t)$ measured on a dilute globular protein solution. Right: The derived D as a function of Q as derived according to Eqn (18).

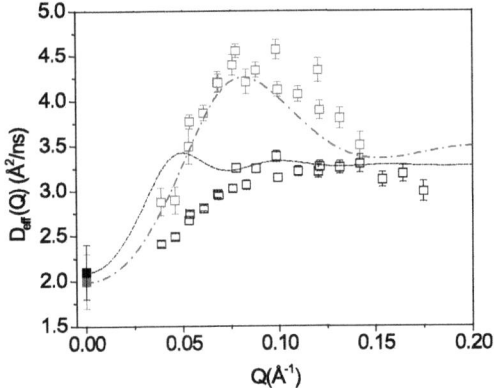

FIGURE 8 NHerf, a flexible multidomain protein [25]. Red (upper curve, partially deuterated) and blue (lower curve, nondeuterated) empty symbols are $D_{eff}(Q)$ values as derived from neutron spin-echo measurements. Full symbols are the diffusion constants for the same samples as measured by PG-NMR.

On the other hand for a flexible object, which shows internal motion on the time scale we are measuring, due to the properties of the Fourier transformation, when Q increases and $1/Q$ becomes comparable to the size of the object, we might start to see the internal motions, potentially including the rotational diffusion. An example is shown on Figure 8, for a flexible protein molecule [25]. At low Q values, we recover the pure center of the mass diffusion as verified by pulsed field gradient NMR, at high Q we start to see an addition contribution (which looks like an increase in the effective $D(Q)$).

In a strict mathematical sense, the $I(Q, t)$ should become a sum of exponentials with Q dependent prefactors. However, experimentally it becomes visible only if the time scales are well separated and the prefactors are sufficiently large. This is the case on Figure 9, where the form fluctuation of microemulsion droplets is much faster than the center of the mass diffusion [26], and due to an interplay of the form factors become particularly visible at a given Q.

We can consider as a limiting case a long polymer, where practically only the internal motions are visible. For a very long chain, the center of the mass diffusion might contribute a negligible decay in the measured time window ($Q^2 Dt \ll 1$) (Figure 10).

Another complication rises with interaction between the objects. As we deal with coherent scattering, at finite concentration and/or with strong, long ranged interaction, the objects do not move any more independently and what we measure is the collective diffusion. In the static small-angle scattering, the measured intensity becomes:

$$I(Q) = NS(Q)P(Q) \tag{19}$$

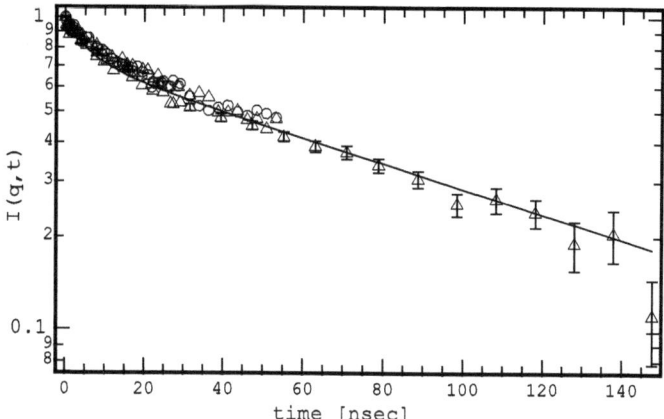

FIGURE 9 $I(Q, t)$ of TDMAO (nonionic surfactant + oil in water) microemulsion droplets in shell contrast at $Q = 0.6$ nm^{-1}. The logarithmic Y scale reveals a first relatively fast (~ 10 ns) decay, identified as shape fluctuation, followed by a slower decay corresponding to the center of the mass diffusion [26].

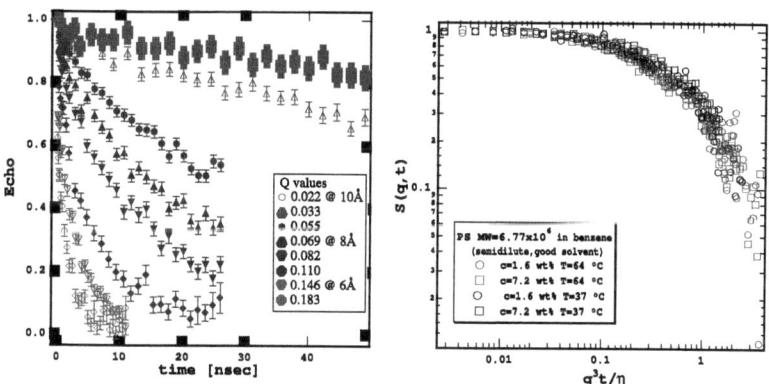

FIGURE 10 Left: $I(Q, t)$ polystyrene (MW $= 6.77 \times 10^6$) semidilute solution in dueterated benzene. The center of the mass diffusion should show negligible decay. Right: Similar curves measured at two temperatures and two concentrations scaled by Q^3/η showing that the Zimm prediction is obeyed. *Data from Ref. [27].*

where $P(Q)$ is the scattered intensity of a single object, N is the number of objects, and $S(Q)$ carries the information about local order which builds up due to the interaction. In $I(Q, t)$ in first approximation this will show up as:

$$I(Q,t) = e^{-Q^2 \frac{Dt}{S(Q)}} \left\langle \sum_{i,j} b_i b_j e^{-i\mathbf{Q} \cdot (\mathbf{r}_j(t) - \mathbf{r}_i(0))} \right\rangle \qquad (20)$$

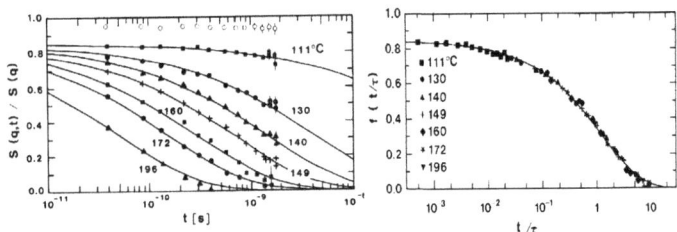

FIGURE 11 Left: $I(Q, t)$ of $Ca_{0.4}K_{0.6}(N_{03})_{1.4}$ glass at different temperatures at $Q = 17$ nm^{-1}. Right: The same curves scaled by the macroscopic viscosity. *Data from Ref. [28]*.

The renormalization of the effective diffusion constant from D to $D/S(Q)$ in comparison to Eqn (18) is the so-called deGennes narrowing. In fact deGennes derived this expression for simple atomic liquids, but it also holds for a solution of colloids.

Dynamics of relatively simple liquids also becomes more complicated when they can reach glass transition without crystallization. The glass transition is characterized by a structure which remains amorphous, but the macroscopic viscosity diverges at a given temperature. NSE can probe this slowing down on a microscopic scale. Interestingly, at first glance the atomic motion seems to slow down with the same rate as the macroscopic viscosity. Here the capability of NSE to measure $I(Q, t)$ as a function of time becomes invaluable. It could be seen immediately that the decay of $I(Q, t)$ is no more a simple exponential but a stretched exponential. Another aspect is that $I(Q, t)$ does not seem to extrapolate to 1 at $t = 0$ (Figure 11).

This means that somewhere between $t = 0$ and our experimental shortest time some other fast process takes place, leading to this drop. This was predicted theoretically by the mode coupling theory and later found also experimentally [28].

1.5.2 NSE Spectroscopy in Magnetism

We have to examine how the specific interaction of the neutron spins with the magnetic moments in the sample influences NSE. A ferromagnetic sample has a magnetic field inside and its return field outside, even worse, most likely will consist of randomly oriented magnetic domains. This will lead to uncontrolled precession of the neutron spins, thus a loss of the spin-echo signal. First we limit our discussion here to the case of paramagnets with some comments on antiferromagnets. As was pointed out, the neutrons will only "see" the microscopic magnetic field $\mathbf{B}(\mathbf{r})$ inside the sample, which enters the scattering cross section formulae by the Fourier transform $\mathbf{B}(\mathbf{Q})$. It can be shown on the basis of Maxwell equations that $\mathbf{B}(\mathbf{Q}) = \mathbf{M}_\perp(\mathbf{Q})$, where $\mathbf{M}_\perp$ is the component of magnetic moment density perpendicular to the scattering vector $\mathbf{Q}$.

If the neutrons propagate in the z direction and the precession plane is in the x, y plane, we can write the beam polarization as

$$\mathbf{P} = P_0 \begin{pmatrix} \cos(\omega t) \\ \sin(\omega t) \\ 0 \end{pmatrix} \qquad (21)$$

If $\mathbf{Q}$ points in the x direction, then the P_x component will be fully spin flipped, thus $P_x => -P_x$, while the y component will be spin flipped with 50% probability and not spin flipped also with 50% probability if the paramagnet is isotropic [21]. After scattering, the polarization can be written as

$$\mathbf{P} = \frac{1}{2} P_0 \begin{pmatrix} \cos(\omega t) \\ \sin(\omega t) \\ 0 \end{pmatrix} + \frac{1}{2} P_0 \begin{pmatrix} -\cos(\omega t) \\ \sin(\omega t) \\ 0 \end{pmatrix} \qquad (22)$$

We should notice that the action of the 180° flipper in the NSE spectrometer is to reverse the precession plane around, e.g., the y axis, thus leaving the y component unchanged and reversing the x component. Scattering from isotropic paramagnet due to the second term in Eqn (22) will give an echo with 50% reduced amplitude and without use of a 180° flipper. The very nice feature here is the fact that without 180° flipper the nuclear scattering will not give any echo, thus there is no need for background correction, and unambiguously only the magnetic scattering will contribute to the echo signal.

We still need to solve the problem, how to determine P_0. Indeed in a simple polarization measurement switching a 180° flipper on-off gives spin up and spin down intensities, and coherent, incoherent, and magnetic scattering will contribute with different signs, and it is not possible to separate P_0 for the magnetic scattering. The way out is the xyx polarization analysis. If instead of measuring up and down while keeping the beam polarization in the z direction we repeat the up, down measurements in the x, y, and z directions, the magnetic contribution thus can be separated [29]. Let 'us illustrate the power of the technique on the $Ho_2Ti_2O_7$ compound [30] (called spin ice), where at low temperature the magnetic structure approaches a frozen state (Figure 12).

All our description above supposed an isotropic paramagnet. It also applies for antiferromagnets if the magnetic domains (or single crystal grains) are randomly oriented. It is not unusual to have antiferromagnetic single crystals. In this case, special care has to be taken whether it is also a single magnetic domain or not and what are the orientation of the magnetic moments relative to Q.

As mentioned, ferromagnetic samples have the problem of depolarizing the neutron beam. With some specific tricks this problem can be partially circumvented. One possibility [21] is turning one component along the precession field before the sample and applying a strong field on the sample, which is

32   Neutron Scattering - Magnetic and Quantum Phenomena

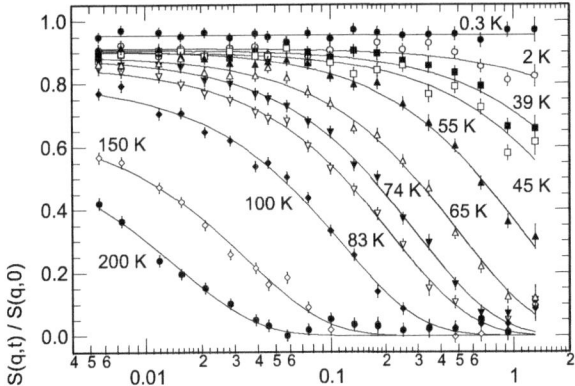

FIGURE 12   Slowing down of $I(Q, t)$ of $Ho_2Ti_2O_7$ which is a topologically frustrated ferromagnet. *Data from Ref. [30]*.

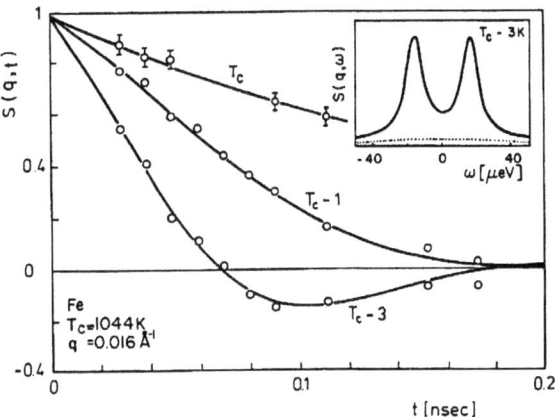

FIGURE 13   Spin dynamics in Fe at and just below the critical temperature [16]. At T_c-3 the magnon peaks are already sufficiently separated and narrow to observe a damped oscillation in $I(Q, t)$.

strong enough to overcome the magnetic field of the randomly oriented domains. This of course might influence the physics to be studied. Another less intrusive method is the intensity modulated neutron spin echo (IMNSE) [11] which is more costly in neutron intensity as in addition the beam has to be polarization analyzed before the sample and repolarized after it with the losses this implies. IMNSE was first used to measure critical magnon dynamics of Fe just under the ferromagnetic transition (Figure 13), where applying of an external strong field would completely change the dynamical behavior.

1.6 NEUTRON SPIN-ECHO FOR ELASTIC SCATTERING AT SMALL ANGLES

Small-angle neutron scattering (SANS) experiments can be performed with the help of the spin-echo labeling techniques. Such approaches are referred to by different acronyms, SERGIS and SESAME are often used examples [31,32]. In general, a SANS setup or a reflectometer strongly resembles a standard diffractometer, however, with a very limited Q range and very accurate angular resolution. Therefore, such facilities are most suited for the study of mesoscopic structures, i.e., from nanometer up to the micron range. To reach the good angular resolution of these setups, the incoming angular beam divergence has to be kept at a minimum in the apparatus. This requires the use of very narrow slits, which again causes a considerable decrease in intensity. This fact makes the intrinsic intensity limited character of neutron sources even more severe. The schematic experimental arrangement of such a conventional setup is shown in Figure 14.

By implementing the spin-echo spin labeling technique we may compensate for this intensity loss, by encoding the scattering information into the polarization of the neutron beam. In this setup the neutrons traverse two identical magnetic field regions, one before and one after the scattering event. The polarization of the out-coming neutron beam will decrease as a result of a scattering process. If these magnetic regions have a particular shape (i.e., flat borders that are inclined with respect to the symmetry axis of the experiment), the remaining polarized fraction of the neutron beam can give information about the scattering distribution. The measured quantity (polarization) is proportional to the one-dimensional projection of the spatial correlation function of the scattering potential along the transverse in-plane direction. Generally, spin-echo methods directly provide the Fourier transform of the scattering function of the sample and are therefore referred to as Fourier methods.

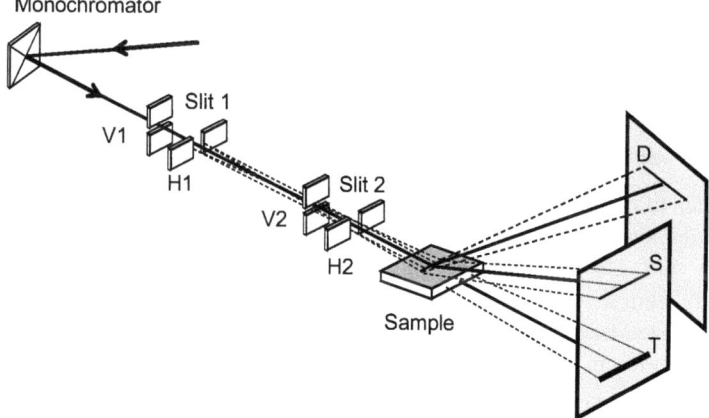

FIGURE 14 Schematic setup of conventional neutron grazing incidence diffraction. With tight horizontal slits the required in-plane resolution can be achieved. As shown in the figure, three types of beams coming off the sample can be distinguished: the specularly reflected beam (S), the off-specularly scattered beam (D), and the transmitted (T) beam. *This figure is taken from Ref. [33].*

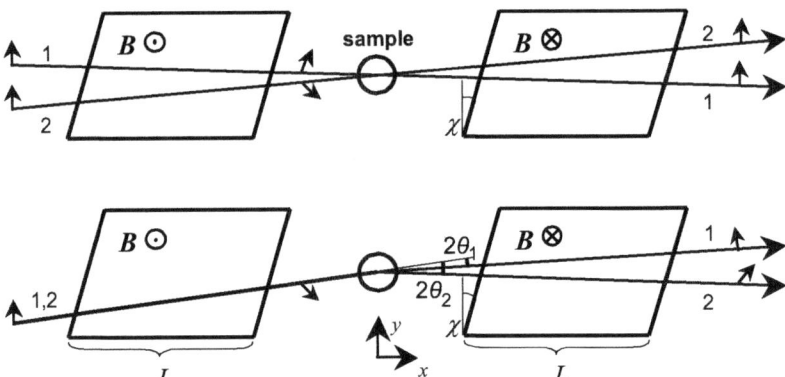

FIGURE 15 Schematic principle of the neutron spin-echo technique for small-angle elastic scattering, taken from Ref. [33]. Two precession regions, which are realized by identical magnetic field regions with inclined borders and opposite field orientation, are situated before and after the sample. For simplicity, the spin polarizer and analyzer are not shown. Top: The polarization state of the neutron spin, represented by the small arrows, is restored after passing through the entire setup if it did not change direction by scattering on the sample. This is independent of the position or direction of the neutron trajectory and the speed of the neutron. Bottom: Scattering of the neutrons by the sample results in a difference in polarization state of the neutrons compared to their incoming polarization. By this the beam polarization is reduced.

The basic principle of the experiment is shown in Figure 15 [33]. The main components of the experiment are the two identical precession regions on either side of the scattering sample, with opposite direction of precession. The neutron beam is polarized before entering the setup and the depolarization of the beam is measured in a detector after a spin analyzer. If the neutron velocity did not change (in magnitude and direction) in the sample region, the path lengths inside the two regions are equal. Thus the total precession angle is zero for all neutron trajectories and the initial neutron beam polarization is restored after the second precession region. This is not the case, if the neutron direction has changed at the sample.

1.6.1 Neutron Beam Polarizers and Analyzers

In order to produce a polarized neutron beam, we need an interaction with a polarized target, e.g., with a magnetized ferromagnet as, for example, Fe, which will efficiently absorb the neutrons with the "wrong" polarization directly from the source (e.g., the research reactor) [34]. The determination of the polarized fraction of the scattered beam can be done in a similar way. The usual spin filters, that are used for the analysis of the scattered neutron beam work only in a relatively narrow angular range since they are able to select the polarized and nonpolarized component near the critical angle of the surface of the filter ($+/-$ a few degrees). One way to overcome this limitation is to scan with the filter orientation, which results in a strong increase in required experimental time. Alternatively, the usage of a ferromagnetic multilayer structure, in which the layers are nearly parallel to each other [35] could eliminate this problem. The advantage of such a unit over a single thin layer polarizer is that its orientation within the experimental setup can be shifted without changing its performance. In part of the experiments, the spin analysis was performed by a magnetized ferromagnetic multilayer, where the local magnetic field was several kGauss. Such a strong magnetic field has to be shielded at the border of the magnetic box, which was done by using a 4-mm thick steel plate with appropriate openings for the neutron beams (total weight ~150 kg).

Another way to analyze neutron beam polarization is to use a spin-dependent nuclear physical process. In this case, there are no angular limitations during the experiments. The only difficulty when using this method is, however, how to polarize, e.g., the usually used ^{3}He gas and to conserve the polarization after polarizing for the time of the experiments [36].

1.6.2 Transport of Polarized Neutron Beams and Spin-Injection Devices

The experiment can be performed by letting the originally linearly polarized neutrons pass through a homogeneous guide field that possesses inclined borders as shown in Figure 15. The neutron beam enters the NSE-range through a current sheet device [37], as shown in Figure 16. These elements allow for a good field transmission between the magnetic spin polarizer and magnetic spin analyzer, both with relatively strong local fields. They can inject (or extract) neutron spins with well defined direction to the approximately zero field NSE region. The blue arrows in Figure 16 show the neutron trajectory leaving the NSE region towards the polarization analyzer before the detector. The red arrow in the plane of the current sheet formed by DC current carrying wires shows the direction of the neutron spin component extracted by this device, which will be guided into the direction of the magnetizing field on the polarization analyzer, defining the direction of its polarization sensitivity.

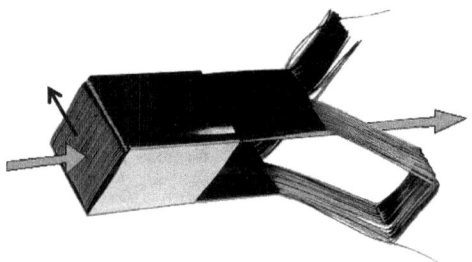

FIGURE 16 Picture of a Forte asymmetric coupling coil. Using this coil, the polarized neutron beam can be implanted into the neutron spin-echo setup with the correct polarization. Such a coil is also used to extract the neutron beam after scattering before the spin analyzer. The thick blue arrow shows the propagation direction of the neutrons in the latter case. The thin red arrow is parallel to the direction of the coil magnetic field close to the beam trajectory. *For more details see Ref. [33].*

The quadratic region of parallel wires at the left end of the coil forms the so-called current sheet. This is a plain sheet of metallic surface with a constant current density. The induced magnetic field of such a geometry is independent of the distance from the sheet. Since during the Larmor precession the magnetic energy is a constant of motion, i.e., the angle of the spin axis to the rotation axis (direction of the local magnetic field) is constant, and since we can say that the current sheet is so thin that the neutrons will traverse it within a very short time, i.e., with practically zero precession angle, the spin of the neutron will not change in the time it takes to travel through the current sheet. The situation is quite different, if the current density is not large enough in the sheet. In such a case, the Larmor precession can be so slow that inevitable nonparallel components of the magnetic field arise. If now the motion of the neutron is fast—similarly to the case of motional narrowing—the polarization is conserved, except if the local magnetic field decreases below the empirical value of 5 Gauss. In a magnetic field above 5 Gauss only a few percentage of the polarization is lost, assuming that the current sheet is ideal or close to being ideal.

Inside the coil, the magnetic field is the sum of the field of the connecting magnetic element (e.g., polarizer) and that of the coil itself, the main component of which is perpendicular to the direction of the current (wires) in the plane of the sheet, as shown by the red arrow in Figure 16 current sheet. Outside the coil, the perpendicular magnetic field component remains, therefore, no depolarization is expected, especially if the neutrons are fast enough. In order to realize this, the coil field needs to be carefully measured and tuned. Magnetic fields **B** < 5 Gauss can be achieved easily, practically with all types of magnetic sources. Usually, the condition that the magnetic field **B** is perpendicular to the current sheet is automatically fulfilled (Maxwell's equations). Details of how to produce such a current sheet coil device are given in Ref. [33].

1.6.3 Precession Region and Magnetic Shielding

In order to generate the required magnetic field regions before and after the scattering region for the spin-echo technique, a similar arrangement is used as in RNSE spectroscopy [12]. The experimental arrangement is shown in Figure 17.

In RNSE, the magnetic field is produced by a set of two coils instead of a single coil, between which the magnetic field value is zero. These coils are thin and their edges form current sheets in order to define the borders of the magnetic field very precisely. In order to achieve the best possible shielding of stray magnetic fields outside the coil assembly, the so-called bootstrap configuration [12] was applied. Here, each single resonance coil is made up of a set of two coils with opposite field directions, as indicated in Figure 17. The realization of this setup is very similar to that of the current sheet coil.

Neutron Optics and Spin Labeling Methods Chapter | 1   37

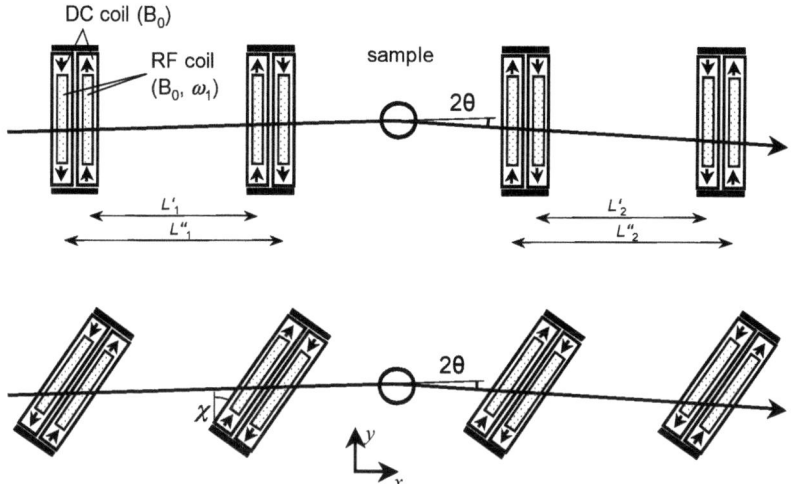

FIGURE 17 Top view of the coil arrangement and corresponding magnetic fields in the resonance neutron spin-echo setup. The bootstrap configuration is realized by two DC-RF coils. Top: Setup, where the field borders are not inclined, corresponding to a spin-echo length of zero. Bottom: Inclined coils result in a nonzero spin-echo length. *More details are given in Ref. [33].*

(cf. Figure 16). First, a fence-like construction is built from brass parts, which is then covered by uniform isolated Cu wire parallel to the base plate. The shorting of the field lines is performed by thin μ-metal foils. In the experiment, first it is verified that with zero guide field ($B = 0$), the neutrons keep their polarization state from the start throughout the whole setup. Then, by changing the guide field ($B > 0$) by an extra coil (Figures 17) the polarization will be reduced.

In such a setup, the scattering angle of the neutrons is labeled the same way the neutron energy change was labeled in NSE, cf. Eqns (11−14). The echo polarization thus becomes the Fourier transform of the scattering function $S(Q)$, i.e., a function of a spin-echo length Y [38], defined analogously to the spin-echo time in Eqn (15).

$$Y = -2m\gamma_L/\pi h \cdot B_0 L \lambda^2 \tan\chi, \quad (23)$$

where L denotes the mean distance of the central plates of the RNSE coil assemblies for each of the two precession devices, B_0 is the magnitude of the DC magnetic field and χ is the inclination angle of the coils to the neutron direction, as indicated in Figure 17.

In addition to the improved magnetic shielding, the bootstrap configuration also provides double the spin-echo length as compared to a setup using single coils, which is advantageous in the study of large-scale structures. More details about the precession devices are given in Ref. [33].

1.6.4 Experimental Results

In the following, we will summarize examples for experimental investigations that have been conducted with this technique to date. The experiments are partly performed in transmission geometry (i.e., spin-echo resolved small-angle neutron scattering) and partly in reflection geometry (spin-echo resolved neutron reflectometry). The experiments described here have all been performed at one of the following machines: the dedicated add-on NSE setup at the EVA reflectometer facility of the Max-Planck-Institut für Metallforschung at the Institut Laue-Langevin in Grenoble (France) [39], which was transferred to the reflectometer N-REX+ at the neutron source Forschungsneutronenquelle "Heinz Maier-Leibnitz" FRM II in Garching (Germany) [40], or at the Second Target Station at the ISIS source of the Rutherford Appleton Laboratory (Oxforshire, UK) [41]. In addition to these machines, a spin-echo resolved SANS setup is in operation at the Delft University of Technology (Delft, Netherlands) [42].

1.6.4.1 Spin-Echo Small-Angle Scattering

The NSE spin labeling method in the transmission geometry was tested on a suspension of spherical polystyrene particles in mostly heavy water, in order to increase the neutron scattering contrast between the solvent and the polystyrene spheres [33]. In this experiment, the scattered fraction of the neutrons on the polystyrene spheres was determined and showed a clear dependence on the sphere size (in the sub-μm range). The known sphere sizes could be very well reproduced by fitting the data to a model describing the scattering on spherical particles in a suspension [43], thereby verifying the power of this method to yield information on the structure of the samples on large length scales up to 450 nm. (Equivalent to Q resolution in the $6 \times 10^{-4} \text{ Å}^{-1}$.) The results are shown in Figure 18.

In a second example, a comparison between the conventional SANS and the spin-echo resolved technique was carried out on a nanoporous aluminum oxide foil. The periodicity of the order of 100 nm was resolved in both measurements; however, the spin-echo data yielded a much higher intensity and information on the relevant length scale directly in real space, as shown in Figure 19. More details can be found in Ref. [33].

1.6.4.2 Spin-Echo Reflectometry

The NSE technique in reflection geometry was verified using an optical diffraction grating [33]. Here the spatial period of 278 nm of the grating structure (nominal value of 3600 lines mm^{-1}) is directly and reliably reproduced in the NSE data, cf. Figure 20. Again, detailed analysis of this result is presented in Ref. [33].

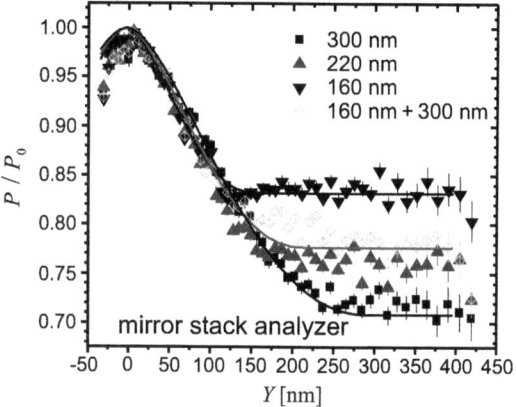

FIGURE 18 Neutron spin-echo small-angle neutron scattering (SANS) experiment on a suspension of polystyrene spheres in heavy water and water in transmission (SANS) geometry. The normalized spin-echo polarization as a function of spin-echo length is clearly correlated with the size of the particles. The experimental values are shown by symbols, the solid lines correspond to the fitted model functions for monodispersed spheres in a dilute solution. *More experimental detail is given in Ref. [33].*

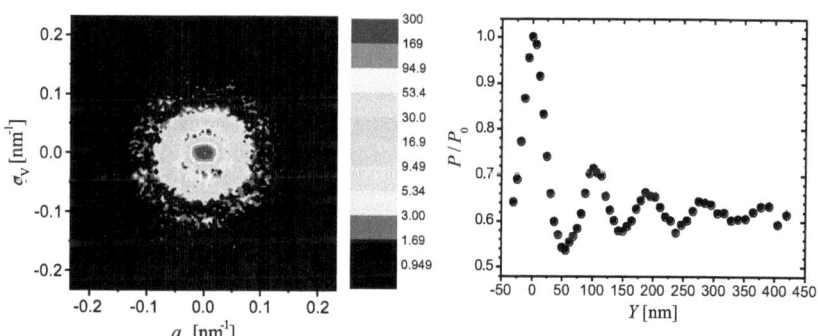

FIGURE 19 Left: Small-angle scattering distribution of an alumina foil with a periodic structure. Right: Data in transmission geometry of the same sample. From the neutron spin-echo data, the real-space periodicity of the pores in the sample can be extracted directly.

Further examples include the study of a self-organized organic layer on a silicon surface. Here a $F_{16}CuP$ (copper hexadecafluorophthalocyanine) layer was covered with a layer of di-indenoperylene (DIP) [44,45]. The investigation of the specular and off-specular reflections of this surface revealed a plateau-like island morphology that the organic layer adopts on the sample [33].

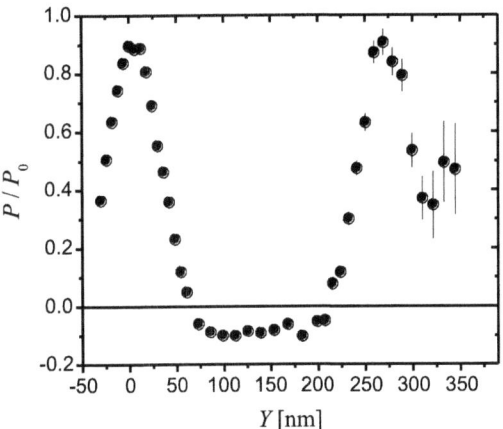

FIGURE 20 Reflection geometry neutron spin-echo data measured on an optical grating (3600 lines/mm). The spatial periodicity (grating constant) can be directly obtained from the spin-echo signal (polarization) as a function of the spin-echo length.

The spin-echo resolved reflectometry technique has also been applied to infer a more complex surface morphology of ultrathin polymer films dewetted from a silicon surface [46]. More recently, the material system P3HT:PCBM (P3HT = poly(3-hexylthiophene-2,5-diyl); PCBM = [6,6]-phenyl-C61-butyric acid methyl ester) has been investigated by this technique in the reflection geometry [47]. This material system is scientifically relevant due to its application in photovoltaic devices. The spin labeling resolved reflectometry data on this system imply that the investigation of buried structures in such thin films can be accessed by this technique, which could be of significant scientific and technological relevance.

REFERENCES

[1] F. Bloch, Phys. Rev. 51 (1937) 994.
[2] D.J. Hughes, M.T. Burgy, Phys. Rev. 81 (1951) 498.
[3] C.G. Shull, E.O. Wollan, W.A. Strauser, Phys. Rev. 81 (1951) 483.
[4] F. Mezei, Acta Phys. Hung 64 (1988) 15−19;
F. Mezei, Physica B 137 (1986) 295.
[5] A. Abragam, G.L. Bacchella, H. Glattli, P. Meriel, M. Pinot, J. Piesvaux, Phys. Rev. Lett. 31 (1973) 776.
[6] c.f. F. Mezei, in: M. Schlenker, et al. (Eds.), Imaging Processes and Coherence in Physics, Heidelberg: Springer Verlag, 1980, pp. 283−295.
[7] F. Mezei, Z. Physik. 255 (1972) 146−160.
[8] F. Mezei (Ed.), Neutron Spin Echo, Lecture Notes in Physics Series, vol. 128, Springer Verlag, Heidelberg, 1980.
[9] F. Mezei, Phys. Rev. Lett. 44 (1980) 1601−1604.

[10] F. Mezei, in: F. Mezei (Ed.), Neutron Spin Echo, Lecture Notes in Physics Series, vol. 128, Springer Verlag, Heidelberg, 1980, pp. 178–192.
[11] B. Farago, Gy. Kali, private communication.
[12] M. Köppe, M. Bleuel, R. Gähler, R. Golub, P. Hank, T. Keller, S. Longeville, U. Rauch, J. Wuttke, Physica B 266 (1999) 75.
[13] W. Häussler, U. Schmidt, G. Ehlers, F. Mezei, Chem. Phys. 292 (2003) 501.
[14] T. Keller, P. Zimmermann, R. Golub, R. Gähler, Physica B 162 (1990) 327.
[15] J.P. Boucher, F. Mezei, L.P. Regnault, J.P. Renard, Phys. Rev. Lett. 55 (1985) 1778–1781.
[16] B. Farago, F. Mezei, Phys. B&C 136 (1986) 100–102.
[17] R. Gähler, R. Golub, T. Keller, Physica B 180 & 181 (1992) 899.
[18] C. Pappas, A. Triolo, R. Kischnik, F. Mezei, Appl. Phys. A 74 (2002) S286.
[19] R. Pynn, J. Phys. E. Sci. Instrum. 11 (1978) 1133;
F. Mezei, in: Neutron Inelastic Scattering, IAEA, Vienna, 1978, pp. 125–134.
[20] F. Li, R. Pynn, J. Appl. Cryst. 47 (2014) 1849.
[21] F. Mezei, in: F. Mezei (Ed.), Neutron Spin Echo, Lecture Notes in Physics Series, vol. 128, Springer Verlag, Heidelberg, 1980, pp. 3–26.
[22] S.P. Bayrakci, T. Keller, K. Habicht, B. Keimer, Science 312 (2006) 1926.
[23] F. Mezei, C. Pappas, T. Gutberlet (Eds.), Neutron Spin Echo, Lecture Notes in Physics Series, vol. 601, Springer Verlag, Heidelberg, 2003.
[24] C.M.E. Zeyen, P.C. Rem, Meas. Sci. Technol. 7 (1996) 782–791.
[25] B. Farago, J.Q. Li, G. Cornilescu, D.J.E. Callaway, Z.M. Bu, Biophys. J. 99 (2010) 3473–3482.
[26] B. Farago, M. Gradzielski, J. Chem. Phys 114 (2001) 10105–10122.
[27] M. Adam, B. Farago, P. Schleger, E. Raspaud, D. Lairez, Macromolecules 31 (1998) 9213.
[28] F. Mezei, W. Knaak, B. Farago, Phys. Rev. Lett. 58 (1987) 571–574.
[29] A.R. Wildes, Rev. Scientific Instruments 70 (1999) 4241–4245 and references therein.
[30] G. Ehlers, A.L. Cornelius, T. Fennell, M. Koza, S.T. Bramwell, J.S. Gardner, J. Phys. Condens. Matter. 16 (2004) S635–S642.
[31] G.P. Felcher, S.G.E. te Velthuis, J. Major, H. Dosch, C. Anderson, K. Habicht, T. Keller, Proc. SPIE 4785 (2002) 164.
[32] sesame.
[33] J. Major, A. Vorobiev, A. Rühm, R. Maier, M. Major, M. Mezger, M. Nülle, H. Dosch, G.P. Felcher, P. Falus, T. Keller, R. Pynn, Rev. Sci. Instrum. 80 (2009) 123903.
[34] S. Masalovich, Nucl. Instrum. Methods Phys. Res. A 581 (2007) 791 and references therein.
[35] P. Falus, A. Vorobiev, T. Krist, Physica B 385–386 (2007) 1149.
[36] J. Bryne, Neutrons, Nuclei and Matter, IOP, Bristol, Philadelphia, 1984.
[37] M.Th. Rekveldt, J. de Physique 32 (1971) C579;
A.I. Okorokov, V.V. Runov, V.I. Volkov, A.G. Gukasov, ZhETF 69 (1976) 590 (Soviet. Phys. JETP 42 (1976) 300);
M. Forte, B.R. Heckel, N.F. Ramsey, K. Green, G.L. Greene, J. Byrne, J.M. Pendlebury, Phys. Rev. Lett. 45 (1980) 2088.
[38] J. Major, H. Dosch, G.P. Felcher, K. Habicht, T. Keller, S.G.E. te Velthuis, A. Vorobiev, M. Wahl, Physica B 336 (2003) 8.
[39] H. Dosch, K. Al Usta, A. Lied, W. Drexel, J. Peisl, Rev. Sci. Instrum. 63 (1992) 5533.
[40] M. Nülle, Spin-Echo Resolved Neutron Scattering from Self-Organised Polymer Interfaces (Ph.D. thesis), MPI für Metallforschung, Stuttgart, Germany, 2010, http://www.fkf.mpg.de/647110/05FRM-II-Group.

[41] R.M. Dalgliesh, S. Langridge, J. Plomp, V.O. de Haan, A. Aÿvan Well, Phys. B Condens. Matter. 406 (2011) 2346.
[42] R. Andersson, L.F. van Heijkamp, I.M. de Schepper, W.G. Bouwman, J. Appl. Cryst. 41 (2008) 868.
[43] O. Glatter, O. Kratky, Small angle X-Ray scattering (Academic, London, 1982), in: L. Rayleigh (Ed.), Proc. R. Soc. London, Ser. A, vol. 84, 1911, p. 25.
[44] D.G. de Oteyza, E. Barrena, S. Sellner, J.O. Ossó, H. Dosch, J. Phys. Chem. B 110 (2006) 16618.
[45] D.G. de Oteyza, E. Barrena, M. Ruiz-Oses, I. Silanes, B.P. Doyle, J.E. Ortega, A. Arnau, H. Dosch, Y. Wakayama, J. Phys. Chem. C 112 (2008) 7168.
[46] A. Vorobiev, J. Major, H. Dosch, P. Mueller-Buschbaum, P. Falus, G.P. Felcher, S.G.E. te Velthuis, J. Phys. Chem. B 115 (2011) 5754.
[47] A.J. Parnell, R.M. Dalgliesh, R.A.L. Jones, A.D.F. Dunbar, Appl. Phys. Lett. 102 (2013) 073111.

Chapter 2

Quantum Phase Transitions

Sara Haravifard,[1,2,3,*] Zahra Yamani[4] and Bruce D. Gaulin[5,6,7]

[1]Department of Physics, Duke University, Durham, North Carolina, USA; [2]The James Franck Institute and Department of Physics, The University of Chicago, Chicago, Illinois, USA; [3]Advanced Photon Source, Argonne National Laboratory, Argonne, Illinois, USA; [4]Canadian Neutron Beam Centre, Chalk River Laboratories, Chalk River, Ontario, Canada; [5]Department of Physics and Astronomy, McMaster University, Hamilton, Ontario, Canada; [6]Canadian Institute for Advanced Research, Toronto, Ontario, Canada; [7]Brockhouse Institute for Materials Research, McMaster University, Hamilton, Ontario, Canada
*Corresponding author: E-mail: haravifard@phy.duke.edu

Chapter Outline

2.1 Introduction	44	
2.1.1 Classical Phase Transitions	45	
2.1.2 Continuous Phase Transitions and Critical Behavior	46	
2.1.3 Quantum Critical Scaling	48	
2.1.4 Quantum Critical Point	49	
2.1.5 Quantum Critical Region	50	
2.2 Experimental Techniques	51	
2.2.1 General Principles of Neutron Scattering	51	
2.2.2 Neutron Scattering Cross Sections	52	
2.2.3 Correlation and Scattering Functions	54	
2.2.4 Magnetic Cross Section	55	
2.2.5 Instruments	56	
2.2.5.1 Triple-Axis Spectrometers	57	
2.2.5.2 Time-of-Flight Spectrometers	58	
2.3 Extreme Environmental Conditions	59	
2.3.1 Cryogenics	59	
2.3.1.1 Helium Closed-Cycle Refrigerator	59	
2.3.1.2 Liquid Helium Bath Cryostats	59	
2.3.1.3 Cryogen-Free Systems	60	
2.3.1.4 Helium-3 Sorption System	62	
2.3.1.5 Helium-3/Helium-4 Dilution Refrigerators	63	
2.3.2 High Magnetic Field	64	
2.3.3 High Pressure	68	
2.3.3.1 Hydrostatic Cells (Piston–Cylinder Devices)	69	
2.3.3.2 Large-Volume (Clamped) Cells	69	
2.3.3.3 Opposed Anvil Cells	70	
2.4 Quantum Phase Transitions in Spin Dimer Systems	71	
2.4.1 Spin Dimer Systems	71	
2.4.2 TlCuCl$_3$	73	
2.4.3 Field-Induced QPT in TlCuCl$_3$	73	
2.4.4 Pressure-Induced QPT in TlCuCl$_3$	79	

Experimental Methods in the Physical Sciences, Vol. 48. http://dx.doi.org/10.1016/B978-0-12-802049-4.00002-6
Copyright © 2015 Elsevier Inc. All rights reserved.   43

2.5	Quantum Phase Transitions in		2.6.4 CeT(In$_{1-x}$M$_x$)$_5$	107
	J$_{eff}$ = 1/2 Pyrochlore Magnets	87	2.6.5 CeRhIn$_5$	109
	2.5.1 XY Pyrochlore Magnets	87	2.6.6 CeCoIn$_5$	111
	2.5.2 Er$_2$Ti$_2$O$_7$	88	2.6.7 CeIrIn$_5$	112
	2.5.3 Spin Excitations in Er$_2$Ti$_2$O$_7$	90	2.7 Quantum Phase Transitions in	
	2.5.4 Yb$_2$Ti$_2$O$_7$	93	Itinerant Magnets	112
	2.5.5 Spin Excitations in		2.7.1 Weak Itinerant	
	Yb$_2$Ti$_2$O$_7$	95	Ferromagnets	112
2.6	Quantum Phase Transitions in		2.7.2 MnSi	115
	Heavy Fermions	99	2.7.2.1 Applied Pressure	118
	2.6.1 Heavy Fermions	99	2.7.3 URu$_2$Si$_2$	120
	2.6.1.1 Spin-Density-Wave QC at a		2.7.3.1 Chemical Pressure (Doping)	122
	Conventional QCP	100	2.7.3.2 Applied Pressure	125
	2.6.1.2 Local QC	100	2.8 Quantum Phase Transitions in	
	2.6.2 Cerium-Based Heavy-Fermion System	101	Transverse Field Ising Systems	126
	2.6.3 CeCu$_{6-x}$Au$_x$	102	2.8.1 Transverse Field Quantum Ising Model	126
	2.6.3.1 Chemical Pressure (Doping)	102	2.8.2 CoNb$_2$O$_6$	130
	2.6.3.2 Applied Pressure	105	2.9 Closing Remarks	138
	2.6.3.3 Applied Magnetic Field	106	Acknowledgments	138
			References	139

2.1 INTRODUCTION

The distinguishing characteristic of a phase transition is the sudden change in one or more physical properties. Studies of quantum magnetism consider systems in which the effects of quantum fluctuations, as well as thermal fluctuations, must be considered. Our most familiar conception of a phase transitions is that driven by fluctuations caused by tuning the temperature close to some critical value T_c, known as a classical phase transition (CPT). As the sample temperature is raised above T_c, thermal fluctuations, whose scale is controlled by $k_B T$ where k_B is the Boltzmann constant, destroy the order and drive the system across the phase transition. Conversely, thermal fluctuations decrease as the temperature is dropped and eventually freeze out as $T \to 0$. Quantum fluctuations, on the other hand, continue to live on at zero temperature and thus, under certain conditions, trigger the system to encounter a phase transition between different quantum phases at $T = 0$. This transition at zero temperature is known as a quantum phase transition (QPT), where the underlying quantum fluctuations are determined by Heisenberg's uncertainty principle. The amplitude of these quantum fluctuations can be tuned by varying an external parameter in the Hamiltonian governing the system, such as applied magnetic field, pressure, or doping level of chemical, which eventually results in development of a new preferable lower-energy ground state.

2.1.1 Classical Phase Transitions

The degree of ordering in a system is the outcome of competition between interactions that enforce ordering to minimize internal energy and thermal fluctuations that increase disorder to maximize entropy. For many systems this results in the formation of different macroscopic phases. The transformation of a system from one of its phases into another is referred to as a CPT. A CPT happens when the free energy of the system or one of its derivatives is discontinuous. The transitions between solid, liquid, and gaseous phases of water, or the transition between the ferromagnetic and paramagnetic (PM) phases of magnetic materials at the Curie point, are familiar examples of such transitions. For detailed discussions on phase transitions, see Refs [1,2].

According to the modern classification, phase transitions are divided into two broad categories: first-order or discontinuous transitions and second-order or continuous transitions.

If there is a finite discontinuity in one or more of the first derivatives of the appropriate thermodynamic potential, the transition is first-order or discontinuous. For example, this can be a discontinuity in the magnetization in a magnetic system. Figure 1 illustrates neutron diffraction data collected at the 1/2 (0,1,3) Bragg reflection of $KMnF_3$, exhibiting temperature dependence typical of a first-order phase transition. An additional characteristic of first-order phase transitions is the existence of latent heat. This means that during such a transition the system either absorbs or releases a fixed amount of energy and, since energy cannot be instantaneously transferred between the system and its environment, first-order transitions are associated with "mixed-phase regimes" in which some parts of the system have completed the transition and others have not. A good example is freezing of water during which the temperature remains constant as the heat is transferred into the environment and the system remains in a mixed phase of fluid and solid until the entire structure has transitioned.

The second class of phase transitions is second-order or continuous phase transitions. These are the transitions in which the first derivatives of thermodynamic potentials are continuous but the second derivatives are discontinuous or infinite. Figure 2 shows the temperature-dependent behavior of neutron diffraction measurements of NbO_2 collected at the superlattice reflection (5/4,5/4,3/2), demonstrating neither discontinuity nor any hysteresis in the intensity, thus signifying the second-order nature of the semiconducting-to-metallic phase transition. The key feature of second-order or continuous phase transitions

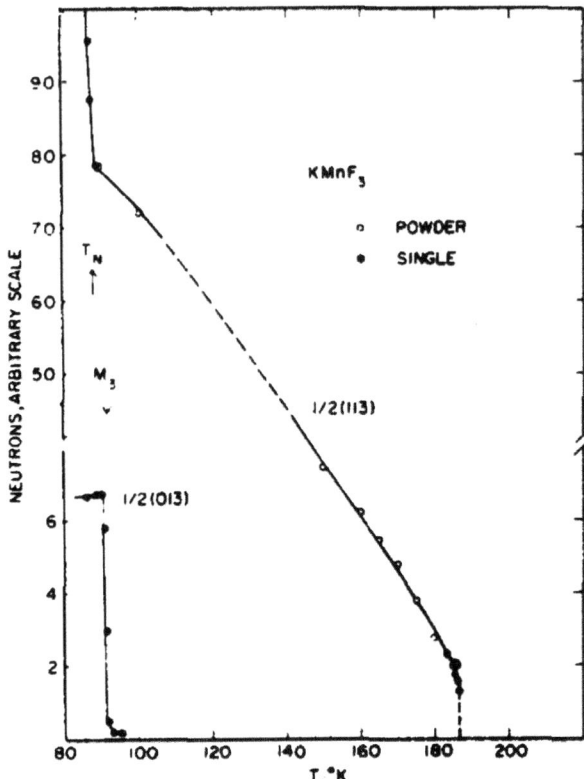

FIGURE 1 The temperature dependence of neutron intensity of the 1/2 (0,1,3) Bragg reflection, displaying a first-order phase transition in KMnF$_3$. *Reprinted with permission from Ref. [3].*

is the existence of fluctuating microregions of both phases near the critical point. The characteristic length of these fluctuating regions diverges at the critical point. For further discussions on neutron scattering techniques used to study the critical phenomena, see Ref. [5].

2.1.2 Continuous Phase Transitions and Critical Behavior

A continuous phase transition can be characterized by an order parameter, a thermodynamic quantity that is zero in one phase (disordered) but has a finite value in another phase (ordered). For instance, in a ferromagnetic phase transition, the total magnetization is an order parameter.

Although the thermodynamic average of an order parameter is zero in the disordered phase, its fluctuations are nonzero. The spatial correlations of the order parameter fluctuations become long-ranged, as the system approaches the critical point at which it goes through the phase transition; eventually the

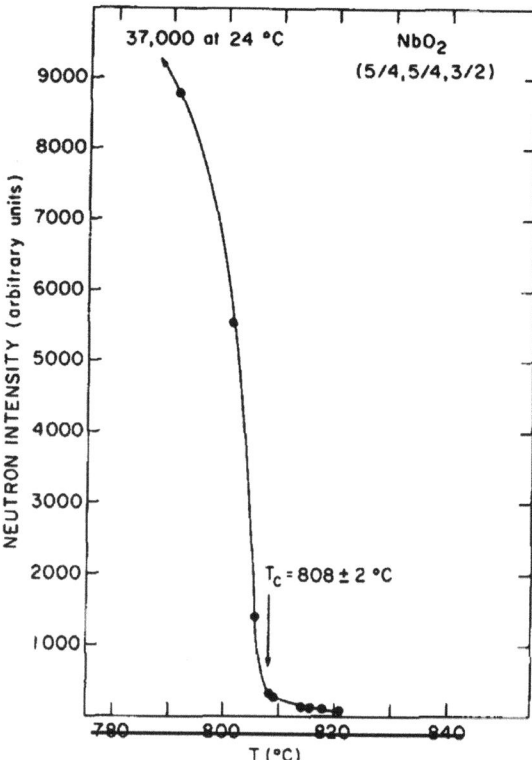

FIGURE 2 Neutron diffraction intensity collected at the superlattice reflection (5/4,5/4,3/2) shows a continuous second-order transition in NbO₂ as it passes from a semiconducting to metallic conductivity phase. *Reprinted with permission from Ref. [4].*

typical length scale of the order parameter fluctuations, known as the correlation length ξ, diverges as:

$$\xi \propto |t|^{-\nu} \tag{1}$$

where t is some dimensionless measure of the distance from the critical point (in the case where the transition occurs at a nonzero temperature T_c, t equals $|T - T_c|/T_c$) and ν is the correlation length critical exponent [6].

Since the correlation length is the only relevant length scale near the critical point, if one rescales all lengths in the system by an arbitrary factor b while simultaneously adjusting the external tuning parameter x such that the correlation length maintains its initial value, the physical properties of a system should remain unchanged. This is known as the homogenous form of the free-energy density f near the critical point:

$$f(t,x) = b^{-d} f\left(tb^{\frac{1}{\nu}}, xb^{y_x}\right) \tag{2}$$

TABLE 1 Commonly Used Critical Exponents [13]

Thermodynamic Quantity	Critical Exponent	Relation
Specific heat	α	$c \propto \|t\|^{\alpha}$
Order parameter	β	$m \propto (-t)^{\beta}$
Susceptibility	γ	$\chi \propto \|t\|^{-\gamma}$
Critical isotherm	δ	$m \propto x^{1/\delta}$
Correlation length	ν	$\xi \propto \|t\|^{-\nu}$
Correlation time	z	$\xi_{\tau} \propto \xi^{z}$

where d is the space dimensionality, y_x is critical exponent associated with the external tuning parameter x, which, for example, could be external magnetic field or pressure (for in-depth discussions, see Refs [7–10]).

Other thermodynamic quantities can also be extracted by differentiating the free energy, giving rise to various critical exponents. Common critical exponents used in magnetism are shown in Table 1.

However, not all exponents are independent from each other [11,12]. The scaling relations shown in Eqn (3) link the four thermodynamic exponents α, β, γ, δ.

$$2 - \alpha = 2\beta + \gamma = \beta(\delta + 1) \tag{3}$$

$$2 - \alpha = d\nu \tag{4}$$

$$\gamma = (2 - \eta)\nu \tag{5}$$

Equations (4) and (5) are referred to as hyperscaling relations. Equation (4), which is also known as Josephson's identity, is only valid for system dimension $d \leq 4$. In continuous QPTs, critical exponents are the same for all classes of phase transitions which may occur in various physical systems based on their system dimension. This remarkable feature is known as universality [9,13].

2.1.3 Quantum Critical Scaling

Continuous phase transitions at $T = 0$ are called QPTs, where thermal fluctuations have vanished and quantum fluctuations demanded by Heisenberg's uncertainty dominate. In continuous phase transitions, depending on the values of the external tuning parameters, the order parameter either is zero or has a finite value and a variation of external parameters can switch the order parameter on or off in the ground state of the system. At finite temperature a phase transition is driven by thermal fluctuations, whereas in QPT at $T=0$ some other nonthermal external parameters such as pressure, magnetic field, or chemical doping tune the quantum fluctuations of the system and thus drive it into the phase transition. In this case the corresponding density operator $e^{-\beta\mathcal{H}}$ resembles a time evolution operator $e^{-i\tau\mathcal{H}}$, where $\tau \equiv -i\beta$ and $\beta = 1/k_B T$, and accordingly shows a fundamental difference between CPT and QPT: in QPT not only the correlation length in the spatial dimension d diverges when $T \to 0$ but also the correlation along the direction of imaginary time $i\tau = 1/k_B T$ extends to infinity as the temperature goes to zero [14]. Thus at zero temperature the imaginary time acts as an additional dimension, and the homogeneity law for the free-energy density can be rewritten as:

$$f(t,x) = b^{-(d+z)} f\left(tb^{\frac{1}{\nu}}, xb^{y_x}\right) \quad (6)$$

where z is referred to as the dynamic critical exponent. In general, z may be different from 1 since the length scales in space can be different from the length scale in the imaginary time direction. Comparing Eqn (6) to the homogeneity law obtained for CPT, it is concluded that all the scaling forms are very similar, considering that a QPT in spatial dimension d is equivalent to a CPT in spatial dimension $d + z$. As a result, the hyperscaling relation for QPT is modified as:

$$2 - \alpha = (d + z)\nu \quad (7)$$

As system temperature approaches $T=0$, thermal fluctuations reduce, and consequently quantum fluctuations become more dominant. Although it is not possible to observe a system at $T=0$, it is conceivable to reach low enough temperatures where quantum fluctuations are more prominent than thermal fluctuations. Similar to Eqn (1), the correlation length along the imaginary time dimension is given as:

$$\xi_\tau \propto |t|^{-\nu z} \quad (8)$$

and thus as T approaches zero and the correlation length along spatial dimensions diverges, so does the correlation length along the imaginary time dimension.

2.1.4 Quantum Critical Point

The quantum critical point (QCP) is the point where the system goes through the QPT at zero temperature. Although the QPT occurs at $T=0$, its effects can be observable in the vicinity of the QCP. The system behavior near QCP is the result of competing thermal fluctuations, as a function of thermal energy scale $k_B T$, and quantum fluctuations, as a function of quantum energy scale $\hbar\omega$. The crossover between classical and quantum behavior occurs when the correlation time reaches $1/k_B T$. Since the correlation time, the shortest time allowed by the quantum mechanics for the system to return to equilibrium after an external arbitrary perturbation, is finite, the crossover temperature is nonzero. This allows for the experimental observation of nonthermally induced QPT behavior at a small but nonzero temperature below the crossover temperature [15].

2.1.5 Quantum Critical Region

In the vicinity of QCP at $x = x_c$, where x_c represents the critical value of the tuning parameter at QPT, at lengths smaller than ξ, the ground state wave function has an entangled critical form, whereas it has a noncritical form at longer lengths. Furthermore, at finite temperatures, the system has an additional characteristic length $\hbar\omega/k_B T$, the so-called de Broglie wavelength of excitations at x_c. For $\xi < \hbar\omega/k_B T$ the ground state wave function has a noncritical form where thermal fluctuations are suppressed. In this case the ground state is thermally excited as temperature rises. For $\xi > \hbar\omega/k_B T$, however, the wave function is in quantum critical region, and thermal fluctuations act directly on the quantum entangled ground state.

There are two types of phase diagrams: with or without a long-range ordered phase at finite temperatures. These phase diagrams are shown in Figure 3 as function of temperature versus tuning parameter x. Upon rising temperature the thermal fluctuations destroy the ordering as the system crosses the phase boundary. As x increases, the quantum fluctuations become more important, and the system will be in a quantum disordered state. If a long-range ordered phase at finite temperature exists (as shown in the left panel of Figure 3), the system will become ordered at low enough temperatures depending on the magnitude of tuning parameter x. At higher temperature the long-range order will be destroyed by thermal fluctuations. By varying the tuning parameters x (magnetic field, pressure, or chemical doping) at low

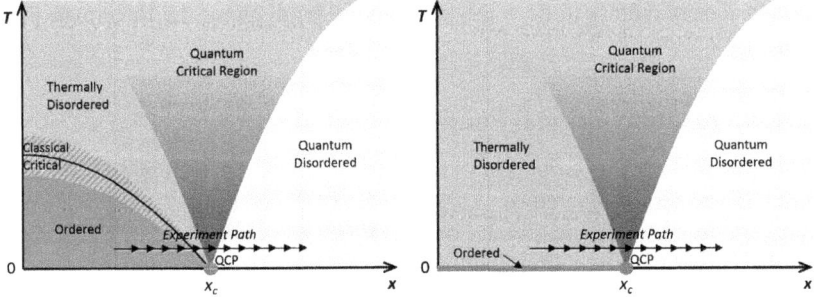

FIGURE 3 Phase diagram for systems with (left) and without (right) a long-range ordered state at finite temperatures. *Adapted from Ref. [12].*

enough temperature, one can observe a crossover from the classical region into the quantum critical region in the quantum disordered phase. If no long-range ordered phase exists in the system at finite temperature as shown in the right panel of Figure 3, one will not see an actual phase transition in the experiment, but instead a sharp crossover as the system passes the quantum critical region, turning more prominent as the temperature approaches zero, and the system gets closer to the QCP.

The fan shape region in the middle of the phase diagrams shows the quantum critical region, where the thermal and quantum fluctuations are of the same order of magnitude and are both important. The region has a characteristic fan shape corresponding to the fact that ξ diverges as $x \to x_c$. Thus surprisingly with increasing temperature, the influence of the quantum criticality (QC) also increases and even expands beyond the QCP at $T=0$. Describing the dynamics of the complex critical state and developing a theory that explains QC has been the subject of intensive theoretical and experimental research. For a comprehensive discussion on QPT, see Ref. [14].

2.2 EXPERIMENTAL TECHNIQUES

Systems exhibiting QPTs are expected to show unusual dynamics controlled by external parameters such as applied magnetic fields, pressure, and chemical doping that tune the quantum fluctuations. The exotic characteristic of these systems can be probed with neutron scattering techniques under extreme environmental conditions. In this section, we briefly review the key aspects of neutron scattering methods typically used to investigate quantum magnetic systems. In the following section, we summarize the major advancements in extreme sample environmental techniques developed for neutron scattering experiments and mainly used to study QPTs. These discussions are intended to assist readers to better navigate through the more involved topics discussed in subsequent sections. An in-depth discussion on the fundamentals of neutron scattering is available in Ref. [16], and those readers who are interested in learning more about the important features of neutron scattering experimental techniques typically used in the field may consult Ref. [17].

2.2.1 General Principles of Neutron Scattering

Neutron scattering is a powerful tool to probe dynamic and static properties of condensed matter at microscopic levels. The energies of cold and thermal neutrons are of the order of microscopic excitations in condensed matter, and the wavelengths of cold and thermal neutrons are comparable to the intermolecular distances. Since neutrons have no electrical charge, there is no Coulomb interaction between them and the nuclei of the sample, and therefore they can easily and deeply penetrate the material.

Neutron scattering events are described by means of energy and momentum transfer. A neutron with incident momentum p_i and incident wave vector k_i has incident energy of:

$$E_i = \frac{p_i^2}{2m} = \frac{\hbar^2 k_i^2}{2m} \qquad (9)$$

After interactions with the sample, the neutron scatters to the direction 2θ, with a momentum p_f, a wave vector k_f, and energy E_f. As with any particle scattering technique, energy and the momentum conservation are the two basic principles of neutron scattering. Based on energy conservation the energy transfer of the incident and scattered neutrons is defined as follows:

$$\hbar\omega = E_f - E_i = \frac{\hbar^2}{2m}\left(k_f^2 - k_i^2\right) \qquad (10)$$

Eventually, momentum conservation makes it possible to define the scattering vector $\mathbf{Q}$:

$$\hbar\mathbf{Q} = \mathbf{p}_f - \mathbf{p}_i = \hbar\mathbf{k}_f - \hbar\mathbf{k}_i \qquad (11)$$

The magnitude of the scattering vector $\mathbf{Q}$ is related to the incident and scattered neutron energies and to the scattering angle 2θ as follows:

$$\mathbf{Q}^2 = \frac{2m}{\hbar^2}\left(E_i + E_f - 2\sqrt{E_i E_f}\cos 2\theta\right) \qquad (12)$$

Both the sign and the magnitude of the energy transfer are used to classify the neutron scattering event.

Neutron scattering is considered elastic when $\Delta\hbar\omega = E_f - E_i = 0$, i.e., the neutrons do not change their energy in the scattering process. If neutrons either gain ($\Delta\hbar\omega > 0$) or lose ($\Delta\hbar\omega < 0$) energy in the scattering process, the scattering is called inelastic.

2.2.2 Neutron Scattering Cross Sections

The quantity measured in a neutron scattering experiment is the double differential cross section, $d^2\sigma/d\Omega dE_f$, which gives the proportion of neutrons with an incident energy E_i scattered into a solid angle element $d\Omega$ with an energy between E_f and $E_f + dE_f$. The geometry of the scattering experiment is shown in Figure 4.

An incident neutron with a wave vector $\mathbf{k_i}$ is scattered into a state with wave vector $\mathbf{k_f}$. Before and after the interaction with the neutron, the sample can be described by the quantum states λ_i and λ_f, respectively. The probability that the combined state of the neutron and the sample makes the transition from the initial state $|\mathbf{k_i}\lambda_i\rangle$ to the final state $|\mathbf{k_f}\lambda_f\rangle$ is given by Fermi's golden rule:

$$\sum_{\mathbf{k_f} \text{ in } d\Omega} W_{\mathbf{k_i},\lambda_i \to \mathbf{k_f},\lambda_f} = \frac{2\pi}{\hbar}\rho_{\mathbf{k_f}}|\langle \mathbf{k_f}\lambda_f|V|\mathbf{k_i}\lambda_i\rangle|^2 \qquad (13)$$

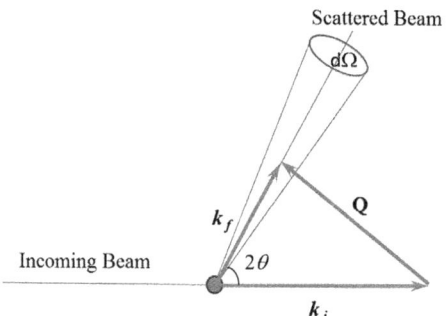

FIGURE 4 The schematic of neutron scattering geometry.

where ρ_{k_f} is the density of the final states of the neutron described by k_f in $d\Omega$ and V is the interaction potential between the nuclei in the sample and the neutron. Using the Born approximation for the cross section, the double differential cross section is given by:

$$\frac{d^2\sigma}{d\Omega dE_f} = \frac{k_f}{k_i} \sum_{\lambda_i,\lambda_f} p_{\lambda_i} |\langle \mathbf{k_f}\lambda_f|V|\mathbf{k_i}\lambda_i\rangle|^2 \delta(\hbar\omega + E_i - E_f) \quad (14)$$

where the δ-function is included to ensure the energy conservation with the neutron energy transfer $\hbar\omega$ and the initial and final sample energies E_i and E_f, respectively. In this equation, p_{λ_i} is the probability that the initial state of the sample is λ_i. This probability is given by Boltzmann distribution:

$$p_{\lambda_i} = \frac{e^{-E_i/k_B T}}{Z}$$

and

$$Z = \sum_{\lambda_i} e^{-E_i/k_B T} \quad (15)$$

k_B is the Boltzmann constant, Z is the partition function of the sample, and T is the sample temperature. After further manipulations, the double differential scattering cross section can be written as:

$$\frac{d^2\sigma}{d\Omega dE_f} = \frac{1}{2\pi\hbar} \frac{k_f}{k_i} \sum_{j,i} \overline{b_j b_i} \int \left\langle e^{-i\mathbf{Q}\cdot\mathbf{R}_i(0)} e^{-i\mathbf{Q}\cdot\mathbf{R}_j(t)} \right\rangle e^{-iEt/\hbar} dt \quad (16)$$

where b_j and b_i are the scattering lengths of the jth and ith nuclei, respectively, $\mathbf{R}_i(0)$ is the position operator of the ith nucleus at time zero, $\mathbf{R}_j(t)$ is the position operator of the jth nucleus at time t, and $\langle\rangle$ denotes a thermal average.

The average $\overline{b_j b_i}$ for cases $j = i$ and $j \neq i$ is given as $\overline{b_j b_i} = \overline{b}^2$ and $\overline{b_j b_i} = \overline{b^2}$, respectively. Accordingly, Eqn (16) can be written as a sum of two components for cases $j = i$ and $j \neq i$, as described below:

$$\frac{d^2\sigma}{d\Omega dE_f} = \frac{1}{2\pi\hbar} \frac{k_f}{k_i} \overline{b}^2 \sum_{j,i}^{j \neq i} \int \left\langle e^{-i\mathbf{Q}\cdot\mathbf{R}_i(0)} e^{-i\mathbf{Q}\cdot\mathbf{R}_j(t)} \right\rangle e^{-iEt/\hbar} dt$$
$$+ \frac{1}{2\pi\hbar} \frac{k_f}{k_i} \overline{b^2} \sum_{j,i}^{j = i} \int \left\langle e^{-i\mathbf{Q}\cdot\mathbf{R}_i(0)} e^{-i\mathbf{Q}\cdot\mathbf{R}_i(t)} \right\rangle e^{-iEt/\hbar} dt \quad (17)$$

This leads to the introduction of *coherent* and *incoherent* scattering, related to the terms of the sum in Eqn (17), respectively. The coherent scattering arises from interference effects and would be the scattering if all the nuclei of any element had the same scattering length b. The incoherent scattering, on the other hand, does not arise from interference effects and is related to the distribution or deviation of scattering length from the mean value $\overline{b}$ and therefore is proportional to $\overline{b^2} - \overline{b}^2$. The coherent and the incoherent scattering cross section into all directions can be defined as follows:

$$\sigma_c = 4\pi \overline{b}^2$$
$$\sigma_i = 4\pi \left(\overline{b^2} - \overline{b}^2 \right) \quad (18)$$

Subsequently, the double differential cross section can be rewritten as:

$$\frac{d^2\sigma}{d\Omega dE} = \frac{1}{8\pi^2\hbar} \frac{k_f}{k_i} \sigma_c \sum_{j,i} \int \left\langle e^{-i\mathbf{Q}\cdot\mathbf{R}_i(0)} e^{-i\mathbf{Q}\cdot\mathbf{R}_j(t)} \right\rangle e^{-iEt/\hbar} dt$$
$$+ \frac{1}{8\pi^2\hbar} \frac{k_f}{k_i} \sigma_i \sum_{j} \int \left\langle e^{-i\mathbf{Q}\cdot\mathbf{R}_j(0)} e^{-i\mathbf{Q}\cdot\mathbf{R}_j(t)} \right\rangle e^{-iEt/\hbar} dt \quad (19)$$
$$= \left(\frac{d^2\sigma}{d\Omega dE} \right)_c + \left(\frac{d^2\sigma}{d\Omega dE} \right)_i$$

The coherent scattering contains information about the correlation between the positions of different nuclei and the collective excitations in a sample. The incoherent scattering component can provide information about the individual nuclei and single-particle excitations in the sample. More detailed discussions are given in Refs [16–19].

2.2.3 Correlation and Scattering Functions

The space–time correlation function describes the position of nuclei in space and time [20] and is given for N nuclei as:

$$G(r,t) = \frac{1}{N} \sum_{i,j}^{N} \langle \delta\{r + \mathbf{R}_i(0) - \mathbf{R}_j(t)\} \rangle \quad (20)$$

The space Fourier transformation of the space–time correlation function $G(r,t)$ is called the intermediate scattering function:

$$I(\mathbf{Q},t) = \int_{-\infty}^{\infty} G(\mathbf{r},t)e^{i\mathbf{Q}\cdot\mathbf{r}}d\mathbf{r} = \frac{1}{N}\sum_{i,j}\left\langle e^{-i\mathbf{Q}\cdot\mathbf{R}_i(0)}e^{-i\mathbf{Q}\cdot\mathbf{R}_j(t)}\right\rangle \quad (21)$$

The time Fourier transform of the intermediate scattering function leads to the scattering function (also called as the dynamic structure factor), which provides information on the sample states as a function of energy and momentum:

$$S(\mathbf{Q},E) = \frac{1}{2\pi\hbar}\int_{-\infty}^{\infty} I(\mathbf{Q},t)e^{-iEt/\hbar}dt \quad (22)$$

Both $I(\mathbf{Q},t)$ and $S(\mathbf{Q},E)$ can be divided into a coherent and an incoherent part. The relationship between the coherent and incoherent scattering functions and the double differential scattering cross section can be written as:

$$\frac{d^2\sigma}{d\Omega dE} = \frac{\sigma_i}{4\pi\hbar}\frac{k_f}{k_i}NS_i(\mathbf{Q},E) + \frac{\sigma_c}{4\pi\hbar}\frac{k_f}{k_i}NS_c(\mathbf{Q},E) \quad (23)$$

2.2.4 Magnetic Cross Section

The interaction between the magnetic moment of a neutron μ_n and the electrons inside the scattering system originates from the Zeeman interaction of the neutron with the magnetic field distribution inside the sample arising from the spin and orbital angular momenta of unpaired electrons. The magnetic moment of a neutron is given by $\mu_n = -\gamma\mu_N\sigma_n$, where $\gamma = 1.913$ is the gyromagnetic ratio of the neutron and $\mu_N = 5.051 \times 10^{-27}$ J T^{-1} is the nuclear magneton. If p denotes the electron momentum operator and R the distance vector measured from this electron, then by introducing the unit vector $\hat{\mathbf{R}} = \mathbf{R}/|\mathbf{R}|$ the total Zeeman interaction with the field produced by this electron can be derived from electromagnetic theory:

$$V_{\text{mag}}(\mathbf{r}) = \frac{\mu_0}{2\pi}\gamma\mu_N\mu_B\sigma_n\cdot\left(\nabla\times\left(\frac{s\times\hat{\mathbf{R}}}{|\mathbf{R}|^2}\right) + \frac{1}{\hbar}\frac{p\times\hat{\mathbf{R}}}{|\mathbf{R}|^2}\right) \quad (24)$$

where $\sigma_n = 2s_n$ is the Pauli spin operator and s is the operator for the electron spin.

The first term originates from the field created by the magnetic moment associated with the electronic spin angular momentum while second term comes from the orbital angular momentum of electronic charges. The motion of these charges may be viewed as current elements and hence contributes to the field distributions as described by the law of Biot and Savart.

Substituting this electromagnetic potential into double partial differential cross section, one derives with the equation for the magnetic scattering differential cross section [20].

$$\left(\frac{d^2\sigma}{d\Omega dE_f}\right)_{mag} = \frac{1}{\hbar}\frac{k_f}{k_i}\left(\frac{\gamma r_0 g}{2}\right)^2 |F(\mathbf{Q})|^2 e^{-2W} \sum_{\alpha,\beta} \left(\delta_{\alpha\beta} - \widehat{Q}_\alpha \widehat{Q}_\beta\right) S^{\alpha\beta}(\mathbf{Q}, E)$$
(25)

The scattering function $S^{\alpha\beta}(\mathbf{Q},E)$ is then given by:

$$S^{\alpha\beta}(\mathbf{Q}, E) = \frac{1}{2\pi} \sum_{\mathbf{R},\mathbf{R}'} e^{-i\mathbf{Q}\cdot(\mathbf{R}-\mathbf{R}')} \int_{-\infty}^{\infty} e^{-iEt/\hbar} \left\langle S_{\mathbf{R}}^\alpha(0) S_{\mathbf{R}'}^\beta(t) \right\rangle dt$$
(26)

A useful property of $S(\mathbf{Q},E)$ is that it is connected to the imaginary part of the generalized magnetic susceptibility $\chi(\mathbf{Q},E) = \chi'(\mathbf{Q},E) + i\chi''(\mathbf{Q},E)$ through the fluctuation-dissipation theorem:

$$S(\mathbf{Q}, E) = [n(E) + 1]\chi''(\mathbf{Q}, E) = \frac{\chi''(\mathbf{Q}, E)}{1 - e^{-E/k_B T}}$$
(27)

in which the exponential term is called the Bose factor. The probability that a neutron gains an energy E is different from the probability that a neutron loses an energy E in the scattering process. This is due to the fact that it is $e^{-E/k_B T}$ times less probable for the system to be in a higher initial state. Hence, the scattering function has to be corrected for this factor when calculating the susceptibility at different temperatures and energy transfers, and this is exactly what the Bose factor does in Eqn (27).

The imaginary part of the susceptibility $\chi''(\mathbf{Q},E)$ is a very important property of the system and describes how the system responds to external forces. For instance, the magnetic susceptibility of a system characterizes the magnetic response of that system to the magnetic field:

$$M(\mathbf{Q}, E) = \chi''(\mathbf{Q}, E) H(\mathbf{Q}, E)$$
(28)

A more detailed discussion on magnetic neutron scattering is given in Ref. [21]. An in-depth discussion on neutron magnetic scattering is available in Chapter 1.

2.2.5 Instruments

In this section we briefly introduce the two main methods used in condensed matter physics and particularly to investigate magnetic systems: triple-axis and time-of-flight spectrometers. More detailed discussion on various neutron scattering instrumentations is given in Ref. [17].

2.2.5.1 Triple-Axis Spectrometers

A triple-axis spectrometer is an extremely simple machine. It is based only on Bragg's law and simple geometry in reciprocal space. In a triple-axis experiment a beam of neutrons traverses a path through the instrument determined by the settings of three angles θ_M, θ_S, and θ_A. Figure 5 shows a simple schematic of triple-axis spectrometer with a monochromator crystal, located in the beam path from source to sample position, and an analyzer crystal, located in the path from sample to detector. Monochromators and analyzers are either perfect crystals or crystals that have been deformed in a controlled manner to obtain certain characteristic properties. Typical monochromator and analyzer materials are pyrolytic graphite, silicon, and germanium.

At the monochromator and analyzer positions, neutrons are reflected according to Bragg's law:

$$n\lambda = 2d \sin \theta_i \qquad (29)$$

where n is an integer and $i = M, A$ for monochromator and analyzer, respectively. The setting of the angle θ_i causes a family of crystal planes, characterized by their distance d, to diffract exactly those neutrons with wavelengths $\lambda = 2\pi/k$ determined by Eqn (29). Hence, a monochromator transforms a polychromatic beam of neutrons to a beam of neutrons with wave numbers k, $2k$, etc. When this beam hits the sample, the scattered neutrons leave the sample along a distribution of directions and with a distribution of energies and spin directions that are determined by the spin-dependent partial differential scattering cross sections.

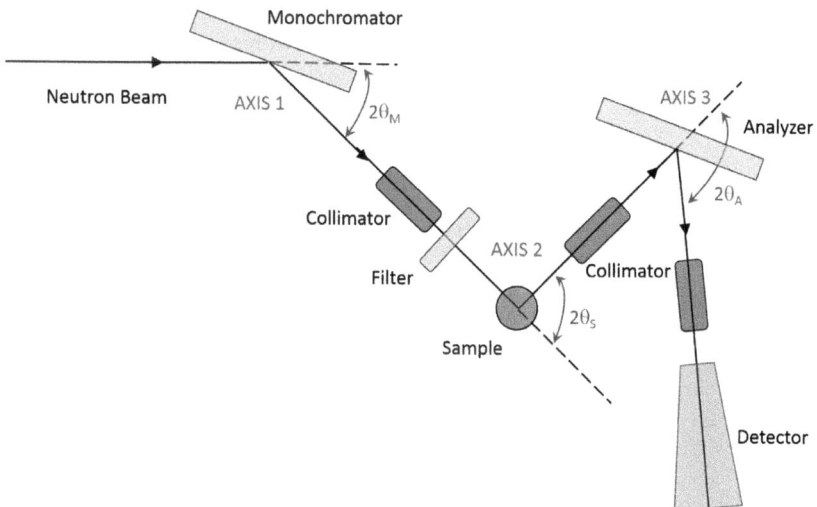

FIGURE 5 Schematic layout of triple-axis spectrometer.

By varying the scattering angle θ_S between $\mathbf{k_i}$ and $\mathbf{k_f}$, the angular distribution can be explored. At the same time, the energy distribution in each given direction can be studied by varying the analyzer angle θ_A. Additional information may be collected by considering not only the change in momentum and energy of the neutron in the scattering process, but also possible changes in its spin state.

2.2.5.2 Time-of-Flight Spectrometers

The time-of-flight method complements the triple-axis spectrometer technique. The triple-axis spectrometer is ideally suited to the study of excitations in oriented samples at specific points in $(\mathbf{Q},E)$ phase space. Time-of-flight instruments, on the other hand, may be used to explore rather large regions of phase space since many detectors simultaneously collect neutrons over a wide range of values of the scattered energy.

Figure 6 illustrates a simple time-of-flight spectrometer. A neutron beam from the reactor is reflected from a monochromator crystal. The monochromatic beam, characterized by its energy E_i and wave vector $\mathbf{k_i}$, is then pulsed by a chopper placed at a known distance L_{CS} from the sample. An array of detectors is arranged at a known fixed distance L_{SD} from the sample, and scattered neutrons arrive at the detectors at times determined by their scattered energies E_f. The time of flight of a neutron from the chopper is given by:

$$t_{CD} = t_{CS} + t_{SD} = \tau_0 L_{CS} + \tau L_{SD} \qquad (30)$$

where t_{CS} and t_{SD} are the times of flight of the neutron from chopper to sample and from sample to detector, respectively; t_0 and t are the reciprocal velocities of the neutron before and after scattering, respectively. If the initial energy E_i is known, then using t_{CD} and the final energy E_f, the energy transfer (i.e., $\hbar\omega = E_i - E_f$) may be determined. Given the angle between the incident and scattered neutron wave vectors, the wave vector transfer (i.e., $\mathbf{Q} = \mathbf{k_i} - \mathbf{k_f}$) can also be calculated.

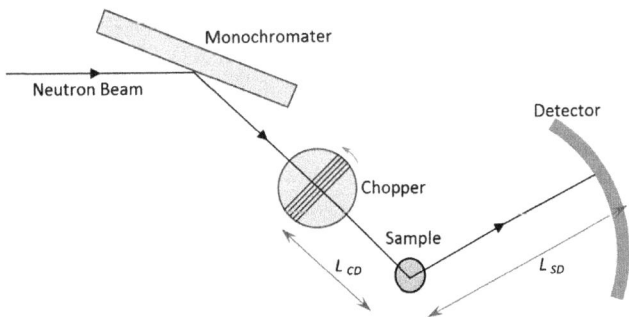

FIGURE 6 Schematic layout of a simple time-of-flight spectrometer.

2.3 EXTREME ENVIRONMENTAL CONDITIONS

Systems exhibiting QPTs are expected to show unusual dynamics controlled by external parameters, such as applied magnetic fields, pressure, and chemical doping, which tune the quantum fluctuations and can be probed with neutron scattering techniques under extreme environmental conditions. The three most common neutron scattering extreme environmental conditions employed to study magnetic systems are cryogenic temperatures, high magnetic fields, and pressure. As discussed in earlier sections, low temperature is an essential requirement to study QPTs, therefore, experiments investigating such phenomena require cryogenic temperatures in combination with high magnetic fields or high pressure to tune the amplitude of quantum fluctuations. A useful review of sample environments is given in Ref. [22].

2.3.1 Cryogenics

As explained earlier a QPT can be investigated near a QCP at low enough temperatures. There are a few different cryogenic systems commonly used to reach such temperatures. Here we present a brief introduction of the low temperature techniques used in neutron scattering experiments; a more in-depth review of low temperature physics is given in Ref. [23].

2.3.1.1 Helium Closed-Cycle Refrigerator

Typical closed-cycle refrigerator (CCR) cryostats (see Figure 7) are either direct cooling or indirect cooling. In direct cooling CCR cryostats, the sample is attached directly to the cold head, which results a very short sample cooling time. On the other hand, in indirect cooling CCR cryostats, commonly known as cryogen-free cryostats, the sample is held in a helium exchange gas chamber. It takes longer time for the sample to cool down in indirect cooling cryostats, but it is much faster to change the sample whereas the direct cooling cryostats require the entire unit to be removed from the vacuum chamber. This also increases the probability of moisture condensation on the cold head or the sample. Recent advancement in the designs, material, and manufacturing of CCR systems has improved the achievable temperature from ~ 10 K to less than 4 K. Extended temperature range CCR systems can achieve base temperatures below 2 K [24].

2.3.1.2 Liquid Helium Bath Cryostats

The liquid helium bath cryostat designed at ILL in 1974 [26], commonly known as orange cryostat (see Figure 8), has been the basis for most cryostats used for neutron scattering experiments globally. Orange cryostats get their popular name from their external color of orange, which has been their exterior paint color since their earliest designs. In this type of cryostat, the sample is attached to the end of a sample stick and is suspended in low-pressure helium

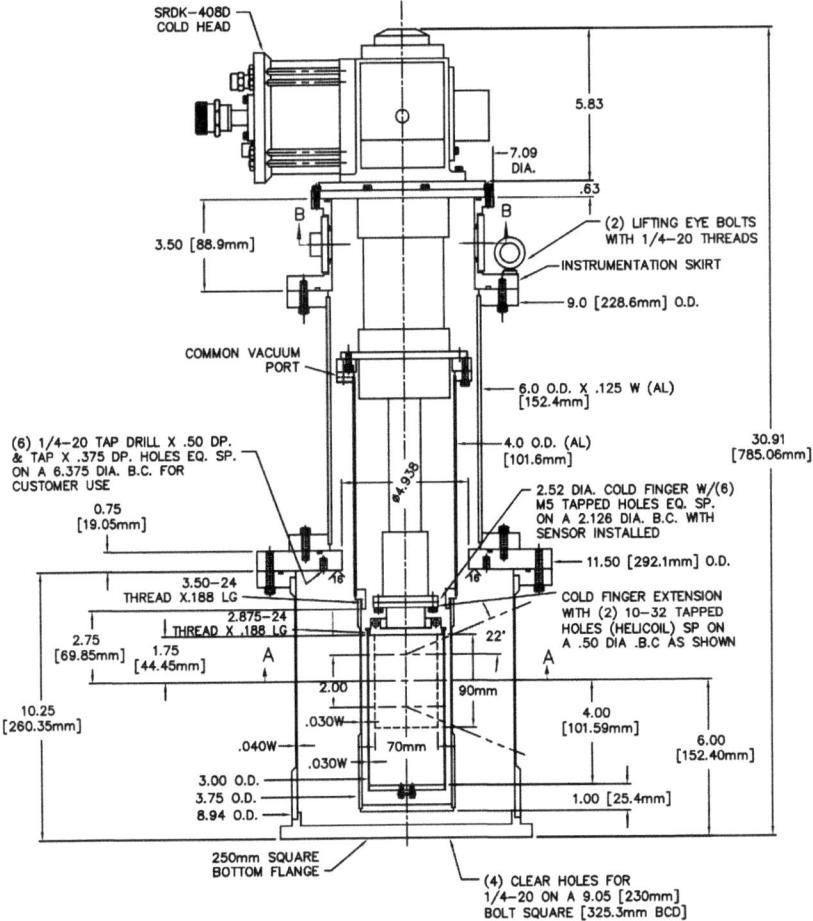

FIGURE 7 Special closed-cycle refrigerator model SHI-4 for neutron scattering by Janis Research Co. used at the NIST Center for Neutron Research for the 4–325 K temperature range. *Reprinted with permission from Ref. [25].*

exchange gas inside the inner vacuum chamber. The sample temperature is controlled by adjusting the heater current and the flow of helium through the helium cold valve. Liquid helium bath cryostats can operate in temperature range between 1.5 and 300 K.

2.3.1.3 Cryogen-Free Systems

The increase in demand for liquid helium along with problems in global supply has raised significant concerns about the availability and affordability of low-temperature experiments with high liquid helium consumption in conventional cryostats mentioned earlier, which work on the basis of

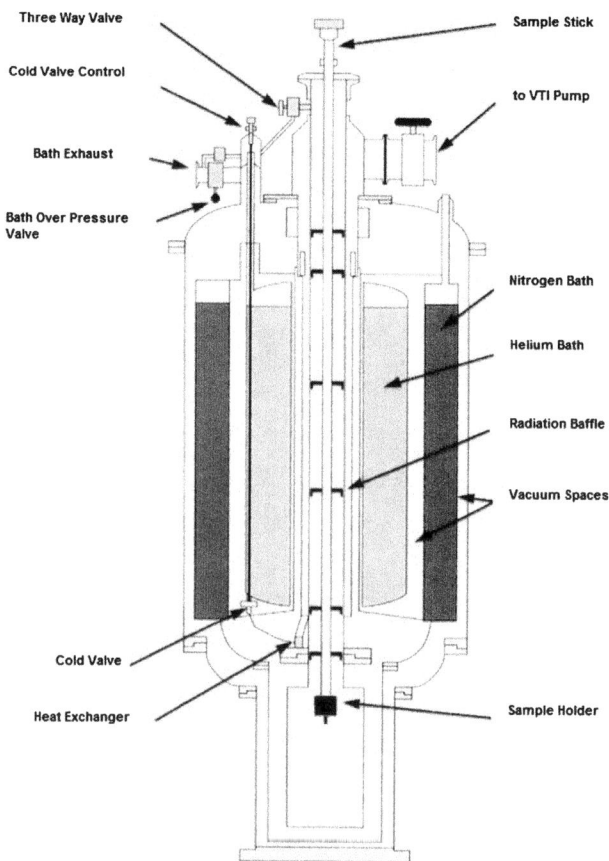

FIGURE 8 Schematic of Orange Cryostat used for neutron scattering experiments. *Courtesy of Oak Ridge National Laboratory, U.S. Department of Energy [27].*

evaporating liquid helium. Currently, there are two options to reduce the consumption of liquid helium. One option is utilizing recondensing systems to recondense the evaporating liquid helium and return it to the cryostat in a closed-cycle; the other option is "dry systems," also known as cryogen-free systems, which do not contain any liquid helium and operate by utilizing the cooling power of the cold head.

The most successful example of cryogen-free systems is the pulse tube refrigerator (PTR). The PTR works on the basis of compression and expansion of helium gas by utilizing oscillating pressure inside the pulse tube (see Figure 9). Figure 10 illustrates schematic of the PTR top-loading system used at ISIS manufactured by AS Scientific Products Ltd [29]. The system consists of (1) the outer vacuum chamber, (2) the top-loading insert, (3) the PTR, (4) the infrared shield, (5) the thermal link between PTR and

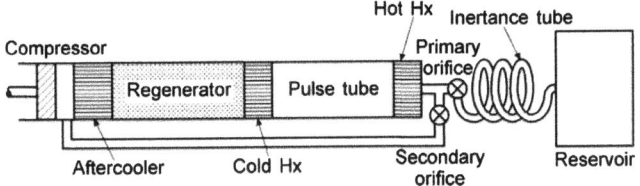

FIGURE 9 Schematic of double-inlet pulse tube refrigerator. *Reprinted with permission from Ref. [28].*

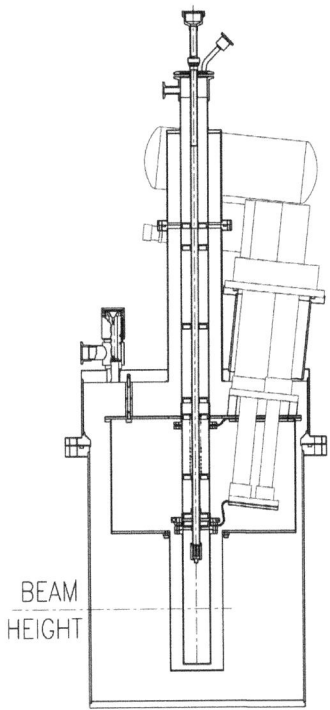

FIGURE 10 Schematic of top-loading pulse tube refrigerator used for neutron scattering. *Reprinted with permission from Ref. [29].*

the insert base flange, and (6) the sample stick. The design provides lower maintenance cost and reduced vibration.

2.3.1.4 Helium-3 Sorption System

Helium-3 sorption systems work on the principle of adsorption, a physical phenomenon in which a gas is trapped on a material surface and retained for a finite period. There are two types of adsorption: chemisorption and physisorption. The former refers to mechanism in which chemical bonds are

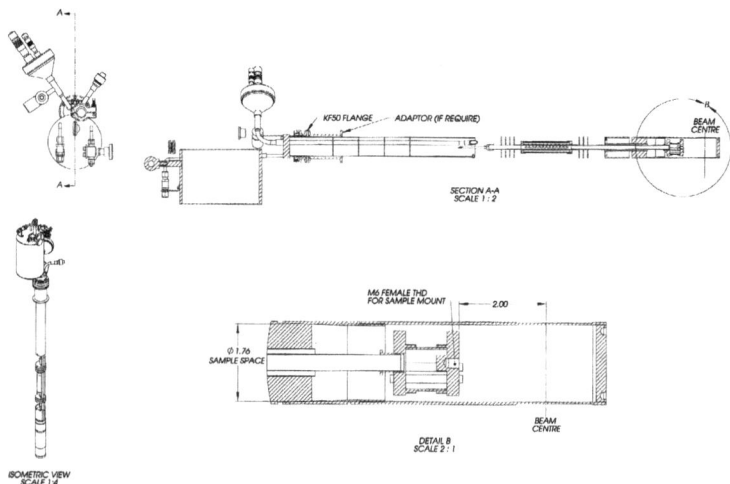

FIGURE 11 Heliox Helium-3 insert by Oxford Instruments used at Oak Ridge National Laboratory. *Courtesy of Oak Ridge National Laboratory, U.S. Department of Energy [30].*

formed by electron transfer; the latter relies on van der Waals' forces. Adsorption is used as the last stage of a cooling system—after a precooling stage such as in a liquid bath cryostat—by creating an evaporating pump and therefore reducing the temperature. He-3 sorption systems can be used for cooling down from 3 K to 300 mK temperatures. Figure 11 shows a helium-3 insert used for neutron scattering experiments at Oak Ridge National Laboratory.

2.3.1.5 Helium-3/Helium-4 Dilution Refrigerators

The helium-3/helium-4 dilution refrigerator insert is the only means of reaching stable base temperatures below 300 mK. ^{3}He-^{4}He dilution refrigerators work on a mixture of ^{3}He-^{4}He based on the idea published by Heinz London in 1951 [31]. However, it was not until 10 years later in 1962 that a dilatation refrigerator system was proposed [32]. In 1965, the first ever refrigerator based on London's idea was built and reached a temperature of 220 mK [33]. Modern dilution refrigerators can cool down to temperatures of about 5 mK with the lowest temperature obtained at ~ 2 mK [34]. In neutron scattering, however, the minimum temperature attainable are about a few tens of millikelvin due to the limitations in use of certain materials for sample cells.

The dilution refrigerator works on the principle of evaporation of liquids. At temperatures below the triple point, the mixture of ^{3}He-^{4}He will separate into two liquid phases separated by a phase boundary as shown in Figure 12 [35]. Inside the dilution refrigerator, a mixture of ^{3}He and ^{4}He is separated into two phases inside the mixing chamber. The mixture condensates into two distinct phases: one only ^{3}He and the other ^{4}He with a small fraction of ^{3}He.

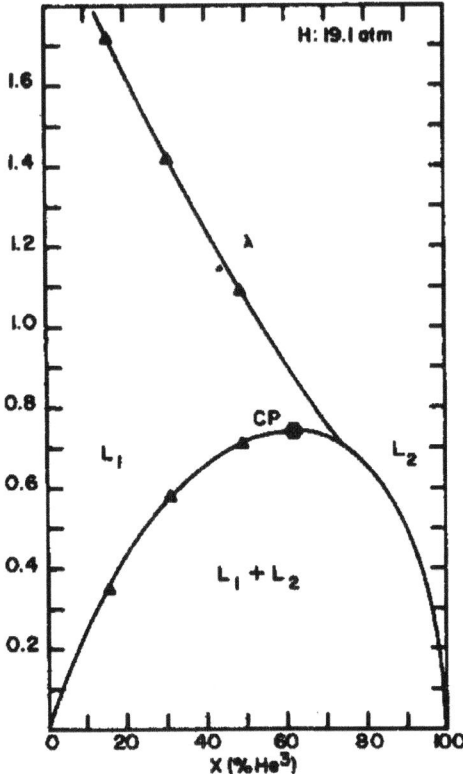

FIGURE 12 $T-X$ phase diagram of ^{3}He-^{4}He at 19.1 atm reported by Tedrow and Lee [35]. L_1 and L_2 represent the ^{4}He-rich and ^{3}He-rich liquid phases, respectively, X denotes the concentration of ^{3}He. *Reprinted with permission from Ref. [35].*

By pumping ^{3}He out of the ^{4}He-rich phase, ^{3}He atoms are forced into the ^{4}He-rich phase through an endothermal process, thus cooling down the system. The atoms lost in the ^{3}He phase are replenished by pumping them back in a closed circulation. Figure 13 presents a drawing of a typical Oxford Instruments dilution refrigerator insert used for neutron scattering experiments at Oak Ridge National Laboratory.

2.3.2 High Magnetic Field

A high magnetic field applied to the sample cooled down to extremely low temperature is an essential tuning parameter in studying QPTs. There are three types of high magnetic fields used in neutron scattering facilities: (1) continuous or steady high magnetic fields, (2) pulsed high magnetic fields, and (3) hybrid high magnetic fields.

Quantum Phase Transitions Chapter | 2   65

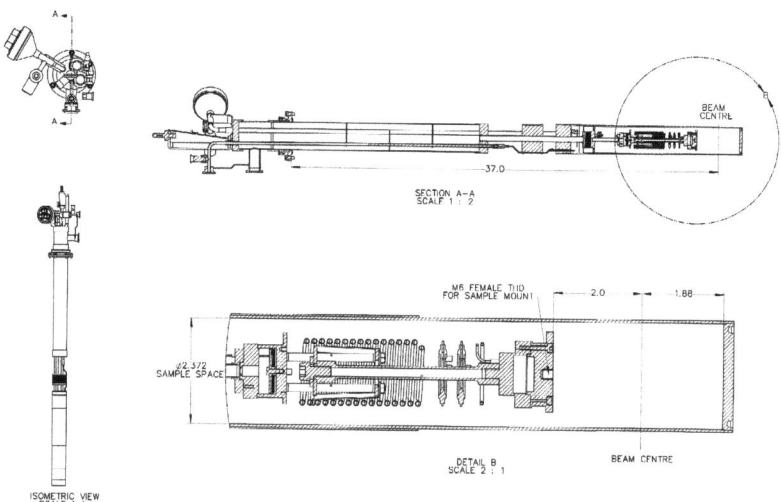

FIGURE 13   Dilution fridge insert designed by Oxford Instruments for Orange Cryostat used in neutron scattering experiments and capable to reach base temperature <50 mK. *Courtesy of Oak Ridge National Laboratory, U.S. Department of Energy [30].*

The most challenging obstacle to overcome is combining the restricted geometries of high magnets and extremely low temperature environments. This is accomplished in superconducting cryomagnets. Since the high magnetic fields of cryomagnet can interact with the magnetic components of the instrument, it is important for the neutron instrument to be made of nonmagnetic material components.

In the conventional design, the high magnetic field is supplied by two superconducting coils placed with a common axis close to each other and with a small gap to allow for the incoming and scattered neutron beams. The magnetic field in this configuration is perpendicular to the incoming neutron beam and is known as vertical field configuration. Figure 14 illustrates a schematic view of a vertical cryomagnetic system designed for neutron scattering experiments. The force generated between the two superconducting coils in this type of cryomagnets will be in the range of tens of tons, and therefore in this design the maximum magnetic field achievable is technically limited [36]. The world record for maximum steady vertical magnetic field for neutron scattering experiments is 17.5 T achieved by Helmholtz Centre Berlin (HZB, formerly Hahn-Meitner Institute) [37,38].

For higher magnetic fields, a one-solenoid configuration is used with its axis in the horizontal direction along the incoming neutron beam. This arrangement is known as horizontal configuration. This configuration significantly limits the range of neutron scattering angles accessible. Tapering the

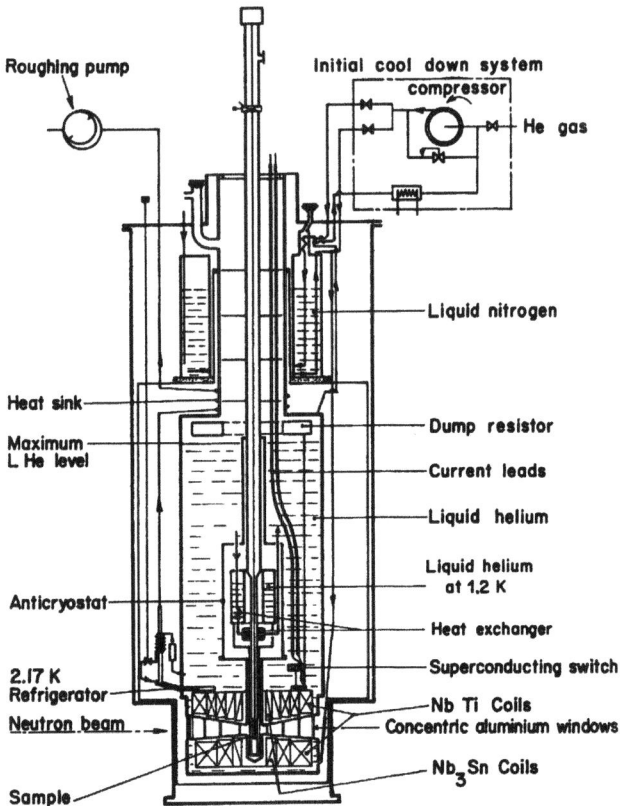

FIGURE 14 Schematic view of a vertical cryomagnetic system designed for neutron scattering experiments with operating temperature range of 50 mK–300 K and maximum continuous magnetic field intensity of 10 T. *Reprinted with permission from Ref. [36].*

openings on both sides of the solenoid axis can achieve scattering angles of about 30° for forward and backward scattered beams. In order to overcome this reduced angular access, a very broad band of neutron wavelengths is used. By using series-connected hybrid magnets, the horizontal configuration allows for steady magnetic fields of above 30 T [39]. Currently, High Magnetic Field Facility at HZB in collaboration with National High Magnetic Field Laboratory in Florida is building a new hybrid magnet for neutron scattering [40]. The series-connected hybrid magnet combines resistive insert coils with an exterior superconducting solenoid enabling it to reach magnetic fields between 26 and 32 T, providing the strongest steady magnetic fields available for neutron scattering experiments worldwide. Figure 15 displays schematic cross section of the hybrid magnet, illustrating the superconducting coil and resistive coil.

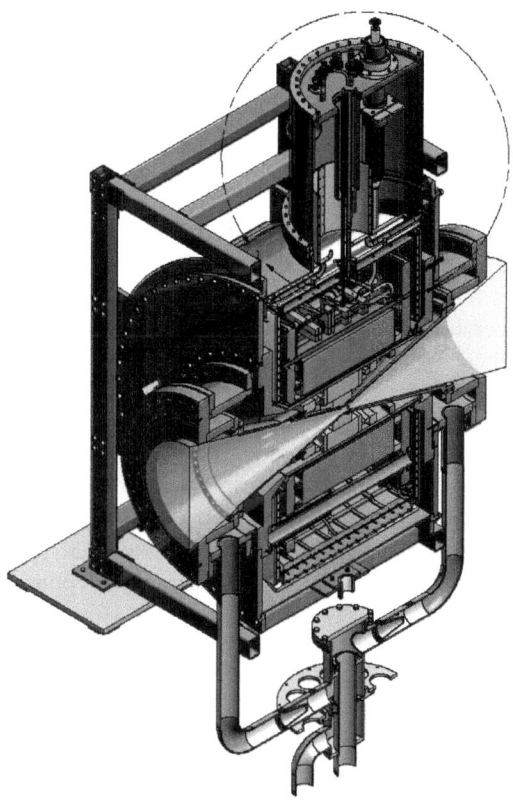

FIGURE 15 Schematic cross section of a series-connected hybrid magnet built by NHMFL for HZB specified for neutron scattering experiments under extreme conditions. *Reprinted with permission from Ref. [40].*

Pulsed resistive magnets can produce higher magnetic fields but only for a fraction of a second. The solenoid-type pulsed magnets consist of a stack of Bitter disks (alternating conductor and insulator disks with a large number of small holes for cooling water, designed by F. Bitter in 1936 [41]) [42]. The neutrons strike the sample parallel to the magnetic field, and the scattered neutrons can be measured in a small angle. The split-type pulsed magnets consist of two coils with a small gap between them. In the split-type pulsed magnets, neutrons enter, perpendicular to the magnetic field, and the scattered neutrons are detected at a single point in the same plane [43]. In comparison with a continuous magnetic field, which allows for all incoming neutrons to be used in the experiment, the pulsed magnets, although capable of reaching higher magnetic fields, allow only for use of a small fraction of the neutrons, which limits their application in inelastic neutron scattering experiments.

2.3.3 High Pressure

Pressure can alter distances between atoms and molecules, which results in a change of atomic and molecular interactions and can therefore be used as a tuning parameter; however, this is very difficult to achieve considering the magnitude of interaction forces between atoms and molecules. Even a very small change in atomic and molecular distances will require application of pressures of up to several gigapascals (GPa). This can be achieved in pressure cells. There are ongoing efforts to make pressure cells capable of reaching higher pressures for neutron scattering.

In low-temperature magnetic neutron scattering experiments, only cells made from limited number of materials can be used that are nonmagnetic, have adequate strength and ductility in extremely low temperatures, and are relatively "transparent" to neutrons (Table 2) [45]. Cells made of materials with lower strength will be larger for the same maximum pressure compared to a stronger material which means they cannot cool down as much and operate at the same extremely low temperatures as the smaller cell. On the other hand, if the cell is made from strong material with low neutron transparency, the scattered neutrons will have a higher noise ratio in comparison with more transparent materials. Similarly, cells made with larger volume, which can hold larger size of sample material, have higher thermal load and are limited in how low they can be cooled down (Table 3).

Currently, the type of high-pressure cells that can be used in neutron scattering experiments at extremely low temperatures are not capable of reaching the same high pressures accessible in X-ray scattering experiments, so considering that X-ray and neutron scattering measurements complement one another, there is a degree of deficiency in high-pressure studies using X-ray scattering techniques. In this section, an introduction to some commonly used techniques in high-pressure neutron scattering experiments is provided. More in-depth review of high-pressure neutron scattering techniques can be found in Ref. [46].

TABLE 2 Mechanical Properties and Neutron Transparency of Materials for Tensile Stress Components at Room Temperature [44]

Material	Tensile Strength (GPa)	Neutron Transparency (%)
A7075-T6	0.505	68
Ti-5Al-2.5Sn ELI	0.628	66
Be-Cu	1.18	7

TABLE 3 Mechanical Properties of Materials for Compressive Stress Components at Room Temperature [44]

Material	Bending Strength (GPa)	Young Modulus (GPa)	Compressive Strength (GPa)
Al_2O_3	0.735	380–400	2.75
ZrO_2	1.57	200	1.85
Si_3N_4	1.08	310	3.5
Diamond	–	910–1250	8.2–16.5

High-pressure cells used in neutron scattering experiments can be divided into the following categories.

2.3.3.1 Hydrostatic Cells (Piston–Cylinder Devices)

Hydrostatic cells work on the basis of pressure from a piston inside a cylinder. Hydrostatic cells made from beryllium copper (Be-Cu) alloy typically have a maximum pressure of less than 1.5 GPa for single-wall cells and 3.0 GPa for double-wall cells. Figure 16 shows a piston–cylinder device.

These high-pressure cells typically utilize compressed helium gas as the pressure medium inside thick cylinders made of neutron-friendly materials such as aluminum alloys or Ti-Zr alloys. Considering He $P-T$ phase diagram, pressure is applied at a temperature in which helium is in gas state and then the temperature of the cell, and the sample is reduced to freeze the helium at the applied pressure. This means every time the pressure needs to be adjusted the cell and the sample must be warmed up to the temperatures above helium-freezing temperature before the helium pressure medium can be repressurized at the new pressure and cooled down again.

2.3.3.2 Large-Volume (Clamped) Cells

Large-volume cells include multianvil cells and toroidal anvil presses. These cells have a large sample volume and therefore can improve signal-to-noise ratio, but because of their sizes they also have a larger thermal load and thus cannot be used in experiments requiring extremely low temperatures.

Clamped cells typically have a barrel-shaped alumina (Al_2O_3) cylindrical core in which there is a small aluminum sample capsule located. The pressure inside the sample capsule is produced by applying external pressure on a pair of tungsten carbide (W-C) pistons inside the cylindrical core (see Figure 17).

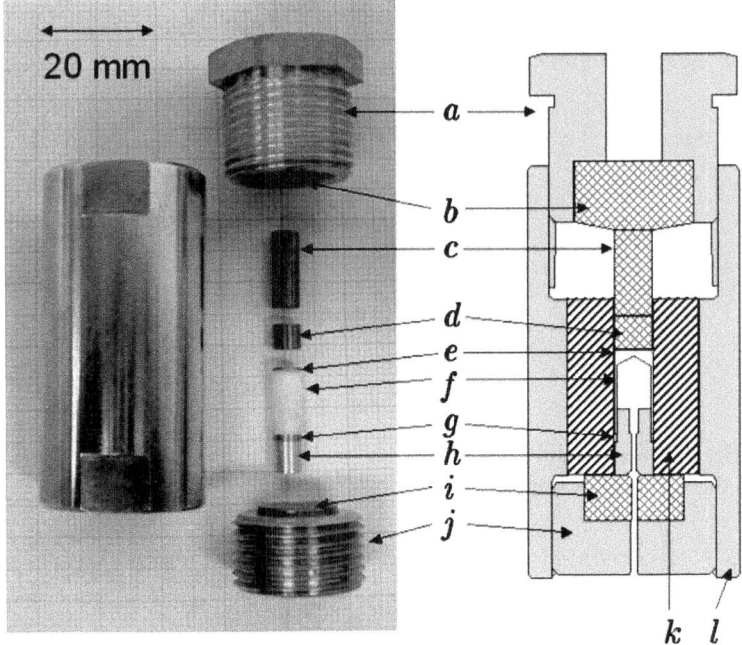

FIGURE 16 Piston—Cylinder Device with schematic cross-section view. (a) Be-Cu locknut; (b) W-C pressure-transmitting pad; (c) W-C piston; (d) W-C short piston; (e) Be-Cu sealing ring; (f) Teflon capsule; (g) Be-Cu sealing ring; (h) Be-Cu flange with electrical feedthrough; (i) W-C pressure retention pad; (j) Be-Cu locknut; (k) Ni-Cr-Al inner cylinder; and (l) Be-Cu outer shell. *Reprinted with permission from Ref. [47].*

2.3.3.3 Opposed Anvil Cells

Opposed anvil cells include diamond anvil cells, sapphire anvil cells, and Paris—Edinburgh presses. In diamond anvil cells, pressure is controlled by a membrane. These cells are compact and therefore have small thermal load, making them suitable for low-temperature experiments. Their small size allows their use in cryostat or dilution refrigerator; however the small size of the cell also means small sample volume which results in weaker signal-to-noise ratio. Figure 18 shows a schematic view of a diamond anvil cell.

Sample sizes required for neutron scattering experiments are significantly larger than those required for X-ray experiments. As a result, large volume cells such as Paris—Edinburgh presses are more commonly used in neutron experiments. These cells can reach pressures approximately 8—10 GPa at temperatures of about 4 K. At higher temperatures, higher pressures—as high as 30 GPa—can be achieved (Table 4).

ISIS is planning a new neutron time-of-flight diffractometer optimized for extreme environment studies of materials, called exceed. It will be providing access to high pressures of above 50 GPa using diamond anvil cells in

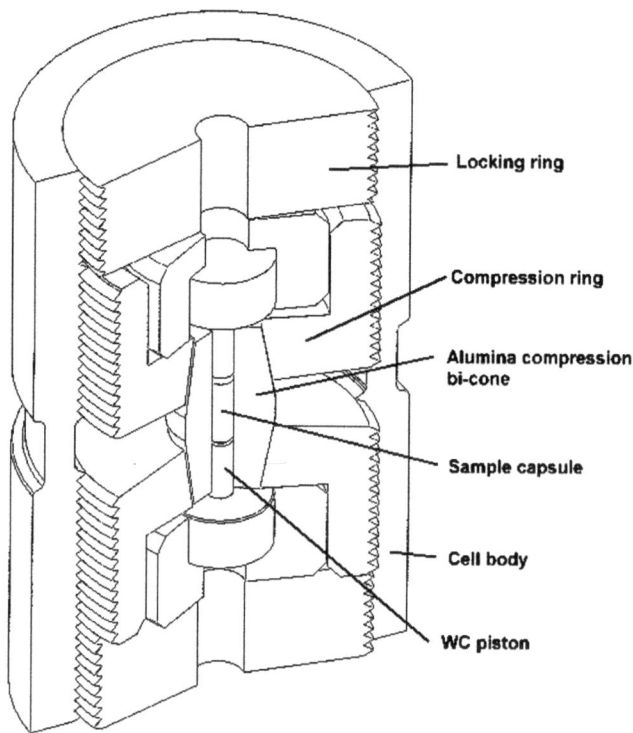

FIGURE 17 Schematic of the 2.5 GPa McWhan Clamped cell. *Reprinted with permission from Ref. [22].*

combination with extremely low temperatures in millikelvin and high pressures in high magnetic fields of up to 10 T [52].

2.4 QUANTUM PHASE TRANSITIONS IN SPIN DIMER SYSTEMS

2.4.1 Spin Dimer Systems

Along with mass and charge, spin is one of the fundamental properties of elementary particles. Interactions between particle spins and the interplay between spin and charge give rise to a rich variety of collective phenomena, making spin materials a fruitful laboratory for studying the onset of ordering, dimensional crossovers, and other questions related to QPTs. Novel collective quantum phenomena were recently observed in magnetic spin dimer systems. Magnetism in these systems consists of spin-1/2 ions where strong antiferromagnetic exchange interactions between pairs of $S = 1/2$ spins lead to a ground state that is a product of singlets with excited triplet states. The comprehensive microscopic characterization of the triplet excitations is used

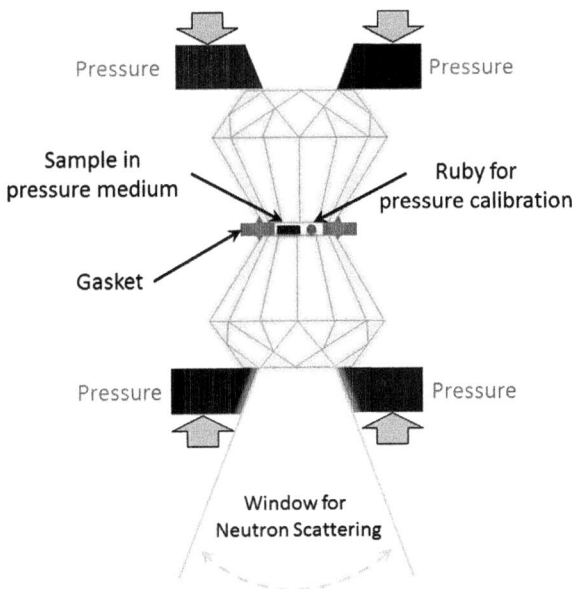

FIGURE 18 Schematic view of the opposed diamond anvil assembly in a Diamond Anvil Cell. Sample is inside the pressure medium encapsulated in a thin metal gasket and is squeezed between the two anvils.

TABLE 4 Maximum Pressure and Sample Volume and the Corresponding Minimum Temperature Achievable at Some User Facilities Globally

Type	Maximum Pressure (GPa)	Corresponding Minimum Temperature (K)	Sample Chamber Volume (mm^3)
Hydrostatic (NIST) [48]	1	1.5	1500
Diamond anvil cell (ISIS) [49]	20	300	45
McWhan clamped (ILL) [50]	2.5	1.5	35
Paris–Edinburgh (ORNL) [51]	10	85	87

as a probe to determine the underlying magnetic interactions and the mechanism behind the unconventional properties and QPTs in dimer systems. Among dynamical experimental techniques, neutron scattering and in particular inelastic neutron scattering offers momentum resolved information on the singlet and triplet states through the transition matrix elements of the spin operator, offering a quantitative privileged testing ground for theoretical predictions. A simple, yet compelling featured article on this subject can be found in Ref. [53]. This section focuses on providing a more detailed insight into model spin dimer system TlCuCl$_3$. However, there are several other similar quantum magnetic systems such as SrCu$_2$(BO$_3$)$_2$ [54–58], Cu(NO$_3$)$_2\cdot$5/2H$_2$O [59,60], PHCC [61,62], Ba$_3$Cr$_2$O$_8$ [63,64] that have inspired extensive theoretical and experimental investigations which may be of interest to readers.

2.4.2 TlCuCl$_3$

Novel collective quantum phenomena in spin dimer systems were recently realized in magnetic isolators ACuCl$_3$ (A = Tl, K), which are based on a crystalline network of dimers formed by two-coupled $S = 1/2$ Cu^{2+} ions. The spins are described by the Hamiltonian shown in Eqn (31) [57].

$$H = \sum_{i<j} J_{ij} \mathbf{S}_i \cdot \mathbf{S}_j \tag{31}$$

where $\mathbf{S}_i$ is the spin-1/2 operator for site i and $J_{ij} \geq 0$ is the antiferromagnetic Heisenberg exchange, characterizing the spin coupling between the Cu^{2+} spins. The dimer ground state is a singlet with total spin $S = 0$, separated by an energy gap from the excited triplet state with total spin $S = 1$.

TlCuCl$_3$ exemplifies QPTs in dimer antiferromagnets. Monoclinic TlCuCl$_3$ crystallizes in the space group P2$_1$/c, containing planer dimers of Cu$_2$Cl$_6$ in which Cu^{2+} ions carry spin-1/2 and are stacked on top of one another to form infinite double chains parallel to the crystallographic axis a and located at the corners and center of the unit cell in the $b-c$. The Cu-Cu dimers are separated by Ti$^+$ ions [65,66]. Figure 19 demonstrates the crystal structure of TlCuCl$_3$. Neutron scattering experiments performed on TlCuCl$_3$ verified that the QPT in the dimer systems could be induced by applied magnetic field and hydrostatic pressure. Below we review neutron scattering experiments performed on TlCuCl$_3$, confirming the field- and pressure-induced QPTs in this system.

2.4.3 Field-Induced QPT in TlCuCl$_3$

In $S = 1/2$ spin dimer system, TlCuCl$_3$ has a nonmagnetic singlet ground state with excitation gap $\Delta/k_B \sim 7.5$ K [65]. The dynamical spin properties of TlCuCl$_3$ were extensively studied by inelastic neutron scattering [67], showing well-defined triplet waves and confirming the dimer origin of spin excitations

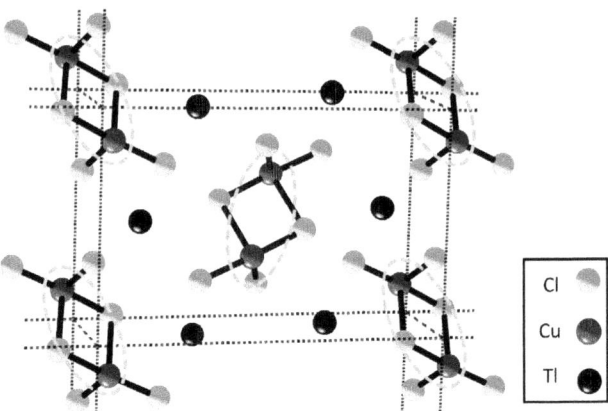

FIGURE 19 Schematic view of TlCuCl$_3$ crystal structure. Dashed ellipsis marks the Cu-Cu dimers.

in this system. The underlying spin-coupling exchange interactions were identified from the observed scattered neutron energy dispersion, which is reported to be dispersive in all directions of the reciprocal space, ratifying the three-dimensional (3D) nature of the correlations. Figure 20 illustrates the typical results of the inelastic neutron profiles of TlCuCl$_3$ at zero magnetic field and as function of temperature. The observed wave vector dependence (upper panel) and the temperature renormalization (middle and lower panels) dependence confirm the singlet−triplet nature of the magnetic excitations. The energy dispersion of the excitations along the main directions of the reciprocal lattice was extracted from the inelastic neutron scattering and is summarized in Figure 21.

We now turn to magnetic properties of spin dimer systems in an applied magnetic field. At zero external magnetic field, pairs of spin-1/2 ions are antiferromagnetically coupled, and the ground state of these materials is simply a product of singlets with excited $S = 1$ ($S_z = +1, 0, -1$) triplets and separated by a finite energy gap Δ. Application of a magnetic field does not alter the nonmagnetic singlet state but causes the Zeeman splitting of the three exited triplet states, which results in linear reduction of the gap energy to the lowest $S_z = +1$ branch. The gap ultimately vanishes at the critical magnetic field $H_c = \Delta/(g\mu_B)$ where g is the effective gyromagnetic factor of the electron spin and μ_B is the Bohr magneton (see Figure 22). At the critical field H_c, each dimer unit is preferably in a renormalized triplet state $S_z = +1$ with a finite net magnetic moment. For QPTs exhibiting 3D nature, the interdimer magnetic exchange interactions provide the 3D coupling, and thus at fields above H_c magnetic long-range ordering is expected. This QPT is characterized by Bose−Einstein condensation (BEC) of the triplet excitations into the singlet ground state by mapping triplet states to a dilute Bose gas [68−70].

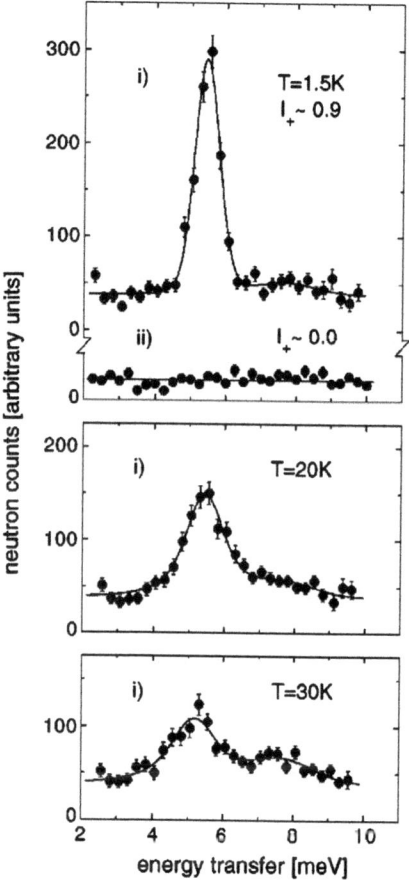

FIGURE 20 Inelastic neutron scattering profiles of TlCuCl$_3$ measured at (i) (1.35,0,0) and (ii) (0,0,3.15), as functions of temperature. *Reprinted with permission from Ref. [67].*

Similarly in TlCuCl$_3$ application of magnetic field results in the Zeeman splitting of the triplet modes which reduces the gap energy. At $H_c \sim 6$ T the lowest triplet branch $S_z = +1$ energetically falls below the value of the nonmagnetic singlet ground state, resulting in long-range ordering of triplets at H_c. The intrinsic parameters of TlCuCl$_3$ make it possible to access the critical region microscopically, and dynamic probes like neutron scattering provide valuable tools to study the QPTs in this system. Neutron scattering measurements performed on TlCuCl$_3$ confirmed the field-induced QPT and the 3D magnetic ordering at $H > H_c$, consistent with BEC of the triplet states [71–74].

Figure 23 demonstrates the summary of inelastic neutron scattering (INS) data collected for TlCuCl$_3$ (squares) at well-resolved reciprocal lattice points and

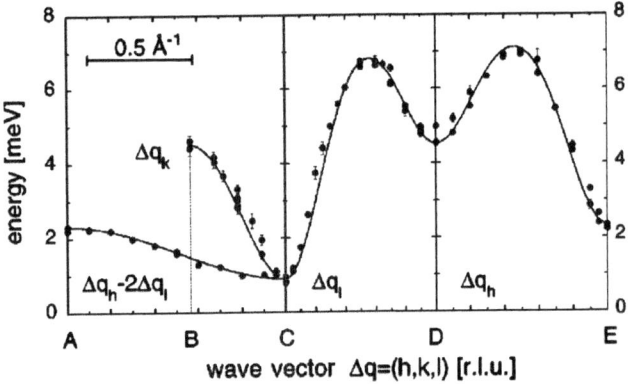

FIGURE 21 Reported energy dispersion of the magnetic excitation modes in TlCuCl$_3$ at $T = 1.5$ K. The results were extracted from inelastic neutron scattering data collected at the relevant directions of reciprocal space and arranged in a reduced scheme representation. Lines are fits to the three-dimensional dimer network corresponding to a Heisenberg model in the strong coupling limit. *Reprinted with permission from Ref. [67].*

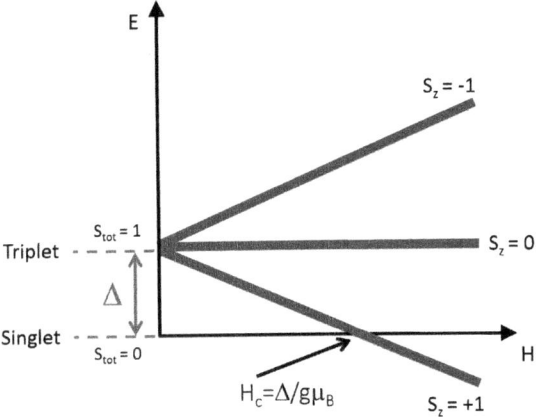

FIGURE 22 Energy spectrum of a spin dimer system, composed of pairs of $S = 1/2$ magnetic ions in an external magnetic field. In zero magnetic field, a finite energy gap Δ separates singlet and the excited triplet states $S_z = +1, 0, -1$. Application of external magnetic field separates the three triplet states due to Zeeman effect such that $S_z = +1$ band crosses the nonmagnetic singlet ground state at $H_c = \Delta/(g\mu_B)$, closing the excitation gap and prompting the formation of magnetic long-range ordering. This quantum phase transition is described by Bose–Einstein condensation of the triplet excitations.

at $H < H_c$ up to 5.5 T [71]. The results verify that external magnetic field does not affect the energy of $S_z = 0$ mode whereas the energy of the $S_z = \pm 1$ modes follows the Zeeman term.

As the triplets undergo BEC at ordering wave vector $\mathbf{Q}_0$, the transverse spin components also form long-range order at the same wave vector. The

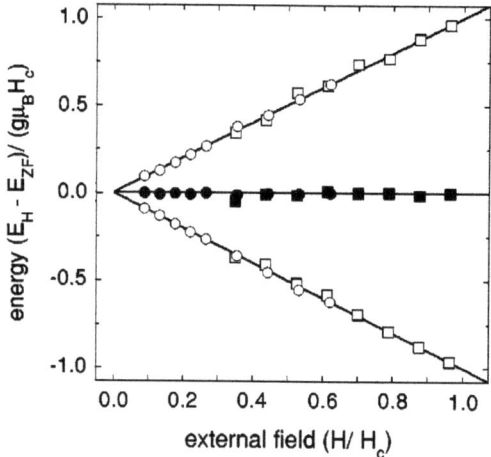

FIGURE 23 Summary of the progressive Zeeman splitting of the triplet modes reported for KCuCl₃ (circles) and TlCuCl₃ (squares) at $T = 1.5$ K and up to 14 and 5.5 T, respectively. *Reprinted with permission from Ref. [71].*

transverse magnetization per site, $m_\perp$, is expressed by $m_\perp = g\mu_B \sqrt{n_c/2}$ where n_c is the condensate density, and therefore the magnetic Bragg peak intensity at Q_0 is proportional to the triplet condensate density n_c. Elastic neutron scattering experiments performed on TlCuCl₃ confirmed the appearance of magnetic Bragg reflections corresponding to the transverse spin ordered state, in agreement with the long-range ordering of the triplets in the BEC state [72]. Figure 24 shows magnetic Bragg reflections observed at $H > H_c$ at 12 T and at 1.9 K. The spin structure shown in Figure 25 is extracted from the nuclear and magnetic Bragg peaks collected at multiple reflections at a field above H_c; the shaded area represents the chemical unit cell. The results are consistent with the long-range field-induced transverse Néel ordering. The phase transition temperature $T_N(H)$ and field $H_N(T)$ are summarized in Figure 26, where the phase boundary can be expressed by the power law as predicted by the BEC theory in quantum magnets, underlying the 3D nature of the observed QPT.

The BEC of triplet excitations is also used to describe the dynamic properties of the magnetic excitations at $H > H_c$. A gapless linear mode, corresponding to the Goldstone mode of the ordered phase, is expected to emerge above the quantum critical phase. The spin dynamics of TlCuCl₃ at fields above H_c was investigated using INS. Three samples of the neutron energy scan data measured at $H > H_c$ phase are presented in Figure 27 [73]. Two sharp transitions are observed in the inelastic neutron profiles up to $H = 12$ T, above which an additional low-lying signal emerges from the edge of the elastic channel (marked by arrow).

78 Neutron Scattering - Magnetic and Quantum Phenomena

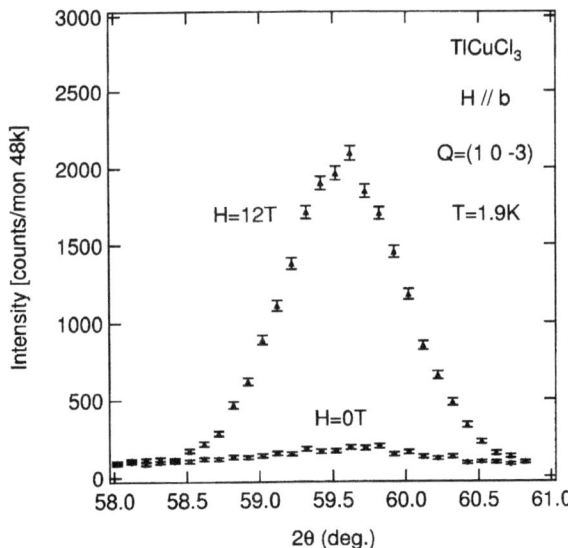

FIGURE 24 $\theta-2\theta$ scans for $\mathbf{Q} = (1,0,-3)$ magnetic Bragg peak, equivalent to those for the lowest magnetic excitation at zero field, measured at $H = 0$ and 12 T at 1.9 K. *Reprinted with permission from Ref. [72].*

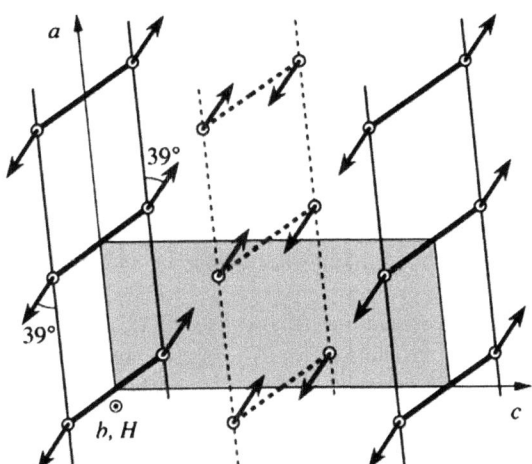

FIGURE 25 Spin structure in the magnetic field-induced ordered phase of TlCuCl$_3$. The external field is applied along the b-axis. The double chains located at the corner and the center of the chemical unit cell in the $b-c$ plane are represented by solid and dashed lines, respectively. The shaded area is the chemical unit cell in the $a-c$ plane. *Reprinted with permission from Ref. [72].*

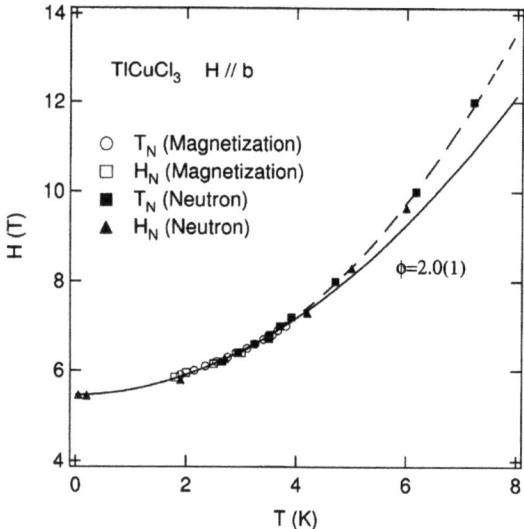

FIGURE 26 The phase boundary in TlCuCl$_3$ determined from the results of temperature (closed rectangles) and field (closed circles) scans of the magnetic Bragg peak intensity. Open circles and rectangles denote the transition points determined from the previous magnetization measurements. The dashed line is a guide for the eyes. The solid line denotes power-law fit as predicted by the magnon Bose–Einstein condensation theory. *Reprinted with permission from Ref. [72].*

Figure 28 presents a summary of the evolution of the magnetic excitations for the Zeeman ($H < H_c$) and the high field ($H > H_c$) regimes extracted from convoluted resolution-limited fits to the collected INS data shown in Figure 27. At $H > H_c$ the excitation spectrum exhibits a dramatic change in nature: $S_z = +1$ Zeeman is absent at finite energies as it remains gapless, while the $S_z = 0$ and $S_z = -1$ Zeeman modes are characteristically renormalized. The magnetic neutron spectra measured under the same experimental conditions at $H = 0$ (spin singlet state) and $H = 14$ T (field-induced ordered state) are compared in Figure 29. The coexistence of the lower- and higher-lying excitations is detected at the field-induced ordered state, illustrating a Goldstone mode with soundlike dispersion. The results are explained within the framework of the BEC of triplet excitations and in agreement with a system with a spontaneously broken XY symmetry in a plane perpendicular to the applied field, verifying the anticipated field-driven QPT.

2.4.4 Pressure-Induced QPT in TlCuCl$_3$

The application of pressure is another method of controlling the quantum magnetism in spin dimer systems. In this section we review the spin dynamics of TlCuCl$_3$ across the pressure-induced QPT, investigated by means of elastic and inelastic neutron scattering. First, we only consider the changes in the

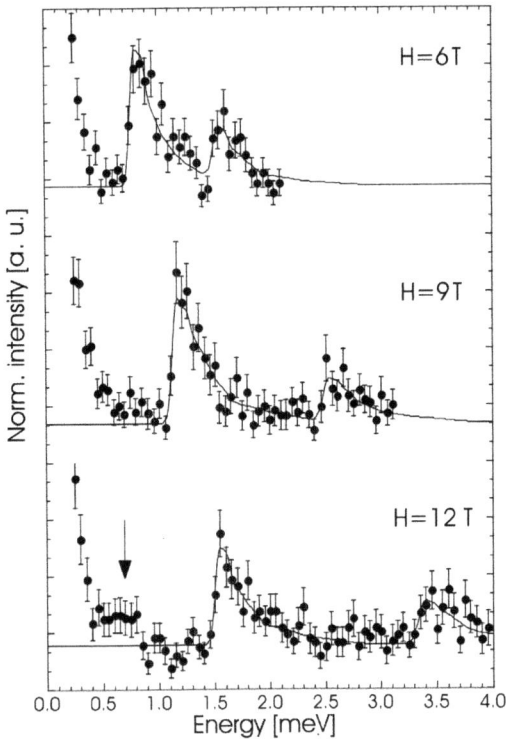

FIGURE 27 Observed spectra in TlCuCl$_3$ collected at $T = 1.5$ K and at $\mathbf{Q} = (0,4,0)$. Arrow marks a weak signal that emerges at $H = 12$ T at the edge of the elastic line. *Reprinted with permission from Ref. [73].*

ground states of the electrons at zero temperature, and later we briefly discuss the thermal excitations, QC, and the $T > 0$ phase diagram.

The relative strength of the exchange interactions defines the nature of the ground state. A key feature of TlCuCl$_3$ is that it has a dominant antiferromagnetic (AF) exchange coupling ($J > 0$), which connects the two Cu^{2+} $S = 1/2$ moments. As a result the spins pair into dimers and form singlet bonds $(|\uparrow\downarrow\rangle - |\downarrow\uparrow\rangle)/\sqrt{2}$, giving rise to a nonmagnetic quantum paramagnet ground state. Figure 30 represents a simple model of a dimer AF across the phase diagram as function of tuning parameter x. The ellipses shown in Figure 30 represent the singlet dimers at ambient pressure; the solid and dashed lines denote the intradimer (J) and the weaker interdimer (J') exchange interactions, respectively, where $J' = J/x$ with $x \geq 1$. The ground state of TlCuCl$_3$ at ambient pressure can be recognized in terms of dimer AF at $x = \infty$ regime, where the ground state of each dimer is a rotationally invariant singlet, and the ground state of the full system is a product over such singlet valence bonds. At $x = 1$, the ground state has AF Néel ordering with the spins polarized in a

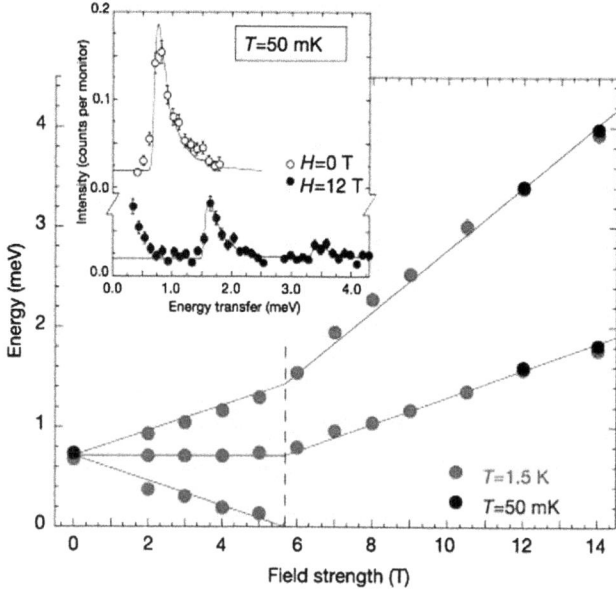

FIGURE 28 Magnetic field dependence of the excitation energies measured in TlCuCl$_3$ at the Bragg point $Q = (0,4,0)$, and fixed $T = 1.5$ K (red (gray in print versions) symbols) and $T = 50$ mK (blue (black in print versions) symbols). The data points are extracted from the least-squares fits to the neutron scattering spectra (inset), curves reflect a Zeeman model. The quantum critical field $H = H_c$ is denoted by the dashed boundary. *Reprinted with permission from Ref. [74].*

staggered spatial pattern, as illustrated in Figure 30. In this case, individual spins have definite orientations, and thus the symmetry of the spin rotations has been broken. The spin-rotation symmetry is restored at $x = x_c$.

The exchange pathways defined by the interatomic bond lengths and angles determine the magnetic interactions between the spin moments. On the other hand, the ground state of a spin system strongly depends on the relative sign and strength of these exchange interactions. For TlCuCl$_3$ applied pressure alters these interatomic pathways and thus inversely tunes x, with a continuous QPT to AF Néel ordering anticipated at x_c. Magnetization as well as neutron diffraction measurements confirmed the pressure-induced long-range AF ordering in TlCuCl$_3$ [75,76]. Figure 31 shows neutron diffraction θ−2θ scans for $P = 1.48$ GPa collected at various temperatures in TlCuCl$_3$, which clearly illustrate the increase in magnetic Bragg reflection intensity with decreasing temperature. The magnetic Bragg reflections were observed at reciprocal points equivalent to those with lowest magnetic excitation energy at ambient pressure—i.e., at $Q = (h,0,l)$ with integer h and odd l. Since the relative ratio of J and J' governs the singlet−triplet excitation gap, it can be deduced that application of pressure enhances J' relative to J, which leads to reduction of

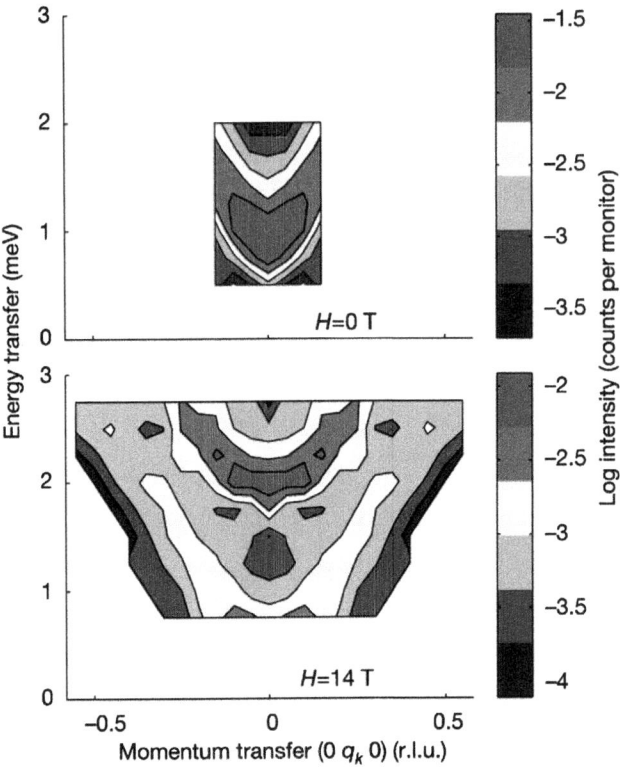

FIGURE 29 Contour plot of the magnetic inelastic neutron scattering intensity measured in TlCuCl$_3$. Goldstone mode with linear dispersion around $\mathbf{K}_0$ (corresponding to $q_k = 0$) is detected. *Reprinted with permission from Ref. [74].*

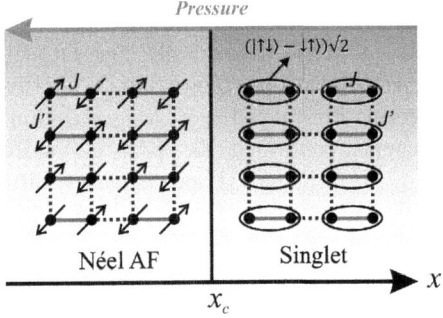

FIGURE 30 Schematic representation of the phase diagram of a simple AF spin dimer system. TlCuCl$_3$ demonstrates similar quantum phase transition with pressure acting as inverse tuning parameter. *Adapted from Ref. [53].*

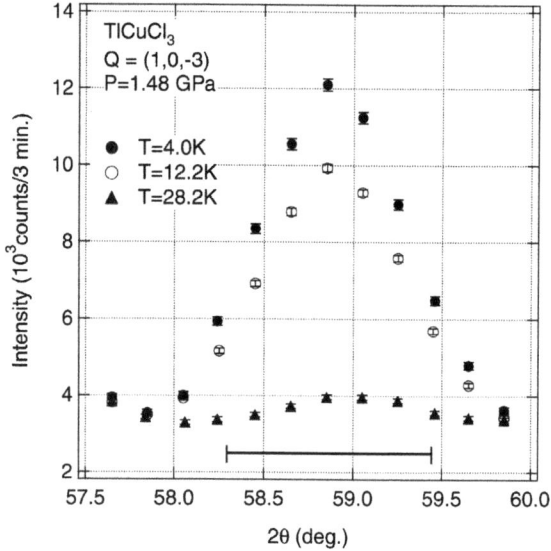

FIGURE 31 θ–2θ scans for the $Q = (1,0,-3)$ reflection measured as function temperature at $P = 1.48$ GPa in TlCuCl$_3$. Reprinted with permission from Ref. [76].

the spin gap corresponding to the lowest excitation energy and eventually to its complete collapse, which then enables the long-range magnetic ordering.

Applied pressure eventually closes the singlet–triplet excitation gap, and at $P = P_c$ the triplet ($S = 1$) components $S_z = +1, 0, -1$ can condense into the singlet ($S = 0$) ground state, whereas in the field-induced QPT (see Section 2.4.3) the degeneracy of the triplet state is lifted at $H = H_c$, and only the lowest energy branch of the triplet states ($S_z = +1$) closes its distinct spin gap and condenses into the singlet state. Considering that the ground state of the pressure-induced Néel phase is the staggered spin configuration (see Figure 30), it is anticipated to detect two classes of excitations: two low-lying transverse (T) spin-wave excitation modes of a conventional well-ordered magnet, resulted from the slow rotation of spin orientations in space (Goldstone), as well as one longitudinal (L) fluctuation mode of the weakly ordered moments (Higgs boson). Furthermore, application of pressure tunes the strength of the interdimer coupling (J) which facilitates the delocalization of triplets, enabling them to hop to a neighboring dimer site, which then leads to dispersion of these excitations.

The excitations of TlCuCl$_3$ have been characterized in detail by elastic and inelastic neutron scattering experiments with continuous pressure control through the QPT. Elastic neutron diffraction measurements were performed to observe the long-range magnetic ordering at $P > P_c$, and the pressure dependence of Néel temperature was extracted from the representative temperature dependence of the identified magnetic Bragg peaks. The pressure

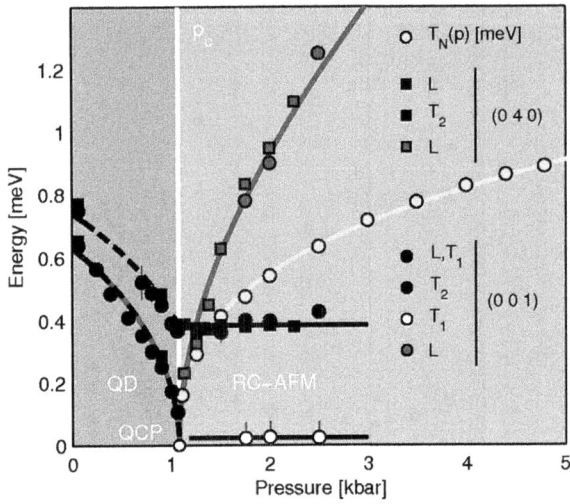

FIGURE 32 Summary of INS results for the gaps of all three triplet excitations as functions of pressure at $T = 1.85$ K. The blue (gray in print versions) region represents the spin singlet phase (Quantum Disordered, QD), while the red (light gray in print versions) region represents the pressure-induced long-range ordered phase (Renormalized Classical Spin Wave AF, RC-AFM). Modes L and T_1 are degenerate within experimental resolution at $P < P_c$. Red (dark gray in print versions) symbols show the longitudinal mode L at $P > P_c$. Solid and dashed lines are theoretical power-law fits. Data for $T_N(P)$ are reprinted from Ref. [77]. QCP, quantum critical point. *Reprinted with permission from Ref. [78].*

dependence of the singlet–triplet gap energy $\Delta(P)$ has been measured using INS. Figure 32 shows a summary of neutron scattering results measured across the QPT as function of pressure at $T = 1.85$ K [78]. Green symbols denote $T_N(P)$ extracted from neutron diffraction data [77], while L and T modes are extracted from INS results collected at $\mathbf{Q} = (0,0,1)$ and $(0,4,0)$, in order to give access to all three spatial direction and thus cover both modes for $P > P_c$ Néel ordered phase [78]. A combined power-law fit of the elastic and inelastic results to $T_N \propto (P - P_c)^\beta$ and $\Delta \propto (P_c - P)^\alpha$, respectively, yields $P_c \sim 1.07$ kbar [77]. Figure 33 shows the typical pressure dependence of INS spectra for the spin excitations in TlCuCl$_3$; the two resolved excitations L and T_2 are marked, indicating that the spin-wave excitations of the ordered phase are accompanied by a pressure-dependent longitudinal mode which shows no sign of divergent decay at QPT, while the gap scales with the Néel temperature and the ordered moment.

We now turn to the experimental implications of QCP at $P = P_c$. As it was discussed in Section 2.1.5, despite the fact that QPT occurs at zero temperature, the transition leaves a clear fingerprint at nonzero temperatures in the vicinity of the QCP. Therefore, here we need to consider the influence of finite

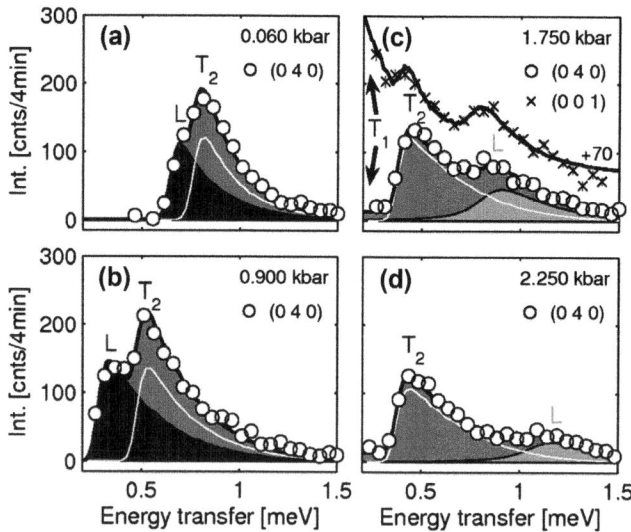

FIGURE 33 INS spectra showing the triplet excitations at $T = 1.85$ K and $\mathbf{Q} = (0,4,0)$ for 0.06 (a), 0.9 (b), 1.75 (c) and 2.25 kbar (d) across the QPT. Complementary data taken at $\mathbf{Q} = (0,0,1)$ are shown in (c). *Reprinted with permission from Ref. [78].*

temperatures on the ground state as it goes through the QPT. Figure 34 demonstrates the schematic $x-T$ phase diagram of TlCuCl$_3$. At finite temperatures and for $x \ll x_c$ spin-wave excitations are induced by thermal fluctuations, distorting the Néel AF ordering, whereas for $x \gg x_c$, spin singlet

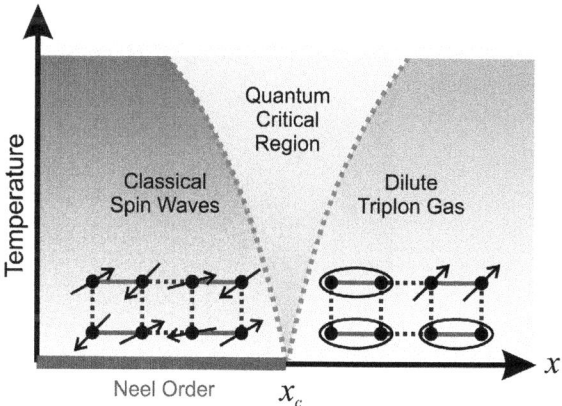

FIGURE 34 Schematic $x-T$ phase diagram of TlCuCl$_3$. Quantum critical region extends to nonzero temperature. *Adapted from Ref. [53].*

dimers are thermally excited to triplons. QC emerges in the intermediate state in the vicinity of x_c, where thermal fluctuations compete with quantum fluctuations, and the ground state wave function has entangled critical form.

As it was noted earlier, applied pressure serves as the inverse of the tuning parameter x and thus at finite temperatures it can be used to directly and continuously control the thermal and quantum fluctuations, yielding comprehensive thermodynamic and mesoscopic information necessary to untie the effects of classical and quantum phenomena close to QCP. High-resolution INS measurements have been successful in achieving such a control, by measuring the magnetic excitation spectrum across the entire quantum critical phase diagram as a function of pressure and temperature. Figure 35 presents INS results mapping the evolution of the spin dynamics of TlCuCl$_3$ throughout the quantum critical phase diagram as a function of pressure and temperature. The results indicate that although thermal and quantum fluctuations operate independently close to QCP, surprisingly they have similar effect on the magnetically ordered phase, accounting for the opening of the excitation gaps and melting the magnetically ordered phase.

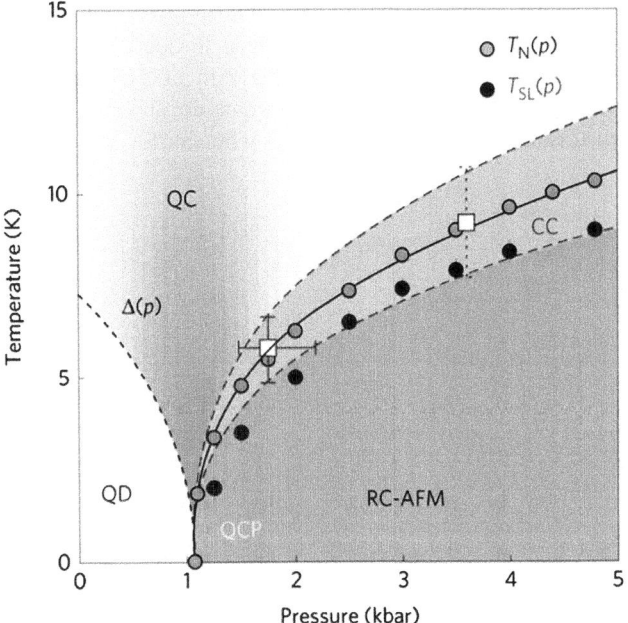

FIGURE 35 Complete experimental phase diagram, showing quantum disordered (QD), quantum critical (QC), classical critical (CC), and renormalized classical (RC-AFM) phases. The dashed lines denote energy scales marking crossovers in behavior. *Reprinted with permission from Ref. [79].*

2.5 QUANTUM PHASE TRANSITIONS IN $J_{\text{eff}} = 1/2$ PYROCHLORE MAGNETS

2.5.1 XY Pyrochlore Magnets

Pyrochlore magnets have been a playground for the physics of geometrical frustration as the A-site of cubic pyrochlores of the form $A_2B_2O_7$ form a network of corner-sharing tetrahedra, shown in Figure 36 [80]. The family of rare earth titanates, $R_2B_2O_7$, has been of particular interest, as a range of R^{3+} ions can occupy the A-site, and the B-site of the pyrochlore structure is occupied by nonmagnetic Ti^{4+} [81].

As rare earth ions reside at the bottom of the periodic table, where spin−orbit coupling is very strong, the magnetism associated with the R^{3+} ions is described by a total angular momentum J, calculated using Hund's rules for a rare earth ion in isolation. The R^{3+} ion is imbedded in the crystalline lattice, and crystalline electric field (CEF) effects will split the $(2J + 1)$-fold degeneracy of the rare earth ion and produce a ground state magnetic moment characteristic of the CEF eigenfunctions that make up the ground state. The anisotropy of the rare earth magnetic moments and the J_{eff} nature of these moments are then determined by the nature of the eigenfunctions that make up the ground state.

Cubic rare earth titanate pyrochlores often display local Ising anisotropy, with magnetic moments constrained to point directly into or out of the tetrahedra on which they reside. This occurs in rare earth pyrochlores related to classical spin ice physics [82], such as $Ho_2Ti_2O_7$ [83,84] and $Dy_2Ti_2O_7$ [85]. However, in $Er_2Ti_2O_7$ and $Yb_2Ti_2O_7$, the CEF effects pick out ground state

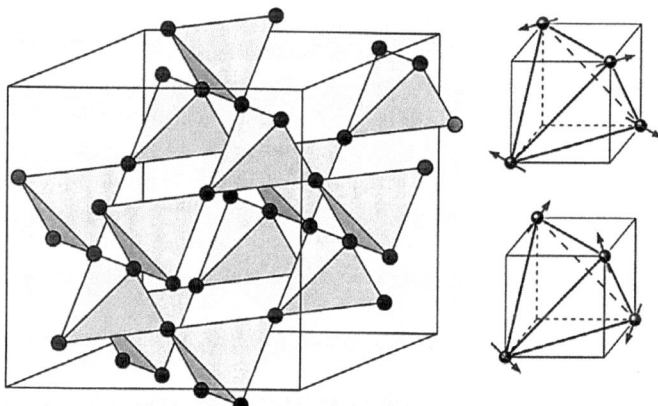

FIGURE 36 The cubic pyrochlore lattice, a network of corner-sharing tetrahedra formed by the rare earth sites in $Er_2Ti_2O_7$ and $Yb_2Ti_2O_7$. The left panel shows the spin configurations taken up by Er^{3+} moments in the ψ_2 and ψ_3 ground states. $Er_2Ti_2O_7$ displays the ψ_2 ground state below $T_N = 1.2$ K. *Reprinted with permission from Ref. [80].*

doublets with eigenfunctions made up primarily of $m_J = \pm 1/2$. This gives local anisotropy to the magnetic moments which is XY or planar, rather than Ising-like, and an effective $J = 1/2$ quantum spin description [86].

A classical description of XY magnetic moments decorating the pyrochlore lattice is shown in the bottom panels of Figure 36. Here we show "spins" that can lie in any direction within the local XY plane. This local XY plane is the plane perpendicular to the local Ising direction, which itself points directly into or out of the tetrahedra. The particular spin configurations illustrated in Figure 36 are those appropriate to the so-called ψ_2 and ψ_3 noncollinear ground states of the XY antiferromagnet on a pyrochlore lattice [87]. In fact the ψ_2 ground state is known to be the ordered structure that describes $Er_2Ti_2O_7$ at low temperatures [88].

2.5.2 $Er_2Ti_2O_7$

$Er_2Ti_2O_7$ is described as an antiferromagnetically coupled XY antiferromagnet on a cubic pyrochlore lattice [89]. Its Curie–Weiss constant which characterizes the sign and strength of its average interactions is ~ 22 K, indicating antiferromagnetic interactions. The low-temperature phase diagram for $Er_2Ti_2O_7$ in the presence of a (1,1,0) magnetic field is shown in the top panel Figure 37. This phase diagram is derived from heat capacity measurements that are shown in the bottom panel of Figure 37. Clearly, the low temperature heat capacity in zero or small fields shows a strong C_p anomaly, indicating a phase transition to an ordered state at $T_N \sim 1.2$ K in zero field. However, we also see that this ordered phase can be destroyed at the lowest temperatures by application of a relatively modest magnetic field applied along the 110 direction—hence a magnetic field induced QPT at $H_c \sim 1.7$ T.

Low-temperature neutron scattering measurements, using both time-of-flight techniques with the disk chopper spectrometer (DCS) instrument at NIST, as well as triple axis measurements, using FLEX at the Helmholz Zentrum Berlin, were used to study the low field magnetic structure and the nature of the QPT at 1.7 T [90]. A map of the elastic scattering within the (H,H,L) scattering plane of $Er_2Ti_2O_7$ is shown in the left panel of Figure 38. The noncollinear ψ_3 magnetic structure that $Er_2Ti_2O_7$ displays at low temperature is a $Q = 0$ antiferromagnetic structure, meaning that the spin configuration of each tetrahedron is the same, as illustrated in the right panel of Figure 36. The magnetic Bragg peaks are also allowed nuclear peaks. Of these, the (2,2,0) Bragg peak has the largest ratio of magnetic to nuclear scattering at low temperatures, and variation of this (2,2,0) Bragg intensity at $T = 0.03$ K as a function of (1,1,0) magnetic field is shown in the right panel of Figure 38.

The resulting elastic scattering order parameter for the QPT shows an interesting $\sim 25\%$ growth at the lowest magnetic fields, followed by a continuous fall off of the order parameter as one pushes through the QPT near $H_c \sim 1.7$ T. The initial growth in the (2,2,0) magnetic Bragg scattering is

Quantum Phase Transitions Chapter | 2 **89**

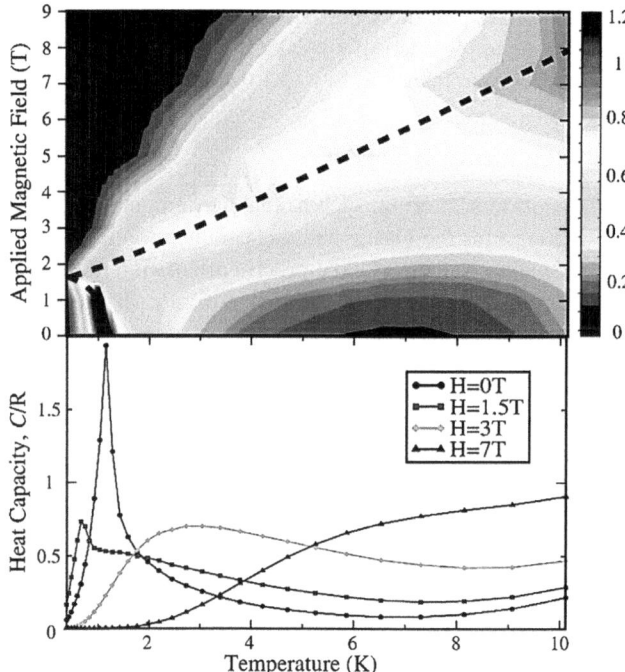

FIGURE 37 The top panel shows the magnetic field, in the (1,1,0) direction, versus temperature phase diagram of $Er_2Ti_2O_7$, showing the (1,1,0) field-induced quantum phase transition at ~ 1.7 T. The bottom panel shows the C_p data from which the phase diagram was determined. *Reprinted with permission from Ref. [90].*

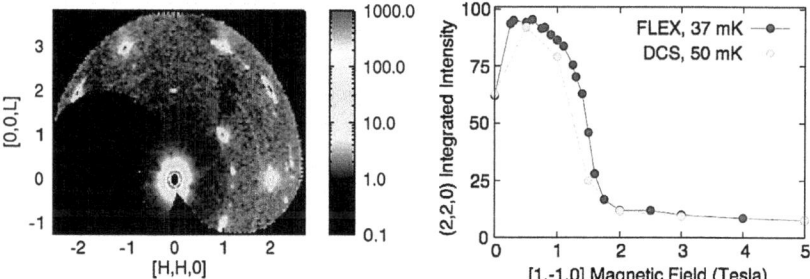

FIGURE 38 The left panel shows a map of the elastic scattering in the (H,H,L) plane of reciprocal space for $Er_2Ti_2O_7$ at $T = 0.05$ K. The right panel shows low-temperature magnetic Bragg intensities at the (2,2,0) position which shows the low field increase, and quantum phase transition at $\sim H = 1.7$ T.

associated with the fact that the ψ_3 ordered state displays six domains, two of which are chosen by application of a weak (1,1,0) field, and this results in the increase of this order parameter at low (1,1,0) fields.

2.5.3 Spin Excitations in $Er_2Ti_2O_7$

Time-of-flight neutron scattering, such as that carried out using DCS at NIST, simultaneously measures the elastic and inelastic neutron scattering spectrum, at wave vectors that are selected by the precise orientation of the single crystal to the incident neutron beam and the detector coverage. By rotating the single crystal about a vertical axis, a large range of wave vectors can be surveyed, along with a dynamic range in energy that is selected by the energy of the monochromatic incident neutrons. In this way, broad surveys of elastic scattering in reciprocal space, as shown in the left panel of Figure 38, and surveys of the inelastic scattering along different high-symmetry directions can be simultaneously probed.

Figure 39 shows one such energy versus direction in reciprocal space slice for $Er_2Ti_2O_7$, mostly at low temperatures, $T = 0.03$ K, and as a function of (1,1,0) magnetic field. The direction of reciprocal space chosen for this plot is the (2,2,L) direction, which is illustrative of the spin waves going into the strong (2,2,0) magnetic Bragg peak, and how they are affected as we push through the QPT as a function of field.

Figure 39(a) and (b) shows this inelastic spectrum in zero field above ($T = 2$ K) and well below the zero-field phase transition to the ψ_3 state. Clearly, one sees the formation of at least two bands of spin-wave excitations, one of which appears to be a Goldstone mode, going soft at (2,2,0) in the ordered phase. This would normally be expected for an ordered system with a continuous symmetry, as expected for such an XY antiferromagnet. In fact, much higher-resolution measurements (not shown here) reveal the Goldstone mode is indeed gapped with a small finite energy excitation at the (2,2,0) zone center [91]. This is an expected consequence of the ground state selection by the "order by disorder" mechanism believed to be relevant for $Er_2Ti_2O_7$ [87].

The finite field inelastic measurements in Figure 39(b–h) show the disruption to these spin waves on application of the (1,1,0) magnetic field, and in particular the qualitative change in the spin-wave spectrum as one pushes through the QPT at $H_c \sim 1.7$ T. The Goldstone-like modes are disrupted even for the smallest applied fields employed, $H = 0.5$ T in Figure 39(c), a likely consequence of domain selection within the ψ_3 state [87]. However, at $H = 1.5$ T, near but below H_c, one observes a new form for the spin waves emerging, which is then gapless and well developed at $H = 2$ T. This form of the spin-wave spectrum is then lifted up in energy in an approximately linear form on further increasing the (1,1,0) magnetic field beyond $H_c \sim 1.7$ T.

At magnetic fields beyond the QPT, the ground state is now a variation of a polarized paramagnet. Consequently, we can now reliably apply linear

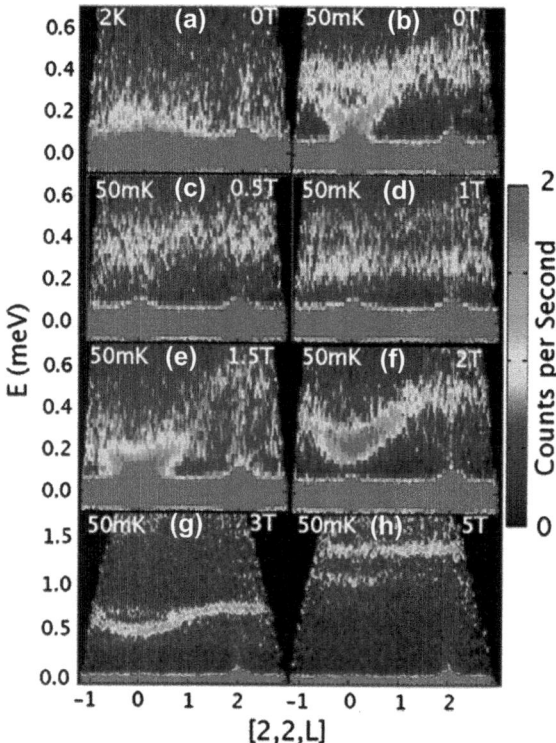

FIGURE 39 The inelastic neutron scattering spectrum along the (2,2,L) direction in reciprocal space is shown for $Er_2Ti_2O_7$ at $T = 2$ K(a) and $T = 0.05$ K (b–h). Goldstone-like modes are seen to evolve to new soft modes at the quantum phase transition, which are subsequently lifted to higher energies at fields beyond the quantum critical point. *Reprinted with permission from Ref. [90].*

spin-wave theory to this high field ground state and extract the underlying microscopic spin Hamiltonian for $Er_2Ti_2O_7$. This type of analysis is shown in Figure 40. The top panel of Figure 40 shows measurements of the $H = 3$ T high field spin-wave dispersion and intensities along seven different directions in reciprocal space, while the bottom row shows the spin-wave spectrum calculated using linear spin-wave theory for the same seven directions. Actually, the five leftmost panels for both the experiment and the theory in Figure 40 correspond to the application of a (1,1,0) magnetic field, while the two rightmost panels show the same measurements for a (1,1,1) magnetic field. The linear spin-wave theory utilized the most general form of anisotropic exchange possible for this system, which is predicated on the point group symmetry of the rare earth site. This results in a 3 × 3 exchange matrix which describes the coupling of each (x,y,z) component of spin on the four neighboring sublattices defined by the vertices of the tetrahedral. Of these nine possible exchange terms, only four are distinct due to the symmetry at the rare

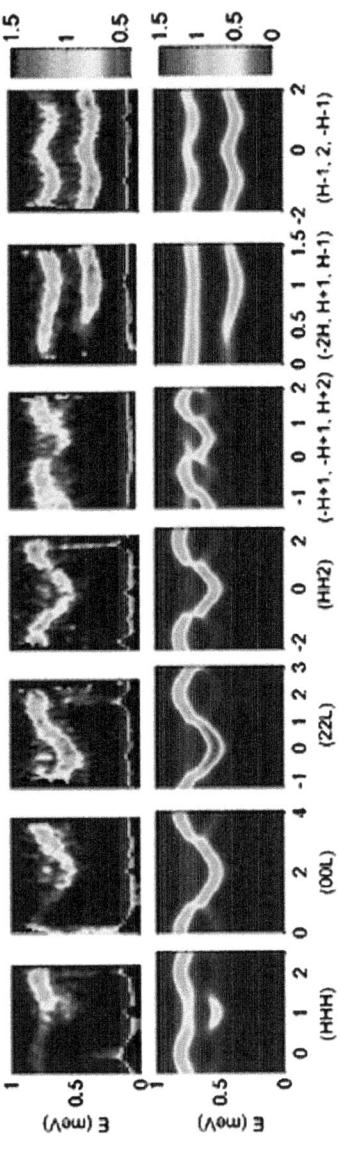

FIGURE 40 The top row shows inelastic neutron scattering spectra along different direction in reciprocal space for $Er_2Ti_2O_7$ at $T = 0.05$ K and $H = 3$ T. The five leftmost panels show data for magnetic field applied along the (1,1,0) direction, while the two rightmost panels show data with the magnetic field along (1,1,1). The bottom row shows theory calculations using anisotropic exchange, which determines the microscopic spin Hamiltonian for $Er_2Ti_2O_7$. *Reprinted with permission from Ref. [87].*

earth site. The spin Hamiltonian can then be written in the following, more conventional form, where we now employ local (x,y,z) basis at each of the four sublattice moment sites:

$$H = \sum_{\langle ij \rangle} \left\{ J_{zz} S_i^z S_j^z - J_\pm \left(S_i^+ S_j^- + S_i^- S_j^+ \right) + J_{++} \left[\gamma_{ij} S_i^+ S_j^+ + \gamma_{ij}^* S_i^- S_j^- \right] \right. \\ \left. + J_{z\pm} \left[S_i^z \left(\zeta_{ij} S_j^+ + \zeta_{ij}^* S_j^- \right) + i \leftrightarrow j \right] \right\} \quad (32)$$

where S^z, S^+, and S^- are the usual spin operators with respect to local z and transverse directions, acting on the effective $S = 1/2$ degrees of freedom relevant to Er^{3+} moments. By fitting the calculated spin-wave spectra to the inelastic measurements, we could estimate the exchange parameters in millielectron volt as:

$$J_{zz} = -2.5 \times 10^{-2} \pm 1.8 \times 10^{-2}, \quad J_\pm = 6.5 \times 10^{-2} \pm 7.5 \times 10^{-3},$$
$$J_{\pm\pm} = 4.2 \times 10^{-2} \pm 5.0 \times 10^{-3}, \quad J_{z\pm} = -8.8 \times 10^{-3} \pm 1.5 \times 10^{-2}$$

and the best fit comparison between experiment and theory is shown in Figure 40 [87].

These results show the largest near-neighbor exchange coupling to be between local transverse components of moment (that is, $J_\pm$ is the strongest). With a microscopic spin Hamiltonian in hand, the selection of the ψ_2 state, as opposed to the ψ_3 state, at low temperatures is now understood as a consequence of the anisotropic exchange and an order-by-disorder mechanism (although this has recently also be proposed as having an energetic origin as well [92]).

The phase diagram for $Er_2Ti_2O_7$ can also be calculated within mean field theory using the microscopic Hamiltonian and this is shown in Figure 41. Remarkably, this shows almost perfect agreement between the location of the field-induced QPT at $H_c \sim 1.7$ T from mean field theory and from the experimental determination for $Er_2Ti_2O_7$. The agreement between the calculated and measured zero-field T_c for $Er_2Ti_2O_7$ is less impressive, but the mean field T_c still comes within approximately a factor of two of the measured T_c. This indicates that fluctuations and frustration are not particularly strong in $Er_2Ti_2O_7$ at low temperatures, as mean field theory provides a reasonably accurate determination of the relevant phase transitions. This should be contrasted with the situation relevant to $Yb_2Ti_2O_7$, to be discussed next.

2.5.4 $Yb_2Ti_2O_7$

$Yb_2Ti_2O_7$ can be thought of as the ferromagnetic counterpart to $Er_2Ti_2O_7$. Once again, CEF effects give rise to a Kramer's doublet ground state for the $J = 7/2$ Yb^{3+} moments that correspond primarily to $m_J = \pm 1/2$, and a $J_{eff} = 1/2$ quantum moment with XY anisotropy. A Curie−Weiss analysis of

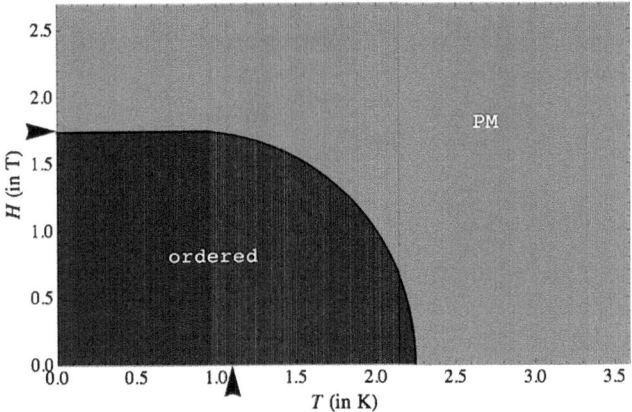

FIGURE 41 The mean field phase diagram obtained for $Er_2Ti_2O_7$ in a (1,1,0) magnetic field is shown, where the blue (dark gray in print versions) region denotes the ordered phase and the green (gray in print versions) region represents the paramagnetic phase (PM). Experimental determinations for H_c and T_c are indicated with black arrows. Note the almost perfect agreement of the calculated quantum phase transition at $H_c \sim 1.7$ T with the measured value. The zero-field agreement between the calculated and measured T_c is less good, but still within a factor of 2. Reprinted with permission from Ref. [93].

$Yb_2Ti_2O_7$'s susceptibility, however, gives net ferromagnetic interactions and a Curie−Weiss constant of $\sim +0.6$ K [94]. The combination of net ferromagnetic exchange interactions and the quantum $J_{\text{eff}} = 1/2$ moments give rise to the possibility that $Yb_2Ti_2O_7$ could be a candidate for a quantum variant of the spin ice problem that has been vigorously studied in the classical spin ice pyrochlore magnets $Ho_2Ti_2O_7$ and $Dy_2Ti_2O_7$, and this, in and of itself, has generated much interest in $Yb_2Ti_2O_7$.

At low temperatures and in zero magnetic field, $Yb_2Ti_2O_7$ is known to display a sharp C_p anomaly at "T_c" ~ 0.26 K, which appears to be a phase transition to an ordered state. However, the nature of this ordered state is not clear and has been strongly debated over the course of the last 5 years. Weak ferromagnetic Bragg peaks have been reported below T_c, and measurements have reported a finite magnetization at low temperatures, but this is also combined with extensive diffuse magnetic scattering at low temperatures, and no sharp, spin-wave excitations at low temperatures, consistent with what would be expected of a conventional ferromagnetic long-range ordered state. More intriguing is that the "T_c" measured by low-temperature heat capacity in $Yb_2Ti_2O_7$ is sample dependent with powder samples grown by solid-state synthesis showing the sharpest C_p anomalies at the highest T_c's, ~ 0.26 K. Single crystal samples grown from the melt display broader C_p anomalies at lower temperatures, and a crystallographic study of powder samples and crushed single crystals shows that while the powder samples tend to be stoichiometric $Yb_2Ti_2O_7$, the single crystals are "lightly stuffed" with a small proportion ($\sim 2\%$) of excess Yb occupying the Ti sublattice [95].

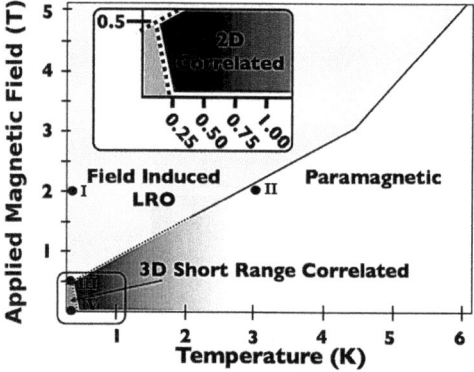

FIGURE 42 The magnetic field, in the (1,1,0) direction, versus temperature phase diagram of $Yb_2Ti_2O_7$, showing the (1,1,0) field-induced quantum phase transition at ~ 0.5T, between field induced long range order (LRO) and correlated phases. This phase diagram is based on diffuse and inelastic neutron scattering measurements as described in the text. *Reprinted with permission from Ref. [94].*

The phase diagram measured for single crystal $Yb_2Ti_2O_7$ in a (1,1,0) magnetic field is shown in Figure 42. The phases at low field, below ~0.5 T at any temperature are disordered, in the sense that the vast majority of the magnetic spectral weight is distributed in diffuse scattering, as opposed to Bragg scattering, which covers all of reciprocal space. These regions are shown in shaded blue and pink in Figure 42, wherein the diffuse scattering is loosely organized into rods characteristic of two-dimensionally correlated systems (blue (dark gray in print versions)), and this scattering bunches together as quasi-Bragg peaks at the lowest temperature and at low field (the pink (gray in print versions) region). However, at all temperatures for fields less than $H = 0.5$ T, no sharp spin waves are observed, and correspondingly we identify these phases as disordered.

2.5.5 Spin Excitations in $Yb_2Ti_2O_7$

Figure 43 shows the spin excitation spectrum of $Yb_2Ti_2O_7$, mostly at $T = 0.03$ K, as measured with the DCS time-of-flight spectrometer. The (H,H,H) direction in reciprocal space is denoted relative to Brillouin zone boundaries in Figure 43(a), while Figure 43(b) and (c) shows the inelastic spectrum along (H,H,H) in zero magnetic field at $T = 4$ and 0.03 K. One can see that even though these temperatures are well above and below "T_c" ~ 0.26 K, both data sets appear as quasielastic diffuse scattering typical of disordered paramagnets, which indeed $Yb_2Ti_2O_7$ is at $T = 4$ K in Figure 43(b).

However, as a (1,1,0) magnetic field is applied at $T = 0.03$ K, one sees that the quasielastic diffuse scattering becomes increasingly organized into sharp

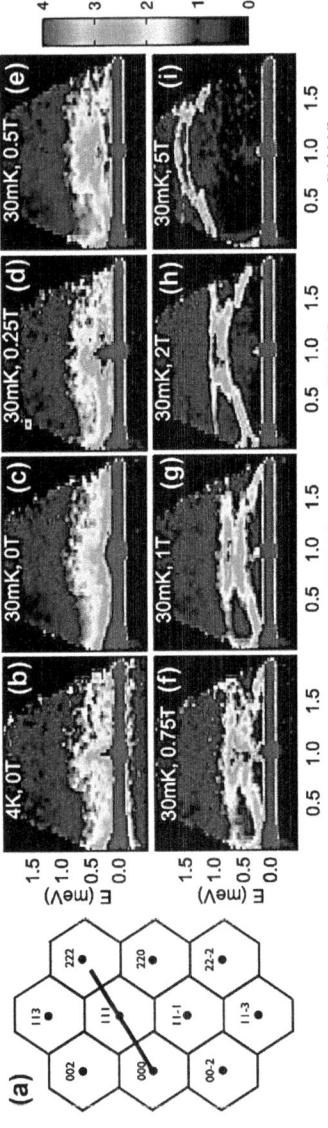

FIGURE 43 The inelastic neutron scattering spectrum along the (H,H,H) direction, as indicated in the leftmost panel (a), in reciprocal space is shown for $Yb_2Ti_2O_7$ at $T = 4K$ (b) and $T = 30mK$ at 0 (c), 0.25 (d), 0.5 (e), 0.75 (f), 1 (g), 2 (h), and 5 T (i). Well-defined spin-wave excitations only exist above the $H = 0.5$ T quantum phase transition in this frustrated magnet at low temperatures. *Reprinted with permission from Ref. [94].*

spin waves at high fields, above $H_c \sim 0.5$ T. We therefore associate the phase above $H_c \sim 0.5$ T with a field-polarized ferromagnetic phase, indicated in green on the $Yb_2Ti_2O_7$ phase diagram shown in Figure 42. This regime of the phase diagram at all temperatures is characterized by sharp and dispersive spin waves of which those shown in Figure 43(f–i) are typical. One clearly observes pronounced structure in both the spin-wave dispersions and intensities at high field, including easily identified gaps in the spin-wave dispersion when the spin-wave wave vectors cross Brillouin zone boundaries as demarked in Figure 43(a).

As was the case for the high field polarized state of $Er_2Ti_2O_7$, we can use the very sharp and structured spin-wave dispersion and intensity to model the microscopic spin Hamiltonian in $Yb_2Ti_2O_7$ using linear spin-wave theory and anisotropic exchange [87]. The general form of the spin Hamiltonian employed is the same one, Eqn (32), as was used for $Er_2Ti_2O_7$. The top and third from the top row of Figure 44 show neutron scattering measurements of the spin excitation spectrum of $Yb_2Ti_2O_7$ at $T = 0.03$ K for five different high-symmetry directions. The top row shows these spectra for a (1,1,0) magnetic field of $H = 5$ T, while the third from the top row shows the corresponding data for $H = 2$ T. The second from the top and bottom rows of Figure 44 show the corresponding high field spin waves as calculated within linear spin-wave theory. Note that both of these fields are in the field polarized regime, about the QPT at $H_c \sim 0.5$ T, and the overall excellent description of the inelastic neutron scattering with the high field linear spin-wave calculation.

This fit of the $H = 5$ T neutron data involved setting the relative intensity of the calculated spin-wave spectra to the measured spin-wave intensity. That is done at a single Q and energy position, and is thereafter not an adjustable parameter in the fit. The comparison between theory and experiment at $H = 2$ T is therefore parameter-free, as the exchange terms in the microscopic spin Hamiltonian and the overall intensity scale have already been fixed in fitting the $H = 5$ T data set. The four symmetry-allowed terms in the microscopic spin Hamiltonian, Eqn (32), are (in millielectron volt):

$J_{zz} = 0.17 \pm 0.04$, $J_{\pm} = 0.05 \pm 0.01$, $J_{\pm\pm} = 0.05 \pm 0.01$, $J_{z\pm} = -0.14 \pm 0.01$

In contrast to the case for $Er_2Ti_2O_7$, the largest term in $Yb_2Ti_2O_7$'s Hamiltonian is J_{zz}, the term which couples together local Ising-like terms on neighboring sites. This ferromagnetic term is the same term which is relevant to classical spin ice physics, as occurs in $Ho_2Ti_2O_7$ and $Dy_2Ti_2O_7$. However, for $Yb_2Ti_2O_7$, the spin Hamiltonian also contains substantial terms which couple transverse, or XY, components of moment, and which therefore support strong quantum fluctuations. Consequently, $Yb_2Ti_2O_7$ is proposed as a quantum spin ice candidate system.

With the microscopic spin Hamiltonian in hand for $Yb_2Ti_2O_7$, its mean field phase diagram can be calculated in much the same way as $Er_2Ti_2O_7$'s

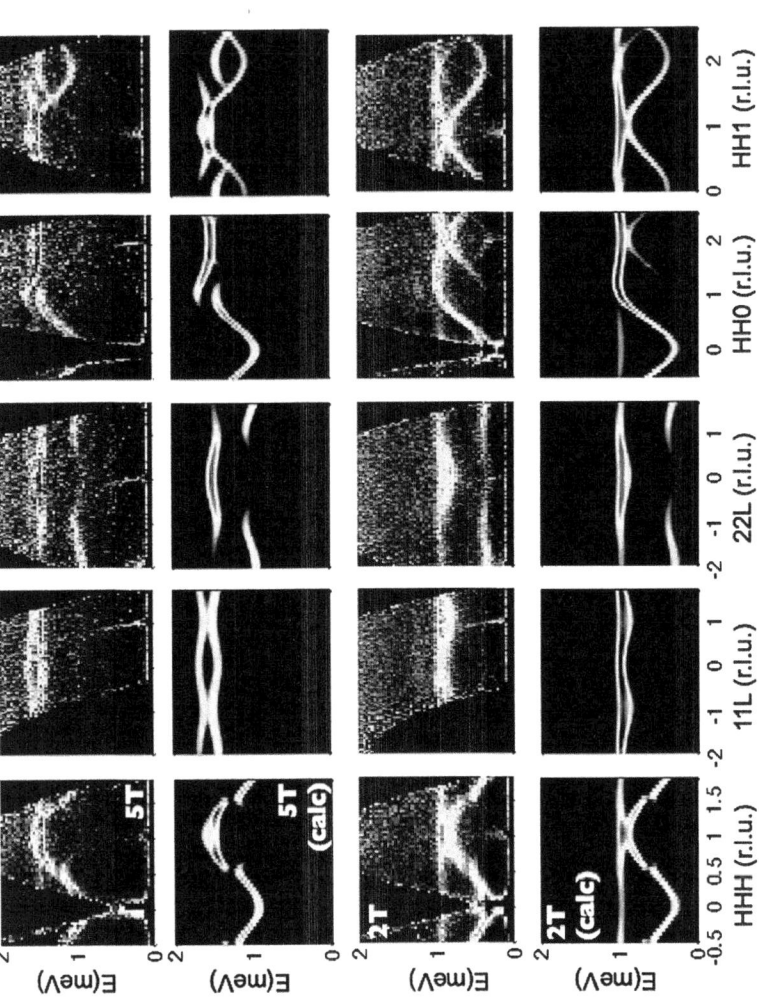

FIGURE 44 The top and third row from the top show inelastic neutron scattering spectra along different direction in reciprocal space for $Yb_2Ti_2O_7$ at $T = 0.03$ K and at $H = 5$ and 2 T, respectively. The second from the top and bottom row shows the corresponding theory calculations using anisotropic exchange, again at $H = 5$ and 2 T, respectively, and these determine the microscopic spin Hamiltonian for $Yb_2Ti_2O_7$. Reprinted with permission from Ref. [86].

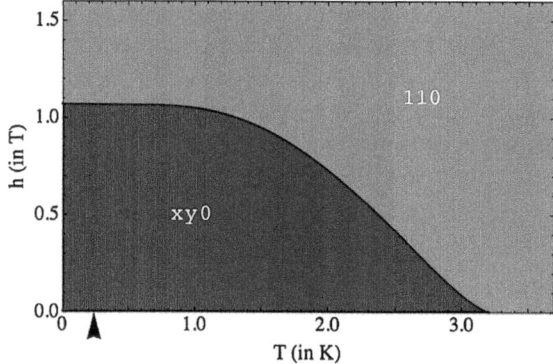

FIGURE 45 Field versus temperature phase diagram obtained from a mean-field analysis. The blue (dark gray in print versions) region denotes a region where the total magnetization lies in the xy plane, and the green (gray in print versions) region is the paramagnetic phase; the two zones are separated by a continuous transition. *Reprinted with permission from Ref. [86].*

was. The result is shown in Figure 45. The experimentally measured QPT at $H_c \sim 0.5$ T is roughly within a factor of two of the mean field $H_c \sim 1.1$ T. However, the mean field T_c is ~ 3 K, and it is ~ 12 times that measured experimentally, taking the C_p anomaly at 0.26 K, shown as the black arrowhead in Figure 45, as the experimental measure. The fact that the mean field analysis provides such a relatively poor description of the zero-field phase behavior in $Yb_2Ti_2O_7$ clearly demonstrates the importance of fluctuations; both quantum mechanical and those originating from geometrical frustration.

2.6 QUANTUM PHASE TRANSITIONS IN HEAVY FERMIONS

2.6.1 Heavy Fermions

Discovery of heavy-fermion and cuprate superconductivity in the 1980s has generated significant interest in the field of strongly correlated electron materials. In forefront of research in this field are efforts to establish how electrons are organized and to ascertain whether there are universal principles that control various classes of these materials. One such principle is QC [96] which facilitates a continuous phase transition at absolute zero temperature between distinct ground states of matter as a function of a nonthermal tuning parameter such as magnetic field, pressure, or doping range. QPT is at the heart of the physics of strongly correlated electronic systems. Experimental and theoretical studies of QPTs remain of extensive current interest as electronic correlations and many-body competing interactions in such systems (especially when close to the border of magnetism) give rise to novel finite-temperature properties such as unconventional superconductivity and non-Fermi liquid (NFL) behavior [97,98].

The presence of unconventional elementary excitations belonging to the quantum critical ground state leads to deviations from the standard Landau theory of Fermi liquids (FL). The coexistence of these with the standard fluctuations of the electron—hole quasiparticles makes understanding of unconventional superconductors and heavy fermions challenging [99]. There are two main categories of QPTs realized in these systems.

2.6.1.1 Spin-Density-Wave QC at a Conventional QCP [100]

The phase transition occurs by the usual spin-density-wave (SDW) instability of the Fermi surface. In this picture, the local moments do not play an essential role in the phase transition, and magnetism simply acts as a source of fluctuations within the Landau QC. The fermionic heavy quasiparticles remain intact, and only the bosonic magnetic fluctuations become critical. The Hertz—Millis—Moriya (HMM) spin fluctuation theory considers such QPTs [100] where fluctuations of the AF moment (the order parameter) solely control the phase transition and govern the FL regime. In this approach, it is assumed that at the QCP the quasiparticles undergo a singular scattering due to the vast presence of low-lying magnetic fluctuations. The NFL properties observed in the vicinity of most heavy-fermion QCPs are believed to be the result of this divergence of the critical fluctuations at the QCP. Despite the success of the standard HMM theories to describe the 3D antiferromagnetic (AF) metals, they fail to account for certain experimental results [101] in ferromagnetic and two-dimensional (2D) antiferromagnet materials due to the presence of both order parameter fluctuations and electron—hole fluctuations.

2.6.1.2 Local QC

A dramatically different scenario is the local QC featuring a new type of QCP beyond the Landau framework [102]. It assumes a breakdown of the quasiparticles at the QCP by coupling to the quantum critical magnetic fluctuations. The local moments survive at all finite temperatures close to the critical point. At the QCP, they become critically quenched and produce the critical magnetic fluctuations that destabilize the Fermi sea and change the Fermi volume. In this type of QCP, the quasiparticles have a composite nature due to the hybridization of local moments and conduction electrons leading to NFL electronic excitations in addition to the fluctuations of the magnetic order parameter. Despite recent developments [103], a full theoretical treatment is still lacking and distinction between the two classes of materials is actively pursued experimentally and theoretically. Furthermore, the effects of disorder often introduced by chemical doping add complexity in such studies [99].

Over the past couple of decades, rare earth intermetallic heavy fermions, transition-metal intermetallic weak ferromagnets, and transition-metal oxides have become important testing grounds to explore QCPs and the breakdown of FL theory in general. In this section, we focus primarily on heavy fermions and transition-metal intermetallic weak ferromagnets. In this section we present a review on $CeCu_{6-x}M_x$ and the so-called Ce-115 system as prominent examples of heavy fermions to showcase the properties of QPT as seen with neutron scattering technique. For other heavy fermion examples, such as $CeCu_2Si_2$, $YbRh_2Si_2$, $CePd_2Si_2$, the reader is referred to Ref. [99]. A comprehensive review of the high transition temperature copper oxides is provided in Chapter 3 of this volume. Neutron scattering has played a crucial role in determining properties of QPTs including magnetic properties such as ordering temperature, correlation lengths, and spin dynamics as a function of a nonthermal parameter such as doping, pressure, or magnetic field. In fact the evidence for the failure of the order-parameter-fluctuation picture for QC first came from inelastic neutron scattering measurements [101] in $CeCu_{6-x}Au_x$ heavy-fermion system considered below.

2.6.2 Cerium-Based Heavy-Fermion System

Heavy-fermion metals with f-electrons possess a large effective electron mass as a result of the large ratio of the on-site Coulomb repulsive interaction to the kinetic energy. Strong electronic correlations lead to integer valence occupancy with the f-electron band in its Mott-insulating state [104–106]. Thus, for energies less than the Mott gap, the f-electrons effectively behave as localized magnetic moments. Two principle energy scales determine [104–106] the properties of these materials: the on-site Kondo screening (T_K) and the intersite Ruderman–Kittel–Kasuya–Yosida (RKKY) exchange magnetic coupling (T_{RKKY}). Below the Kondo temperature, the localized f-electrons start to hybridize with the itinerant d-electrons forming a larger and heavier Fermi surface. The local moments interact with each other through the RKKY exchange coupling which promotes magnetic ordering, and they could also interact with the conduction electrons via an AF Kondo exchange coupling which favors a Kondo-screened singlet ground state. The central question is then the interplay between the local moments and the conduction-electron bands which determines the magnetic ordering temperature T_N and characterizes the QPT between a PM heavy FL and an antiferromagnetic (AF) metal. In real systems, however, the presence of multiple energy scales (such as spin, charge, orbital, and lattice degrees of freedom) causes more complexity [99,105].

Since QPTs between distinct ground states arise as a result of competing interactions in quantum many-body systems, heavy-fermion systems with naturally competing interactions (lattice of localized magnetic moments and a band of conduction electrons), provide a fertile setting for QPTs. Theoretically, QCPs in heavy fermion (HF) metals have mostly been treated by applying the HMM theory of order parameter fluctuations [100]. In this "conventional" approach it is assumed that the heavy, "composite" charge carriers keep their integrity at the QCP, i.e., they exist on either side of it. The reader is referred to Ref. [99] for the prototype example of itinerant SDW-type $CeCu_2Si_2$. Here we consider $CeCu_{6-x}Au_x$ and $CeRhIn_5$ as examples of the new type of QC (local) characterized with the destruction of the Kondo effect and the disintegration of the composite fermions.

2.6.3 $CeCu_{6-x}Au_x$

$CeCu_{6-x}Au_x$ is one of the most prominent examples of heavy fermions displaying a local QPT [106]. The parent compound, $CeCu_6$, has an orthorhombic structure [107] (space group Pnma, Figure 46) at room temperature with a second-order structural phase transition at $T_s \sim 230$ K to a monoclinic structure (space group P$2_1/c$ and monoclinic angle of $\sim 91.58°$). The lattice degrees of freedom are not influenced by the 4f-electrons since $LaCu_6$ also has a similar structural phase transition [106]. $CeCu_6$ is a PM FL metal with an exceptionally large electronic contribution to its specific heat and no sign of superconductivity. A key question in understanding the QC in $CeCu_{6-x}Au_x$ is whether the AF order near the QCP originates from an FL-like state or from local moments [101]. The QPT from FL to AF metal is suggested [108] to be the result of a competition between the screening of the Ce 4f moments in a crystal-field split $^2F_{5/2}$ doublet ground state by conduction electrons (Kondo effect) and the tendency toward AF order via the RKKY coupling between the Ce moments.

2.6.3.1 Chemical Pressure (Doping)

Neutron scattering experiments performed on $CeCu_6$ show that incommensurate AF correlations develop below 10 K with a long-range AF order setting

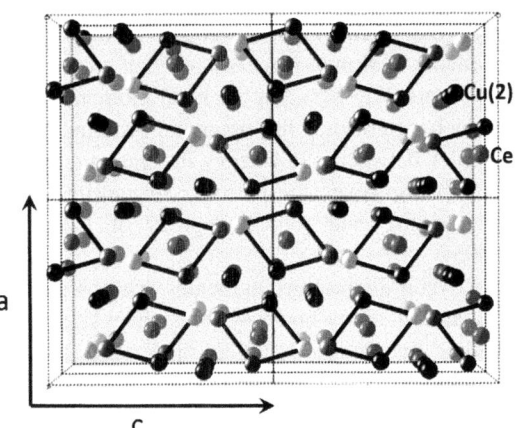

FIGURE 46 Crystal structure of $CeCu_6$ along b-axis, where red (gray in print versions) and blue (dark gray in print versions) circles denote Ce and Cu(2) ions, respectively, and the dashed lines represent the orthorhombic unit cell.

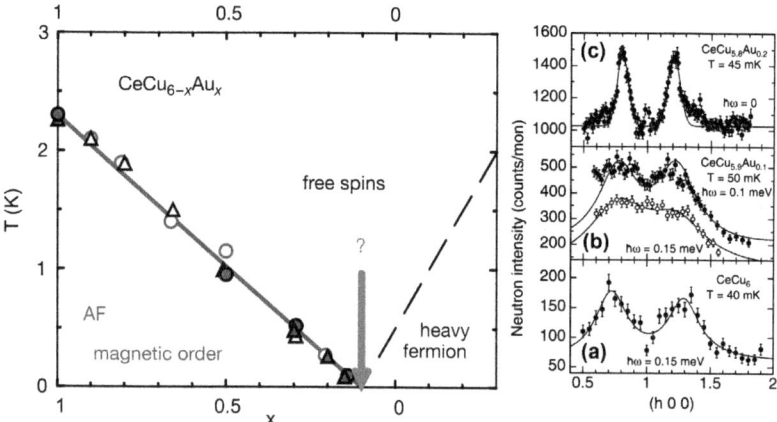

FIGURE 47 (Left) The phase diagram for CeCu$_{6-x}$Au$_x$ as a function of Au concentration, x, [101]. With increasing x the system crosses a quantum critical point at $x_c = 0.1$ and acquires an AF-ordered magnetic phase for $x > x_c$. (Right) (a–c) **Q** scans along (1,0,0) in CeCu$_{6-x}$Au$_x$ at different energy transfers E at $T = 40$–50 mK. Solid lines represent fits to the data with Lorentzian ($x = 0$; 0.1) and Gaussian line shape ($x = 0.2$). *Reprinted with permission from Ref. [109].*

in only at extremely low temperatures (below 2 mK) [106]. Doping CeCu$_6$ with gold atoms, CeCu$_{6-x}$Au$_x$, beyond a critical concentration $x_c \sim 0.1$, however, induces [106] long-range AF order at finite Neel temperatures (T_N). The phase diagram as a function of Au concentration is shown in Figure 47 (left). For $x \leq 1$, Au occupies exclusively the Cu(2) sites in the orthorhombic CeCu$_6$ structure [106]. At the critical concentration x_c where T_N is driven to zero, CeCu$_{6-x}$Au$_x$ exhibits many pronounced deviations from the FL behavior [106] in its physical properties including resistivity, specific heat, and susceptibility.

Neutron scattering measurements show that the AF ordering temperature rises linearly with x for $x > x_c$. At the QCP, the magnetic Bragg peaks broaden into ridges which connect into a "butterfly"-shaped critical line [106], as shown in Figure 48. Similar to T_N, low-temperature staggered moment sets in gradually beyond x_c establishing the existence of an AF QCP.

At the critical doping, $x_c = 0.1$, inelastic neutron scattering measurements have revealed the presence of 2D magnetic critical fluctuations associated with the QCP [109], see Figure 47 (right panel). The critical fluctuations are coupled to quasiparticles with 3D dynamics (precursors to the 3D ordering for $x > 0.1$). The anomalous critical exponents [101] at the QCP measured by inelastic neutron scattering on CeCu$_{5.9}$Au$_{0.1}$ also point to the fluctuations being 2D, despite the underlying electronic system being 3D. Similarly, such theory can accurately predict the dependencies of T_N on the Au concentration and pressure. In addition to the presence of 2D critical dynamics which is prerequisite for the validity of the local criticality model [102], the unusual E/T scaling of

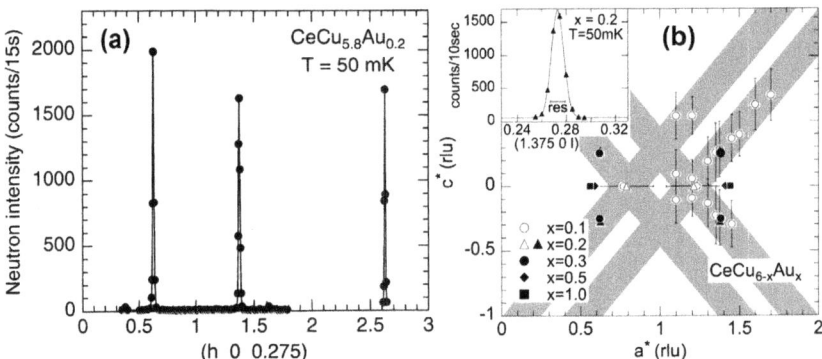

FIGURE 48 Neutron scattering results for $CeCu_{6-x}Au_x$. (a) Resolution-limited magnetic Bragg reflections for $CeCu_{5.8}Au_{0.2}$ corresponding to an incommensurate magnetic ordering wave vector $Q = (0.625,0,0.27)$. (b) Position of the dynamic correlations $x = 0.1$ ($E = 0.1$ meV, $T = 100$ mK) and magnetic Bragg peaks $0.2 < x < 1.0$ in the a^*–c^* plane in $CeCu_{6-x}Au_x$. Open symbols for $x = 0.2$ represent short-range-order peaks. Vertical and horizontal bars indicate the Lorentzian linewidths for $x = 0.1$. The four shaded rods are related by the orthorhombic symmetry ignoring the small monoclinic distortion. The inset shows a resolution limited magnetic Bragg peak for $x = 0.2$ at $T = 50$ mK via elastic scan at $(1.375,0,l)$. *Reprinted with permission from Ref. [106].*

dynamical susceptibility $\chi''(Q,\omega)$ observed [101] with neutron scattering (Figure 49) is also consistent with local criticality providing strong evidence that the local criticality is playing a crucial role in the observed NFL behavior.

The unusual magnetic dynamics in $CeCu_{6-x}Au_x$, observed near the $x = x_c \approx 0.1$, can be understood in terms of a critical Kondo breakdown (local QC).

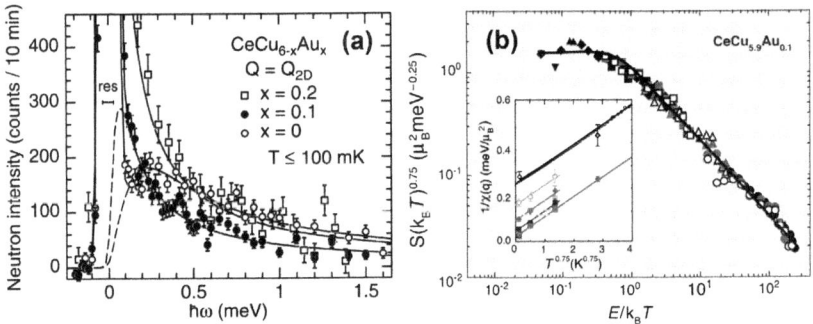

FIGURE 49 (a) Energy scans in $CeCu_{6-x}Au_x$ at $T < 100$ mK performed on top of the rodlike structure at critical $Q = Q_{2D}$ values [109]. (b) E/T scaling of the dynamic susceptibility measured at different wave vectors specified in the inset in $CeCu_{5.9}Au_{0.1}$. The different symbols represent data taken at different magnetic wave vectors. The temperature and energy exponent is fractional ($= 0.75$). (Inset) The inverse of the bulk magnetic susceptibility, $1/\chi(q = 0) \equiv H/M$, and that of the static susceptibility at other wave vectors derived from the dynamical spin susceptibility through the Kramers–Kronig relation. *Reprinted with permission from Ref. [110].*

The local Kondo quasiparticles are destroyed by coupling to the quantum critical magnetic fluctuations (local QC) [102], and the Kondo screening scale (Kondo temperature T_K) is reduced [106] and finally vanished at QCP, x_c. In the HMM type approach for a 3D SDW QPT, however, the Kondo scaling (Kondo temperature T_K) remains finite even at the QCP. Moreover, the HMM model predicts the critical slowing down of the spin fluctuations only at Q_{AF}, whereas the data indicate that magnetic fluctuations become critical in energy for all momentums (local criticality).

2.6.3.2 Applied Pressure

Since Au has larger radius than Cu, doping with Au results in an expansion of the lattice (commonly referred to as "negative chemical pressure") leading to a reduction in the hybridization between the Ce 4f-electrons and the conduction electrons, i.e., weakening of the Kondo screening. In fact, magnetic order is suppressed, and a nonmagnetic ground state is established for $x > x_c$ by application of large enough hydrostatic pressure with a linear suppression of T_N with pressure, corresponding to the its linear dependence to x [108]. Neutron scattering measurements under applied pressure indicate that pressure tuning of $CeCu_{6-x}Au_x$ with $x > x_c$ toward the QCP results in a recovery of the magnetic ordering wave vector of corresponding $CeCu_{6-x}Au_x$ with lower Au content and the same ordering temperature (see Figures 50 and 51). The transport and thermodynamic measurements also indicate that concentration and pressure equally tune the QCP.

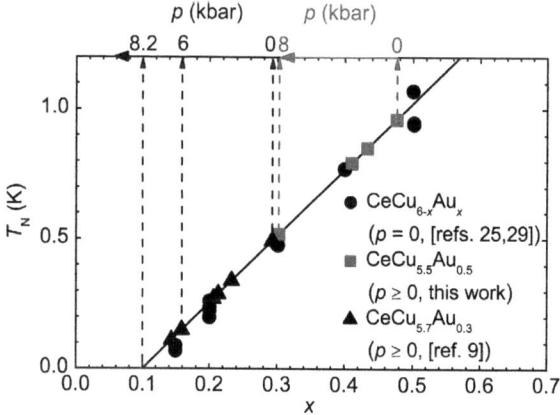

FIGURE 50 The Néel temperature T_N of $CeCu_{6-x}Au_x$ at ambient pressure versus Au concentration x (bottom x-axis) together with the pressure dependence (top x-axis) for $x = 0.3$ and $x = 0.5$ doping concentrations. In order to compare the effects of Au doping and hydrostatic pressure, T_N was mapped onto x by virtue of the linear fit. *Reprinted with permission from Ref. [108].*

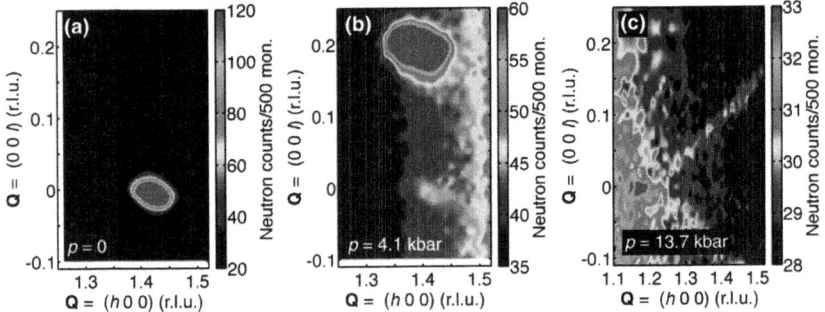

FIGURE 51 Color-coded neutron intensity map of the reciprocal $(h,0,l)$ plane in CeCu$_{5.5}$Au$_{0.5}$ for pressures $P = 0$ (a), 4.1 (b) and 13.7 (c) kbar at $T < 110$ mK showing magnetic Bragg peaks at different Q positions. The color scales are adjusted individually for each pressure to make also weak peaks visible. Note the different scale for the horizontal $(h,0,0)$ axis in (c). *Reprinted with permission from Ref. [111].*

2.6.3.3 Applied Magnetic Field

Although both pressure- and doping-induced QCPs in CeCu$_{6-x}$Au$_x$ show the characteristics of local QC, the field-induced QCP has the properties of an SDW QCP. A detailed comparison of pressure- and field-tuned QCPs on the same system CeCu$_{5.8}$Au$_{0.2}$ [112] demonstrated that field, as opposed to pressure, drives the system to a 3D SDW QCP. Inelastic neutron scattering measurements under applied magnetic field determine $E/T^{1.5}$ scaling corresponds to the HMM scenario [100] and confirm that the magnetic field tuning is different from the concentration and pressure tuning of QCP [112] (see Figure 52). It is an open question whether the apparent similarity to the

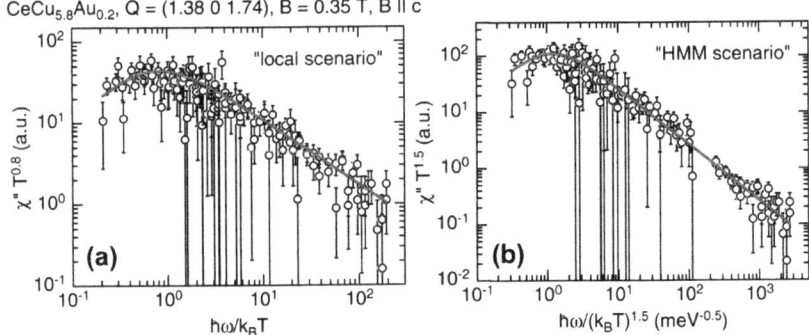

FIGURE 52 Scaling plots of the imaginary part of the dynamic susceptibility $\chi''(Q,\omega,T)$ at $Q = (1.38, 0, 1.74)$, $B = 0.35$ T $(B||c)$ in CeCu$_{5.8}$Au$_{0.2}$ plotted as $\chi''(T^\alpha) = f(\omega/T^\beta)$ with (a) $\alpha = 0.8$, $\beta = 1.0$ and (b) $\alpha = 1.5$, $\beta = 1.5$. Fits indicate that the Hertz-Millis-Moriya scenario describes the data in applied field better than local quantum criticality. *Reprinted with permission from Ref. [112].*

spin fluctuation theory represents crossover phenomena or the approach to a new QCP. It must be noted that near a QCP many different effects can come into play, including local Kondo screening, lattice coherence and Fermi volume collapse, quantum critical fluctuations, and possible dimensional reduction and disorder all making it challenging to unambiguously discern a class of QC [99,105]. How lattice disorder, structural transitions, and other thermoelectric properties affect and are related to the QPTs and the associated NFL behavior remains an open question in the physics of heavy-fermion systems.

Similar weakening of the 4f conduction-electron hybridization (due to lattice expansion) can be induced [106] by other dopants in $CeCu_{6-x}M_x$ with M = Ag ($x_c = 0.2$), Pd ($x_c = 0.05$), and Pt ($x_c = 0.1$). Again the system is driven toward a QCP beyond which long-range AF order appears analogous to the doping with M = Au as discussed above. Remarkably, the specific heat coefficient of all of these different systems is very similar showing a characteristic logarithmic dependence over nearly two decades in temperature down to 50 mK. This suggests that NFL behavior observed at the respective x_c in different systems does not depend on which dopant M is used, and for the different $CeCu_{6-x}M_x$ systems a common origin for the quantum critical state at x_c is likely.

2.6.4 $CeT(In_{1-x}M_x)_5$

The so-called Ce-115 heavy-fermion compounds with chemical formula $CeTIn_5$ (T = Co, Rh, Ir) have generated great interest due to their NFL behavior, field-induced magnetism, proximity to AF order and magnetic QCPs, the coexistence of unconventional superconductivity and magnetism under pressure by chemical substitution or in a magnetic field, and their relatively high transition temperatures [113,114]. While $CeCoIn_5$ and $CeIrIn_5$ exhibit superconductivity, $CeRhIn_5$ exhibits a long-range AF order at ambient pressure with superconductivity only emerging above a critical pressure. The crystal structure of these compounds is tetragonal (space group 123) derived from cubic $CeIn_3$ [115] and consists of nearly 2D $CeIn_3$ layers, separated by intercalated TIn_2 layers along the c-axis as shown in Figure 53. This layered structure reinforces a 2D character for the interactions between the cerium ions. For spin fluctuation-mediated superconductivity, it has been shown that a reduction of the dimensionality favors superconductivity [116]. This seems to be supported by the strong increase of T_c, from ~ 0.2 K in $CeIn_3$ to ~ 2.3 K in the two-layer $CeCoIn_5$ compound (see Figure 54).

An investigation of the phase diagram shown in Figure 54 indicates that an increase of the T_c goes together with an increase of the c/a ratio of the lattice parameters [116]. The fact that $CeIrIn_5$ has an even larger unit cell volume than $CeRhIn_5$ but it does not show any AF order suggests that perhaps the entire series of $CeMIn_5$ is superconducting and T_c increases in a linear fashion

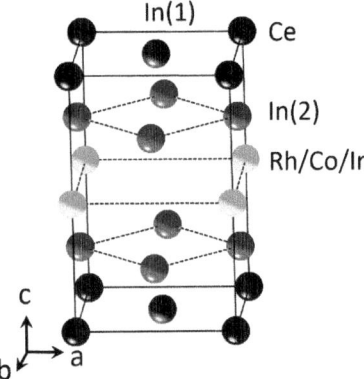

FIGURE 53 Crystal structure of CeTIn$_5$ (T = Rh, Co, Ir).

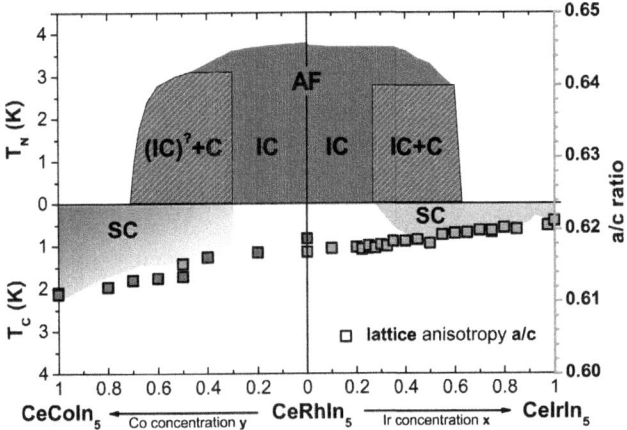

FIGURE 54 Phase diagram of CeRh$_{1-x}$Ir$_x$In$_5$ and CeRh$_{1-y}$Co$_y$In$_5$ showing the magnetically ordered and superconducting regions as function of the Ir and Co concentration. The incommensurate antiferromagnetic ordered phases (IC AF) and the commensurate ones (C AF) are colored dark gray and hatched gray, respectively, and the regions of superconductivity (SC) are marked light gray. The colored squares show the lattice anisotropy $a = c$ (right scale) for these systems. *Reprinted with permission from Ref. [116].*

from Ir to Rh to Co but slight changes in electronic structure of the Rh system favors an AF ground state instead of a superconducting one. We will mainly consider the properties of CeRhIn$_5$ and CeCoIn$_5$ compounds as CeIrIn$_5$ compound is the least studied show casing evidence for possible QPTs from neutron scattering measurements as a function of doping, magnetic field, and applied pressure in these materials.

2.6.5 CeRhIn$_5$

CeRhIn$_5$ compound is not superconducting at ambient pressure [114]. Instead it orders antiferromagnetically at $T_N = 3.8$ K with a spiral magnetic structure along the c-axis and a relatively large magnetic moment of ~ 0.8 μ_B. The ordered moment at ambient pressure is a substantial fraction of the moment expected (0.92 μ_B) in the crystal field doublet ground state of Ce^{3+} suggesting it mainly has 4f-localized character, albeit with small hybridization of the 4f-electron with ligand states.

Application of hydrostatic pressure drives [117] the AF CeRhIn$_5$ compound toward an instability and incipient superconductivity at a pressure of about 0.6 GPa where T_N exhibits a maximum. T_N is reduced nonmonotonically while T_c increases and crosses T_N at $p_c^* \sim 1.75$ GPa. With further increase of pressure, magnetic order disappears and only superconductivity survives with a maximum T_c of 2.3 K (Figure 55). Strong deviations from the FL behavior are observed over a wide pressure range [114]. The pressure-induced transition to superconductivity appears to be of first order thus thought to mask the presence of any potential magnetic QCP [113,114].

Neutron diffraction measurements under pressure have indicated [117] that initially on cooling below T_N, magnetic scattering intensity appears at the same wave vector as the propagation wave vector at ambient pressure (Q_1) but with further cooling intensity at this position decreases, and scattering instead grows at a slightly different wave vector (Q_2) (see Figure 55). Temperature at which scattering becomes finite at Q_2 corresponds to the onset of transition to superconductivity observed with resistivity, whereas a bulk transition to superconductivity occurs at the temperature where scattering at Q_1 vanishes. It is unclear whether the change in the magnetic structure allows the bulk superconductivity or whether superconductivity drives the change in the magnetic structure. Further neutron scattering

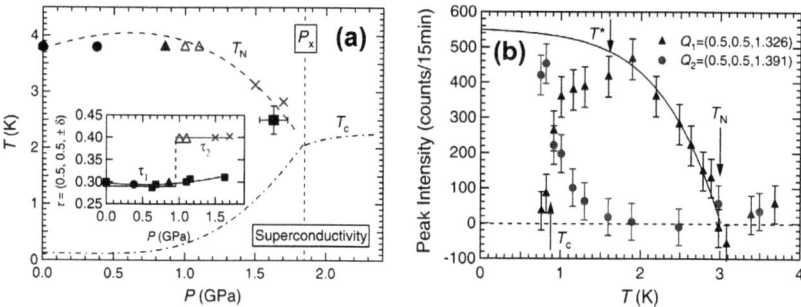

FIGURE 55 (a) T–P phase diagram for CeRhIn$_5$. The dash-dotted and broken lines correspond to T_c and T_N, respectively. (b) T-dependence of the peak intensities at $Q_1 = (0.5,0.5,1.326)$ (triangles) and $Q_2 = (0.5,0.5,1.391)$ (circles). *Reprinted with permission from Ref. [117].*

experiments under pressure and to temperatures less than bulk T_c are necessary to clarify these questions and to look for the spin-resonance excitations.

As mentioned, once pressure exceeds p_c^*, evidence for magnetic order disappears, however, with an applied magnetic field, magnetic order reappears in the superconducting state [113] (see Figure 56). The nature of the field-induced order in CeRhIn$_5$ remains undetermined, however, the de Haas−van Alphen results [118] suggest a jump in the Fermi surface and a divergence in the effective mass indicating that local QC induces superconductivity in CeRhIn$_5$. The field-induced order extends into the normal state with no evidence for a Fulde−Ferrell−Larkin−Ovchinnikov state [113]. Further studies of the AF/superconducting phase boundary under pressure have provided evidence for the presence of a quantum critical line between a phase of microscopically coexistence with superconductivity and a purely superconducting phase [113]. A quantum tetracritical point where four phases (a phase of d-wave superconductivity coexisting with magnetic order, a purely d-wave superconducting state, a PM phase, and an antiferromagnetically ordered phase) meet is suggested [114] to exist if the line of field-induced transitions persists at $T \to 0$. The presence of such quantum tetracritical point in CeRhIn$_5$ has not been experimentally confirmed.

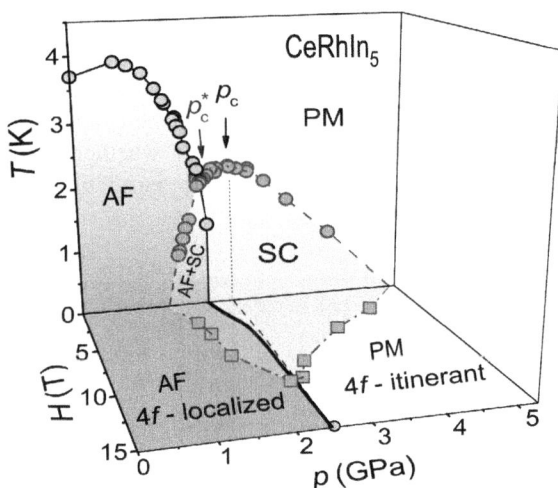

FIGURE 56 The combined temperature, pressure, and field H phase diagram of CeRhIn$_5$ with antiferromagnetic (blue (dark gray in print versions)), superconducting (yellow (light gray in print versions)), and coexistence AF + SC (green (gray in print versions)) phases. The thick black line in the $H-p$ plane indicates the proposed line where the Fermi surface changes from 4f "localized" (small Fermi surface and topology comparable to LaRhIn$_5$) to 4f "itinerant" (large Fermi surface as in CeCoIn$_5$). PM, paramagnetic; SC, superconductivity. *Reprinted with permission from Ref. [113].*

2.6.6 CeCoIn$_5$

The heavy-fermion CeCoIn$_5$ superconductor displays [114] intriguing NFL behavior in the normal state above $T_c = 2.3$ K all consistent with close proximity to an AF QCP. Superconductivity in this system emerges T_c from heavy low-energy excitations related to the AF QCP that arise at high temperature. The development of a coherent heavy quasiparticle band in CeCoIn$_5$ is evidenced by the observation of a drop in the electrical resistivity around 50 K followed by a T-linear resistivity at lower temperature (above T_c), a behavior that has been associated with the proximity to the QCP [114]. Other properties such as specific heat, thermal expansion coefficient, and nuclear spin relaxation rate also exhibit NFL behavior.

Initial neutron scattering measurements on CeCoIn$_5$ identified [114] strong magnetic fluctuations with relatively large moment (0.6 μ_B) at the commensurate wave vector (0.5,0.5,0.5) even though no static magnetic order is observed in zero applied field. Upon cooling below T_c, the magnetic spectral weight is shifted to higher energies creating a spin resonance at 0.6 meV, believed to be similar to the resonance excitation seen in cuprate superconductors. More recent detailed inelastic neutron scattering experiments (see Figure 57) provide [119] clear evidence that the spin-resonance mode is in fact incommensurate with the lattice and has Ising nature. The incommensurate peak position corresponds to the propagation vector of the adjacent field-induced static magnetic ordered phase (the so-called Q-phase). Moreover, the direction of the magnetic moment fluctuations appears also the same direction of the ordered magnetic moments in the Q-phase. Similar symmetry

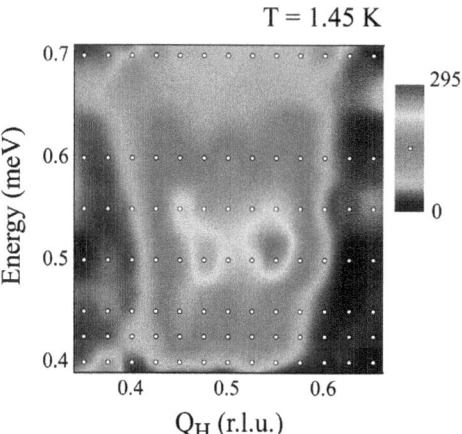

FIGURE 57 Color-coded intensity plot of the INS spectra of CeCoIn$_5$ as a function of Q_H and E for $\mathbf{Q} = (Q_H, Q_H, 0.5)$ at 1.45 K. The empty circles indicate positions where data were collected. *Reprinted with permission from Ref. [119].*

observed for both the resonance mode, and the Q-phase supports a scenario where the static order is realized by a condensation of the magnetic excitation.

In a recent study where different techniques (neutron scattering, muon spin rotation, and X-ray absorption spectroscopy) were combined with the effects of hydrostatic pressure, and equivalent "negative" pressure (doping with Cd) has revealed [120] strong evidence for the presence of a putative pressure-induced QCP at $p_c \simeq 0.4$ GPa (see Figure 58). Neutron scattering measurements in this study confirm the presence of commensurate long-range antiferromagnetic order with a magnetic moment of ~ 0.4 μ_B for a sample with $T_c \approx T_N$. The results suggest that the magnetic order of itinerant character coexist with bulk superconductivity and that the suppression of the antiferromagnetic order appears to be driven by a modification of the bandwidth/carrier concentration. This indicates that both hydrostatic and chemical pressure strongly affect the electronic structure and consequently the interplay between superconductivity and magnetism.

2.6.7 CeIrIn$_5$

CeIrIn$_5$ is least studied among the Ce-115 series, however, it has similarly intriguing properties as other members. It exhibits [114] bulk superconductivity below 0.4 K with a resistive transition temperature ranging between 1.2 and 1.4 K. Similar to CeCoIn$_5$ and CeRhIn$_5$, superconductivity in CeIrIn$_5$ develops from a heavy-fermion normal state and arising from pairing of very heavy quasiparticles [114]. Measurements of several physical properties reveal NFL behavior suggesting the proximity of CeIrIn$_5$ to an AF QCP.

Application of pressure increases the superconducting transition temperature eventually reaching the essentially pressure-independent resistive T_c at a pressure of 3 GPa [121]. With further increase of the applied pressure, T_c decreases. This pressure response, plotted in Figure 59, is reminiscent of that in CeCoIn$_5$ (see Figure 58). Recent NMR measurements in an applied pressure in a Cd-doped sample, CeIr(In$_{0.925}$Cd$_{0.075}$)$_5$, reveal [121] that similar with CeCoIn5, doping with Cd induces AF in zero applied pressure. It is concluded that the antiferromagnetism emerges locally around Cd dopants. An application of pressure suppresses the magnetism suggesting the existence of a pressure-induced QCP hidden by the presence of superconductivity. Neutron scattering measurements under applied pressure and for different Cd doping are required to confirm or deny the validity of this speculation.

2.7 QUANTUM PHASE TRANSITIONS IN ITINERANT MAGNETS

2.7.1 Weak Itinerant Ferromagnets

Although there is clear evidence for AF QCPs, the existence of an FM QCP still remains an open question. Recent substantial experimental and theoretical efforts have indicated that depending on the strength of quenched disorder, the

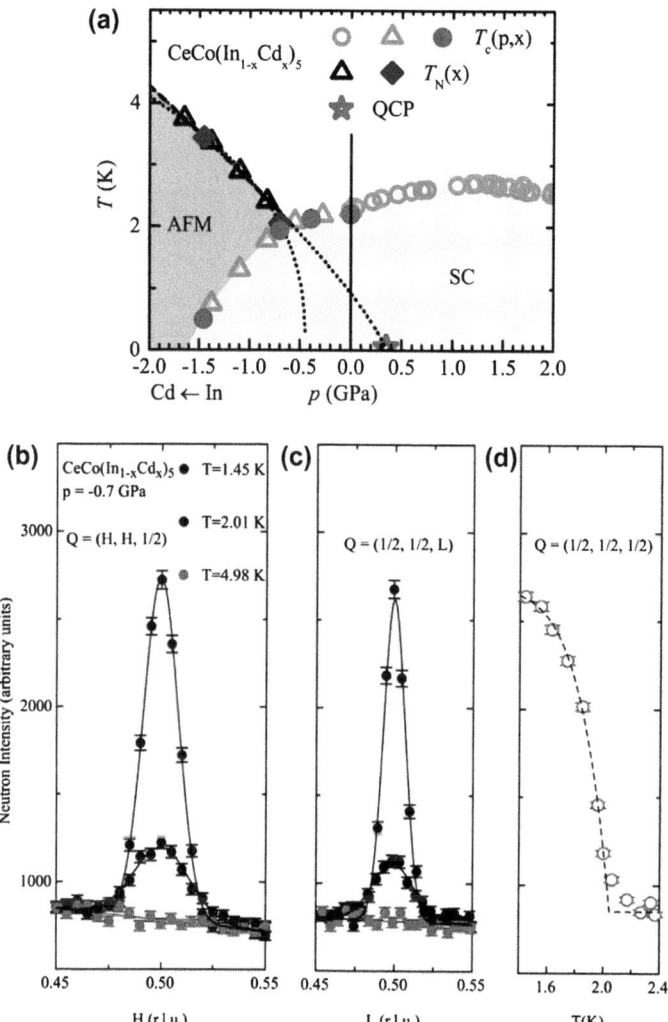

FIGURE 58 (a) The phase diagram of $CeCo(In_{1-x}Cd_x)_5$ with the equivalent "negative" chemical pressures (empty blue (dark gray in print versions) and orange (light gray in print versions) triangles) and positive hydrostatic pressures (empty orange (light gray in print versions) circles). The red (gray in print versions) star represents the position of the putative quantum critical point (QCP). (b–d) Scans in the reciprocal space of the neutron diffraction intensity of $CeCo(In_{1-x}Cd_x)_5$, performed along the lines $\mathbf{Q} = (H,H,0.5)$ in (b) and $\mathbf{Q} = (0.5,0.5,L)$ in (c). Reciprocal lattice units (r.l.u.) are used as coordinates of the reciprocal space. The $CeCo(In_{1-x}Cd_x)_5$ sample has a doping x corresponding to a negative pressure of $p = -0.7$ GPa. (d) Temperature dependence of the neutron diffraction intensity measured at $\mathbf{Q} = (0.5,0.5,0.5)$. SC, superconducting. *Reprinted with permission from Ref. [120].*

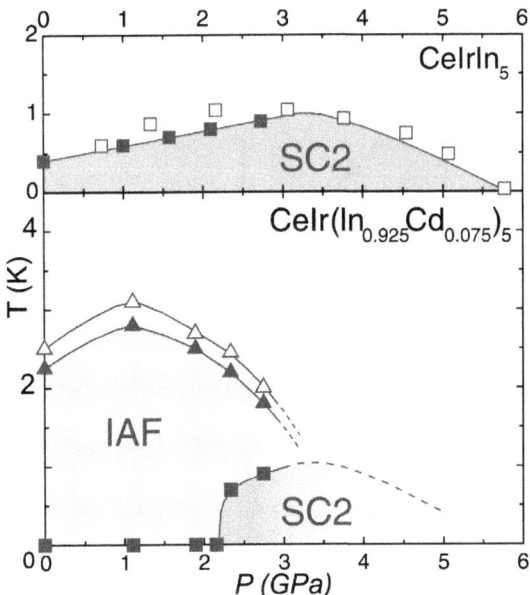

FIGURE 59 The pressure (P)–temperature (T) phase diagram for pure CeIrIn$_5$ (top) and CeIr(In$_{0.925}$Cd$_{0.075}$)$_5$ (bottom). The red (gray in print versions) dome in bottom diagram, represents the inhomogeneous AF ordered phase (IAF). Similar to CeCoIn5, the pressure-induced quantum critical point is hidden by the presence of superconductivity (SC). *Reprinted with permission from Ref. [121].*

ferromagnetic quantum transitions can be classified in different categories [122]. For pure systems, there is a discontinuous, or first-order, QPT from the ferromagnetic to a PM ground state as a function of the control parameter (hydrostatic pressure or uniaxial stress, composition, or an external magnetic field). QPTs in disordered materials are often continuous or second order with QPT singularities in the PM phase near the transition. Transition from the ferromagnetic state to other ground states such as antiferromagnetic or SDWs is also observed in some systems [122].

Instead of displaying a QCP, many FM systems undergo a disorder QPT where a line of second-order transitions at relatively high temperatures is separated from a line of first-order ones at low temperatures by a tricritical point [122]. Theory indicates [105,122] that this is because a 3D itinerant FM QCP is inherently unstable toward either a first-order phase transition or an inhomogeneous magnetic phase such as modulated or textured structures. Indeed several stoichiometrically clean magnetic transition-metal compounds show NFL behavior close to the FM instability but the transition changes to first order [105]. A large variety of materials (including 3d-electrons transition metals and 4f- and 5f-electron systems) display a first-order transition occurs across their phase diagram. Here we consider the QPT is MnSi as an example of FM QPT in a clean system with a first-order transition. We then consider the QPT in Re-doped URu$_2$Si$_2$ as an example of systems with presence of disorder.

2.7.2 MnSi

The intermetallic compound MnSi is a weak 3d itinerant helimagnet [123] that crystallizes in the cubic B20-type noncentrosymmetric structure represented by $P2_13$ space group and with a lattice constant of 4.558 Å [124] (see Figure 60). The simple crystal structure, the high quality of single crystals in combination with its remarkable physical properties such as anomalous resistivity NFL behavior [123,125] over a large region of the phase diagram has made MnSi the subject of intense experimental and theoretical studies over the past several decades.

At ambient pressure and zero applied magnetic field, MnSi undergoes a magnetic transition at $T_c = 29.5$ K to a helical magnetic structure [123]. Polarized and small angle neutron scattering experiments have confirmed [126,127] that in the absence of external magnetic fields, the magnetic structure consists of four left-handed spiral (see Figure 60) domains oriented along four <111> directions with a long helix pitch of approximately 180 Å

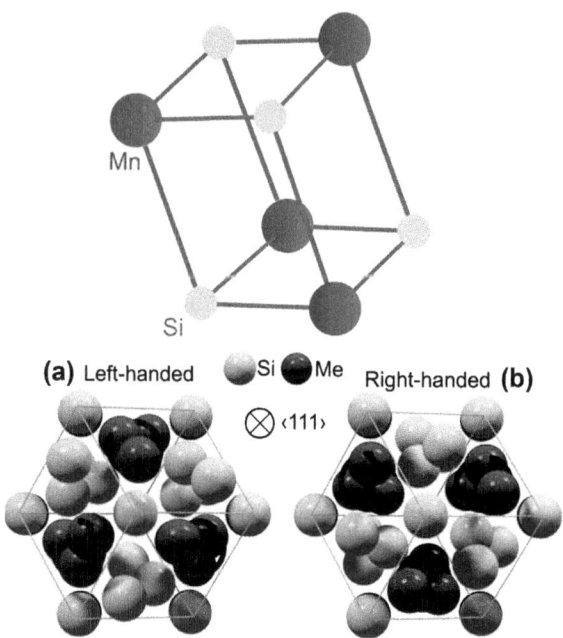

FIGURE 60 (Top) The primitive unit cell of MnSi contains four formula units [124]. Four Mn and four Si ions are situated at the symmetry-related sites: (u,u,u), (12 + u,12 − u,ū), (12 − u,ū,12 + u), and (ū,12 + u,12 − u). (Bottom) The chirality of the Si sublattice shown in panel (a) is left-handed for the left-handed crystal structure ($u_{Mn} = 0.135$ and $u_{Si} = 0.845$). It is right-handed in panel (b) for the right-handed crystal structure ($u_{Mn} = 0.865$ and $u_{Si} = 0.155$). Note that the Mn sublattice shows opposite chirality as indicated by the black lines. *Reprinted with permission from Ref. [128].*

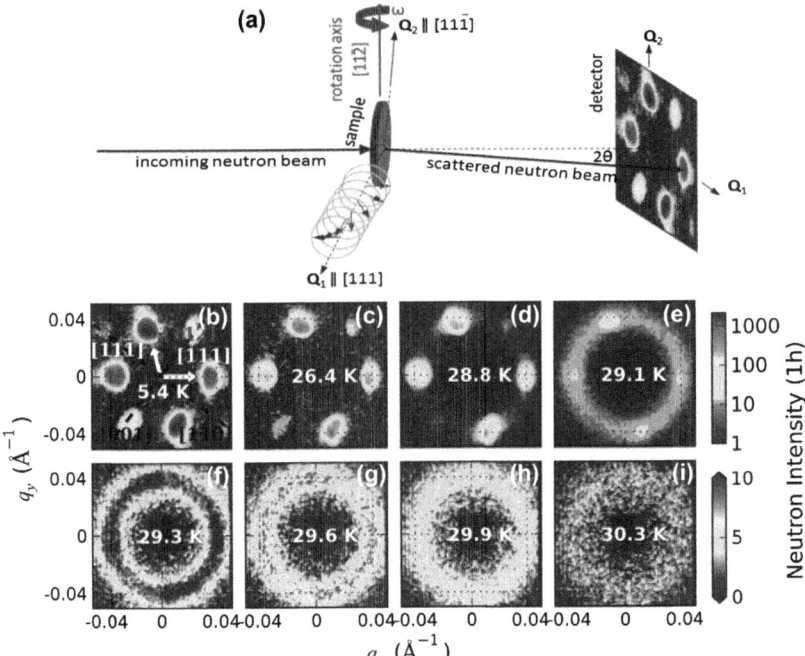

FIGURE 61 (a) The small-angle neutron scattering setup used to study the temperature dependence of magnetic correlations, where the neutron beam scatters off the magnetic helix. Turning the sample around the rotation axis parallel to the (1,1,2) zone axis (red (gray in print versions) arrow) the sample can be rocked through the Bragg condition for the magnetic helix. With this sample orientation the magnetic Bragg condition can be fulfilled for the two magnetic propagation vectors $\mathbf{Q}_1 = (1,1,1)$ and $\mathbf{Q}_2 = (1,1,1)$ (white arrows in panel (b)) corresponding to two of the four possible helical Q-domains. Below T_c (panels (b–d)) discrete magnetic Bragg spots corresponding to the helical order are visible, whereas above T_c (panels (f–i)) the magnetic intensity spreads out over a sphere in reciprocal space. *Reprinted with permission from Ref. [126].*

(see Figure 61). Despite the helical magnetic structure in the absence of external magnetic fields, MnSi is widely regarded as weak itinerant ferromagnet. The long wavelength of the helix allows one to approximate the system as a ferromagnet. In addition, the spontaneous magnetic moment of 0.4 μ_B is considerably smaller compared to the effective PM moment of 2.2 μ_B estimated from the Curie-Weiss (CW) law of the susceptibility in the PM regime [123].

The intimate interplay between three fundamental interactions at different energy scales [123] controls the magnetic properties of MnSi. The strongest interaction is ferromagnetic exchange interaction between the Mn ions which favors a parallel alignment of magnetic moments ($\mu = 0.4\ \mu_B$) in the layers. The weaker Dzyaloshinskii–Moriya spin–orbit interaction, g_{so}, nonzero as a result of noncentrosymmetric crystal structure, tends to align the neighboring spins perpendicular to one another causing the spins twist into a helix.

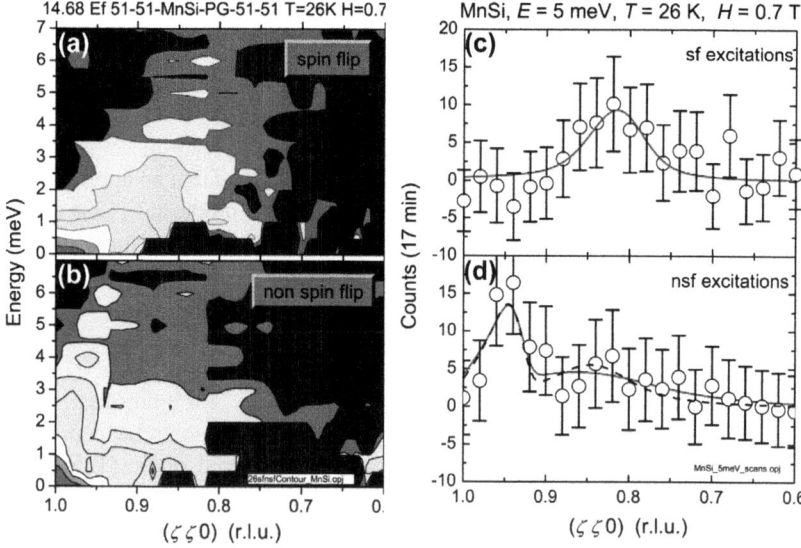

FIGURE 62 Contour maps of MnSi measured with polarized neutron inelastic scattering along the [110] direction at $T = 26$ K and in a field $B = 0.7$ T for (a) spin-flip (SF) and (b) non-spin-flip (NSF) components. The data clearly indicate the spin-wave branch merging into the Stoner continuum. The contours of the NSF scattering (b) show the steep phonon branch as well as the longitudinal fluctuations, which extend to high energy transfers. Constant energy scans for $E = 5$ meV at $T = 26$ K and $B = 0.7$ T for a Stoner excitation near (0.8,0.8,0) for SF (c) and NSF (d) components. *Reprinted with permission from Ref. [129].*

The pitch periodicity is much larger than the lattice spacing and reflects the small coupling constant of this spin—orbit interaction which causes the helical order. Finally the weakest energy scale is the higher-order spin—orbit interaction, g_{so}^2, induced by the cubic CEFs, which pins the helix propagation vector along the $<111>$ crystallographic directions. The typical size of the magnetic domains in the ordered state is 10^4 Å.

Inelastic neutron scattering measurements on MnSi have revealed [129,130] the presence of three types of magnetic excitations characteristic of weak itinerant ferromagnets (see Figure 62): the single-particle spin excitations (weakly temperature-dependent Stoner continuum), the collective spin waves below the Stoner boundary which merge tangentially into the Stoner continuum, and the low-energy PM excitations (critical scattering). It is suggested that the lack of inversion symmetry in MnSi not only leads to a helical magnetic structure in zero external field but additionally to the observed chirality in the magnetic excitation spectrum in the ferromagnetic phase [131]. Moreover, critical chiral magnetic fluctuations have been observed above T_c in the PM phase [132]. The maxima of the critical scattering lie on a sphere with a radius that corresponds to the helix pitch in the helimagnetic phase.

The presence of chiral fluctuations even outside the helimagnetic phase suggests that chirality in MnSi is a key player in determining its complex properties.

2.7.2.1 Applied Pressure

The interest in MnSi was renewed when it was discovered that an application of modest hydrostatic pressure has profound effect on its properties [133] driving the system toward a first-order QPT. The application of hydrostatic pressure suppresses T_c to zero at a critical pressure $p_c = 14.6$ kbar. The pressure and temperature phase diagram is shown in Figure 63. The transition temperature to the helimagnetic order, T_c, is reduced with increasing hydrostatic pressure, resulting in first-order phase transition for $p^* \leq p \leq p_c$ ($p^* \sim 12$ kbar, $p_c = 14.6$ kbar) [133]. Moreover, for pressure range $p_c \leq p \leq p_0$ ($p_0 \sim 21$ kbar), neutron scattering measurements have provided evidence for the presence of an unusual partially ordered phase [125] below a characteristic temperature T_0 which is not apparent in susceptibility or resistivity measurements. As seen in Figure 64, while the helimagnetic order is

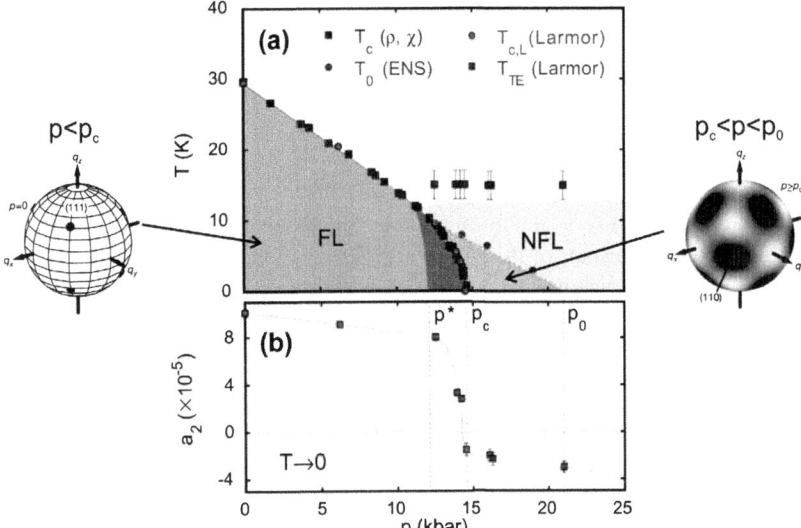

FIGURE 63 (a) Phase diagram of MnSi as a function of pressure. (b) Extrapolated zero-temperature variation of lattice constant related to the applied pressure. The blue (gray in print versions) shading indicates the regime of Fermi liquid (FL) behavior, where dark blue (dark gray in print versions) shading shows the regime of phase segregation seen in μSR. The green (lighter gray in print versions) shading represents the regime of non-Fermi liquid (NFL) resistivity, where dark green (light gray in print versions) shading indicates the regime of partial order. The transition temperature $T_{c,L}$ is obtained from lattice constant, and the crossover temperature T_{TE} represents the appearance of lattice contraction, both extracted from neutron Larmor diffraction data. *Reprinted with permission from Refs [125,133].*

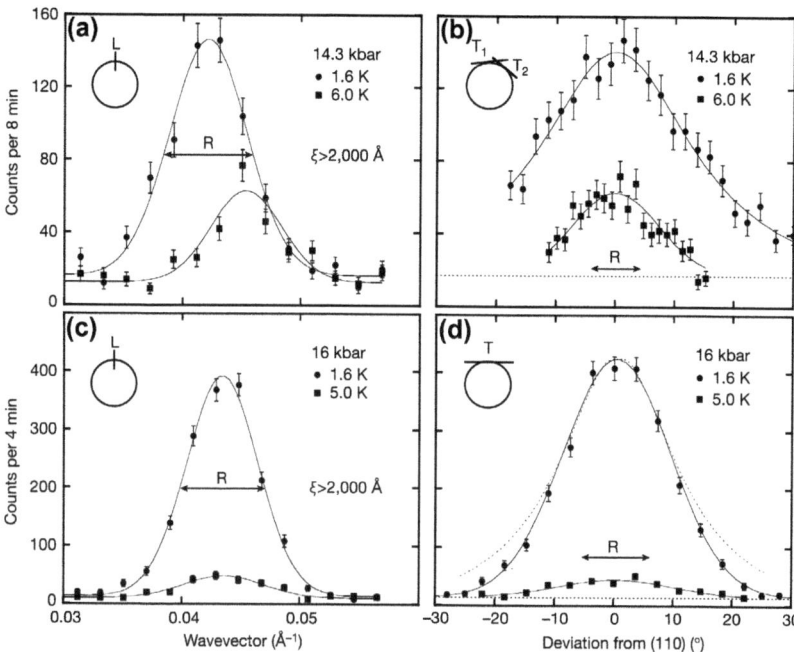

FIGURE 64 Longitudinal (L) and transverse (T) scans of the magnetic neutron scattering intensity with respect to (1,1,0). Data are shown for 14.3 kbar (a and b), just below p_c, and 16 kbar (c and d), well above p_c. The direction of the scans with respect to a cross section of the spherical surface is depicted in the top-left corner of each panel. The arrow marked R shows the instrumental resolution. *Reprinted with permission from Ref. [133].*

characterized by resolution-limited peaks corresponding to spiral wave vectors $\sim 0.037\ \text{Å}^{-1}$ parallel to the <111> directions, in the partially ordered phase the scattering signal spreads diffusely over a sphere of radius $\sim 0.043\ \text{Å}^{-1}$, weakly preferring the (1,1,0) directions over the (1,1,1) and (1,0,0) directions. The intensity distribution with a broad maximum at (1,1,0) suggests that the direction of the helix, fixed by crystal field interactions below p_c, is unlocked above p_c.

Even though MnSi is well described by a weakly ferromagnetic FL model at low temperatures and ambient pressure, transport measurements [133] have shown that resistivity exhibits an NFL behavior at the pressures larger than p_c. NFL behavior in MnSi occurs over a very large temperature range (three orders of magnitude from a few millikelvin to a crossover temperature of approximately 12 K). Furthermore, the suppression of magnetism under hydrostatic pressure over a wide pressure range suggests the existence of a new phase of matter [133]. As the FL phase of MnSi vanishes at the same pressure p_c where the partial magnetic order appears, the depinning of the helical order may be regarded as a good driving candidate for the FL to NFL transition.

However, the partial magnetic order is not observed for the complete NFL temperature range but only below a crossover temperature T_0 that vanishes at the pressure p_0. In addition, the Muon-spin rotation (µSR) experiments indicate [133] that below p_c the helical order occurs in a decreasing volume fraction for $T_c \to 0$, as represented in the dark blue shaded area in Figure 63. The abrupt drop of the zero-field muon spin precession frequency at p_c supports the first-order nature of the magnetic phase transition at P_c.

The profound effect of applied pressure on the different properties of MnSi has resulted in a key question: Is the transition from the FL to the NFL regime driven by a QCP? In general for a 3D metal, such as MnSi, the breakdown of FL theory is only expected at QCPs [98]. In MnSi, the fact that $T_c \to 0$ at P_c was used to argue that the phase transition is only "weakly" first order and therefore it can be explained in terms of a ferromagnetic QCP [134] resulting in the NFL behavior. However, this explanation is in contradiction with the observed signatures of a first-order transition in the ac-susceptibility, and the existence of the magnetic order above P_c which clearly suggests only the weakest energy scale (the cubic anisotropy energy) is suppressed.

Recent thermal expansion measurements conducted [125] under pressure by means of high-precision neutron Larmor diffraction show no singularity at P_c, demonstrating that the transition cannot be explained by a QCP (see Figure 63). This is because the thermal expansion is the associated variable to the control applied pressure that is used to tune the system across the FL to NFL phase transition. Hence, it provides a unique handle to investigate the nature of the transition. Similar results for the transition at P_0 also suggest that the NFL phase is instead a novel metallic state far from QC [125]. High-precision neutron Larmor diffraction is a relatively new technique [135] based on neutron resonance spin echo. In this technique each single neutron is marked by a Larmor precession phase. This phase is independent of the Bragg angle or the velocity of the single neutron and only depends on the lattice spacing d, thus allowing for a determination of lattice spacing with very high accuracy ($\Delta d/d \sim 10^{-6}$).

2.7.3 URu_2Si_2

The uranium-based intermetallic compound, URu_2Si_2, is a heavy-fermion superconductor ($T_c = 1.5$ K) [136] with a simple tetragonal structure (space group I4 mmm) as shown in Figure 65. Interest in this material grew substantially in 1984, when specific heat measurements clearly indicated the presence of yet another second-order phase transition at a higher temperature of $T_0 = 17.5$ K to a "hidden-order" phase. The small antiferromagnetic moment of 0.03 μ_B that develops at a commensurate antiferromagnetic wave vector (1,0,0) below ~ 17 K cannot explain the large specific heat jump at this phase transition [136]. Antiferromagnetism, therefore, cannot be the main cause of the hidden order. The system appears to have condensed into a new

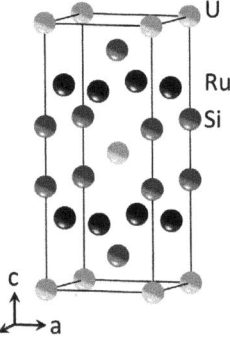

FIGURE 65 The crystal structure of URu_2Si_2 is depicted. It is body-centered tetragonal with space group I4 mmm.

phase of matter for which the order parameter and associated symmetries differ from conventional expectations. Detailed neutron scattering measurements of magnetic excitations found two distinct modes: one at the commensurate antiferromagnetic wave vector, $Q_C = (1,0,0)$ and the other at the incommensurate wave vector, $Q_{IC} = (1.4,0,0)$ [137] (see Figure 66). Below T_0, both modes appear much more strongly, with a gap opening for the excitations at the incommensurate wave vector.

Neutron scattering experiments indicate [138] that the crystal fields and orbital currents cannot be considered as a source of hidden order. Extensive search by means of neutron scattering measurements for the presence of such orbital currents revealed no sign of scattering predicted by the model

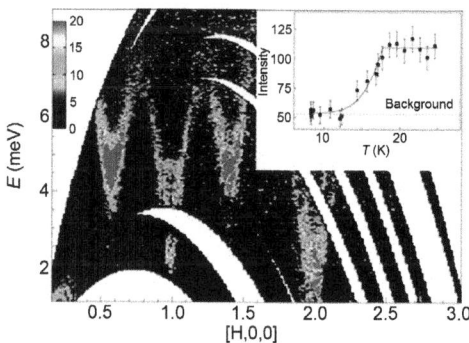

FIGURE 66 Excitation spectrum of URu_2Si_2 in the $(H,0,0)$ plane at $T = 1.5$ K (where H and L are reciprocal lattice vectors a^* and c^*, L is integrated from -0.12 to 0.12). Note the minima at the AF zone center $(1,0,0)$ and incommensurate positions $(1 \pm 0.4,0,0)$. The feature at $(2,0,0)$ is due to phonons. The inset shows how the incommensurate excitations become gapped through the transition by counting at the point $(0.6,0,0)$ at 0.25 meV transfer on a triple-axis spectrometer. *Reprinted with permission from Ref. [137].*

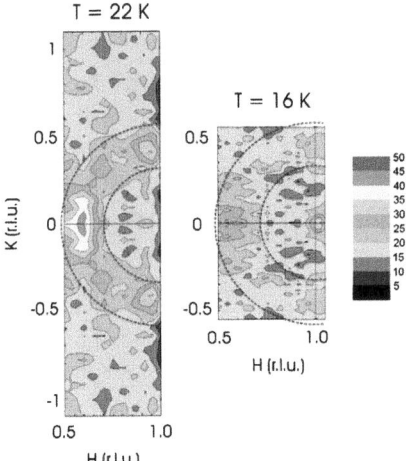

FIGURE 67 Contour plot of scattering in the $(H,K,0)$ plane at $T = 22$ K and $T = 16$ K with 0.25 THz (~ 1 meV) energy transfer showing the formation of a ring of quasielastic scattering forms in the $(H,K,0)$ plane centered at an incommensurate radius 0.4 from the $(1,0,0)$ AF Bragg position. Similar searches in the $(H,K,0)$ and $(H,0,L)$ planes in the elastic channel revealed no sign of scattering predicted by the orbital model. *Reprinted with permission from Ref. [138].*

(see Figure 67). The spins fluctuating above the 17 K transition are centered on the incommensurate wave vector $(1 \pm \delta, 0, 0)$ with $\delta = 0.4$. Emerging from that wave vector is a high-velocity cone of strongly damped gapless excitations that extend over a finite region of the Brillouin zone in this precursor phase to hidden order. They provide evidence of itinerant rather than localized spins.

Furthermore, neutron scattering study of magnetic excitations above T_0 show [137] that these gapless spin fluctuations give rise to a term in the specific heat that closely accounts for the magnitude of the giant-specific heat linear in T that was previously attributed entirely to electrons (see Figure 68). In addition, the decrease of the specific heat below T_0 is now understood to arise from the formation of a spin gap below T_0. More recently, the excitations have been interpreted as the response of itinerant spins to Fermi surface nesting, similar to that of chromium, and specific nesting vectors were proposed [139]. Despite this considerable progress, the symmetry of the order parameter that condenses in the hidden-order phase remains unknown.

2.7.3.1 Chemical Pressure (Doping)

Substitution of Re, Tc, or Mn for Ru atoms suppresses the hidden-order phase establishing a ferromagnetic ground state, instead, beyond a certain dopant concentration [140]. Here we consider $URu_{2-x}Re_xSi_2$ as it has been studied in more details among different types of doping. For Re doping the system develops an ferromagnetic (FM) ground state for $x > 0.15$ but it appears that

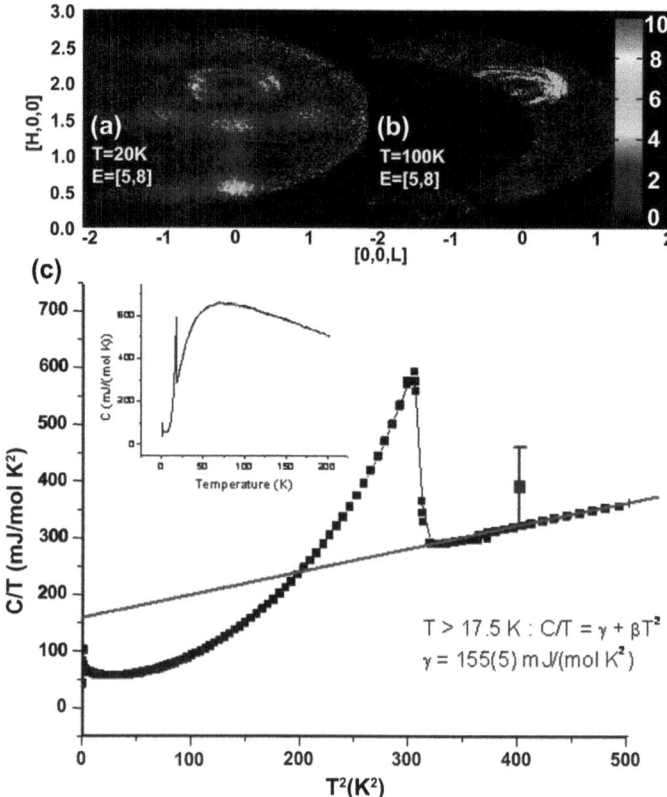

FIGURE 60 Cuts in the $(H,0,L)$ scattering plane integrated from 5 to 8 meV at 20 K (a) and 100 K (b). The incommensurate scattering disappears at 100 K (indicated by a circle) just above the "coherence" temperature where a signature of heavy-quasiparticle formation is seen in the specific heat (inset in c). (c) The C/T as a function of T^2 shows the 17.5 K anomaly. The linear portion of the specific heat, γ, is calculated with the solid line to be 155(5) mJ mol^{-1} K^{-2}. The blue (gray in print versions) data point is the calculation of $\gamma = 220(70)$ mJ mol^{-1} K^{-2} from the spin fluctuations observed at 20 K. *Reprinted with permission from Ref. [137].*

the exact critical concentration is hard to determine [140]. T_c increases monotonically with increasing x and reaches a maximum of almost 40 K at $x = 0.8$, above which the material does not remain in a single phase. Neutron scattering experiments show clear evidence for the existence of FM long-range order for $x = 0.8$ [141]. Figure 69 shows the phase diagram as a function of Re doping.

In a recent detailed neutron scattering study [142] of URu$_{2-x}$Re$_x$Si$_2$ with $x = 0.1$ (close to critical doping concentration), no elastic incommensurate scattering was found at the wave vector (1.4,0,0) nor at the commensurate antiferromagnetic point (1,0,0). The latter is in contrast to the parent material, where the weak elastic scattering at the commensurate point is believed to

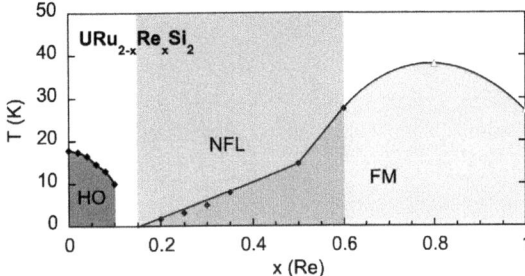

FIGURE 69 Magnetic phase diagram of $URu_{2-x}Re_xSi_2$ showing the antiferromagnetic (hidden order) and ferromagnetic phases. *Reprinted with permission from Ref. [140].*

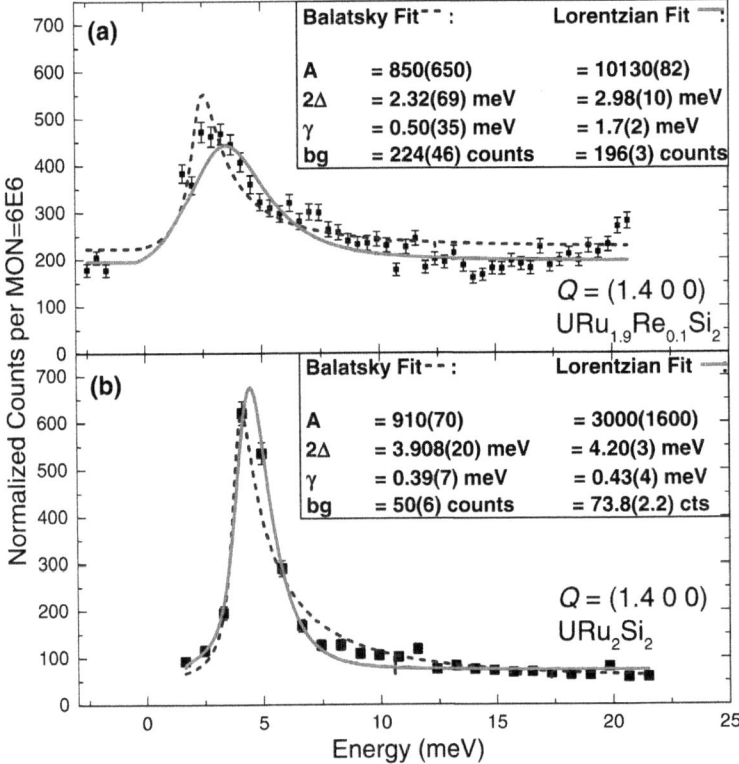

FIGURE 70 Incommensurate fluctuations (a) for $URu_{1.9}Re_{0.1}Si_2$ at 2 K and (b) for pure URu_2Si_2 at 3 and 5 K (combined data). For the Re doping, the nesting gap energy, is reduced to 60% of its value in the pure system. The observed increased spectral width suggest that the fluctuations are highly damped by doping. *Reprinted with permission from Ref. [142].*

arise from a minority AF phase [136]. As seen in Figure 70 the dynamic magnetic excitation spectrum of $URu_{1.9}Re_{0.1}Si_2$ measured with neutron scattering as a function of temperature also exhibits an incommensurate spin gap at low temperatures similar to that in pure URu_2Si_2 at the same incommensurate wave vector. However, doping has reduced the gap value. These measurements also indicate while the intensity at the incommensurate wave vector is reduced. Doping also increases the spectral width indicating that fluctuations are damped. The slowing of fluctuations is more dramatic at the commensurate hidden-order wave vector where the lifetime is so short that the characteristic energy barely gives a peak in the spectrum. It may also signify the destruction of perfect nesting by charge impurities. It appears that toward the FM QCP induced by Re doping, the gapped incommensurate fluctuations remain robust while the commensurate spin fluctuations (related to hidden order) are significantly weakened and slowed down [142].

As expected for close proximity to a QPT, pronounced NFL behavior has been observed [122] in several thermoelectric properties such as specific heat, the electrical resistivity, and dynamical magnetic susceptibility for a large concentration range $0.15 < x < 0.8$. Transport measurements indicate that the system is highly disordered as evident from the residual resistivity [140]. Difficulties in defining the precise critical concentration have hampered efforts in determining the critical exponents [140]. It is also suggested that the presence of strong disorder and interplay between remnants of the hidden-order and the induced ferromagnetism is behind the unusual behavior near the onset of ferromagnetism [140].

2.7.3.2 Applied Pressure

Extensive studies of the properties of URu_2Si_2 under applied pressure, including neutron scattering measurements under hydrostatic pressure, have provided important clues on the nature of the superconductivity, magnetism, and the hidden order in this material. As seen in Figure 71, these studies show [143] that applied pressure initially strengthens the hidden order and increases T_0. This is in marked contrast from other perturbations such as chemical pressure as discussed in the case of Re doping (or applied magnetic fields) that tend to destroy the hidden-order phase. Neutron scattering measurements under an applied pressure reveal that beyond a critical pressure, $P_c \sim 0.8$ GPa, standard Néel AF ordered phase emerges from within the hidden-order dome. The pressure-induced phase diagram is shown in Figure 71. Since this region has a larger antiferromagnetic moment (0.4 μ_B) than the hidden-order region, it has been called the large moment antiferromagnetic (LMAF) phase.

The spin correlations appear qualitatively similar between the LMAF phase and the hidden-order phase. This suggests that the hidden-order phase is closely related to the conventional antiferromagnetic phase that arises with the application of hydrostatic pressure beyond P_c. Moreover, from the fact that the end point of the superconducting phase boundary extrapolates to P_c, one can

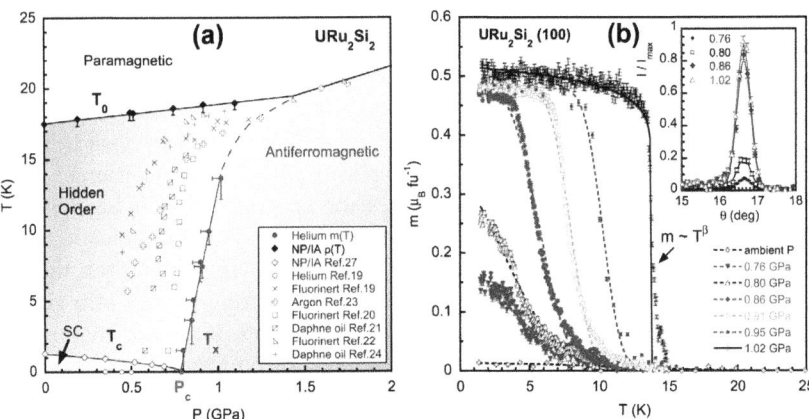

FIGURE 71 (a) Pressure-temperature phase diagram of URu_2Si_2 comparing various pressure media. T_0 initially increases before an AF phase emerges beyond 0.8 GPa. It is suspected that superconductivity and antiferromagnetism meet at a multicritical quantum point, P_c. (b) Temperature dependence of the AF moment under hydrostatic pressure measured by elastic neutron scattering. Rocking scans at 1.5 K normalized to the intensity at 1.02 GPa are shown in the inset. The sharp onset and weak slope in the ordered phase suggests the transition is discontinuous. *Reprinted with permission from Ref. [143].*

conclude that the superconducting pairing energy scale goes to zero exactly at the onset of LMAF phase. Thus there is an intimate relationship between superconducting, LMAF and hidden-order phases in URu_2Si_2.

2.8 QUANTUM PHASE TRANSITIONS IN TRANSVERSE FIELD ISING SYSTEMS

2.8.1 Transverse Field Quantum Ising Model

Although the underlying physics of QPTs are complex and challenging, there are few relatively simple models which can demonstrate the fundamentals of this rich phenomena. Among them, Ising magnet in a transverse external magnetic field is the simplest quantum spin model. Choosing the z-axis to be the Ising axis, the corresponding Hamiltonian is given by:

$$H = -\sum_{\langle ij \rangle} J_{ij}^{zz} S_i^z S_j^z + \Gamma \sum_i S_i^x \qquad (33)$$

where S_i^z and S_i^x are the z and x components of the spin at lattice site i, respectively. $J_{ij}^{zz} > 0$ is the longitudinal exchange coupling along z, and Γ is an effective transverse magnetic field along x, perpendicular to the Ising axis. The first term in the Hamiltonian describes the magnetic interaction between the nearest-neighbor spins, and the second term corresponds to the transverse magnetic field, which forces the spins to align along x. In the absence of

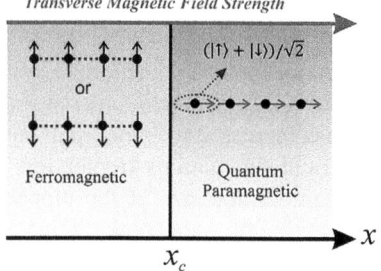

FIGURE 72 Schematic phase diagram of spin Ising model system as function of tuning parameter x (transverse magnetic field) at zero temperature. For $x < x_c$ the ground state is an ordered ferromagnet; while for $x > x_c$ the ground state is a quantum paramagnet which is a superposition of spin-up and spin-down states. *Adapted from Ref. [53].*

transverse magnetic field, when $\Gamma = 0$ the model reduces to the recognized classical Ising model. In the classical limit and at zero temperature when there are no correlations with other degrees of freedom, such as phonons, all possible spin configurations are dynamically stable and all spins are parallel.

Figure 72 presents a schematic phase diagram of the spin Ising model system at zero temperature and as a function of the tuning parameter x; in this case x depicts the transverse magnetic field strength which disrupts the order. In order to further appreciate the influence of the transverse magnetic field, the second term of Eqn (33) can be rewritten in the following form:

$$\Gamma \sum_i S_i^x = \Gamma \sum_i \left(S_i^+ + S_i^- \right) \quad (34)$$

where S_i^+ and S_i^- are the spin-flip operators at site i. This expression clearly demonstrates that the transverse field prompts quantum spin tunneling between the spin-up and spin-down states. These spin flips are in fact the same quantum fluctuations that melt the magnetic order at Γ_c and induce QPT.

When transverse field is increased above the critical field $\Gamma_c \equiv x_c$, the system undergoes a QPT into a PM state. Contrary to the classical paramagnet which can be described in terms of spin fluctuations between the up and down states, quantum paramagnet ground state is a superposition of the spin-up and spin-down states, where the spins will eventually point in the direction of the applied field at the extreme high limit [12]. The spectral density of these quantum fluctuations can be characterized by an energy scale using the Jordan–Wigner transformation, mapping the spins to noninteracting spinless fermions [14], where $\Delta(\Gamma)$ corresponds to the gap energy between the ground state and the first excited state, and $\Delta(\Gamma) \to 0$ as $\Gamma \to \Gamma_c$. The change in the ground state as a function of Γ is accompanied by a corresponding change in the nature of the low-lying excitations and thus a change in the associated energy gap $\Delta(\Gamma)$. Recent advancements in neutron scattering techniques have enabled researchers to tune the ground state by applying magnetic field and thus detecting such energy excitations [144].

As noted above and according to Eqn (33), for $\Gamma = 0$ and $J_{ij}^{zz} > 0$ all spins are aligned parallel forming a ferromagnetic ordered state at zero temperature. The moment, temperature is increased to some small but finite value, the order is destroyed and the thermal fluctuations trigger few spins to flip into the opposite direction. The number and size of the flipped region expands with rising temperature, randomizing the spin configuration and resulting in the fall of the total magnetization of the system. Eventually at a critical temperature the magnetization vanishes and the system becomes PM. Figure 73 shows the schematic phase diagram of the Ising model system in transverse magnetic field as function of temperature. The intermediate regime marked in Figure 73 corresponds to the quantum critical region close to the QCP, where the system displays the intensely coupled dynamics of complex entangled quantum excitations.

Despite its simplicity the experimental realization of the transverse field Ising model systems is quite rare. To explore the underlying physics of QPT in spin Ising model system, several key components are required. The main challenge is to find an Ising chain compound with sufficiently small exchange interaction J of a few millielectron volt, that can be matched within the limits of achievable laboratory magnetic fields in order to reach the QCP and tune the QPT (1 meV ~ 10 T). Additionally in real-world materials, the exchange interactions are truly 3D and thus there are inevitable interchain couplings, which introduce more complex dynamics into the system, thus finding a material with well-isolated spin chains along a strong easy axis is a significant factor. Remarkably, LiHoF$_4$ and CoNb$_2$O$_6$ are two of the exceptional prot10typical Ising system compounds which their QCP can be experimentally accessed.

LiHoF$_4$ is a nearly ideal 3D Ising ferromagnet. Although LiHoF$_4$ is not a low-dimensional system, remarkably, it does host a rich collection of

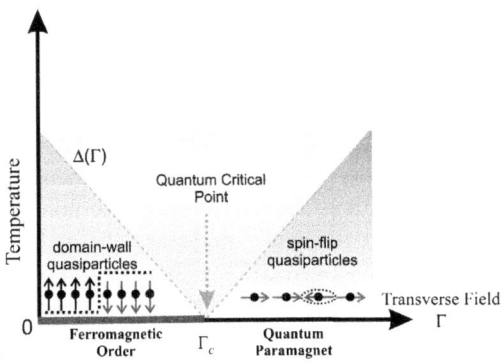

FIGURE 73 Schematic phase diagram of spin Ising model system at finite temperatures and in transverse magnetic field. *Adapted from Ref. [144]*.

collective quantum phenomena, ranging from quantum annealing, to entanglement and coherently oscillating spin clusters, to tunneling of single moments and domain walls. In LiHoF$_4$, magnetic moments are carried by Ho^{3+} ions, for which the magnetic interaction is weak and dominated by the classical dipole−dipole interactions. Random substitution of nonmagnetic Y for Ho in LiHoF$_4$ tunes the classical quenched disorder, resulting in a variety of quantum mechanical phases as function of dilution: ranging from ferromagnetic, to spin glass, and to antiglass (spin-liquid like) phase (for more reading, see Refs [145−151]). Additionally, application of transverse magnetic field can independently tune quantum fluctuations, inducing QPTs and novel criticality phenomena in this system [152]. The Hamiltonian of LiHoF$_4$ includes terms for the crystal field, the hyperfine interactions, the Zeeman effect, and the classical dipole−dipole interactions, in addition to the nearest-neighbor exchange interactions.

The magnetic properties of LiHoF$_4$ have been studied through a sequence of experimental techniques [153,154]. The $H-T$ phase diagram of LiHoF$_4$, determined from susceptibility measurements, is shown in Figure 74 [152]. The phase boundary separates the PM and ferromagnetic phases. At zero field, pure LiHoF$_4$ orders ferromagnetically at temperatures below 1.53 K, whereas the ordering temperature reduces under applied transverse magnetic field, approaching zero at critical field of 49.3 kOe as system crosses to the PM phase. The solid line through the data shown in Figure 74 corresponds to the mean field description used to analyze the data, which incorporates the known hyperfine interaction coupling of the Ho nuclear spins and the electrons, as well as the real crystal field parameters for the electrons attached to the Ho^{3+} ions.

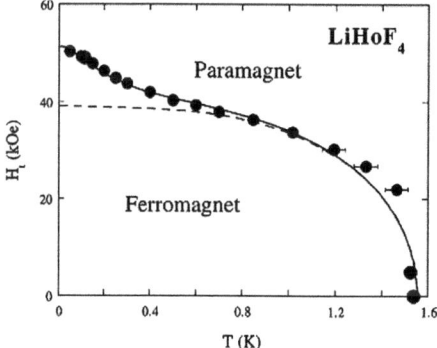

FIGURE 74 Phase diagram of LiHoF$_4$ extracted from susceptibility measurements as a function of temperature and transverse magnetic field. Dashed line is mean field theory including only the electronic degrees of freedom, solid line is full mean field theory including nuclear moments as well. *Reprinted with permission from Ref. [152].*

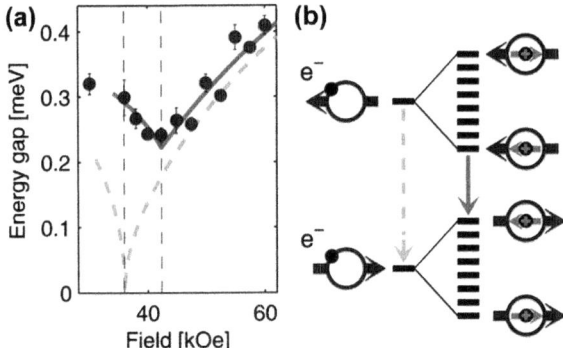

FIGURE 75 (a) Map of the energy spectrum of LiHoF$_4$ as a function of transverse magnetic field. The solid and dashed lines represent calculated energies obtained with and without hyperfine coupling, respectively. (b) Schematic cartoon of the effect of hyperfine coupling of electronic (blue (dark gray in print versions)) and nuclear (red (gray in print versions)) energy levels as transverse field approaches QCP. *Reprinted with permission from Ref. [155].*

Neutron spectroscopy techniques performed in applied transverse magnetic field have been used to study the dynamics of LiHoF$_4$ through the QPT [155]. The results confirm the electronic mode softening expected for QPT in simple Ising model system. The results, however, reveal that at the critical field the mode reaches a finite energy due to the hyperfine coupling to the nuclear spins that act as a spin bath with local degrees of freedom. The experimental results are in good agreement with the theoretical calculations, as plotted in Figure 75. Further studies have suggested that the nuclear spin bath has dramatic effects on the phase diagram and excitation spectrum of LiHoF$_4$ near the QCP by controlling the length scale over which the excitations could be entangled [156].

While LiHoF$_4$ is a 3D Ising model system, CbNb$_2$O$_6$ is an excellent real-world realization of the one-dimensional (1D) spin Ising model. The low dimensionality of CbNb$_2$O$_6$ makes it an ideal system to investigate the QPT phenomena in the transverse field Ising model systems. In the section below we briefly review the characteristics of QPT in this compound. A comprehensive review of QPTs in the transverse field spin Ising models is provided in Ref. [157].

2.8.2 CoNb$_2$O$_6$

CoNb$_2$O$_6$ crystallizes in the orthorhombic columbite structure with space group Pbcn. The Co^{2+} magnetic ions are arranged in Co–O layers, and due to the crystal-field effects from the distorted CoO$_6$ environment, their spins align parallel or antiparallel into zigzag chains through Co–O–Co superexchange interactions along a preferred crystalline axis in the a–c plane, forming an ordered 1D spin chain. The crystal structure of CoNb$_2$O$_6$ is shown in Figure 76.

Quantum Phase Transitions Chapter | 2 131

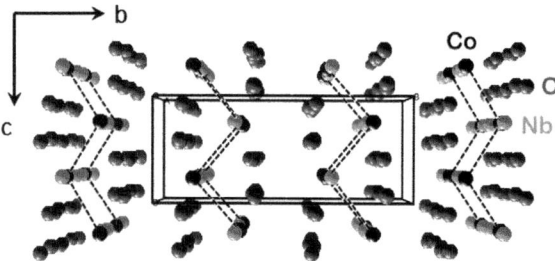

FIGURE 76   Crystal structure of $CoNb_2O_6$.

In addition to the dominant intrachain ferromagnetic exchange couplings, there are also weak interchain antiferromagnetic coupling in $CoNb_2O_6$, thus each Co chain is weakly coupled to its neighbors. At low temperatures, the interplay between the effective interchain interactions and the single ion anisotropy of Co^{2+} results in the formation of rich 3D magnetic ordering. A number of powder and single-crystal neutron diffraction experiments were performed to resolve the magnetic structure of $CoNb_2O_6$. Initial neutron diffraction measurements revealed that at low temperatures in zero field, $CoNb_2O_6$ magnetically orders in a sinusoidally modulated incommensurate structure at 2.95 K, and into a noncollinear antiferromagnetic structure at 1.97 K [158,159]. Figure 77 represents neutron diffraction data collected at magnetic Bragg reflection in the 3D ordered state [160]. The effect of the interplay between the interchain interactions and single-ion anisotropy has been investigated by measuring the deviation of the magnetic Bragg scattering function from the delta function. Further high-resolution powder neutron diffraction investigations of the low-temperature magnetic properties of $CoNb_2O_6$ have confirmed the two previously reported ordered states and have further revealed the existence of another magnetic phase with different propagation vector for temperatures below 1.97 K [161]. Moreover, neutron diffraction experiments performed in finite magnetic fields present the magnetic phase diagrams for several field directions [162,163].

Recent high-resolution inelastic neutron scattering measurements performed on single crystal of $CoNb_2O_6$ have revealed a dramatic change in the fundamental characteristics of spin excitations in the ordered phase and the PM phase [144,164]. Considering that the direction of the spin alignment z is in $a-c$ crystallographic plane, in order to tune the critical point, a transverse magnetic field has been applied parallel to the b-axis, perpendicular to the local Ising axis. Since in $CoNb_2O_6$ the exchange interaction J is small ($J \sim 1.94$ meV), at critical field of 5.5 T, the long-range 3D magnetic ordered phase is suppressed in a continuous QPT. The field-dependence profile of the 3D magnetic Bragg peak is shown in Figure 78.

Inelastic neutron scattering simulates excitations in the spin chain. At transverse fields below QCP, each neutron flips one of the spins in the chain,

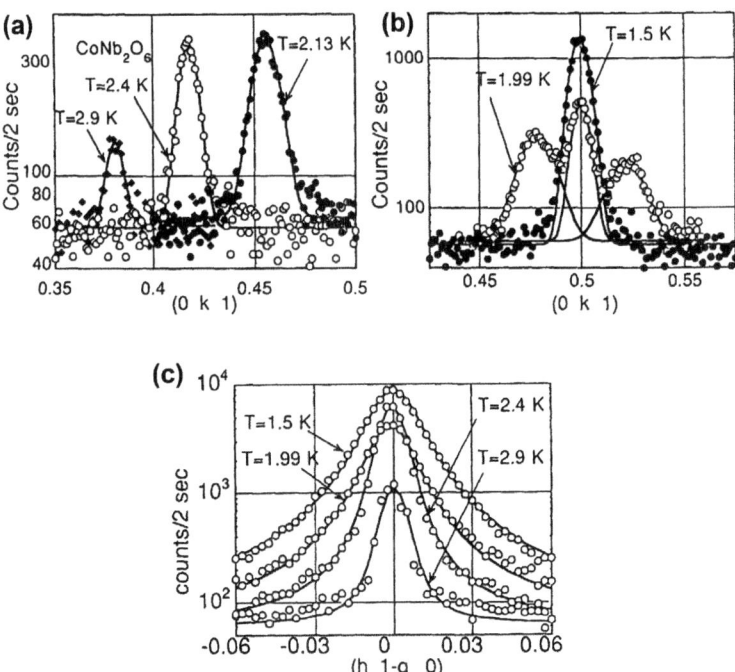

FIGURE 77 Neutron diffraction results showing the temperature dependence of the profile of the magnetic Bragg reflections of $CoNb_2O_6$ along (0,k,1) for $0.35 < k < 0.5$ (a) and $0.45 < k < 0.55$ (b) and along (h,1-q,0) (c). Reprinted with permission from Ref. [160].

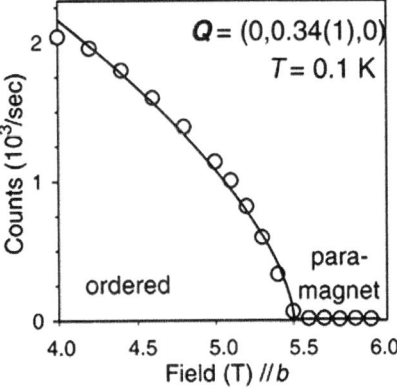

FIGURE 78 Intensity of magnetic Bragg peak collected as function of applied transverse field. Reprinted with permission from Ref. [144].

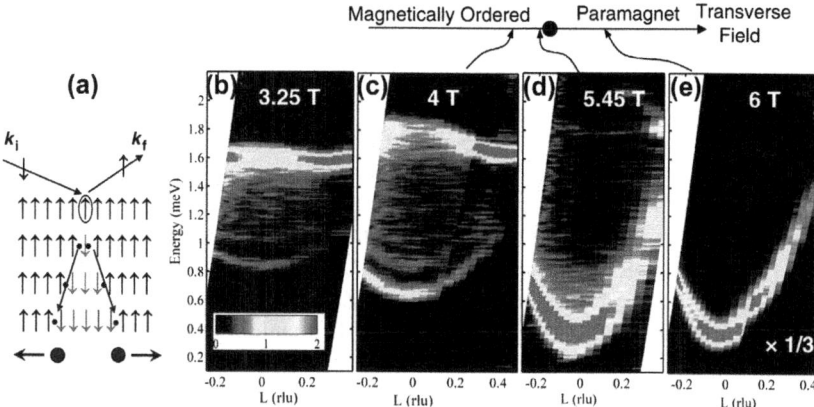

FIGURE 79 (a) Cartoon of a neutron spin-flip scattering below the critical field. Each neutron flips one spin in the chain which produces two domain walls, creating a pair of independently propagating kinks in a ferromagnetically ordered chain. (b–e) Color maps of spin excitations in the vicinity of the critical field as a function of wave vector along the chain. *Reprinted with permission from Ref. [144].*

producing two domain walls, or kinks, which may propagate away from one another, as shown schematically in Figure 79(a). Thus for $\Gamma < \Gamma_c$ neutrons scatter by creating a pair of kinks, interpolating between the two spin-up and spin-down ground states. For $\Gamma > \Gamma_c$, however, the magnetic order is suppressed, and the system is in a PM state. In this case the inelastic neutron scattering excitations consist of spin-flip quasiparticles (see Figure 73). Figure 79(b–e) shows the results of neutron scattering measurements as function of energy and L for applied transverse fields in the vicinity of QCP, where L represents the wave vector of excitations.

The continuum spectrum presented in Figure 79(b) and (c) corresponds to the ordered phase for $\Gamma < \Gamma_c$ and is characteristic of the formation of kink pairs. Transverse magnetic field directly tunes the kinetic energy of the kinks, causing them to hop, which consequently results in the increase of the continuum bandwidth and the decrease in its gap energy. The signature of these excitations is observed in the neutron data: the spectrum at high energies is attributed to the kinetically bound pair kinks traveling in the same direction and the one at low energies corresponds to the dispersion curve of a single kink. The separation between the two spectrums is the widest for $L = 0$ (the kinks have equal momentums in opposite directions) and decreases for $L > 0$ (kinks have different momentums). Figure 79(e) shows the inelastic neutron scattering results for $\Gamma > \Gamma_c$, demonstrating only a single sharp mode with quadratic dispersion, typical of the excitations associate with only one quasiparticle (the flipped spin) in the PM phase. Figure 79(d) shows the spectrum close to QCT and illustrates characteristics of both phases, where both continuum and discrete spectra are present.

We now turn to the effect of natural interactions between the adjunct 1D Co^{2+} Ising chains in $CoNb_2O_6$ crystal. This intrinsic weak interchain coupling provides a small longitudinal magnetic field in the system in the form of added perturbations in the 1D Ising model Hamiltonian, which gives rise to the rich structure of bound states [165–167]. At QCP the quantum critical state should have a gapless continuum of states, however, this additional longitudinal field stabilizes the bound states by creating a series of kinks, mutating them into the quantum resonances which are a signature of the underlying emergent symmetries that govern the physics of the QCP.

Twenty years ago, theorist Alexander Zamolodchikov proposed that the spectrum in the vicinity of the QCP of the 1D quantum Ising chain in a weak longitudinal field is governed by the so-called Lie group E8 symmetry, where a group of eight bound-state quasiparticles (kinks) emerge and the energies of the two lower bound states approach the specific "golden ratio" [168]. Despite the fact that the exceptional Lie group E8 has a very complex mathematical structure, amazingly it can be verified with neutron scattering. Even at zero applied transverse field, neutron scattering produces pairs of kinks. In the ordered phase and in the presence of the longitudinal field, kink propagation disturbs the interchain interactions. As the kink separation grows, the domain between them grows, and accordingly due to the coupling between the neighboring chains, the energy cost to create larger domains grows as well. As a consequence of this energy disadvantage, a type of string tension between kinks arises which repels their separation. This "kink confinement" is ascribed to the creation of discrete bound states of the kink pairs.

Figure 80 presents the results of inelastic neutron scattering measurements of $CoNb_2O_6$ at zero applied transverse field as a function of temperature. Figure 80(a) demonstrates a gapped continuum scattering at the ferromagnetic zone center ($L = 0$), corresponding to the unbound kink pair excitations above the ordering temperature (2.95 K) at $T = 5$ K. Figure 80(b) shows that the continuum splits into a sequence of sharp discrete modes as the temperature cools to 40 mK, below the ordering temperature, and drives the system deep in the magnetically ordered phase. Figure 80(c) and (d) presents model calculations compiled for zero applied transverse field for the two temperature regions below and above the ordering temperature, showing quadratic dispersion for the excitation modes. Figure 80(e) shows energy scan cuts of the color maps presented in Figure 80(a) and (b), at the ferromagnetic zone center, revealing five sharp modes for the magnetically ordered phase. The instrumental resolution convoluted simulations of dynamical correlations for the ordered phase are shown in Figure 80(f). The results are in good agreement with the experimentally observed spectrum shown in Figure 80(b). A cartoon of the "confined kink" region is sketched in Figure 80(e). The arrows mark the linear string tension, associated with the interchain couplings, that confines kinks into bound states. A comparison of the observed and calculated bound-state energies for the ordered phase is plotted in Figure 80(h).

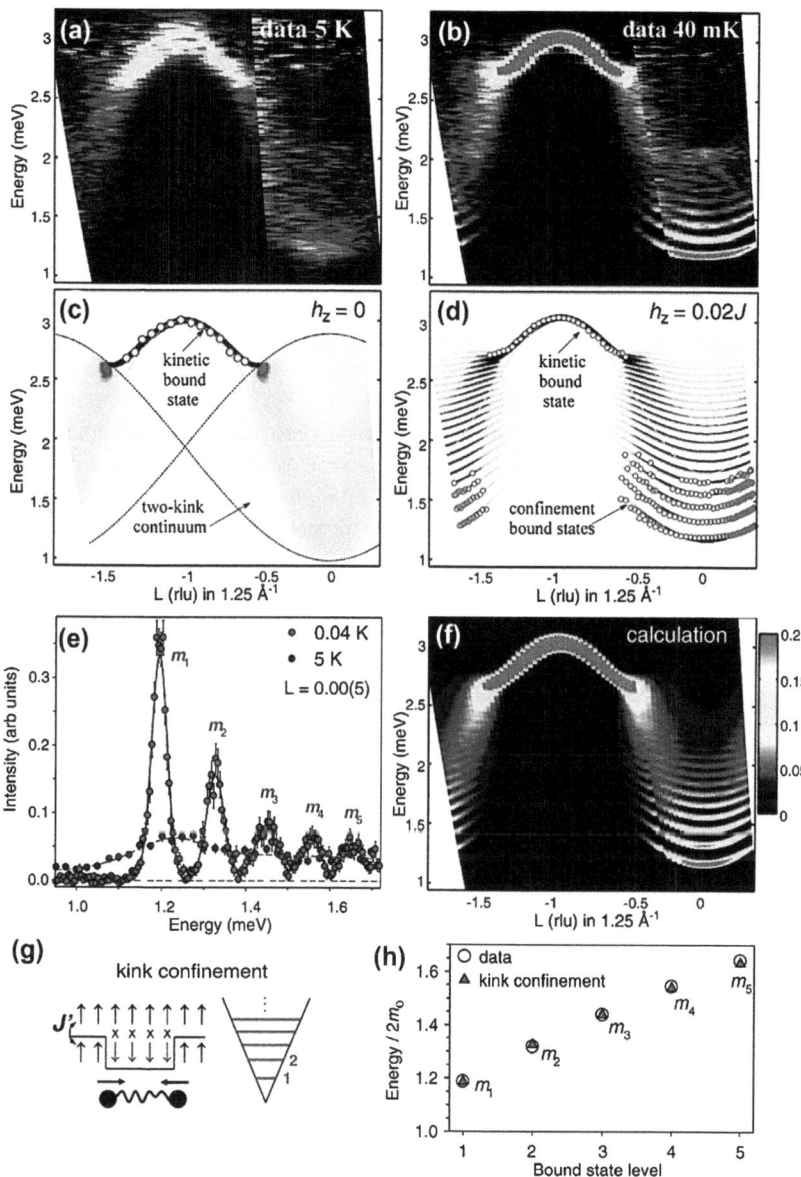

FIGURE 80 Zero-field spin excitations in $CoNb_2O_6$. Inelastic neutron scattering results collected for paramagnet (a) and magnetically ordered phases (b). Model calculations compiled for paramagnet (c) and magnetically ordered (d) phases. Energy scan cuts of the data shown in (e) at the zone center. (f) Simulation calculations of the dynamical correlations performed for magnetically ordered phase. (g) Schematic cartoon depicting the "kink confinement." (h) Comparison between the calculated and observed bound-state energies. *Reprinted with permission from Ref. [144].*

We now turn to the influence of QPT at finite transverse magnetic field, discussing how these discrete modes evolve as transverse field approaches Γ_c. According to the E8 Lie group, it is anticipated that only the eight lowest energy states of the kink bound quasiparticles remain stable at QCP, whereas all others decay. As it was discussed above, the longitudinal field associate with the interchain coupling stabilizes the eight bound states and generates an energy gap. Figure 81(a) and (b) shows the neutron energy scans at the antiferromagnetic zone center just below QCP, resolving two out of eight kink bound-state energy gaps (m_1 and m_2) within the available experimental resolution. Figure 81(c) illustrates the softening of m_1 and m_2 gaps as $\Gamma \to \Gamma_c$. It is reported that the m_2 mode was not detected at fields above 5 T. Figure 81(d) demonstrates how the ratio of m_1 and m_2 modes evolves with increasing field and approaches the golden mean value of $(1 + \sqrt{5})/2$, exactly as predicted by the E8 group theory for the ratio of the two lowest quantum resonances.

Moreover, comprehensive inelastic neutron scattering measurements of $CoNb_2O_6$ have investigated the dispersion relations of the magnetic excitations at transverse fields above QCP and have revealed the effects of interchain couplings [169]. The results confirm the spin-flip nature of the quasiparticle excitations in the PM phase, reporting a main sharp resolution-limited dispersive mode at low energies. A trace of an additional, much weaker, mode associated with buckling of the magnetic chains and alternating rotation

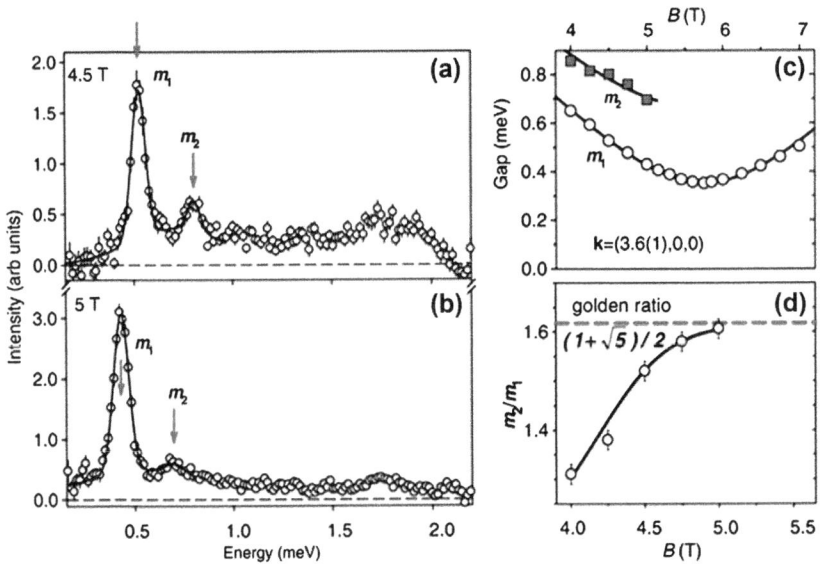

FIGURE 81 Energy scans at the zone center just below the critical field, at 4.5 T (a) and 5 T (b). (c) Softening of the energy gaps as field approaches quantum critical point. (d) The ratio of m_1 and m_2 modes approaches the golden mean value, as predicted by E8 symmetry. *Reprinted with permission from Ref. [144].*

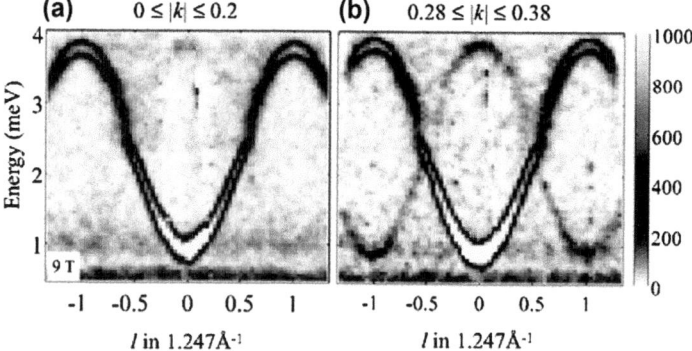

FIGURE 82 Dispersion relations of the magnetic excitations in paramagnetic phase collected at transverse field above the critical field. For small interchain wave-vector k (a), a single dispersive mode is visible, and for finite k (b), a second less intense mode is observed. *Reprinted with permission from Ref. [169].*

of the Ising axes have also been reported (see Figure 82). Modulations of the energy dispersion in the planes normal and along the chain direction were resolved for fields above QCP and interpreted in terms of linear spin-wave model and quantum renormalization effects for PM phase.

Motivated by these results, follow-up high-resolution inelastic neutron scattering experiments have been performed on $CoNb_2O_6$ further investigating the origin of the single-particle dispersion and the anomalous broadening effects at intermediate energies (see Figure 83) [170]. The results suggest that the observed anomalous broadening spreads over a narrow range of energies at small transverse magnetic fields. Using the single-particle dispersion fits to the high-resolution inelastic neutron scattering data presented in Figure 83 it has been suggested that the longitudinal field, including the spontaneous decay of the single-particle mode into two-particle states, is responsible for the detected anomalous broadening.

Recent neutron diffraction experiments under uniaxial pressure applied along the isosceles direction in $CoNb_2O_6$ lattice suggest that the pressure-tuned

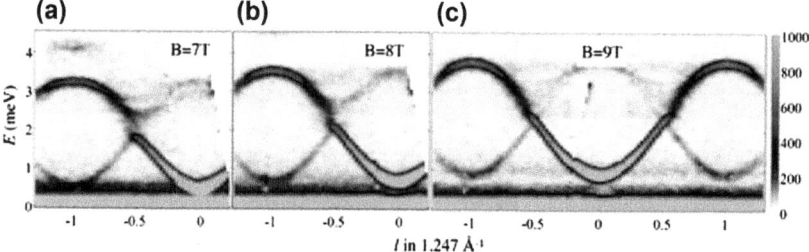

FIGURE 83 High-resolution inelastic neutron scattering data for single-particle dispersion in the paramagnetic phase with momentum aligned along the Ising chain, measured at (a) B = 7, (b) 8, and (c) 9 T. *Reprinted with permission from Ref. [170].*

anisotropic deformations significantly modifies the spin fluctuations and thus increases the nearest-neighbor to next-nearest-neighbor interaction ratio [171]. Future experiment is in progress to explore the effect of such changes and their impact on QPT in spin Ising model systems. Moreover, higher-resolution neutron scattering instruments are being built in order to resolve all the eight excitation masses predicted by emergent E8 symmetry and possibly the anticipated corresponding gap.

2.9 CLOSING REMARKS

The variation of external parameters changes the characteristics of a system, resulting in phase transitions. We encounter with phase transitions in our daily lives: the thermal motion of water molecules destroys the crystal structure of the ice, as it melts and transits into the liquid phase. In thermal phase transitions the order, which exists at lower temperatures, is destroyed by thermal fluctuations. QPTs, however, occur at zero temperature where thermal fluctuations are fully suppressed, and the transformation in ground state is governed by quantum fluctuations rooted in the Heisenberg uncertainty principle and is induced by variations in some nonthermal tuning parameters such as pressure, magnetic field, or chemical doping. The signatures of QPTs can expand to finite temperatures and thus can be investigated experimentally, as long as the quantum fluctuations dominate the thermal fluctuations. In such a scenario, the collective quantum fluctuations and their interplay with thermal fluctuations at the quantum critical region can drastically modify the materials characterizations even at finite temperatures and provide a route for novel electronic and magnetic phenomena. Neutron scattering techniques, particularly the recent advancements in the field of extreme sample environmental conditions, have played a vital role in exploring such emergent states of matter by offering a tunable tool to directly probe the mesoscopic dynamics in the vicinity of QPTs. In this chapter we presented few examples of experimental observations of the induced QPTs in spin magnetic systems. One of the most challenging open questions remaining is the inclusive theoretical description of the experimentally observed phenomena and a verification of the previously proposed interpretations. We hope this review promotes future experimental and theoretical studies on this subject.

ACKNOWLEDGMENTS

This work was supported in part by Duke University; the University of Chicago through NSF grant No. DMR-1206519 and the US Department of Energy Office of Basic Energy Sciences grant No. DE-FG02-99ER45789; Argonne National Laboratory through the US Department of Energy Office of Basic Energy Sciences under contract No. NEAC02-06CH11357; and by NSERC of Canada.

REFERENCES

[1] H.E. Stanley, Introduction to Phase Transitions and Critical Phenomena, Oxford University Press, Oxford, 1987.
[2] H. Nishimori, G. Ortiz, Elements of Phase Transitions and Critical Phenomena, Oxford University Press, Oxford, 2010.
[3] G. Shirane, V.J. Minkiewicz, A. Lisz, Solid State Commun. 8 (23) (1970) 1941.
[4] S.M. Shapiro, J.D. Axe, G. Shirane, P.M. Raccah, Solid State Commun. 15 (2) (1974) 377.
[5] M.F. Collins, Magnetic Critical Scattering, Oxford University Press, Oxford, 1989.
[6] J.J. Binney, N.J. Dowrick, A.J. Fisher, M. Newman, The Theory of Critical Phenomena: An Introduction to the Renormalization Group, Oxford University Press, Oxford, 1992.
[7] P. Chaikin, T. Lubensky, Principles of Condensed Matter Physics, Cambridge University Press, Cambridge, 1995.
[8] N. Goldenfeld, Lectures on Phase Transitions and Renormalization Group, Addison-Wesley, Reading, Massachusetts, 1992.
[9] K.G. Wilson, Phys. Rev. B 4 (9) (1971) 3174.
[10] M.E. Fisher, Rev. Mod. Phys. 46 (4) (1974) 597.
[11] J. Cardy, Scaling and Renormalization in Statistical Physics, Cambridge University Press, Cambridge, 1996.
[12] M. Vojta, Rep. Prog. Phys. 66 (12) (2003) 2069.
[13] T. Vojta, Quantum phase transitions, in: K.H. Hoffmann, M. Schreiber (Eds.), Computational Statistical Physics, Springer, Berlin, 2002, pp. 211–226.
[14] S. Sachdev, Quantum Phase Transitions, second ed., Cambridge University Press, Cambridge, 2011.
[15] S. Sachdev, Nat. Phys. 4 (2008) 173.
[16] D.L. Price, F. Fernandez-Alonso, An introduction to neutron scattering, in: F. Fernandez-Alonso, D.L. Price (Eds.), Experimental Methods in the Physical Sciences, first ed.Neutron Scattering, vol. 44, Academic Press, 2013, pp. 1–135.
[17] M. Arai, Experimental techniques, in: F. Fernandez-Alonso, D.L. Price (Eds.), Experimental Methods in the Physical Sciences, Neutron Scattering, vol. 44, Academic Press, 2013, pp. 245–316.
[18] S.W. Lovesey, The Theory of Neutron Scattering from Condensed Matter, vol. 2, Clarendon Press, 1986.
[19] G.L. Squires, Introduction to the Theory of Thermal Neutron Scattering, Dover Publications, 1997.
[20] L. Van Hove, Phys. Rev. 95 (1) (1954) 249.
[21] T. Chatterji, Magnetic neutron scattering, in: T. Chatterji (Ed.), Neutron Scattering from Magnetic Materials, Elsevier Science, 2005, pp. 1–24.
[22] I.F. Bailey, Z. Kristallogr. 218 (2) (2003) 84.
[23] C. Enss, S. Hunklinger, Low-temperature Physics, Springer-Verlag, Berlin, Heidelberg, 2005.
[24] M. De Palma, F. Thomas, S. Pujol, New 1.8 K Cryogen Free Cryostat Available to the Users, ILL Annual Report, 2000, p. 89.
[25] https://www.ncnr.nist.gov/equipment/displex_files/BLCCR.pdf (last accessed on 20.06.15).
[26] D. Brochier, ILL Tech Report 77, 1974.
[27] http://neutrons2.ornl.gov/equipment/category.cfm?category=Liquid%20Helium%20Cryostats (last accessed on 20.04.15).

[28] R. Radebaugh, Development of the Pulse Tube Refrigerator as an Efficient and Reliable Cryocooler, Proc Inst Refrigeration, London, 1999–2000.
[29] B. Evans, R. Down, J. Keeping, O. Kirichek, Z. Bowden, Meas. Sci. Technol. 19 (3) (2008) 034018.
[30] http://neutrons-old.ornl.gov/equipment/ (last accessed 11.5.2015).
[31] H. London, in: Proc. 2nd Int. Conf. Low Temp. Phys., 1951. Oxford.
[32] H. London, G.R. Clarke, E. Mendoza, Phys. Rev. 128 (1962) 1992.
[33] P. Das, R. DeBruyn Ouboter, K.W. Taconis, in: Proc. 9th Int. Conf. Low Temp. Phys., 1965. London.
[34] G.J. Frossati, Phys. Colloq. 30 (C6) (1978) 1578–1589.
[35] P.M. Tedrow, D.M. Lee, Phys. Lett. 9 (2) (1964) 130.
[36] J. Rossat-Mignod, J. Olivier, P. Burlet, L.P. Regnault, J.M. Effantin, et al., Rev. Phys. Appl. 19 (9) (1984) 783.
[37] F. Mezei, BENSC Operation, Hahn-Meitner-Institut Berlin Annual Report, 2005.
[38] P. Smeibidl, F. Mezei, M. Meissner, K. Prokes, B. Schroder-Smeibidl, M. Steiner, Phys. B 329 (2003) 1666.
[39] M. Steiner, D.A. Tennant, P. Smeibidl, J. Phys. Conf. Ser. 51 (2006) 470.
[40] P. Smeibidl, A. Tennant, H. Ehmler, M. Bird, J. Low Temp. Phys. 159 (2010) 402.
[41] F. Bitter, Rev. Sci. Instrum. 7 (1936) 482.
[42] Y. Nakagawa, S. Miura, A. Hoshi, K. Nakajima, H. Takano, Jpn. J. Appl. Phys. 22 (6) (1983) 1020.
[43] M.D. Bird, IEEE Trans. Appl. Supercond. 16 (2) (2006) 1676.
[44] F. Amita, A. Onodera, Rev. Sci. Instrum. 69 (7) (1998) 2738.
[45] M.K. Jacobsen, C.J. Ridley, A. Bocian, O. Kirichek, P. Manuel, D. Khalyavin, M. Azuma, J.P. Attfield, K.V. Kamenev, Rev. Sci. Instrum. 85 (2014) 043904.
[46] S. Klotz, Techniques in High Pressure Neutron Scattering, CRC Press, 2012.
[47] H. Taniguchi, S. Takeda, R. Satoh, A. Taniguchi, H. Komatsu, K. Satoh, Rev. Sci. Instrum. 81 (2010) 033903.
[48] https://www.ncnr.nist.gov/equipment/Pressure.html (last accessed on 21.06.15).
[49] http://www.isis.stfc.ac.uk/sample-environment/high-pressure-and-gas-handling8450.html (last accessed on 21.06.15).
[50] https://www.ill.eu/instruments-support/sample-environment/equipment/high-pressures/clamped-cells/ (last accessed on 21.06.15).
[51] https://neutrons.ornl.gov/snap/capabilities (last accessed on 21.06.15).
[52] http://www.isis.stfc.ac.uk/instruments/exeed/sample-environment/exeed-sample-environment8273.html (last accessed on 21.06.15).
[53] S. Sachdev, B. Keimer, Phys. Today 64 (2) (2011) 29.
[54] H. Kageyama, M. Nishi, N. Aso, K. Onizuka, T. Yosihama, K. Nukui, K. Kodama, K. Kakurai, Y. Ueda, Phys. Rev. Lett. 84 (2000) 5876.
[55] B.D. Gaulin, S.H. Lee, S. Haravifard, J.P. Castellan, A.J. Berlinsky, H.A. Dabkowska, Y. Qiu, J.R.D. Copley, Phys. Rev. Lett. 93 (2004) 267202.
[56] S. Haravifard, A. Banerjeeb, J.C. Langa, G. Srajer, D.M. Silevitch, B.D. Gaulin, H.A. Dabkowska, T.F. Rosenbaum, Proc. Natl. Acad. Sci. U.S.A. 109 (7) (2011) 2286.
[57] S. Haravifard, A. Banerjee, J. van Wezel, D.M. Silevitch, A.M. dos Santos, J.C. Lang, E. Kermarrec, G. Srajer, B.D. Gaulin, H.A. Dabkowska, et al., Proc. Natl. Acad. Sci. U.S.A. 111 (40) (2014) 14372.
[58] S. Haravifard, S.R. Dunsiger, S. El Shawish, B.D. Gaulin, H.A. Dabkowska, M.T.F. Telling, T.G. Perring, J. Bonča, Phys. Rev. Lett. 97 (2006) 247206.

[59] M.W. van Tol, L.S.J.M. Henkens, N.J. Poul, Phys. Rev. Lett. 27 (1971) 739.
[60] K.M. Diederix, J.P. Groen, L.S.J.M. Henkens, T.O. Klaassen, N.J. Poulis, An experimental study on the magnetic properties of the singlet ground-state system in $Cu(NO_3)_2 \cdot 2.5H_2O$: II. The long-range ordered state, Phys. B+C 94 (1) (1978) 9.
[61] M.B. Stone, I. Zaliznyak, D.H. Reich, C. Broholm, Phys. Rev. B 64 (2001) 144405.
[62] M.B. Stone, I.A. Zaliznyak, T. Hong, C.L. Broholm, D.H. Reich, Nature 440 (2006) 187.
[63] T. Nakajima, H. Mitamura, Y. Ueda, J. Phys. Soc. Jpn. 75 (5) (2006) 054706.
[64] A.A. Aczel, Y. Kohama, M. Jaime, K. Ninios, H.B. Chan, L. Balicas, H.A. Dabkowska, G. M. Luke, Bose-Einstein condensation of triplons in $Ba_3Cr_2O_8$, Phys. Rev. B 79 (2009) 100409(R).
[65] A. Oosawa, M. Ishii, H. Tanaka, J. Phys. Condens. Matter. 11 (1999) 265.
[66] W. Shiramura, K. Takatsu, H. Tanaka, K. Kamishima, M. Takahashi, H. Mitamura, T. Goto, J. Phys. Soc. Jpn. 66 (1997) 1900.
[67] N. Cavadini, G. Heigold, W. Henggeler, A. Furrer, H.U. Güdel, K. Krämer, H. Mutka, Phys. Rev. B 63 (2001) 172414.
[68] S. Wessel, S. Haas, Phys. Rev. B 62 (2000) 316.
[69] T. Giamarchi, C. Rüegg, O. Tchernyshyov, Nat. Phys. 4 (2008) 198.
[70] V. Zapf, M. Jaime, C.D. Batista, Rev. Mod. Phys. 86 (2014) 563.
[71] N. Cavadini, C. Rüegg, A. Furrer, H.U. Güdel, K. Krämer, H. Mutka, P. Vorderwisch, Phys. Rev. B 65 (2002) 132415.
[72] H. Tanaka, A. Oosawa, T. Kato, H. Uekusa, Y. Ohashi, K. Kakurai, A. Hoser, J. Phys. Soc. Jpn. 70 (4) (2001) 939.
[73] C. Rüegg, N. Cavadini, A. Furrer, K. Krämer, H.U. Güdel, P. Vorderwisch, H. Mutka, Appl. Phys. A 74 (2002) 840.
[74] C. Rüegg, N. Cavadini, A. Furrer, H.U. Güdel, K. Krämer, H. Mutka, A. Wildes, K. Habicht, P. Vorderwisch, Nature 423 (2003) 62.
[75] H. Tanaka, K. Goto, M. Fujisawa, T. Ono, Y. Uwatoko, Phys. B 329–333 (2003) 697.
[76] A. Oosawa, M. Fujisawa, T. Osakabe, K. Kakurai, H. Tanaka, J. Phys. Soc. Jpn. 72 (5) (2003) 1026.
[77] C. Rüegg, A. Furrer, D. Sheptyakov, T. Strässle, K.W. Krämer, H.U. Güdel, L. Mélési, Phys. Rev. Lett. 93 (2004) 257201.
[78] C. Rüegg, B. Normand, M. Matsumoto, A. Furrer, D. McMorrow, K. Krämer, H.U. Güdel, S. Gvasaliya, H. Mutka, M. Boehm, Phys. Rev. Lett. 100 (2008) 205701.
[79] P. Merchant, B. Normand, K.W. Krämer, M. Boehm, D.F. McMorrow, C. Rüegg, Nat. Phys. 10 (2014) 373.
[80] C. Lacroix, P. Mendels, F. Mila, Introduction to frustrated magnetism, in: Springer Series in Solid-State Sciences, Springer, Heidelberg, 2011.
[81] J. Gardner, M. Gingras, J. Greedan, Rev. Mod. Phys. 82 (2010) 53.
[82] S. Bramwell, M. Gingras, Science 294 (2001) 1495.
[83] M.J. Harris, S.T. Bramwell, D.F. McMorrow, T. Zeiske, K.W. Godfrey, Phys. Rev. Lett. 79 (1997) 2554.
[84] J.P. Clancy, J.P.C. Ruff, S.R. Dunsiger, Y. Zhao, H.A. Dabkowska, J.S. Gardner, Y. Qiu, J.R.D. Copley, T. Jenkins, B.D. Gaulin, Phys. Rev. B 79 (2009) 014408.
[85] A. Ramirez, A. Hayashi, R. Cava, R. Siddardthan, B. Shastry, Nature 399 (1999) 333.
[86] K.A. Ross, L. Savary, B.D. Gaulin, L. Balents, Phys. Rev. X 1 (2011) 021002.
[87] L. Savary, K.A. Ross, B.D. Gaulin, J.P.C. Ruff, L. Balents, Phys. Rev. Lett. 109 (2012) 167201.
[88] A. Poole, A.S. Wills, J. Lelièvre-Berna, J. Phys. Condens. Matter. 19 (2007) 452201.

[89] J.D.M. Champion, M.J. Harris, P.C.W. Holdsworth, A.S. Wills, G. Balakrishnan, S.T. Bramwell, E. Cizmar, T. Fennell, J.S. Gardner, J. Lago, et al., Phys. Rev. B 68 (2003) 020401.
[90] J.P.C. Ruff, J.P. Clancy, A. Bourque, M.A. White, M. Ramazanoglu, J.S. Gardner, Y. Qiu, J.R.D. Copley, M.B. Johnson, H.A. Dabkowska, et al., Phys. Rev. Lett. 101 (2008) 147205.
[91] K.A. Ross, Y. Qiu, J.R.D. Copley, H.A. Dabkowska, B.D. Gaulin, Phys. Rev. Lett. 112 (2014) 057201.
[92] B. Javanparast, A.G.R. Day, Z. Hao, M.J.P. Gingras, Phys. Rev. B 91 (2015) 174424.
[93] L. Savary, L. Balents, Inventors, 2015. Private Communication.
[94] K.A. Ross, J.P.C. Ruff, C.P. Adams, J.S. Gardner, H.A. Dabkowska, Y. Qiu, J.R.D. Copley, B.D. Gaulin, Phys. Rev. Lett. 103 (2009) 227202.
[95] K.A. Ross, T. Proffen, H.A. Dabkowska, J.A. Quilliam, L.R. Yaraskavitch, J.B. Kycia, B. D. Gaulin, Phys. Rev. B 86 (2012) 174424.
[96] P. Coleman, A.J. Schofield, Nature 433 (2005) 226.
[97] G. Stewart, Rev. Mod. Phys. 73 (2001) 797.
[98] A.J. Schofield, Contemp. Phys. 40 (1999) 95.
[99] L. Carr, Understanding Quantum Phase Transitions, CRC Press, 2010.
[100] T. Moriya, Y. Takahashi, Spin Fluctuations in Itinerant Electron Magnetism, Citeseer, 1985.
[101] A. Schroder, G. Aeppli, R. Coldea, M. Adams, O. Stockert, H. Lohneysen, E. Bucher, R. Ramazashvili, P. Coleman, Nature 407 (2000) 351.
[102] Q. Si, S. Rabello, K. Ingersent, J.L. Smith, Nature 413 (2001) 6858.
[103] M.A. Metlitski, S. Sachdev, Phys. Rev. B 82 (2010) 075127.
[104] P. Gegenwart, Q. Si, F. Steglich, Nat. Phys. 4 (2008) 186.
[105] P. Gegenwart, F. Steglich, C. Geibel, M. Brando, Euro Phys. J. Spec. Top. 224 (2015) 975.
[106] H. Lohneysen, A. Rosch, M. Vojta, P. Wolfle, Rev. Mod. Phys. 79 (2007) 1015.
[107] M.L. Vrtis, J.D. Jorgensen, D.G. Hinks, J. Solid State Chem. 84 (1990) 93.
[108] A. Hamann, O. Stockert, V. Fritsch, K. Grube, A. Schneidewind, H. Lohneysen, Phys. Rev. Lett. 110 (2013) 096404.
[109] O. Stockert, H. Lohneysen, W. Schmidt, M. Enderle, M. Loewenhaupt, J. Low Temp. Phys. 161 (2010) 55.
[110] Q. Si, F. Steglich, Science 329 (2010) 1161.
[111] V. Fritsch, O. Stockert, C.L. Huang, N. Bagrets, W. Kittler, C. Taubenheim, B. Pilawa, S. Woitschach, Z. Huesges, S. Lucas, et al., Euro Phys. J. Spec. Top. 224 (2015) 997.
[112] O. Stockert, M. Enderle, H. Lohneysen, Phys. Rev. Lett. 99 (2007) 237203.
[113] G. Knebel, D. Aoki, J. Flouquet, C. R. Phys. 12 (2011) 542.
[114] D.J. Thompson, Z. Fisk, J. Phys. Soc. Jpn. 81 (2011) 011002.
[115] E. Moshopoulou, Z. Fisk, J. Sarrao, J. Thompson, J. Solid State Chem. 158 (2001) 25.
[116] T. Willers, F. Strigari, Z. Hu, V. Sessi, N.B. Brookes, E.D. Bauer, J.L. Sarrao, J. Thompson, A. Tanaka, S. Wirth, et al., Proc. Natl. Acad. Sci. U.S.A. 112 (2015) 2384.
[117] N. Aso, K. Ishii, H. Yoshizawa, T. Fujiwara, Y. Uwatoko, G.F. Chen, N.K. Sato, K. Miyake, J.Phys. Soc. Jpn. 78 (2009) 073703.
[118] O. Stockert, F. Steglich, Annu. Rev. Condens. Matter. Phys. 2 (2011) 79.
[119] S. Raymond, G. Lapertot, Phys. Rev. Lett. 115 (2015) 037001.
[120] L. Howald, E. Stilp, P. Dalmas de Réotier, A. Yaouanc, S. Raymond, C. Piamonteze, G. Lapertot, C. Baines, H. Keller, Sci. Rep. 5 (2015) 12528.

[121] M. Yashima, N. Tagami, S. Taniguchi, T. Unemori, K. Uematsu, H. Mukuda, Y. Kitaoka, Y. Ota, F. Honda, R. Settai, et al., Phys. Rev. Lett. 109 (2012) 117001.
[122] M. Brando, D. Belitz, F. Grosche, T. Kirkpatrick, arXiv preprint, 2015 arXiv:1502.02898.
[123] S.M. Stishov, A.E. Petrova, Phys. Usp. 54 (2011) 1117.
[124] T. Jeong, W. Pickett, Phys. Rev. B 70 (2004) 075114.
[125] C. Pfleiderer, P. Boni, T. Keller, U. Roler, A. Rosch, Science 316 (2007) 1871.
[126] M. Janoschek, M. Garst, A. Bauer, P. Krautscheid, R. Georgii, P. Boni, C. Pfleiderer, Phys. Rev. B 87 (2013) 134407.
[127] M. Ishida, Y. Endoh, S. Mitsuda, Y. Ishikawa, M. Tanaka, J. Phys. Soc. Jpn. 54 (1985) 2975.
[128] S. Grigoriev, D. Chernyshov, V. Dyadkin, V. Dmitriev, E.V. Moskvin, D. Lamago, Th Wolf, D. Menzel, J. Schoenes, S. Maleyev, H. Eckerlebe, Phys. Rev. B 81 (2010) 12408.
[129] P. Boni, B. Roessli, K. Hradil, J. Phys. Condens. Matter. 23 (2011) 254209.
[130] Y. Ishikawa, G. Shirane, J. Tarvin, M. Kohgi, Phys. Rev. B 16 (1977) 4956.
[131] G. Shirane, R. Cowley, C. Majkrzak, J. Sokoloff, B. Pagonis, C. Perry, Y. Ishikawa, Phys. Rev. B 28 (1983) 6251.
[132] B. Roessli, P. Boni, W. Fischer, Y. Endoh, Phys. Rev. Lett. 88 (2002) 237204.
[133] C. Pfleiderer, D. Reznik, L. Pintschovius, H. Lohneysen, M. Garst, A. Rosch, Nature 427 (2004) 227.
[134] C. Pfleiderer, G. McMullan, S. Julian, G. Lonzarich, Phys. Rev. B 55 (1997) 8330.
[135] F. Mezei, C. Pappas, T. Gutberlet, Neutron Spin Echo Spectroscopy: Basics, Trends and Applications, vol. 601, Springer, 2003.
[136] J.A. Mydosh, P.M. Oppeneer, Philos. Mag. 94 (2014) 3642.
[137] C. Wiebe, J. Janik, G. MacDougall, G. Luke, J. Garrett, H. Zhou, Y.J. Jo, L. Balicas, Y. Qiu, J. Copley, et al., Nat. Phys. 3 (2007) 96.
[138] C. Wiebe, G. Luke, Z. Yamani, A. Menovsky, W. Buyers, Phys. Rev. B 69 (2004) 132418.
[139] J. Janik, H. Zhou, Y. Jo, L. Balicas, G. MacDougall, G. Luke, J. Garrett, K. McClellan, E. Bauer, J. Sarrao, et al., J. Phys. Condens. Matter. 21 (2009) 192202.
[140] N.P. Butch, M.B. Maple, J. Phys. Condens. Matter. 22 (2010) 164204.
[141] M. Torikachvili, L. Rebelsky, K. Motoya, S. Shapiro, Y. Dalichaouch, M. Maple, Phys. Rev. B 45 (1992) 2262.
[142] T. Williams, Z. Yamani, N. Butch, G. Luke, M. Maple, W. Buyers, Phys. Rev. B 86 (2012) 235104.
[143] N.P. Butch, J.R. Jeffries, S. Chi, J.B. Leao, J.W. Lynn, M.B. Maple, Phys. Rev. B 82 (2010) 060408.
[144] R. Coldea, D.A. Tennant, E.M. Wheeler, E. Wawrzynska, D. Prabhakaran, M. Telling, K. Habicht, P. Smeibidl, K. Kiefer, Science 327 (2010) 177.
[145] D.H. Reich, B. Ellman, J. Yang, T.F. Rosenbaum, G. Aeppli, Phys. Rev. B 42 (1990) 4631.
[146] S. Tabei, M.J. Gingras, Y.J. Kao, P. Stasiak, J.Y. Fortin, Phys. Rev. Lett. 97 (2006) 237203.
[147] D.M. Silevitch, D. Bitko, J. Brooke, S. Ghosh, G. Aeppli, T.F. Rosenbaum, Nature 448 (2007) 567.
[148] M. Schechter, Phys. Rev. B 77 (2008) 020401(R).
[149] J. Brooke, T.F. Rosenbaum, G. Aeppli, Nature 413 (2001) 610.
[150] J. Brooke, D. Bitko, T.F. Rosenbaum, G. Aeppli, Science 284 (1999) 779.
[151] J.O. Piatek, B. Dalla Piazza, N. Nikseresht, N. Tsyrulin, I. Živković, K.W. Krämer, M. Laver, K. Prokes, S. Mataš, N.B. Christensen, et al., Phys. Rev. B 88 (2013) 014408.
[152] D. Bitko, T.F. Rosenbaum, G. Aeppli, Phys. Rev. Lett. 77 (1996) 940.
[153] J.A. Griffin, M. Huster, R.J. Folweiler, Phys. Rev. B 22 (1980) 4370.

[154] W. Wu, D. Bitko, T.F. Rosenbaum, G. Aeppli, Phys. Rev. Lett. 71 (1993) 1919.
[155] H.M. Rønnow, R. Parthasarathy, J. Jensen, G. Aeppli, T.F. Rosenbaum, D.F. McMorrow, Science 308 (2005) 389.
[156] H.M. Rønnow, J. Jensen, R. Parthasarathy, G. Aeppli, T.F. Rosenbaum, D.F. McMorrow, C. Kraemer, Phys. Rev. B 75 (2007) 54426.
[157] A. Dutta, G. Aeppli, B.K. Chakrabarti, U. Divakaran, T.F. Rosenbaum, D. Sen, Quantum Phase Transitions in Transverse Field Spin Models: From Statistical Physics to Quantum Information, first ed., Cambridge University Press, 2015.
[158] I. Maartense, I. Yaeger, B.M. Wanklyn, Solid State Commn. 21 (1977) 93.
[159] W. Scharf, H. Weitzel, I. Yaeger, I. Maartense, B.M. Wanklyn, J. Magn. Magn. Mater. 13 (1979) 121.
[160] S. Mitsuda, K. Hosoya, T. Wada, H. Yoshizawa, T. Hanawa, M. Ishikawa, K. Miyatani, K. Saito, K. Kohn, J. Phys. Soc. Jpn. 63 (10) (1994) 3568.
[161] P.W.C. Sarvezuk, E.J. Kinast, C.V. Colin, M.A. Gusmão, J.B.M. da Cunha, O. Isnard, J. Appl. Phys. 109 (2011) 07E160.
[162] C. Heid, H. Weitzel, P. Burlet, M. Bonnet, W. Gonschorek, T. Vogt, J. Norwig, H. Fuess, J. Magn. Magn. Mater. 151 (1995) 123.
[163] C. Heid, H. Weitzel, P. Burlet, M. Winkelmann, H. Ehrenberg, H. Fuess, Physica B 234–236 (1997) 574.
[164] B.G. Levi, Phys. Today 63 (3) (2010) 13.
[165] B.M. McCoy, T.T. Wu, Phys. Rev. D 18 (1978) 1259.
[166] G. Delfino, G. Mussardo, Nucl. Phys. B 455 (1995) 724.
[167] S. Lee, R.K. Kaul, L. Balents, Nat. Phys. 6 (2010) 703.
[168] A.B. Zamolodchikov, Int. J. Mod. Phys. A 4 (1989) 4235.
[169] I. Cabrera, J.D. Thompson, R. Coldea, D. Prabhakaran, R.I. Bewley, T. Guidi, J.A. Rodriguez-Rivera, C. Stock, Phys. Rev. B 90 (2014) 14418.
[170] N.J. Robinson, F.H.L. Essler, I. Cabrera, R. Coldea, Phys. Rev. B 90 (2014) 174406.
[171] S. Kobayashi, S. Hosaka, H. Tamatsukuri, T. Nakajima, S. Mitsuda, K. Prokeš, K. Kiefer, Phys. Rev. B 90 (2014) 060412(R).

Chapter 3

High-Temperature Superconductors

Yu Song* and Pengcheng Dai
Department of Physics and Astronomy, Rice University, Houston, Texas, USA
Corresponding author: E-mail: Yu.Song@rice.edu

Chapter Outline

3.1 Introduction 145
3.2 Hole-Doped Cuprate Superconductors 147
  3.2.1 Magnetic Order and Spin Waves in the Parent Compounds 147
  3.2.2 Evolution of Magnetic Excitations upon Doping 151
  3.2.3 Neutron Scattering versus RIXS 156
  3.2.4 The "Resonance" Mode in the Superconducting State 158
3.3 Iron-Based Superconductors 160
  3.3.1 Introduction, Compounds, and Phase Diagrams 160
  3.3.2 Antiferromagnetically Ordered Metallic Parent Compounds 162
  3.3.3 Evolution of Magnetic Order and Magnetic Excitations with Doping 168
  3.3.4 Persistence of High-Energy Magnetic Excitations 176
  3.3.5 The Resonance Mode in Iron-Based Superconductors 179
  3.3.6 Magnetism in $A_xFe_{2-y}Se_2$ (A = Alkali Metal or Tl) Compounds 182
  3.3.7 Polarization Dependence of Low-Energy Magnetic Excitations 186
3.4 Summary and Outlook 190
Acknowledgments 193
References 193

3.1 INTRODUCTION

Understanding the microscopic origin of high-temperature superconductivity in cuprate and iron-based superconductors is at the forefront of condensed matter physics. Since the parent compounds of cuprate and iron-based superconductors are antiferromagnets with long-range magnetic order and superconductivity arises by electron- or hole-doping into the parent compounds, the key question to be answered is how magnetism is related to superconductivity. The relevance of magnetism is indisputable, evidenced by proximity of superconductivity to magnetic order in cuprates, iron pnictides and heavy fermion superconductors and the presence of intense magnetic excitations in the superconducting state obviating from the paradigm of conventional BCS superconductors. However, the exact role of magnetism and whether magnetic excitations play the role of pairing glue analogous to phonons in conventional superconductors is still not resolved. Neutron scattering has played a key role toward our current understanding of magnetism in high-temperature superconductors and related compounds, here we review key results and recent progress in cuprate and iron-based superconductors.

The discovery of superconductivity in the cuprate $La_{2-x}Ba_xCuO_4$ [1] took the physics community by surprise. It completely goes against the rules set out by Matthias [2] for finding higher T_c, the weak electron−phonon coupling in these materials suggested a non-phonon pairing mechanism (hence "unconventional" superconductivity in contrast to phonon mediated "conventional" superconductivity; alternatively, unconventional superconductivity may be defined by its sign-changing superconductivity order parameter on different parts of the Fermi surface). Neutron diffraction soon uncovered the parent compound La_2CuO_4 to be a Mott insulator with antiferromagnetic order [3]. This finding highlighted the relevance of magnetism and set the stage for research in high-temperature superconductivity where neutron scattering played a key role. With the discovery of superconductivity above 77 K in $YBa_2Cu_3O_{6+x}$ [4], cuprates became bona fide high-temperature superconductors. Recent developments in neutron scattering instrumentation and single-crystal growth techniques allowed comprehensive studies of magnetic excitations in the cuprates, despite the small magnetic moment of Cu^{2+} and the large bandwidth of magnetic excitations (in magnetically ordered cuprate parent compounds, this corresponds to the spin-wave bandwidth which is $\sim 2J$, J being the nearest-neighbor exchange coupling). The fact that cuprates are two-dimensional (2D) systems coupled with time-of-flight neutron spectroscopy with large area detectors allowed magnetic excitations throughout the Brillouin zone to be mapped efficiently. While there is still no consensus regarding the exact mechanism of superconductivity in the cuprates, a universal picture for the evolution of spin excitations has emerged for different families of hole-doped cuprates. The spin waves in the parent compounds and their evolution in doped cuprates have been reviewed in Refs [5−10]. Here key results and more recent results are reviewed.

The discovery of superconductivity in iron pnictides [11] presents a new challenge. While also based on 2D structural motifs, the multiband nature makes this system even more challenging to understand and each family of materials appears unique. Unlike the cuprates, even the nature of the magnetism in the parent compounds of the iron pnictides is perplexing. While spin waves have been mapped out in the parent compounds of iron-pnictide superconductors, interpretations involving either itinerant electrons or local moments have been presented [12]. Compared to hole-doped cuprates, significant differences exist among different families of iron-based superconductors, both in their electronic structure and spin-excitation spectra. Neutron scattering has made significant contributions toward understanding the magnetism in iron pnictides and has been reviewed in Refs [12–14]. Here we will review neutron scattering results in the iron pnictides/chalcogenides focusing on systematic single-crystal studies on systems derived from $BaFe_2As_2$, NaFeAs, and FeTe. Alkali metal iron chalcogenides present another intriguing case of unconventional superconductivity where superconductivity can be induced with the suppression of antiferromagnetic order [15]. While the jury is still out on which phase is responsible for superconductivity and how the insulating phase with block antiferromagnetic order is related to superconductivity, neutron scattering has played an essential role in determining the nature of these systems. The physics of this specific system has been reviewed in Ref. [16], here we will survey the neutron scattering results so far and discuss their significance.

Polarized neutron scattering allows the polarization dependence of magnetic excitations and the ordered magnetic moment to be studied [17]. Typically, such a setup requires a guide field, and in the case of superconductors the Meissner effect tends to affect the polarization of the neutrons. Development of techniques allowing the sample to be in zero-field allowed detailed studies of polarization dependence of magnetic excitation in both the cuprates and iron-based superconductors. By tuning the polarization direction of the neutron, it is further possible to measure the magnetic response polarized along each crystallographic direction. Here we review the principles used in such experiments and some results.

With improvements of energy resolution below 0.1 eV, resonant inelastic X-ray scattering (RIXS) has demonstrated comparable capability in measuring high-energy spin excitations in both cuprates [18] and iron pnictides [19]. With RIXS instruments having even better energy resolution scheduled to come on line, RIXS is undoubtedly challenging neutron scattering's monopoly in measuring magnetic excitations resolving both energy and momentum. Since RIXS and neutron scattering probe different regions of reciprocal space, these two techniques provide complementary information in the study of high-temperature superconductors. RIXS results on the cuprates and iron pnictides and comparison with neutron scattering are discussed.

3.2 HOLE-DOPED CUPRATE SUPERCONDUCTORS
3.2.1 Magnetic Order and Spin Waves in the Parent Compounds

For the purpose of neutron scattering, compounds based on La_2CuO_4 and $YBa_2Cu_3O_{6+x}$ are most studied due to availability of large single crystals. La_2CuO_4 orders magnetically below $T_N \approx 320$ K [20]. The ordered moments are collinear and lie in the CuO_2 planes oriented at 45° from the Cu—O bond directions (Figure 1(a)) [21]. In the case of $YBa_2Cu_3O_{6+x}$, the actual doping per Cu^{2+} in the CuO_2 planes p does not have a simple relation to x, but for samples with $x \leq 0.15$, p is sufficiently small that it can be regarded as the parent phase [22]. $YBa_2Cu_3O_{6.05}$ orders below $T_N \approx 410$ K with magnetic moments lying in the CuO_2 planes parallel to the Cu—O bond directions (Figure 1(b)) [23]. In both systems, the magnetic unit cells have the same size along c as the structural unit cells shown in Figure 1, whereas in the CuO_2 plane the size is doubled. Therefore, magnetic Bragg peaks occur at (1/2,1/2,1) and equivalent positions. In the case of $YBa_2Cu_3O_{6+x}$, the bilayer structure gives rise to a sinusoidal modulation of the magnetic structure factor along L (reciprocal lattice units are defined as $(H, K, L) = \left(\frac{q_x a}{2\pi}, \frac{q_y b}{2\pi}, \frac{q_z c}{2\pi}\right)$ where the momentum transfer $\mathbf{Q} = (q_x, q_y, q_z)$ is in units of $Å^{-1}$). Since both cuprates and iron-based superconductors are quasi-2D, the L component of $\mathbf{Q}$ is sometimes dropped resulting in notation of the form (H,K).

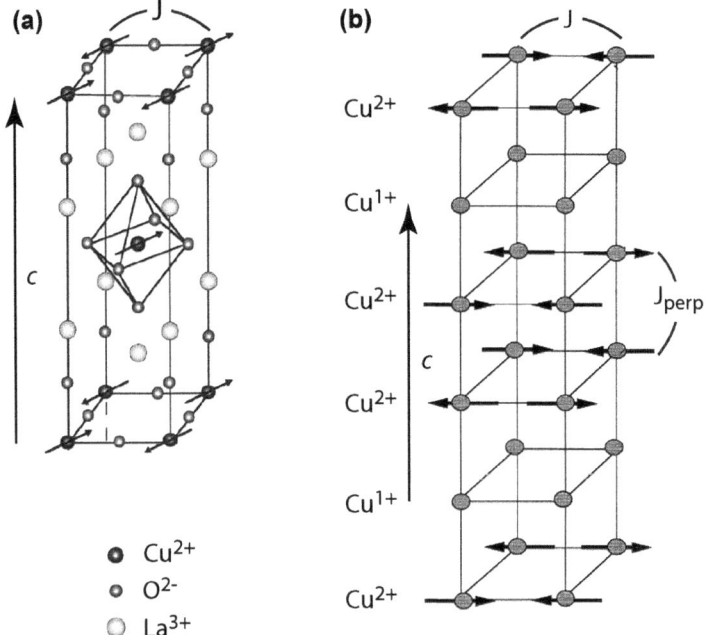

FIGURE 1 Magnetic structure for La_2CuO_4 [21] and $YBa_2Cu_3O_{6+x}$ with $x < 0.4$ [23]. (a) Magnetic structure of La_2CuO_4 [21]. *Adapted with permission from Ref. [21].* (b) The magnetic structure of $YBa_2Cu_3O_{6+x}$ with $x < 0.4$, only Cu atoms are shown, the structure consists of magnetic Cu^{2+} layers and nonmagnetic Cu^{1+} layers. *Adapted from Ref. [23].* Lengths of the unit cells along c axis are shown in the figure by vertical arrows.

The magnetic interactions in cuprate parent compounds are dominated by the superexchange between in-plane nearest-neighbors J which is on the order of 100 meV [7]. Magnetic interactions over longer distances, while much weaker than J, are still significant compared to the superconductivity pairing energies. To obtain these interactions, it is necessary to obtain the dispersion of spin waves along the zone boundary. In La_2CuO_4 this was accomplished and the dispersion of spin waves along the zone boundary is interpreted as due to a cyclic exchange term J_c (58(4) meV) that is almost half of J (143(2) meV) at 10 K. Within this interpretation, the next-nearest (J') and next-next-nearest (J'') interactions are 5% of J_c [20,24] (Figure 2(a) and (b)). In addition, it was found that spin waves at the (1/2,0) zone boundary position show strong damping compared to another zone boundary position (1/4,1/4) [20] (Figure 2(c)). This is surprising as the energy of the spin waves at the zone boundaries ($\sim 2J$) is much smaller than the charge-transfer gap in La_2CuO_4. The authors argued that

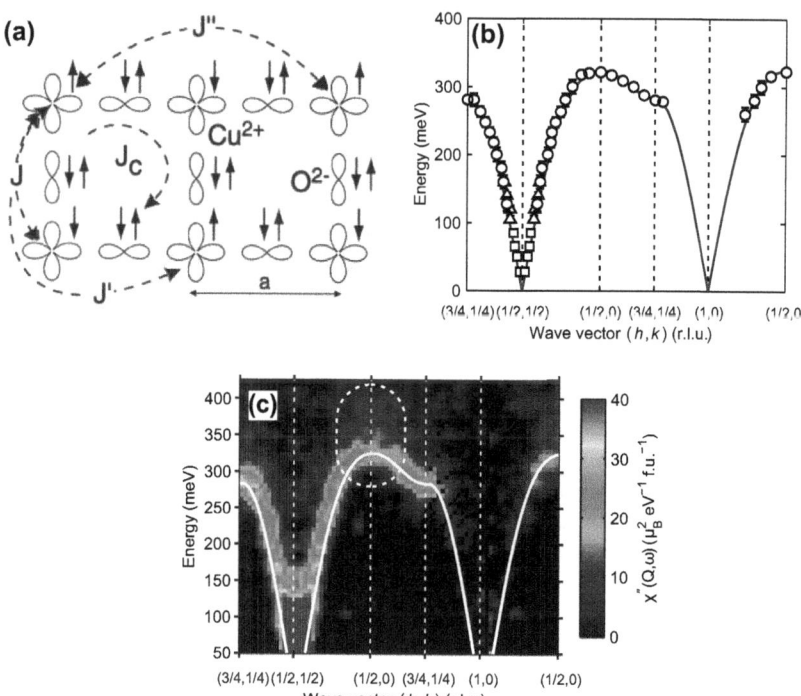

FIGURE 2 Spin waves in La_2CuO_4. (a) The exchange couplings for La_2CuO_4. Adapted with permission from Ref. [24]. The dispersion (b) and intensity (c) of spin waves in La_2CuO_4. Adapted with permission from Ref. [20]. The spin waves are damped near (1/2,0) despite the spin-wave energies being much smaller than the charge-transfer gap in La_2CuO_4, as can be seen in (c) inside the dashed ellipse.

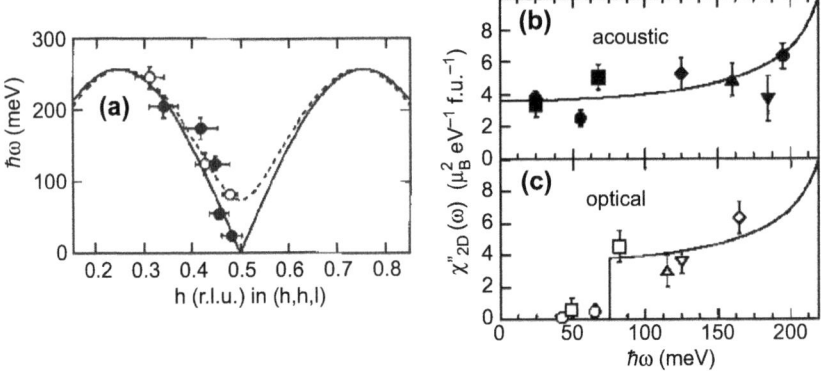

FIGURE 3 Spin waves in $YBa_2Cu_3O_{6.15}$. Adapted with permission from Ref. [25]. (a) The spin-wave dispersions of the acoustic and optical spin-wave modes. The closed symbols and the solid line correspond to the acoustic branch, the open symbols and the dotted line refer to the optical spin-wave branch. The local susceptibilities for the acoustic (b) and optical (c) spin-wave branches. Above ~100 meV, the acoustic and optical branches become almost identical.

the damped zone boundary spin waves at (1/2,0) decay into a two-spinon continuum which has a lower energy at (1/2,0) than (1/4,1/4).

Compared to La_2CuO_4 where the couplings between CuO_2 layers are weak, the parent phase of $YBa_2Cu_3O_{6+x}$ consists of coupled CuO_2 bilayers. This doubling of CuO_2 planes results in an optical spin-wave branch not present in monolayer La_2CuO_4. Measurements of both the acoustic and the optical spin-wave branches enabled a determination of the in-plane nearest-neighbor coupling $J = 125(5)$ meV and the out-of-plane nearest-neighbor coupling $J_{perp} = 11(2)$ meV (Figure 3) [25]. So far there is no report of spin-wave dispersion along the zone boundary in $YBa_2Cu_3O_{6+x}$ and it is not clear if the cyclic exchange and damping of spin waves at (1/2,0) are unique to La_2CuO_4 or common features of cuprate parent compounds.

Due to quantum fluctuations, both the spin-wave energies and the intensities will be renormalized compared to values derived from the classical large-S limit. Neutron scattering cannot directly measure the spin-wave energy renormalization factor Z_c, instead $Z_c = 1.18$ is assumed to extract the exchange couplings [24,25]. The spin-wave intensity renormalization factors Z_d for La_2CuO_4 is 0.40(4) [20] and for $YBa_2Cu_3O_{6.15}$ $Z_\chi = 0.4(1)$ ($Z_d = Z_\chi Z_c$) has been reported [25], these values are smaller than $Z_d \sim 0.6$ obtained in theoretical studies [26]. Quantum fluctuations, therefore, increase the spin-wave bandwidth and decrease spin-wave intensities. The total moment squared obtained in La_2CuO_4 is $\langle M_{tot}^2 \rangle = 1.9(3)$ μ_B^2, compared to 3 μ_B^2 expected for $S = 1/2$ [20]. The reduction of the observed total moment may be due to the presence of magnetic excitations at higher energies that is not included in the estimate, another possible cause is hybridization of Cu 3d orbitals with O p orbitals causing a marked drop of magnetic intensity as found in Sr_2CuO_3 [27].

Such hybridization effects will be manifested in the magnetic form factor, while it has long been established that the Cu^{2+} magnetic form factor is anisotropic [28], only in recent works [29] hybridization effects are taken into account when considering the magnetic form factor.

2D spin correlations survive above T_N in both La_2CuO_4 and $YBa_2Cu_3O_{6+x}$, a result of the dominant nearest-neighbor J being much larger than the interlayer coupling and 3D magnetic ordering is bottlenecked by the weaker interlayer coupling [7]. It is interesting to note that persistence of 2D spin correlations above T_N is also found in $BaFe_2As_2$ [30] and $CeRhIn_5$ [31], the parent compounds of iron-based superconductors, and heavy fermion superconductors.

3.2.2 Evolution of Magnetic Excitations upon Doping

Upon doping by holes, the commensurate antiferromagnetic order in the parent compound is very quickly suppressed at $p \sim 0.02$ as shown in the phase diagram of $La_{2-x}Sr_xCuO_4$ (Figure 4) [8,32]. The long-range commensurate order

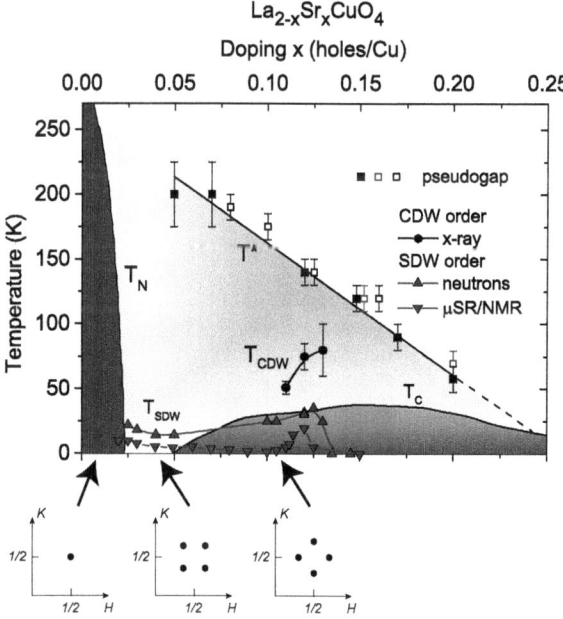

FIGURE 4 The phase diagram of $La_{2-x}Sr_xCuO_4$. *Adapted with permission from Ref. [32].* Schematics for low-energy magnetic excitations are shown at the bottom of the figure for the commensurate long-range ordered, the spin-glass (SG), and the superconducting (SC) regime. The low-energy excitations in the commensurate long-range ordered phase are commensurate spin waves, in the SG phase they form quartets split along the (110) direction and in the SC phase they form quartets split along the Cu—O bond directions ((100) direction). In untwinned samples the quartet in the SG phase becomes two peaks. CDW, charge density waves; SDW, spin density waves.

is supplanted by incommensurate magnetic order in the doping range $p \sim 0.02$ to ~ 0.055 corresponding to a spin-glass phase. For doping levels higher than $p \sim 0.055$, superconductivity appears. At $p = 0.125$ there is a slight dip in the superconductivity dome and optimal superconductivity is achieved at $p \sim 0.16$ (this is termed optimal doping, compounds below and above this doping level are described as underdoped and overdoped, respectively). For $p \sim 0.25$, superconductivity disappears giving way to a Fermi liquid metallic state.

In the spin-glass phase, the incommensurate magnetic order and excitations are split along the (110) direction at 45° to the Cu—O bonds. Upon just entering the superconducting state, the incommensurate magnetic order split along the (110) direction rotates by 45° becoming parallel to (100) and then disappears at higher doping levels [33], while the low-energy excitations persist (low-energy excitations stem from the magnetic ordering wave vector and also rotate by 45° along with the magnetic ordering wave vector near the boundary between superconductivity and the spin-glass regime). Despite the rotation by 45° of the incommensurate magnetic order and low-energy excitations, the splitting scales linearly with doping until saturating around optimal doping (Figure 5(a)) [8,33]. Both the transitions from long-range commensurate order to incommensurate order split along (110), and the rotation of incommensurate magnetic order involves samples with phase separation, suggesting both transitions are first order in nature. In contrast, a quantum critical point seems to be involved with the disappearance of the incommensurate magnetic order split along (100), where application of a magnetic field pushes this quantum critical point to higher doping levels [34–36] (Figure 5(b)).

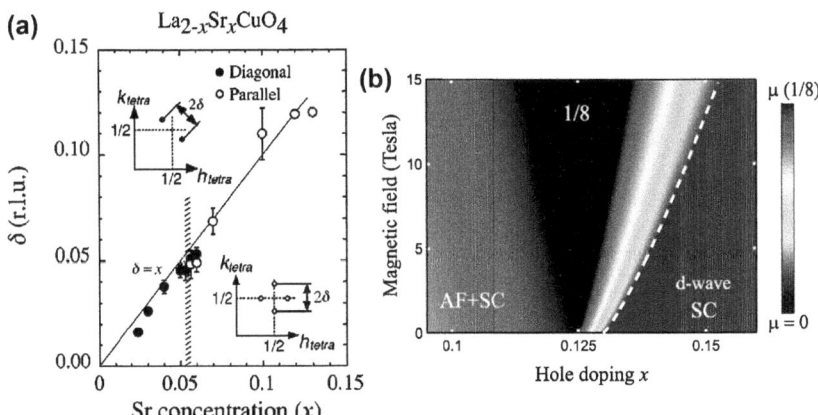

FIGURE 5 (a) The splitting of incommensurate peaks in $La_{2-x}Sr_xCuO_4$. Despite the rotation by 45° of the splitting direction, the splitting itself increases linearly with $p \approx \delta$. *Adapted with permission from Ref. [33].* (b) The disappearance of incommensurate order along (100) in $La_{2-x}Sr_xCuO_4$ involves a quantum phase transition, the critical point can be driven to higher doping levels with the application of a magnetic field. *Adapted with permission from Ref. [36].* SC, superconducting.

A very similar phase diagram with incommensurate magnetic excitations rotating from split along (110) to split along (100) is also found in $Bi_{2+x}Sr_{2-x}CuO_{6+y}$ [37], demonstrating that the complexity of the phase diagram outlined above is not unique to $La_{2-x}(Sr,Ba)_xCuO_4$. In $YBa_2Cu_3O_{6+x}$, there is no phase with incommensurate magnetic ordering split along (110). When long-range magnetic order in the parent phase is suppressed with increasing doping, incommensurate magnetic order split along (100) sets in [38]. In $La_{2-x}Sr_xCuO_4$ with incommensurate order split long (110) and $YBa_2Cu_3O_{6+x}$ with incommensurate excitations split along (100), it has been found in untwinned samples that the low-energy incommensurate magnetic peaks display twofold rotational symmetry (Figure 6) [38–40]. Given that the

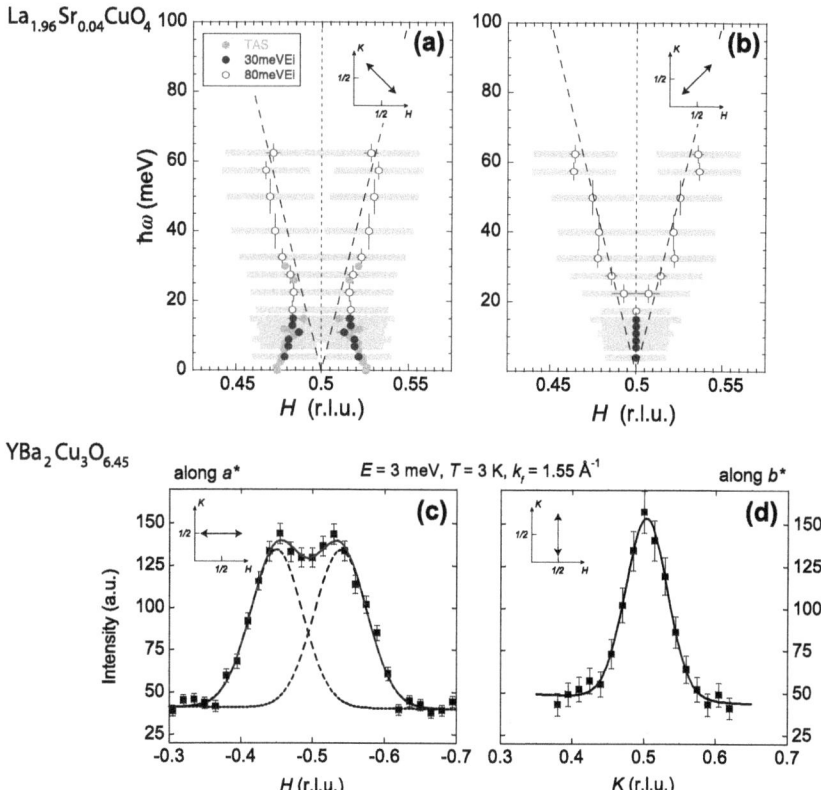

FIGURE 6 (a and b) The twofold rotational symmetry of low-energy excitations in untwinned $La_{1.96}Sr_{0.04}CuO_4$, along the two directions at 45° relative to Cu−O bonds. Clear differences can be seen at low energies, whereas at high energy they both exhibit dispersions similar to La_2CuO_4 (dotted line). *Adapted with permission from Ref. [39].* (c and d) In untwinned $YBa_2Cu_3O_{6.45}$, twofold rotational symmetry is observed at low energies, however, the symmetry axes are along Cu−O bonds, rotated by 45° compared to $La_{1.96}Sr_{0.04}CuO_4$. *Adapted with permission from Ref. [38].*

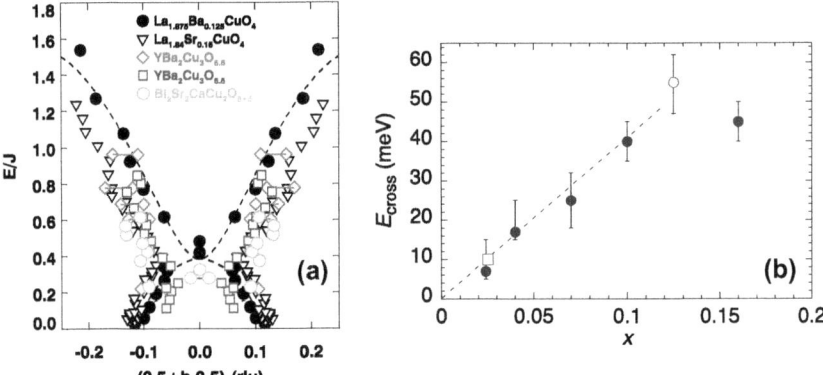

FIGURE 7 (a) The universal hourglass dispersion found in different families of cuprates scaled by J of the parent compound. The dashed line corresponds to $La_{1.875}Ba_{0.125}CuO_4$. It is clear that doping holes causes a softening of high-energy excitations compared to the parent compounds which have bandwidths $\sim 2J$. Adapted with permission from Ref. [8]. (b) The energy E_{cross} at which the upward dispersion and the downward dispersion meet scales with doping, similar to the splitting of low-energy incommensurate excitations. Adapted with permission from Ref. [42].

incommensurability decreases with increasing temperature similar to order parameters, these observations may be associated with electronic nematicity [41]. While we adopt the unit cells in Figure 1, it should be noted that in $La_{2-x}Sr_xCuO_4$ there are small structural distortions that cause (110) and (1−10) to become inequivalent, in $YBa_2Cu_3O_{6+x}$ Cu−O chains cause (100) and (010) to become inequivalent.

The low-energy excitations in hole-doped cuprates emanating from incommensurate positions disperse inward toward (1/2,1/2) until some energy E_{cross} above which it starts to disperse out resembling spin waves in the corresponding parent compounds although with a slightly reduced effective J. Taken together, these features form the well-known universal hourglass dispersion [8] (Figure 7(a)). The inward dispersion has a velocity that is more or less independent of doping, resulting in E_{cross} also increasing linearly with doping and saturating around optimal doping similar to the splitting of incommensurate magnetic order (Figure 7(b)) [42]. A remarkable achievement toward the understanding of hole-doped cuprates is establishing that these behaviors are universal for different families of hole-doped cuprates. The hourglass dispersion has been established in both $La_{2-x}(Sr,Ba)_xCuO_4$, $YBa_2Cu_3O_{6+x}$, and $Bi_2Sr_2CaCu_2O_{8+\delta}$ [8]. The linear dependence of incommensurate splitting with doping has been established for $YBa_2Cu_3O_{6+x}$ and $Bi_{2+x}Sr_{2-x}CuO_{6+y}$ in addition to $La_{2-x}Sr_xCuO_4$ [37]. It should be noted that the splitting is typically determined from elastic magnetic peaks, but in compounds where no magnetic order is found, it is determined from the splitting of low-energy magnetic excitations.

While it is generally accepted that the upward dispersing excitations above E_{cross} are spin-wave-like excitations, there are mainly two approaches for understanding the inward dispersion below E_{cross}. One possibility is these excitations are due to the dynamic version of static stripes found in $La_{2-x}Ba_xCuO_4$ [7]. The phase diagram of $La_{2-x}Ba_xCuO_4$ is similar to $La_{2-x}Sr_xCuO_4$, however, superconductivity is strongly suppressed at $p = 0.125$ where both charge- and spin-stripe orders are found for this doping. Doped holes segregate into stripes resulting in charge ordering and form antiphase domain boundaries for the magnetically ordered regions, resulting in incommensurate elastic magnetic peaks split along Cu—O bond directions. The charge stripe is therefore correlated with spin-stripe ordering. This is in agreement with the observation that the charge-ordering temperature T_{CO} and the spin-ordering temperature T_{SO} are both enhanced near $p = 0.125$ and both compete with superconductivity [32]. While charge ordering and stripe ordering are also found in $La_{2-x}Sr_xCuO_4$ [32], structural distortions in $La_{2-x}Ba_xCuO_4$ making 100 and 010 inequivalent allow such ordering to be much stronger causing the stronger suppression of superconductivity at $p = 0.125$ in $La_{2-x}Ba_xCuO_4$ [9]. These results suggest that the inward dispersion emanating from incommensurate positions in cuprates is due to fluctuations of spin stripes associated with charge stripes [9]. The similar magnetic excitation spectra found for $YBa_2Cu_3O_{6.45}$, $La_{1.875}Ba_{0.125}CuO_4$, and $Bi_2Sr_2CaCu_2O_{8+\delta}$ also supports this scenario [43—45].

In $La_{2-x}Ba_xCuO_4$, static stripes are found and superconductivity is strongly suppressed at $p = 0.125$, whereas in $La_{2-x}Sr_xCuO_4$ and $YBa_2Cu_3O_{6+x}$ T_c, superconductivity is only mildly suppressed. Recently charge ordering and fluctuations have been observed in underdoped $YBa_2Cu_3O_{6+x}$ [46,47] which competes with superconductivity. Upon application of a magnetic field static charge ordering can be induced [48] or enhanced [46] concomitant with the suppression of superconductivity, demonstrating the competition between charge ordering and superconductivity [46]. The charge ordering in $YBa_2Cu_3O_{6+x}$ is found to arise from charge stripes which break fourfold rotational symmetry of the lattice, similar to stripes in $La_{2-x}Ba_xCuO_4$ [49]. Therefore, electronic orders breaking fourfold rotational symmetry that compete with superconductivity seem to be another unifying theme for the hole-doped cuprates. It is worthwhile to note that the detected magnetism in these proposed stripes display 2D rather than 1D magnetism because the magnetic moments from different spin stripes are in fact correlated with each other.

Given the metallic nature of doped cuprates, the hourglass dispersion can also be tackled from the perspective of itinerant electrons by interpreting the low-energy magnetic excitations as particle-hole excitations [50,51]. This approach is appealing since it provides a link between the iron pnictides and the cuprates. However, observation of an hourglass dispersion in insulating $La_{5/3}Sr_{1/3}CoO_4$ suggests itinerant electrons are not necessary and the hourglass dispersion should be ascribed to fluctuating stripes [52]. Since spin stripes are

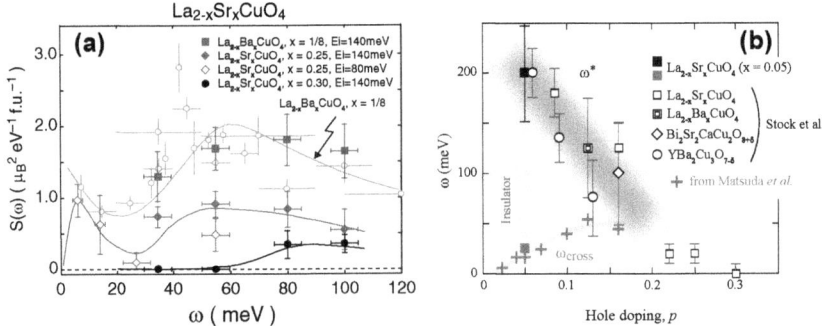

FIGURE 8 Evolution of magnetic spectral weight with doping. (a) Integrated magnetic intensity for $La_{2-x}Sr_xCuO_4$; with increasing doping, the intensity is strongly suppressed [55]. (b) The energy at which χ'' falls to half of the parent compound closely follows the pseudogap energy, E_{cross} is also plotted (ω_{cross} in the figure) [56]. Adapted with permission from Refs [8,55].

based on charge stripes, the observation of an hourglass dispersion and incommensurate magnetic order in the absence of charge-stripe order in $La_{1.6}Sr_{0.4}CoO_4$ led the authors to argue that neither the stripe scenario nor band effects are connected to the hourglass dispersion [53], but rather nanoscopic phase separation is responsible [54]. More research is needed to settle the origin of the hourglass spectrum and its relationship to superconductivity.

With increasing hole-doping, the overall magnetic intensity is suppressed, and the effective coupling is reduced resulting in softened and weaker spin excitations [55] (Figure 8(a)). Interestingly, by tracking the energy at which the magnetic signal falls to half of that of the parent compound, it was found that this energy tracks the well-known pseudogap energy very well (Figure 8(b)) [56]. The suppression of magnetic spectral weight agrees with Raman scattering measurements of bimagnons [57].

3.2.3 Neutron Scattering versus RIXS

In RIXS, the scattering process is similar to inelastic neutron scattering, an incident photon is scattered exchanging both energy and momentum with the sample. By tuning the incident photon energy to match the absorption edge of specific elements the inelastic scattering cross sections can be significantly enhanced, this in addition allows for element selectivity. The RIXS inelastic cross section is sensitive to magnetic, charge, and orbital degrees of freedom, while versatile this also means the RIXS cross section is more complicated than that for neutron scattering. Compared to neutron scattering RIXS only needs very small sample volumes, but since incident photons are typically $\sim$keV it is challenging to resolve excitations in solids which are typically ~ 0.1 eV [58]. Recent advances in RIXS instrumentation have allowed measurements of spin excitations in cuprates with reasonable energy resolution.

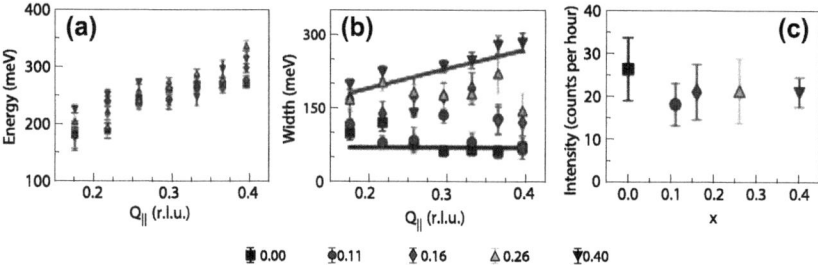

FIGURE 9 The evolution of magnetic excitations in $La_{2-x}Sr_xCuO_4$ with doping seen by resonant inelastic X-ray scattering measured along (1,0) direction at $(Q_\|,0)$. (a) The dispersions show no/small doping dependence. (b) The energy widths of magnetic excitations become much broader for doped cuprates. (c) The averaged magnetic intensity does not change with doping. *Adapted with permission from Ref. [59].*

The spin excitations in $La_{2-x}Sr_xCuO_4$ were found to neither soften nor weaken in terms of intensity even up to the overdoped regime with $p = 0.2$ (Figure 9), compared to spin waves in La_2CuO_4 [59]. The persistence of spin excitations was also found in many other hole-doped cuprate superconductors [18,60]. These results seem to contradict the observations of neutron scattering, where significant reduction of magnetic spectral weight is observed for doped cuprates [56]. An important difference between neutron scattering and RIXS is that neutron scattering measures near (1/2,1/2), RIXS measures near (0,0). In the antiferromagnetically ordered parent compounds, spin waves emanating from (0,0) and (1/2,1/2) have identical dispersion but the structure factor near (1/2,1/2) is much larger. When holes are doped resulting in the destruction of antiferromagnetic order, (0,0) and (1/2,1/2) are no longer equivalent. Therefore, the persistence of spin excitations with doping near (0,0) does not necessarily contradict the suppression of spin excitations near (1/2,1/2). However, since the magnetic spectral weight is concentrated near (1/2,1/2), the integrated magnetic spectral weight will still be strongly suppressed with doping despite the persistence of spin excitations near (0,0).

Also it should be noted that, whereas the neutron scattering cross section is simple and well understood, the RIXS cross section is considerably more complicated including contributions from magnetic, charge, and orbital degrees of freedom [58]. Therefore, it would be ideal if the RIXS results can be reproduced with neutron scattering especially for the doped compounds where, due to the suppression of antiferromagnetic order, magnetic excitations near (0,0) are no longer equivalent to those near (1/2,1/2). However, so far no such results have been reported, possibly because the structure factor is too small near (0,0) for neutron scattering.

Compared to neutron scattering the sample volume required for RIXS is much smaller, and it has even been possible to measure spin excitations in heterostructures consisting of single La_2CuO_4 layers [61]. In addition, RIXS is element sensitive, making it an ideal probe to study compounds where more than one element could be magnetic. With several new RIXS instruments with even better energy resolution coming on line in the near future, RIXS is poised to become a very powerful probe in measuring spin excitations in addition to neutron scattering.

3.2.4 The "Resonance" Mode in the Superconducting State

Upon cooling below T_c, a "resonance" mode has been found in many superconducting cuprates. Loosely speaking, the resonance mode refers to strong enhancements of collective magnetic excitations at certain wave vectors and energy transfers due to superconductivity. Experimentally, the mode is observed by subtracting the neutron scattering intensity in the normal state just above T_c from the signal in the superconducting state (Figure 10(a) and (b)). In addition, the resonance mode responds to the onset of superconductivity. In hole-doped cuprates, the energy of the resonance mode does not change as a function of temperature while its intensity resembles the superconductivity order parameter (Figure 10(c)). Formation of the resonance mode is typically associated with

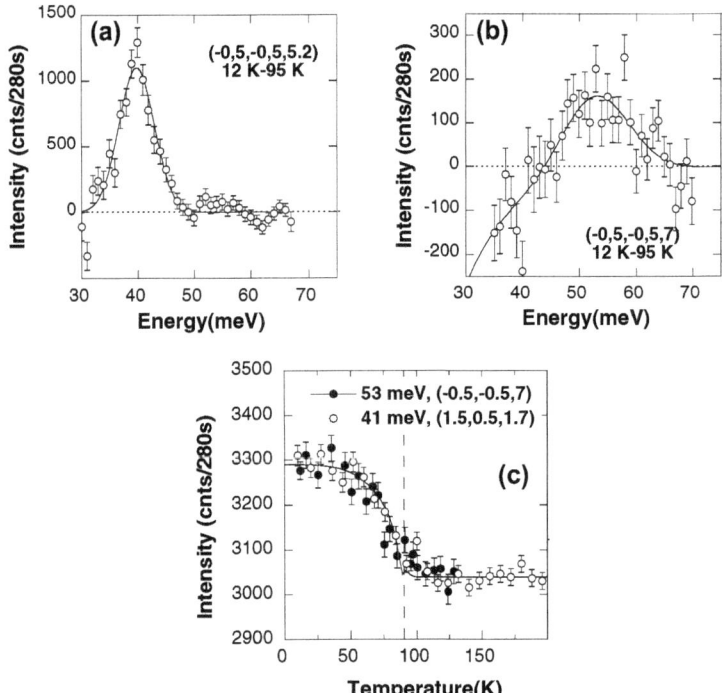

FIGURE 10 The resonance mode seen in $YBa_2Cu_3O_{6.85}$, obtained by subtracting the signal just above T_c from the signal well below T_c. Adapted with permission from Ref. [78]. (a) The odd mode. (b) The even mode. (c) Temperature dependence of both the odd and even modes scaled together.

the opening of a spin gap at energies below the resonance, where upon entering the superconducting state magnetic spectral weight shifts from inside the spin gap to the resonance mode. The resonance mode in hole-doped cuprates was initially discovered in YBa$_2$Cu$_3$O$_{6+x}$ [62] and was later found in La$_{2-x}$Sr$_x$CuO$_4$ [63], Bi$_2$Sr$_2$CaCu$_2$O$_{8+\delta}$ [64], Tl$_2$Ba$_2$CuO$_{6+\delta}$ [65], and HgBa$_2$CuO$_{4+\delta}$ [66]. Similar enhancements were also observed in electron-doped cuprates [67], iron-based superconductors [68], and heavy fermion superconductors [69–71]. It has been argued that the energy of the resonance mode scales with T_c [72] or the superconducting gap [73]. The spin gap in YBa$_2$Cu$_3$O$_{6+x}$ due to superconductivity is found to scale with T_c [74]. In the cuprates, iron pnictides, and heavy fermion superconductors, the change in magnetic exchange energy in the superconducting state due to the resonance mode is sufficient to account for the superconductivity condensation energy [71,75,76].

In the case of La$_{2-x}$Sr$_x$CuO$_4$, the resonance occurs at incommensurate positions below E_{cross}, the spin gap is small, and the magnetic excitations above T_c are already strong unlike in other hole-doped cuprates [63,77]. Interestingly in bilayer cuprates YBa$_2$Cu$_3$O$_{6+x}$ [78] and Bi$_2$Sr$_2$CaCu$_2$O$_{8+\delta}$ [79], two resonance modes, one with odd symmetry and another with even symmetry in the bilayers, are found (Figure 10(a) and (b)), similar to the acoustic and optical spin waves found in the parent phase of YBa$_2$Cu$_3$O$_{6+x}$ [25]. The average energy of the two modes has been argued to correspond to the energy of the resonance in single-layer cuprates [80], the average has been also used by Yu et al. [73] to arrive at the universal relationship between the resonance energy and the superconducting gap. In underdoped YBa$_2$Cu$_3$O$_{6+x}$, the resonance mode was found to survive above T_c in the pseudogap state [81].

In slightly underdoped YBa$_2$Cu$_3$O$_{6+x}$, the resonant mode forms an hourglass dispersion [78], similar to the excitations in the normal state of La$_{1.875}$Ba$_{0.125}$CuO$_4$ [44]. One way to unify the magnetic excitations in different systems is to consider them to be dynamic stripes as described above, which is essentially a local-moment approach. Such a description is a natural extension of the spin waves in the insulating parent compounds. Alternatively, an entirely itinerant approach appropriate for optimal doped and overdoped cuprates has been widely studied [82,83]. In this picture, the resonance is a spin-exciton inside the particle-hole continuum, and spin fluctuations measured by neutron scattering at the wave vector $\mathbf{Q}$ are favored when $\mathbf{Q}$ connects Fermi surfaces (at $\mathbf{k}$ and $\mathbf{Q} + \mathbf{k}$) with opposite superconductivity order parameters ($\Delta(\mathbf{k} + \mathbf{Q}) = -\Delta(\mathbf{k})$). In the case of cuprates, this is satisfied at $\mathbf{Q} = (1/2,1/2)$ which connects "hot spots" near the Fermi surface with opposite superconductivity order parameters. The hourglass dispersion can be reproduced by calculating the threshold of the particle-hole continuum which puts an upper limit on the resonance energy (Figure 11) [80,82]. It has been suggested that this interpretation of the resonance mode serves as a "common thread" linking different families of unconventional superconductors [83]. The observation of a "Y"-shaped normal state response in YBa$_2$Cu$_3$O$_{6+x}$ suggests that this approach is more appropriate for

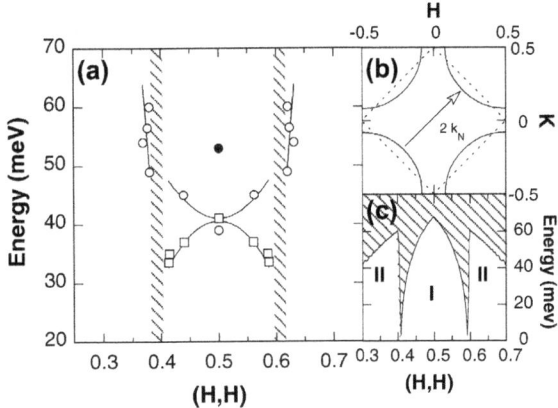

FIGURE 11 Resonance modes in $YBa_2Cu_3O_{6.85}$. Adapted with permission from Ref. [78]. (a) Experimentally observed dispersion of the resonance, open (full) symbols represent resonance modes with odd (even) symmetry. (b) Typical Fermi surface found in hole-doped cuprates. (c) Electron-hole spin-flip continuum, the downward dispersion is enforced by the continuum going to zero at the border between regions I and II.

$YBa_2Cu_3O_{6+x}$ than dynamic stripes [84]. This approach, however, cannot account for the hourglass magnetic excitations in $La_{1.875}Ba_{0.125}CuO_4$, where stripe order is clearly a better description. The enhancement of magnetic excitations in $La_{2-x}Sr_xCuO_4$ is interpreted as a reduction in damping rather than a spin-exciton [85]. Such a mechanism accounts for the enhancement of phonon lifetimes and intensities in conventional superconductors [86]. Recently, a "Y"-shaped magnetic response in both the superconducting and the normal state was uncovered in $HgBa_2CuO_{4+\delta}$ [87], representing a departure from the universal hourglass dispersion.

The dichotomy of theoretical descriptions of magnetic excitations in doped cuprates has been a long-standing issue in the field. It is clear that cuprates become less correlated with increasing doping and for a large part of the phase diagram reside in the intermediate coupling regime. The difficulty with reconciling the itinerant and the local moment approaches could be the lack of a suitable description of magnetism in the intermediate coupling regime. Similar difficulties are all the more apparent in the iron-based superconductors, where even how to describe the magnetism in the parent compounds is not entirely clear.

3.3 IRON-BASED SUPERCONDUCTORS

3.3.1 Introduction, Compounds, and Phase Diagrams

While the physics of superconducting cuprates remains to be elucidated, the discovery of high-temperature superconductivity in iron pnictides [11] opened up a whole new perspective to tackle the problem. The discovery of antiferromagnetic order in the parent compound highlights the relevance of magnetism in superconductivity [88] similar to the cuprates. However, in stark contrast to the cuprates, the parent compounds of iron pnictides are metallic, and the presence of nested electron and hole Fermi surfaces suggests that the magnetism may be of itinerant origin [2]. This initial observation raises the question whether superconductivity in the two systems can have the same origin. Unlike the parent compounds of the cuprates, the parent compounds of the iron-based superconductors reside in the intermediate coupling regime, similar to doped cuprates [89]. It is not clear whether the iron pnictides are also close to a Mott-insulating phase as theoretically proposed [90].

Iron-based superconductors that are well studied can be divided into five families, iron pnictides 1111 (e.g., $LaFeAsO_{1-x}F_x$), 122 (e.g., $Ba(Fe_{1-x}Co_x)_2As_2$), 111 (e.g., $NaFe_{1-x}Co_xAs$), iron chalcogenides (e.g., $FeTe_{1-x}Se_x$), and alkali metal iron chalcogenides (e.g., $K_xFe_{2-y}Se_2$). The parent compounds of iron pnictides (e.g., LaFeAsO, $BaFe_2As_2$, and NaFeAs) are metallic paramagnets with tetragonal crystal structure at room temperature. As the temperature is lowered, the crystal structure changes to orthorhombic at the structural transition temperature T_S, and then long-range antiferromagnetic order develops at T_N. Although the structural transition either precedes or coincides with the magnetic transition, it is believed to be driven by electronic degrees of freedom [91]. The parent compounds of the iron pnictides adopt collinear magnetic order (stripe order) [88], the magnetic structure of FeTe is instead bicollinear magnetic order (double-stripe order) [92]. The insulating alkali metal iron chalcogenide is made up of antiferromagnetically aligned ferromagnetic blocks with four Fe atoms (block checkerboard antiferromagnetic order) [93].

Starting from the parent compound of the iron pnictides such as $BaFe_2As_2$, superconductivity can be induced by applying pressure, electron-doping, hole-doping, and even isovalent substitution [94,95]. For example, starting from $BaFe_2As_2$, superconductivity can be induced by introducing alkali metals at Ba sites; Co, Ni, Ru, or Rh at Fe sites; and P at As sites. Curiously, while doping Cr or Mn at Fe sites suppresses the magnetic order, superconductivity is not induced. The phase diagrams for $Ba_{1-x}K_xFe_2As_2$ and $Ba(Fe_{1-x}Co_x)_2As_2$ are shown in Figure 12(a), superconductivity is induced in both cases by doping either holes or electrons. However, T_S and T_N split with Co doping while they remain coincident for K doping. There is also a region where superconductivity coexists with antiferromagnetic order in both cases. The phase diagrams for other iron pnictides are similar to Figure 12(a). For NaFeAs and LaFeAsO, T_S and T_N are already split in the parent phase, whereas LiFeAs is paramagnetic with tetragonal crystal structure at all temperatures. The phase diagram for $FeTe_{1-x}Se_x$ is shown in Figure 12(b), doping Se in FeTe suppresses magnetic ordering and results in a glassy magnetic state at low dopings before inducing bulk superconductivity. There is no coexistence of long-range magnetic order and superconductivity. The end member FeSe also superconducts at 8 K, and it has also been proposed that Fe_4Se_5 is the parent compound for

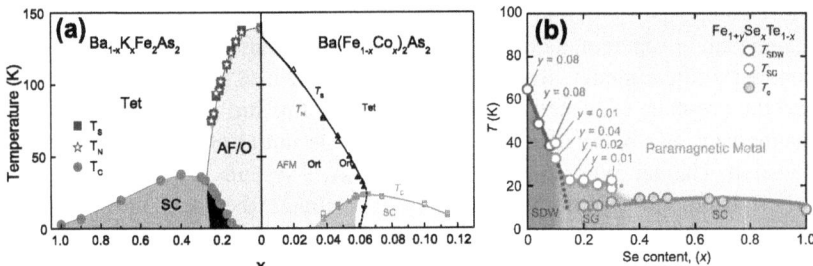

FIGURE 12 (a) Phase diagram of $Ba_{1-x}K_xFe_2As_2$ and $Ba(Fe_{1-x}Co_x)_2As_2$. T_S and T_N split with Co doping, but remains coincident for K doping [14]. (b) Phase diagram of $FeTe_{1-x}Se$, a glassy magnetic phase separates long-range magnetic order from bulk superconductivity. *Adapted with permission from Ref. [229].*

iron chalcogenides instead of FeTe [96]. The case of alkali metal iron chalcogenides is more complicated due to phase separation. While a phase diagram where superconductivity is induced with the suppression of antiferromagnetic order has been proposed [15], the exact phase responsible for superconductivity is still under debate [16]. Sections 3.3.2−3.3.5 will focus on magnetism in the iron pnictides and iron chalcogenides. Magnetism in alkali-metal iron chalcogenides is discussed in Section 3.3.6. Results from polarized neutron scattering experiments are discussed in Section 3.3.7.

3.3.2 Antiferromagnetically Ordered Metallic Parent Compounds

At room temperature, the parent compounds of iron pnictides adopt a tetragonal structure as shown in Figure 13(a) and (b), while for 122 compounds the chemical unit cell consists of two FeAs layers; 11, 111, and 1111 compounds have only one FeAs/FeTe(Se) layer in the chemical unit cell. Upon cooling, the tetragonal crystal structure changes to orthorhombic, this transition is first order for $SrFe_2As_2$ and $CaFe_2As_2$ but is second order for $BaFe_2As_2$ and NaFeAs [97,98]. The orthorhombic unit cell is rotated by 45° compared to the tetragonal unit cell and is twice as large in the *ab* plane. Stripe magnetic order appears in the orthorhombic phase with spins oriented along the longer in-plane orthorhombic axes (a_O) when temperature is further lowered (Figure 13(d)). In the notation of the orthorhombic chemical unit cell, the magnetic order appears at $(1,0,1)_O$ and equivalent wave vectors for 122 compounds and $(1,0,0.5)_O$ for 111 and 1111 compounds. The magnetic transition is first order for $BaFe_2As_2$, $SrFe_2As_2$, and $CaFe_2As_2$, but is second order for NaFeAs [97,98]. Upon electron doping, both the structural and magnetic transitions become second order. In the discussion of magnetic scattering, we typically adopt the orthorhombic notation and use subscripts to indicate which notation is used. In the orthorhombic notation, magnetic scattering occurs near $(1,0)_O$, while in tetragonal

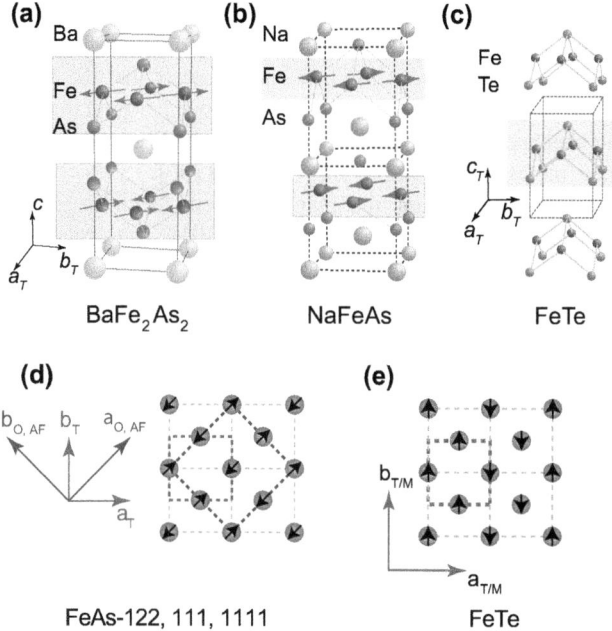

FIGURE 13 The crystal structures of (a) BaFe$_2$As$_2$, (b) NaFeAs, and (c) FeTe. The dashed lines indicate the tetragonal chemical unit cell, and arrows are ordered moments in the magnetically ordered state. (d) In-plane magnetic structure of iron pnictides including 122, 111, and 1111 compounds. The green (dark gray in print versions) dashed box is the tetragonal chemical unit cell, and the dashed magenta (gray in print versions) box is the orthorhombic chemical unit cell. The ordered moments are aligned along the a_O direction. (e) In-plane magnetic structure for FeTe, the green (dark gray in print versions) dashed line is the chemical unit cell for both the tetragonal and monoclinic phases.

notation this position becomes $(0.5, 0.5)_T$. In systems that are tetragonal and in orthorhombic systems due to twinning, magnetic signal is also seen at $(0,1)_O$.

FeTe is also tetragonal at room temperature, below ~70 K the system orders magnetically accompanied by a structural transition. Below the structural transition temperature FeTe changes to a weakly monoclinic structure, with $a_M > b_M$ and $\beta < 90°$. The magnetic order in FeTe has a bicollinear structure as shown in Figure 13(e), in this unit cell magnetic order occurs at $(0.5, 0)_T$, and the spins are oriented along b_M. In FeTe samples an additional complicating factor is the presence of excess Fe, a small amount of which is necessary to stabilize the structure, therefore, the actual chemical composition is Fe$_{1+x}$Te. For $x < 0.12$ the system adopts the commensurate magnetic order in Figure 13(e) with a weakly monoclinic crystal structure. For $x > 0.12$ the magnetic order instead becomes incommensurate, characterized by the wave vector $(0.5 - \delta, 0, 0.5)_T$ and the crystal structure becomes orthorhombic [99]. For FeTe$_{1-x}$Se$_x$, with increasing Se doping the amount of excess Fe decreases, and in superconducting samples excess Fe is no longer an issue of concern (Figure 12(b)). In stark contrast to the parent compounds of iron pnictides, Fermi-surface nesting is absent in FeTe and suggests that the magnetic order originates from local moments. However, it has been argued that the magnetic order in the absence of Fermi-surface nesting in FeTe can still be reproduced with itinerant models involving d_{xz}/d_{yz} orbitals [100].

In contrast to the cuprates, the magnetic form factor in iron pnictides is isotropic and agrees with that of metallic iron [101]. The ordered moments in the parent compounds of iron pnictides are smaller than 1 μ_B. Such values are smaller than the predictions from density functional theory (DFT), but calculations combining with dynamical mean field theory (DMFT) yields a much better agreement [102]. This highlights the fact that iron pnictides reside in the intermediate coupling regime and that neither a fully local nor a fully itinerant picture is sufficient to describe the underlying physics. Alternatively, it has been proposed that the small-ordered moment observed by neutron diffraction is due to local structural variations [103]. The ordered moment size in FeTe is ~ 2 μ_B, suggesting it is a more localized system compared to the iron pnictides [92].

Similar to the cuprates, measurements of high-energy spin waves can be done with the incident neutron beam parallel to the c-axis on chopper time-of-flight instruments. In such a geometry, L cannot be independently varied but is determined by incident energy, energy transfer, and momentum transfer in the ab plane. If the magnetism in iron pnictides is truly 2D as in the cuprates, this method can unequivocally obtain the complete magnetic response for all energy transfers. If the magnetism is 3D, and the magnetic signal is sharp in reciprocal space, it is possible to choose energy transfers that correspond to either zone center or the zone boundary along L and determine if there is significantly L dependence. This was done for nearly optimal-doped Ba(Fe$_{0.935}$Co$_{0.065}$)$_2$As$_2$ and no L dependence was found for $E > 10$ meV, demonstrating the magnetic excitations are essentially L independent for this compound [104]. For the parent compounds, the spin-anisotropy gaps depend on L. For BaFe$_2$As$_2$, the spin-anisotropy gap is 10 meV for the magnetic zone center ($L = 1$) and 20 meV for the zone boundary ($L = 0$) [105]. For NaFeAs, the gap size is 5 meV for the magnetic zone center ($L = 0.5$) and 7 meV for the zone boundary ($L = 0$) [106]. For LaFeAsO, the gap is ~ 10 meV for both the magnetic zone center ($L = 0.5$) and zone boundary ($L = 0$) [107]. Since the L dependence of the spin-anisotropy gap is essentially determined by interlayer coupling J_c [107], these results suggest that the magnetism in LaFeAsO is more 2D than NaFeAs, which in turn is more 2D than BaFe$_2$As$_2$. It should also be noted that these results are obtained using triple-axis spectrometers where L can be independently varied, whereas estimates from time-of-flight measurements with $\mathbf{k_i}$ parallel to c-axis are not sensitive to J_c [108,109].

Spin waves in the entire Brillouin zone have been mapped out for BaFe$_2$As$_2$ [108], CaFe$_2$As$_2$ [110], SrFe$_2$As$_2$ [111], and NaFeAs [109]. As can be seen in Figure 14(e), the spin waves emanate from $(1,0)_O$ and disperse

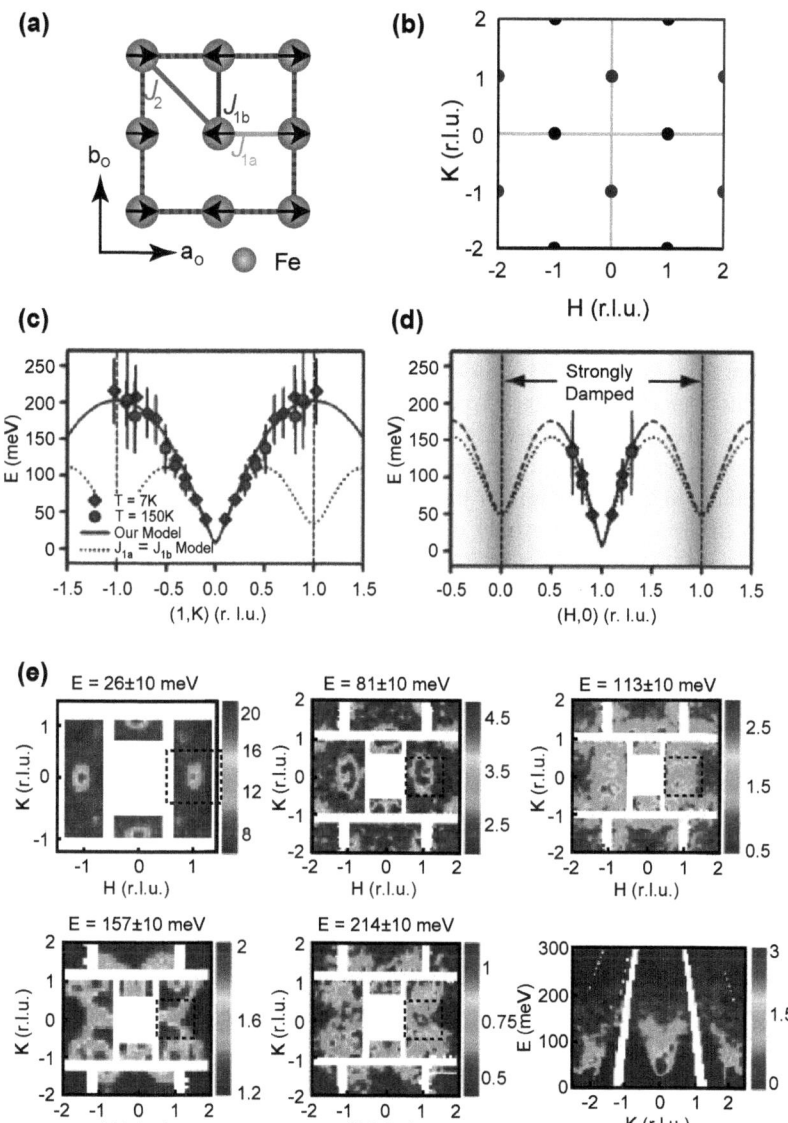

FIGURE 14 Spin waves in $BaFe_2As_2$. *Adapted with permission from Ref. [108]*. (a) The in-plane magnetic structure of $BaFe_2As_2$ and definition of effective exchange couplings. (b) The magnetic Bragg peaks in reciprocal space, the blue (dark gray in print versions) points are Bragg peaks for the structure in (a), and the red (gray in print versions) points are from domains rotated by 90° from (a), the notation in reciprocal space corresponds to the orthorhombic chemical unit cell in (a) enclosed in the green (light gray in print versions) box. (c) The dispersion of spin waves along the transverse direction centered at (1,0,L), the solid line represents a model with $J_{1a} \neq J_{1b}$, and the dashed line is the model with $J_{1a} = J_{1b}$. Similarly, the dispersion along the longitudinal direction centered at (1,0,L) is shown in (d). (e) Constant-E slices of time-of-flight data for $BaFe_2As_2$, the spin waves emanate from (1,0) and disperse toward (1,1) with a bandwidth over 200 meV.

toward $(1,1)_O$, the spin-wave velocity along the longitudinal direction is higher than that along the transverse direction. The results can be interpreted using Heisenberg models with effective exchange interactions (Figure 14(a)). For BaFe$_2$As$_2$, the obtained exchange couplings are $SJ_{1a} = 59(2)$ meV, $SJ_{1b} = -9(1)$ meV, and $SJ_2 = 14(1)$ meV. It should be noted that SJ can be directly obtained by fitting spin waves measured by neutron scattering, and only when S is known as in the case of parent compounds of cuprates, J can be determined. In the case of iron-based superconductors, SJ is typically reported. Within the Heisenberg model, the larger spin-wave velocity along the longitudinal direction than the transverse direction is due to $J_{1a} + J_{1b} > 0$ [94]. Anisotropic nearest-neighbor exchange with $J_{1a} \neq J_{1b}$ is also necessary to account for the observed spin-wave dispersion as shown in Figure 14(c), whereas a model with isotropic exchange interactions does not correctly account for the magnetic band top being at $(1,1)_O$. Given the metallic nature of the parent compounds, spin waves are expected to be damped by particle-hole excitations in large portions of the Brillouin zone in contrast to La$_2$CuO$_4$ for which the spin waves are only damped possibly by interactions with spinon pairs near $(1/2,0)$ [20]. As can be seen in Figure 14(e), the spin waves at low energies near $(1,0)_O$ are well defined, and with increasing energy they become more diffuse. This can be understood by calculating particle-hole excitation spectra which is weak near $(1,0)_O$ [112]. Experimentally, damping of spin waves is accounted for by using an empirical form [108–110]. Alternatively, the spin waves can also be interpreted using a Heisenberg model with isotropic J_1 and J_2 by adding a biquadratic coupling of the form $-K(S_i \cdot S_j)^2$ [113–115]. Compared to the Heisenberg model with anisotropic exchange, this model can also account for the magnetic excitations in the tetragonal paramagnetic state [30,111].

The spin waves can also be understood using itinerant models. For SrFe$_2$As$_2$, it was argued that the magnetic excitations are better explained by a five-band itinerant mean-field model [111]. However, calculations using the random-phase approximation (RPA) within a purely itinerant model to calculate the spin excitations cannot reproduce the local magnetic susceptibility in BaFe$_2$As$_2$ (see Section 3.3.4), whereas DMFT can qualitatively reproduce the local susceptibility [116], stressing the importance of electronic correlations in understanding the magnetism in iron pnictides. Compared to the 122 parent compounds, the magnetic excitations in NaFeAs are qualitatively the same except that their bandwidth is much narrower suggesting that NaFeAs is more correlated [102,109]. CaFe$_2$As$_2$ presents an intriguing case where under external pressure (or internal strain field due to FeAs precipitates) both the magnetic order and spin waves completely disappear as the system enters the so-called collapsed tetragonal phase where the c-axis lattice parameter is significantly reduced [117]. NMR [118] and DFT + DMFT [119] calculations suggest that in the collapsed tetragonal phase, electron correlations are suppressed, and CaFe$_2$As$_2$ becomes a good metal with suppressed spin correlations.

High-Temperature Superconductors Chapter | 3 167

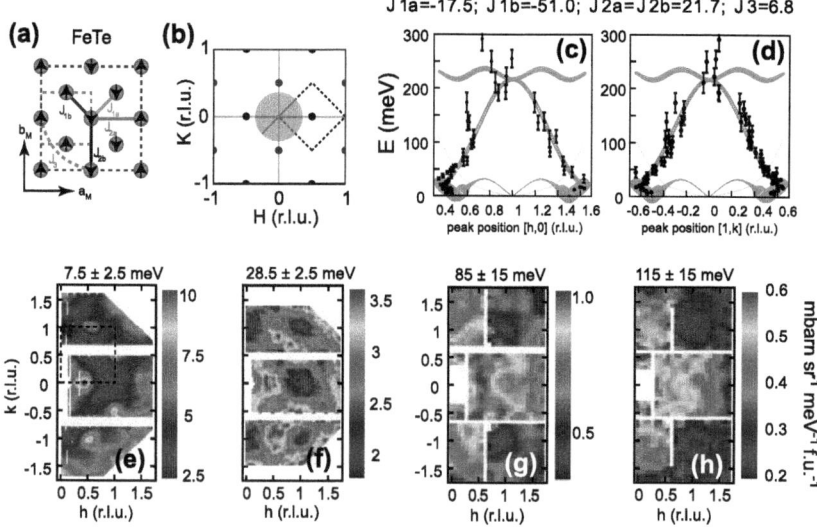

FIGURE 15 Spin waves in FeTe. *Adapted with permission from Ref. [120].* (a) The in-plane collinear commensurate magnetic structure of FeTe and definition of effective exchange couplings. (b) The magnetic Bragg peaks in reciprocal space, the blue (dark gray in print versions) points are Bragg peaks for the structure in (a), and the red (gray in print versions) points are from domains rotated by 90° from (a). The notation in reciprocal space corresponds to the monoclinic chemical unit cell in (a) enclosed in the green (light gray in print versions) box. (c) and (d) show the dispersion extracted from the data, the gray circles are the dispersion expected from the Heisenberg model using parameters above panels (c) and (d). (e–h) show constant energy slices of spin waves in FeTe. Above 85 meV, the excitations become diffuse.

Spin waves in FeTe have also been mapped out for samples with 5% [120] and 10% [121] excess Fe. The magnetic excitations can be fit with a Heisenberg model with anisotropic nearest-neighbor exchange interactions (Figure 15(a)) [120]. However, the spin waves become very diffuse above 85 meV (Figure 15). In another experiment, the magnetic excitations in FeTe were interpreted as correlations of liquidlike spin plaquettes [121]. Interestingly, the effective moment was found to increase with increasing temperature suggesting that local moments become entangled with itinerant electrons at low temperatures via a Kondo-type mechanism. It should be noted that excess Fe significantly affects low-energy magnetic excitations [122]. FeTe with $\sim 6\%$ excess Fe orders at $(1/2,0,1/2)_T$, and spin waves emanating from this wave vector have a 7 meV spin gap, while samples with 14% excess Fe exhibit gapless magnetic excitations stemming from the incommensurate magnetic ordering wave vector $(0.38,0,1/2)_T$. While the low-energy spectral weight in $Fe_{1.14}Te$ is stronger than $Fe_{1.06}Te$, the total moment obtained by considering both elastic and inelastic spectral weight are similar [123]. The commensurate and incommensurate magnetism compete with each other as demonstrated in $Fe_{1.08}Te$, below T_N the magnetism is commensurate with a spin gap while above T_N incommensurate magnetic excitations are seen [124].

3.3.3 Evolution of Magnetic Order and Magnetic Excitations with Doping

Starting from the parent compounds of the iron pnictides, the ordered moments become smaller, and T_N decreases with increasing doping. The orthorhombicity $\delta = (a_O - b_O)/(a_O + b_O)$ for the tetragonal-to-orthorhombic structural transition becomes smaller and T_S also decreases. Therefore, it becomes challenging to study the evolution of the magnetic order and the structural transition near the region where magnetic order and superconductivity coexists (Figure 12(a)), especially with powder samples. Therefore, although the phase diagrams for 1111 compounds were the first to be mapped out [94], the majority of detailed studies in the region where superconductivity coexists with antiferromagnetic order have been performed on doped 122 single crystals. Doping Co or Ni at Fe sites in $BaFe_2As_2$ splits the magnetic and the structural transition while both transitions are driven toward lower temperatures (Figure 12(a)). Increasing doping further, the magnetic order changes from long-range commensurate to transverse-split short-range incommensurate in a first-order fashion (Figure 16(c)) [125,126]. The incommensurate magnetic order was argued to be a hallmark of magnetism driven by a spin-density-wave instability [125], recent results from NMR and neutron resonant spin-echo measurements [127–129] instead suggest that the incommensurate short-range magnetic order is due to formation of a cluster spin-glass phase. In this scenario, the disappearance of short-range magnetic order with increasing doping corresponds to the disappearance of volume fraction for the spin-glass phase [128,129]. In the underdoped region of the phase diagram where superconductivity coexists with magnetic order, it was found that superconductivity competes with magnetic order, whether it is long-range commensurate (Figure 16(a)) or short-range incommensurate order (Figure 16(b)) [126,129,130]. Since the long-range magnetic order from the parent compound disappears through a first-order transition to the short-range incommensurate order phase, there is no quantum critical point in the conventional sense [131]. Application of a magnetic field suppresses superconductivity and enhances magnetic order for both commensurate long-range order [132] and incommensurate short-range order (Figure 16(c)) [126]. However, there is no report of field-induced magnetic order in a sample which does not order under zero field as in $La_{2-x}Sr_xCuO_4$ [35]. Compared to Co and Ni, doping Cu in $BaFe_2As_2$ does not induce incommensurate magnetic order near the putative quantum critical point [133]. Although similar to the former cases, long-range magnetic order also disappears through a phase with short-range magnetic order without a quantum critical point.

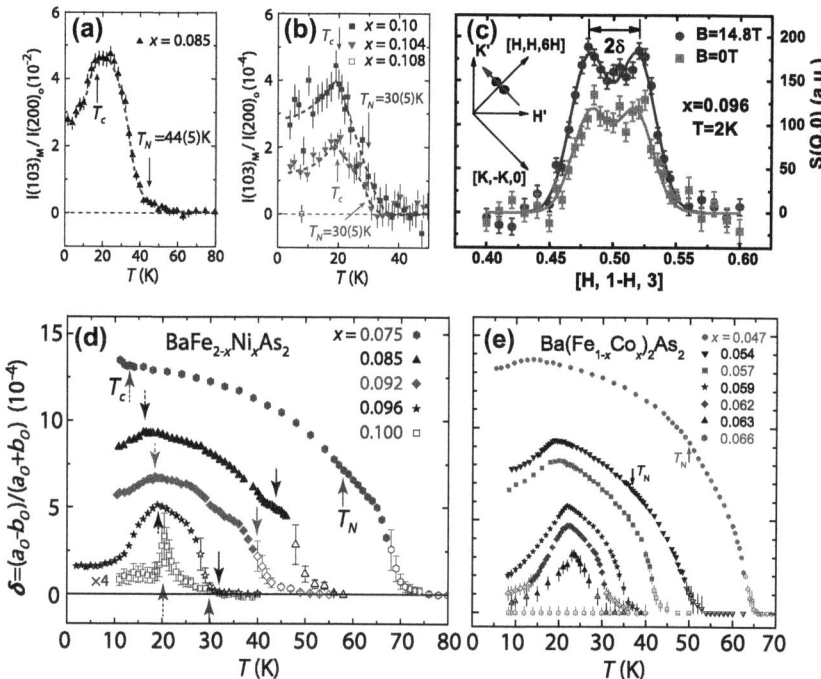

FIGURE 16 Evolution of magnetic order and structural distortion in $BaFe_{2-x}Ni_xAs_2$ and $Ba(Fe_{1-x}Co_x)_2As_2$. (a) The magnetic order parameter in $BaFe_{2-x}Ni_xAs_2$ with $x = 0.085$ which shows long-range commensurate magnetic order, below T_c the magnetic order is suppressed [131]. (b) The magnetic order parameter for $BaFe_{2-x}Ni_xAs_2$ with $x = 0.1$, 0.104, and 0.108. Samples $x = 0.1$ and 0.104 exhibit short-range incommensurate magnetic order, and no magnetic ordering is seen in the sample with $x = 0.108$ [131]. (c) Effect of an applied magnetic field on the short-range incommensurate magnetic order in $BaFe_{2-x}Ni_xAs_2$ with $x = 0.096$. The inset shows the geometry of the experimental setup [126]. (d) Orthorhombicity for $BaFe_{2-x}Ni_xAs_2$, clear suppression of orthorhombicity is seen below T_c, but does not reenter the tetragonal state [131]. (e) Orthorhombicity for $Ba(Fe_{1-x}Co_x)_2As_2$, reenters tetragonal state for samples with $x = 0.063$ [134]. *Adapted with permission from Refs [126,131,134].*

In the underdoped region where superconductivity coexists with antiferromagnetic order in the orthorhombic state, the onset of superconductivity suppresses orthorhombicity (Figure 16(d) and (e)) [131,134]. In the case of $Ba(Fe_{1-x}Co_x)_2As_2$ for critical doping ($x = 0.063$), the system changes from tetragonal to orthorhombic at T_S and then reenters the tetragonal state below T_c. In the case of $BaFe_{2-x}Ni_xAs_2$, similar behavior is seen but no reentrance behavior was observed, even though for samples with $x = 0.096$ and 0.1 the orthorhombicity becomes temperature independent at the lowest temperatures measured (Figure 16(d)) [131]. The suppression of orthorhombicity is suggested to be a result of magnetoelastic coupling [134], therefore it is a secondary effect linked to the competition between magnetic order and superconductivity.

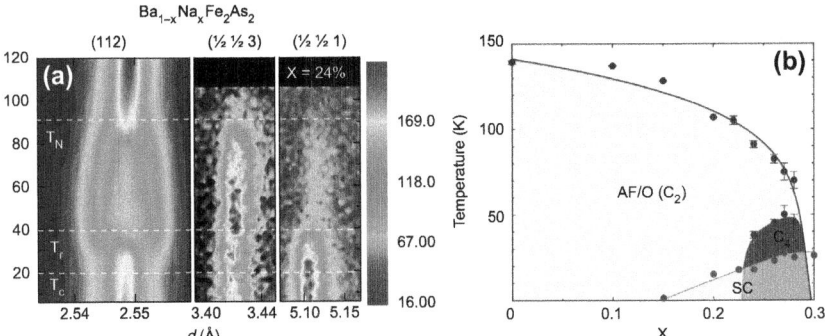

FIGURE 17 Reentrant tetragonal phase in $Ba_{1-x}Na_xFe_2As_2$. *Adapted with permission from Ref. [135]*. (a) Temperature scans of the structural peak (1,1,2) and magnetic peaks (1/2,1/2,3) and (1/2,1/2,1) in $Ba_{1-x}Na_xFe_2As_2$ with $x = 0.24$. The reentry temperature is marked as T_r, the magnetic peaks show significant change in intensity below T_r suggesting possible spin reorientation. (b) The phase diagram for $Ba_{1-x}Na_xFe_2As_2$ showing reentrant phase with C_4 rotational symmetry with magnetic order, the reentry temperature occurs above T_c. SC, superconducting.

Doping holes in $BaFe_2As_2$ by substituting Ba with K and Na also induces superconductivity while suppressing magnetic order, however, the magnetic ordering transition temperature T_N and the structural transition temperature T_S do not split with doping, and superconductivity is achieved at a much higher doping level with a superconductivity dome that spans a much larger region (Figure 12(a)). For $Ba_{1-x}Na_xFe_2As_2$, reentry into the tetragonal state was also found (Figure 17) [135]. In contrast to $Ba(Fe_{1-x}Co_x)_2As_2$, the reentry occurs above T_c. The reentry to the tetragonal state is possibly accompanied by a reorientation of magnetic moments (Figure 17(a)) [135]. The presence of magnetic order in tetragonal $Ba_{1-x}Na_xFe_2As_2$ favors the presence of a magnetic Ising-nematic state [91].

Doping Mn and Cr at Fe sites in $BaFe_2As_2$ does not split the magnetic and structural transitions nor induces superconductivity. In the phase diagram for $Ba(Fe_{1-x}Cr_x)_2As_2$, while the stripe magnetic order from $BaFe_2As_2$ is suppressed at low Cr concentrations, checkerboard magnetic order (characterized by the ordering wave vector $(1,1)_O$) appears for doping levels $x > 0.3$ (Figure 18(a)) [136]. The phase diagram for $Ba(Fe_{1-x}Mn_x)_2As_2$ is similar [137,138]. For Mn doping levels $x > 0.1$, it was found that the system enters a Griffiths-type phase before the stripe magnetic order associated with $BaFe_2As_2$ is completely suppressed. Therefore, Mn dopants act as strong local magnetic impurities [138]. In $Ba(Fe_{1-x}Mn_x)_2As_2$ with $x = 0.075$ which exhibit only stripe magnetic order, Neel-type magnetic excitations at $(1,1,L)$ are seen suggesting such magnetic fluctuations compete with stripe magnetic fluctuations [139]. Isovalent doping P at As sites or Ru at Fe sites in $BaFe_2As_2$ also suppresses magnetic order. In these cases, the structural transition temperature T_S remains coincident with the magnetic

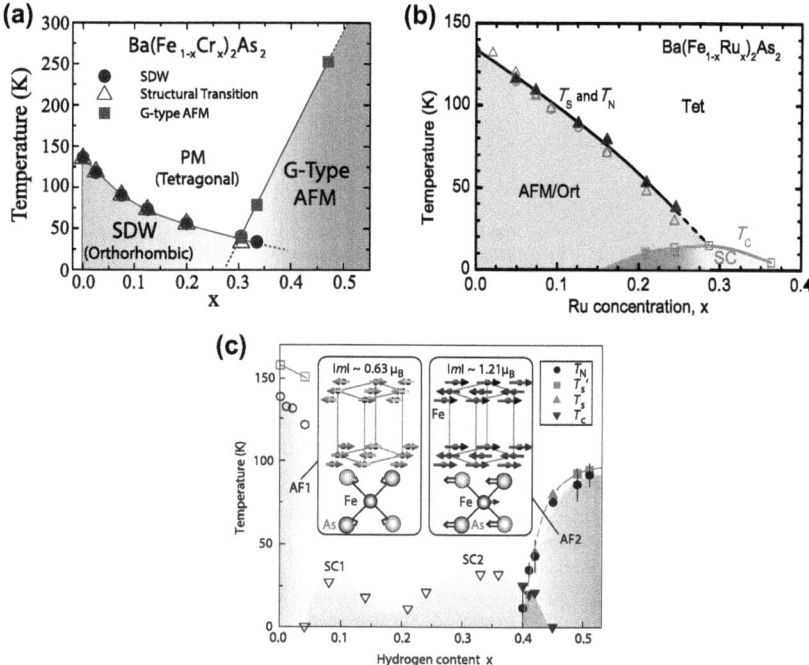

FIGURE 18 Magnetic phase diagram of Cr-doped [136] and Ru-doped [230] BaFe$_2$As$_2$ and H-doped LaFeAsO [143]. (a) In Ba(Fe$_{1-x}$Cr$_x$)$_2$As$_2$, stripe magnetic order is suppressed with Cr doping up to ~30% and then G-type magnetic order appears [136]. (b) In Ba(Fe$_{1-x}$Ru$_x$)$_2$As$_2$, T_N and T_S remain coincident for all doping levels, and superconductivity is induced at ~20% Ru doping [230]. (c) In LaFeAsO$_{1-x}$H$_x$, two superconductivity domes associated with two magnetically ordered phases are observed, the two magnetically ordered end compounds have different magnetic structures and order moments [143]. *Adapted with permission from Refs [136,143,230].*

transition temperature T_N, and superconductivity is induced for ~20% dopants. BaFe$_2$(As$_{1-x}$P$_x$)$_2$ is especially interesting because of the numerous reports of a quantum critical point [140] in this system [141]. However, results from single-crystal neutron diffraction suggest magnetic order in this system also disappears in a weakly first-order fashion [142]. An interesting case occurs when electron dopant H is introduced into LaFeAsO, superconductivity forms two domes and for doping levels ~50%, another magnetically ordered phase appears (Figure 17(c)) [143]. The second magnetically ordered phase also exhibits stripe order, but compared to LaFeAsO it is ferromagnetic along c-axis and the moments are perpendicular to the stripe direction. The two superconductivity domes derived from two separate ordered phases highlight the connection between magnetism and superconductivity.

For spin excitations, upon doping Ni in BaFe$_2$As$_2$ the immediate effect is to suppress the spin-anisotropy gap and, therefore, the c-axis coupling present in the parent compound [144]. The low-energy magnetic excitations centered at

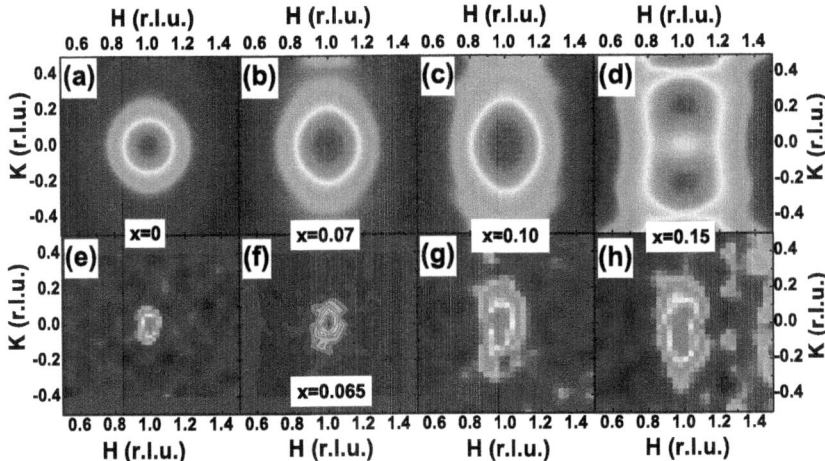

FIGURE 19 Evolution of low-energy magnetic excitations in BaFe$_{2-x}$Ni$_x$As$_2$ at $E = 8$ meV. Adapted with permission from Ref. [145]. (a–d) random-phase approximation calculations. (e–h) Neutron scattering measurements. The qualitative agreement suggests the low-energy excitations can be ascribed to itinerant electrons.

$(1,0)_O$ broaden along both the longitudinal and the transverse directions but with a higher rate along the transverse direction. The evolution of low-energy magnetic excitations with doping can be qualitatively reproduced by calculations using the RPA favoring an itinerant origin of low-energy magnetic excitations (Figure 19) [145]. Remarkably, doping K at Ba site in BaFe$_2$As$_2$ results in the low-energy magnetic excitations becoming longitudinally elongated at optimal doping (Figure 20(a)) [146], in agreement with weak coupling model predictions [147]. Further K doping results in two split peaks situated at $(1 \pm \delta, 0)$ in overdoped Ba$_{1-x}$K$_x$Fe$_2$As$_2$ [148] and KFe$_2$As$_2$ (Figure 20(b)) [149].

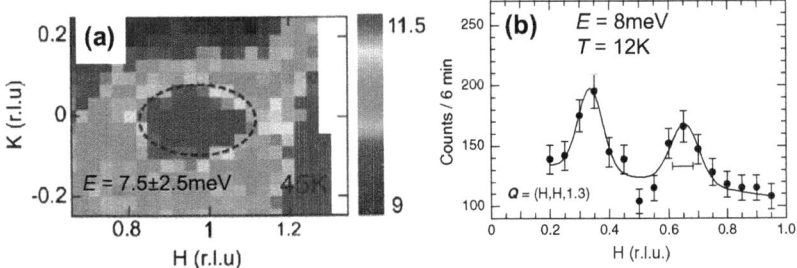

FIGURE 20 Evolution of low-energy magnetic excitations in Ba$_{1-x}$K$_x$Fe$_2$As$_2$. (a) In optimal doped Ba$_{0.67}$K$_{0.33}$Fe$_2$As$_2$, the low-energy magnetic excitations become longitudinally elongated, rotated by 90° compared to electron-doped BaFe$_2$As$_2$. Adapted with permission from Ref. [146]. (b) In KFe$_2$As$_2$, magnetic excitations split into two peaks along the longitudinal direction. Adapted with permission from Ref. [149].

Interestingly in $SrCo_2As_2$, magnetic excitations are found to be centered at $(1,0)_O$ but are longitudinally elongated [150] resembling K-doped rather than Co-/Ni-doped $BaFe_2As_2$ [145]. It is not clear how such an evolution occurs as Co gradually replaces Fe in $SrFe_2As_2$.

Doping Se in FeTe suppresses static magnetic order and causes magnetic spectral weight to be transferred from $(0.5,0)_T$ to $(0.5,0.5)_T$ near optimal doping [151,152]. Therefore, in superconducting $FeTe_{1-x}Se_x$ compounds, magnetic excitations are situated at the same wave vector as in the iron pnictides. Coexistence of magnetic excitations at $(0.5,0)_T$ and near $(0.5,0.5)_T$ is found over a wide doping range for $FeTe_{1-x}Se_x$. With increasing Se content, magnetic excitations at $(0.5,0)_T$ are suppressed while those at $(0.5,0.5)_T$ are enhanced [153−155]. In moderately Se-doped $FeTe_{1-x}Se_x$, low-energy magnetic excitations are found at $(0.5,0)_T$ while at higher energies the spectral weight transfers to incommensurate peaks transversely split around $(0.5,0.5)_T$ [153,156]. Increasing Se to ~50%, the magnetic response becomes commensurate at $(0.5,0.5)_T$ [157]. The magnetic excitations in $FeTe_{1-x}Se_x$ are compared for nonsuperconducting doping $x = 0.27$ and superconducting doping $x = 0.49$ in Figure 21 [156]. For $E = 6$ meV, the magnetic excitations in $FeTe_{0.73}Se_{0.27}$ are centered at $(0.5,0)_T$ (Figure 21(a)) but at higher energies become transversely split incommensurate peaks centered around $(0.5,0.5)_T$ (Figure 21(b)). In $FeTe_{0.51}Se_{0.49}$, the corresponding magnetic excitations are transversely elongated commensurate peaks centered at $(0.5,0.5)_T$ (Figure 21(f) and (g)). However, the high-energy magnetic excitations are similar in the two compounds (Figure 21(c−e) and (h−k)). These results suggest that the appearance of superconductivity in $FeTe_{1-x}Se_x$ is accompanied by the appearance of low-energy magnetic excitations around $(0.5,0.5)_T$. The appearance of superconductivity and magnetic fluctuations at $(0.5,0.5)_T$ in superconducting $FeTe_{1-x}Se_x$ is significant since this corresponds to the nesting wave vector [158], suggesting that despite the different magnetic ground state of FeTe and parent compounds of iron pnictides, similar magnetic fluctuations are found in superconducting compounds of both systems [154]. In Cu-doped $Fe_{1-x}Cu_xTe_{0.5}Se_{0.5}$, low-energy magnetic excitations are enhanced and resistivity of the system resembles an insulator rather than a metal. However, it is not clear whether the enhanced magnetic excitations are due to a redistribution of high-energy spectral weight or an enhanced magnetic moment [159].

LiFeAs represents an oddity among iron-pnictide parent compounds as it does not order magnetically and instead becomes superconducting below $T_c = 18$ K. Low-energy magnetic excitations in LiFeAs form transversely split incommensurate peaks (Figure 22(a)) [160,161], similar to electron overdoped 122 compounds. While the absence of static magnetic order in LiFeAs differs from other parent compounds, the low-energy magnetic excitations agree with Fermi-surface measurements using angle-resolved photoemission spectroscopy (ARPES) [162]. Given the Fermi surface of LiFeAs, two possible nesting

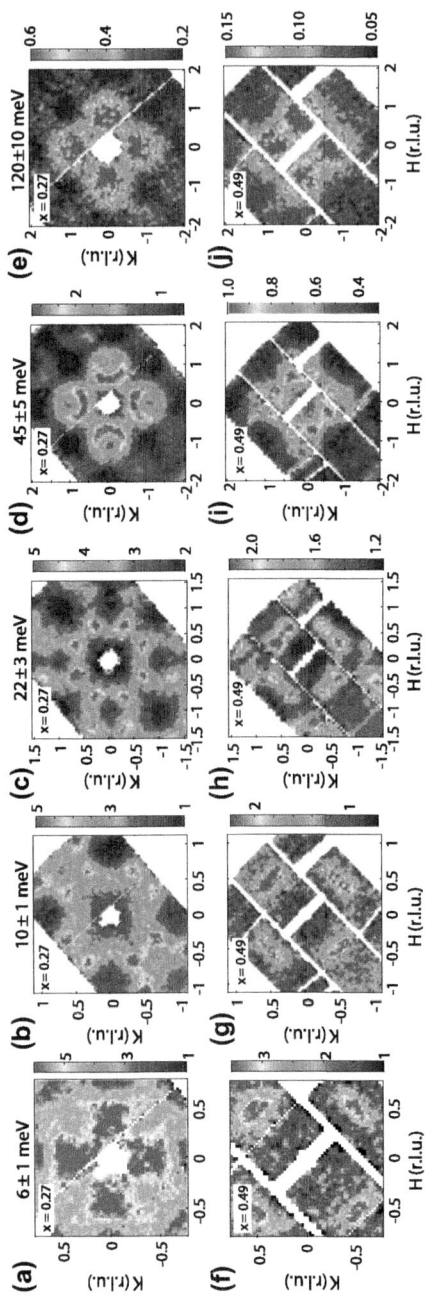

FIGURE 21 Comparison of magnetic excitations in FeTe$_{1-x}$Se$_x$ with $x = 0.27$ and $x = 0.49$. Adapted with permission from Ref. [156]. (a–e) Constant-E slices for FeTe$_{1-x}$Se$_x$ with $x = 0.27$. (f–j) Identical slices for $x = 0.49$.

High-Temperature Superconductors **Chapter | 3** **175**

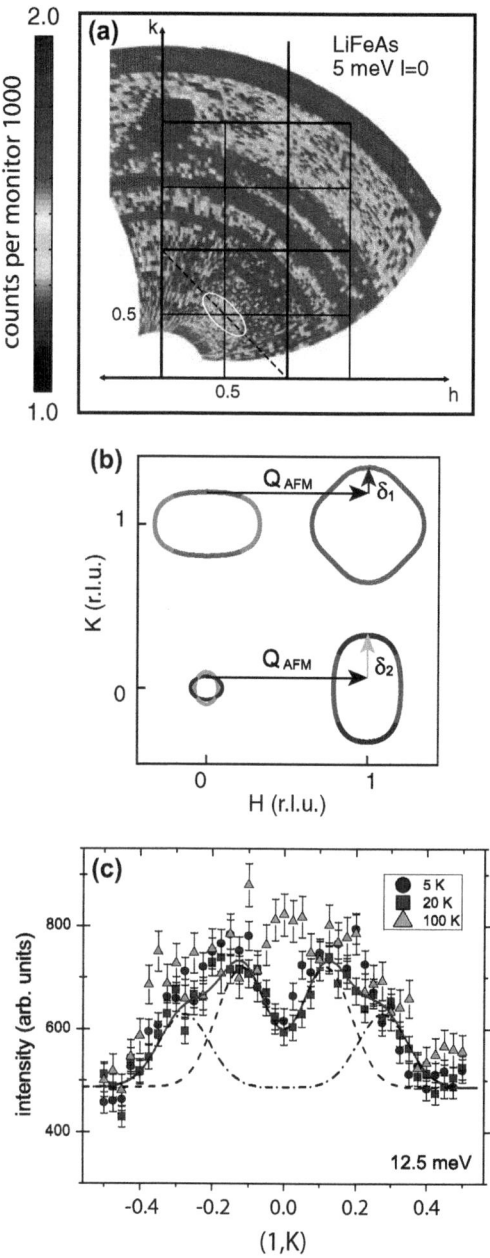

FIGURE 22 Magnetic excitations in LiFeAs. (a) Transverse split incommensurate peaks found in LiFeAs [161]. Here, tetragonal notation is used where $(0.5,0.5)_T$ correspond to $(1,0)_O$ in orthorhombic notation. (b) Fermi surface of LiFeAs from angle-resolved photoemission spectroscopy where two possible nesting vectors are possible [162]. (c) The magnetic excitations can be described by two pairs of incommensurate peaks, suggesting both nesting conditions in (b) give rise to magnetic fluctuations [163]. *Adapted with permission from Refs [161,163].*

wave vectors are possible (Figure 22(b)) and signature for two pairs of incommensurate split peaks are seen experimentally from inelastic neutron scattering (Figure 22(c)) [163].

3.3.4 Persistence of High-Energy Magnetic Excitations

The systematic evolution of magnetic excitations with electron doping has been studied for BaFe$_{2-x}$Ni$_x$As$_2$ (Figure 23) [76,108,116,164]. At low energies, sharp spin waves in BaFe$_2$As$_2$ become ellipses elongated along the transverse direction while high-energy magnetic excitations remain similar. For BaFe$_{1.7}$Ni$_{0.3}$As$_2$, low-energy magnetic excitations are completely suppressed below ~50 meV (Figure 23(u−v)) [76] while high-energy magnetic excitations persist. The persistence of high-energy magnetic excitations in

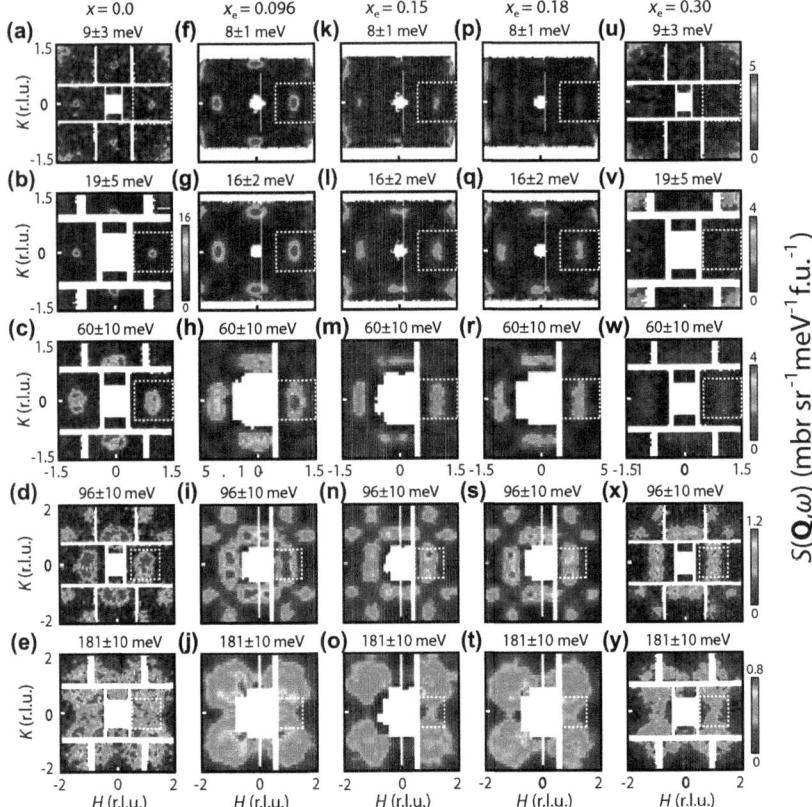

FIGURE 23 Evolution of high-energy magnetic excitations in BaFe$_{2-x}$Ni$_2$As$_2$. (a−e) BaFe$_2$As$_2$ [108]. (f−j) BaFe$_{1.904}$Ni$_{0.096}$As$_2$ [164]. (k−o) BaFe$_{1.85}$Ni$_{0.15}$As$_2$ [164]. (p−t) BaFe$_{1.82}$Ni$_{0.18}$As$_2$ [164]. (u−y) BaFe$_{1.7}$Ni$_{0.3}$As$_2$ [76].

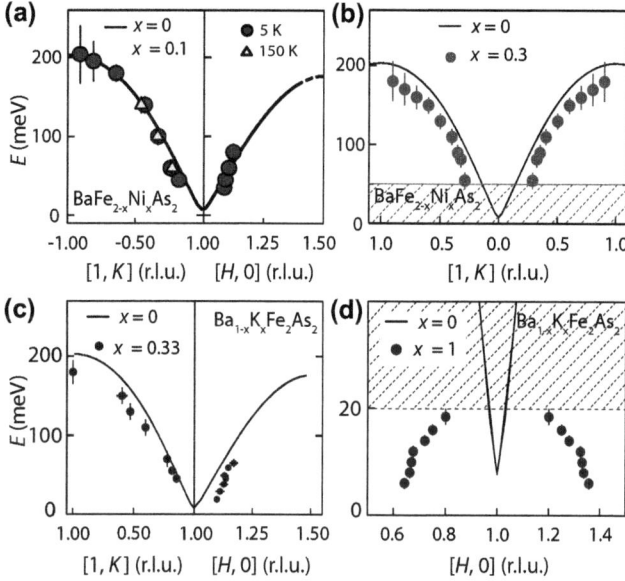

FIGURE 24 Doping dependence of the dispersion of magnetic excitations in $BaFe_2As_2$ compared with undoped $BaFe_2As_2$ (solid black line). (a) $BaFe_{1.9}Ni_{0.1}As_2$ [116]. (b) $BaFe_{1.7}Ni_{0.3}As_2$ [76]. (c) $Ba_{0.67}K_{0.33}Fe_2As_2$ [76]. (d) KFe_2As_2 [76].

$BaFe_{2-x}Ni_xAs_2$ contrasts with what was observed in hole-doped cuprates, where with increasing doping magnetic excitations are strongly suppressed (Figure 8(b)). The dispersion of magnetic excitations in $BaFe_{2-x}Ni_xAs_2$ with $x = 0.1$ and 0.3 only slightly softens compared to $BaFe_2As_2$ (Figure 24(a) and (b)). A similar persistence of magnetic excitations is also seen in $Ba_{0.67}K_{0.33}Fe_2As_2$ (Figure 24(c)) but in KFe_2As_2 only low-energy magnetic excitations are seen (Figure 24(d)). Based on these results, it was argued that in electron-/hole-doped $BaFe_2As_2$, low-energy magnetic excitations originate from nested Fermi surfaces, whereas high-energy magnetic excitations are due to short-range local magnetic correlations [76]. In this picture, the persistence of high-energy magnetic excitations can be understood since they are not sensitive to doping, and the large spin gap in $BaFe_{1.7}Ni_{0.3}As_2$ is due to the suppression of itinerant magnetism. It is still not clear how the large spin gap in $BaFe_{1.7}Ni_{0.3}As_2$ opens with increasing doping: Is the spectral weight below ~ 50 meV gradually suppressed or does the gap increase in size with increasing doping? More work is needed to clarify this issue.

To quantitatively trace the evolution of magnetic spectral weight with doping, it is useful to calculate the local susceptibility, defined as

$$\chi''(E) = \int \chi''(Q,E) dQ \bigg/ \int dQ,$$

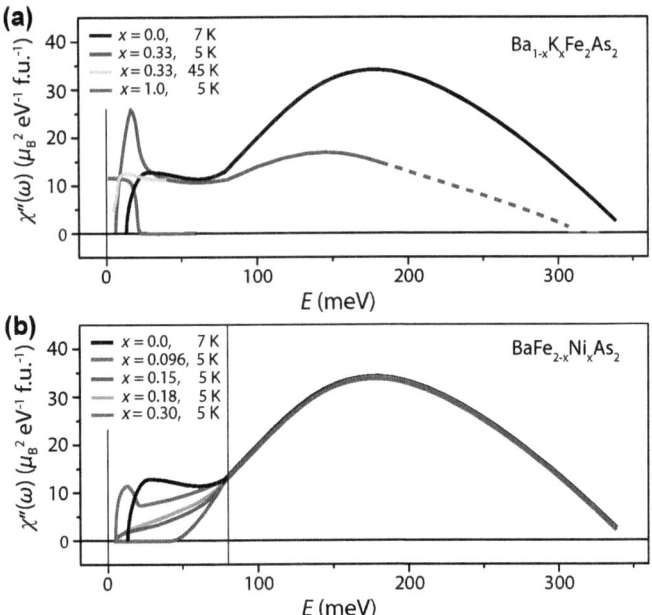

FIGURE 25 The local susceptibility for $Ba_{1-x}K_xFe_2As_2$ and $BaFe_{2-x}Ni_xAs_2$ shown in units of μ_B^2/eV/formula unit [14]. (a) $Ba_{1-x}K_xFe_2As_2$. (b) $BaFe_{2-x}Ni_xAs_2$.

where in practice the integration should be carried over an area large enough such that the average is equivalent to the average over the whole reciprocal space [104]. The local susceptibility therefore represents the overall strength of magnetic fluctuations and by integrating the local susceptibility over energy one obtains the total fluctuating moment through

$$\langle m^2 \rangle = \frac{3\hbar}{\pi} \int \frac{\chi''(E)dE}{1 - \exp\left(-\frac{\hbar\omega}{k_BT}\right)}.$$

The local susceptibility has been studied using time-of-flight neutron spectroscopy for $Ba_{1-x}K_xFe_2As_2$ and $BaFe_{2-x}Ni_xAs_2$ [14], and results are summarized in Figure 25. For $Ba_{1-x}K_xFe_2As_2$, the low-energy local susceptibility is significantly modified with doping. For optimal doped $Ba_{0.67}K_{0.33}Fe_2As_2$, there is a strong resonance mode in the superconducting state ($T_c = 38$ K). As in the cuprates, while the resonance is the most prominent feature in the magnetic excitation spectrum, it accounts for only a small part of the total spectral weight. The effect of K doping on high-energy magnetic excitations seems to be a gradual suppression of spectral weight, which strongly contrasts with Ni doping where high-energy magnetic excitations persist. The total fluctuating moment for $BaFe_2As_2$ is estimated to be ~ 3.6 μ_B^2/Fe [30] and with Ni doping it gradually decreases to ~ 2.8 μ_B^2/Fe in $BaFe_{1.7}Ni_{0.3}As_2$ [164]. The fluctuating

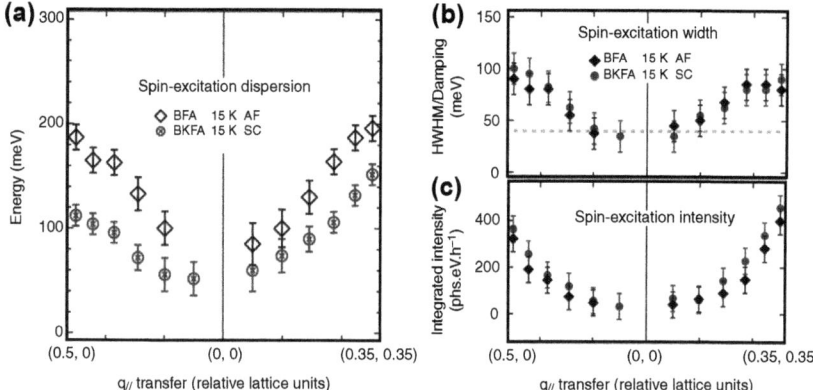

FIGURE 26 Magnetic excitations in BaFe$_2$As$_2$ (BFA) and Ba$_{0.6}$K$_{0.4}$Fe$_2$As$_2$ (BKFA) seen by resonant inelastic X-ray scattering [19]. (a) The dispersion of magnetic excitations in the two compounds. (b) The widths of excitations, the dashed line represents instrument resolution. (c) The intensity of magnetic excitations. *Adapted with permission from Ref. [19].*

moment in Ba$_{0.67}$K$_{0.33}$Fe$_2$As$_2$ and KFe$_2$As$_2$ reduces to $\sim 1.7\ \mu_B^2/$Fe and $0.1\ \mu_B^2/$Fe, respectively [76], within the energy range when data are currently available.

The persistence of high-energy magnetic excitations is also seen by RIXS in Ba$_{0.6}$K$_{0.4}$Fe$_2$As$_2$ (Figure 26) [19]. Compared to neutron scattering results [76], RIXS also sees a softening of magnetic excitations with K doping (Figure 26(a)) but the persistence of intensity with K doping disagrees with neutron scattering results (Figure 26(c)). This discrepancy is similar to what was observed in the cuprates, where neutrons saw significant suppression of spectral weight but RIXS saw persistence of magnetic intensity. Similar to the case of the cuprates, neutron scattering measures magnetic signals near $(1,0)_O$ and RIXS measures near $(0,0)$ (magnetic spectral weight is concentrated near $(1,0)_O$), dispersion of spin waves are identical for these two points in the magnetically ordered parent compounds but with the suppression of magnetic order these two points are no longer equivalent. Neutron scattering and RIXS therefore provide complementary information on how spin waves evolve from magnetically ordered parent compounds to magnetic excitations in the superconducting compounds.

3.3.5 The Resonance Mode in Iron-Based Superconductors

Soon after the discovery of superconductivity in the iron pnictides, a resonance mode was found in Ba$_{1-x}$K$_x$Fe$_2$As$_2$ [68]. Later, it was confirmed in many other iron pnictides [161,165–170]. The resonance-mode intensity tracks the superconductivity order parameter and forms in the superconducting state by shifting spectral weight from lower energies so a spin gap also opens in the superconducting state. In the electron-underdoped regime where superconductivity and antiferromagnetic order coexist, the static order is suppressed when entering the superconducting state so the corresponding spectral weight also shifts to the resonance mode [130,165]. The resonance mode in underdoped iron pnictides centered at $(1,0)_O$ disperses both along c [171] and b [172], but is sharply peaked along a. The dispersion along L becomes weaker with increasing doping, but remains dispersive in overdoped compounds [168,173]. The dispersion of the resonance mode is due to 3D spin correlations in the antiferromagnetically ordered parent compounds. Moving away from antiferromagnetic order well past optimal doping, the resonance becomes dispersionless (Figure 27(a)) [168]. Alternatively, the dispersion of the resonance in doped

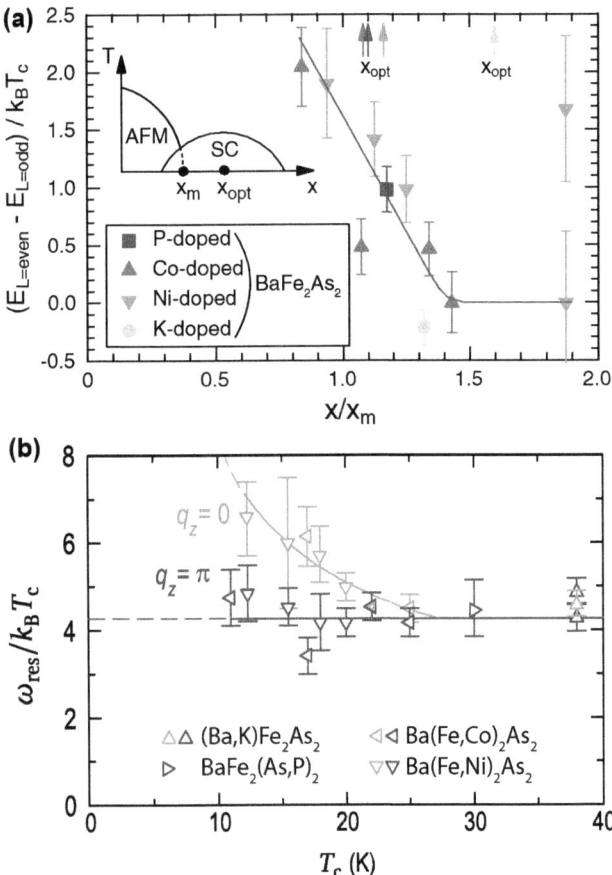

FIGURE 27 Dispersion along c of the resonance mode in doped BaFe$_2$As$_2$. (a) Dispersion along c becomes weaker when moving further away from BaFe$_2$As$_2$ [168]; x_m is the critical doping level where magnetic order disappears. (b) Alternatively, the dispersion along c may depend on T_c. In samples with smaller T_c, the dispersion becomes stronger [174]. *Adapted with permission from Refs [168,174].*

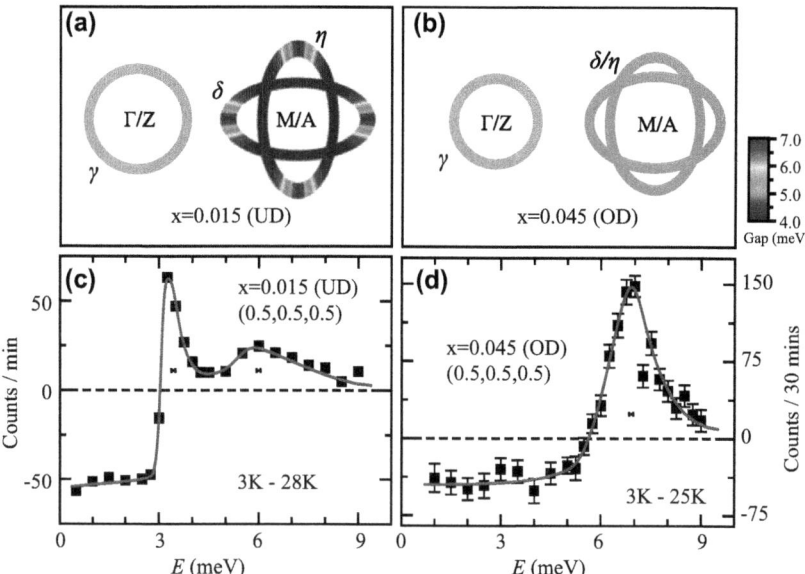

FIGURE 28 Double-resonance mode in underdoped NaFe$_{0.985}$Co$_{0.015}$As and single-resonance mode in overdoped NaFe$_{0.955}$Co$_{0.045}$As [170] seen in neutron scattering experiments. (a) Schematic diagram of the anisotropic superconducting gap as indicated by the color scale in underdoped NaFe$_{0.985}$Co$_{0.015}$As. (b) Schematic of the isotropic superconducting gap in overdoped NaFe$_{0.955}$Co$_{0.045}$As. (c) Double resonance found in underdoped NaFe$_{0.985}$Co$_{0.015}$As. (d) Single resonance found in overdoped NaFe$_{0.955}$Co$_{0.045}$As. *Adapted with permission from Ref. [170].*

BaFe$_2$As$_2$ has been argued to be related to T_c, with samples with lower T_c exhibiting a stronger dispersion (Figure 27(b)) [174]. With increasing doping, the resonance splits along the transverse direction for BaFe$_{1.85}$Ni$_{0.15}$As$_2$ [145].

While the resonance in iron pnictides is mostly interpreted as a spin-exciton with $s\pm$ pairing symmetry where the superconductivity order parameter changes sign on different Fermi surfaces [2,83], the broad resonance mode observed in 122 compounds [146,175] led to the proposals of orbital-mediated $s++$ superconductivity (the superconductivity order parameter maintains the same sign on different Fermi surfaces) in the iron pnictides [176,177]. The sharp resonance found in both overdoped (Figure 28(d)) [169] and underdoped NaFe$_{1-x}$Co$_x$As (Figure 28(c)) [170] ruled out the $s++$ scenario since the $s++$ scenario can only account for broad resonance modes. Intriguingly, in underdoped NaFe$_{1-x}$Co$_x$As, two resonance modes were found [170]. The origin of the double resonance is still under debate. It may be due to anisotropy of the superconducting gap for underdoped NaFe$_{1-x}$Co$_x$As which becomes isotropic for overdoped NaFe$_{1-x}$Co$_x$As (Figure 28(a) and (b)) [178] or the two resonances actually occur at different wave vectors (one at $(1,0)_O$ and the other at

$(0,1)_O$) but due to twinning they are simultaneously observed [179]. Inelastic neutron scattering experiments on detwinned samples are needed to determine the origin of the double resonance.

A resonance mode has also been reported in superconducting LiFeAs [161]. Similar to the normal-state magnetic excitations, the resonance mode is also incommensurate. Neutron resonance modes are also found in superconducting FeTe$_{1-x}$Se$_x$ [152,157,180]. While the normal-state magnetic excitations can be transversely split centered at $(0.5,0.5)_T$, the resonance mode is commensurate at $(0.5,0.5)_T$ (Figure 29) [157].

3.3.6 Magnetism in A_xFe$_{2-y}$Se$_2$ (A = Alkali Metal or Tl) Compounds

The report in 2010 of superconductivity in $K_{0.8}$Fe$_2$Se$_2$ with $T_c \sim 30$ K initiated research on A_xFe$_{2-y}$Se$_2$ [181]. Initial neutron diffraction data suggested the coexistence of superconductivity with block antiferromagnetic order with the composition $A_{0.8}$Fe$_{1.6}$Se$_2$ (Figure 30(a)) [93]. Later, it became clear that superconducting A_xFe$_{2-y}$Se$_2$ samples are phase separated [182–184], and that block antiferromagnetism and superconductivity originate from different phases [185–188]. The parent phases have been proposed to be either $A_{0.8}$Fe$_{1.6}$Se$_2$ with block antiferromagnetic order (245 phase, Figure 30(a)) [93,189,190] or AFe$_{1.5}$Se$_2$ with stripe antiferromagnetic order and rhombus iron vacancy (122 phase, Figure 30(b)) [187], whereas the superconducting phase has the structure of the 122 phase but with disordered iron vacancies and no magnetic order [187,188]. Fe vacancies in $A_{0.8}$Fe$_{1.6}$Se$_2$ order at $T_S = 500-578$ K and Fe moments order at a slightly lower temperature $T_N = 471-559$ K with an ordered moment of ~ 3.3 μ_B/Fe (Figure 30 (a)) [190]. Magnetic moments in KFe$_{1.5}$Se$_2$ were found to order at $T_N \sim 280$ K with an ordered moment ~ 2.8 μ_B/Fe [187].

While insulating K_xFe$_{2-y}$Se$_2$ samples are made up entirely of the 245 phase, semiconducting and superconducting samples also contain a significant portion of the 122 phase. The 122 phase in the semiconducting sample orders magnetically but the 122 phase in superconducting samples do not. Furthermore, the magnetic order in the 122 phase resembles the magnetic order in the parent phase of iron pnictides (Figure 30(b)). Based on these results, it was suggested that the 122 phase with magnetic order is the actual parent compound from which superconductivity is derived [187]. It is worth noting that a similar phase separation of 245 and 122 phases is seen in Rb$_x$Fe$_{2-y}$S$_2$. However, superconductivity is absent in this system [191].

Spin waves have been mapped out for block-ordered Rb$_{0.8}$Fe$_{1.6}$Se$_2$ (Figure 31(a–d)) [192], (Tl,Rb)$_{0.8}$Fe$_{1.6}$Se$_2$ [193], and K$_2$Fe$_4$Se$_5$ [194]. These results were analyzed using linear spin-wave theory. Two optical branches (Figure 31(a) and (b)) in addition to an acoustic branch are observed

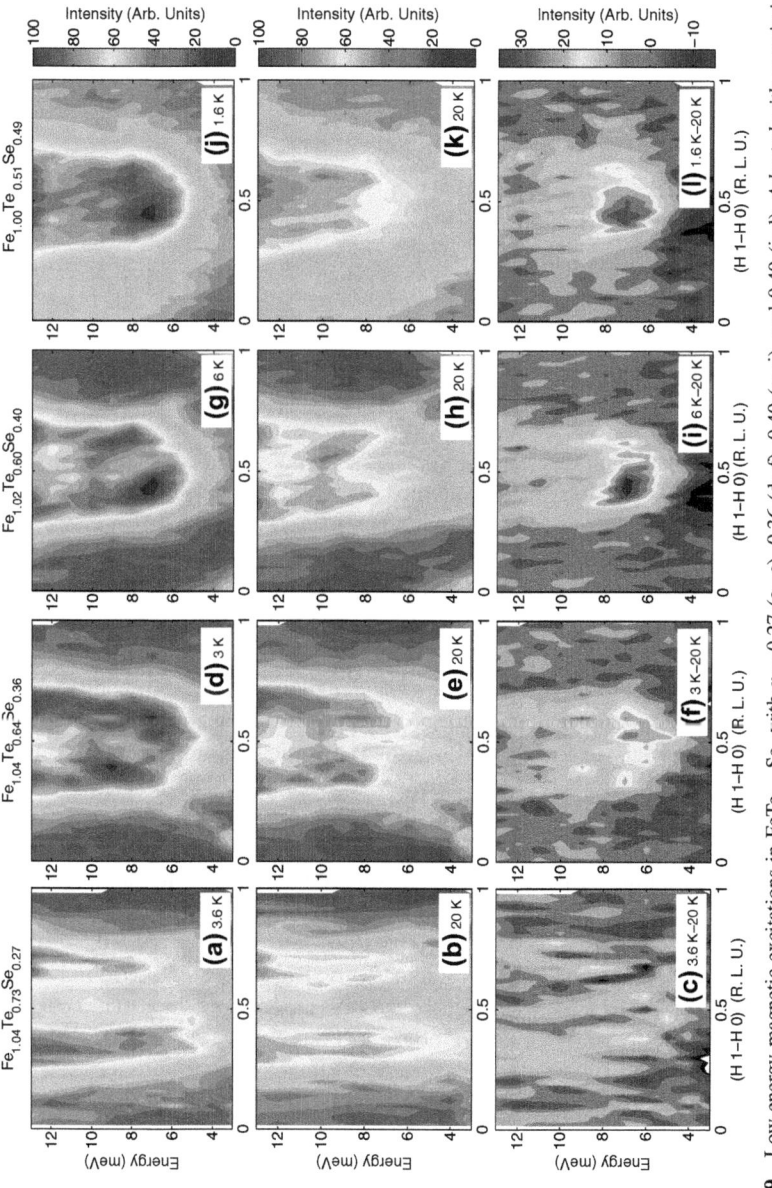

FIGURE 29 Low-energy magnetic excitations in FeTe$_{1-x}$Se$_x$ with $x = 0.27$ (a–c), 0.36 (d–f), 0.40 (g–i), and 0.49 (j–l). Adapted with permission from Ref. [157]. The samples with $x = 0.27$ and 0.36 show filamentary superconductivity. For $x = 0.40$, $T_c = 13$ K and for $x = 0.49$, $T_c = 14$ K. While the response at 20 K may be commensurate or incommensurate, 3–20 K data are commensurate suggesting the resonance mode is commensurate.

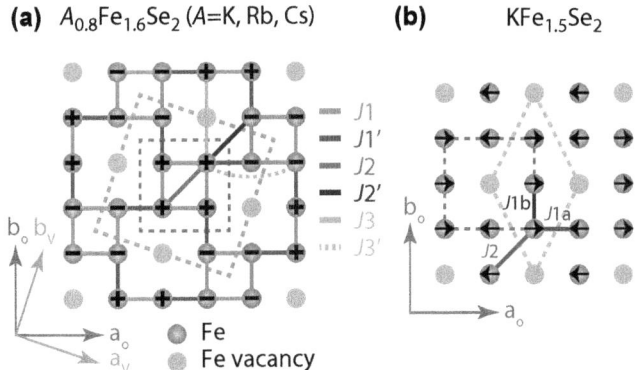

FIGURE 30 Fe vacancy order and magnetic order in (a) $A_2Fe_4Se_5$ (A = K, Rb, Cs) and (b) $KFe_{1.5}Se$ [14]. The gray dashed box in (a) is the structural and magnetic unit cell for the $A_2Fe_4Se_5$ phase.

(Figure 31(c)), which can also be seen by examining the local susceptibility (Figure 31(d)). The fluctuating moment is found to be $\langle m^2 \rangle \sim 16\, \mu_B^2/\text{Fe}$. Considering that the ordered moment is $\sim 3\, \mu_B/\text{Fe}$, the total moment can be determined from

$$M_{\text{total}}^2 = M_{\text{ord}}^2 + \langle m^2 \rangle,$$

resulting in a total moment $M_{\text{tot}}^2 \sim 25\, \mu_B^2/\text{Fe}$, consistent with $24\, \mu_B^2/\text{Fe}$ expected for Fe with $S = 2$ within experimental error [192]. In addition, the study of the temperature dependence of magnetic excitations in $Rb_{0.8}Fe_{1.6}Se_2$ found that contrary to $BaFe_2As_2$ [30], the magnetic excitations do not persist above T_N [195]. Magnetic excitations arising from the 122 phase have also been studied (Figure 31(e–h)). In this case, the spin waves form cones stemming from $(1,0)_O$ similar to the parent compounds of the iron pnictides [196]. While this evidence favors a superconducting state arising from doping of the magnetically ordered 122 phase, in $A_xFe_{2-y}Se_2$ the resonance mode associated with superconductivity does not appear at $(0.5, 0.5)_T$ as in iron pnictides. Instead, the resonance in $Rb_xFe_{2-y}Se_2$ is found near $(0.5, 0.25)_T$ at $E \sim 14$ meV (Figure 32(a)) [197], corresponding to the nesting wave vector between two electron pockets (Figure 32(c)) [198]. This finding is later confirmed by similar measurements in related systems [198–201]. The resonance is elongated along the transverse direction and does not disperse along c. Also, the resonance mode does not show the symmetry of the 245 phase clearly indicating it does not arise from the blocked ordered phase [198,199]. Since the resonance is believed to arise from quasiparticle excitations that are enhanced when the superconducting gaps connected by the resonance wave vector have opposite signs [83], the Fermi surface of

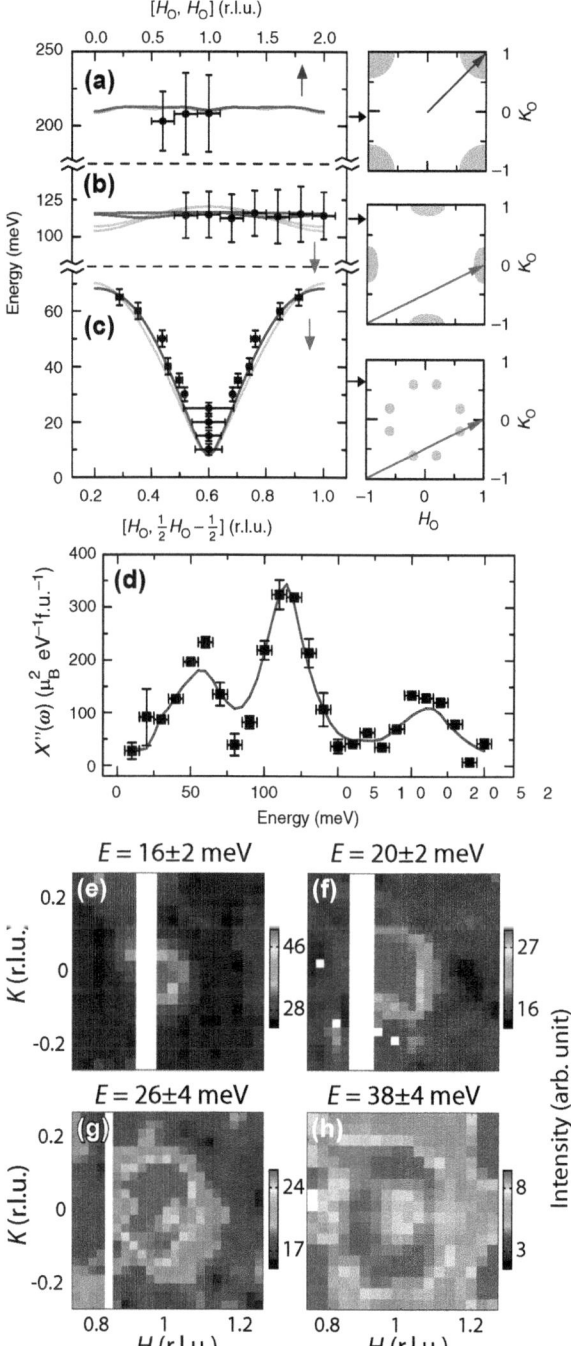

FIGURE 31 Spin waves in the 245 and 122 phases of $A_xFe_{2-y}Se_2$. (a and b) Optical and (c) acoustic branches of spin waves in $Rb_{0.8}Fe_{1.6}Se_2$ [192]. The blue (dark gray in print versions) solid lines are for a spin wave model with $J_3 > 0$, and pink (light gray in print versions) solid lines are for $J_3 = 0$. The dispersion curves are along directions shown to the right of each panel. (d) Local susceptibility of $Rb_{0.8}Fe_{1.6}Se_2$ [192]. The solid line is the calculated local susceptibility based on a spin wave model. (e–h) Constant-E slices of spin waves from the 122 phase of $K_xFe_{2-y}Se_2$ [196]. *Adapted with permission from Refs [192,196].*

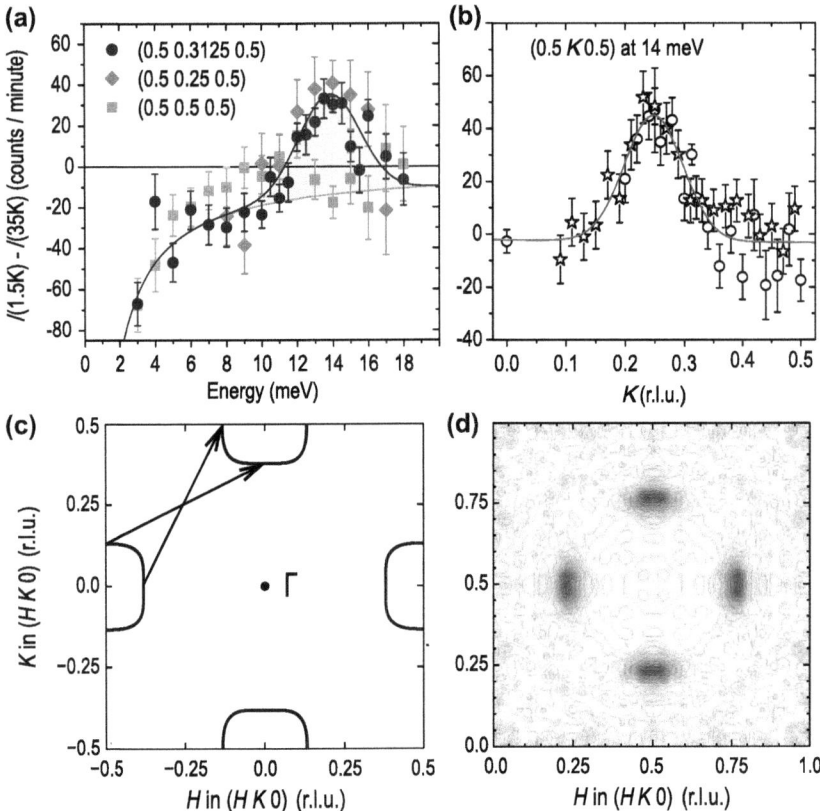

FIGURE 32 The resonance mode in $A_x Fe_{2-y} Se_2$. (a) 1.5–35 K data for constant-$\mathbf{Q}$ scan of the resonance mode in $Rb_x Fe_{2-y} Se_2$ with $T_c = 32$ K [197]. (b) 1.5–35 K data for constant energy scan of the resonance mode [197]. (c) Possible nesting wave vector that connects the two electron Fermi surfaces corresponding to the resonance wave vector [198]. (d) H-K map at $E = 15$ meV for $Rb_x Fe_{2-y} Se_2$ [198]. *Adapted with permission from Refs [197,198].*

$A_x Fe_{2-y} Se_2$ (Figure 32(c)) suggests that superconductivity has d-wave symmetry rather than $s\pm$ as in the superconducting iron pnictides.

3.3.7 Polarization Dependence of Low-Energy Magnetic Excitations

By polarizing the incident neutron beam and analyzing the scattered neutrons with a particular spin direction, it becomes possible to study the polarization dependence of magnetic excitations [17]. It is typically useful to choose three such directions x, y, and z, to be parallel and perpendicular to the scattering vector $\mathbf{Q}$ in the scattering plane and perpendicular to the scattering plane, respectively. Since the polarization of the magnetic signal has to be perpendicular to $\mathbf{Q}$ and the neutron-polarization direction for spin-flip scattering, the spin-flip scattering cross sections are [202]:

$$\sigma_{xx}^{SF} \sim M_y + M_z + B^{SF}$$
$$\sigma_{yy}^{SF} \sim M_z + B^{SF}$$
$$\sigma_{zz}^{SF} \sim M_y + B^{SF}$$

M_i is the magnetic signal polarized along i and B^{SF} is the background for spin-flip scattering. Similarly for non-spin-flip scattering the cross sections are:

$$\sigma_{xx}^{NSF} \sim B^{NSF}$$
$$\sigma_{yy}^{NSF} \sim M_y + B^{NSF}$$
$$\sigma_{zz}^{NSF} \sim M_z + B^{NSF}$$

B^{NSF} is the non-spin-flip background. By measuring all three cross sections (either spin-flip or non-spin-flip), M_y and M_z can be unequivocally determined. Doing the same at an equivalent wave-vector position in another Brillouin zone allows M_a, M_b, and M_c to be determined when effects of instrumental resolution can be estimated [106,202–204].

In optimal doped $YBa_2Cu_3O_{6+x}$, polarized neutron scattering experiments found that the normal state response and the resonance mode are isotropic in spin space (i.e., $M_y = M_z$), but low-energy magnetic excitations in the superconducting state are anisotropic with magnetic excitations polarized along c being suppressed [205]. Another advantage of using polarized neutrons in $YBa_2Cu_3O_{6+x}$ is that it allows the magnetic signal to be separated from phonons which overlap with the magnetic signal. However, so far there are a limited number of such studies in the cuprates possibly due to the weak magnetic signal of the cuprates and reduced flux for polarized experiments.

In $BaFe_2As_2$, significant spin anisotropy was found [204,206]. The spin waves polarized along c have a smaller gap than those polarized in-plane, suggesting that it costs more energy to rotate spins in the ab-plane than perpendicular to it [206]. More interestingly, longitudinal magnetic excitations that cannot be explained using spin-wave theory are also observed, highlighting the contribution of itinerant electrons to magnetism in the iron pnictides (Figure 33(a–b)) [204]. Similar anisotropy is also found in NaFeAs, but with smaller spin-anisotropy gaps (Figure 33(c–d)) [106]. Upon electron doping in $BaFe_2As_2$ and NaFeAs, a similar anisotropy is found to persist even when antiferromagnetic order is suppressed [202,203,207,208], and magnetic excitations become isotropic only well on the overdoped side of the phase diagram [202,209]. For $Ba_{1-x}K_xFe_2As_2$, significant anisotropy exists in optimal and overdoped samples [210,211]. Remarkably, using polarized neutron scattering, it was found that the broad resonance mode in near-optimal doped $Ba(Fe_{0.94}Co_{0.06})_2As_2$ consists of two parts, a low-energy component polarized along c and a higher-energy isotropic component (Figure 34(a–b)) [208].

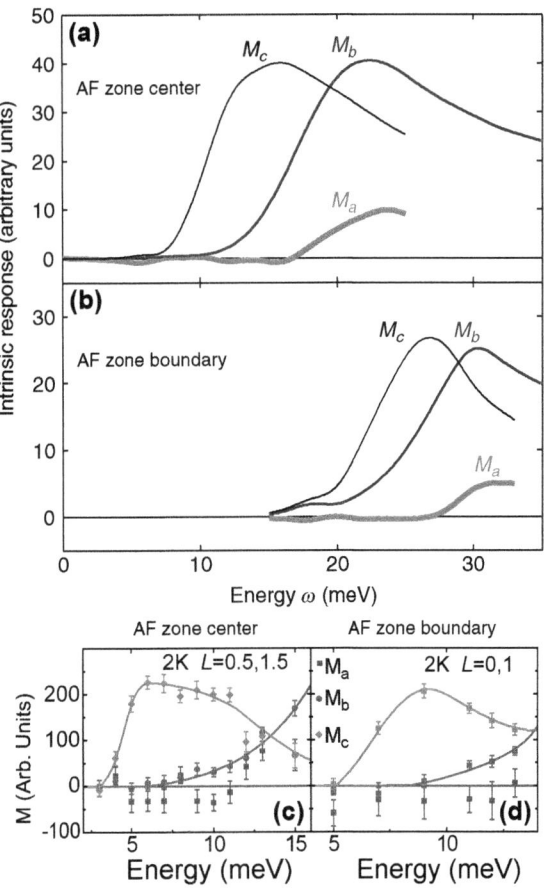

FIGURE 33 Polarization of spin waves in $BaFe_2As_2$ [204] and NaFeAs [106]. (a) The spin waves in $BaFe_2As_2$ at the magnetic zone center. (b) The spin waves in $BaFe_2As_2$ at the zone boundary along c. (c) The spin waves in NaFeAs at the magnetic zone center. (d) The spin waves in NaFeAs at the zone boundary along c. *Adapted with permission from Refs [106,204].*

Similar behavior is also seen in optimal (Figure 34(c–d)) [210] and slightly overdoped $Ba_{1-x}K_xFe_2As_2$ [211]. In underdoped $NaFe_{0.985}Co_{0.015}As$ with double resonances coexisting with antiferromagnetic order [170], the lower-energy resonance is anisotropic, but the resonance at higher energy is isotropic [202]. These findings suggest that the double resonance in $NaFe_{0.985}Co_{0.015}As$ [170] is actually a common feature of iron pnictides, but in other systems the two modes are too broad and merge into one in unpolarized neutron scattering experiments. The first resonance mode is dispersive along c and anisotropic, while the second resonance is isotropic in spin space and forms a flat band along c [170,202]. In-plane spin anisotropy in near-optimal doped $BaFe_{1.904}Ni_{0.096}As_2$ exists at $E = 3$ meV even in the paramagnetic tetragonal

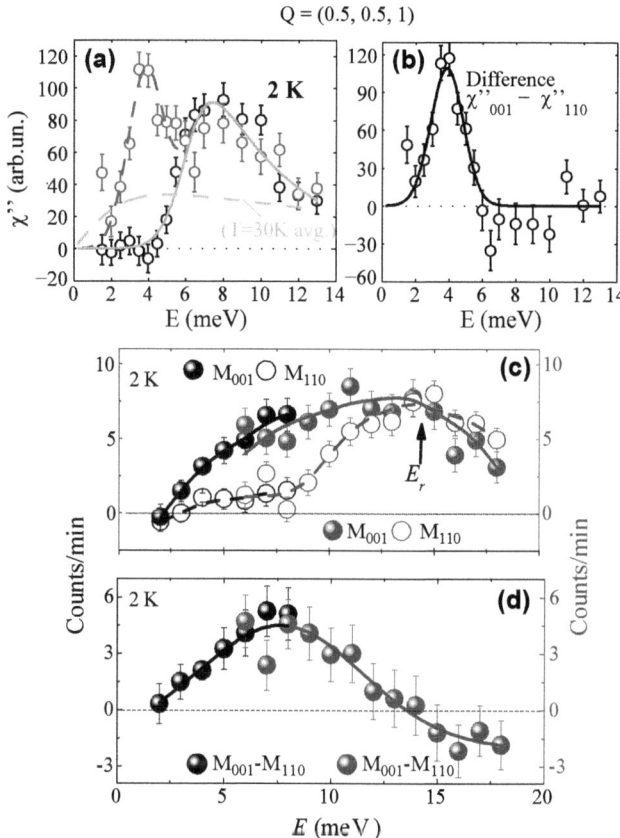

FIGURE 34 Polarization of magnetic excitations in near-optimal doped Ba(Fe$_{0.94}$Co$_{0.06}$)$_2$As$_2$ [208] and optimal-doped Ba$_{0.67}$K$_{0.33}$Fe$_2$As$_2$ [210]. (a) In-plane and out-of-plane polarized magnetic excitations for Ba(Fe$_{0.94}$Co$_{0.06}$)$_2$As$_2$ at 2 K. (b) The difference along the two directions. (c) Similar in-plane and out-of-plane polarized magnetic excitations for Ba$_{0.67}$K$_{0.33}$Fe$_2$As$_2$ at 2 K. (d) The difference along the two directions. *Adapted with permission from Refs [208,210].*

state but is enhanced in the superconducting state, while magnetic excitations at $E = 7$ meV are isotropic at all temperatures (Figure 35) [203]. This finding suggests that the anisotropy of the first resonance could be caused by electronic anisotropy or orbital ordering already present in the normal state. Weak spin anisotropy is also found in superconducting FeTe$_{1-x}$Se$_x$, with the in-plane component slightly larger than the out-of-plane component [212,213].

So far, there are few theoretical studies on the polarization of spin excitations in iron pnictides to explain the experimental observations. While the energy scale of spin anisotropy is small compared to exchange couplings in the parent compounds, spin anisotropy has an energy scale comparable to the resonance mode in many compounds and may be an important ingredient

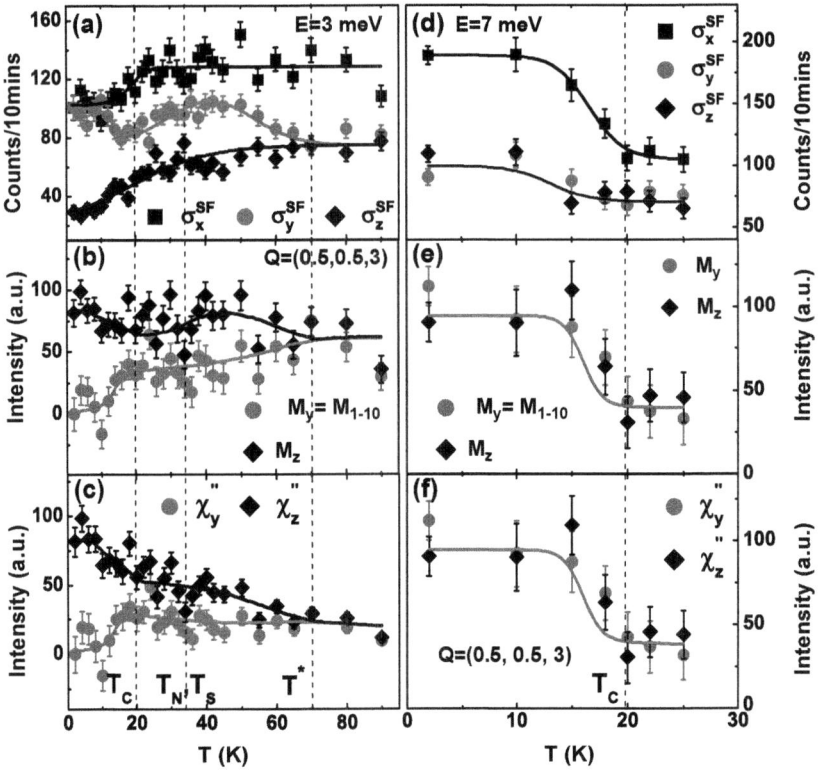

FIGURE 35 Polarization of magnetic excitations in near-optimal doped BaFe$_{1.904}$Ni$_{0.096}$As$_2$ [203]. (a) Cross sections for x, y, and z at (0.5,0.5,3) $E = 3$ meV. (b) M_y and M_z. (c) Susceptibility χ_y'' and χ_z'' which is related to M_y and M_z by the Bose factor. (d–f) Similar plots as (a–c) but for $E = 7$ meV.

for understanding the origin of superconductivity. Further experimental and theoretical efforts are needed to understand the polarization of low-energy magnetic excitations in both the cuprates and iron-based superconductors.

3.4 SUMMARY AND OUTLOOK

Superconductivity in both the cuprates and iron-based superconductors is derived from antiferromagnetically ordered parent compounds. However, whereas the parent compounds of cuprates are half-filled, single-band Mott insulators, the parent compounds of iron pnictides have multiple bands derived from several orbitals and are metallic. In both cases, strong magnetic excitations with a wide magnetic bandwidth are observed by neutron scattering. Whereas in hole-doped cuprates and iron pnictides, the magnetic excitations are suppressed with increasing doping, high-energy magnetic excitations persist in electron-doped iron pnictides. In the superconducting state, resonance modes are found in both systems. While the resonance mode is the most prominent feature in the superconducting state, it accounts for a small fraction of the entire magnetic spectral weight, but nevertheless it is able to account for the superconductivity condensation energy.

An electronic nematic state that exists in iron pnictides has been the focus of considerable research [91]. However, due to structural twinning, limited progress has been made from a neutron scattering perspective. Neutron diffraction has been carried out on Ba(Fe$_{1-x}$Co$_x$)$_2$As$_2$ [214,215] and NaFeAs [98] by applying uniaxial pressure to detwin samples and found that uniaxial pressure can affect both T_S and T_N. Inelastic neutron scattering has also been done on detwinned BaFe$_{2-x}$Ni$_x$As$_2$ under uniaxial pressure and found that anisotropy between $(1,0)_O$ and $(0,1)_O$ persists even above T_N, favoring a spin-nematic scenario for electronic nematicity [216]. By studying the spin—spin correlation length between T_S and T_N in twinned samples, anomalies in spin excitations between T_S and T_N also suggest a magnetic scenario for electronic nematicity [217]. Studies of magnetic excitations over the entire Brillouin zone using detwinned samples and in situ removal of uniaxial strain are needed to clarify the connection between electronic nematicity and magnetism in iron pnictides.

The strength of electronic correlations for the parent compounds of iron pnictides is in fact similar to superconducting cuprates [89]. Further, superconductivity can be induced in iron pnictides in almost every way that suppresses the antiferromagnetic ordering [94], whereas in the cuprates the route is much more specific. Based on these general results, an alternate view of the iron pnictides has been proposed (Figure 36) [218]. In this picture, the iron pnictides are also derived from a Mott insulating phase with strong electronic correlations when all Fe orbitals are half-filled ($n = 5$, corresponding to doping one hole to each Fe in BaFe$_2$As$_2$ with $n = 6$) and each band individually becomes Mott insulating [218—220]. For slightly higher fillings, the system is tuned to a selective Mott insulating state with different orbitals having different degrees of localization. The parent compounds then correspond to 20% doping, which in the cuprates is slightly overdoped (Figure 4). The strength of electronic correlations is found to increase from the electron-doped to the hole-doped side of the phase diagram for BaFe$_2$As$_2$ [218,221], culminating in KFe$_2$As$_2$ ($n = 5.5$) with a large mass enhancement close to a selective Mott transition [222]. In this picture, the magnetic order in the parent compounds corresponds to competing phases similar to stripe order in La$_{2-x}$Ba$_x$CuO$_4$, which naturally accounts for why superconductivity can be induced in numerous ways by suppressing magnetic order, and the similar strength of electronic correlations found in superconducting cuprates and parent compounds of iron pnictides. While this view is certainly tantalizing by putting the cuprates and iron pnictides on the same footing, experimental evidence for a Mott insulating phase with half-filled Fe orbitals would be desirable.

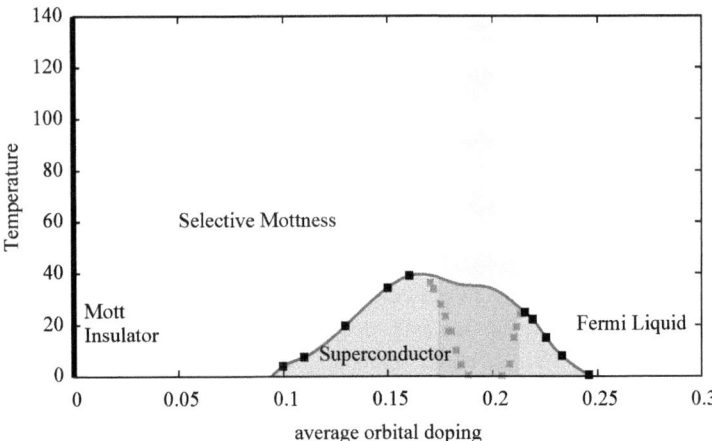

FIGURE 36 Proposal of unified phase diagram for iron pnictides and cuprates [218]. Zero doping correspond to a yet-to-be-discovered Mott insulating phase with half-filled ($n = 5$) Fe orbitals. 10% doping corresponds to KFe_2As_2 with $n = 5.5$, and 20% doping corresponds to $BaFe_2As_2$ ($n = 6$). In this picture, the magnetic order in $BaFe_2As_2$ is then viewed as a competing phase with superconductivity similar to the formation of stripes in the cuprates. *Adapted with permission from Ref. [218].*

For research on the cuprates, one challenge would be to have one model system to be well characterized by different probes, since currently most scanning tunneling microscopy and ARPES studies have examined different compounds compared to those well characterized by neutron scattering. Combining neutron scattering and ARPES results, it was argued that spin fluctuations have sufficient strength to mediate superconductivity in $YBa_2Cu_3O_{6+x}$ [223]. Polarization dependence studies of magnetic excitations are also interesting to study anisotropies in spin space. Further, it has the potential to unveil broad magnetic features that are difficult to observe using unpolarized neutrons [205].

There are many systems that are not well studied by neutron scattering due to sample limitations, such as $Bi_2Sr_2Ca_{n-1}Cu_nO_{2n+4+\delta}$ (BSCCO), 1111 iron pnictides, and FeSe. When samples of the necessary size become available, it will be highly interesting to carry out neutron scattering experiments on these systems. In addition, newer systems such as $Ca_{10}(Pt_3As_8)(Fe_2As_2)_5$ and $Ca_{10}(Pt_4As_8)(Fe_2As_2)_5$ [224,225], $Ca_{1-x}(Pr,La)_xFeAs_2$ (112) [226,227] and $La_{0.4}Na_{0.6}Fe_2As_2$ [228] are also of great interest for neutron scattering. To understand the role of magnetism in high-temperature superconductors, neutron scattering plays and will continue to play a key role. To further the current understanding, new materials, combination of different probes, and theory will be important.

ACKNOWLEDGMENTS

We acknowledge help from X. Lu in preparing some of the figures. The materials synthesis and neutron scattering efforts on the hole-doped 122 and 111 family of materials are supported by the U.S. DOE, Office of Basic Energy Sciences, under Contract No. DE-SC0012311. The neutron scattering work on the electron-doped 122 family of materials is supported by U.S. NSF, DMR-1362219. The combination of RPA calculation and neutron scattering is supported by U.S. NSF, DMR-1436006 and DMR-1308603. Part of the materials synthesis at Rice University is supported by the Robert A. Welch Foundation Grants No. C-1839.

REFERENCES

[1] J.G. Bednorz, K.A. Müller, Z. Phys. B 64 (1986) 189.
[2] I.I. Mazin, Nature 464 (2010) 183.
[3] D. Vaknin, S.K. Sinha, D.E. Moncton, D.C. Johnston, J.M. Newsam, C.R. Safinya, et al., Phys. Rev. Lett. 58 (1987) 2802.
[4] M.K. Wu, J.R. Ashburn, C.J. Torng, P.H. Hor, R.L. Meng, L. Gao, et al., Phys. Rev. Lett. 58 (1987) 908.
[5] M.A. Kastner, R.J. Birgeneau, G. Shirane, Y. Endoh, Rev. Mod. Phys. 70 (1998) 897.
[6] R.J. Birgeneau, C. Stock, J.M. Tranquada, K. Yamada, J. Phys. Soc. Jpn. 75 (2006) 111003.
[7] J.M. Tranquada, in: J.R. Schrieffer, J.S. Brooks (Eds.), Handbook of High-Temperature Superconductivity, Springer, New York, 2007, pp. 257–298.
[8] M. Fujita, H. Hiraka, M. Matsuda, M. Matsuura, J.M. Tranquada, S. Wakimoto, et al., J. Phys. Soc. Jpn. 81 (2012) 011007.
[9] J.M. Tranquada, AIP Conf. Proc. 1550 (2013) 114.
[10] J.M. Tranquada, G. Xu, I.A. Zaliznyak, J. Magn. Magn. Mater. 350 (2014) 148.
[11] Y. Kamihara, T. Watanabe, M. Hirano, H. Hosono, J. Am. Chem. Soc. 130 (2008) 3296.
[12] P. Dai, J. Hu, E. Dagotto, Nat. Phys. 8 (2012) 709.
[13] M.D. Lumsden, A.D. Christianson, J. Phys. Condens. Matter. 22 (2010) 203203.
[14] P. Dai, Rev. Mod. Phys. 87 (2015) 855.
[15] M.-H. Fang, H.-D. Wang, C.-H. Dong, Z.-J. Li, C.-M. Feng, J. Chen, et al., Europhys. Lett. 94 (2011) 27009.
[16] E. Dagotto, Rev. Mod. Phys. 85 (2013) 849.
[17] F. Fernandez-Alonso, D.L. Price, in: F. Fernandez-Alonso, D.L. Price (Eds.), Neutron Scattering — Fundamentals, Experimental Methods in the Physical Sciences, vol. 44, Academic Press, Amsterdam, 2013, pp. 1–136.
[18] M.P.M. Dean, J. Magn, Magn. Mater. 376 (2015) 3.
[19] K.-J. Zhou, Y.-B. Huang, C. Monney, X. Dai, V.N. Strocov, N.-L. Wang, et al., Nat. Commun. 4 (2013) 1470.
[20] N.S. Headings, S.M. Hayden, R. Coldea, T.G. Perring, Phys. Rev. Lett. 105 (2010) 247001.
[21] Y.S. Lee, R.J. Birgeneau, M.A. Kastner, Y. Endoh, S. Wakimoto, K. Yamada, et al., Phys. Rev. B 60 (1999) 3643.
[22] R. Liang, D.A. Bonn, W.N. Hardy, Phys. Rev. B 73 (2006) 180505.
[23] P. Burlet, J.Y. Henry, L.P. Regnault, Phys. C (Amsterdam, Neth.) 296 (1998) 205.
[24] R. Coldea, S.M. Hayden, G. Aeppli, T.G. Perring, C.D. Frost, T.E. Mason, et al., Phys. Rev. Lett. 86 (2001) 5377.

[25] S.M. Hayden, G. Aeppli, T.G. Perring, H.A. Mook, F. Doğan, Phys. Rev. B 54 (1996) 6905.
[26] J. Lorenzana, G. Seibold, R. Coldea, Phys. Rev. B 72 (2005) 224511.
[27] A.C. Walters, T.G. Perring, J.-S. Caux, A.T. Savici, G.-D. Gu, C.-C. Lee, et al., Nat. Phys. 5 (2009) 867.
[28] S. Shamoto, M. Sato, J.M. Tranquada, B.J. Sternlieb, G. Shirane, Phys. Rev. B 48 (1993) 13817.
[29] Z. Xu, C. Stock, S. Chi, A.I. Kolesnikov, G. Xu, G. Gu, et al., Phys. Rev. Lett. 113 (2014) 177002.
[30] L.W. Harriger, M. Liu, H. Luo, R.A. Ewings, C.D. Frost, T.G. Perring, et al., Phys. Rev. B 86 (2012) 140403.
[31] W. Bao, G. Aeppli, J.W. Lynn, P.G. Pagliuso, J.L. Sarrao, M.F. Hundley, et al., Phys. Rev. B 65 (2002) 100505.
[32] T.P. Croft, C. Lester, M.S. Senn, A. Bombardi, S.M. Hayden, Phys. Rev. B 89 (2014) 224513.
[33] M. Fujita, K. Yamada, H. Hiraka, P.M. Gehring, S.H. Lee, S. Wakimoto, et al., Phys. Rev. B 65 (2002) 064505.
[34] E. Demler, S. Sachdev, Y. Zhang, Phys. Rev. Lett. 87 (2001) 067202.
[35] B. Khaykovich, S. Wakimoto, R.J. Birgeneau, M.A. Kastner, Y.S. Lee, P. Smeibidl, et al., Phys. Rev. B 71 (2005) 220508.
[36] J. Chang, Ch Niedermayer, R. Gilardi, N.B. Christensen, H.M. Rønnow, D.F. McMorrow, et al., Phys. Rev. B 78 (2008) 104525.
[37] M. Enoki, M. Fujita, T. Nishizaki, S. Iikubo, D.K. Singh, S. Chang, et al., Phys. Rev. Lett. 110 (2013) 017004.
[38] D. Haug, V. Hinkov, Y. Sidis, P. Bourges, N.B. Christensen, A. Ivanov, et al., New J. Phys. 12 (2010) 105006.
[39] M. Matsuda, G.E. Granroth, M. Fujita, K. Yamada, J.M. Tranquada, Phys. Rev. B 87 (2013) 054508.
[40] V. Hinkov, D. Haug, B. Fauqué, P. Bourges, Y. Sidis, A. Ivanov, et al., Science 319 (2008) 597.
[41] E. Fradkin, S.A. Kivelson, M.J. Lawler, J.P. Eisenstein, A.P. Mackenzie, Annu. Rev. Condens. Matter Phys. 1 (2010) 153.
[42] M. Matsuda, J.A. Fernandez-Baca, M. Fujita, K. Yamada, J.M. Tranquada, Phys. Rev. B 84 (2011) 104524.
[43] S.M. Hayden, H.A. Mook, P. Dai, T.G. Perring, F. Doğan, Nature 429 (2004) 531.
[44] J.M. Tranquada, H. Woo, T.G. Perring, H. Goka, G.D. Gu, G. Xu, et al., Nature 429 (2004) 534.
[45] G. Xu, G.D. Gu, M. Hücker, B. Fauqué, T.G. Perring, L.P. Regnault, et al., Nat. Phys. 5 (2009) 642.
[46] J. Chang, E. Blackburn, A.T. Holmes, N.B. Christensen, J. Larsen, J. Mesot, et al., Nat. Phys. 8 (2012) 871.
[47] G. Ghiringhelli, M. Le Tacon, M. Minola, S. Blanco-Canosa, C. Mazzoli, N.B. Brookes, et al., Science 337 (2012) 821.
[48] T. Wu, H. Mayaffre, S. Krämer, M. Horvatić, C. Berthier, W.N. Hardy, et al., Nature 477 (2011) 191.
[49] R. Comin, R. Sutarto, E.H. da Silva Neto, L. Chauviere, R. Liang, W.N. Hardy, et al., Science 347 (2015) 1335.
[50] M. Norman, Phys. Rev. B 63 (2001) 092509.
[51] T. Das, R.S. Markiewicz, A. Bansil, Phys. Rev. B 85 (2012) 064510.
[52] A.T. Boothroyd, P. Babkevich, D. Prabhakaran, P.G. Freeman, Nature 471 (2011) 341.

[53] Y. Drees, D. Lamago, A. Piovano, A.C. Komarek, Nat. Commun. 4 (2013) 2449.
[54] Y. Drees, Z.W. Li, A. Ricci, M. Rotter, W. Schmidt, D. Lamago, et al., Nat. Commun. 5 (2014) 5731.
[55] S. Wakimoto, K. Yamada, J.M. Tranquada, C.D. Frost, R.J. Birgeneau, H. Zhang, Phys. Rev. Lett. 98 (2007) 247003.
[56] C. Stock, R.A. Cowley, W.J.L. Buyers, C.D. Frost, J.W. Taylor, D. Peets, et al., Phys. Rev. B 82 (2010) 174505.
[57] S. Sugai, H. Suzuki, Y. Takayanagi, T. Hosokawa, N. Hayamizu, Phys. Rev. B 68 (2003) 184504.
[58] L.J.P. Ament, M. van Veenendaal, T.P. Devereaux, J.P. Hill, J. van den Brink, Rev. Mod. Phys. 83 (2011) 705.
[59] M.P.M. Dean, G. Dellea, R.S. Springell, F. Yakhou-Harris, K. Kummer, N.B. Brookes, et al., Nat. Mater. 12 (2013) 1019.
[60] M. Le Tacon, G. Ghiringhelli, J. Chaloupka, M. Moretti Sala, V. Hinkov, M.W. Haverkort, et al., Nat. Phys. 7 (2011) 725.
[61] M.P.M. Dean, R.S. Springell, C. Monney, K.J. Zhou, J. Pereiro, I. Božović, et al., Nat. Mater. 11 (2012) 850.
[62] J. Rossat-Mignod, L.P. Regnault, C. Vettier, P. Bourges, P. Burlet, J. Bossy, et al., Phys. C (Amsterdam, Neth.) 185–189 (1991) 86.
[63] T.E. Mason, A. Schröder, G. Aeppli, H.A. Mook, S.M. Hayden, Phys. Rev. Lett. 77 (1996) 1604.
[64] H.F. Fong, P. Bourges, Y. Sidis, L.P. Regnault, A. Ivanov, G.D. Gu, et al., Nature 398 (1999) 588.
[65] H. He, P. Bourges, Y. Sidis, C. Ulrich, L.P. Regnault, S. Pailhès, et al., Science 295 (2002) 1045.
[66] G. Yu, Y. Li, E.M. Motoyama, X. Zhao, N. Barišić, Y. Cho, et al., Phys. Rev. B 81 (2010) 064518.
[67] S.D. Wilson, P. Dai, S. Li, S. Chi, H.J. Kang, J.W. Lynn, Nature 442 (2006) 59.
[68] A.D. Christianson, E.A. Goremychkin, R. Osborn, S. Rosenkranz, M.D. Lumsden, C.D. Malliakas, et al., Nature 456 (2008) 930.
[69] N.K. Sato, N. Aso, K. Miyake, R. Shiina, P. Thalmeier, G. Varelogiannis, et al., Nature 410 (2001) 340.
[70] C. Stock, C. Broholm, J. Hudis, H.J. Kang, C. Petrovic, Phys. Rev. Lett. 100 (2008) 087001.
[71] O. Stockert, J. Arndt, E. Faulhaber, C. Geibel, H.S. Jeevan, S. Kirchner, et al., Nat. Phys. 7 (2011) 119.
[72] S. Li, P. Dai, Front. Phys. 6 (2011) 429.
[73] G. Yu, Y. Li, E.M. Motoyama, M. Greven, Nat. Phys. 5 (2009) 873.
[74] P. Dai, H.A. Mook, R.D. Hunt, F. Doğan, Phys. Rev. B 63 (2001) 054525.
[75] H. Woo, P. Dai, S.M. Hayden, H.A. Mook, T. Dahm, D.J. Scalapino, et al., Nat. Phys. 2 (2006) 600.
[76] M. Wang, C. Zhang, X. Lu, G. Tan, H. Luo, Y. Song, et al., Nat. Commun. 4 (2013) 2874.
[77] N.B. Christensen, D.F. McMorrow, H.M. Rønnow, B. Lake, S.M. Hayden, G. Aeppli, et al., Phys. Rev. Lett. 93 (2004) 147002.
[78] S. Pailhès, Y. Sidis, P. Bourges, V. Hinkov, A. Ivanov, C. Ulrich, et al., Phys. Rev. Lett. 93 (2004) 167001.
[79] L. Capogna, B. Fauqué, Y. Sidis, C. Ulrich, P. Bourges, S. Pailhès, et al., Phys. Rev. B 75 (2007) 060502.
[80] Y. Sidis, S. Pailhès, V. Hinkov, B. Fauqué, C. Ulrich, L. Capogna, et al., C. R. Phys. 8 (2007) 745.

[81] P. Dai, H.A. Mook, S.M. Hayden, G. Aeppli, T.G. Perring, R.D. Hunt, et al., Science 284 (1999) 1344.
[82] M. Eschrig, Adv. Phys. 55 (2006) 47.
[83] D.J. Scalapino, Rev. Mod. Phys. 84 (2012) 1383.
[84] V. Hinkov, P. Bourges, S. Pailhès, Y. Sidis, A. Ivanov, C.D. Frost, et al., Nat. Phys. 3 (2007) 780.
[85] D.K. Morr, D. Pines, Phys. Rev. B 61 (2000) 6483.
[86] P.B. Allen, V.N. Kostur, N. Takesue, G. Shirane, Phys. Rev. B 56 (1997) 5552.
[87] M.K. Chan, C.J. Dorow, L. Mangin-Thro, Y. Tang, Y. Ge, M.J. Veit et al., http://arxiv.org/abs/1402.4517 (last accessed 20.07.15).
[88] C. de la Cruz, Q. Huang, J.W. Lynn, J. Li, W. Ratcliff II, J.L. Zarestky, et al., Nature 453 (2008) 899.
[89] D.N. Basov, R.D. Averitt, D. van der Marel, M. Dressel, K. Haule, Rev. Mod. Phys. 83 (2011) 471.
[90] Q. Si, E. Abrahams, Phys. Rev. Lett. 101 (2008) 076401.
[91] R.M. Fernandes, A.V. Chubukov, J. Schmalian, Nat. Phys. 10 (2014) 97.
[92] W. Bao, Y. Qiu, Q. Huang, M.A. Green, P. Zajdel, M.R. Fitzsimmons, et al., Phys. Rev. Lett. 102 (2009) 247001.
[93] W. Bao, Q. Huang, G. Chen, M.A. Green, D. Wang, J. He, et al., Chin. Phys. Lett. 28 (2011) 086104.
[94] D.C. Johnston, Adv. Phys. 59 (2010) 803.
[95] P.C. Canfield, S.L. Bud'ko, Annu. Rev. Condens. Matt. Phys. 1 (2010) 27.
[96] T.-K. Chen, C.-C. Chang, H.-H. Chang, A.-H. Fang, C.-H. Wang, W.-H. Chao, et al., Proc. Natl. Acad. Sci. U.S.A. 111 (2014) 63.
[97] M.G. Kim, R.M. Fernandes, A. Kreyssig, J.W. Kim, A. Thaler, S.L. Bud'ko, et al., Phys. Rev. B 83 (2011) 134522.
[98] Y. Song, S.V. Carr, X. Lu, C. Zhang, Z.C. Sims, N.F. Luttrell, et al., Phys. Rev. B 87 (2013) 184511.
[99] E.E. Rodriguez, C. Stock, P. Zajdel, K.L. Krycka, C.F. Majkrzak, P. Zavalij, et al., Phys. Rev. B 84 (2011) 064403.
[100] M.-C. Ding, H.-Q. Lin, Y.-Z. Zhang, Phys. Rev. B 87 (2013) 125129.
[101] W. Ratcliff II, P.A. Kienzle, J.W. Lynn, S. Li, P. Dai, G.F. Chen, et al., Phys. Rev. B 81 (2010) 140502.
[102] Z.P. Yin, K. Haule, G. Kotlair, Nat. Mater 10 (2011) 932.
[103] J.L. Niedziela, M.A. McGuire, T. Egami, Phys. Rev. B 86 (2012) 174113.
[104] C. Lester, J.-H. Chu, J.G. Analytis, T.G. Perring, I.R. Fisher, S.M. Hayden, Phys. Rev. B 81 (2010) 064505.
[105] J.T. Park, G. Friemel, T. Loew, V. Hinkov, Y. Li, B.H. Min, et al., Phys. Rev. B 86 (2012) 024437.
[106] Y. Song, L.P. Regnault, C. Zhang, G. Tan, S.V. Carr, S. Chi, et al., Phys. Rev. B 88 (2013) 134512.
[107] M. Ramazanoglu, J. Lamsal, G.S. Tucker, J.-Q. Yan, S. Calder, T. Guidi, et al., Phys. Rev. B 87 (2013) 140509.
[108] L.W. Harriger, H.Q. Luo, M.S. Liu, C. Frost, J.P. Hu, M.R. Norman, et al., Phys. Rev. B 84 (2011) 054544.
[109] C. Zhang, L.W. Harriger, Z. Yin, W. Lv, M. Wang, G. Tan, et al., Phys. Rev. Lett. 112 (2014) 217202.
[110] J. Zhao, D.T. Adroja, D.-X. Yao, R. Bewley, S. Li, X.F. Wang, et al., Nat. Phys. 5 (2009) 555.

High-Temperature Superconductors **Chapter | 3 197**

[111] R.A. Ewings, T.G. Perring, J. Gillett, S.D. Das, S.E. Sebastian, A.E. Taylor, et al., Phys. Rev. B 83 (2011) 214519.
[112] E. Kaneshita, T. Tohyama, Phys. Rev. B 82 (2010) 094441.
[113] A.L. Wysocki, K.D. Delashchenko, V.P. Antropov, Nat. Phys. 7 (2011) 485.
[114] D. Stanek, O. Sushkov, G.S. Uhrig, Phys. Rev. B 84 (2011) 064505.
[115] R. Yu, Z. Wang, P. Goswami, A.H. Nevidomskyy, Q. Si, E. Abrahams, Phys. Rev. B 86 (2012) 085148.
[116] M. Liu, L.W. Harriger, H. Luo, M. Wang, R.A. Ewings, T. Guidi, et al., Nat. Phys. 8 (2012) 376.
[117] J.H. Soh, G.S. Tucker, D.K. Pratt, D.L. Abernathy, M.B. Stone, S. Ran, et al., Phys. Rev. Lett. 111 (2013) 227002.
[118] Y. Furukawa, B. Roy, S. Ran, S.L. Bud'ko, P.C. Canfield, Phys. Rev. B 89 (2014) 121109.
[119] S. Mandal, R.E. Cohen, K. Haule, Phys. Rev. B 90 (2014) 060501.
[120] O.J. Lipscombe, G.F. Chen, C. Fang, T.G. Perring, D.L. Abernathy, A.D. Christianson, et al., Phys. Rev. Lett. 106 (2011) 057004.
[121] I.A. Zaliznyak, Z. Xu, J.M. Tranquada, G. Gu, A.M. Tsvelik, M.B. Stone, Phys. Rev. Lett. 107 (2011) 216403.
[122] C. Stock, E.E. Rodriguez, M.A. Green, P. Zavalij, J.A. Rodriguez-Rivera, Phys. Rev. B 84 (2011) 045124.
[123] C. Stock, E.E. Rodriguez, O. Sobolev, J.A. Rodriguez-Rivera, R.A. Ewings, J.W. Taylor, et al., Phys. Rev. B 90 (2014) 121113.
[124] D. Parshall, G. Chen, L. Pintschovius, D. Lamago, Th Wolf, L. Radzihovsky, et al., Phys. Rev. B 85 (2012) 140515.
[125] D.K. Pratt, M.G. Kim, A. Kreyssig, Y.B. Lee, G.S. Tucker, A. Thaler, et al., Phys. Rev. Lett. 106 (2011) 257001.
[126] H. Luo, R. Zhang, M. Laver, Z. Yamani, M. Wang, X. Lu, et al., Phys. Rev. Lett. 108 (2012) 247002.
[127] B. Farago, J. Major, F. Mezei, Neutron optics and spin labeling methods, in: Felix Fernandez-Alonso, David L. Price (Eds.), Neutron Scattering - Magnetic and Quantum Phenomena, in: Thomas Lucatorto, Albert C. Parr, Kenneth Baldwin (Eds.), Experimental Methods in the Physical Sciences, vol. 48, Elsevier, Amsterdam, 2015, pp. 1–42.
[128] A.P. Dioguardi, J. Crocker, A.C. Shockley, C.H. Lin, K.R. Shirer, D.M. Nisson, et al., Phys. Rev. Lett. 111 (2013) 207201.
[129] X. Lu, D.W. Tam, C. Zhang, H. Luo, M. Wang, R. Zhang, et al., Phys. Rev. B 90 (2014) 024509.
[130] D.K. Pratt, W. Tian, A. Kreyssig, J.L. Zarestky, S. Nandi, N. Ni, et al., Phys. Rev. Lett. 103 (2009) 087001.
[131] X. Lu, H. Gretarsson, R. Zhang, X. Liu, H. Luo, W. Tian, et al., Phys. Rev. Lett. 110 (2013) 257001.
[132] M. Wang, H. Luo, M. Wang, S. Chi, J.A. Rodriguez-Rivera, D. Singh, et al., Phys. Rev. B 83 (2011) 094516.
[133] M.G. Kim, J. Lamsal, T.W. Heitmann, G.S. Tucker, D.K. Pratt, S.N. Khan, et al., Phys. Rev. Lett. 109 (2012) 167003.
[134] S. Nandi, M.G. Kim, A. Kreyssig, R.M. Fernandes, D.K. Pratt, A. Thaler, et al., Phys. Rev. Lett. 104 (2010) 057006.
[135] S. Avci, O. Chmaissem, J.M. Allred, S. Rosenkranz, I. Eremin, A.V. Chubukov, et al., Nat. Commun. 5 (2014) 3845.

[136] K. Marty, A.D. Christianson, C.H. Wang, M. Matsuda, H. Cao, L.H. VanBebber, et al., Phys. Rev. B 83 (2011) 060509.
[137] M.G. Kim, A. Kreyssig, A. Thaler, D.K. Pratt, W. Tian, J.L. Zarestky, et al., Phys. Rev. B 82 (2010) 220503.
[138] D.S. Inosov, G. Friemel, J.T. Park, A.C. Walters, Y. Texier, Y. Laplace, et al., Phys. Rev. B 87 (2013) 224425.
[139] G.S. Tucker, D.K. Pratt, M.G. Kim, S. Ran, A. Thaler, G.E. Granroth, et al., Phys. Rev. B 86 (2012) 020503.
[140] S. Haravifard, B. Gaulin, Z. Yamani, Quantum phase transitions, in: Felix Fernandez-Alonso, David L. Price (Eds.), Neutron Scattering - Magnetic and Quantum Phenomena, in: Thomas Lucatorto, Albert C. Parr, Kenneth Baldwin (Eds.), Experimental Methods in the Physical Sciences, vol. 48, Elsevier, Amsterdam, 2015, pp. 43−144.
[141] T. Shibauchi, A. Carrington, Y. Matsuda, Annu. Rev. Condens. Matter Phys. 5 (2014) 113.
[142] D. Hu, X. Lu, W. Zhang, H. Luo, S. Li, P. Wang, et al., Phys. Rev. Lett. 114 (2015) 157002.
[143] M. Hiraishi, S. Iimura, K.M. Kojima, J. Yamaura, H. Hiraka, K. Ikeda, et al., Nat. Phys. 10 (2014) 300.
[144] L.W. Harriger, A. Schneidewind, S. Li, J. Zhao, Z. Li, W. Lu, et al., Phys. Rev. Lett. 103 (2009) 087005.
[145] H. Luo, Z. Yamani, Y. Chen, X. Lu, M. Wang, S. Li, et al., Phys. Rev. B 86 (2012) 024508.
[146] C. Zhang, M. Wang, H. Luo, M. Wang, M. Liu, J. Zhao, et al., Sci. Rep. 1 (2011) 115.
[147] J.T. Park, D.S. Inosov, A. Yaresko, S. Graser, D.L. Sun, Ph. Bourges, et al., Phys. Rev. B 82 (2010) 134503.
[148] J.-P. Castellan, S. Rosenkranz, E.A. Goremychkin, D.Y. Chung, I.S. Todorov, M.G. Kanatzidis, et al., Phys. Rev. Lett. 107 (2011) 177003.
[149] C.H. Lee, K. Kihou, H. Kawano-Furukawa, T. Saito, A. Iyo, H. Eisaki, et al., Phys. Rev. Lett. 106 (2011) 067003.
[150] W. Jayasekara, Y. Lee, A. Pandey, G.S. Tucker, A. Sapkota, J. Lamsal, et al., Phys. Rev. Lett. 111 (2013) 157001.
[151] H.A. Mook, M.D. Lumsden, A.D. Christianson, S.E. Nagler, B.C. Sales, R. Jin, et al., Phys. Rev. Lett. 104 (2010) 187002.
[152] Y. Qiu, W. Bao, Y. Zhao, C. Broholm, V. Stanev, Z. Tesanovic, et al., Phys. Rev. Lett. 103 (2009) 067008.
[153] S. Chi, J.A. Rodriguez-Rivera, J.W. Lynn, C. Zhang, D. Phelan, D.K. Singh, et al., Phys. Rev. B 84 (2011) 214407.
[154] T.J. Liu, J. Hu, B. Qian, D. Fobes, Z.Q. Mao, W. Bao, et al., Nat. Mater 9 (2010) 716.
[155] Z. Xu, J. Wen, G. Xu, Q. Jie, Z. Lin, Q. Li, et al., Phys. Rev. B 82 (2010) 104525.
[156] M.D. Lumsden, A.D. Christianson, E.A. Goremychkin, S.E. Nagler, H.A. Mook, M.B. Stone, et al., Nat. Phys. 6 (2010) 182.
[157] A.D. Christianson, M.D. Lumsden, K. Marty, C.H. Wang, S. Calder, D.L. Abernathy, et al., Phys. Rev. B 87 (2013) 224410.
[158] A. Subedi, L. Zhang, D.J. Singh, M.H. Du, Phys. Rev. B 78 (2008) 134514.
[159] J. Wen, S. Li, Z. Xu, C. Zhang, M. Matsuda, O. Sobolev, et al., Phys. Rev. B 88 (2013) 144509.
[160] M. Wang, M. Wang, H. Miao, S.V. Carr, D.L. Abernathy, M.B. Stone, et al., Phys. Rev. B 86 (2012) 144511.
[161] N. Qureshi, P. Steffens, Y. Drees, A.C. Komarek, D. Lamago, Y. Sidis, et al., Phys. Rev. Lett. 108 (2012) 117001.

[162] S.V. Borisenko, V.B. Zabolotnyy, D.V. Evtushinsky, T.K. Kim, I.V. Morozov, A.N. Yaresko, et al., Phys. Rev. Lett. 105 (2010) 067002.
[163] N. Qureshi, P. Steffens, D. Lamago, Y. Sidis, O. Sobolev, R.A. Ewings, et al., Phys. Rev. B 90 (2014) 144503.
[164] H. Luo, X. Lu, R. Zhang, M. Wang, E.A. Goremychkin, D.T. Adroja, et al., Phys. Rev. B 88 (2013) 144516.
[165] A.D. Christianson, M.D. Lumsden, S.E. Nagler, G.J. MacDougall, M.A. McGuire, A.S. Sefat, et al., Phys. Rev. Lett. 103 (2009) 087002.
[166] S. Chi, A. Schneidewind, J. Zhao, L.W. Harriger, L. Li, Y. Luo, et al., Phys. Rev. Lett. 102 (2009) 107006.
[167] J. Zhao, C.R. Rotundu, K. Marty, M. Matsuda, Y. Zhao, C. Setty, et al., Phys. Rev. Lett. 110 (2013) 147003.
[168] C.H. Lee, P. Steffens, N. Qureshi, M. Nakajima, K. Kihou, A. Iyo, et al., Phys. Rev. Lett. 111 (2013) 167002.
[169] C. Zhang, H.-F. Li, Y. Song, Y. Su, G. Tan, T. Netherton, et al., Phys. Rev. B 88 (2013) 064504.
[170] C. Zhang, R. Yu, Y. Su, Y. Song, M. Wang, G. Tan, et al., Phys. Rev. Lett. 111 (2013) 207002.
[171] D.K. Pratt, A. Kreyssig, S. Nandi, N. Ni, A. Thaler, M.D. Lumsden, et al., Phys. Rev. B 81 (2010) 140510.
[172] M.G. Kim, G.S. Tucker, D.K. Pratt, S. Ran, A. Thaler, A.D. Christianson, et al., Phys. Rev. Lett. 110 (2013) 177002.
[173] M. Wang, H. Luo, J. Zhao, C. Zhang, M. Wang, K. Marty, et al., Phys. Rev. B 81 (2010) 174524.
[174] D.S. Inosov, J.T. Park, A. Charnukha, Y. Li, A.V. Boris, B. Keimer, et al., Phys. Rev. B 83 (2011) 214520.
[175] M.D. Lumsden, A.D. Christianson, D. Parshall, M.B. Stone, S.E. Nagler, G.J. MacDougall, et al., Phys. Rev. Lett. 102 (2009) 107005.
[176] S. Onari, H. Kontani, M. Sato Phys. Rev. B 81 (2010) 060504.
[177] S. Onari, H. Kontani, Phys. Rev. B 84 (2011) 144518.
[178] Q.Q. Ge, Z.R. Ye, M. Xu, Y. Zhang, J. Jiang, B.P. Xie, et al., Phys. Rev. X 3 (2013) 011020.
[179] W. Lv, A. Moreo, E. Dagotto, Phys. Rev. B 89 (2014) 104510.
[180] D.N. Argyriou, A. Hiess, A. Akbari, I. Eremin, M.M. Korshunov, J. Hu, et al., Phys. Rev. B 81 (2010) 220503.
[181] J. Guo, S. Jin, G. Wang, S. Wang, K. Zhu, T. Zhou, et al., Phys. Rev. B 82 (2010) 180520.
[182] A. Ricci, N. Poccia, G. Campi, B. Joseph, G. Arrighetti, L. Barba, et al., Phys. Rev. B 84 (2011) 060511.
[183] R.H. Yuan, T. Dong, Y.J. Song, P. Zheng, G.F. Chen, J.P. Hu, et al., Sci. Rep. 2 (2012) 221.
[184] F. Chen, M. Xu, Q.Q. Ge, Y. Zhang, Z.R. Ye, L.X. Yang, et al., Phys. Rev. X 1 (2011) 021020.
[185] W. Li, H. Ding, Z. Li, P. Deng, K. Chang, K. He, et al., Phys. Rev. Lett. 109 (2012) 057003.
[186] W. Li, H. Ding, P. Deng, K. Chang, C. Song, K. He, et al., Nat. Phys. 8 (2012) 126.
[187] J. Zhao, H. Cao, E. Bourret-Courchesne, D.-H. Lee, R.J. Birgeneau, Phys. Rev. Lett. 109 (2012) 267003.
[188] S.V. Carr, D. Louca, J. Siewenie, Q. Huang, A. Wang, X. Chen, et al., Phys. Rev. B 89 (2014) 134509.
[189] M. Wang, M. Wang, G.N. Li, Q. Huang, C.H. Li, G.T. Tan, et al., Phys. Rev. B 84 (2011) 094504.

[190] F. Ye, S. Chi, W. Bao, X.F. Wang, J.J. Ying, X.H. Chen, et al., Phys. Rev. Lett. 107 (2011) 137003.
[191] M. Wang, W. Tian, P. Valdivia, S. Chi, E. Bourret-Courchesne, P. Dai, et al., Phys. Rev. B 90 (2014) 125148.
[192] M. Wang, C. Fang, D.-X. Yao, G. Tan, L.W. Harriger, Y. Song, et al., Nat. Commun. 2 (2011) 580.
[193] S. Chi, F. Ye, W. Bao, M. Fang, H.D. Wang, C.H. Dong, et al., Phys. Rev. B 87 (2013) 100501.
[194] Y. Xiao, S. Nandi, Y. Su, S. Price, H.-F. Li, Z. Fu, et al., Phys. Rev. B 87 (2013) 140408.
[195] M. Wang, X. Lu, R.A. Ewings, L.W. Harriger, Y. Song, S.V. Carr, et al., Phys. Rev. B 87 (2013) 064409.
[196] J. Zhao, Y. Shen, R.J. Birgeneau, M. Gao, Z.-Y. Lu, D.-H. Lee, et al., Phys. Rev. Lett. 112 (2014) 177002.
[197] J.T. Park, G. Friemel, Y. Li, J.-H. Kim, V. Tsurkan, J. Deisenhofer, et al., Phys. Rev. Lett. 107 (2011) 177005.
[198] G. Friemel, J.T. Park, T.A. Maier, V. Tsurkan, Y. Li, J. Deisenhofer, et al., Phys. Rev. B 85 (2012) 140511.
[199] G. Friemel, W.P. Liu, E.A. Goremychkin, Y. Li, J.T. Park, O. Sobolev, et al., Europhys. Lett. 99 (2012) 67004.
[200] M. Wang, C. Li, D.L. Abernathy, Y. Song, S.V. Carr, X. Lu, et al., Phys. Rev. B 86 (2012) 024502.
[201] A.E. Taylor, R.A. Ewings, T.G. Perring, J.S. White, P. Babkevich, A. Krzton-Maziopa, et al., Phys. Rev. B 86 (2012) 094528.
[202] C. Zhang, Y. Song, L.-P. Regnault, Y. Su, M. Enderle, J. Kulda, et al., Phys. Rev. B 90 (2014) 140502.
[203] H. Luo, M. Wang, C. Zhang, X. Lu, L.-P. Regnault, R. Zhang, et al., Phys. Rev. Lett. 111 (2013) 107006.
[204] C. Wang, R. Zhang, F. Wang, H. Luo, L.P. Regnault, P. Dai, et al., Phys. Rev. X 3 (2013) 041036.
[205] N.S. Headings, S.M. Hayden, J. Kulda, N. Hari Babu, D.A. Cardwell, Phys. Rev. B 84 (2011) 104513.
[206] N. Qureshi, P. Steffens, S. Wurmehl, S. Aswartham, B. Büchner, M. Braden, Phys. Rev. B 86 (2012) 060410.
[207] O.J. Lipscombe, L.W. Harriger, P.G. Freeman, M. Enderle, C. Zhang, M. Wang, et al., Phys. Rev. B 82 (2010) 064515.
[208] P. Steffens, C.H. Lee, N. Qureshi, K. Kihou, A. Iyo, H. Eisaki, et al., Phys. Rev. Lett. 110 (2013) 137001.
[209] M. Liu, C. Lester, J. Kulda, X. Lu, H. Luo, M. Wang, et al., Phys. Rev. B 85 (2012) 214516.
[210] C. Zhang, M. Liu, Y. Su, L.-P. Regnault, M. Wang, G. Tan, et al., Phys. Rev. B 87 (2013) 081101.
[211] N. Qureshi, C.H. Lee, K. Kihou, K. Schmalzl, P. Steffens, M. Braden, Phys. Rev. B 90 (2014) 100502.
[212] P. Babkevich, B. Roessli, S.N. Gvasaliya, L.-P. Regnault, P.G. Freeman, E. Pomjakushina, et al., Phys. Rev. B 83 (2011) 180506.
[213] K. Prokeš, A. Hiess, W. Bao, E. Wheeler, S. Landsgesell, D.N. Argyriou, Phys. Rev. B 86 (2012) 064503.
[214] C. Dhital, Z. Yamani, W. Tian, J. Zeretsky, A.S. Sefat, Z. Wang, et al., Phys. Rev. Lett. 108 (2012) 087001.

[215] C. Dhital, T. Hogan, Z. Yamani, R.J. Birgeneau, W. Tian, M. Matsuda, et al., Phys. Rev. B 89 (2014) 214404.
[216] X. Lu, J.T. Park, R. Zhang, H. Luo, A.H. Nevidomskyy, Q. Si, et al., Science 345 (2014) 657.
[217] Q. Zhang, R.M. Fernandes, J. Lamsal, J. Yan, S. Chi, G.S. Tucker, et al., Phys. Rev. Lett. 114 (2015) 057001.
[218] L. de' Medici, G. Giovannetti, M. Capone, Phys. Rev. Lett. 112 (2014) 177001.
[219] H. Ishida, A. Liebsch, Phys. Rev. B 81 (2010) 054513.
[220] T. Misawa, K. Nakamura, M. Imada, Phys. Rev. Lett. 108 (2012) 177007.
[221] M. Nakajima, S. Ishida, T. Tanaka, K. Kihou, Y. Tomioka, C.-H. Lee, et al., J. Phys. Soc. Jpn. 83 (2014) 104703.
[222] F. Hardy, A.E. Böhmer, D. Aoki, P. Burger, T. Wolf, P. Schweiss, et al., Phys. Rev. Lett. 111 (2013) 027002.
[223] T. Dahm, V. Hinkov, S.V. Borisenko, A.A. Kordyuk, V.B. Zabolotnyy, J. Fink, et al., Nat. Phys. 5 (2009) 217.
[224] N. Ni, J.M. Allred, B.C. Chan, R.J. Cava, Proc. Natl. Acad. Sci. U.S.A. 108 (2011) 1019.
[225] S. Kakiya, K. Kudo, Y. Nishikubo, K. Oku, E. Nishibori, H. Sawa, et al., J. Phys. Soc. Jpn. 80 (2011) 093704.
[226] H. Yakita, H. Ogino, T. Okada, A. Yamamoto, K. Kishio, T. Tohei, et al., J. Am. Chem. Soc. 136 (2014) 846.
[227] N. Katayama, K. Kudo, S. Onari, T. Mizukami, K. Sugawara, Y. Sugiyama, et al., J. Phys. Soc. Jpn. 82 (2013) 123702.
[228] J.-Q. Yan, S. Nandi, B. Saparov, P. Čermák, Y. Xiao, Y. Su, et al., Phys. Rev. B 91 (2015) 024501.
[229] N. Katayama, S. Ji, D. Louca, S. Lee, M. Fujita, T.J. Sato, et al., J. Phys. Soc. Jpn. 79 (2010) 113702.
[230] M.G. Kim, D.K. Pratt, G.E. Rustan, W. Tian, J.L. Zarestky, A. Thaler, et al., Phys. Rev. B 83 (2011) 054514.

Chapter 4

Magnetic Structures

V. Ovidiu Garlea* and Bryan C. Chakoumakos
Quantum Condensed Matter Division, Oak Ridge National Laboratory, Oak Ridge, Tennessee, USA
Corresponding author: E-mail: garleao@ornl.gov

Chapter Outline

- 4.1 Introduction 204
- 4.2 The Beginnings of Magnetic Structure Determination Using Neutron Diffraction 206
- 4.3 Fundamentals of Magnetic Scattering 211
  - 4.3.1 Scattering Amplitude from Magnetic Order 211
  - 4.3.2 The Magnetic Propagation Vector Formalism 215
    - 4.3.2.1 Magnetic Structures with $k = 0$ or $\tau/2$ 218
    - 4.3.2.2 Spin Density Wave or Sine Wave Structures 219
    - 4.3.2.3 Helical Magnetic Structures 221
  - 4.3.3 Symmetry Considerations in Magnetic Structures 222
    - 4.3.3.1 Shubnikov Space Groups 223
    - 4.3.3.2 Magnetic Superspace Groups 227
    - 4.3.3.3 Representation Analysis 228
- 4.4 Practical Aspects in Determination of Magnetic Structures 231
  - 4.4.1 Steps in the Determination of Magnetic Structures 231
  - 4.4.2 Limitations of Neutron Scattering for Complex Magnetic Structures 234
- 4.4.3 Standard Description for Magnetic Structures 235
- 4.5 Hierarchy of Magnetic Structures (Important Examples) 235
  - 4.5.1 Three-Dimensional Networks 237
    - 4.5.1.1 Corner-Sharing Tetrahedral Lattices in Pyrochlore Oxides 237
    - 4.5.1.2 Interpenetrated Tetrahedra and Diamond Lattices in Spinel Compounds 242
    - 4.5.1.3 Diamond-Lattice in Spinel Systems 246
    - 4.5.1.4 Edge-Sharing Tetrahedra in Double Perovskites 248
  - 4.5.2 Layered Structures 250
    - 4.5.2.1 Square-Planar Metal-Oxygen/ Halogen Layers 251
    - 4.5.2.2 Square Lattices of Metal Atoms Coordinated by Pnictogen/ Chalcogen Atoms 253

Experimental Methods in the Physical Sciences, Vol. 48. http://dx.doi.org/10.1016/B978-0-12-802049-4.00004-X
Copyright © 2015 Elsevier Inc. All rights reserved.
Battelle Memorial Institute LLC, Operator of Oak Ridge National Laboratory, Contract No. DE-AC05-00OR22725

4.5.2.3 Combined Square-Planar Metal-Oxygen and Metal-Pnictogen Lattices 256	4.5.3.3 Effective One-Dimensionality Caused by Frustrated Interchain Interactions 270
4.5.2.4 Planes of Edge-Sharing Equilateral Triangles 258	4.5.3.4 Field-Induced Magnetic Order in One-Dimensional Quantum Systems 272
4.5.2.5 Planes of Edge-Sharing Nonequilateral Triangles 260	
4.5.2.6 Planar Kagomé Lattices, Based on Corner-Sharing Triangles 262	4.6 Disordered Magnetic Structures 274
4.5.2.7 Layers of Honeycomb Lattices 264	4.6.1 Diffuse Scattering from Disordered Alloys 275
4.5.3 Quasi-One-Dimensional Lattices 266	4.6.2 Diffuse Scattering from Systems with Reduced Dimensionality 276
4.5.3.1 Long-Range Magnetic Order in Simple Chain-Based Structures 266	4.6.3 Diffuse Magnetic Scattering in Frustrated Magnets 278
	4.7 Concluding Remarks 281
4.5.3.2 Triangular-Based Magnetic Chains 268	Acknowledgments 281
	References 282

4.1 INTRODUCTION

A magnetic state in a solid, whether ordered or disordered, is the net result of the competing influences, thermal energy tending to randomize moments versus some quantum mechanical coupling tending to order moments. Our discussion is predicated upon the assumption that unpaired electron-spin density is localized, i.e., a discrete magnetic moment can be associated with an atom or ion in a solid, leading to well-defined (atom centered) magnetic moments and well-defined Bragg reflections arising when these atom-centered moments are arranged on a lattice. Solids in which some or all of the electrons are itinerant, as in most metals, are not included here. The electron is the primary carrier of magnetism, having an intrinsic angular momentum ("spin") that leads to an intrinsic magnetic moment (the Bohr magneton, $\mu_B = 9.284 \times 10^{-24}$ J/T). Much smaller contributions to the magnetic moment of the atom arise from the orbital motion of the electrons and the nuclear spin. An atom has a net moment when a d- or f-electron shell is incompletely filled, the spin and orbital momenta of the electrons in the shell do not exactly cancel.

An elementary review of the origins and various types of magnetic order in solids is given by Hurd [1].

Interest in magnetic structures is greater than ever, and even though major neutron scattering facilities have come and gone over time, the number of neutron powder and single-crystal diffractometers available to researchers now is greater than any time in history. The number and complexity of the magnetic structures also is growing, and as a result, standards for magnetic crystallographic information files (CIF) have been internationally agreed upon, to aid in the uniform description of magnetic structures and to facilitate the use of databases and crystallographic software tools.

Magnetic structure analysis began with the birth of neutron diffraction, and in itself constitutes one of the unrivaled applications of neutron scattering. The development of this tool has been driven, like many measurement and analysis schemes, by the advent and advancement of computers. Symmetry analysis and least-squares fitting of diffraction data is not computationally demanding given today's computer power, but computers make the computations and visualization easy and accessible to anyone with a laptop computer. What constitutes a magnetic structure is the nuclear crystal structure with the addition of the moment size and directions for each of the unique magnetic ions. Because the magnetic ions can adopt a lattice that is the same or larger than the parent nuclear lattice, the propagation vector relating the magnetic lattice in terms of the nuclear lattice is also given.

Magnetic structure analysis historically has been determined by trial and error, comparing powder or single-crystal diffraction peak intensities calculated from intuitive models to the measured intensities of magnetic peaks. Depending on the experience of the investigator and the degree of complexity of the models, the trial and error approach has been successful in a great many cases. Symmetry analysis, introduced nearly 50 years ago, has only surpassed the trial and error approach in recent years. Symmetry analysis provides useful constraints on possible models and can allow for a more complete appraisal of all possible solutions. Even so, limited amounts of data, particularly from powder diffraction, occasionally will not guarantee a unique solution.

The vast majority of magnetic structures are determined using neutron powder diffraction, but to a lesser extent, when single-crystals are available, neutron single-crystal diffraction is used. Neutron single-crystal diffraction necessarily constrains the number of model solutions to a greater extent, given the three-dimensional nature of the data. Optional enhancements such as polarization analysis are used to unambiguously separate the magnetic from the nuclear scattering, which is crucial in some cases. In addition to the temperature control necessary to record data above and below the magnetic ordering temperature, the application of an applied magnetic field can be advantageous to align moments in a particular direction. Large-scale magnetic structures can be studied with small-angle neutron scattering, and neutron polarized reflectometry is the method of choice for magnetic multilayer heterostructures. An in-depth description of the aforementioned experimental techniques is given in Ref. [2].

Our focus and approach here primarily addresses the great diversity of long-range ordered states that are possible, ranging from simple ferro- and antiferromagnetic orders to the more complex helical and conical spiral orders, commensurate and incommensurate. Disordered and glassy states are also mentioned, but in much less detail. We are purposely omitting here the more exotic quantum ground states such as spin-liquid and spin-nematic states, which are more appropriately covered in Chapter 2 by S. Haravifard, B. Gaulin, and Z. Yamani.

This chapter is organized into five major sections: the beginnings of magnetic structure determination using neutron diffraction, the fundamentals of magnetic scattering, practical aspects in determining magnetic structures, a hierarchy of magnetic structures with important examples, and an overview of disordered magnetic structures. Our goal is to expose the reader to the current state of how magnetic structures are determined using neutron diffraction and how they are described. We limit ourselves to the analysis of powder and single-crystal diffraction data, recognizing that magnetism in thin-film heterostructures and large-scale magnetic structures, are covered elsewhere in this volume, e.g., Chapter 6 by B.P. Toperberg and H. Zabel.

4.2 THE BEGINNINGS OF MAGNETIC STRUCTURE DETERMINATION USING NEUTRON DIFFRACTION

The understanding and application of neutron diffraction to the study of magnetic structures developed soon after the discovery of the neutron by Chadwick in 1932 [3]. Demonstrations of neutron diffraction from crystalline materials were made by Mitchell and Powers in 1936 [4] and von Halban and Preiswerk [5], using Ra-Be sources in the pre-reactor days. The theoretical basis of neutron diffraction from magnetic ions was put forth by Bloch [6] and he suggested three applications: (1) measurement of the magnetic scattering at small angles to determine the magnetic moment of the neutron; (2) production of polarized neutrons by transmission through magnetized iron, as a result of the different transmissions for neutrons with spins oriented parallel and antiparallel to the magnetization; and (3) measurement of magnetic scattering factors for ions to determine their spatial distribution. Schwinger [7] immediately pointed out a correction to Bloch's formulation in noting that only the atomic moment components in the plane perpendicular to the scattering vector are effective in scattering neutrons. An important conclusion from Bloch's work was that the magnetic scattering cross section was comparable in size to the nuclear scattering cross section. Halpern and Johnson published two notes in 1937 that expanded the theoretical basis outlined by Bloch and Schwinger to include not only elastic scattering, but also the cases in which the neutron interacts with the sample [8]. These suggestions triggered several experimental efforts to determine the magnetic moment and spin state of the neutron [9−11] because these properties were uncertain or not exactly known at the time. These resulted in the magnetic moment being about 2 nuclear magnetons with a spin state of 1/2. The first quantitative determination of the neutron magnetic moment ($\mu_n = 1.935 \pm 0.02$ absolute nuclear magnetons) was reported by Alvarez and Bloch using a magnetic resonance method [12]. In 1939, Halpern and Johnson published their seminal work on magnetic scattering of neutrons [13], which has served as the basis for all subsequent theoretical and experimental work on magnetic scattering and polarized beam production, as noted by Moon [14], who gives an excellent account of this early history with particular reference to the development of neutron polarization methods and their use; Moon was Shull's first graduate student at M.I.T. Halpern's group continued to develop the theoretical treatment of neutron scattering from magnetic materials [15,16], but much of the parallel experimental verification and development could not be profitably undertaken until more powerful sources of neutrons became available from nuclear reactors. At New York University, while Halpern and his colleagues were laying the theoretical groundwork for magnetic neutron scattering, Shull was also there doing his graduate work (1937−1941) toward a PhD. Although, Shull's thesis research had been in a different area of physics, he was well familiar with the work on magnetic neutron scattering [17], and this knowledge no doubt provided incentive for his later work on neutron diffraction.

Higher intensity neutron sources were realized from the first permanent nuclear reactors constructed as part of the Manhattan Project (1939−1946), and these allowed systematic study of nuclear and magnetic scattering using neutrons. The early neutron scattering work at Oak Ridge, Tennessee, was performed at the X-10 Pile (later known as the Oak Ridge Graphite Reactor), which operated from November 1943 to November 1963. In May 1944, a heavy water reactor, CP-3, was started at the original site of Argonne National Laboratory, south of Chicago, Illinois, and operated until 1950, then it was dismantled and rebuilt as CP-3' and operated until 1954. Even though these reactors were primarily used for the needs of the Manhattan Project, groups working at Oak Ridge and Chicago undertook measurements that demonstrated the future utility of neutron diffraction and how to produce polarized neutron beams. E.O. Wollan, L.B. Borst, and W.H. Zinn were all able to observe neutron diffraction in 1944 using these two reactors. Subsequently, an extended period of work at Oak Ridge by Wollan and Shull, who joined Wollan's group in 1946, laid the foundations of neutron diffraction and magnetic scattering. An excellent account of this early history is given by Ref. [18], and recounted in Shull's 1994 Nobel Prize acceptance speech. The other noteworthy reactor completed just after the Manhattan project ended was

NRX, a heavy water reactor at the Chalk River Laboratories in Deep River, Ontario, Canada, that began operating in 1947 and continued until 1993. Nearly all of the Nobel Prize-winning research by Brockhouse and Shull was conducted at the NRX and X-10 pile, respectively.

Shull worked with Wollan in Oak Ridge from June 1946 to 1955, afterward moving to M.I.T. During this relatively short period, it is truly amazing how much these two and their colleagues accomplished. They had exclusive and unencumbered access to the Oak Ridge Graphite Reactor, which operated in a near continuous mode. The research field was virgin ground, but they had to essentially build the first two-axis neutron spectrometer, and then systematically build an understanding of neutron scattering amplitudes and cross sections that would quantitatively account for the observed intensities in the powder diffraction pattern. M.K. Wilkinson, who joined the nascent neutron scattering group at Oak Ridge in 1950 as Shull's first postdoctoral fellow, provides a detailed historical account of the group's early accomplishments [19]. Perhaps the biggest discovery in terms of magnetism was the experimental confirmation of Néel's theory of antiferromagnetism; see Refs [20,21]. Extra Bragg peaks were observed in the powder diffraction pattern of MnO that showed the ordered magnetic cell was larger than the chemical cell (Figure 1). Antiferromagnetism was also confirmed in FeO, CoO, NiO, and hematite-type α-Fe_2O_3 [22]. The ferrimagnetic structure of magnetite was also confirmed. With magnetite in an applied magnetic field they actually controlled the direction of the magnetization relative to the scattering vector and used almost the entirely magnetic (1,1,1) reflection to provide another test that the form of the magnetic interaction proposed by Schwinger, Halpern, and Johnson was correct, i.e., the intensity would vary as $\sin^2\alpha$, where α is the

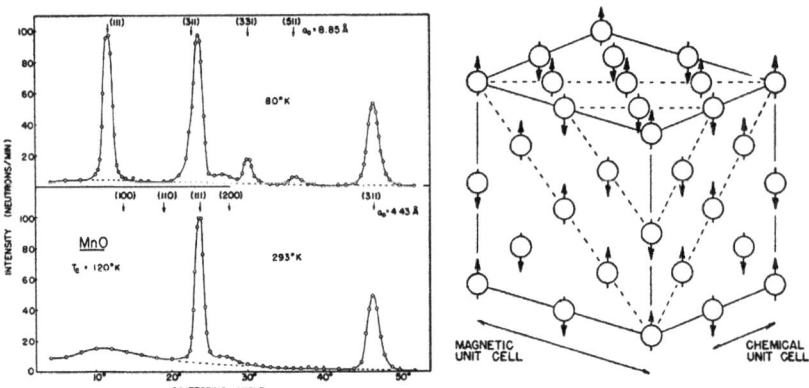

FIGURE 1 Neutron powder diffraction patterns for MnO at 80 and 293 K. The structure drawing shows the antiferromagnetic structure below its Néel temperature of 120 K, with the magnetic cell dimension twice as long as the chemical cell dimension. *Reproduced with permission from Ref. [22].*

angle between the magnetization and the scattering vector [23]. This early work as summarized by Wilkinson [19], included the first neutron scattering investigations of the magnetic properties of rare earth oxides [24] and rare earth metals [25]; investigations of antiferromagnetism in manganous fluoride and some isomorphous compounds [26]; a determination of the ferrimagnetic structure of the intermetallic compound Mn_2Sb [27]; investigations of the magnetic structure in 3d-transition metals [28] and alloys [29]; investigations of the high temperature and critical scattering from iron [30,31]; and a detailed analysis of the magnetic structures of the perovskite-type compounds [$(1-x)$ La, xCa]MnO_3 [32]. This highly cited paper has become a classic reference for the colossal magnetoresistance (CMR) manganites (Figure 2), and their proposed scheme for the different magnetic structures is still widely used. Other nuclear reactors were being built at universities and other national laboratories around the world by the late 1950s, and the instruments and methods to do neutron diffraction developed at Oak Ridge were being duplicated to form their own neutron scattering research programs.

It was evident early on that polarized neutron beams would have some utility for a variety of purposes. Already by the early 1950s, three methods had been experimentally proven for producing polarized neutron beams: (1) transmission through magnetized iron, (2) total reflection from magnetized mirrors, and (3) Bragg scattering from ferromagnetic single crystals; see Refs

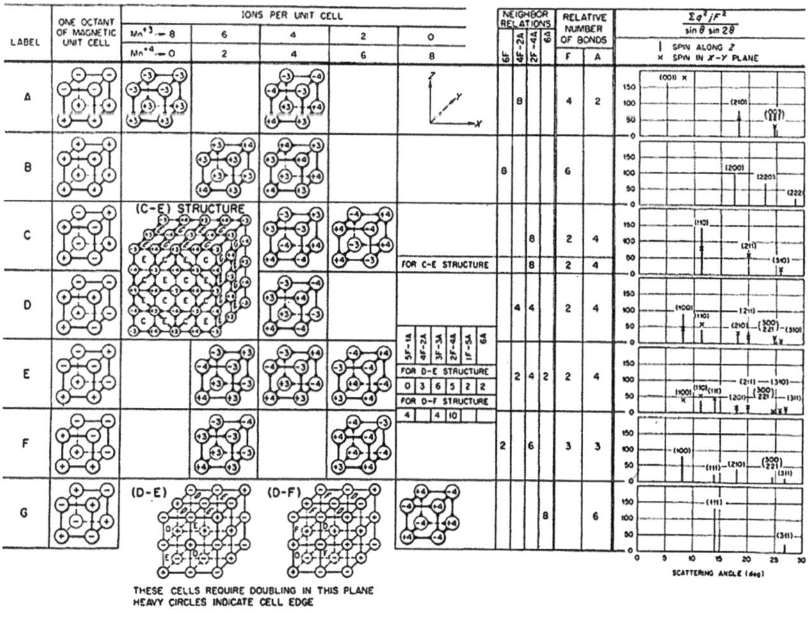

FIGURE 2 Scheme of magnetic structures of the perovskite manganites devised by Wollan and Koehler [32], cited over 1900 times. *Reproduced with permission from Ref. [32].*

[14,33]. The widespread use of polarized neutron beams to unambiguously separate the magnetic and nuclear scattering was launched by the clear and practical paper on polarization analysis by Moon, Riste, and Koehler [34]. Moon [14] summarized this now classic paper by the following: in planning experiments and understanding experimental results, it is almost sufficient to remember that (1) neutrons will flip their spins when scattered by spin components perpendicular to the neutron polarization and will not flip their spin when scattered by spin components parallel to the neutron polarization, (2) neutrons are not scattered by atomic spin components parallel to the scattering vector, and (3) neutrons do not flip their spin in coherent nuclear scattering. When the neutron polarization is along the scattering vector, all magnetic scattering is spin-flip (SF), and all nuclear coherent scattering is non-spin-flip (NSF). For more information of polarized neutron scattering, the reader is referred to Chapter 1 of Ref. [2].

Given that magnetic scattering arises from the interaction of the neutron magnetic moment with the magnetic moments of the unpaired electrons in the outer orbitals of certain ions, typically ions of the transition metals, lanthanides, actinides, and certain molecules, the Fourier transform of the spatial distribution of these unpaired electrons constitutes the magnetic form factor [35]. Magnetic form factors resemble X-ray form factors, but they decay more rapidly with scattering angle because neutron magnetic scattering is only from the outer electrons. Early on, measurements of magnetic form factors were undertaken to understand their different contributions [36]. The main contribution is from pure spin of the free ion, but this can be modified by the electron density overlap between neighboring atoms and by orbital contributions to the total angular momentum.

A sufficient number of magnetic crystal structures had been reported by neutron diffraction in the first 10 years (from 1946) that review articles on the subject began appearing as early as 1959 [37]. The structural models for the first 25 years or so were determined by trial and error, and by hand calculation, because mainframe computers did not come into widespread use until the early 1950s, and personal computers were not popular until the 1980s. The representation analysis approach was practically developed in 1968 [38], and Rietveld's whole pattern fitting was introduced in 1969 [39], but the early Rietveld method computer programs, such as DBWS [40], did not include magnetic scattering nor offer the group theoretical tools that are available today. The computational tools to make the analysis of magnetic structures relatively easy (e.g., GSAS [41], FullProf Suite [42], JANA [43], SARAh [44,45]) did not come into being until the 1990s. The growth in the number of magnetic structures reported annually had been steadily increasing up to 2003 with marked increases in 1980 and 1990 likely associated with introduction of computers and computational tools (Figure 3). These numbers also reflect the number of operational neutron sources and diffractometers available that could

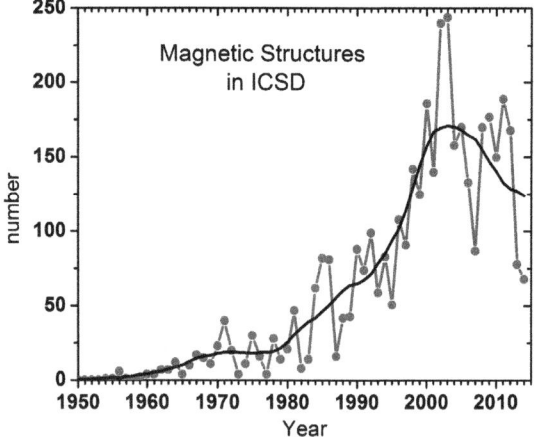

FIGURE 3 Number of entries in the Inorganic Crystal Structure Database that reports magnetic structure. Blue (dark gray in print versions) line is a 10-year smoothed average.

provide the raw data needed for magnetic structure analysis, so the flattening or down-turn in the numbers since 2003 might be linked in part to the permanent closures of the HFBR facility at Brookhaven National Laboratory in 1996 and the IPNS facility at Argonne National Laboratory in 2008, and the gradual ramp-ups of the SNS (United States) and JPARC (Japan). The expectation is that the number of magnetic structures determined annually will again increase as the newer neutron scattering facilities mature, and the dissemination of user-friendly computational tools continues, because the discovery of novel magnetic structures is unceasing, driven by their own interest, their relationships to unconventional superconductors, the needs for better permanent magnets, and the desire to develop spintronic technologies.

4.3 FUNDAMENTALS OF MAGNETIC SCATTERING

4.3.1 Scattering Amplitude from Magnetic Order

We will start this section with a brief review of the important expressions describing the scattering of neutrons from crystalline magnetic systems, which are useful for our discussion throughout this chapter. For a more detailed description, the reader is referred to the classic textbooks on neutron scattering [2,46–49].

The general concept of the differential cross section for neutron scattering has been introduced in Chapter 1 of the volume "Neutron Scattering— Fundamentals" [2], as well as in this volume in Chapter 2 by S. Haravifard, B. Gaulin, and Z. Yamani, and Chapter 3 by Y. Song and P. Dai. Our focus will be directed here to the nearly static arrangement of magnetic moments with time-independent correlations, for which the differential scattering cross section can be approximated as purely elastic and defined as:

$$\left(\frac{d\sigma}{d\Omega}\right)_{el} = (\gamma r_0)^2 |\langle \mathbf{D}_\perp(\mathbf{Q})\rangle|^2 \qquad (1)$$

where $\gamma = -1.913$ is the gyromagnetic ratio of the neutron, $r_0 = 2.8179$ fm is the classical electron radius. The term $\mathbf{D}_\perp(\mathbf{Q})$ represents the projection of the time-independent magnetic interaction vector operator, $\mathbf{D}(\mathbf{Q})$, on a plane perpendicular to the neutron scattering vector $\mathbf{Q}$ (i.e., $\mathbf{D}_\perp(\mathbf{Q}) = \hat{\mathbf{Q}} \times \mathbf{D} \times \hat{\mathbf{Q}}$), where $\hat{\mathbf{Q}}$ represents a unit vector along $\mathbf{Q}$. We define here the scattering cross section as an extensive quantity, as given in Refs [46–48], but it can be also formulated as an intensive quantity (i.e., divided by the number of magnetic centers, $1/N_m$) as in Refs [2,49]. A schematic view of the scattering geometry and the relevant projection of the interaction vector are shown in Figure 4. This angular dependence of the magnetic scattering allows the determination of both the amplitude and orientation of the magnetic moment. The magnetic interaction operator is related to the Fourier transform of the total magnetization density of the scattering system:

$$\mathbf{D}(\mathbf{Q}) = -\mathbf{M}(\mathbf{Q})/2\mu_B = -\frac{1}{2\mu_B} \int \mathbf{M}(r) e^{i\mathbf{Q}\cdot r} dr \qquad (2)$$

For electrons obeying Russell–Saunders (*LS*) coupling, localized on atoms at position $\mathbf{R}_j$, it can be expressed as:

$$\langle \mathbf{D}(\mathbf{Q})\rangle = \sum_j f_j(\mathbf{Q}) \frac{1}{2} \langle \mathbf{\mu}_j\rangle e^{i\mathbf{Q}\cdot \mathbf{R}_j} \qquad (3)$$

where μ_j stands for the magnetic moment of the jth atom in units of Bohr magnetons defined as $\mu = g\mathbf{S}$ for spin-only ($\mathbf{S} = $ spin operator) contributions, $\mu = g\mathbf{J}$ for the contribution of total angular momentum ($\mathbf{J}$), or some effective

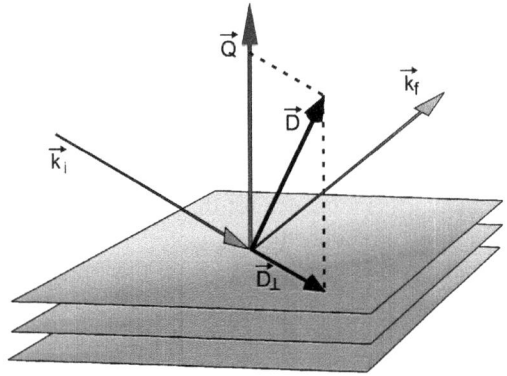

FIGURE 4 Projection of the magnetic interaction vector $\mathbf{D}$ on the diffraction plane, perpendicular to $\mathbf{Q}$.

spin operator in the case of scattering from ions with partially quenched orbital angular momentum. The term $f_j(\mathbf{Q})$ denotes the magnetic form factor, a normalized scalar function ($f_j(0) = 1$) which represents the Fourier transform of the moment distribution over the volume of the atom and describes the $\mathbf{Q}$-dependence of the scattering amplitude from a single ion. When the magnetization arises from electrons of a single open shell with the orbital quantum number l, the magnetic form factor can be expressed as a sum of radial integrals $\langle j_{2k}(Q) \rangle = \int j_{2k}(Q \cdot r)|R_l(r)|^2 r^2 dr$ with $k = 0,1,2,\ldots,l$, where $R_l(r)$ is the radial part of the electronic wave function, and j_{2k}'s are the spherical Bessel functions. In the limit of small momentum transfers $\mathbf{Q}$, the electron density can be treated as spherical in the so-called dipole approximation, and the form factor is defined by:

$$f_j(\mathbf{Q}) = \frac{g_S}{g} \langle j_0(Q) \rangle + \frac{g_L}{g} [\langle j_0(Q) \rangle + \langle j_2(Q) \rangle], \quad (4)$$

where g, g_S, g_L are the Lande g-factors associated with total, intrinsic electronic spin, and orbital, respectively. Although, almost exclusively used in analyzing neutron diffraction data, the dipole approximation does not account for the anisotropy of the magnetic form factors that can be significant for ions with only one or two unpaired electrons [50–52]. When orbital angular momentum is quenched by the crystalline fields, g_L in the above formula can be replaced by "$g - 2$" and the magnetic form factor is given by:

$$f_j(\mathbf{Q}) = \langle j_0(Q) \rangle + \left(1 - \frac{2}{g}\right) \langle j_2(Q) \rangle \quad (5)$$

Another special case is encountered for the rare earths and actinides where the spin–orbit coupling is larger than crystal field splitting and the magnetic moment, in the Russell–Saunders approximation, is determined by the total angular momentum $\mathbf{J} = \mathbf{L} + \mathbf{S}$. Within this approximation, the construction of atomic levels is based on the assumption that the orbital angular momenta of electrons combine to give the total orbital momentum $\mathbf{L}$ and their spins give the total spin $\mathbf{S}$. In such situations, the magnetic form factor in the dipole approximation can be expressed as [53]:

with
$$f_j(\mathbf{Q}) = \langle j_0(Q) \rangle + c_2 \langle j_2(Q) \rangle \quad (6)$$

$$c_2 = \frac{2}{g_J} - 1 = \frac{J(J+1) + L(L+1) - S(S+1)}{3J(J+1) - L(L+1) + S(S+1)}$$

The radial integrals for most of the known magnetic ions have been calculated numerically using ab initio methods such as self-consistent Hartree–Fock and Dirac–Fock calculations [54–57]. They have also been empirically approximated by the sum of the three Gaussians: $\langle j_0 \rangle = A \cdot \exp(-a\zeta^2) + B \cdot \exp(-b\zeta^2) + C \cdot \exp(-c\zeta^2) + D$ and $\langle j_{2k>0} \rangle = \zeta^2 \cdot$

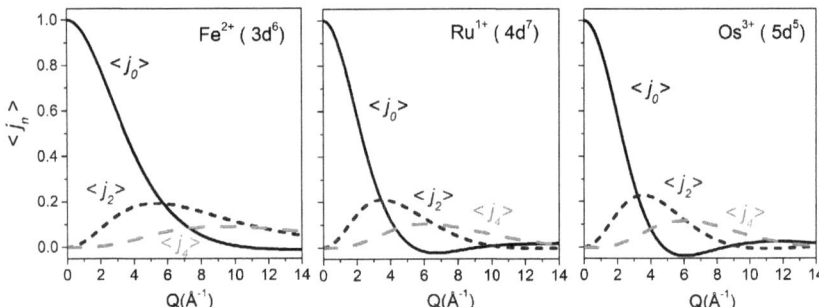

FIGURE 5 The dependence of $\langle j_{2k}\rangle$, $k = 0,1,2$ radial integrals on the momentum transfer for typical magnetic 3d, 4d, and 5d ions, calculated using the three Gaussian approximation.

$(A' \cdot \exp(-a'\zeta^2) + B' \cdot \exp(-b'\zeta^2) + C' \cdot \exp(-c'\zeta^2) + D')$, with $\zeta = Q/4\pi = \sin(\Theta)/\lambda$. The coefficients corresponding to the most common magnetic ions with 3d, 4d, 4f, and 5f electronic configurations have been tabulated by Brown [35], and for the 5d ions by Kobayashi et al. [58]. Figure 5 shows a comparison of the first two radial integrals ($k = 0,1,2$) for selected magnetic ions with 3d, 4d, and 4f electrons, calculated using the aforementioned empirical expressions. As readily visible in the figure, the form factor decreases more rapidly with Q when magnetization density originates from electrons with more expanded shells.

If we consider a periodic magnetic order and define the time-dependent position vector for the jth magnetic moment as $\mathbf{R}_j = \mathbf{R}_l + \mathbf{r}_j + \mathbf{u}_j$, where $\mathbf{R}_l$ refers to the center of the magnetic unit cell, $\mathbf{r}_j$ is the relative position of the magnetic moment inside the unit cell, and $\mathbf{u}_j$ is the instantaneous displacement from the equilibrium position, Eqn (3) becomes:

$$\langle \mathbf{D}(\mathbf{Q})\rangle = \sum_l \sum_j f_j(\mathbf{Q}) \frac{1}{2}\langle \mathbf{\mu}_j\rangle e^{i\mathbf{Q}\cdot(\mathbf{R}_l+\mathbf{r}_j)} e^{-W_j(\mathbf{Q})} = \frac{1}{|\gamma r_0|}\sum_l \mathbf{F}_\mathbf{M}(\mathbf{Q})\langle\mathbf{\mu}_j\rangle e^{i\mathbf{Q}\cdot\mathbf{R}_l} \quad (7)$$

where $W_d(\mathbf{Q}) = \frac{1}{2}\langle(\mathbf{Q}\cdot\mathbf{u}_j)^2\rangle$ is the Debye–Waller factor, and $\mathbf{F}_\mathbf{M}(\mathbf{Q})$ is the magnetic unit cell structure factor defined as:

$$\mathbf{F}_\mathbf{M}(\mathbf{Q}) = \frac{|\gamma r_0|}{2}\sum_j f_j(\mathbf{Q})\langle\mathbf{\mu}_j\rangle e^{i\mathbf{Q}\cdot\mathbf{r}_j} e^{-W_j(\mathbf{Q})} \quad (8)$$

The constant $|\gamma r_0|/2$ that is often identified in various textbooks as $p = 2.696$ fm gives a sense of the magnitude for the magnetic scattering in comparison with the nuclear scattering characterized by a coherent scattering length $|b_c|$ varying from 0.44 to 12.6 fm. In a more general case of a magnetic unit cell containing distinct atoms located in nonequivalent sites (s), j becomes the label of atoms inside the site s and the summation in Eqn (8) extends over both indexes j and s to give:

$$\mathbf{F_M(Q)} = p\sum_{j,s} f_s(\mathbf{Q})\langle \boldsymbol{\mu}_j^s \rangle e^{i\mathbf{Q}\cdot\mathbf{r}_j^s} e^{-W_j^s(\mathbf{Q})} \qquad (9)$$

To define the Bragg condition for the magnetic scattering, we introduce the reciprocal lattice vector $\boldsymbol{\tau} = h\mathbf{a}^* + k\mathbf{b}^* + l\mathbf{c}^*$ ($\boldsymbol{\tau} = 2\pi\mathbf{H}$, in crystallographic notation) where $\mathbf{a}^* = 2\pi(\mathbf{b}\times\mathbf{c})/\upsilon_0$, $\mathbf{b}^* = 2\pi(\mathbf{c}\times\mathbf{a})/\upsilon_0$, and $\mathbf{c}^* = 2\pi(\mathbf{a}\times\mathbf{b})/\upsilon_0$ are reciprocal unit cells vectors, and $\upsilon_0 = \mathbf{a}(\mathbf{b}\times\mathbf{c})$ is the volume of the unit cell. Bragg's law can be expressed conveniently as $\mathbf{Q} = \boldsymbol{\tau}$. However, for a more general situation when the periodicity of the magnetic order does not coincide with the nuclear cell, the Bragg condition for the magnetic scattering becomes $\mathbf{Q} = \boldsymbol{\tau}_\mathbf{M} = \boldsymbol{\tau} \pm \mathbf{k}$, where $\mathbf{k}$ is called the propagation vector or wave vector of the magnetic structure. As it will be detailed in the following section, the $\mathbf{k}$ vector defines the magnetic moment of any atom by associating it to that of a related atom in the nuclear cell (zeroth cell) using a phase relation. Returning our attention to the expression of the differential scattering cross section, the summation over all lattice points gives:

$$\left|\sum_l e^{i\mathbf{Q}\cdot\mathbf{R}_l}\right|^2 = N\frac{(2\pi)^3}{\upsilon_0}\sum_{\boldsymbol{\tau}_\mathbf{M}} \delta(\mathbf{Q} - \boldsymbol{\tau}_\mathbf{M}) \qquad (10)$$

$\boldsymbol{\tau}_\mathbf{M}$ is the reciprocal lattice vector of the magnetic structure, and N represents the number of unit cells, and Eqn (1) becomes:

$$\left(\frac{d\sigma}{d\Omega}\right)_{el} = N\frac{(2\pi)^3}{\upsilon_0}\sum_{\boldsymbol{\tau}_\mathbf{M}} \delta(\mathbf{Q} - \boldsymbol{\tau}_\mathbf{M})|\mathbf{F_{M\perp}}(\boldsymbol{\tau}_\mathbf{M})|^2 \qquad (11)$$

where $\mathbf{F_{M\perp}} = \hat{\mathbf{Q}}\times\mathbf{F_M}\times\hat{\mathbf{Q}} = \mathbf{F_M} - (\mathbf{F_M}\cdot\hat{\mathbf{Q}})\hat{\mathbf{Q}}$ defines the component of the magnetic structure factor perpendicular to $\mathbf{Q}$.

4.3.2 The Magnetic Propagation Vector Formalism

The vector nature of the magnetic moments makes it possible to form magnetic orders for which periodicities are not described by the same unit cell as the nuclear (chemical) lattice. However, a relation between moment orientations of equivalent magnetic atoms located in different nuclear unit cells can always be established by means of a "propagation vector" or "wave vector" $\mathbf{k}$ (as depicted in Figure 6) [59,60]. Given the periodic nature of magnetic structures, the moment distribution $\boldsymbol{\mu}_{lj}$ associated with the Bravais lattice j can be expressed by a Fourier series expansion, as follows:

$$\boldsymbol{\mu}_{lj} = \sum_\mathbf{k} \mathbf{S}_{\mathbf{k}j} e^{i\mathbf{k}\cdot\mathbf{R}_l} \qquad (12)$$

FIGURE 6 Correlation of the magnetic moment on atom j within unit cell "l" to that in the "zeroth" unit cell as a function of the propagation vector, **k**, and the lattice vector $\mathbf{R}_l$.

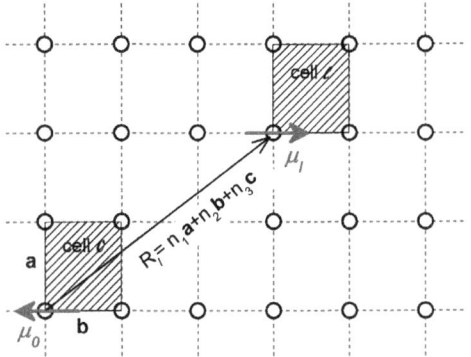

where **k** vectors are confined to the first Brillouin zone of the nuclear unit cell, $\mathbf{R}_l = n_1\mathbf{a} + n_2\mathbf{b} + n_3\mathbf{c}$, ($n_i \in \mathbb{Z}$) denotes the translation vector of the lth unit cell with respect to the zeroth cell, and $\mathbf{S}_{\mathbf{k}l}$'s are complex vectors expressed in the same units as magnetic moments (Bohr magnetons) and have components defined with respect to the unitary frame of the conventional unit cell ($\mathbf{a}/a$, $\mathbf{b}/b$, $\mathbf{c}/c$).

While the majority of magnetic structures involve only a single propagation vector, it is important to realize that there are cases when several **k** vectors can operate concomitantly to form, the so-called, multi-**k** structures (note the summation over **k** in Eqn (12)). In this context, it is useful to bring into discussion the effect of the symmetry elements $g = \{h,\mathbf{t}\}$ of a given space group, $\mathbf{G}_0$, on a particular **k**-vector. When the propagation vector corresponds to a symmetry element, some symmetry operations g of the group $\mathbf{G}_0$ will leave the vector **k** unchanged (or transformed to an equivalent vector, $g\mathbf{k} = \mathbf{k}' = \mathbf{k} + \mathbf{H}$), and those g's form the subgroup $\mathbf{G}_\mathbf{k}$ known as the wave vector group or the "little group." The action of the remaining g's will generate a set of nonequivalent propagation vectors ($\mathbf{k}' \neq \mathbf{k} + \mathbf{H}$), which is termed the "star of **k**." Each individual vector $\mathbf{k}_\mathbf{L}$ belonging to this set is defined as the "arm" of the star, and their total number, $l_\mathbf{L}$, cannot exceed the number of elements of $\mathbf{G}_0$. It is therefore natural to expect the existence of magnetic structures where several arms $\mathbf{k}_\mathbf{L}$ of a star are involved. Distinguishing between single-**k** and multi-**k** cases can pose a significant experimental challenge, as several domains in the crystal also may be ordered according to different single arms (**K**-domains) that give exactly the same contribution to the magnetic scattering as the multi-**k** structure. In such situations, only scattering experiments on single crystals under external uniaxial stress or magnetic field, which may favor the population of a certain **K**-domain over another, can remove the ambiguity [61−63]. Among the simplest examples of indistinguishable multi-**k** structures are those associated with the stars of the wave vectors $\mathbf{k} = (0,0,1/2)$ for the primitive cubic Bravais lattice, shown in

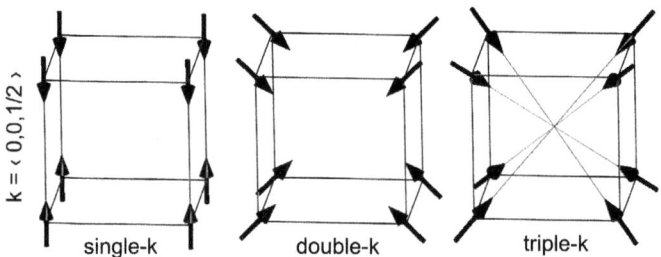

FIGURE 7 Possible multi-k magnetic structures associated with the stars of the propagation vector $\mathbf{k} = (0,0,1/2)$ for the primitive cubic Bravais lattice.

Figure 7. In this case, several magnetic configurations can be defined including a collinear single-k structure with $\mathbf{k} = (0,0,1/2)$, or $(1/2,0,0)$, or $(0,1/2,0)$, and noncollinear structures with double-k, e.g., $\mathbf{k}_1 = (0,0,1/2)$ and $\mathbf{k}_2 = (1/2,0,0)$; or triple-k, $\mathbf{k}_1 = (0,0,1/2)$, $\mathbf{k}_2 = (1/2,0,0)$, and $\mathbf{k}_3 = (0,1/2,0)$. Experimental realization of multi-k magnetic structures can be found in the uranium monopnictides and UO_2 [61–63]. It is generally accepted that the stabilization of multi-k structures requires in addition to single-ion anisotropy, a competition with either anisotropic bilinear or multipolar exchange interactions. This supports the fact that these magnetic structures are more often encountered in rare earth and actinide compounds than in 3d compounds, due to the anisotropic interactions originating from strong hybridization of f electrons with band electrons [63]. In some circumstances, multi-k magnetic structures can also include $\mathbf{k}_L$ arms belonging to different stars. In these structures, the experimental observation of each contributing $\mathbf{k}$ is less challenging because the magnetic reflections will occur at different $\mathbf{Q}$ positions. Such magnetic structures can occur when high-order exchange interactions give rise to noncollinear magnetic structures composed from both ferromagnetic and antiferromagnetic components, e.g., HoP [64] and $DyMn_6Ge_6$ [65] compounds, or in systems with competing magnetic ground states, e.g., Tm_3Ge_4 [66]. Also common is the magnetic ordering according to multiple wave vectors of systems with more than one type of magnetic atom that are weakly coupled, e.g., $PrCo_2P_2$ [67], or display site-dependent anisotropic single-ion magnetism, e.g., $SrHo_2O_4$ [68]. A comprehensive discussion of the multi-k structure problem is given by Rossat-Mignod in Ref. [60].

Special attention must also be paid to the role of propagation vector $\mathbf{k}$ in the Fourier expansion [Eqn (12)]. Excepting for the cases when $\mathbf{k} = (0,0,0)$ or $\tau/2$ (where $\tau = 2\pi\mathbf{H}$ is a reciprocal lattice vector), a center of inversion in the crystal does not keep the $\mathbf{k}$ invariant but it is changed to the nonequivalent arm $-\mathbf{k}$. Both $\mathbf{k}$ and $-\mathbf{k}$ contribute to the star of $\mathbf{k}$, and Eqn (12) can be rewritten as a summation over the total number of pairs $(\mathbf{k},-\mathbf{k})$:

$$\mu_{lj} = \sum_{\mathbf{k}'} \left(\mathbf{S}_{\mathbf{k}j} e^{i\mathbf{k}\mathbf{R}_l} + \mathbf{S}_{-\mathbf{k}j} e^{-i\mathbf{k}\mathbf{R}_l} \right) \quad (13)$$

To obtain a real magnetic moment μ_{lj}, the Fourier components must obey the equality $\mathbf{S}_{-\mathbf{k}j} = \mathbf{S}^*_{\mathbf{k}j}$ where "*" denotes the complex conjugate. Even in the cases where the vector $-\mathbf{k}$ is not present in the star of $\mathbf{k}$, a complex conjugate term has to be added to enable a real moment. Therefore, in crystals without a center of inversion, the contribution from the two different stars $\{\mathbf{k}\}$ and $\{-\mathbf{k}\}$ will be required. The structures where the pair $(\mathbf{k},-\mathbf{k})$ is required to produce real moments should not be confused with the multi-$\mathbf{k}$ structures.

For simplicity, we will restrict ourselves in the following discussion to cases that involve only a single propagation vector $\mathbf{k}$ and its inverse $-\mathbf{k}$. Using the formalism of propagation vectors given in Eqn (12), one can now give a generic classification of the possible magnetic structures. When the vector $\mathbf{k}$ can be expressed as a rational part of a principal vector of the reciprocal lattice, the corresponding magnetic structure is termed "commensurate." In other words, the magnetic cell is a simple multiple of the nuclear cell, as found in the majority of known ferromagnets, antiferromagnets, and ferrimagnets. A different category of magnetic structures is generated by $\mathbf{k}$ vectors that cannot be defined as rational parts of a vector of reciprocal lattice, leading to a magnetic cell that is "incommensurate" with the nuclear cell.

4.3.2.1 Magnetic Structures with $\mathbf{k} = 0$ or $\tau/2$

The simplest types of commensurate magnetic structures are those associated with the propagation vector $\mathbf{k} = (0,0,0)$. The Fourier components need to be real and can be identified in this case with the true magnetic moment: $\mu_{lj} = \mathbf{S}_{0j}e^{i0\cdot\mathbf{R}_l} = \mathbf{S}_{0j} = \mu_{0j}$. The magnetic and nuclear unit cells are of the same size. In terms of scattering cross section (defined by Eqn (11)), the magnetic contributions occur at the nodes of the nuclear reciprocal lattice, $\mathbf{Q} = \tau_M = \tau \pm \mathbf{k} = \tau$, as displayed in Figure 8. For a Bravais lattice (single magnetic atom per primitive cell) $\mathbf{k} = (0,0,0) = 0$ implies a ferromagnetic alignment of the magnetic moments. In contrast, for non-Bravais lattices this case can result in any of ferromagnetic, ferrimagnetic, or antiferromagnetic structures. Furthermore, when the crystallographic unit cell contains more than one magnetic site, even fairly complex noncollinear spin arrangements can be stabilized by the single-ion anisotropy or anisotropic exchange interactions [60,69]. Special attention is required for centered cells, which include atoms defined by centering vector translations, i.e., $\mathbf{t}_c = (1/2,1/2,1/2)$ for a body-centered cell. In this case, the $\mathbf{k} = (0,0,0)$ is not equivalent to $\mathbf{k} = (0,0,1)$, as the former will give a ferromagnet ($\mu_{0t} = \mathbf{S}_0 e^{i0\cdot 1/2} = \mu_0$), while the latter will produce an antiferromagnet ($\mathbf{S}_0 e^{i\cdot 2\pi\cdot 1/2} = -\mu_0$).

The magnetic orderings with the propagation vector corresponding to a high-symmetry point of the surface of the Brillouin zone, $\mathbf{k} = \tau/2$, represent another class of simple commensurate structures. The $\tau/2$ symmetry points for each of the 14 Bravais lattices are reproduced from Ref. [60] in Table 1. This propagation vector gives rise to an alternating sequence of magnetic moments

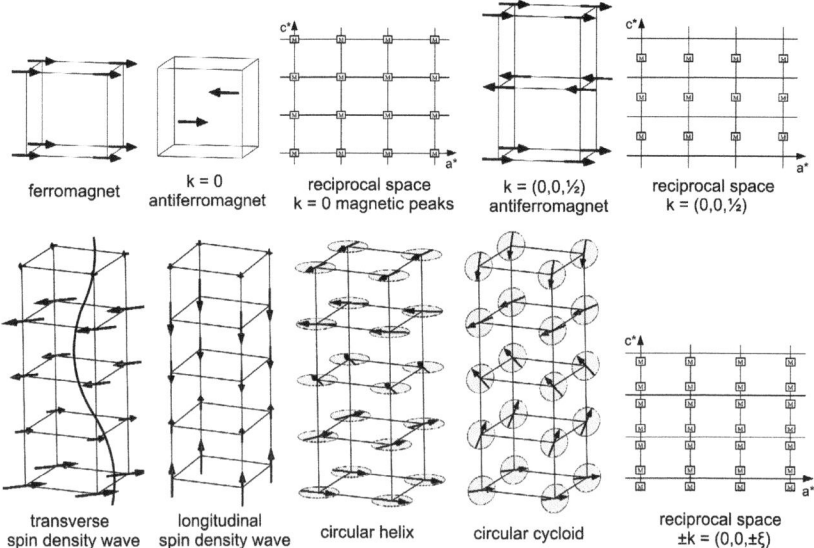

FIGURE 8 Different types of magnetic structures and the positions of the associated magnetic reflections relative to the nodes of the nuclear reciprocal lattice. Drawings of the magnetic structures were made using FPStudio software [81].

pointing in opposite directions when moving along the **k** direction: $\mu_{lj} = \mathbf{S}_{0j} e^{i\pi \cdot \mathbf{H} \cdot \mathbf{R}_l} = \mathbf{S}_{0j} = (-1)^l \mu_{0j}$. The magnetic cell could become twice, four, or eight times larger than the nuclear cell and the magnetic reflection will occur at half integer positions, $Q = \tau_M = \tau/2$, as depicted in Figure 8.

For the most general situation, ($\mathbf{k} \neq 0$ or $\tau/2$), the magnetic-moment distribution is given by Eqn (13), where, as discussed above, the pairs ($\mathbf{k}, -\mathbf{k}$) need to be considered to produce a real moment. If we narrow the calculation to a single pair of vectors, **k** and −**k**, the equation becomes: $\mu_{lj} = \mathbf{S}_{\mathbf{k}j} e^{i\mathbf{k}\mathbf{R}_l} + \mathbf{S}_{-\mathbf{k}j} e^{-i\mathbf{k}\mathbf{R}_l}$. Expanding the two exponentials and applying the condition $\mathbf{S}_{-\mathbf{k}j} = \mathbf{S}_{\mathbf{k}j}^*$, it leads to the expression:

$$\mu_{lj} = 2\mathrm{Re}(\mathbf{S}_{\mathbf{k}j})\cos(\mathbf{k}\mathbf{R}_l) + 2\mathrm{Im}(\mathbf{S}_{\mathbf{k}j})\sin(\mathbf{k}\mathbf{R}_l) \quad (14)$$

Depending upon the explicit form of the complex Fourier coefficients one can conveniently generate several types of magnetic structures:

4.3.2.2 Spin Density Wave or Sine Wave Structures

The Fourier coefficients takes the form: $\mathbf{S}_{\mathbf{k}j} = \frac{1}{2} S_{0j} \widehat{\mathbf{u}}_j e^{i\varphi_j}$, where S_{0j} is the amplitude of the moment in the zeroth cell, $\widehat{\mathbf{u}}_j$ is a unit vector ($|\widehat{\mathbf{u}}| = 1$) giving the polarization direction of the moment, and φ_j is a phase factor. This yields a magnetic moment with a cosine modulation: $\mu_{lj} = 2\mathrm{Re}(\mathbf{S}_{\mathbf{k}j}) \cos(\mathbf{k}\mathbf{R}_l) = S_{0j} \cos(\mathbf{k}\mathbf{R}_l + \varphi_j)\widehat{\mathbf{u}}_j$. Depending on the polarization direction with

TABLE 1 Symmetry Points of the Brillouin Zones of the 14 Bravais Lattices

Bravais Lattice		Symmetry Points k = H/2				
Cubic	P	(0,0,1/2)	(1/2,1/2,0)	(1/2,1/2,1/2)		
	I	(0,0,1)	(1/2,1/2,0)			
	F	(0,0,1)	(1/2,1/2,1/2)			
Hexagonal	P	(0,0,1/2)	(1/2,0,0)	(1/2,0,1/2)		
	R	(0,0,1/2)	(1/2,1/2,0)	(1/2,1/2,1/2)		
Tetragonal	P	(0,0,1/2)	(1/2,0,0)	(1/2,1/2,0)	(1/2,0,1/2)	(1/2,1/2,1/2)
	I ($c > a$)	(0,0,1)	(1/2,1/2,0)	(1/2,0,1/2)		
	I ($c < a$)	(1,0,0)	(1/2,1/2,0)	(1/2,0,1/2)		
Orthorhombic	P	(0,0,1/2)	(1/2,1/2,0)	(1/2,1/2,1/2)		
	C ($a > b$)	(0,0,1/2)	(1,0,0)	(1/2,1/2,0)	(1,0,1/2)	(1/2,1/2,1/2)
	F ($c > a > b$)	(0,0,1)	(1,0,0)	(1/2,1/2,1/2)		
	I ($c > a > b$)	(0,0,1)	(1/2,1/2,0)	(1/2,0,1/2)		
Monoclinic	P	(0,0,1/2)	(0,1/2,0)	(1/2,0,1/2)	(0,1/2,1/2)	(1/2,1/2,1/2)
	B ($c > a$)	(0,1/2,0)	(0,0,1)	(1/2,0,1/2)	(0,1/2,1)	(1/2,1/2,1/2)
Triclinic	P	(0,0,1/2)				

Reproduced from Ref. [60].

respect to the propagation vector, the spin-density waves (SDWs) can be either longitudinal ($\widehat{\mathbf{u}}_j \parallel \mathbf{k}$) or transverse ($\widehat{\mathbf{u}}_j \parallel \mathbf{k}$). Schematic representations of these structures are shown in Figure 8. Examples of sine wave structures are those found in the elementary rare earth metals like Er, Tm, Nd, and Pr [70–75]. It is quite typical for these magnetic structures to remain stable only for a narrow temperature range below the ordering temperature. Upon decreasing temperature, the tendency of moments to saturate may lead to the instability of the modulation that results in the development of harmonics of $\mathbf{k}$ in addition to the fundamental component. For instance, in Er and Tm metals higher-order harmonics ($2\mathbf{k}$, $3\mathbf{k}$, $5\mathbf{k}$) appear with decreasing temperature [72,73]. Other expected behaviors of SDW structures with incommensurate $\mathbf{k}$ vectors are their evolution into helical structures (discussed below) that allow uniform moments, or the shifting of the wave vector toward commensurate values [60,76–78]. In this context, the commensurate structures with $\mathbf{k} = \tau/4$ or $\tau/6$ can be listed under the same category of spin-wave modulated structures, although they may appear as standard antiferromagnetic structures. To clarify this aspect, let us examine the case of wave vector $\mathbf{k} = (0,0,1/4)$ for the primitive cubic Bravais lattice. The moment distribution in this situation can be described by $\boldsymbol{\mu}_l = S_0 \cos\left(\frac{2\pi}{4}\mathbf{R}_l + \varphi\right)\widehat{\mathbf{u}}_j$. For a choice of phase $\varphi = 0$, the magnetic structure consists of $(S_0,0,-S_0,0)$ sequence, while for $\varphi = (2n+1)\pi/4$ the sequence becomes $(S_0/\sqrt{2}, S_0/\sqrt{2}, -S_0/\sqrt{2}, -S_0/\sqrt{2})$. Structures with the $(+,+,-,-)$ sequence are common to rare earth compounds with strong anisotropic exchange interaction as in the actinide monopnictides [63].

4.3.2.3 Helical Magnetic Structures

The Fourier coefficients for this class of magnetic structures are written as complex vectors $\mathbf{S}_{\mathbf{k}j} = \frac{1}{2}S_{0j}(\widehat{\mathbf{u}}_j + i p_j \cdot \widehat{v}_j)e^{i\varphi_j}$ where $\widehat{\mathbf{u}}_j$ and $\widehat{\mathbf{v}}_j$ are orthogonal unitary vectors ($\widehat{\mathbf{u}}_j \cdot \widehat{\mathbf{v}}_j = 0$; $|\widehat{\mathbf{u}}| = |\widehat{\mathbf{v}}| = 1$), and φ_j is the phase factor and p_j is a weighting parameter between the two components. The moment distribution for the sublattice j is given in this case by $\boldsymbol{\mu}_{lj} = S_{0j}[\cos(\mathbf{k}\mathbf{R}_l + \varphi_j)\widehat{\mathbf{u}}_j + p\sin(\mathbf{k}\mathbf{R}_l + \varphi_j)\widehat{\mathbf{v}}_j]$. This indicates that the moment direction rotates within the plane defined by $\widehat{\mathbf{u}}_j$ and $\widehat{\mathbf{v}}_j$ vectors, by an angle defined by $\mathbf{k}(\mathbf{R}_l - \mathbf{R}_{l'})$ when moving from the unit l' to l. If the propagation vector is perpendicular to this plane, the magnetic structure corresponds to a *helix* or *spiral* (see Figure 8). When the $\mathbf{k}$ vector is contained in the $(\widehat{\mathbf{u}}_j, \widehat{\mathbf{v}}_j)$ plane, the structure corresponds to a *cycloid*. Furthermore, the value of the parameter p, gives the so-called "envelope" or the projection of the plane given by the two spin components. For $p = 1$, all j moments will have the same magnitude and the helical structure will have a "circular envelope," whereas for $p \neq 1$ the moment values will be distributed on an ellipse giving an "elliptical envelope." Since on each atomic position the magnetic moment can reach its saturated value, the helical magnetic structures have an increased stability at low temperatures as compared to the sine wave structures. Examples of helical structures are encountered in the elements Ho, Tb, and Dy [75,79].

A distinct case of helical structure is that defined by the commensurate propagation vector $\mathbf{k} = (1/3, 1/3, 0)$, which in a hexagonal structure yields the well-known 120° or triangular spin structure [80]. In a hexagonal Bravais lattice, the moments can be expressed by $\mu_l = S_0 \left[\cos\left(\frac{2\pi}{3}\mathbf{R}_l\right)\mathbf{u} + \sin\left(\frac{2\pi}{3}\mathbf{R}_l\right)\mathbf{v}\right]$, with $\mathbf{u}$ chosen as $[1,0,0]$ and $\mathbf{v}$ orthogonal to $\mathbf{u}$, i.e., $\mathbf{v} = 1/\sqrt{3}\,[1,2,0]$. This generates the following spin arrangement in the basal plane: $\mu_{(00)}\|[1,0,0], \mu_{(10)}\|[0,1,0], \mu_{(11)}\|[-1,-1,0], \mu_{(20)}\|[-1,-1,0]$, and so forth, all with the same magnitude S_0.

For either spin-wave or helical structures, the magnetic reflections will be located at positions associated with both $\mathbf{k}$ and $-\mathbf{k}$ vectors giving rise to satellites surrounding each of the reciprocal lattice points: $(h,k,l)^+$ at $\mathbf{Q} = \tau + \mathbf{k}$, and $(h,k,l)^-$ at $\mathbf{Q} = \tau - \mathbf{k}$ (see Figure 8). In case of contributions from the harmonics of $\mathbf{k}$, additional reflections will occur at positions corresponding to fractions of $\mathbf{k}$.

4.3.3 Symmetry Considerations in Magnetic Structures

To derive a magnetic structure from neutron diffraction data one needs to first identify the propagation vector $\mathbf{k}$ by indexing the magnetic reflections, and to determine the coupling between the Fourier components $\mathbf{S}_{kj}$, their directions and magnitudes, by fitting the integrated intensities to a certain model. The latter task can pose a serious challenge when dealing with a large number of magnetic ions (j) per unit cell, and in that case group theory can play an important role. It is reasonable to expect that the pattern of magnetic interactions of a system retains most of the symmetry properties of the crystallographic structure, and thus by considering the symmetry of the crystal one should be able to systematically generate the possible couplings between Fourier components $\mathbf{S}_{kj}$.

For symmetry considerations a magnetic structure can be represented by an axial vector function $\mathbf{S}(\mathbf{r})$, which is defined for each of the magnetic atoms of the crystal. Under the action of symmetry elements g of a certain group, the function $\mathbf{S}(\mathbf{r})$ may either remain invariant or be transformed into another function $\mathbf{S}'(\mathbf{r})$. This leads to the existence of two different approaches for describing the symmetry of magnetic structures: the magnetic symmetry approach, which in essence consists in defining the set of operations that leave the function $\mathbf{S}(\mathbf{r})$ invariant, and the "representative" analysis approach where one specifies $\mathbf{S}'(\mathbf{r})$ that result from a given $\mathbf{S}(\mathbf{r})$ under all symmetry operations. The first approach, introduced in the 1950s, is based on the conventional symmetry groups to which a nonspace operation $1'$ of spin reversal (aka time-reversal) has been added to produce the so-called magnetic Shubnikov groups [82—88]. This approach has been limited to the description of the invariance symmetry properties for commensurate magnetic structures. To further extend the invariance approach to incommensurate structures, the superspace formalism has recently been employed to produce magnetic superspace groups [89—92]. The alternative approach, representation analysis, is more general and consists on the decomposition of the magnetic configurations into basis functions transforming according to different irreducible representations (IRs) of the wave vector group G_k. This approach proposed in 1958 by Dzyaloshinsky [93,94] and further developed by Bertaut [38,95,96] is now the most widely used to solve both commensurate and incommensurate magnetic structures. We will present next some of the most relevant aspects of the aforementioned approaches, for the most part, following the notations from Ref. [97]. For more thorough reviews of the magnetic symmetry topic, the readers are referred to the book of Izyumov [59,98], as well as to the recent articles by Rodríguez-Carvajal and Bourée [97], Wills [99], Chapon [100], and Perez-Mato et al. [91,92].

4.3.3.1 Shubnikov Space Groups

To describe the invariance of magnetic structure, a new "spin-reversal" operator (aka antisymmetry, antiidentity, or time-inversion) that defines the current loop type symmetry of an axial vector is used. The antiidentity operation was introduced by Heesch [101], but magnetic symmetry is usually termed Shubnikov symmetry, after the crystallographer Shubnikov who rediscovered the antisymmetry concept as a way to expand the classical symmetry groups. This antisymmetry operator, identified by $1'$, can be combined with any conventional symmetry operator h to form a new "primed" operator h'. The effect of such symmetry operators on a magnetic moment described as axial vectors associated with a current loop is exemplified in Figure 9. It is important to emphasize that the magnetic moment μ_j is

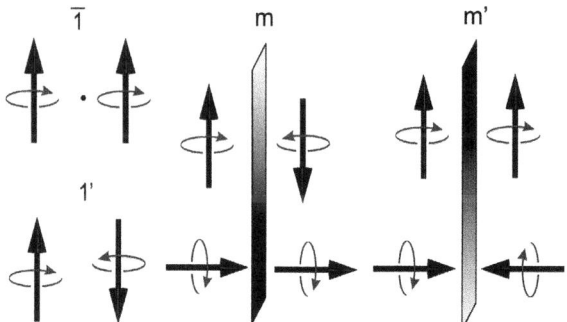

FIGURE 9 Transformations of magnetic moments described as axial vectors associated with a current loop, under the action of inversion, spin-reversal, mirror, and antimirror operators.

transformed only by the rotational part of the operator $g = \{h|t\}$. The resulting moment μ'_j can be mathematically expressed as:

$$\mu'_j = g\mu_j = \det(h)\delta h\mu_j \tag{15}$$

where the determinant $\det(h)$ describes the current loop type symmetry, while the term δ takes the value 1 for unprimed symmetry elements and -1 for the primed ones. The position in the zeroth cell of the transformed moment changes according to the equation:

$$\mathbf{r}'_j = g\mathbf{r}_j = \{h|t\}\mathbf{r}_j = h\mathbf{r}_j + \mathbf{t} = \mathbf{r}_i + \mathbf{a}_{gj} \tag{16}$$

The vector $\mathbf{a}_{gj}$ is called the "returning vector" because it links the transformed position $\mathbf{r}'_j$ outside the zeroth cell to a symmetry equivalent $\mathbf{r}_i$ inside the zeroth cell. For the operator g to be a magnetic symmetry operator, it should leave the moment invariant such as $\mu'_j = \mu_i$. Following the above formulas, notice that the moment is not reversed by the inversion operation $\overline{1}$ (see Figure 9). Furthermore, the mirror operator m leaves the moment invariant only if it aligns perpendicular to the plane of the mirror, whereas the m' leaves invariant only the moment lying within the mirror.

By adding the spin reversal operator to any of the standard rotational operators the number of crystallographic point groups increases from 32 to 122. If one defines the time-reversal group as formed by two elements $\Theta = \{1, 1'\}$, a magnetic group **M** can be obtained as a subgroup of the direct product of **R** with the crystallographic group **G**: $\mathbf{M} \subset \mathbf{G} \otimes \Theta$. The magnetic point groups can be classified into three types. The first type is made by those identical to the 32 crystallographic point groups, not involving the $1'$ operation, termed "single-color" or "colorless." Note that the group nomenclature uses the analogy between the concept of spin-reversal and color change, however, as noted above, color and spin differ in the way the regular group operations act upon them.

The second type, named "gray" groups, consist of the 32 groups containing symmetry elements h in both pure and prime forms, from the construction $\mathbf{M} = \mathbf{G} \cup \mathbf{G}1'$. The presence of a spin-reversal operator in each such group precludes nonzero magnetic moments, and consequently they are also known as "paramagnetic" groups. The third type involves unprimed elements of the subgroup **H** of index 2 of **G** (**H** is a so-called "halving subgroup" of **G**), and the remaining operators **G\H** (read **G** "not" **H**) that are being primed, from the construction $\mathbf{M} = \mathbf{H} \cup (\mathbf{G}\backslash\mathbf{H})1'$. These resulting 58 magnetic groups are called "black-white" or Heesch groups. A lucid presentation on how to expand the crystallographic point groups to include antisymmetry operations is given by Boisen [102]. Because **H** is a halving subgroup, $\mathbf{M} = \mathbf{H} \cup (\mathbf{G}\backslash\mathbf{H})1'$ has the same number of symmetry elements as **G**, and exactly half are antisymmetry operations.

It is quite evident that not all of the colorless and black-white magnetic point groups defined above can be realized in a magnetically ordered system. In that sense, a point group is called admissible if all its operators leave at least one spin component invariant. There are 31 admissible magnetic point groups that are listed together with their admissible moment direction in Table 2 [85,97].

To derive the magnetic lattices, the concept of translation group is generalized to the case of Shubnikov symmetry by considering the antitranslation operation $\mathbf{t}' = \mathbf{t}1'$. Note that this concept replaces in a way the propagation vector formalism defined in the previous section, causing a limitation of the Shubnikov symmetry to the commensurate structures with $\mathbf{k} = (0,0,0)$ and $\tau/2$. The single color magnetic lattices coincide with the 14 conventional Bravais lattices, while the "paramagnetic" lattices do not need to be considered because in that case the crystal is not magnetically ordered. The derivation of black-white translation groups can be done in a similar way as done for the magnetic point groups, by using the subgroups of index 2, $\mathbf{H_L}$ of

TABLE 2 Magnetic Point Groups and Corresponding Moment Directions That Enable magnetic Order

Admissible Magnetic Point Groups					Admissible Moment Directions
1	$\bar{1}$				Any direction
$2'$	$2'/m'$	$m'm2'$			Perpendicular to the twofold axis, and to the unprimed plane for $m'm2'$
m'					Any direction within the plane
m					Perpendicular to the plane
$m'm'm$					Perpendicular to the unprimed plane
$2'2'2$					Along the unprimed axis
2	$2/m$	$m'm'2$			Along the twofold axis
4	$\bar{4}$	$4/m$	$42'2'$		Along the fourfold axis
$4m'm'$	$\bar{4}2m'$	$4/mm'm'$			Along the fourfold axis
3	$\bar{3}$	$32'$	$3m'$	$\bar{3}m'$	Along the threefold axis
6	$\bar{6}$	$6/m$	$62'2'$		Along the sixfold axis
$6m'm'$	$\bar{6}m'2'$	$6/mm'm'$			Along the sixfold axis

Reproduced from Ref. [97].

translation group T: $\mathbf{M_L} = \mathbf{H_L} + (\mathbf{T} - \mathbf{H_L})1'$. These result in 34 black-white Bravais lattices which are listed in Refs [85,98].

In direct correlation with the magnetic point groups and translational groups, one can obtain a total of 1651 Shubnikov space groups that consist of 230 single-color, 230 paramagnetic, and 1191 black-white groups. The latter type of magnetic space groups can in turn be grouped in two categories: 674 Shubnikov groups of the "first kind" where the subgroup of translation is the same as that of the space group, and 517 of "second kind" where the translation subgroup contains antitranslations leading to primitive magnetic unit cells larger than the primitive crystal cells. Such magnetic lattices correspond to those defined by the propagation vector $\mathbf{k} = \tau/2$.

Representations for space-group symmetry operations can be given by the following 4 × 4 matrix,

$$\begin{bmatrix} h_{11} & h_{12} & h_{13} & t_1 \\ h_{21} & h_{22} & h_{23} & t_2 \\ h_{31} & h_{32} & h_{33} & t_3 \\ 0 & 0 & 0 & \pm 1 \end{bmatrix}$$

where

$$R = \begin{bmatrix} h_{11} & h_{12} & h_{13} \\ h_{21} & h_{22} & h_{23} \\ h_{31} & h_{32} & h_{33} \end{bmatrix}$$

is the point-group operation, and the vector $\mathbf{t}$ embodies the location and translation of the space-group operation. The ± 1 in the (4,4) entry denotes either a regular operation (+1) or an antisymmetry operation (−1). The meaning of the vector $\mathbf{t}$ is not always obvious by inspection, yet a simple recipe for constructing and interpreting space group symmetry operations is given in Ref. [103]. The various shorthand notations for space-group symmetry operations, and those adopted in the International Tables, are given by Litvin and Kopský (2011).

There are two notations used in the literature for describing magnetic space groups following Belov−Neronova−Smirnova (BNS) [84] and Opechowski−Guccione (OG) [85]. Both notations are identical for the major part of magnetic space groups except for the second kind black-white magnetic space groups. A list of all magnetic space groups using the OG notation has been compiled by Litvin [87], followed by a reinterpretation in terms of the BNS notation by Grimmer [88]. Other excellent resources are the ISOTROPY Software Suite [104] and Bilbao Crystallographic Server [105−107] that host databases and programs related to crystallographic and magnetic symmetry.

4.3.3.2 Magnetic Superspace Groups

The superspace approach implemented by de Wolff et al. [108] to describe periodic perturbations can be applied to the modulated commensurate or incommensurate magnetic structures that cannot be dealt with using the classical Shubnikov space groups. In addition, it also offers an elegant way to connect the magnetic symmetry of the modulated phase with any tensor property of the crystal, such as ferroelectricity [90,91,109]. Note that the coupling between magnetic and polarization order parameters is largely covered in the Chapter 5 of this volume.

In this approach, each magnetic moment j at position $\mathbf{r}_j$ inside the unit cell l is associated with a modulation function defined along a continuous coordinate $x_4 = \mathbf{k}(\mathbf{R}_l + \mathbf{r}_j)$:

$$\mu_j(x_4) = \mathbf{M}_j e^{ix_4} + \mathbf{M}_j^* e^{-ix_4} \quad (17)$$

One can note from comparing the above equation with the previous definition of moment distribution in Eqn (12) that the $\mathbf{M}_{kj}$ coefficients are related to the Fourier components $\mathbf{S}_{kj}$ by a phase factor that depends on the atom position $\mathbf{r}_j$ inside the unit cell, $\mathbf{S}_{kj} = \mathbf{M}_{kj} \exp(i\mathbf{k} \cdot \mathbf{r}_j)$.

Extension of symmetry invariance to modulated magnetic structures comes down to defining a set of symmetry operators of a superspace group that constrains the form of the modulation functions associated with symmetry-related atoms within the unit cell. Any symmetry operation $\{h,\delta|\mathbf{t}\}$ of the gray magnetic space group of the basic structure changes the modulation functions by an operation-specific translation τ such that the new modulation function becomes $\mu'_j(x_4) = \mu_j(x_4 + \tau)$. It means that the original structure can be recovered by performing an additional translation τ along the internal coordinate x_4. In consequence, the symmetry operators of a magnetic superspace group will have to be defined in a $(3 + 1)$ mathematical space as $\{h,\delta|\mathbf{t},\tau\}$, where the ordinary magnetic space group operations are followed by the translation τ. If $\{h,\delta|\mathbf{t},\tau\}$, belongs to a $(3 + 1)$-dimensional superspace group, the action of h on the propagation vector $\mathbf{k}$ transforms this vector to either $\mathbf{k}$ or $-\mathbf{k}$. For an incommensurate propagation vector, the symmetry relation between the magnetic modulation functions associated with two magnetic atoms related by the operation $\{h|\mathbf{t}\}\ \mathbf{r}_i = \mathbf{r}_j + \mathbf{R}_l$, is

$$\mu_j(\delta_k x_4 + \tau + \mathbf{k} \cdot \mathbf{t}) = \delta \det(h) h\ \mu_i(x_4) \quad (18)$$

where δ_k being either $+1$ or -1 depending on whether h keeps the wave vector invariant or transforms it into $-\mathbf{k}$ [91]. If the magnetic modulations are accompanied by structural modulations, these are described by corresponding atomic modulation functions $\mathbf{u}(x_4)$ constrained by the same superspace group:

$$\mathbf{u}_j(\delta_k x_4 + \tau + \mathbf{kt}) = h\ \mathbf{u}_i(x_4) \quad (19)$$

The symbols of the (3 + 1)-dimensional magnetic superspace groups are similar to the well-established symbols of the nonmagnetic superspace groups tabulated in the International Tables of Crystallography, Vol. C [110]. They include the standard symbol of a gray Shubnikov group, followed by the propagation vector $\mathbf{k} = (\alpha,\beta,\gamma)$ with α, β, γ denoting the irrational components of $\mathbf{k}$, and the intrinsic translations τ associated with each point-group generator. The intrinsic internal translations are represented by lowercase letters $(0, s = 1/2, t = 1/3, q = 1/4, h = 1/6)$ in a one-line symbol, e.g., $Pbnm1'(0,\beta,0)$ s00s. Recent implementation of superspace formalism in computer programs JANA2006 [111] and ISODISTORT [104] allows for the automatic calculation of possible magnetic superspace symmetries for any gray space group and propagation vector. More in depth information on the magnetic superspace approach and specific illustrative examples can be found in the article of Perez-Mato et al. [91,92].

4.3.3.3 Representation Analysis

In the representation analysis method, the magnetic structures are described in terms of the basis functions of IRs of the space group of the propagation vector $\mathbf{G_k}$. This approach is applicable in equal measure to both commensurate and incommensurate magnetic structures. To emphasize the completeness of this approach, E.F. Bertaut stated in one of his introductory articles [96] that: "representation analysis is not only labeling for classifying a structure, but consists mainly of the search for the structure before it is known and of the discussion of the interactions which might explain the final structure model." Indeed, for the past two decades this has been the method most frequently employed for solving new magnetic structures. The literature describing this approach is quite abundant, and we recommend the reviews by Bertaut [38,95,96], Izyumov [59,98], Wills [99], Ballou [112], and Rodríguez-Carvajal [42], just to name a few.

We recall that the group $\mathbf{G_k}$ is formed by a set of symmetry operators that leave the propagation vector invariant, such that $\mathbf{G_k} = \{g \in \mathbf{G} | g\mathbf{k} = \mathbf{k} + \mathbf{H}\}$. The operations of all symmetry elements $g = \{h|\mathbf{t_h}\}$ of the $\mathbf{G_k}$ group on the Fourier components $\mathbf{S}_{\mathbf{k}j\alpha}$ (with $\alpha = 1,2,3$ or x,y,z, and j indexing the symmetry-equivalent points $j = 1,...,n$ inside the unit cell) can be described using a large matrix named the magnetic representation Γ_{mag}. Realize that the atoms occupying nonequivalent sites (different Wyckoff positions) do not mix under the symmetry operators and need to be treated separately. The rearrangement (permutation) of the labels of equivalent atoms under the action of all symmetry operators can be represented by a "permutation representation" Γ_{perm}, whereas the transformations of the axial vector components (α to α') can be represented by an "axial vector representation" Γ_{axial}. The permutation representation is formed by matrices of order n, while the axial representation consists of matrices of maximum dimension 3. Given the presence of two independent effects of the symmetry operations, the magnetic representation can be expressed as the direct product of those two representations: $\Gamma_{mag} = \Gamma_{perm} \times \Gamma_{axial}$. The magnetic representation matrix will have a dimension $3n$ and its components are determined as:

$$\Gamma_{jj'\alpha\alpha'}(g) = e^{i\mathbf{k}\cdot\mathbf{a}_{gj}} \det(h)\, h_{\alpha\alpha'} \delta_{j',gj} \tag{20}$$

where $\delta_{j',gj}$ is the Kronecker delta and has the value 1 when the operator g transforms the atom j into the symmetry-equivalent j', and 0 otherwise. The phase factor $\mathbf{ka}_{gj}$ is required to relate an atomic position shifted outside the zero*th* cell under the action of a symmetry operation, to a translationally equivalent position inside the zeroth cell, i.e., $\mathbf{a}_{gj}$ is the so-called returning vector defined in Eqn (16). The determinant $\det(h)$ gives the sense of the current loop for an axial vector under the action of a rotational operator h.

The magnetic representations are generally reducible and can be decomposed into IRs:

$$\Gamma_{mag} = \sum_{v} n_v \Gamma_v \tag{21}$$

where n_v is the number of times an IR Γ_v appears in the magnetic representation, which can be calculated from the formula:

$$n_v = \frac{1}{n(\mathbf{G_k})} \sum_{g \in \mathbf{G_k}} \chi_{\Gamma_{mag}}(g) \chi_{\Gamma_v}(g)^* \tag{22}$$

Here, n_v represents the order of the group $\mathbf{G_k}$, and $\chi_\Gamma(g)$ are the characters of Γ_v and Γ_{mag} representations. These characters can be computed from the characters of the associated permutation and axial representations, $\chi_\Gamma(g) = \chi_{\Gamma_{perm}}(g) \times \chi_{\Gamma_{axial}}(g)$, which, in turn, are given by the number of position labels remaining unchanged and the traces of the rotational matrices, respectively.

Once the IRs Γ_v are identified, their basis vectors ψ_v can be calculated using the projection operator technique that consists in using axial unit vectors ϕ_α, e.g., $\phi_1 = (1,0,0)$, $\phi_2 = (0,1,0)$, and $\phi_2 = (0,0,1)$, as test functions and projecting from them the part that transforms according to each IR Γ_v. The projection operator formula that will give the atomic component of the basis vector $\psi_v^{j\lambda}$ is:

$$\psi_v^{j\lambda} = \sum_{g \in \mathbf{G_k}} \Gamma_v^{\lambda*}(g) e^{i\mathbf{ka}_{gj}} \det(h) h_{\alpha\alpha'} \delta_{j',gj} \phi_\alpha \tag{23}$$

where $\lambda = 1,\ldots,d_v$, defines the elements of the matrix Γ_v^* of dimension d_v. The formula is applied for each equivalent position j of the crystallographic site, to each element λ of the matrix Γ_v^*. The Fourier components $\mathbf{S}_{\mathbf{k}j}$ that describe the magnetic structure can be expressed as a linear combination of the basis vectors:

$$\mathbf{S}_{\mathbf{k}j} = \sum_{\lambda n} C_v^{\lambda n} \psi_v^{j\lambda n} \tag{24}$$

with v indicating the active IR and n indicating the number of times the IRs appear in the decomposition $n = 1,...,n_v$. The coefficients $C_v^{\lambda n}$, called "mixing coefficients," represent the free parameters of the magnetic structure used to refine the orientation and magnitude of the Fourier components. In the case that more than one IR is involved in the magnetic phase transition, Eqn (23) will contain an additional summation over v. However, according to Landau theory [113], a second-order transition to a magnetically ordered state involves a single IR (single v), which puts a stringent limit on the number of possible magnetic models and the number of parameters that are involved in the magnetic structure refinement. In practice, magnetic structures with magnetic sublattices (single crystallographic site) defined by mixed basis vectors of two independent IRs are rare, but not unseen [59]. These can occur in magnetic states associated with a first-order transition, or resulting from several successive second-order transitions each involving a different representation. Another scenario is that of slightly distorted structures where the energy difference of possible magnetic states (classified according to different IRs of the distorted structure) is small and appears as degenerate levels.

In the case that the magnetic structure involves multiple magnetic sites (s), the calculation of the basis vectors [Eqn (23)] has to be repeated for each index j of a particular site s. The resulting magnetic configurations depend on the relative strength of the intrasite and intersite interactions and can involve one or multiple IRs. The situation of two magnetic sites (A and B sublattices) and the three resulting distinguishing magnetic configurations has been discussed by Rosat-Mignod in Ref. [60]. If the intersite interaction is dominant ($I_{AB} \geq I_A, I_B$) both sites order at a single critical temperature and the basis vectors describing the coupling between the two sublattices must belong to the same IR. The second situation involves one dominant intrasite interaction ($I_A > I_{AB} > I_B$), where two distinct phase transitions could take place but the nonnegligible intersite coupling might determine a polarization of site B by site A, implying that the basis vectors of each site belong to the same representation. The third case corresponds to weak intersite coupling ($I_A \geq I_B > I_{AB}$), in which case the two sites behave independently, each displaying different ordering temperatures and noncoupled basis vectors.

A slight adjustment of the representation analysis described above is required for modulated structures where $\mathbf{k}$ is not kept invariant ($-\mathbf{k} \neq \mathbf{k} + \mathbf{H}$). One recalls that both $\mathbf{k}$ and $-\mathbf{k}$ are required to recover real magnetic moments, but since the little group $G_\mathbf{k}$ excludes the symmetry operations which transform $\mathbf{k}$ in $-\mathbf{k}$, the coupling between the two groups of Fourier components is lost. To restore this coupling, by taking advantage of the existing symmetry operations which transform $\mathbf{k}$ in $-\mathbf{k}$ in the symmetry analysis, one needs to either use the "spin inversion" operator and consider all symmetry elements of the paramagnetic space group as done by Bertaut [96], or to make use of the operator "conjugation" and employing the Wigner corepresentations instead of the usual representations [114–116]. Corepresentations can be constructed from Kovalev's tables [114].

Conveniently, the use of representation/corepresentation analysis is now also facilitated by computer programs that have been developed to calculate the irreducible matrix representations and the associated basis vectors. The IRs can be calculated ab initio (e.g., the KAREP program [117]) or taken from Kovalev tables [114]. Among the most frequently used computer programs are SARAh [44,45], MODY [118], BasIreps [81], and ISOTROPY [104]. A suite of programs is also available in the Bilbao Crystallographic Server [105–107]. An extensive list of examples illustrating the use of representation analysis can be found in Refs [59] and [97].

As a final remark we would like to point out that a relation between the representation theory and magnetic Shubnikov groups is given by the Niggli–Indenbom theorem which states that "the magnetic space groups correspond to real one-dimensional irreducible representations of conventional crystallographic space groups." This emphasizes the generality of the representational theory approach over the Shubnikov magnetic groups, by offering a wider framework capable of describing not only commensurate but also complex incommensurate magnetic structures.

4.4 PRACTICAL ASPECTS IN DETERMINATION OF MAGNETIC STRUCTURES

4.4.1 Steps in the Determination of Magnetic Structures

In the quest towards determining the magnetic structure of a material, one has to begin with a well-defined structural model, including the lattice parameters, crystal symmetry, atomic positions and occupancies, as well as associated atomic displacement parameters. In addition, other factors contributing to the scattering intensity such as sample absorption, extinction parameters, and sample twinning have to be well defined. In this sense, neutron data collected in the paramagnetic state is a typical starting point. It is also prerequisite that the sample is well characterized macroscopically by magnetization and specific-heat measurements to acquire a basic understand of the magnetic phase diagram. The presence of magnetic impurities in the sample can mislead the proper indexing of the magnetic peaks. It is possible that depending on the details of the structure factors involved, the magnetic scattering may be stronger than the nuclear and, therefore, the magnetic peaks of an impurity may only be seen at low temperature.

Once the aforementioned considerations are addressed and neutron diffraction data of the magnetically ordered material becomes available, four basic steps need to be undertaken:

- Identify the propagation vector **k**.
- Explore the symmetry-allowed couplings between the Fourier components S_{kj} inside the unit cell by performing symmetry analysis.

- Select the best physically meaningful models that are compatible with neutron and magnetization data.
- Refine the direction and the amplitude of the Fourier components.

The identification of the propagation vectors **k** and the proper indexing of the magnetic reflections may be seen as the most important, and potentially, challenging. This is particularly problematic when dealing with incommensurate magnetic order and only powder diffraction data are available, due to the limited information caused by powder averaging. For simple commensurate propagation vectors, the indexing can be done intuitively, or by successive trials, or using graphical methods as described by Roisat-Mignod [60]. Computer-generated stereograms showing circles produced by the intersection of pairs of spheres with radii $\mathbf{d}*(=h\mathbf{a}* + k\mathbf{b}^* + l\mathbf{c}* \pm \tau)$ drawn about different reciprocal-lattice points were used in 1974 by Knapp et al. [119] to index neutron powder diffraction peaks from spiral antiferromagnets. A more advanced computer program (DISPIRAL) capable of exploring commensurate and incommensurate trial propagation vectors following grid searches was later developed by Wilkinson [120]. At present, computer programs that allow searching for the **k** vectors by performing grid searches or nonlinear procedures are part of data analysis packages, e.g., K-SEARCH program included in the Fullprof Suite [81]. However, such routines encounter difficulties for situations where only few magnetic peaks are available, or where several symmetry-unrelated **k** vectors are involved in the order. To address this problem, another method has been recently introduced by Wills [121] that is based on reverse Monte Carlo (RMC) refinement of moment orientations for trial propagation vectors corresponding to the different high-symmetry points, lines, and planes in the Brillouin zone of the crystal structure before the ordering transition. This procedure has been implemented in the program SARAh [44,45] through a metaprogram structure with the Rietveld refinement program FullProf [81]. Regardless of the method used, the solutions for incommensurate wave vectors obtained from powder diffraction data are often not unique and additional single-crystal studies can be pivotal in the further modeling of magnetic structure. In contrast to powder measurements, single-crystal data leave no ambiguities in the wave vector determination especially when using modern instruments, with large position sensitive detectors (PSDs) or quasi-Laue diffractometers that allow the three-dimensional mapping of reciprocal space.

Once the propagation vector is determined, the next step is to perform a symmetry analysis of the possible couplings between the Fourier components $\mathbf{S}_{kj}$ inside the unit cell. This analysis will significantly limit the number of trial structures to those that are compatible with the propagation vector and crystal space group before the phase transition [122]. Such symmetry constraints will not only make the process of magnetic structural determination more rigorous but it will also allow reducing the necessary number of free parameters to be refined. Details of the approaches used for describing symmetry of the magnetic structure, namely the representation/corepresentation analysis and magnetic space groups and superspace groups, are given in Section 4.6.3. A number of computer programs have been developed that allow the unspecialized user to perform these calculations automatically, by providing the space group, magnetic atom positions and the propagation vector. At present, the most frequently used programs for carrying out representation and corepresentation analysis are SARAh [44,45], BasIreps [81], and MODY [118]. Conveniently, the SARAh and BasIreps programs are integrated with common refinement codes to enable the rapid examination of the compatibility of magnetic structure models with experimental data. The superspace formalism is implemented in JANA2006 [111] and the ISOTROPY Software Suite [104]. A large number of resources for symmetry analysis are also available at the Bilbao Crystallographic Server [105–107]. A review of the symmetry-based computational tools for magnetic crystallography is given in Ref. [92]. Some of the aforementioned programs are also capable of integrating the results of the group theory (i.e., representation and corepresentation analysis) with the magnetic Shubnikov group and superspace group descriptions.

The remaining two steps in magnetic-structure determination can be described together within the overall goal of establishing the direction and the amplitude of the Fourier components after the symmetry-allowed models have been identified. A selection of suitable models can be made by a trial-and-error process. However, a careful inspection of the magnetic Bragg intensities done beforehand can be instrumental in reducing the list of reasonable possibilities. One should always remember that the magnetic scattering is only given by the components of the moments that are perpendicular to the scattering vector and, therefore, the presence of systematic absences can help to deduce the main direction of the magnetic moments. Additionally, specific sets of systematic absences or extinction rules caused by symmetry elements can help to determine the appropriate magnetic space group. The computer program MAGNEXT [106] available from the Bilbao Crystallographic Server can be used to obtain the list of the magnetic space groups compatible with a particular set of observed systematic absences, or to list all systematic absences for any chosen Shubnikov magnetic space group.

At last, the refinement of the directions and magnitudes of the magnetic moments from a specific model can be performed by using one of the available software packages: FULLPROF [42,81], JANA [43,111], GSAS [41], and RIETAN-2000 [123]. Besides the conventional least-squares refinement, alternative algorithms such as RMC and simulated annealing have been introduced to avoid (false) local minima, commonly associated with a large number of parameters and limited experimental observations. In FULLPROF, the simulated annealing (SAnn) method can be used on either integrated intensities or profile intensities. For complex magnetic structures where multiple solutions cannot be distinguished by traditional unpolarized neutron diffraction methods, the use of polarized neutrons with spherical polarization analysis is needed [49,124].

4.4.2 Limitations of Neutron Scattering for Complex Magnetic Structures

Neutron powder diffraction is the simplest method to use at first and often is sufficient for solving a magnetic structure. The limitations arising for powder data are due to powder averaging, i.e., only the modulus of the scattering vector is measured and not the direction, so it can be sometimes difficult to correctly assign the propagation vector. In addition, the directions of the moments within the basal plane of a uniaxial crystal (trigonal, tetragonal, or hexagonal) cannot be determined and no orientational information at all can be obtained from crystals with cubic symmetry. Domain information is also lost. Sample size requirements are much larger than that for X-ray diffraction, but the minimum sample size for neutron diffraction continues to shrink with increases in source brightness and improved neutron optics and detectors (see Chapters 2 and 3 of Ref. [2]).

In contrast, neutron single-crystal diffraction allows reciprocal space exploration in three dimensions and offers much greater sensitivity to small magnetic moments, leaves no ambiguities in the propagation vectors, and allows separation of **k**-domains from different sets of magnetic reflections. Neutron single-crystal diffraction when combined with polarization analysis is the most powerful technique to determine complex magnetic structures. The limitations arising for single-crystal data are due to orientation domains when present, that act in the same way as powder averaging; it can be difficult to properly model the domain populations. Sample size requirements (~ 0.5 mm^3) for neutron scattering measurements are still large compared to X-ray single-crystal diffraction. Extinction effects can be a problem as well when crystals are too perfect or ideally imperfect [125,126]. Primary and secondary extinctions manifest by reduced intensities in crystals affected by multiple scattering events. They are difficult to predict and correct before the observed and calculated intensities can be compared. As such, it is essential that good characterization of the single crystal's extinctions be done in the paramagnetic state.

A general limitation of neutron diffraction is the lack of sensitivity to extremely small moments, as compared to some other more sensitive methods like muon spin spectroscopy [127]. Multi-**k** structures and multiple single-**k** structures will be indistinguishable and will require applied magnetic fields and/or uniaxial stress to unravel them. A more general shortcoming is the well-known crystallographic phase problem, here as it applies to the magnetic Bragg peaks. See the example in Section 4.3.2 for the case of wave vector

$\mathbf{k} = (0,0,1/4)$ for the primitive cubic Bravais lattice. The moment distribution can be described by $\mu_l = S_0 \cos\left(\frac{2\pi}{4}\mathbf{R}_l + \varphi\right)\hat{\mathbf{u}}_j$. For a choice of phase $\varphi = 0$, the magnetic structure consists of a $(S_0, 0, -S_0, 0)$ sequence, whereas for $\varphi = (2n+1)\pi/4$ the sequence becomes $(S_0/\sqrt{2}, S_0/\sqrt{2}, -S_0/\sqrt{2}, -S_0/\sqrt{2})$.

4.4.3 Standard Description for Magnetic Structures

As the interest in magnetic structures continues to grow and the development of user-friendly analysis tools enables more investigators to generate structure solutions, the need for a standard description of magnetic structure becomes evident. The lack of such uniform description has made it impossible to date to organize the existing magnetic structures into a database and the transfer of information has been largely ineffective. On the other hand, many of the early reported structures obtained by trial and error without symmetry considerations are incomplete or even incorrect. To address such concerns, a Commission on Magnetic Structures of the International Union of Crystallography [128] was established in 2011. The commission's mandate covers the following: (1) establish standards for the description and dissemination of magnetic structures; (2) develop CIF standards for magnetic structures and promote their use in crystallographic software; (3) develop a database for magnetic structures based on the sharing of magnetic CIF files; (4) cooperate in establishing and maintaining standards of common interest, such as magnetic symmetry-group tables, magnetic nomenclature, and magnetic form factor data. H.T. Stokes and B.J. Campbell extended the CIF dictionaries and produced in early 2014 a first mCIF format [104] that includes the most important aspects of a magnetic structure dictionary: the magnetic space group number based on D. Litvin's tables [129], the propagation vector, crystal and magnetic symmetry operators, as well as the explicit magnetic cell (according to the crystal cell and $\mathbf{k}$-vector) with all the magnetic atoms and the associated moments (in Bohr magneton units) with their projections onto the possibly nonorthogonal axes of the unit cell. This standard format is now in the process of being adopted by all the analysis packages. J.M. Perez-Mato and colleagues have started an online database of magnetic structures based on this new magnetic mCIF standard at the Bilbao Crystallographic Server.

4.5 HIERARCHY OF MAGNETIC STRUCTURES (IMPORTANT EXAMPLES)

While a classification of the magnetic structures can be obtained using propagation-vector criteria, as given in Section 4.3, a more physically sound hierarchy has to take into account the various terms contributing to the spin Hamiltonian that governs the magnetic properties of the system.

An effective spin Hamiltonian can primarily be expressed as a sum of single-spin terms, such as magnetic field coupling $(\mathcal{H}_\mathbf{H} = -\mathbf{H}\sum_i g_i \mu_B \mathbf{S}_i)$ and single-ion anisotropy $(\mathcal{H}_{an} = -\frac{1}{2}\sum_i D S_{iz}^2)$, and two-spin terms which comprise isotropic exchange interactions $(\mathcal{H}_{ex} = -\sum_{ij} J_{ij} \mathbf{S}_i \cdot \mathbf{S}_j)$, anisotropic interactions $(\mathcal{H}_{an-ex} = -\sum_{ij} J_{ij}^{xy}(S_{ix} \cdot S_{jx} + S_{iy} \cdot S_{jy}) + J_{ij}^z(S_{iz} \cdot S_{jz}))$, antisymmetric Dzyaloshinskii–Moriya exchange $(\mathcal{H}_{DM} = -\sum_{ij} \mathbf{D}_{ij} \mathbf{S}_i \times \mathbf{S}_j)$, and dipolar interactions $\left(\mathcal{H}_{dip} = \sum_{ij} \frac{(g\mu_B)^2}{r_{ij}^3}[3(\mathbf{r}_{ij} \cdot \mathbf{S}_i)(\mathbf{r}_{ij} \cdot \mathbf{S}_j) - \mathbf{S}_i \cdot \mathbf{S}_j]\right)$. Three or four-spin potentials are more rarely encountered.

The most important contribution to the Hamiltonian comes from the isotropic exchange interaction term which has a bilinear form $(\mathcal{H}_{ex} = -\sum_{ij} J_{ij} \mathbf{S}_i \cdot \mathbf{S}_j)$ and is invariant under rotations in spin space. The exchange interactions can occur through various mechanisms and may extend beyond first neighbors along any of the directions of the crystal structure. A long-range magnetically ordered state can be achieved through optimum balance of all the involved exchange interactions, with the effective dimensionality of the spin lattice playing an important role. Here, we refer to dimensionality as a measure of the number of spatial dimensions presenting relevant exchange couplings. In general, the dimensionality of magnetic correlations is determined by the underlying crystal structure. As such, a two-dimensional magnetic lattice is typically associated to a spatially anisotropic crystal structure that consists of sheets of atoms that are separated by a weak bonding, whereas a quasi-one-dimensional magnetic lattice is formed in crystal structures containing infinite chains of atoms with strong intra-chain, and negligible inter-chain bonding.

Besides the isotropic exchange interaction, all other terms of the spin Hamiltonian can be present as perturbations, which nevertheless in certain conditions may become decisive in the selection of the magnetic ground state. All perturbative terms, excluding the dipolar interactions, are anisotropic and depend directly on the local rotation symmetry of the magnetic ion. For instance, the magnetic-field coupling $\mathcal{H}_\mathbf{H} = -\mathbf{H}\sum_i g_i \mu_B \mathbf{S}_i$ is dependent on the direction of the field, $\mathbf{H}$, as well as on the g-tensor of site i, which follows the rotational symmetry of the site. Other anisotropic terms like the single-ion anisotropy, the Dzyaloshinskii–Moriya and the anisotropic exchange are due to spin-orbit coupling and can put rigorous constraints on the orientation of magnetic moments.

With most of the magnetic ground states being realized primarily through the competition between multiple isotropic interactions (nearest-neighbors (NN), or next-nearest-neighbors (NNN) interactions), it seems that the most suitable classification of magnetic structures should be based on the effective dimensionality of the magnetic lattice. In the following, we will present a series of examples of magnetic orderings occurring in three-dimensional, layered, and quasi-one-dimensional systems. We will concentrate here on long-range ordered structures, and will not expand on the other possible exotic disordered magnetic states such as the spin ice, spin glass, and spin liquid. The drawings of the magnetic structures presented in this section have been produced using the VESTA [130], Fpstudio [81], and Diamond [131] programs. Some of the magnetic structures have been obtained from the magnetic structure database at the Bilbao Crystallographic Server [105,106].

4.5.1 Three-Dimensional Networks

Among the most-studied crystalline systems with uniformly distributed interactions, one can enumerate the perovskites, double perovskites, pyrochlores, spinels, and garnets. All of these crystal structure types have extremely diverse chemistries, and in each case the high-symmetry parent structures are cubic. The perovskite family with the general formula ABX_3 is without doubt the most popular due to a large variety of physical properties and important potential applications. The cubic perovskite structure consists of BX_6 octahedra, which share corners infinitely in all three dimensions, and A ions located in the 12-fold cavities in between B-site polyhedra. The magnetic ions are assembled to form a simple cubic lattice. Pioneering work on characterization and classification of the magnetic orderings encountered in perovskites was conducted by Wollan and Koehler in 1955 [32]. That work has formed the basis for further exploration of magnetic structures not only for perovskite-related structures but also for other types of structures with similar magnetic-ordering schemes. To date, the magnetic structures of perovskites have been extensively reviewed in many textbooks, including those focused on neutron scattering by Izyumov [59] and Chatterji [49]. Due to space constraints, we will not discuss here the perovskites, but rather concentrate on structures in which the magnetic ions form triangles or tetrahedra, as shown in Figure 10. The tetrahedron represents the quintessential framework for the three-dimensional frustrated magnetic lattices and the materials based on this type of geometry are expected to display complex magnetic orders [132].

4.5.1.1 Corner-Sharing Tetrahedral Lattices in Pyrochlore Oxides

The pyrochlore oxides with the general formula $A^{3+}_2B^{4+}_2O_7$, where A is typically a rare earth ion and B a transition metal, consist of two distinct interpenetrating lattices (A,B) of corner-sharing tetrahedra. In the presence of strong geometrically frustrated NN antiferromagnetic interactions, the NNN exchange interactions, antisymmetric Dzyaloshinskii–Moriya exchange, dipolar interactions, and magnetoelastic couplings are all relevant in the development of the spin–spin correlations and lead to the formation of a broad diversity of unconventional magnetic and thermodynamic behaviors. In

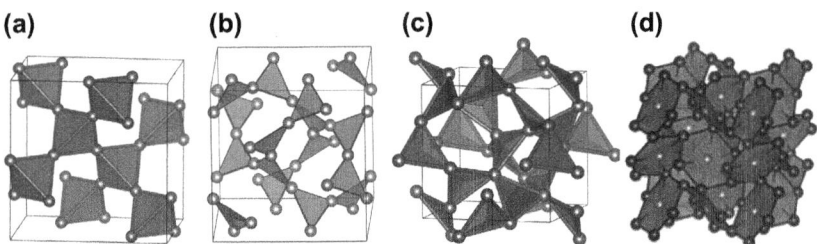

FIGURE 10 Near-neighbor connectivities in some cubic structures that can promote magnetic frustration: (a) spinel B-site or pyrochlore B-site or Laves Cu_2MgCu-sites; (b) garnet A-site; (c) β-Mn Mn_2-sites; (d) bixbyite M-sites.

addition to the long-range magnetic orderings, which will be highlighted here, the other possible magnetic ground states exhibited by pyrochlores include the spin-liquid, spin-glass, or disordered spin-ice states. The properties of these materials have been widely discussed in numerous review articles such as those by Subramanian et al. [133], Greedan [134], and Gardner et al. [135]. The majority of the magnetically ordered states found in pyrochlores are defined by a propagation vector $\mathbf{k} = (0,0,0)$. Representational analysis calculations can be performed for each of the magnetic sites to determine the symmetry-allowed magnetic configurations that can result from a second-order magnetic phase transition, given the $Fd\bar{3}m$ crystal symmetry and the aforementioned magnetic wave vector. Calculations carried out using the program SARAh-coRepresentational Analysis [136] indicate that the magnetic representation of the crystallographic site A^{3+} occupying the $16d$ crystallographic site can be decomposed in terms of four IRs: $\Gamma_{mag}(A) = 1\Gamma_3^{(1)} + 1\Gamma_5^{(2)} + 1\Gamma_7^{(3)} + 2\Gamma_9^{(3)}$. The IRs are labeled according to the numbering scheme of Kovalev [114], with the superscript representing the order of the representation. The associated basis vectors are listed in Table 3 and are depicted in Figure 11, for the four symmetry-related ions that form a tetrahedron. The remaining A^{3+} positions are related to these four by the four cell-centering translations. Similar basis vectors can be generated for the crystallographic site $16c$, occupied by the B^{4+} ions.

As we will see in the following, several of these possible magnetic configurations are encountered in the pyrochlores studied to date, their selection depending on the nature of the perturbative terms contributing to the spin-Hamiltonian. As the magnetic phase transitions in pyrochlores have been shown to be of the first order, one could also expect orderings according to several IRs. It is however commonly found that the Fourier components of magnetic moments are defined as linear combinations of basis vectors that belong to a single representation, meaning that the terms which drive the transition to being of first order are relatively weak.

TABLE 3 The Basis Vectors (BV) Corresponding to the 16d Sites of the Pyrochlore Structure, Defined by Space Group $Fd\bar{3}m$ and Propagation Vector k = (0,0,0)

IR	BV	Atom 1 (1/2,1/2,1/2)	Atom 2 (1/2,1/4,1/4)	Atom 3 (1/4,1/2,1/4)	Atom 4 (1/4,1/4,1/2)
Γ_3	ψ_1	(1,1,1)	(1,-1,-1)	(-1,1,-1)	(-1,-1,1)
Γ_5	ψ_2	(1,-1,0)	(1,1,0)	(-1,-1,0)	(-1,1,0)
	ψ_3	(1,1,-2)	(1,-1,2)	(-1,1,2)	(-1,-1,-2)
Γ_7	ψ_4	(0,-1,1)	(0,1,-1)	(0,1,1)	(0,-1,-1)
	ψ_5	(1,0,-1)	(-1,0,-1)	(-1,0,1)	(1,0,1)
	ψ_6	(-1,1,0)	(1,1,0)	(-1,-1,0)	(1,-1,0)
Γ_9	ψ_7	(1,1,0)	(-1,1,0)	(1,-1,0)	(-1,-1,0)
	ψ_8	(0,0,1)	(0,0,1)	(0,0,1)	(0,0,1)
	ψ_9	(0,1,1)	(0,-1,-1)	(0,-1,1)	(0,1,-1)
	ψ_{10}	(1,0,0)	(1,0,0)	(1,0,0)	(1,0,0)
	ψ_{11}	(1,0,1)	(-1,0,1)	(-1,0,-1)	(1,0,-1)
	ψ_{12}	(0,1,0)	(0,1,0)	(0,1,0)	(0,1,0)

IR, irreducible representation; BV, basis vectors.

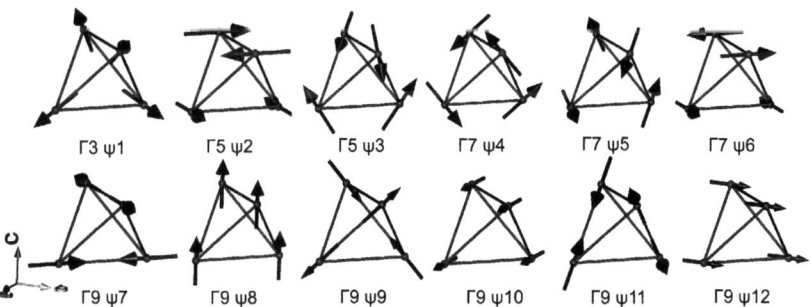

FIGURE 11 The magnetic structures according to various basis vectors of irreducible representations associated with the space group $Fd\bar{3}m$ and propagation k = (0,0,0).

As a first example, we consider the $Gd_2Sn_2O_7$, for which the Gd sublattice (A^{3+} site) orders magnetically below a temperature of 1 K. Analysis of neutron diffraction measurements carried out by Wills et al. [137] have revealed a magnetic structure with Gd moments parallel to the tetrahedral

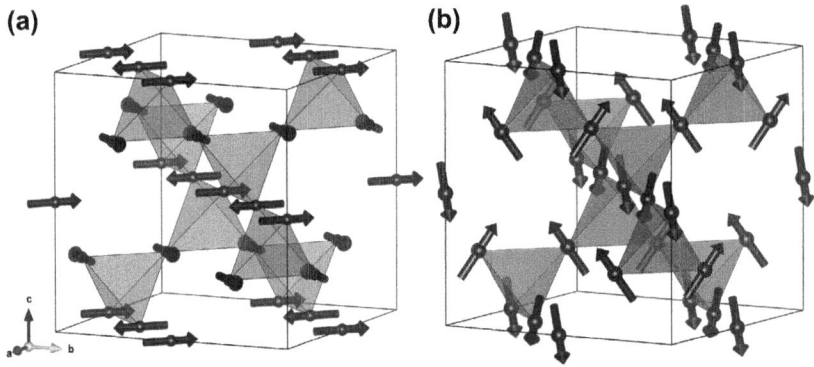

FIGURE 12 Magnetic structures of (a) $Gd_2Sn_2O_7$ and (b) $Er_2Ti_2O_7$ described by the basis vectors of the irreducible representations Γ_7 and Γ_5, respectively.

edges, as displayed in Figure 12(a). This model is defined by using the basis vector ψ_5 of the IR Γ_7 and is similar to that predicted theoretically by Palmer and Chalker [138] for a Heisenberg pyrochlore antiferromagnet with dipolar interactions. A quite different magnetic behavior is found in the analogous titanate $Gd_2Sn_2O_7$, which displays two magnetic transitions, at ~ 0.7 and 1 K, to structures with the ordering vector $\mathbf{k} = (1/2,1/2,1/2)$ [139,140]. The difference in the magnetic structures observed in these similar Gd compounds is likely due to a different relative magnitude in the two types of third-neighbor super-exchange interactions on the pyrochlore lattice [137]. This difference emphasizes the extreme sensitivity of the magnetic ground state to small adjustments in the spin—spin interactions driven by small changes in composition in these highly frustrated materials. Another kind of magnetic structure has been reported for $Er_2Ti_2O_7$, which orders below a Néel temperature of 1.173 K with the propagation vector $\mathbf{k} = (0,0,0)$ [141]. $Er_2Ti_2O_7$ has been proposed as a realization of the XY pyrochlore antiferromagnet with dipolar interactions, with the Er^{3+} moments lying perpendicular to the $<111>$ local axes. The refinement of the magnetic structure indicated that only the two basis vectors ψ_2 or ψ_3 of the IR Γ_5 or a superposition of both is consistent with the observed magnetic intensities. The refined moment of 3.0 μ_B at 50 mK shows a large deviation from the 9.59 μ_B expected from the free-ion term, suggesting the presence of large quantum fluctuations. To confirm the magnetic structure model, and determine the ratio of basis-vector contributions within the Γ_5 representation, a spherical neutron polarimetry experiment was undertaken [141]. The results revealed that the ordered state is described almost exclusively by the ψ_3 basis vector, and further evidences $Er_2Ti_2O_7$ as a good model $<111>$ XY pyrochlore antiferromagnet. The corresponding spin configuration of $Er_2Ti_2O_7$ is shown in Figure 12(b).

The ruthenate perovskite $Ho_2Ru_2O_7$ is a typical example of coexistence of two magnetic sublattices since both Ru^{4+} and Ho^{3+} ions are magnetically active. Neutron scattering measurements on this compound have revealed two transitions to long-range ordered states at 95 and 1.4 K. The first transition involves just the Ru^{4+} ions, while the second involves both Ru and Ho sublattices [142]. The neutron diffraction pattern measured at 20 K has been well described by an ordering of the Ru moments according to the IR Γ_9, which has six associated basis vectors. A linear combination of two of them (e.g., ψ_7 and ψ_8) is needed to describe a canted spin-ice-like arrangement (i.e., "two-in-two-out" spin configuration canted off the local <111>) to produce a predominant ferromagnetic component along the c-direction. Diffraction data measured below 1.4 K were fitted by assuming that both the Ru and Ho moments order according to the representation Γ_9, with the refined moments of 1.8(6) μ_B for the Ru and 6.3(2) μ_B for the Ho. Both magnetic sublattices display a canted spin-ice-like configuration with the uncompensated spin-projections pointing in opposite directions, as illustrated in Figure 13.

Considering the delicate balance between competing energy scales in these materials, intuition suggests that an applied magnetic field may significantly alter their magnetic ground state. Many of the pyrochlore oxides have been subjected to an applied magnetic field and in most cases dramatic effects have been observed. For instance, new metastable states, magnetization plateaus, and slow spin dynamics have been observed in the spin-ice systems $Ho_2Ti_2O_7$ and $Dy_2Ti_2O_7$ by applying small fields along the [0,0,1] or $[1,\bar{1},0]$ directions. Ferromagnetic-like magnetic structures have been induced in the spin liquid $Tb_2Ti_2O_7$, with the details of the spin configuration depending strongly on the direction of the applied field. In this context, it is convenient to look at the pyrochlore lattice as formed by two sets of orthogonal magnetic chains: one running parallel to the [1,1,0] direction and the second running along $[1,\bar{1},0]$, referred to as α and β chains, respectively. A magnetic field applied parallel to the [1,1,0] direction will have a different effect on the spins located on the two chains. For describing the induced magnetic order, a subgroup of the $Fd\bar{3}m$ space group, such as $Fdd2$, needs to be used that allows splitting of the 16d site into two different subsets corresponding to the α and β chains. In the case of $Tb_2Ti_2O_7$, a magnetic field of 1 T gives rise at 0.3 K to a local spin-ice order

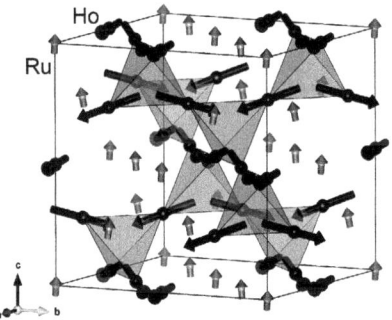

FIGURE 13 Magnetic structure of the $Ho_2Ru_2O_7$ pyrochlore at 1.5 K.

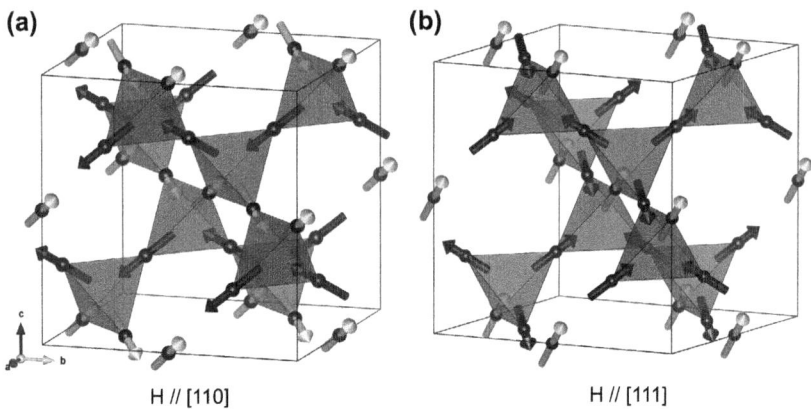

FIGURE 14 Field-induced magnetic ordering in Tb$_2$Ti$_2$O$_7$ pyrochlore, with the external magnetic field oriented along the [1,1,0] (a) and [1,1,1] (b) directions.

(two-in-two-out) but with Tb moments of the α and β chains exhibiting different magnitudes [143]. The α moments reach 5.7(3) μ_B, while the β moments remain smaller than 2 μ_B. A view of the magnetic structure is shown in Figure 14(a). The β moments remain perpendicular to the field direction up to about 3 K, above which they reorient abruptly along the field. An antiferromagnetic-like structure with propagation vector $\mathbf{k} = (0,0,1)$ has also been observed above applied fields of 2 T and below 2 K.

Just as the applied field along the [1,1,0] direction induces a magnetic breakup of the cubic pyrochlore into a set of quasi-one-dimensional chains, a [1,1,1] field leads to a decomposition of the pyrochlore lattice into weakly coupled kagomé planes with their normal along the [1,1,1] direction. The field applied along the [1,1,1] leads to a magnetic structure with lower symmetry than the underlying crystal structure and, therefore, none of the four representations predicted for the $Fd\bar{3}m$ with $\mathbf{k} = (0,0,0)$ were compatible with the measured single-crystal neutron diffraction data. To model the experimental data, a rhombohedral symmetry $R\bar{3}m$ that leads to a splitting of the Tb 16d position into two nonequivalent 3b and 9e positions had to be considered [144]. The proposed magnetic structure model consists of a three-in-one-out/one-in-three-out spin arrangement defined by the IR Γ_3 of the $R\bar{3}m$ group with $\mathbf{k} = (0,0,0)$, equivalent to the magnetic space group $R\bar{3}m'$. The H||[1,1,1] field-induced magnetic order in Tb$_2$Ti$_2$O$_7$ is depicted in Figure 14(b).

4.5.1.2 Interpenetrated Tetrahedra and Diamond Lattices in Spinel Compounds

The AB$_2$X$_4$ spinel structure is chemically versatile and can be found among transition metal oxides, sulfides, and selenides. The B cations are octahedrally coordinated by six X anions and form the same kind of corner-sharing tetrahedral lattice as found in pyrochlores. The A site forms a diamond lattice where each magnetic atom has four magnetic neighbors. The choice of metal cations and their distribution between the A- and B-sites give rise to an extremely tunable magnetic system with a large variety of magnetic ground states. Among the most studied in the recent years are the Cr- and V-based spinels that exhibit complex electronic and magnetic properties [145]. The chromate spinels ACr_2O_4 with a magnetically neutral A-site exhibit simultaneous lattice distortion and magnetic order due to the spin−lattice coupling. The nature of the lattice distortions and the corresponding magnetic order depend strongly on the nature of A-site ions. Diffraction studies showed that $HgCr_2O_4$ distorts to an orthorhombic *Fddd* structure with a magnetic order given by a wave vector **k** = (1,0,0), $CdCr_2O_4$ orders in a tetragonal $I4_1/amd$ and incommensurate **k** =(1,δ,0), whereas $ZnCr_2O_4$ becomes tetragonal $I\bar{4}m2$ with the magnetic propagation vector (1,0,0) [145]. The spinel vanadates, $A^{2+}V^{3+}_2O_4$, with V^{3+} ($3d^2$) cation having an orbital triplet degree of freedom, provide excellent model systems for studying the role of orbital ordering in geometrically frustrated antiferromagnets. In contrast with the chromates, these materials feature a sequence of two phase transitions: a first transition involving a structural distortion of the VO_6 octahedra and a consequent partial lifting of the orbital degeneracy, followed at lower temperature by a transition to an antiferromagnetic state. ZnV_2O_4, CdV_2O_4, and MgV_2O_4 have been the most thoroughly investigated compounds of this series [145]. ZnV_2O_4 exhibits a cubic-to-tetragonal lattice distortion at 50 K and a magnetic long-range order at $T_N = 40$ K. The crystal symmetry of the tetragonal structure of ZnV_2O_4 was found to be $I4_1/amd$ and the low-temperature antiferromagnetic structure of the vanadium sublattice was described by a propagation vector **k** = (0,0,1) [146]. The magnetic moments are aligned parallel to the *c*-axis and the magnitude of the ordered magnetic moment was refined as 0.65(5) μ_B per V^{3+} ion, which is much smaller than the expected value of $S = 1$ spin indicating the existence of large spin fluctuations caused by strong magnetic frustration. The magnetic structure of ZnV_2O_4 is shown in Figure 15(a).

Replacing the atom at the A-site by a magnetic species increases the complexity of the magnetic phase diagram. In MnV_2O_4, where Mn^{2+} is in a $3d^5$ high spin configuration $S = 5/2$ with quenched orbital angular momentum, neutron scattering measurements showed the existence of two consecutive magnetic transitions, first a transition to a collinear ferrimagnetic state with **k** = (0,0,0), followed at a slightly lower temperature by a second transition to a noncollinear state resulting from development of antiferromagnetic components in the *ab* plane. The second transition accompanied by a tetragonal distortion to an orbitally ordered tetragonal phase described by the space group $I4_1/a$ [147]. The FeV_2O_4 spinel is another interesting material, which contains on the tetrahedrally coordinated A-site Fe^{2+} ($3d^6$) cations with high-spin $S = 2$ and twofold orbital degeneracy. X-ray and neutron diffraction experiments revealed a cascade of structural transitions with decreasing temperature, from

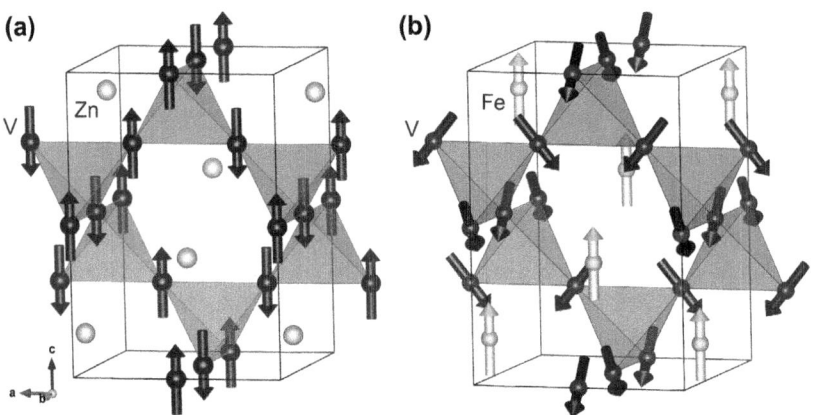

FIGURE 15 (a) Collinear antiferromagnetic structure of ZnV_2O_4 spinel and (b) the canted ferrimagnetic structure of FeV_2O_4 spinel below 60 K.

cubic $Fd\bar{3}m$ to high-temperature tetragonal $I4_1/amd$, to face-centered orthorhombic $Fddd$, and to low-temperature tetragonal $I4_1/amd$ [148]. Similar to MnV_2O_4, the magnetic structure exhibits an evolution from a collinear to canted ferrimagnetic state that coincides with the lowest temperature structural transition. The canting from the axis defined by the Fe^{2+} moments angle jumps abruptly from zero to $\sim 55° \pm 4°$ at the orthorhombic-to-tetragonal transition at ~ 60 K. This noncollinear ferrimagnetic structure is illustrated in Figure 15(b). The spin directions imposed by the $I4_1/amd$ space group results in a "two-in–two-out" structure similar to the ordered spin-ice state [148].

The spin configuration in FeV_2O_4 is slightly different from what has been assumed for the ordered state of MnV_2O_4 where the in-plane spins orientation cannot be uniquely assigned. Representation analysis performed for the V-site of the tetragonal space group $I4_1/a$ and $\mathbf{k} = (0,0,0)$ indicate a decomposition of the magnetic representation into four IRs $\Gamma_{mag}(V) = 3\Gamma_1^{(1)} + 3\Gamma_3^{(1)} + 3\Gamma_5^{(1)} + 3\Gamma_7^{(1)}$, while for $I4_1/amd$ there are five IRs $\Gamma_{mag}(V) = 1\Gamma_1^{(1)} + 2\Gamma_3^{(1)} + 1\Gamma_5^{(1)} + 2\Gamma_7^{(1)} + 3\Gamma_9^{(2)}$. The labeling of these representations is according to the Kovalev scheme [114]. Among these, only one of each decomposition allows a ferrimagnetic alignment of A = Mn/Fe and V sublattices. The basis vectors of these representations are listed in Table 4.

It is notable that the in-plane V moments for MnV_2O_4 are described by two basis vectors that constrain their relative phases to form an orthogonal configuration. The moments can point to any in-plane direction according to the degree of mixing of the two basis vectors as shown in Figure 16. The exact direction of the magnetic moments in this case cannot be uniquely determined using neutron diffraction. In contrast, the direction of the moments in FeV_2O_4 is fixed by a single in-plane basis vector to form a two-in-two-out configuration.

TABLE 4 Basis Vectors of Irreducible Representation Γ_1 (Magnetic Group $I4_1/a$) and Γ_3 (Magnetic Group $I4_1/am'd'$) of the Space Groups $I4_1/a$ and, Respectively, $I4_1/amd$ with Propagation Vector k = (0,0,0)

	MnV$_2$O$_4$ Γ_1 (Shubnikov $I4_1/a$)			FeV$_2$O$_4$ Γ_3 (Shubnikov $I4_1/am'd'$)	
V Site	$\psi 1$	$\psi 2$	$\psi 3$	$\psi 1$	$\psi 2$
V$_1$ (0,0,1/2)	(1,0,0)	(0,1,0)	(0,0,1)	(0,1,0)	(0,0,1)
V$_2$ (0,1/2,1/2)	(−1,0,0)	(0,−1,0)	(0,0,1)	(0,−1,0)	(0,0,1)
V$_3$ (1/4,3/4,1/4)	(0,1,0)	(−1,0,0)	(0,0,1)	(1,0,0)	(0,0,1)
V$_4$ (3/4,3/4,1/4)	(0,−1,0)	(1,0,0)	(0,0,1)	(−1,0,0)	(0,0,1)

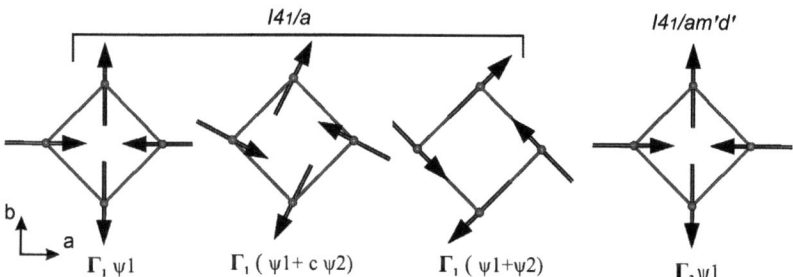

FIGURE 16 Symmetry-allowed magnetic structures for tetragonal phases of MnV$_2$O$_4$ and FeV$_2$O$_4$ spinel compounds. While both materials exhibit an orthogonal arrangement of the spins on each tetrahedron, the orientation of the moments in MnV$_2$O$_4$ is not constrained to a particular direction.

Another intricate series of ferrimagnetic spinels, with both A and B sublattices occupied by magnetic ions, is formed by the chromates ACr$_2$O$_4$ with A = Mn, Co, Ni, and Fe. The lattices of Mn, Co, and Fe compounds [149] are cubic at all temperatures, while NiCr$_2$O$_4$ [150] displays a tetragonal lattice below 310 K which at lower temperatures ($T_s \sim$ 30 K) distorts to an orthorhombic structure defined by the $Fddd$ space group. The magnetic structure of the tetragonal phase of NiCr$_2$O$_4$ has been reported to be collinear ferrimagnetic with the spins aligned along the [1,1,0] direction, as shown in Figure 17(a). At the structural transition, near 30 K, new transverse antiferromagnetic components that can be described with a propagation vector k = (0,0,1) start to develop [150]. Similarly, the Mn, Co, and Fe compounds order magnetically in two sequences comprising a first transition to a collinear ferrimagnetic state followed at low temperatures by the appearance of spiral components. In particular, CoCr$_2$O$_4$ orders at 97 K to a collinear ferrimagnetic

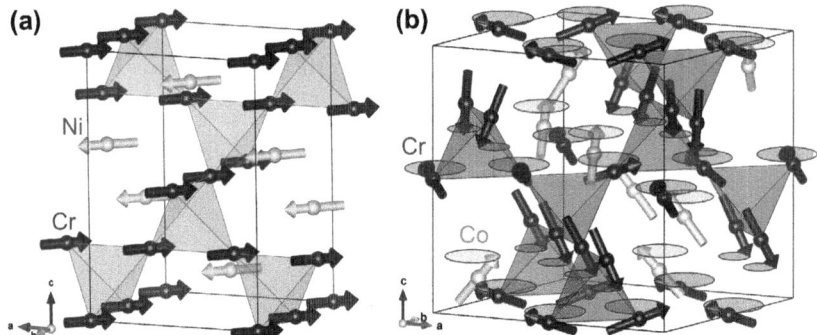

FIGURE 17 (a) Collinear ferrimagnetic structure of tetragonal $NiCr_2O_4$ spinel at temperatures between 75 and 30 K. (b) Ferrimagnetic-spiral magnetic structure of $CoCr_2O_4$ spinel below 25 K.

structure with the spins aligned along the [0,0,1] direction. The transition to a ferrimagnetic spiral structure occurs below 25 K, with the spiral components that lie within the (0,0,1) plane described by a propagation vector $\mathbf{k} = (0.62, 0.62, 0)$. Under the action of this propagation vector, the Cr site is split into two distinct sublattices (two orbits) corresponding to the two orthogonal chains α and β defined in the case of the pyrochlores. The two sites are not related by the symmetry operators of the wave vector space $\mathbf{G_k}$, and thus are magnetically independent. The mean values of cone angles of ferrimagnetic spiral order at 8 K have been determined as 48° for Co, 71° for the Cr ions of the α chain, and $-152°$ for the Cr ions of the β chain. The spiral structure of $CoCr_2O_4$ is illustrated in Figure 17(b).

4.5.1.3 Diamond-Lattice in Spinel Systems

In the case when only the A-site of the spinel lattice is occupied by magnetic atoms, new physics has been shown to occur arising from the frustrating effects of competing exchange interactions. The diamond-lattice form by the A cations can be seen as composed of two interpenetrating face-centered cubic (fcc) sublattices, with each ion having 4 NN and 12 NNN. The competition between NN (J_1) and NNN (J_2) exchange interactions can suppress the long-range magnetic order, leading to highly degenerate spin-liquid ground states. For a critical ratio $J_2/J_1 > 1/8$, theory predicts a large ground-state degeneracy consisting of spin spirals whose propagation wave vectors lie on a two-dimensional surface in momentum space. The degeneracy of the spin-liquid phase could be lifted by thermal or quantum effects to enable a magnetic ordering transition via the "order-by-disorder" mechanism [151]. Several strongly frustrated materials in this class have been recently the focus of intensive theoretical and experimental studies, particular attention being devoted to the oxide $CoAl_2O_4$ [152–155] and to the sulfide spinels $MnSc_2S_4$ and $FeSc_2S_4$ [156,157].

$CoAl_2O_4$ has been identified as a good model system to study unconventional ground states induced by the magnetic frustration of the diamond lattice, but the experimental results were somewhat sample dependent. The emergence of strong diffuse scattering observed in most samples has been explained as a consequence of order-by-disorder, an order-by-quenched-disorder or a kinetically inhibited order mechanism [152,155]. Long-range magnetic order to a collinear antiferromagnetic state below 9.8 K has also been reported [153]. The magnetic structure consists of an antiparallel alignment of the spins associated with the two interpenetrating fcc sublattices. The spins are pointing along the [0,0,1] direction with a measured magnitude of only 1.9(5) μ_B/Co, lower than 3 μ_B/Co expected for Co^{2+} ($S = 3/2$, $g = 2$) [153]. Recently, it has been acknowledged that not only the frustration due to the J_2/J_1 ratio but also the presence of Co/Al site disorder come into play in the observed magnetic behavior of this material [154].

The sulfide spinels $FeSc_2S_4$ and $MnSc_2S_4$ are two other highly celebrated materials for the presence of strong frustration effects in their magnetic properties. The long-range magnetic order is absent in $FeSc_2S_4$ down to 50 mK. Its magnetic frustration parameter $f = \Theta_{CW}/T_N$ (where Θ_{CW} is the Curie–Weiss temperature) is of the order of 1000, one of the largest values reported to date [157] and it has been characterized as a spin-orbital liquid below 45 K. In contrast, $MnSc_2S_4$ displays long-range order at $T_N = 2.3$ K, and bears the characteristics of a spin liquid above T_N. Neutron scattering studies revealed the presence of an incommensurate magnetic modulation around $T_{N1} = 2.3$ K, which shifts on cooling to become a locked-in commensurate structure with propagation vector $\mathbf{k} = (3/4,3/4,0)$ below $T_{N2} = 1.9$ K. The magnetic structure at $T = 1.5$ K has been modeled as a cycloid with the moments confined to the ab plane [156]. A drawing of the proposed magnetic structure is shown in Figure 18.

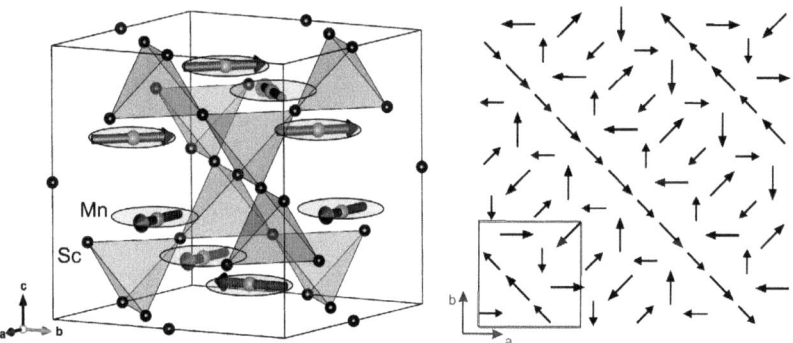

FIGURE 18 Left panel: Spiral magnetic structure of $MnSc_2S_4$ defined by a commensurate propagation vector $\mathbf{k} = (3/4,3/4,0)$. A view of the structure in the $a-b$ plane is shown in the right panel.

4.5.1.4 Edge-Sharing Tetrahedra in Double Perovskites

The magnetic lattices consisting of edge-sharing tetrahedra represent another three-dimensional framework that is highly sought after from the magnetic frustration perspective. This lattice can be found in the double perovskites $A_2BB'O_6$ when one of the B-type sites forming a frustrated fcc lattice is occupied by magnetic ions. The presence of chemical order of B and B' cations offers a wider range of magnetic exchange interactions in comparison to the simple ABO_3 perovskite structure. Dominant NN interactions along superexchange B'-O-O-B' pathways are expected to stabilize an antiferromagnetic order between successive (0,0,1) ferromagnetic planes as depicted in the left panel of Figure 19. This order can be described by a propagation vector $\mathbf{k} = (0,0,1)$ of the fcc lattice and is often referred to as "type I" spin ordering [158]. NNN exchange interactions along a pathway of the form B-O-A-O-B, where A is typically a diamagnetic alkaline earth or lanthanide cation, can also come into play to stabilize two other types of antiferromagnetic structures (Figure 19). If the strength of this latter interaction is dominant, it leads to type II order with spins alternating along two directions. A magnetic structure of type III, defined by a propagation vector $\mathbf{k} = (1/2,1/2,1/2)$ is obtained in the compounds where the two types of interactions are comparable in strength [158]. There is a vast amount of research on double perovskites containing a single magnetic site B', and all three antiferromagnetic structures described above have been experimentally realized. For instance, Sr_2LuRuO_6 and Ba_2YRuO_6 [158] order magnetically in a type I structure; Ca_2WMnO_6, Sr_2WMnO_6, and Sr_2MoMnO_6 [159] exhibit type II structures; and Ba_2LaRuO_6 [160] orders in a type III structure. On the other hand, the effect of geometrical frustration in the B' lattice has been evidenced by the absence of long-range magnetic order down to 2 K in Ba_2LnSbO_6 and Sr_2LnSbO_6 (Ln = Dy, Ho, and Gd) [161]. While strong frustration is indeed expected in ideally ordered double perovskites with negligible NNN interactions, this should be relieved significantly when large structural distortions are present. This is not so in the case of the highly distorted double perovskites La_2NaRuO_6 and La_2NaOsO_6. Neutron powder diffraction measurements for La_2NaRuO_6

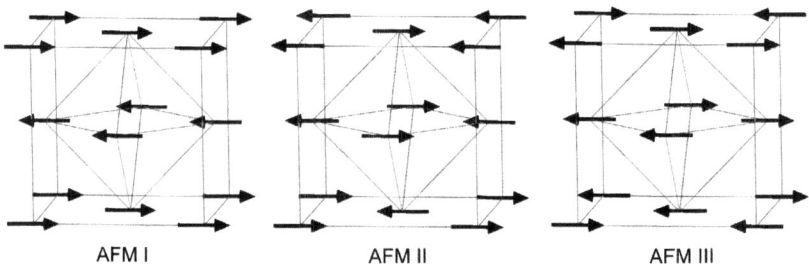

FIGURE 19 Magnetic structure models expected for the B-site in double perovskites $A_2BB'O_6$.

indicate an incommensurate magnetic ground state with a propagation vector $\mathbf{k} = (0,0,0.091)$, and a drastically suppressed ordered moment for La_2NaOsO_6 for which no magnetic Bragg peaks have been detected [162]. Such behavior suggests a different mechanism for magnetic frustration as arising from competition of inequivalent NN and NNN exchange interactions.

If more than one magnetic species is contained in the double perovskite lattice, the interplay between intra-lattice frustration and the inter-lattice interaction can lead to a wide variety of remarkable properties and magnetic states including ferromagnetism, ferrimagnetism, antiferromagnetism, spin-liquids, and spin-glasses. Ferromagnetism above room temperature, CMR, and half-metallic behavior have been reported for BB′ = FeMo-based double perovskites [163]. The Os-based compound Sr_2FeOsO_6 is an antiferromagnetic semiconductor with two magnetic phase transitions at 140 and 67 K [164], while the system Sr_2CrOsO_6 is a ferrimagnetic insulator with one of the highest Curie temperatures of 725 K [165] within this class of materials.

Here we provide details of the magnetic structures reported for the Sr_2CoOsO_6 semiconductor. The effective exchange coupling between cobalt and osmium sites is weak and the two sublattices are found to exhibit different ground states and spin dynamics. Two independent neutron diffraction studies performed for this compound revealed two antiferromagnetic phases with transition temperatures $T_N = 108$ and 67 K [166,167]. The first transition is accompanied by a structural distortion in the ab plane from tetragonal $I4/m$ symmetry to monoclinic $I2/m$ symmetry. While both studies agree on the overall low-temperature static-spin configuration, a point of dispute is the ordering sequence of the Os and Co moments. The first antiferromagnetic ordered phase below 108 K is associated by Morrow et al. [166] to an ordering of the Os moments with a propagation vector $\mathbf{k}_1 = (1/2,1/2,0)$, while the Co ions are thought to only order below 67 K in an antiferromagnetic structure with a propagation vector $\mathbf{k}_2 = (1/2,0,1/2)$. Alternately, Yan et al. [167] propose for the first ordered state, a magnetic structure consisting of both Co and Os partially ordered moments aligned within the ab plane. In the second antiferromagnetic phase, the Co moments freeze into a new spin-canted state toward the c-direction while Os static moments remain almost unchanged. A randomly canted state for the Os moments is also suggested. The static magnetic order at base temperature is reproduced in Figure 20.

The next example illustrates the case where both A- and B′-sites are occupied by magnetic ions. Magnetic structures of double perovskites Nd_2NaRuO_6 and Nd_2NaOsO_6 have been recently reported [168]. These materials exhibit large room-temperature distortions to a monoclinic $P2_1/n$ structure and present distinctively different magnetic structures. A neutron diffraction study revealed that the Nd and B′ spins order at the same temperature, 14 K for B′ = Ru and 16 K for B′ = Os, with commensurate propagation vectors: $\mathbf{k} = (0,0,0)$ for the Ru system and $\mathbf{k} = (1/2,1/2,0)$ for the Os analog. In Nd_2NaRuO_6, the Ru sublattice orders in a type I

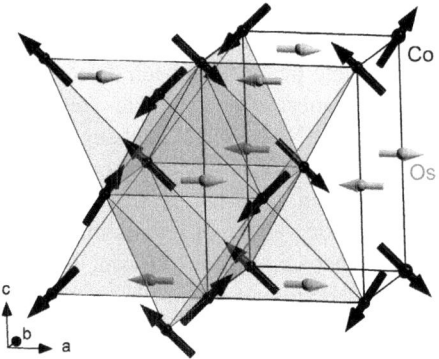

FIGURE 20 Low-temperature magnetic ordering of Co and Os sublattices in Sr_2CoOsO_6.

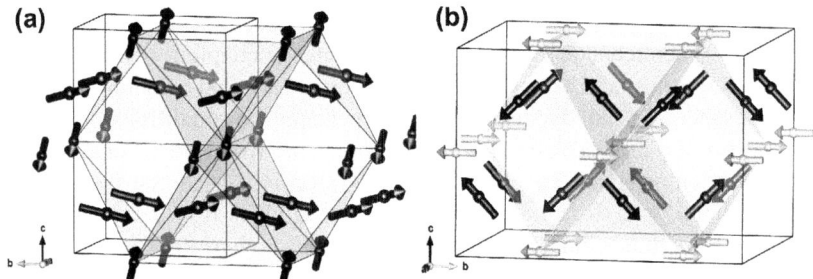

FIGURE 21 Magnetic structures with (a) $\mathbf{k} = (0,0,0)$ and (b) $\mathbf{k} = (1/2,1/2,0)$ proposed for the monoclinic double perovskites Nd_2NaRuO_6 and Nd_2NaOsO_6, respectively.

antiferromagnetic structure with the moments lying in the ac plane and forming ferromagnetic planes stacked antiferromagnetically along the c-axis (Figure 21). The Nd moments are arranged in a canted ferromagnetic configuration along the b-axis. Overall, the system presents net magnetization along the b-direction. In the case of Nd_2NaOsO_6, the Os moments are confined to the b-axis and their directions alternate in both the basal plane and along the c-direction to form a type II antiferromagnetic configuration. In contrast to the Ru analog, the Nd magnetic sublattice in Nd_2NaOsO_6 is antiferromagnetic and no net magnetization is associated with the system.

4.5.2 Layered Structures

The particularity of layered structured magnetic systems became evident in the 1960s after the neutron scattering studies of K_2NiF_4 single crystals lead Birgeneau et al. [169] to the conclusion that this system is a realization of a two-dimensional Heisenberg antiferromagnet model. The coupling between NNN layers was found to be smaller than the intraplane exchange by at least three orders of magnitude. Earlier neutron scattering experiments performed by Plumier [170] revealed the onset of long-range order within the square NiF_2 layer but not between the layers. The two-dimensional nature of the system manifested experimentally by the presence of reciprocal lattice rods or "Bragg ridges" rather than Bragg peaks in the c^*-direction [169]. Even with a long-range order established between layers, the quasi-two-dimensional character of the exchange interactions placed the layered systems in a distinct category. Soon after the investigation of K_2NiF_4, quasi-two-dimensional behavior was found in other isostructural materials including K_2MnF_2, K_2CoF_2, and Rb_2CoF_2. Renewed interest in layered perovskite-type compounds has occurred at intervals of about 20 years, largely as a result of the discovery of superconductivity in the cuprate oxides and more recently in the FeAs pnictides. More information on the topic of superconductivity can be found in Chapter 3. In addition to the square lattices, the two-dimensional triangular-based lattices have also been the subject of intense research since they represent well-suited models for exploring geometrically frustrated magnetism. Several variants of triangular packings with varying degrees of geometric frustration can be distinguished, including edge-shared triangular, kagomé, and honeycomb lattices. In the following, we will highlight selected magnetic structures that have been encountered in layered systems containing either square- or triangular-based lattices.

4.5.2.1 Square-Planar Metal-Oxygen/Halogen Layers

The layered perovskite containing square-planar metal oxide layers are largely known as the Ruddlesden—Popper phases and are defined by the general formula $A_2A'_{n-1}B_nX_{3n+1}$ where n is the number of perovskite blocks that are separated by an A_2 layer. An extended family of layered perovskites include those containing separating motifs of rock-salt Bi_2O_2 layer $\{Bi_2O_2\}$-$\{A_{(n-1)}B_2X_7\}$ or alkali metal $M^{+1}A_{(n-1)}B_nX_{(3n+1)}$ known as Aurivillius and Dion—Jacobson phases, respectively. Most of the magnetic structures found in A_2BX_4 compounds are similar to that found in K_2NiF_4 [170] and consist of an antiferromagnetic alignment of the NN spins within the basal plane. The square antiferromagnetic layers are stacked such that the spins of NNN planes along the c-axis are parallel. Such a structure can be described by a propagation vector $\mathbf{k} = (1/2,1/2,0)$ and is illustrated in Figure 22(a). A quite different stacking of the antiferromagnetic layers has been surprisingly found in Ca_2MnO_4 [171]. In this compound, the adjacent moments separated by the c-lattice spacing of 12 Å are coupled antiparallel, instead of parallel, and the magnetic unit cell is doubled with respect to that of K_2NiF_4. The moments are aligned along [0,0,1] direction and their refined magnitude is 2.0 μ_B per Mn^{4+} ion, significantly less than the spin-only value of 3 μ_B. We shall emphasize here that the propagation vector $\mathbf{k} = (1/2,1/2,1/2)$ describing this structure can produce two distinct spin configurations. For a choice of a phase

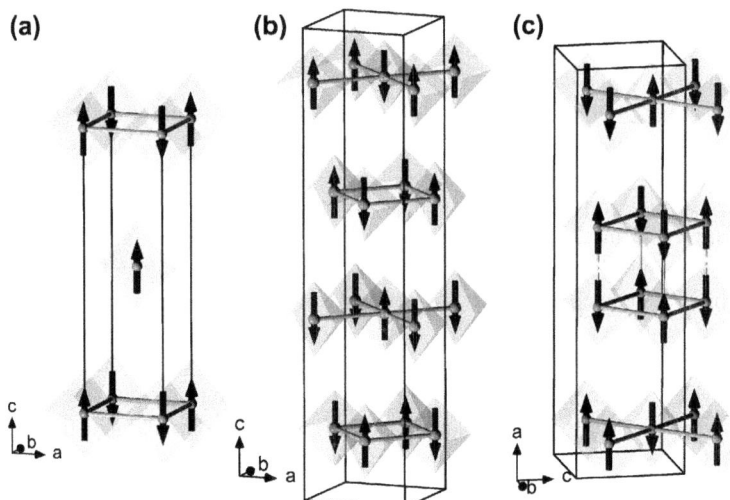

FIGURE 22 Magnetic structures of Ruddlesden–Popper phases. (a) K_2NiF_4, (b) Ca_2MnO_4, and (c) Ca_3MnO_7.

factor zero in the Fourier coefficient $S_{kj} = 1/2 S_{0j} e^{i\varphi_j}$ (see Section 4.3.2), the magnetic moment distribution defined by the cosine modulation $\mu = S_0 \cos(2\pi \mathbf{k} \cdot \mathbf{R}_l + \varphi_j)$ will give rise to a $(S_0, 0, -S_0, 0)$ sequence of magnetic layers along c-axis. According to this model, the ions at the cell center position $(1/2, 1/2, 1/2)$ carry no spin. In contrast, for a phase $\varphi = \pi/4$ the sequence becomes $(S_0/\sqrt{2}, S_0/\sqrt{2}, -S_0/\sqrt{2}, -S_0/\sqrt{2})$. The latter model coincides with that reported for Ca_2MnO_4 [171], although no clear argument was given in support of this model versus the other. Fifteen years after the magnetic structure of Ca_2MnO_4 was reported, a detailed structural investigation on single-crystal samples revealed that its actual crystal structure consists of a supercell ($a = \sqrt{2}a'$, $c = 2c'$) of the presumed $I4/mmm$ cell of the K_2NiF_4-type structure [172]. This superstructure described by the tetragonal space group $I4_1/acd$ results from rotations of the MnO_6 octahedra about their z axes. Using the newly determined crystal structure, the magnetic configuration can be uniquely redefined using a propagation vector $\mathbf{k} = (0,0,0)$ and the basis vector of representation Γ_8 (following Kovalev notation), or the Shubnikov group $I4_1'/a'cd'$. This provides a confirmation for the correctness of the suggested magnetic model. The magnetic structure of Ca_2MnO_4 is displayed in Figure 22(b).

The discovery of large magnetocaloric effects in the bilayered perovskite manganites with chemical compositions $La_{2-2x}A_{2x}Mn_2O_7$ (A = Sr, Ca) has made those some of the most attractive materials of the Ruddlesden–Popper series [173]. We will only describe here the magnetic structure of the parent material $Ca_3Mn_2O_7$. This compound was found to exhibit an octahedral tilting distortion involving a symmetry lowering to the orthorhombic space group $Cmc2_1$ [174]. Magnetic reflections appearing in the neutron diffraction patterns below about 115 K were indexed by either $\mathbf{k} = (0,0,0)$ or $(1,0,0)$ propagation vectors, but the pseudo-tetragonal lattice prevented an unambiguous distinction between these possibilities. In such a situation, the observation of weak ferromagnetism by magnetic susceptibility measurements provided the key information for choosing the $\mathbf{k} = (0,0,0)$ indexing scheme and a magnetic model that allows ferromagnetic components. Representation analysis for the magnetic site Mn ($8b$) of $Cmc2_1$ and $\mathbf{k} = (0,0,0)$ indicates that only three models allow weak ferromagnetism, namely Γ_2 (magnetic space group $Cm'c'2_1$), Γ_3 ($Cmc'2_1'$), and Γ_4 ($Cm'c2_1'$), with FM components along the c-axis, a-axis, and b-axis, respectively. The magnetic structure was solved in the $Cm'c2_1'$ Shubnikov group with a dominant component along a-axis coupled antiferromagnetically with the NNs, and a weak ferromagnetic component along the b-axis. This magnetic structure is shown in Figure 22(c).

4.5.2.2 Square Lattices of Metal Atoms Coordinated by Pnictogen/Chalcogen Atoms

The discovery of high-temperature superconductivity in iron-based materials [175] has resulted in significant interest and motivated the exploration of related materials. The three best-known families are the pnictides based on the ZrCuSiAs-type (tetragonal, $P4/nmm$) or $ThCr_2Si_2$-type ($I4/mmm$) crystal structures, and the monochalcogenides Fe(Se,Te) ($P4/nmm$). These materials are known as the "1111," "122," and "11," respectively, and their common structural feature is the existence of a square lattice of iron atoms tetrahedrally coordinated by either arsenic, selenium, tellurium, or phosphorus anions that are staggered above and below the iron lattice. These slabs are either separated by spacer layers using alkali, alkali-earth, rare earth, or rare earth oxide/fluoride; or simply stacked together as in the 11 series. Similar to the cuprates, high-T_c superconductivity in the iron pnictides emerges when the antiferromagnetic order of parent compounds is suppressed [176,177]. However, the antiferromagnetic state of the parent Fe-based superconductors is different from that of the cuprates. While the cuprates display an NN antiferromagnetism, the iron sublattice undergoes magnetic orderings that consist of stripes of ferromagnetically aligned Fe moments that alternate antiferromagnetically for the 1111 and 122 pnictides, and of double-stripe ordering in chalcogenides, as illustrated in Figure 23. The Fe-based superconductors also have substantial magnetoelastic coupling and the magnetic-ordering transition is preceded or coincides with a lowering of crystal symmetry from tetragonal to orthorhombic.

The layered chalcogenides are structurally the simplest but perhaps the most remarkable among the Fe-based superconductors. Their physical properties markedly change upon the covalent substitution of Se, Te, and S. Furthermore, the nonsuperconducting parent material $Fe_{1+y}Te$ exhibits magnetic properties that are extremely sensitive to nonstoichiometric Fe located at

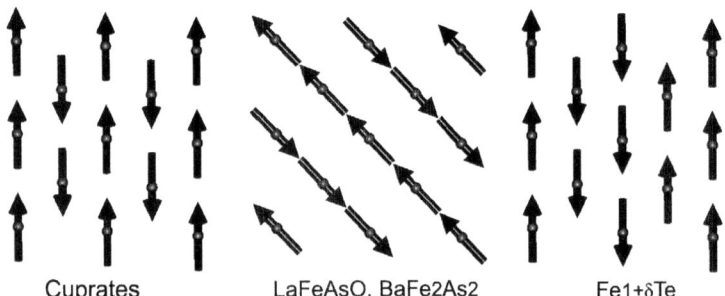

FIGURE 23 Comparison of the in-plane nearest-neighbor antiferromagnetic order in cuprates with the stripe order of ferromagnetically aligned Fe moments in LaFeAsO and BaFe$_2$As$_2$, and the double-stripe order in Fe$_{1+\delta}$Te. *Adapted from Ref. [178]*.

interstitial sites [179]. At low concentrations, $y \lesssim 0.05$, Fe$_{1+y}$Te display a first-order magnetostructural transition from the paramagnetic tetragonal *P4/nmm* phase to monoclinic *P2$_1$/m* with bicollinear antiferromagnetic order defined by the propagation vector $\mathbf{k} = (1/2,0,1/2)$, with moments pointing in the *b*-direction. At high $y \gtrsim 0.12$, the low-temperature phase is orthorhombic *Pnmm*, with an incommensurate helimagnetic spin structure with propagation vector $\mathbf{k} = (0.385,0,1/2)$. In this helical ordering, the moments are confined to the *bc* plane and trace out a circular envelope. For the sample with intermediate doping, the frustration effects of the interstitial Fe decouple different orders and lead to a sequence of transitions [180]. A lattice distortion to a monoclinic structure is closely followed by an incommensurate amplitude-modulated magnetic order that competes with the bicollinear commensurate magnetism. A temperature-dependent hybridization leads to the formation of the ferromagnetic zigzag chains which stabilize the bicollinear order causing a shift of scattering away from an incommensurate position $\mathbf{k} = (\delta,0,1/2)$ toward $(1/2,0,1/2)$. The magnetic structures of different Fe$_{1+y}$Te compounds are shown in Figure 24.

Extending the study to the Co congeners brought into light fascinating magnetic behaviors and rich phase diagrams. For instance, NdCoAsO undergoes three magnetic phase transitions below room temperature. Upon cooling, a ferromagnetic ordering of small Co moments emerges near 69 K followed at 14 K by a spin reorientation into an antiferromagnetic state with small moments induced on the Nd sites [181]. A subsequent magnetic transition occurs at 3.5 K that corresponds to the antiferromagnetic ordering with propagation vector $\mathbf{k} = (0,0,1/2)$ of larger Nd moments. The Co moments in each CoAs layer are ferromagnetically aligned parallel to the tetragonal *a*-axis and are compensated by Nd moments pointing in the opposite direction. The ferromagnetic layers are stacked antiferromagnetically along the *c*-direction giving rise to a magnetic unit cell twice as large as the chemical cell. This magnetic configuration corresponds to the Γ_{10} IR, and the magnetic space

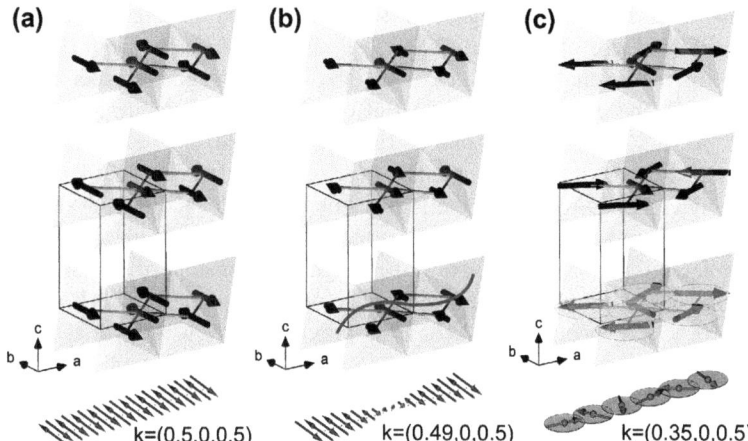

FIGURE 24 Three different magnetic orderings encountered in $Fe_{1+x}Te$: (a) commensurate bicollinear antiferromagnetic order with the moments along the b-direction in FeTe; (b) incommensurate spin-density wave in $Fe_{1.11}Te$; and (c) helical structure with moments confined to the ab plane in $Fe_{1.14}Te$.

group $Pcc'n$. The low-temperature magnetic structure of NdCoAsO is displayed in Figure 25(a).

The rare earth cobalt arsenides, RCo_2As_2 (R = La, Ce, Pr, Nd), exhibit ferromagnetic ordering of Co magnetic moments along the c-axis in the temperature range 60–200 K [182]. At lower temperatures, the rare earth magnetic moments (R = Pr, Nd) order antiparallel to the Co moments

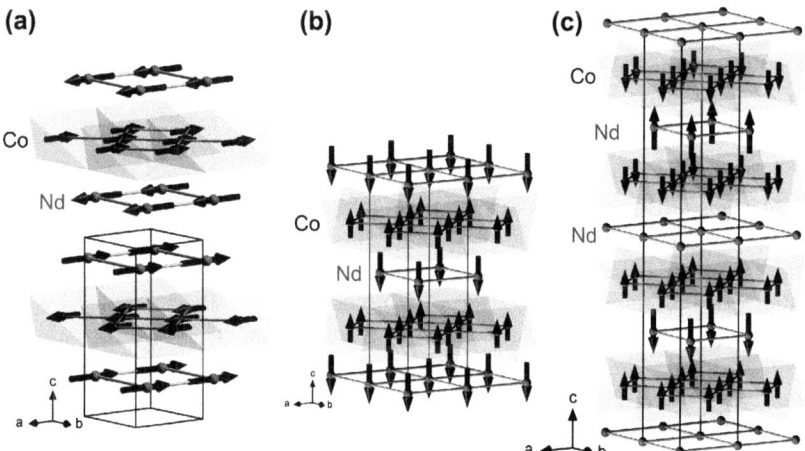

FIGURE 25 Magnetic structures of (a) NdCoAsO antiferromagnet defined by the wave vector $\mathbf{k} = (0,0,1/2)$; (b) $NdCo_2As_2$ with ferrimagnetic arrangement; and (c) $NdCo_2P_2$ consisting of antiferromagnetic order with the Nd sublattice presenting moments only in every other layer.

stabilizing a ferrimagnetic ground state, as shown in Figure 25(b). The magnetic behavior of isostructural RCo_2P_2 is quite different from that established for the arsenates. Only $LaCo_2P_2$ shows ferromagnetic ordering of Co moments in the ab-plane, while in $EuCo_2P_2$ the Co atoms are not magnetic [183]. All the other RCo_2P_2 phases (R = Ce, Pr, Nd, Sm) exhibit antiferromagnetic ordering in the Co sublattice near or above the room temperature and antiferromagnetic ordering in the R sublattice below 20 K [67,183–185]. The Eu moments order in an incommensurate spiral structure with a propagation vector $\mathbf{k} = (0,0,0.85)$ [183], while the magnetic moments of Pr and Nd ions were found to order antiferromagnetically with a commensurate $\mathbf{k} = (0,0,1)$ and (0,0,1/2), respectively [67,185]. As previously discussed, for the case of $\mathbf{k} = (0,0,1/2)$ in a body-centered cell (*I4/mmm*) there are two undistinguishable magnetic structures of Nd moments. In the first model, the moments are ordered in each Nd layer of the unit cell, whereas in the second model the $4f$ moments are present in every other Nd layer, giving the stacking sequence $(0,-,0,+)$. The refined moment values for the second model, shown in Figure 25, were in a much better agreement with the theoretically expected magnetic moment for the ground state of the Nd^{3+} ion, 3.28 μ_B, and it therefore was chosen to be the correct one [185].

Quaternary compositions $La_{1-x}R_xCo_2P_2$, with R = Pr or Nd, exhibit unconventional magnetic behavior, not observed for their ternary congeners. For example, $La_{0.75}Pr_{0.25}Co_2P_2$ undergoes at 167 K a ferromagnetic ordering of Co moments in the ab plane, followed at 66 K by a ferromagnetic ordering of Pr moments parallel to the c-axis, which causes a rotation of the Co moments toward the c-axis in the direction opposite to the Pr moments. This leads to a noncollinear ferrimagnetic structure that switches the direction of the total magnetization from the ab plane to the c-axis. A third magnetic transition has been observed at 35 K, which has been associated with the stabilization of the collinear ferrimagnetic order along the c-axis [184].

4.5.2.3 Combined Square-Planar Metal-Oxygen and Metal-Pnictogen Lattices

In the search for new classes of superconductors, a new series of layered compounds $Sr_2Mn_3Pn_2O_2$ (Pn = pnictogen = P, As, Sb) has been discovered that consist of alternate stacking of square MnO_2 oxide layers and intermetallic Mn_2Pn_2 slabs [186,187]. The two magnetic sublattices formed by two nonequivalent sites, Mn_1 and Mn_2, associated with the two distinct layers (MnO and Mn-Pn, respectively) have tetragonal symmetry. Each of the Mn atoms has four NNs within each plane and four NNs in each of the adjacent planes. The Mn_2 sublattice orders at relatively high temperatures of 340 K (Pn = As) and 300 K (Pn = Sb) in an intra- and interlayer NNs antiferromagnetic structure (G-type) with moments antiparallel to the c-axis. The magnetic order is described by a propagation vector $\mathbf{k} = (0,0,0)$. In contrast, long-range antiferromagnetic order on the Mn_1 sublattice is observed only for $Pn = Sb$ at a much lower temperature, near 65 K. The magnetic configuration is identical to that of K_2NiF_4, described by a wave vector $\mathbf{k} = (1/2,1/2,0)$ and with the moments confined to the MnO_2 plane. In $Sr_2Mn_3As_2O_2$, the Mn_1 sublattice only exhibits a short-range two-dimensional antiferromagnetic ordering below 70 K, evidenced by diffuse scattering with Warren-type line shape [186]. The Warren function, originally introduced to characterize powders of two-dimensional crystallites randomly oriented in space, exhibits an asymmetric profile (sawtooth type) with a sharp increase in intensity at the scattering angle of a two-dimensional Bragg reflection (h,k) and a slow decrease at higher scattering angles [188]. Partial replacement of Mn by Cu in $Sr_2Mn_3As_2O_2$ resulted in a completely different magnetic ordering [187]. Refinement of neutron diffraction data revealed that the Cu is found only in the Mn_2 site, associated with the Mn–As layers in the structure. The transition temperature of this sublattice (Mn_2/Cu) is reduced from 340 to 95 K. In fact, both Mn_1 and Mn_2 sublattices order simultaneously and the NN antiferromagnetism is replaced by ferrimagnetic ordering. The spins within a layer are ferromagnetically aligned with each other, but adjacent ferromagnetic layers are antiferromagnetically aligned. The different magnetic structures observed in $Sr_2Mn_3Pn_2O_2$ series are shown in Figure 26.

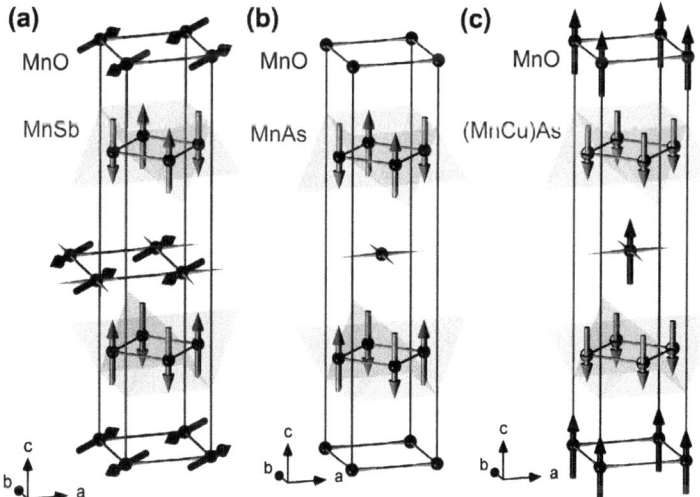

FIGURE 26 Magnetic structures of the $Sr_2Mn_3Pn_2O_2$ series containing MnO_2 and Mn_2Pn_2-type layers. (a) $Sr_2Mn_3Sb_2O_2$ exhibits Mn_1 and Mn_2 sublattices that order in two steps with wave vectors $\mathbf{k} = (0,0,0)$ and $(1/2,1/2,0)$, respectively; (b) $Sr_2Mn_3As_2O_2$ develops magnetic order only in the Mn–As layer (Mn_2 sublattice); (c) $Sr_2Mn_2CuAs_2O_2$ orders in a ferrimagnetic state with antiparallel alignment of Mn_1 and Mn_2 sublattices.

4.5.2.4 Planes of Edge-Sharing Equilateral Triangles

The triangular lattice with antiferromagnetically coupled NN spins is the simplest lattice structure featuring geometrical frustration. In spite of the magnetic frustration, a large variety of magnetic structures can be stabilized by NNN interactions, anisotropy of the orbital configuration, or magnetoelastic coupling. Remarkable properties, such as ferroelectricity, have been found to arise from some of the noncollinear spin arrangements formed in these lattices. The chemistry of the planar lattices consisting of edge-sharing equilateral triangles is diverse, and an excellent review of the most relevant materials has been provided by Cava et al. [132] or Collins and Petrenko [189]. Among the most-studied class of materials of this type are the delafossites with general formula $A^+M^{3+}O_2$, where A^+ = Cu, Ag, Pt, Pd, and M^{3+} is a transition-metal ion or lanthanide. The building blocks for the equilateral triangular magnetic plane and formed by layers of edge-sharing MO_6 octahedra that are stacked alternately with nonmagnetic A^+ layers. The A^+ cations have a dumbbell coordination with two oxygen anions from the planes above and below. Depending on the manner of stacking of the two layers, structural variants with hexagonal (called 2H, space group $P6_3/mmc$) or rhombohedral (called 3R, space group $R\bar{3}m$) symmetry can be formed. The two types of stacking sequences result in the formation of either eclipsed (on top of each other) or staggered (shifted) near-neighbor planes, with the latter also enabling frustrated out-of-plane interactions. Representative for this class of compounds are $CuCrO_2$ and $CuFeO_2$, which exhibit staggered stacking but quite different magnetic ground states. $CuFeO_2$ undergoes successive antiferromagnetic phase transitions: a partially disordered amplitude-modulated magnetic state with $\mathbf{k} = (\sim 0.2, \sim 0.2, 3/2)$ and moments parallel to the c-axis, at $T_{N1} = 14$ K, and a collinear commensurate four-sublattice spin structure with $\mathbf{k} = (1/4, 1/4, 3/2)$ at $T_{N2} = 11$ K [190]. The magnetic transitions are accompanied by second- and first-order structural phase transitions from rhombohedral ($R\bar{3}m$) to monoclinic ($C2/m$) symmetry [191]. In contrast, $CuCrO_2$ exhibits two-dimensional order below 24.2 K followed closely at 23.6 K by a long-range order into a proper screw-helix structure defined by the propagation vector $\mathbf{k} = (0.329, 0.329, 0)$ [192,193]. As illustrated in Figure 7, the magnetic structure model consists of magnetic moments rotating in the $[1,\bar{1},0] - [0,0,1]$ plane orthogonal to the wave vector. A similar magnetic structure has been found for the Ag^+ equivalent, $AgCrO_2$, which orders at 24 K with wave vector $\mathbf{k} = (0.327, 0.327, 0)$ [194]. However, the delafossite $AgFeO_2$, was found to exhibit below 9 K a cycloidal magnetic structure with elliptical envelope instead of the collinear order seen in $CuFeO_2$ [195].

Another popular class of layered materials containing equilateral triangles is that of the α-$NaFeO_2$ rock salt-type structure. The stacking of the triangular planes is staggered, as in Cu^+-and Ag^+-based delafossite compounds, but the nonmagnetic spacing layer consists of octahedrally coordinated alkali or alkaline earth metals. The α-NaFeO$_2$ compound exhibits a sequence of magneto-structural phase transitions starting at $T_{N1} = 11$ K with a long-range ordered SDW with $\mathbf{k}_{hex} = (0.12, 0.12, 3/2)$ and a lowering of structure symmetry from rhombohedral $R\bar{3}m$ to monoclinic $C2/m$ [196]. This incommensurate SDW phase evolves below $T_{N2} = 7.5$ K to a cycloidal multiferroic phase and, finally, below $T_{N3} = 5.5$ K to a commensurate collinear ground state with propagation vector defined in the hexagonal setting as $\mathbf{k} = (-1/4, 1/4, -1)$. The difference between the collinear ground states found in α-NaFeO$_2$ and CuFeO$_2$ has been attributed to different signs of the NN interactions and the predominant contribution of the dipolar interactions to the magnetic anisotropy in the case of α-NaFeO$_2$ [196].

The disilver dioxides Ag$_2$MO$_2$ (M = Ni, Cr, Mn) represent another series related to delafossite-type oxides with staggered triangular lattice planes of MO$_2$ separated by metallic Ag$_2$ layers [197–199]. A structural distortion from rhombohedral to a monoclinic $C2/m$ symmetry occurs in all compounds of this series, but reportedly much smaller in Ag$_2$NiO$_2$ as compared to Ag$_2$MnO$_2$ and Ag$_2$CrO$_2$ [199]. In Ag$_2$NiO$_2$, a long-range magnetic order with $\mathbf{k}_{mono} = (1, 1/3, 1/2)$ in the monoclinic notation, or $\mathbf{k}_{hex} = (1/3, 1/3, 0)$ in the hexagonal lattice, occurs below a Néel temperature of $T_N = 56$ K. The magnetic moments are aligned collinearly in the ac plane and the moment distribution at the lth Ni site follows the expression $m_l = m_0 \cos(2\pi \mathbf{k} \cdot l)$, where $m_0 = (0.31, 0, 0.65)$ μ_B at 5 K. Figure 27(b) shows the commensurate SDW

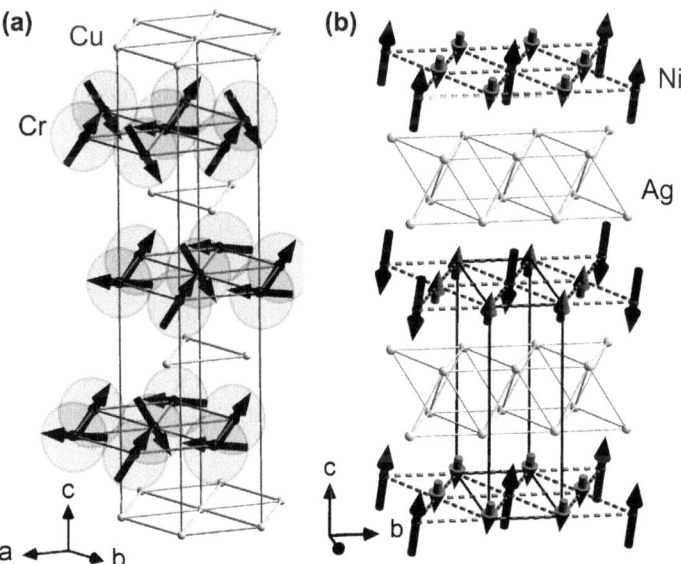

FIGURE 27 (a) Proper screw-helix structure with $\mathbf{k} = (0.329, 0.329, 0)$ in CuCrO$_2$ delafossite; (b) Magnetic structure of Ag$_2$NiO$_2$ consisting of a collinear spin arrangement with $\mathbf{k} = (1, 1/3, 1/2)$.

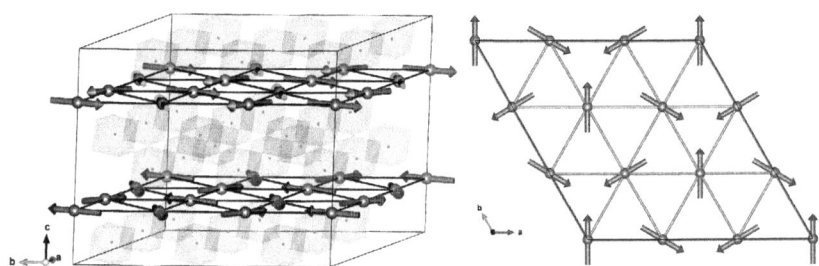

FIGURE 28 Noncollinear 120° structure in the *ab* plane adopted by $Ba_3NiNb_2O_9$.

model in which the magnitude of the ordered moment of one site is half of that of the other crystallographically equivalent site [199]. In contrast, Ag_2MnO_2 has a short-range glassy-like magnetic ground state, while Ag_2CrO_2 exhibits a partially disordered state with a five-sublattice magnetic structure defined by the wave vector $k_{hex} = (1/5,1/5,0)$.

Examples of materials that display eclipsed stacking-type of the equilateral triangles include those with the composition AMX_3 where A is an alkali metal (e.g., Rb, Cs), M is a transition metal and X is a halogen atom (X = Cl, Br, I). These compounds crystallize in the hexagonal space group $P6_3/mmc$ and contain columns of face-sharing MX_6 octahedra arranged in a triangular array and spaced by the large Rb or Cs ions. The magnetic properties of this class have been thoroughly reviewed by Collins and Petrenko [189]. Another classic example is that of the layered compounds VX_2 (X = Cl, Br, I) with the CdI_2 structure ($P\bar{3}m1$ space group), which are typically ordered with a 120° spin structure in the *ac* plane with $k = (1/3,1/3,1/2)$ [200]. Recent experimental efforts to search for new multiferroic materials have brought to light a new series of triangular antiferromagnets that order with chiral 120° spin structures. For instance, $RbFe(MoO_4)_2$ [201] and $RbAg_2Fe[VO_4]_2$ [202], both with low-temperature crystal symmetries $P\bar{3}$, order with an *ab* plane 120° spin structure that propagates with wave vector $k = (1/3,1/3,\xi)$, with $\xi = 0.458$ for the molybdate and $\xi = 0.372$ for the vanadate. Another exciting class of materials that has emerged in recent years is the 6H-perovskites $Ba_3M'M''_2O_9$ [203–205]. Most of the compounds of this type with M' = Ni, Co, Mn, and M'' = Nb or Sb, adopt at low temperatures the noncollinear 120° structure in the *ab* plane, described by a propagation vector $k = (1/3,1/3,1/2)$. This magnetic structure is shown in Figure 28 for the case of $Ba_3NiNb_2O_9$.

4.5.2.5 Planes of Edge-Sharing Nonequilateral Triangles

A particular type of triangular plane is observed in certain layered compounds that undergo Jahn–Teller distortions leading to a crystal structure that maps out a triangular spin lattice with anisotropic interactions. Examples of such compounds are Cs_2CuCl_4 [206], $NaMnO_2$ [207], $NaNiO_2$ [208], or the crednerite $CuMnO_2$ [209–211]. Cs_2CuCl_4 has the magnetic Cu ions arranged in

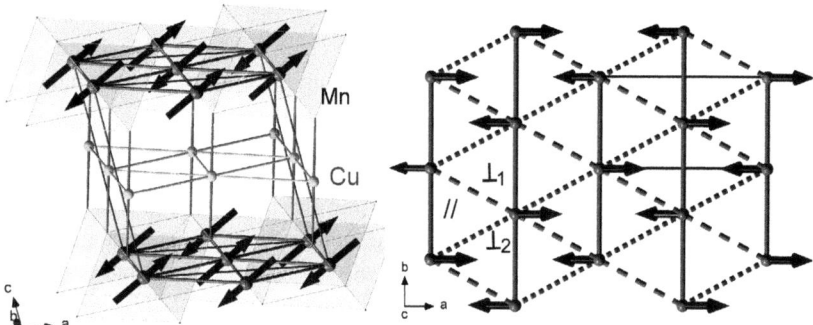

FIGURE 29 Antiferromagnetic order with wave vector $\mathbf{k} = (1/2,1/2,1/2)$ observed in the $CuMnO_2$ crednerite with crystal structure consisting of planes of isosceles triangles.

chains running along the b-axis, which are displaced by $b/2$ with respect to each other to form planes of isosceles triangles. The magnetic ordering was found to set in below $T_N = 0.6$ K and the magnetic structure is cycloidal with spins rotating along the crystallographic b-direction with the ordering wave vector $\mathbf{k} = (0, 0.472, 0)$ [206]. The $NaNiO_2$ and $NaMnO_2$ crystallize in the monoclinic space group $C2/m$ and represent distorted variants of the α-$NaFeO_2$ structure type due to the presence of the Jahn−Teller distortion in the MO_6 octahedra. The magnetic structures of the two congeners are found to be quite different, revealing different degrees of magnetic frustration. The in-plane magnetic exchange scheme can be represented by two distinct exchange integrals: $J_\parallel$, which defines an antiferromagnetic chain parallel to the b-axis, and $J_\perp$, denoting the frustrated couplings along the equivalent isosceles sides (see Figure 29). Unlike the case of $NaNiO_2$, the half-filled t_{2g} Mn^{3+} orbitals along the short Mn−Mn distance of edge-sharing octahedra promote strong antiferromagnetic direct-exchange interactions leading to strong interchain frustration. While in the Ni system, the magnetic moments are ferromagnetically aligned in the NiO_2 layers and antiparallel between adjacent layers along the c-direction (i.e., $\mathbf{k} = (0,0,1/2)$), the Mn system orders with a propagation vector $\mathbf{k} = (1/2,1/2,0)$ with the moments coupled antiferromagnetically in the MnO_2 layer and ferromagnetically between layers. The moments point along the direction of the ferro-ordered $d(3z^2 - r^2)$ orbitals, which are elongated along the z-axis direction, as expected due to the strong anisotropy of Mn^{3+} [207]. The magnetic frustration is lifted by magnetoelastic coupling evidenced by a structural transition to a triclinic cell below $T_N = 45$ K. An even more intriguing behavior has been observed in the case of $CuMnO_2$ crednerite, which has a distorted delafossite structure. This system was also found to exhibit a magnetoelastic stabilization of a long-range magnetic ordering which occurs below 65 K, but the magnetic structure is defined by the propagation vector $\mathbf{k} = (-1/2, 1/2, 1/2)$ and consists of intra- and interlayer antiferromagnetic coupled moments [209,210]. As illustrated in

Figure 29, the MnO$_6$ octahedra undergo a small distortion in the equatorial plane, which produces a segregation of interchain ($\perp$) Mn—Mn bonds into two different components. The influence of spin disorder on the ground-state properties of this system has been explored using hole-doped compounds Cu(Mn$_{1-x}$Cu$_x$)O$_2$ with $0 \leq x \leq 0.07$. It has been found that the spin disorder alters the magnetoelastic characteristics and reduces the effective dimensionality of the magnetic lattice. Upon increasing the doping level, the structural distortion is suppressed and the long-range magnetic order is gradually transformed into a two dimensional order with an intermediate change of the interlayer coupling from antiferromagnetic to ferromagnetic [209].

4.5.2.6 Planar Kagomé Lattices, Based on Corner-Sharing Triangles

The kagomé lattice is derived from the edge-shared triangular lattice by periodic removal of 1/4 of the magnetic sites and consists of equilateral triangles that share corners to form six-triangle rings (see Figure 30). This lattice is considered to be one of the most highly frustrated spin systems with infinitely degenerated ground states [212]. The existing materials approximating the kagomé lattice show a variety of ground states such as spin liquid, short-range order, or long-range order that can be stabilized by further-neighbor and Dzyaloshinskii—Moriya interactions. The closest experimental realizations of the kagomé-lattice antiferromagnet are the volborthite [213], herbertsmithite [214], kapellasite [215], and jarosite series [216—218]. A review of the principal families of kagomé antiferromagnets is provided by Mendels and Wills [132].

Magnetically ordered states are found in the jarosites which are a group of hydroxide-sulfates with the general chemical formula AM$_3$(OH)$_6$(SO$_4$)$_2$, where A is a monovalent cation such as Na, K, Rb, Ag, Tl, NH$_4$ and M is Fe, Cr, or V. The jarosites crystal structure is described by the space group symmetry $R\bar{3}m$ and the magnetic ions are distributed on a single crystallographic site 9d. A detailed analysis of the possible magnetic models

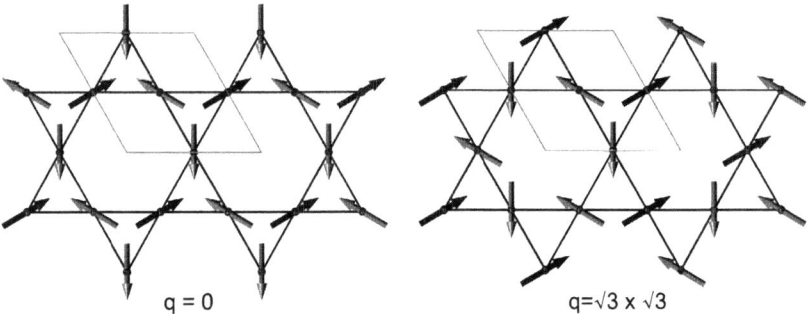

FIGURE 30 Magnetic orders on a classical kagomé lattice with 120° between neighboring spins. The $q = 0$ spin arrangement has uniform vector chirality, whereas the $\sqrt{3} \times \sqrt{3}$ structure consists of an alternative spin arrangement with staggered vector chirality.

compatible with the crystal symmetry and the two experimentally observed propagation vectors $\mathbf{k} = (0,0,0)$ and $(0,0,3/2)$ has been performed by Wills [216]. For both $\mathbf{k}$'s, only three ordering patterns are possible: a 120° structure, an umbrella mode, and a more complex structure given by the basis vector associated with the IRs Γ_2 $(0,0,3/2) = \Gamma_3$ $(0,0,0)$, Γ_4 $(0,0,3/2) = \Gamma_1$ $(0,0,0)$, and Γ_5 $(0,0,3/2) = \Gamma_6$ $(0,0,0)$, respectively. Both propagation vectors point to a "$q = 0$"-type structure with uniform vector chirality, as opposed to the "$q = \sqrt{3} \times \sqrt{3}$" structure with staggered vector chirality predicted to be stabilized by quantum fluctuations [212]. The two types of spin structures are shown in Figure 30. The observation of a $q = 0$ structure indicates the presence of additional magnetic interactions that are beyond the simple NN kagomé model.

The Fe-based jarosites $AFe_3(SO_4)_2(OH)_6$, with A = K, Na, Ag, order at low temperatures into the planar "$q = 0$"-type 120° magnetic structure with an antiferromagnetic stacking of the kagomé layers with the propagation vector $\mathbf{k} = (0,0,3/2)$ (Figure 31). The presence of an intermediate phase with an umbrella structure where the moments are slightly canted out of the kagomé plane has been suggested for the K and Na compounds [217]. The Cr-jarosite $KCr_3(SO_4)_2(OD)_6$ was found to order with a propagation vector $\mathbf{k} = (0,0,0)$, which corresponds to a ferromagnetic stacking of the kagomé layers, in contrast to that seen in the Fe analog.

Two important classes of materials that contain distorted kagomé-type lattices are the langasites, with general formula $Ln_3Ga_5SiO_{14}$ [219–221], and the vanadates $M_3(VO_4)_2$ where M = Ni, Co, and Cu [222–224]. The langasite class has the flexibility that their kagomé lattice planes can be decorated with either rare earth ions (Ln = Nd or Pr) or transition metal ions. The Nd-based langasite, $Nd_3Ga_5SiO_{14}$, has been investigated with neutron scattering down

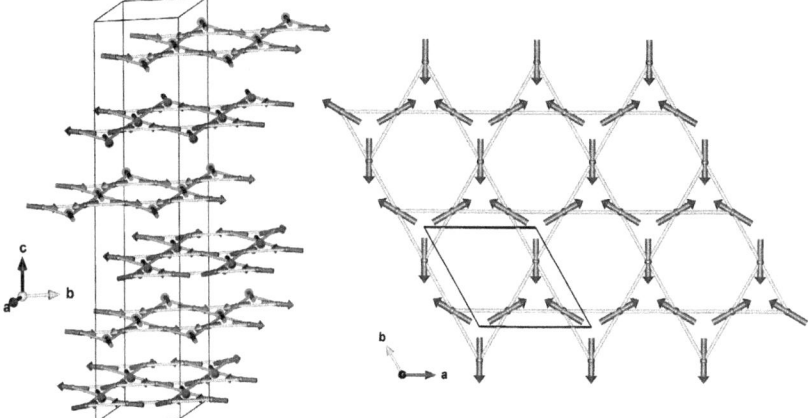

FIGURE 31 Magnetic order in the kagomé lattice of $AFe_3(OH)_6(SO_4)_2$ (A = K, Na, Ag) consisting of a planar 120° arrangement with uniform vector chirality.

to temperatures of 46 mK with no evidence of magnetic long-range order in a zero field but the possibility of a spin-liquid ground state [221]. The isostructural compound $Ba_3NbFe_3Si_2O_{14}$, with the magnetic Fe^{3+} ions forming the corner-sharing triangular lattice, exhibits long-range order below $T_N = 27$ K defined by a wave vector $\mathbf{k} = (0,0,\xi)$, with ξ close to 1/7 [220]. The magnetic structure consists of magnetic moments lying in the ab planes arranged in a 120° configuration that propagates helically along the c-axis with a period of approximately seven lattice parameters. The vanadates $M_3(VO_4)_2$ contain buckled staircase kagomé layers formed by two distinct M sites and feature a complex field-temperature magnetic phase diagram [222–224]. In particular, the system $Ni_3V_2O_8$ has attracted a great deal of attention due to the appearance of ferroelectricity that accompanies one of its magnetic phase transitions. Neutron scattering experiments revealed the existence of four zero-field phase transitions to states that comprise two incommensurate orders, one amplitude modulated and the other helical, followed by a commensurate canted antiferromagnetic order and, finally, by an admixture of commensurate and a cycloidal incommensurate orderings at the lowest temperature [223,224].

4.5.2.7 Layers of Honeycomb Lattices

The layered honeycomb lattices are derived from the edge-sharing triangular planar lattice by removing 1/3 of the magnetic sites in an ordered fashion. A representative class of materials for this type of structure is $BaM_2(XO_4)_2$ (M = Co, Ni and X = P, As, V) [225–228]. These layered compounds crystallize in rhombohedral symmetry $R\bar{3}$, and contain staggered honeycomb layers of edge-sharing MO_6 octahedra spaced by nonmagnetic Ba and XO_4 layers. The Ni-based compounds order magnetically into collinear antiferromagnetic states with the magnetic moments confined to the ab plane. However, the wave vectors that define the ordering sequence of the spins are different: $\mathbf{k} = (1,0,1/2)$ for $BaNi_2(VO_4)_2$, $\mathbf{k} = (0,0,0)$ for $BaNi_2(PO_4)_2$, and $\mathbf{k} = (1/2,0,1/2)$ in the case of $BaNi_2(AsO_4)_2$. In the vanadate and the phosphate variants, each Ni spin is aligned antiferromagnetically with its three NNs within the honeycomb plane, whereas in the arsenate the magnetic ions form ferromagnetic chains that are coupled antiferromagnetically (Figure 32). A far more complicated magnetic structure has been observed in the Co-based compound $BaCo_2(AsO_4)_2$, which appears to exhibit a higher degree of magnetic frustration [225]. In that case the three-dimensional order consists in a periodic stacking of helimagnetic layers defined by the propagation vector $\mathbf{k} = (0.261, 0, -4/3)$. This magnetic structure is shown in Figure 33.

A second class of compounds with honeycomb magnetic lattices is derived from the delafossite-type structure containing a mix of magnetic and nonmagnetic atoms in a 2:1 ratio, to give the formula $Cu_3B'_2B''O_6$ [229]. This class is typified by $Cu_3Ni_2SbO_6$ and $Cu_3Co_2SbO_6$ compounds which present a monoclinic distortion of the hexagonal 6-layer delafossite polytype with space

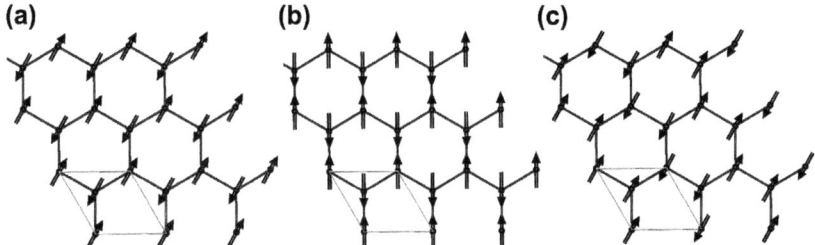

FIGURE 32 Magnetic order configuration of the honeycomb lattices in (a) $BaNi_2(VO_4)_2$, (b) $BaNi_2(PO_4)_2$, and (c) $BaNi_2(AsO_4)_2$.

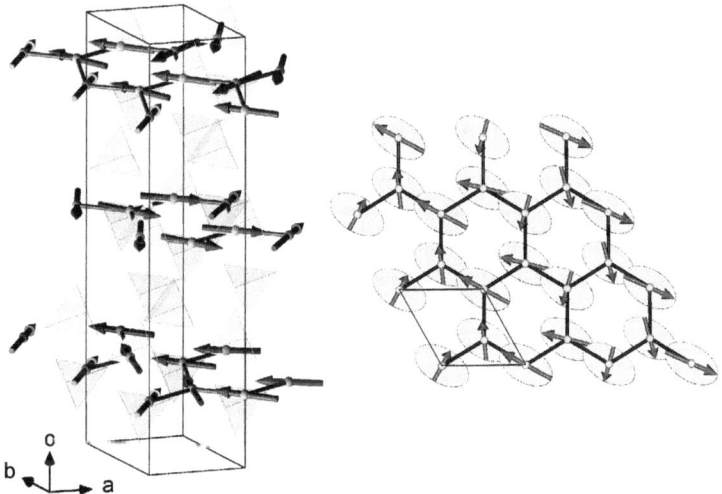

FIGURE 33 Helicoidal magnetic structure of the honeycomb lattice system $BaCo_2(AsO_4)_2$ defined by the propagation vector $\mathbf{k} = (0.26, 0, -4/3)$.

group $C2/c$. Similar monoclinic structures are found for honeycomb-layer compounds derived from the $NaFeO_2$ structures, such as $Na_3Co_2SbO_6$ [230]. The magnetic structures of both $Cu_3Ni_2SbO_6$ and $Cu_3Co_2SbO_6$ compounds are defined by a propagation vector $\mathbf{k} = (1,0,0)$ and consist of ferromagnetic chains along the b-direction, which are antiferromagnetically coupled to their neighboring chains (Figure 34). The coupling between the adjacent honeycomb layers is antiferromagnetic. This type of order resembles that observed in the honeycomb structure $BaNi_2(AsO_4)_2$. The orientation of the magnetic moments in the Co delafossite compound is parallel to the b-axis direction, whereas in the Ni compound the moments are confined to the ac plane and are pointing almost perpendicular to the honeycomb planes. Such magnetic models were described in the frame of the basis vectors of the IR Γ_3 (in Kovalev notation) for Co, and the irreducible Γ_1 for the case of Ni [229].

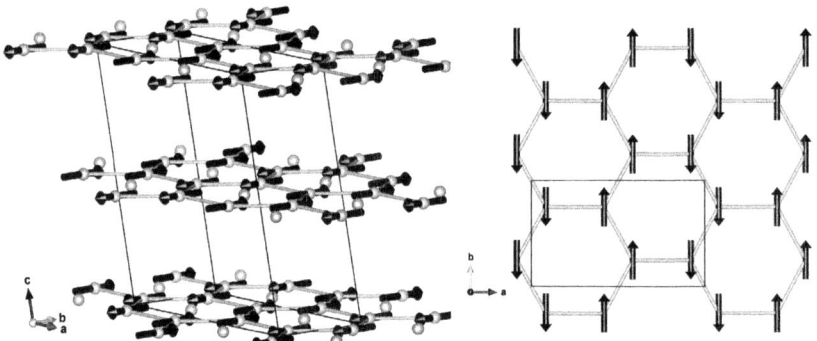

FIGURE 34 Magnetic structure of the honeycomb lattice of $Cu_3Co_2SbO_6$ consisting of ferromagnetic chains, which are antiferromagnetically coupled to their neighboring chains.

4.5.3 Quasi-One-Dimensional Lattices

In the field of low-dimensional magnetism, special emphasis has been devoted to one-dimensional magnets since these systems provide a unique possibility to study ground states of quantum models, and the interplay of quantum and thermal fluctuations. The structure of magnetic system with predominant one-dimensional character is such that strong exchange coupling between magnetic ions is provided only along one crystallographic direction. Although one-dimensional Heisenberg systems are not expected to display long-range order for $T > 0$, we will direct our attention to those quasi-one-dimensional systems where the presence of weak interchain couplings is sufficient to cause three-dimensional magnetic order at finite temperature. The examples presented below comprise inorganic and metal-organic compounds with chain-based structures, systems where competing interactions lead to effective quasi-one-dimensional magnetic lattices, and field-induced magnetic order in one-dimensional quantum systems.

4.5.3.1 Long-Range Magnetic Order in Simple Chain-Based Structures

The alkali-metal pyroxenes AMX_2O_6 ($A = Li^+$, Na^+; $M = Ti^{3+}$, V^{3+}, Cr^{3+}, Mn^{3+}, and Fe^{3+}; $X = Si^{4+}$, Ge^{4+}) have been studied extensively due to their quasi-one-dimensional magnetism and multiferroic phenomena. This class of compounds exhibits a monoclinic crystal structure with the magnetic M^{3+} ions forming zigzag chains running along the crystallographic c-axis. Within the chains, the edge-sharing MO_6 octahedra provide a 90° M-O-M pathway for superexchange interactions. Since the chains are well separated by nonmagnetic XO_4 tetrahedra, the intrachain magnetic interactions are considered to be much stronger than the interchain interactions. The V-based pyroxenes have been considered as promising one-dimensional Haldane-chain systems in spite of the fact that they all undergo long-range antiferromagnetic ordering due to relatively strong interchain exchange interactions [231,232]. The magnetic ground state of Cr-based pyroxenes evolve from long-range antiferromagnetic order for LiCr(Si,Ge)$_2$O$_6$ to ferromagnetic order for NaCrGe$_2$O$_6$. The magnetic structure of LiCrSi$_2$O$_6$ consists of antiferromagnetic chains of Cr that are coupled ferromagnetically, while the magnetic structure of NaCrGe$_2$O$_6$ features ferromagnetic chains and an overall ferromagnetic configuration [233].

A magnetically driven ferroelectricity has been observed in the magnetically ordered state of the iron pyroxene NaFeGe$_2$O$_6$ compound which consists of a cycloid defined by the incommensurate propagation vector $\mathbf{k}$ = (0.335,0,0.081) [234]. The Mn-congener, NaMnGe$_2$O$_6$, undergoes a long-range magnetic ordering transition at T_N = 7 K to an antiferromagnetic state, described by a commensurate propagation wave vector $\mathbf{k}$ = (0,0,1/2). The spin arrangements of the Mn moments along the chain c-axis direction yield an up-up-down-down sequence and ferromagnetic coupling between chains [235]. This magnetic structure is shown in Figure 35(a).

Hybrid materials containing both inorganic and organic components are a goldmine of opportunities to study low-dimensional magnetism [236]. Such materials, often called metal-organic frameworks (MOFs), offer a large number of combinatorial possibilities to form novel one-dimensional systems by connecting low-dimensional inorganic subnetworks through adequately functionalized magnetically inert molecules. Due to their strong low-dimensional character, these materials offer a good selection of quantum magnets with spin-liquid-like ground states, enabling recent breakthroughs in quantum physics [237]. For those quasi-one-dimensional MOFs that order magnetically, transition temperatures are typically extremely low. In addition,

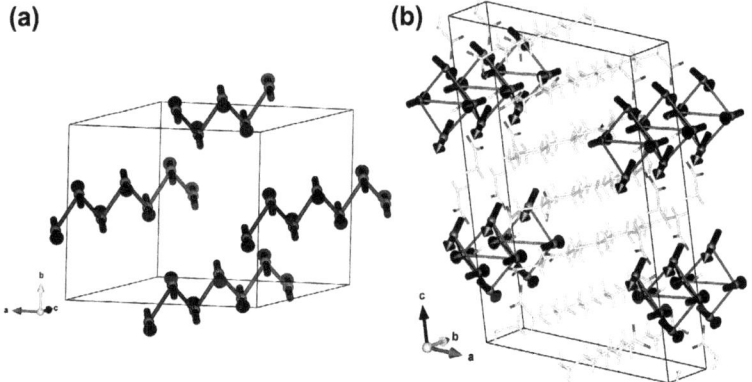

FIGURE 35 Magnetic structures of quasi-one-dimensional systems: (a) NaMnGe$_2$O$_6$ pyroxene ordered antiferromagnetically with a wave vector $\mathbf{k}$ = (0,0,1/2) and (b) the Co-based metal-organic system Co$_4$(OH)$_2$(C$_{10}$H$_{16}$O$_4$)$_3$ that involves a collinear antiferromagnetic sublattice and a canted antiferromagnetic sublattice with uncompensated spin along the c-axis.

the magnetic contribution to the total neutron scattering from these materials is expected to be weak due to the small number of magnetic atoms per unit cell, which makes the measurements more difficult. Another challenge is posed by the presence of hydrogen atoms that are pernicious for neutron experiments due to their large incoherent cross section. Thus, in many cases, hydrogen-bearing samples for neutron experiments must be deuterated. An example of a magnetically ordered hybrid system is the cobalt hydroxy-sebacate, $Co_4(OH)_2(C_{10}H_{16}O_4)_3$, which adopts a $P2_1/c$ monoclinic structure and contains ladders of Co^{2+} ions running along the b-direction, separated from each other by distances of 11 to 21 Å [238]. The one-dimensional subnetwork contains two crystallographically inequivalent magnetic ions, Co(1) and Co(2), that have distorted octahedral and trigonal bipyramidal oxygen environments, respectively. The ladder is constituted on a central zigzag chain of corner-sharing Co(1)O$_6$ octahedra that are sharing oxygen atoms with adjacent Co(2)O$_5$ trigonal bipyramids. The magnetic ordering occurs below 5.4 K and is described by a commensurate propagation vector $\mathbf{k} = (0,0,0)$. The Co(1) moments are aligned along the b-axis in a collinear antiferromagnetic configuration, whereas the Co(2) moments adopt a canted antiferromagnetic configuration with the main direction along b-axis and a ferromagnetic component along the c-axis, as shown in Figure 35(b). The refined values of the magnetic moments were 2.19(7) μ_B and 2.68(8) μ_B for Co(1) and Co(2), respectively. The magnetic structure is not modified upon the application of a magnetic field, but the ordering temperature increases to 13 K at $H = 2$ T due to an apparent strengthening of the interchain correlations [238]. Further examples of quasi-one-dimensional MOFs and their response to externally applied magnetic fields are provided in the Section 6.5.3.4.

4.5.3.2 Triangular-Based Magnetic Chains

One particular type of magnetic chain that has drawn considerable attention from a theoretical standpoint is the so-called sawtooth chain, or delta chain, in which the magnetic ions form zigzag chains of corner-sharing triangles. Such a configuration is one of the prototype examples of highly frustrated lattices and can be seen as derived from a kagomé lattice by removing magnetic sites in periodic lines perpendicular to the kagomé spine direction [132]. Experimental realization of a magnetic sawtooth lattice is rare and only a handful of such compounds are known, including the oxidized delafossite YCuO$_{2.5}$ [239], the olivines [240–243], and the oxy-arsenates $A_2Fe_2O(AsO_4)_2$, with A = K and Rb [244].

The olivines are a large crystal chemical family, with general formula M$_2$XO$_4$, where M is a transition metal ion and X is a group III or IV atom, most commonly Si and Ge. They crystallize in an orthorhombic lattice (space group $Pbnm$) with isolated silicate or germanate tetrahedra, and two symmetrically distinct metal sites (M$_1$ and M$_2$), one forming the central spine of

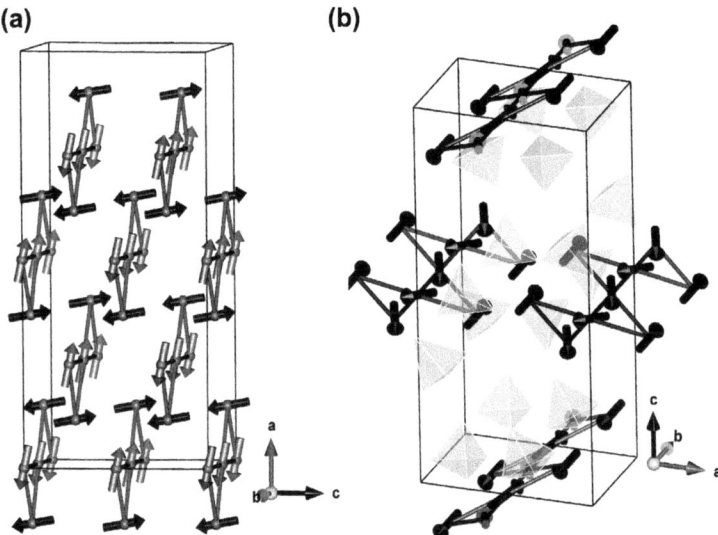

FIGURE 36 (a) Canted antiferromagnetic magnetic structure of Ni_2SiO_4 described by the propagation vector $\mathbf{k} = (1/2,0,1/2)$ and (b) antiferromagnetic order of oxy-arsenate $Rb_2Fe_2O(AsO_4)_2$ with $\mathbf{k} = (0,0,0)$ consisting of ferrimagnetic sawtooth chains coupled antiferromagnetically.

the sawtooth chains and the other one constituting the apex of the triangles zigzagging along the linear chains (Figure 36(a)). The magnetic structures of manganese and iron orthosilicate, Mn_2SiO_4 (with mineral name tephroite) and Fe_2SiO_4 (fayalite), were first investigated in the 1960s by Santoro et al. [241]. Both compounds were reported to order into a collinear antiferromagnetic structure and to rearrange at lower temperature into a partially canted arrangement. The magnetic structure of Ni_2SiO_4 (liebenbergite) determined 40 years later had also revealed evidence of frustrated interactions by a significant canting away from collinearity [240]. The magnetic structure of this compound is defined by a propagation vector $\mathbf{k} = (1/2,0,1/2)$ and has been described using a co-representation formed by the combination of the representations Γ_3 and Γ_7, in Kovalev notation. The magnetic structure of Ni_2SiO_4 is displayed in Figure 36. The moments of the two nonequivalent magnetic sites Ni1 and Ni2 are canted away from the a-axis by 26.26° and 69.42°, respectively. A complex magnetic phase diagram has been recently reported for another member of the olivine family, Mn_2GeO_4, which features both a ferroelectric polarization and a ferromagnetic magnetization [243]. Three different ordered magnetic states have been found below 47 K, comprising a commensurate order with $\mathbf{k}_C = (0,0,0)$ where the sawtooth chains form ferromagnetic layers that are coupled antiferromagnetically, a noncollinear antiferromagnetic order where ferromagnetic layers are alternated with layers containing a noncollinear spin arrangement, and at last, a mixed state where the commensurate order based on ferromagnetic layers coexists with a spiral incommensurate state defined by the propagation vector $\mathbf{k}_{IC} = (0.136, 0.211, 0)$. In addition to the transition-metal-ion olivines, rare earth olivines have been discovered in the form of $ZnLn_2S_4$, with Ln = Er, Tm, and Yb [242]. No long-range magnetic order has been detected in these compounds, and its absence was attributed to the geometrical frustration of the sawtooth triangular chains.

Similar to the olivines, the sawtooth-like chains in the oxy-arsenate $Rb_2Fe_2O(AsO_4)_2$ are formed by corner-sharing isosceles triangles of Fe^{3+} ions occupying two nonequivalent crystallographic sites. The chains are isolated magnetically from each other via diamagnetic $(AsO_4)^{3-}$ units along the a-axis and Rb^+ cations along the c-axis direction. Neutron diffraction measurements indicate the onset of a long-range antiferromagnetic order below 25 K, with the magnetic structure consisting of ferrimagnetic chains coupled antiferromagnetically with each other along the c-axis direction. Within each chain, the Fe_1 moments located at the tip of the sawtooth are aligned collinearly along the b-direction, while the Fe_2 moments are reversely canted by approximately 30°, forming a zigzag pattern in the plane of the triangular chain [244]. This spin structure is shown in Figure 36(b). The magnetic coupling between adjacent chains along the c-direction changes under an applied magnetic field from antiferromagnetic to ferromagnetic to produce an overall ferrimagnetic state.

4.5.3.3 Effective One-Dimensionality Caused by Frustrated Interchain Interactions

One-dimensional magnetic lattices can also arise as a result of a magnetic frustration that can drastically suppress the exchange interactions along two spatial dimensions. This situation is encountered in the family of compounds AR_2O_4 with A = Ba, Sr and R = rare earth [68,245–248]. These compounds crystallize in the $CaFe_2O_4$-type structure, with space group $Pnma$, and consist of three-dimensional honeycomb columns made of zigzag chains which extend along c-axis and give rise to geometric frustration (Figure 37). The zigzag chain structure can be regarded as a realization of the ANNNI (axial NNN Ising) or J_1-J_2 model, where the rungs and legs correspond to NN (J_1) and NNN (J_2) interactions, respectively. Theory predicts for this model a $J_2/J_1 = 1/2$ critical point where the magnetic ground state changes from a simple Néel antiferromagnetic, up-down-up-down, to an up-up-down-down double Néel configuration [249]. Another peculiarity of these compounds is that the rare earth atoms are distributed over two inequivalent octahedrally coordinated sites ($4c$) that give rise to two distinct zigzag chains. The different crystal-field environments of the two rare earth sites have been shown to play an important role in dictating the magnetic properties [68].

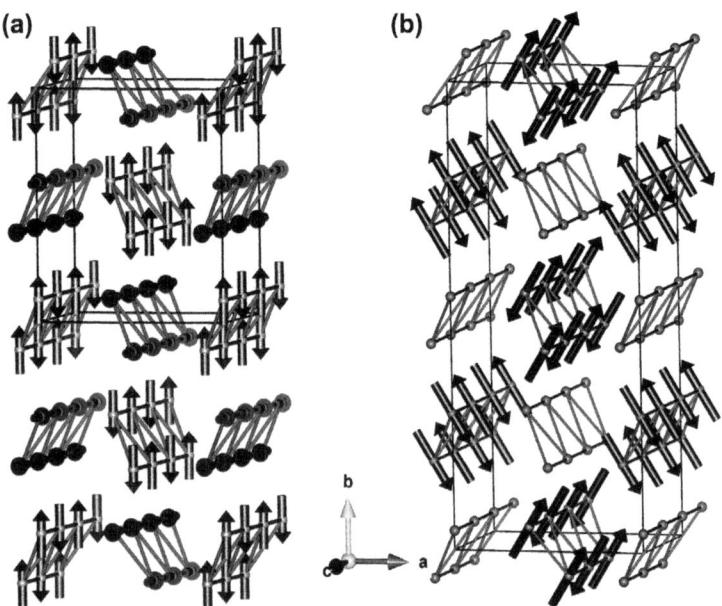

FIGURE 37 Magnetic structures of the quasi-one-dimensional magnetic systems containing zigzag chains. (a) The magnetic ground state of $SrHo_2O_4$ consists of independently ordered ladders with wave vectors $\mathbf{k} = (0,0,0)$ and $(0,0,1/2)$, that feature up-down-up-down and up-up-down-down spin configurations, respectively. (b) The magnetic order in $BaNd_2O_4$ involves only one of the two inequivalent Nd sites and is characterized by a propagation vector $\mathbf{k} = (0,1/2,1/2)$ and up-up-down-down spin configuration.

While the majority of the compounds of this series display short-range magnetic order, the long-range order has been found to arise from only one rare earth site. In $SrEr_2O_4$, an antiferromagnetic long-range order of one of the Er sites appears below 0.75 K. This order is described by a wave vector $\mathbf{k} = (0,0,0)$ and corresponds to a Néel order characteristic of a small J_2, with the moments point along the c-axis direction. The diffuse magnetic scattering coexisting with the long-range magnetic signal has been modeled by a short-range order of the second Er site according to a double-Néel configuration, up-up-down-down, defined by a dominant J_2. A similar magnetic ground state has been observed in $SrHo_2O_4$ below $T_N = 0.66$ K, except that the order of both Ho sites remains short-range correlated. The quasi-long-range ordered magnetic structure of $SrHo_2O_4$ was described with two different propagation vectors, $\mathbf{k}_1 = (0,0,0)$ and $\mathbf{k}_2 = (0,0,1/2)$, each associated with a different Ho site. The $\mathbf{k}_1$ describes the up-down-up-down Néel ordering of Ho1 with a static moment of 6.080(3) μ_B aligned along the c-axis, whereas the $\mathbf{k}_2$ describes the up-up-down-down order of Ho2 moments along the b-axis and a magnitude of 7.740(3) μ_B [68,248]. The magnetic order of $SrHo_2O_4$ is sketched in

Figure 37(a). In contrast to the Er- and Ho-based compounds, $SrDy_2O_4$ does not show any magnetic phase transition down to the lowest available temperatures (~ 20 mK), while $SrYb_2O_4$ features long-range magnetic order at $T_N = 0.92$ K with both inequivalent Yb sites ordering according to the $\mathbf{k} = (0,0,0)$ wave vector. Both sites have magnetic moments confined to the ab plane but feature different directions and magnitudes that are reduced as compared to the full ionic moment. Another well-characterized member of the series is $SrTb_2O_4$ that displays the highest transition temperature to a long-range order of all in the series ($T_N = 4.28(2)$ K) [246]. Long-range order occurs only for one of the sites and the propagation vector is incommensurate with the crystalline lattice, refined as $\mathbf{k} = (0.5924, 0.0059, 0)$. The ordered moment is confined to the bc plane and is modulated in amplitude with a maximum of 1.92(6) μ_B at 0.5 K.

The Ba-based members of this family, BaR_2O_4, have been studied to a lesser extent than the Sr counterparts because only polycrystalline samples were available. A neutron powder diffraction study of $BaNd_2O_4$ revealed a long-range antiferromagnetic ground state characterized by a propagation vector $\mathbf{k} = (0, 1/2, 1/2)$ that arises below $T_N = 1.7$ K from only one of the two Nd sites [247]. Four undistinguishable magnetic structure models can be considered for this system, but they all imply confined moments to the ab plane. One of these models is displayed in Figure 37(b). The magnetic moments are arranged in the sequence up-up-down-down, which is the prediction for a classical Ising J_1-J_2 chain in the large J_2 limit.

4.5.3.4 Field-Induced Magnetic Order in One-Dimensional Quantum Systems

The topic of field-induced magnon condensation in gapped quantum antiferromagnets has recently received a great deal of attention from experimentalists. The most characteristic feature is the destruction of the nonmagnetic spin liquid ground state and the emergence of a long-range antiferromagnetically ordered phase at some critical field H_c. The triplet of low-lying $S = 1$ gap excitations in such systems is subjected to Zeeman splitting by external magnetic fields, and a quantum phase transition to a long-range magnetic ordered phase occurs when the gap for one member of the triplet is driven to zero. To date, the phenomenon has been observed and studied experimentally in numerous one-dimensional materials, including Haldane spin-chains [250–252], bond-alternating $S = 1$ chains [253], $S = 1/2$ spin-ladders and tubes [254,255].

In the prototypical spin ladder material $(CH_3)_2CHNH_3CuCl_3$ (abbreviated as IPA-$CuCl_3$, where IPA denotes isopropyl ammonium), field-induced magnetic order occurs at $H_c = 9.6$ T at a temperature of 50 mK [254]. The spin ladder in this system is built of magnetic $S = 1/2$ Cu^{2+} ions and runs parallel to the a-axis of the triclinic $P\bar{1}$ crystal structure. Exchange interactions along the ladder legs are antiferromagnetic and the pairs of spins on each ladder rung are correlated ferromagnetically. The ordered phase induced by a magnetic field applied along the c-direction is characterized by a commensurate propagation vector $\mathbf{k} = (1/2,0,0)$, and the ordered moments were found to be oriented almost parallel to the crystallographic b-direction. The two spins on each ladder rung are parallel to each other. The refined value of the ordered moment at $H = 12$ T and 50 mK is 0.49(1) μ_B. A rather exotic type of magnetic ordering was observed in the case of the spin-tube material Cu_2Cl_4-$C_4H_8SO_2$ (Sul-Cu_2Cl_4) [255]. This system can be described as an array of four-leg spin tubes with dominant antiferromagnetic NN coupling along the legs and several weaker rung interactions of comparable strength. The tube's legs run along the crystallographic c-axis of the triclinic $P\bar{1}$ structure. An important feature of Sul-Cu_2Cl_4 is a partial geometric frustration of exchange interactions on the tube rungs. The conveniently small energy gap ($\Delta \approx 0.52$ meV) of this system has been overcome by applying a moderate magnetic field, $H_c \approx 4$ T. The neutron scattering experiments performed under magnetic field revealed an incommensurate chiral phase stabilized by geometrically frustrated interactions. The induced ordered state has been described as planar-spiral defined by a propagation vector $\mathbf{k} = (-0.22,0,0.48)$. All spins were assumed to be confined in the plane perpendicular to the direction of the applied field, $H||b$-axis. Figure 38 shows the spiral magnetic structure proposed for Sul-Cu_2Cl_4.

For the $S = 1$ linear-chain antiferromagnets, such as $Ni(C_5D_{14}N_2)_2N_3(PF_6)$ (known as NDMAP) [251,252] and $Ni(C_5H_{14}N_2)_2N_3(ClO_4)$ (NDMAZ) [250], the magnetic anisotropy has a strong impact on the phase transition, and the value of the critical field depends on the relative orientation of the applied field and

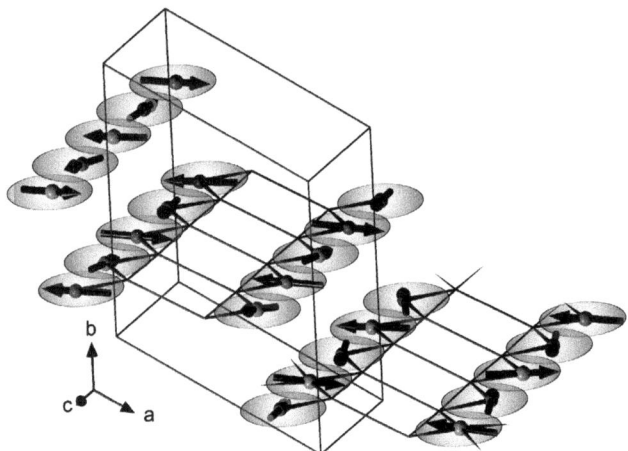

FIGURE 38 Field-induced incommensurate magnetic order in the quantum spin-tube Sul-Cu_2Cl_4 (Cu_2Cl_4-$C_4H_8SO_2$), described as a planar spiral structure with wave vector $\mathbf{k} = (-0.22,0,0.48)$.

the anisotropy tensor. NDMAP features two equivalent sets of Haldane spin chains with noncollinear local anisotropy axes, tilted by 16° relative to the c-axis. Experiments performed in magnetic fields applied along the c-axis (that is, the macroscopic magnetic anisotropy axis) demonstrated the development of a staggered antiferromagnetic long-range order with wave vector $\mathbf{k} = (0,1/2,1/2)$. A weak ferromagnetic tilt of all spins along the field direction was expected but could not be probed. The measured magnetic reflections were modeled by a collinear magnetic structure with staggered moments pointing along the crystallographic a-axis, perpendicular to the applied field. The magnitude of the moment was found to be 0.9(1) μ_B which is about half of the classical sublattice magnetization for an ordered $S = 1$ system. An applied magnetic field along a general direction (14.2° relative to the c-axis, in the ac crystallographic plane) revealed two consecutive field-induced transitions, indicating that long-range ordering occurs in each set of spin-chains independently. As the external field exceeds $H_c = 3.4$ T at 35 mK, one of the chains (type-A) acquires a long-range Néel order while the second chain (type-B) remains in the quantum-disordered gapped phase. In this regime the only magnetized Ni^{2+} ions in NDMAP are located on type-A chains and form a collinear antiferromagnetic structure with propagation vector $\mathbf{k} = (0,1/2,1/2)$, similar to the case of applying the field parallel to the c-axis. Above the second transition, at $H_c = 4.1$ T, the gap in type-B chains closes and the corresponding Ni^{2+} ions acquire static long-range antiferromagnetic spin correlations. The static magnetic moments are located on both A and B sublattices, and a definitive relative alignment between spins of the two sublattices is established via dipolar interactions and/or order-from-disorder fluctuation effects [252].

4.6 DISORDERED MAGNETIC STRUCTURES

Defects and disorder in magnetic materials can give rise to diffuse scattering that can be fruitfully analyzed to determine the atomic magnetization distribution. Many types of magnetic disorder are possible, such as from random occupation on a regular lattice (e.g., dilute metallic alloys), amorphous materials (e.g., bulk metallic glasses), from reduced dimensionality (layers, rods, clusters, etc.), and from frustrated magnetic interactions (e.g., spin ices, spin glasses, and spin liquids). Physically meaningful formalisms to calculate the magnetic diffuse scattering and the underlying atomic magnetization distribution range from mature for some types to rapidly evolving in other cases. In general, nuclear disorder scattering must be first separated from the magnetic disorder scattering, which often requires the use of polarized neutrons and applied magnetic fields. Also, the total scattering will generally have both elastic and inelastic contributions due to static and dynamic magnetic correlations. Multiple scattering and isotopic or nuclear-spin-disorder scattering may also be important corrections. Magnetic diffuse scattering studies benefit from diffractometers with large area detectors, and the ability to isolate the elastic-only contribution, such as enabled by the CORELLI spectrometer [256] at the Spallation Neutron Source, Oak Ridge National Laboratory. In the following, several examples of the occurrence of magnetic diffuse scattering will be presented.

4.6.1 Diffuse Scattering from Disordered Alloys

Inhomogeneities in the distribution of magnetization, such as in a ferromagnetic alloy, give rise to elastic diffuse scattering of neutrons. Shull and Wilkinson [29] noted that the intensity of this scattering depends upon the difference in the magnetic-scattering amplitudes of the two atomic species, and they gave a simple formula for the cross section in the case of an unpolarized incident neutron beam and for an unmagnetized sample with randomly oriented domains,

$$D(\mathbf{Q}) = \frac{2}{3}n(1-n)(f_1 - f_2)^2, \qquad (25)$$

where n and $(1-n)$ are the fractional abundances of the two species and f_1 and f_2 are the magnetic scattering factors. The factor 2/3 arises because of the randomness of the magnetization vector relative to the scattering vector for the unmagnetized sample. Moreover, they showed that analysis of magnetic diffuse scattering combined with magnetization data offers information on the magnetization distribution on an atomic scale.

The physical model for the elastic magnetic diffuse scattering for dilute-to-concentrated binary alloys is more completely stated [257] as

$$|\langle D(\mathbf{Q})\rangle|^2_{\text{diff}} = \sum_{\substack{j\in d \\ j'\in d'}} f_d(\mathbf{Q})f_{d'}(\mathbf{Q})\Delta\mu_j\Delta\mu_{j'}e^{i\mathbf{Q}\cdot(\mathbf{j}'-\mathbf{j})}e^{-(W_d-W_{d'})}, \qquad (26)$$

where $\Delta\mu$ is the difference between the magnetic moment of the impurity and the host atoms, such that

$$\Delta\mu_j = \Delta\mu\cdot p_j + \sum_{j'\neq j}\phi(\mathbf{j}-\mathbf{j}')p_{j'} \qquad (27)$$

If the site is occupied by an impurity atom, $p_j = 1$, otherwise, $p_j = 0$, and $\phi(\mathbf{j} - \mathbf{j}')$ is the effect of the impurity atom on the spin of the host. If one assumes that the magnetic defects associated with each impurity atom are independent of each other, i.e., ignoring the effect of the impurity atom on $f(\mathbf{Q})$, then

$$|\langle D(\mathbf{Q})\rangle|^2_{\text{diff}} = f^2c(1-c)[\Delta\mu + \phi(\mathbf{Q})]^2e^{-2W} \qquad (28)$$

where $\phi(\mathbf{Q})$ is the Fourier transform of $\phi(\mathbf{j} - \mathbf{j}')$. The total magnetic-defect moment per impurity atom, other than that at the impurity site itself, is $\phi(\mathbf{Q} = 0)$.

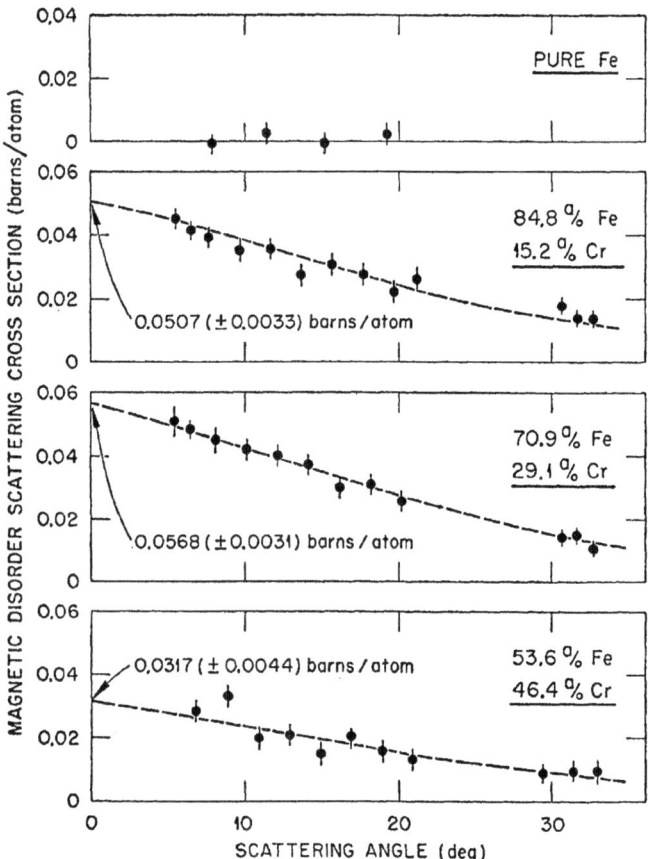

FIGURE 39 Ferromagnetic-disorder scattering for a series of disordered Fe-Cr alloys. The dashed lines are the form factor for pure iron. *Reproduced with permission from Ref. [29].*

Examples of magnetic diffuse scattering studies in alloys date back to the mid-1950s, around the time of the first neutron diffraction experiments. Shull and Wilkinson [29] presented neutron powder diffraction results for a series of alloys of transition elements and examined the appearance of ferromagnetic-disorder scattering as detected by the change in diffuse scattering intensity when the sample was magnetized parallel to the scattering vector. For the example of Fe-Cr (Figure 39), the absolute cross section for pure Fe is much smaller than for the Fe-Cr alloys, for which ferromagnetic-disorder scattering is present, and the intercept of the form factor curves at zero scattering angle are used to evaluate the differences in the magnetic moments of the iron and chromium atoms.

4.6.2 Diffuse Scattering from Systems with Reduced Dimensionality

Progression from a three-dimensional magnetic order down to zero dimensionality, can lead to magnetic diffuse scattering when the successive building

Magnetic Structures Chapter | 4 277

units (e.g., layers, ladders, clusters, etc.) are magnetically disordered with respect to the neighboring building units. A quite interesting example of a system displaying diffuse scattering from low-dimensional magnetic spin correlations is $SrHo_2O_4$ [258]. As discussed in Section 5, the SrR_2O_4 systems consist of magnetic ions arranged in zigzag chains or triangular ladders that run along the c-axis. These chains interconnect to form a distorted honeycomb structure in the ab plane. The low temperature magnetic structure of $SrHo_2O_4$ has been studied on a single crystal sample using unpolarized and polarized neutrons experiments using D7 [259] and D10 [260] at the ILL and WISH [261] at ISIS. $SrHo_2O_4$ displays two distinct types of short-range magnetic order: one characterized by the appearance of broad diffuse-scattering peaks around $\mathbf{k} = (0,0,0)$ positions in the $(h,k,0)$ scattering plane, and a second type characterized by planes of scattering intensity at the $(h,k,\pm l/2)$ positions. These planes are seen as "rods" of scattering intensity in both the $(h,0,l)$ and $(0,k,l)$ planes in reciprocal space at half-integer positions along the l-axis (Figure 40), which suggests that the second type of short-range order present in $SrHo_2O_4$ is principally one-dimensional in nature. The observed lack of coupling between the two types of short-range ordering suggests that each of the crystallographically inequivalent Ho^{3+} sites in $SrHo_2O_4$ contributes to only one of the two coexisting structures. The polarized neutron experiments

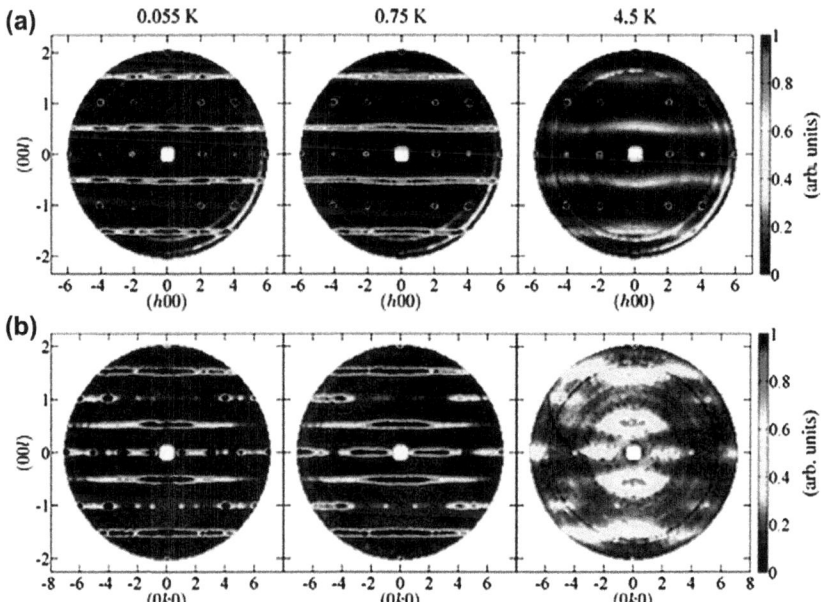

FIGURE 40 Diffuse scattering from $SrHo_2O_4$ at different temperatures, measured using polarized neutrons at the D7 diffractometer, ILL. Two independent components of the scattering function, non-spin-flip and spin-flip channels, were measured in the $(h,0,l)$ (a) and $(0,k,l)$ planes (b), respectively. Rods of scattering intensity are visible at low temperature at $(h,0,1/2)$ and $(0,k,1/2)$. Diffuse magnetic peaks associated with a $\mathbf{k} = (0,0,0)$ structure are also visible in the $(0,k,l)$ plane at 55 mK. *Reproduced with permission from Ref. [258].*

performed using the D7 instrument at ILL were configured to measure two independent components of the scattering function: NSF and SF channels, and allowed to gain more insight about the nature of the diffuse magnetic scattering [258]. In the $(0,k,l)$ measurement, the diffuse scattering appears at $(0,k,1/2)$ only in the SF channel demonstrating that the spin structure has no components parallel to the a-direction, i.e., all of the magnetic moments lie in the bc plane. In the $(h,0,l)$ measurement the diffuse scattering appears at $(h,0,1/2)$ only in the NSF measurement, which indicates that all of the spins lie parallel with the b-axis. The diffuse-scattering peaks associated with the **k** = (0,0,0) magnetic structure are only visible in the SF data for both the $(h,0,l)$ and $(0,k,l)$ scattering planes, which means that all of the spins that give rise to the **k** = (0,0,0) scattering are parallel to the c-axis. This is in agreement with the previously proposed magnetic structure that was established from a refinement performed using powder neutron diffraction data.

4.6.3 Diffuse Magnetic Scattering in Frustrated Magnets

In frustrated magnets, the macroscopic degeneracy of magnetic ground states can lead to the suppression of long-range magnetic order. As such, the magnetic diffuse scattering that replaces the classical magnetic Bragg peaks remains the distinctive signature of a short-range magnetic correlation in the system. Pure β-Mn is a unique example of an elemental quantum spin liquid. Elemental manganese adopts several allotropes depending on temperature and pressure. β-Mn, which is stable in the temperature range, 1000−1373 K, crystallizes in the A13 structure type (cubic $P4_132$, a = 6.315 Å) with two unique Mn positions (Mn$_1$ 8a; Mn$_2$ 12d). Large, high-quality single-crystals have only been grown using Co substitution. The β-Mn structure can tolerate ~40 atomic % Co substitution and neutron diffraction demonstrates that the Co substitution occurs only on the Mn$_1$ sites, which are 40% of the total Mn sites. The magnetic behavior of both the pure and Co-substituted β-Mn is explained by models in which the Mn$_1$ (8a sites) are nonmagnetic. As illustrated in Figure 41, the Mn$_2$ atoms form paddle-wheel arrangements made-up of equivalent equilateral triangles (aka a "hyperkagomé" three-dimensional network), that have been identified as the source for magnetic frustration. Paddison et al. [262] showed that β-Mn$_{0.8}$Co$_{0.2}$ does not order down to 1.5 K and the magnetic diffuse scattering observed from the neutron diffraction of several doped samples is unlikely to result from Co doping but is instead an inherent property of β-Mn. Highly structured diffuse scattering patterns measured at T = 1.5 K are shown in four reciprocal−space planes in Figure 42 (top left panels). The diffuse magnetic scattering modeled by employing RMC refinements and mean-field theory calculations suggested that ferromagnetic correlations between fifth-nearest neighbors J_5 are as strong as NN

Magnetic Structures Chapter | 4 279

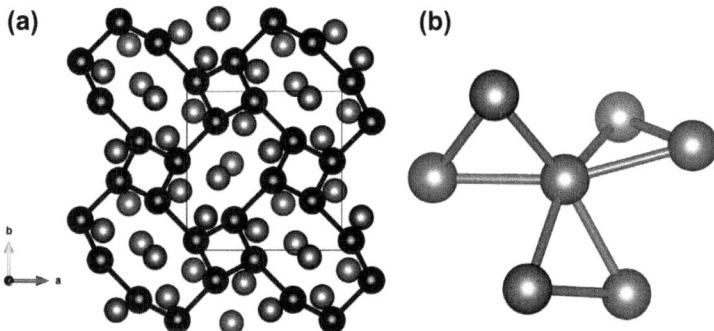

FIGURE 41 (a) β-Mn structure, (1,0,0) projection, showing the Mn_1 (dark) and Mn_2 (light) atoms. Mn_1 forms right-handed spirals running in the <100> directions that are cross-linked to make a network of Mn_1 only comprising the shortest Mn...Mn near-neighbors (shown as connecting rods); (b) The Mn_2 atoms form these paddle-wheel arrangements made-up of equivalent equilateral triangles (aka "hyperkagomé" network), the purported source of the frustration.

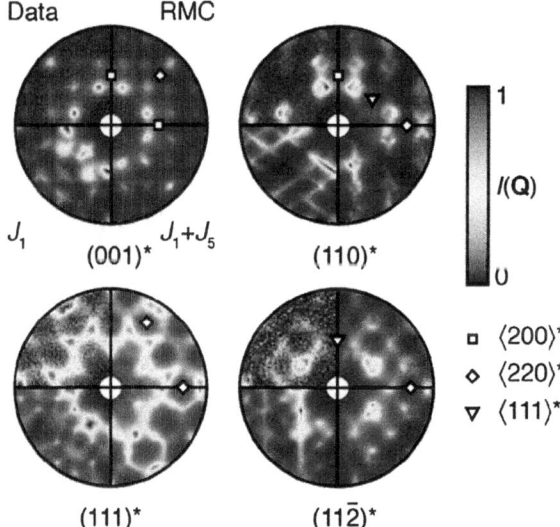

FIGURE 42 Magnetic diffuse scattering patterns for β-$Mn_{0.8}Co_{0.2}$ in four reciprocal lattice planes, clockwise from top left (0,0,1), (1,1,0), (1,1,−2), and (1,1,1). For each, the top left panel shows experimental neutron scattering data at $T = 1.5$ K, the top right panel shows the reverse Monte Carlo (RMC) fit to data, the bottom panel shows calculated scattering using a mean-field theory for a J_1–J_5 model (right), and for antiferromagnetic J_1 interactions only (left). Whereas the J_1-only model is a poor description of the data, both the RMC fit and the J_1–J_5 model give good agreement with experiment. *Reproduced with permission from Ref. [262].*

antiferromagnetic correlations J_1. The data can be fit well using a $J_1 + J_5$ model or RMC, but is inadequately described with a J_1-only model that was previously assumed to explain frustration of single spins on a hyperkagomé lattice (Figure 42).

Spinel is one of the more common structure types (AB_2X_4, cubic, $Fd\bar{3}m$) that can manifest frustration owing to the tetrahedral network of NN B-sites, and is one of many structure types that have interactions in triangular motifs. A good spinel example is $ZnFe_2O_4$, which is often thought of as an antiferromagnet ($T_N = 13$ K); however, magnetic susceptibility data give a Curie–Weiss temperature of about 100 K; therefore, strong spin frustration is also expected. Kamazawa et al. [263] conducted neutron scattering measurements on a single-crystal sample and demonstrated the lack of long-range order in pure $ZnFe_2O_4$, but only short-range order arises because spin correlations cannot develop fully in the presence of geometrical frustration. The measured elastic contour maps in the $(H,K,0)$ and (H,H,L) zones at 15 and 1.5 K are shown in Figure 43. The diffuse scattering follows the magnetic form factor,

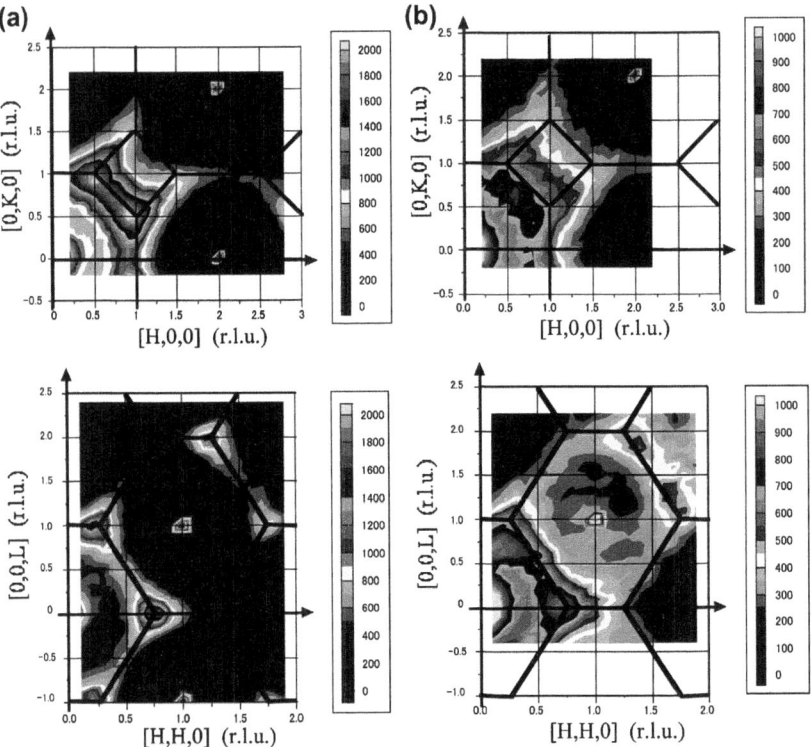

FIGURE 43 Contour maps of magnetic diffuse scattering in $ZnFe_2O_4$ $(H,K,0)$ and (H,H,L) zones at 15 K (a) and 1.5 K (b). The solid lines indicate the Brillouin-zone boundaries of the fcc structure. The diffuse scattering in the $(H,K,0)$ zone of is well explained with the assumptions that the nearest-neighbor interaction is ferromagnetic and the third-neighbor interaction is antiferromagnetic. *Reproduced with permission from Ref. [263].*

and is temperature dependent, which is consistent with the scattering originating from the magnetic contribution of the B atoms. At 15 K, strong diffuse scattering has been found to be distributed along and slightly inside the first Brillouin zone boundaries of the fcc structure with the strongest intensity around $\sim (0.7, 0.7, 0)$, while at 1.5 K, strong diffuse scattering appears at $(1, 1/2, 0)$, $(3/4, 3/4, 0)$, $(1/4, 1/4, 1)$, and symmetrical positions. The observed diffuse pattern and the correlation lengths deduced from diffuse scattering line widths suggest that the magnetic diffuse scattering originates from ferromagnetic coupling of the first-neighbor Fe^{+3} spins at higher temperatures and antiferromagnetic coupling of third-neighbor spins at lower temperatures.

4.7 CONCLUDING REMARKS

The development of user-friendly programs for nonexperts will be key to grow the user community engaged in magnetic structural studies. Already, the adoption of standard descriptors for magnetic structures (e.g., mCIF file format) is enabling magnetic structural databases to become a significant resource. Next-generation neutron scattering facilities like the proposed second target station at the SNS at Oak Ridge (USA), the ESS in Lund (Sweden), and the CSNS (China), where the instruments will be designed to allow measurements of much smaller samples and equipped with polarization capabilities, will have a major positive impact on magnetic structural studies. Also, experimental advances in sample environments will enable more extreme conditions, e.g., higher pressures and higher static and pulsed magnetic fields, which will all directly impact the discovery of more exotic magnetic structures.

The outlook for continued growth of magnetic structure studies is good, owing in large part to the expanding availability of neutron powder diffractometers worldwide, the ever-improving user friendliness of the analysis tools, and the proliferation of magnetic structure determination workshops. But all of these are simply consequences of the demands from the neutron user community. What is really driving the growth is the science. New materials are being synthesized at an unprecedented rate. Our need "to do more with less" is driving technological innovation that requires material property control at the electron level. Fundamental physics and chemistry theories in magnetism are being pushed to new limits. In parallel, the empirical knowledge that will be gained from data mining large numbers of magnetic crystal structures collected together in functional databases is yet to be realized, but likely will follow a similar evolution as that gained from implementation of crystal structure databases.

ACKNOWLEDGMENTS

The authors completed this work while employed at Oak Ridge National Laboratory which is managed by UT-Battelle, LLC for the U.S. Department of Energy, under Contract No. DE-AC05000R22725, and were sponsored by the Scientific User Facilities Division, Office of Basic Energy Sciences, US Department of Energy.

REFERENCES

[1] C.M. Hurd, Contemp. Phys. 23 (1982) 469.
[2] F. Fernandez-Alonso, D.L. Price, Neutron Scattering — Fundamentals, in: Experimental Methods in the Physical Sciences, vol. 44, Elsevier, Amsterdam, 2013.
[3] J. Chadwick, Proc. R. Soc. London A 136 (1932) 692.
[4] D.P. Mitchell, P.N. Powers, Phys. Rev. 50 (1936) 486.
[5] H. von Halban, P. Preiswerk, C. R. Acad. Sci. Paris 203 (1936) 73.
[6] F. Bloch, Phys. Rev. 50 (1937) 259.
[7] J.S. Schwinger, Phys. Rev. 51 (1937) 544.
[8] O. Halpern, M.H.J. Johnson, Phys. Rev. 51 (1937) 992.
[9] O.R. Frisch, H. von Halban, J. Koch, Phys. Rev. 53 (1938) 719.
[10] J.G. Hoffman, M.S. Livingston, H.A. Bethe, Phys. Rev. 51 (1937) 214.
[11] P.N. Powers, Phys. Rev. 54 (1938) 827.
[12] L.W. Alvarez, F. Bloch, Phys. Rev. 57 (1940) 111.
[13] O. Halpern, M.H.J. Johnson, Phys. Rev. 55 (1939) 898.
[14] R.M. Moon, Phys. B 267–268 (1999) 1.
[15] O. Halpern, M. Hamermesh, M.H. Johnson, Phys. Rev. 59 (1941) 981.
[16] O. Halpern, T. Holstein, Phys. Rev. 59 (1941) 960.
[17] [Online]. Available: http://www.nobelprize.org/nobel_prizes/physics/laureates/1994/shull-bio.html (accessed 20.12.14).
[18] T.E. Mason, T.J. Gawne, S.E. Nagler, M.B. Nestor, J.M. Carpenter, Acta Cryst. A 69 (2012) 37.
[19] M.K. Wilkinson, Phys. B 137 (1986) 3.
[20] M. Prévot, D. Dunlop, Phys. Earth Planet. Inter. 126 (2001) 3.
[21] L. Néel, Ann. Phys. 3 (1948) 137.
[22] C. Shull, W. Strauser, E. Wollan, Phys. Rev. 83 (1951) 333.
[23] C. Shull, E. Wollan, W. Strauser, Phys. Rev. 81 (1951) 483.
[24] W. Koehler, E. Wollan, Phys. Rev. 92 (1953) 1380.
[25] W. Koehler, E. Wollan, Phys. Rev. 97 (1955) 1177.
[26] R. Erickson, Phys. Rev. 90 (1953) 779.
[27] M. Wilkinson, N. Gingrich, C. Shull, J. Phys. Chem. Solids 2 (1957) 289.
[28] C. Shull, M. Wilkinson, Rev. Mod. Phys. 25 (1953) 100.
[29] C. Shull, M. Wilkinson, Phys. Rev. 97 (1955) 304.
[30] M. Wilkinson, C. Shull, Phys. Rev. 103 (1956) 516.
[31] H. Gersch, C. Shull, M. Wilkinson, Phys. Rev. 103 (1956) 525.
[32] E. Wollan, W. Koehler, Phys. Rev. 100 (1955) 545.
[33] C.G. Shull, E.O. Wollan, W.C. Koehler, Phys. Rev. 84 (1951) 912.
[34] R. Moon, T. Riste, W. Koehler, Phys. Rev. 181 (1969) 920.
[35] J. Brown, Magnetic form factors, in: International Tables for Crystallography, Volume C: Mathematical, Physical and Chemical Tables, Kluwer Academic Publishers, Dordrecht, 1995, pp. 391–399.
[36] R. Moon, Phys. B 137 (1986) 19.
[37] C.G. Shull, J. Phys. Radium 20 (1959) 169.
[38] E.F. Bertaut, Acta Cryst. A 24 (1968) 217.
[39] H. Rietveld, J. Appl. Crystallogr. 2 (1969) 65.
[40] D. Wiles, R. Young, J. Appl. Crystallogr. 14 (1981) 149.
[41] A. Larson, R. Von Dreele, General Structure Analysis System (GSAS), Los Alamos National Laboratory, Los Alamos, 1994.

[42] J. Rodriguez-Carvajal, Phys. B 192 (1993) 55.
[43] V. Petříček, M. Dušek, JANA98-Crystallographic Computing System, Academy of Sciences of the Czech Republic, Praha, Czech Republic, 1998.
[44] A. Wills, Phys. B 276 (2000) 680.
[45] [Online]. Available: http://www.ucl.ac.uk/chemistry/staff/academic_pages/andrew_wills/ (accessed 25.07.15).
[46] W. Marshall, S.W. Lovesey, Theory of Thermal Neutron Scattering, Clarendon Press, Oxford, 1971.
[47] S.W. Lovesey, Theory of Neutron Scattering from Condensed Matter, Clarendon Press, Oxford, 1984.
[48] G.L. Squires, Introduction to the Theory of Thermal Neutron Scattering, Dover Publications, Cambridge, New York, 1978.
[49] T. Chatterji, Neutron Scattering from Magnetic Materials, Elsevier, Amsterdam, 2006.
[50] I.A. Zaliznyak, S.-H. Lee, Magnetic neutron scattering, in: Y. Zhu (Ed.), Modern Techniques for Characterizing Magnetic Materials, Kluwer Academic, New York, 2004.
[51] S. Shamoto, M. Sato, J.M. Tranquada, B.J. Sternlieb, G. Shirane, Phys. Rev. B 48 (1993) 13817.
[52] J.W. Lynn, G. Shirane, M. Blume, Phys. Rev. Lett. 37 (1976) 154.
[53] D.F. Johnston, Proc. Phys. Soc. (London) 88 (1966) 37.
[54] E. Clementi, C. Roetti, At. Data Nucl. Data Tables 14 (3−4) (1979) 177.
[55] A.J. Freeman, J.P. Desclaux, J. Magn. Magn. Mater. 12 (1979) 11.
[56] J. Descleaux, A. Freeman, J. Magn. Magn. Mater. 8 (1978) 119.
[57] A.D. McLean, R.S. McLean, At. Data Nucl. Data Tables 26 (1981) 197.
[58] K. Kobayash, T. Nagaob, M. Itob, Acta Cryst. A A67 (2011) 473.
[59] Y.A. Izyumov, V.E. Naish, R.P. Ozerov, Neutron Diffraction of Magnetic Materials, Consultants Bureau, Plenum Publishing Corporation, New York, 1991.
[60] J. Rossat-Mignod, in: D.L. Price, K. Skold (Eds.), Neutron Scattering in Condensed Matter Research, Academic Press, Amsterdam, 1986.
[61] P. Burlet, S. Quezèl, J. Rossat-Mignod, Solid State Commun. 55 (1985) 1057.
[62] J. Rossat-Mignod, P. Burlet, S. Quezel, J.M. Effatin, in: Physics of Magnetic Materials, World Scientific, Singapore, 1985, p. 411.
[63] J. Rossat-Mignod, G.H. Lander, P. Burlet, in: G.H. Lander, A.J. Freeman (Eds.), Handbook of the Physics and Chemistry of the Actinides, vol. 1, North-Holland Publ., Amsterdam, 1984.
[64] H.R. Child, M.K. Wilkinson, J.W. Cable, W.C. Koehler, E.O. Wollan, Phys. Rev. 131 (1963) 922.
[65] P. Schobinger-Papamantellos, J. Schefer, J.H.V.J. Brabers, K.H.J. Buschow, J. Alloys Comp. 215 (1994) 111.
[66] O. Zaharko, P. Schobinger-Papamantellos, W. Sikora, C. Ritter, Y. Janssenk, E. Bruck, F.R. de Boer, K.H.J. Buschow, J. Phys. Condens. Matter. 10 (1998) 6553.
[67] M. Reehuis, P.J. Brown, W. Jeitschko, M.H. Möller, T. Vomhof, J. Phys. Chem. Solids 54 (1993) 469−475.
[68] A. Fennell, V.Y. Pomjakushin, A. Uldry, B. Delley, B. Prévost, A. Désilets-Benoit, A.D. Bianchi, R.I. Bewley, B.R. Hansen, T. Klimczuk, R.J. Cava, M. Kenzelmann, Phys. Rev. B 89 (2014) 224511.
[69] A.S. Wills, J. Phys. IV France 11 (2001) 133.
[70] W.C. Koehler (Ed.), Magnetic Properties of Rare Earth Metals, Plenum, New York, 1972, p. 81.
[71] B. Lebech, J. Appl. Phys. 52 (1981) 2019.

[72] J.W. Cable, E.O. Wollan, W.C. Koehler, M.K. Wilkinson, Phys. Rev. 140 (1965) A1896.
[73] W.C. Koehler, J.W. Cable, E.O. Wollan, M.K. Wilkinson, Phys. Rev. 126 (1962) 1672.
[74] R.M. Moon, J.W. Cable, W.C. Koehler, J. Appl. Phys. 33 (1964) 1041.
[75] W.C. Koehler, J. Appl. Phys. 36 (1965) 1078.
[76] M. Kuznietz, P. Burlet, J. Rossat-Mignod, J. Magn. Magn. Mater. 54−57 (1986) 553.
[77] Y. Nambu, L.L. Zhao, E. Morosan, K. Kim, G. Kotliar, P. Zajdel, M.A. Green, Phys. Rev. Lett. 106 (2011) 037201.
[78] C.M. Thompson, J.E. Greedan, V.O. Garlea, R. Flacau, M. Tan, Inorg. Chem. 53 (2) (2014) 1122.
[79] W.C. Koehler, J.W. Cable, M.K. Wilkinson, E.O. Wollan, Phys. Rev. 151 (1966) 414.
[80] A. Yoshimori, J. Phys. Soc. Jpn. 14 (6) (1959) 807.
[81] [Online]. Available: https://www.ill.eu/sites/fullprof/ (accessed 25.07.15).
[82] E.M. Lifshitz, L.D. Landau, Electrodynamics of Continuous Media, Pergamon Press, Oxford, 1960.
[83] A.V. Shubnikov, Symmetry and Anti-symmetry of Finite Figures, Izd. Akad. Nauk SSSR, Moscow, 1951.
[84] N.V. Belov, N.N. Neronova, T.S. Smirnova, Sov. Phys. Crytallogr. 2 (2) (1957) 311.
[85] R. Guccione, W. Opechowski, Magnetic symmetry, in: G.T. Rado, H. Shull (Eds.), Magnetism, vol. II A, Academic Press, New York, 1965, p. 105. Chapter 3.
[86] W. Opechowski, Crystallographic and Metacrystallographic Groups, Elsevier Science Publishers, Amsterdam, 1986.
[87] D.B. Litvin, Acta Cryst. A 64 (2008) 419.
[88] H. Grimmer, Acta Cryst. A 65 (2009) 145.
[89] A. Janssen, T. Janner, Acta Cryst. A 36 (1980) 399.
[90] V. Petříček, J. Fuksa, M. Dušek, Acta Cryst. A 66 (2010) 649.
[91] J.M. Perez-Mato, J.L. Ribeiro, V. Petříček, M.I. Aroyo, J. Phys. Condens. Matter. 24 (2012) 163201.
[92] J. Perez-Mato, S. Gallego, E. Tasci, L. Elcoro, G. de la Flor, M.I. Aroyo, Annu. Rev. Mater. Res. 45 (2015) 217.
[93] I.E. Dzyaloshinsky, J. Phys. Chem. Solids 4 (4) (1958) 241.
[94] I.E. Dzyaloshinsky, Sov. Phys. JETP 5 (1957) 1259.
[95] E.F. Bertaut, J. Appl. Phys. Suppl. 33 (1962) 1138.
[96] E.F. Bertaut, J. Phys. Coll. C1 (32) (1971) 462.
[97] F. Bourée, J. Rodríguez-Carvajal, EPJ Web Conf. 22 (2012) 00010.
[98] Y.A. Izyumov, R.P. Ozerov, Magnetic Neutron Diffraction, Plenum Press, New York, 1970.
[99] A.S. Wills, J. Mater. Chem. 15 (2005) 245.
[100] L.C. Chapon, EPJ Web Conf. 22 (2012) 13.
[101] H. Heesch, Z. Kristallogr. 71 (1929) 95.
[102] M.B. Boisen Jr., Z. Kristallogr. 145 (1977) 197.
[103] M.B. Boisen Jr., G.V. Gibbs, Can. Mineral. 16 (1978) 293.
[104] [Online]. Available: http://iso.byu.edu/iso/isotropy.ph (accessed 01.08.15).
[105] E.S. Tasci, G. de la Flor, D. Orobengoa, C. Capillas, J.M. Perez-Mato, EPJ Web Conf. 22 (2012) 00009.
[106] S.V. Gallego, E.S. Tasci, G. de la Flor, J.M. Perez-Mato, M.I. Aroyo, J. Appl. Cryst. 45 (2012) 1236.
[107] [Online]. Available: http://www.cryst.ehu.es/ (accessed 10.08.15).
[108] P.M. de Wolff, Acta Cryst. A 30 (1974) 777.

[109] I. Urcelay-Olabarria, J.M. Perez-Matto, J.L. Ribeiro, J.L. Garcia-Munoz, E. Ressouche, V. Skumryev, A.A. Mukhin, Phys. Rev. B 87 (2013) 014419.
[110] T. Janssen, A. Janner, A. Looijenga-Vos, P.M. de Wolff, The space-group distribution of molecular organic structures, in: International Tables for Crystallography C, Kluwer Academic Publishers, Dordrecht, 2004, pp. 907−945.
[111] [Online]. Available: http://jana.fzu.cz/ (accessed 25.07.15).
[112] R. Ballou, B. Ouladdiaf, Representation analysis of magnetic structures, in: Neutron Scattering from Magnetic Materials, Elsevier, New York, 2006, pp. 93−152.
[113] E.M. Lifshitz, L.D. Landau, Statistical Physics, Pergamon Press, Oxford, 1959.
[114] O.V. Kovalev, Representations of the Crystallographic Space Groups, second ed., Gordon and Breach, New York, 1993.
[115] J. Schweizer, C. R. Phys. 6 (2005) 375.
[116] L.C. Chapon, P.G. Radaelli, Phys. Rev. B 76 (2007) 054428.
[117] E. Hovestreydt, M. Aroyo, S. Sattler, H. Wondratschek, J. Appl. Cryst. 25 (1992) 544.
[118] W. Sikora, F. Białas, L. Pytlik, J. Appl. Cryst. 37 (2004) 1015.
[119] B.M. Knapp, F. Sinclair, C. Wilkinson, J. Appl. Cryst. 7 (1974) 370−372.
[120] C. Wilkinson, J. Appl. Cryst. 24 (1991) 365.
[121] A. Wills, Z. Kristallogr. Suppl. 30 (2009) 39.
[122] A.S. Wills, Appl. Phys. A 74 (Suppl. 1) (2002) S856.
[123] F. Izumi, T. Ikeda, Mater. Sci. Forum 321−324 (2000) 198.
[124] B. Roessli, P. Böni, Polarized neutron scattering, in: E.R. Pike, P. Sabatier (Eds.), Scattering and Inverse Scattering in Pure and Applied Science, Academic Press, 2002, pp. 1242−1263.
[125] A. Delapalme, Aust. J. Phys. 41 (1988) 383−391.
[126] S. Chandrasekhar, Adv. Phys. 9 (36) (1960) 363.
[127] A. Yaouanc, P. Dalmas de Reotier, Muon Spin Rotation, Relaxation, and Resonance: Applications to Condensed Matter, Oxford University Press, Oxford, 2011.
[128] [Online]. Available: http://magcryst.org/ (accessed 01.08.15).
[129] D. Litvin, Acta Cryst. A57 (2001) 729.
[130] K. Momma, F. Izumi, J. Appl. Crystallogr. 44 (2011) 1272.
[131] [Online]. Available: http://www.crystalimpact.com/diamond (accessed 25.07.15).
[132] C. Lacroix, P. Mendels, F. Mila, Introduction to Frustrated Magnetism, in: Springer Series in Solid-state Sciences, 2010.
[133] M.A. Subramanian, G. Aravamudan, G.V. Subba Rao, Prog. Solid State Chem. 15 (1983) 55.
[134] J.E. Greedan, Magnetic Properties of Nonmetallic Compounds Based on Transition Elements, in: Landolt-Börnstein, New Series, vol. 27, Springer-Verlag, Berlin, 1992, p. 100.
[135] J.S. Gardner, M.J.P. Gingras, J.E. Greedan, Magnetic pyrochlore oxides, Rev. Mod. Phys. 83 (2010) 53.
[136] A. Wills, Phys. B 276−278 (2000) 680.
[137] A.S. Wills, M.E. Zhitomirsky, B. Canals, J.P. Sanchez, P. Bonville, J. Phys. Condens. Matter. 18 (2006) L37.
[138] S.E. Palmer, J.T. Chalker, Phys. Rev. B 62 (2000) 488.
[139] J.D.M. Champion, A.S. Wills, T. Fennell, S.T. Bramwell, J.S. Gardner, M.A. Green, Phys. Rev. B 64 (2001) 140407.
[140] J.R. Stewart, G. Ehlers, A.S. Wills, S.T. Bramwell, J.S. Gardner, J. Phys. Condens. Matter. 16 (2004) L321.

[141] A. Poole, A.S. Wills, E.E. Lelièvre-Berna, J. Phys. Condens. Matter. 19 (452201) (2007).
[142] C.R. Wiebe, J.S. Gardner, S.-J. Kim, G. Luke, A.S. Wills, B.D. Gaulin, J.E. Greedan, I. Swainson, Phys. Rev. Lett. 93 (2004) 076403.
[143] H. Cao, A. Gukasov, I. Mirebeau, P. Bonville, G. Dhalenne, Phys. Rev. Lett. 101 (2008) 196402.
[144] A.P. Sazonov, A. Gukasov, H.B. Cao, P. Bonville, E. Ressouche, C. Decorse, I. Mirebeau, Phys. Rev. B 88 (2013) 184428.
[145] S.-H. Lee, H.H. Takagi, D.D. Louca, M.M. Matsuda, S. Ji, H. Ueda, Y. Ueda, T. Katsufuji, J.-H. Chung, S. Park, S.-W. Cheong, C. Broholm, J. Phys. Soc. Jpn. 79 (1) (2010) 011004.
[146] M. Reehuis, A. Krimmel, N. Büttgen, A. Loidl, A. Prokofiev, Eur. Phys. J. B 35 (2003) 311.
[147] V.O. Garlea, R. Jin, D. Mandrus, B. Roessli, Q. Huang, M. Miller, A.J. Schultz, S.E. Nagler, Phys. Rev. Lett. 100 (2008) 066404.
[148] G.J. MacDougall, V.O. Garlea, A.A. Aczel, H.D. Zhou, S.E. Nagler, Phys. Rev. B 86 (2012) 060414(R).
[149] K. Tomiyasu, J. Fukunaga, H. Suzuki, Phys. Rev. B 70 (2004) 214434.
[150] I. Kagomiya, K. Tomiyasu, J. Phys. Soc. Jpn. 73 (9) (2004) 2539.
[151] D. Bergman, J. Alicea, E. Gull, S. Trebst, L. Balents, Nat. Phys. 3 (2007) 487.
[152] G.J. MacDougall, D. Gout, J.L. Zarestky, G. Ehlers, A. Podlesnyak, M.A. Guire, D. Mandrus, S.E. Nagler, Proc. Natl. Acad. Sci. 108 (2011) 15693.
[153] B. Roy, A. Pandey, Q. Zhang, T.W. Heitmann, D. Vaknin, D.C. Johnston, Y. Furukawa, Phys. Rev. B 88 (2013) 174415.
[154] O. Zaharko, S. Toth, O. Sendetskyi, A. Cervellino, A. Wolter-Giraud, T. Dey, A. Maljuk, V. Tsurkan, Phys. Rev. B 90 (2014) 134416.
[155] L. Savary, E. Gull, S. Trebst, J. Alicea, D. Bergman, L. Balents, Phys. Rev. B 84 (2011) 064438.
[156] A. Krimmel, M. Mücksch, V. Tsurkan, M.M. Koza, H. Mutka, C. Ritter, D.V. Sheptyakov, S. Horn, A. Loidl, Phys. Rev. B 73 (2006) 014413.
[157] V. Fritsch, J. Hemberger, N. Buttgen, E.-W. Scheidt, H.-A. Krug von Nidda, A. Loidl, V. Tsurkan, Phys. Rev. Lett. 92 (2004) 116401.
[158] C. Jones, P.D. Battle, J. Solid State Chem. 78 (1989) 108.
[159] A. Munoz, J. Alonso, M. Casais, M. Martinez-Lope, M. Fernandez-Diaz, J. Phys. Condens. Matter 14 (2002) 8817.
[160] P.D. Battle, J.B. Goodenough, R. Price, J. Solid State Chem. 46 (1983) 234.
[161] H. Karunadasa, Q. Huang, B.G. Ueland, P. Schiffer, R.J. Cava, Proc. Natl. Acad. Sci. 100 (14) (2003) 8097.
[162] A.A. Aczel, P.J. Baker, D.E. Bugaris, J. Yeon, H.-C. zur Loye, T. Guidi, D.T. Adroja, Phys. Rev. Lett. 112 (2014) 117603.
[163] D. Serrate, J.M. de Teresa, M.R. Ibarra, J. Phys. Condens. Matter 19 (2007) 023201.
[164] A.K. Paul, M. Reehuis, V. Ksenofontov, B. Yan, A. Hoser, D.M. Többens, P. Adler, M. Jansena, C. Felser, Phys. Rev. Lett. 111 (2013) 167205.
[165] Y. Krockenberger, K. Mogare, M. Reehuis, M. Tovar, M. Jansen, G. Vaitheeswaran, V. Kanchana, F. Bultmark, A. Delin, F. Wilhelm, A. Rogalev, A. Winkler, L. Alff, Phys. Rev. B 75 (2007) 020404(R).
[166] R. Morrow, R. Mishra, O.D. Restrepo, M.R. Ball, W. Windl, S. Wurmehl, U. Stockert, B. Buechner, P.M. Woodward, J. Am. Chem. Soc. 135 (2013) 18824.
[167] B. Yan, A.K. Paul, S. Kanungo, M. Reehuis, A. Hoser, D.M. Többens, W. Schnelle, R.C. Williams, T. Lancaster, F. Xiao, J.S. Möller, S.J. Blundell, W. Hayes, C. Felser, M. Jansen, Phys. Rev. Lett. 112 (2014) 147202.

[168] A.A. Aczel, D.E. Bugaris, J. Yeon, C. de la Cruz, H.-C. zur Loye, S.E. Nagler, Phys. Rev. B 88 (2013) 014413.
[169] R.J. Birgeneau, H.J. Guggenheim, G. S, Phys. Rev. Lett. 22 (14) (1969) 720.
[170] R. Plumier, E. Legrand, J. Phys. Radium 24 (1963) 741.
[171] D.E. Cox, G. Shirane, R.J. Birgeneau, J.B. Macchesney, Phys. Rev. 188 (2) (1969) 930.
[172] M. Leonowicz, K. Poeppelmeier, J. Longo, J. Solid State Chem. 59 (1985) 71.
[173] T. Kimura, Y. Tomioka, H. Kuwahara, A. Asamitsu, M. Tamura, Y. Tokura, Science 274 (1996) 1698.
[174] M. Lobanov, M. Greenblatt, E.N. Caspi, J.D. Jorgensen, D.V. Sheptyakov, B.H. Toby, C. Botez, P.W. Stephens, J. Phys. Condens. Matter. 16 (2004) 5339.
[175] Y. Kamihara, T. Watanabe, M. Hirano, H. Hosono, J. Am. Chem. Soc. 130 (2008) 3296.
[176] C. de la Cruz, Q.Q. Huang, J.W. Lynn, J. Li, W. Ratcliff II, J.L. Zarestky, H.A. Mook, G.F. Chen, J.L. Luo, N.L. Wang, P. Dai, Nature 453 (2008) 899.
[177] Q. Huang, Y. Qiu, W. Bao, J.W. Lynn, M.A. Green, Y.C. Gasparovic, T. Wu, G. Wu, X.H. Chen, Phys. Rev. Lett. 101 (2008) 257003.
[178] D.J. Singh, Sci. Technol. Adv. Mater. 13 (2012) 054304.
[179] E.E. Rodriguez, C. Stock, P. Zajdel, K.L. Krycka, C.F. Majkrzak, P. Zavalij, M.A. Green, Phys. Rev. B 84 (2011) 064403.
[180] D. Fobes, I.A. Zaliznyak, Z. Xu, R. Zhong, G. Gu, J.M. Tranquada, L. Harriger, D. Singh, V.O. Garlea, M. Lumsden, B. Winn, Phys. Rev. Lett. 112 (2014) 187202.
[181] M.A. McGuire, D.J. Gout, V.O. Garlea, A.S. Sefat, B.C. Sales, D. Mandrus, Phys. Rev. B 81 (2010) 104405.
[182] C.M. Thompson, X. Tan, K. Kovnir, V.O. Garlea, A.A. Gippius, A.A. Yaroslavtsev, A.P. Menushenkov, R.V. Chernikov, N. Büttgen, W. Krätschmer, Y.V. Zubavichus, M. Shatruk, Chem. Mater. 26 (2014) 3825.
[183] M. Reehuis, W. Jeitschko, M.H. Möller, P.J. Brown, J. Phys. Chem. Solids 53 (1992) 687.
[184] K. Kovnir, C.M. Thompson, V.O. Garlea, D. Haskel, Phys. Rev. B 88 (2013) 104429.
[185] C.M. Thompson, K. Kovnir, V.O. Garlea, E.S. Choi, H.D. Zhou, M. Shatruk, J. Mater. Chem. C 2 (2014) 7561.
[186] S.L. Brock, N.P. Raju, J.E. Greedan, S.M. Kauzlarich, J. Alloys Compd. 237 (1996) 9.
[187] R. Nath, V.O. Garlea, A.I. Goldman, D.C. Johnston, Phys. Rev. B 81 (2010) 224513.
[188] E. Warren, Phys. Rev. 59 (1941) 693.
[189] M.F. Collins, O.A. Petrenko, Can. J. Phys. 75 (9) (1997) 605.
[190] S. Mitsuda, N. Kasahara, T. Uno, M. Mase, J. Phys. Soc. Jpn. 67 (1998) 4026.
[191] F. Ye, Y. Ren, Q. Huang, J.A. Fernandez-Baca, P. Dai, J.W. Lynn, T. Kimura, Phys. Rev. B 73 (2006) 220404R.
[192] M. Frontzek, G. Ehlers, A. Podlesnyak, H. Cao, M. Matsuda, O. Zaharko, N. Aliouane, S. Barilo, S.V. Shiryaev, J. Phys. Condens. Matter. 24 (2012) 016004.
[193] M. Soda, K. Kimura, T. Kimura, K. Hirota, Phys. Rev. B 81 (2010) 100406(R).
[194] Y. Oohara, S. Mitsuda, H. Yoshizawa, N. Yaguchi, H. Kuriyama, T. Asano, M. Mekata, J. Phys. Soc. Jpn. 63 (1994) 847−850.
[195] N. Terada, D.D. Khalyavin, P. Manuel, Y. Tsujimoto, K. Knight, P.G. Radaelli, H.S. Suzuki, H. Kitazawa, Phys. Rev. Lett. 109 (2012) 097203.
[196] N. Terada, D.D. Khalyavin, J.M. Perez-Mato, P. Manuel, D. Prabhakaran, A. Daoud-Aladine, P.G. Radaelli, H.S. Suzuki, H. Kitazawa, Phys. Rev. B 89 (2014) 184421.
[197] H. Yoshida, S. Ahlert, M. Jansen, Y. Okamoto, J. Yamaura, Z. Hiroi, J. Phys. Soc. Jpn. 77 (2008) 074719.
[198] M. Matsuda, C. de la Cruz, H. Yoshida, M. Isobe, R.S. Fishman, Phys. Rev. B 85 (2012) 144407.

[199] H. Nozaki, M. Mansson, B. Roessli, V. Pomjakushin, K. Kamazawa, Y. Ikedo, H.E. Fischer, T.C. Hansen, H. Yoshida, Z. Hiroi, J. Sugiyama, J. Phys. Condens. Matter. 25 (2013) 286005.
[200] H. Kadowaki, K. Ubukoshi, K. Hirakawa, J.L. Martinez, G. Shirane, J. Phys. Soc. Jpn. 56 (1987) 4027.
[201] M. Kenzelmann, G. Lawes, A.B. Harris, G. Gasparovic, C. Broholm, A.P. Ramirez, G.A. Jorge, M. Jaime, S. Park, Q. Huang, A. Shapiro, L.A. Demianets, Phys. Rev. Lett. 98 (2007) 267205.
[202] N.E. Amuneke, J. Tapp, C.R. de la Cruz, M. Möller, Chem. Mater. 26 (2014) 5930−5935.
[203] J. Hwang, E.S. Choi, F. Ye, C.R. Dela Cruz, Y. Xin, H.D. Zhou, P. Schlottmann, Phys. Rev. Lett. 109 (2012) 257205.
[204] M. Lee, E.S. Choi, X. Huang, J. Ma, C.R. Dela Cruz, M. Matsuda, W. Tian, Z.L. Dun, S. Dong, H.D. Zhou, Phys. Rev. B 90 (2014) 224402.
[205] Y. Doi, Y. Hinatsu, K. Ohoyama, J. Phys. Condens. Matter. 16 (2004) 8923.
[206] R. Coldea, D.A. Tennant, R.A. Cowley, D. McMorrow, B. Dorner, Z. Tylczynski, J. Phys. Condens. Matter. 8 (1996) 7473.
[207] M. Giot, L.C. Chapon, J. Androulakis, M.A. Green, P.G. Radaelli, A. Lappas, Phys. Rev. Lett. 99 (2007) 247211.
[208] C. Darie, P. Bordet, S. de Brion, M. Holzapfel, O. Isnard, A. Lecchi, J.E. Lorenzo, E. Suard, Eur. Phys. J. B 43 (2005) 159.
[209] V.O. Garlea, A.T. Savici, R. Jin, Phys. Rev. B 83 (2011) 172407.
[210] F. Damay, M. Poienar, C. Martin, A. Maignan, J. Rodriguez-Carvajal, G. André, J.P. Doumerc, Phys. Rev. B 80 (2009) 094410.
[211] N. Terada, Y. Tsuchiya, H.I Kitazawa, T. Osakabe, N. Metoki, N. Igawa, K. Ohoyama, Phys. Rev. B 84 (2011) 064432.
[212] S. Sachdev, Phys. Rev. B 45 (1992) 12377.
[213] Z. Hiroi, M. Hanawa, N. Kobayashi, M. Nohara, H. Takagi, Y. Kato, M. Takigawa, J. Phys. Soc. Jpn. 70 (2001) 3377.
[214] M. Shores, E. Nytko, B. Bartlett, D. Nocera, J. Am. Chem. Soc. 127 (2005) 13462.
[215] B. Fåk, E. Kermarrec, L. Messio, B. Bernu, C. Lhuillier, F. Bert, P. Mendels, B. Koteswararao, F. Bouquet, J. Ollivier, A.D. Hillier, A. Amato, R.H. Colman, A.S. Wills, Phys. Rev. Lett. 109 (2012) 037208.
[216] A.S. Wills, Phys. Rev. B 63 (2001) 064430.
[217] T. Inami, S. Maegawa, M. Takano, J. Magn. Magn. Mat. 177−181 (1998) 752.
[218] M. Nishiyama, T. Morimoto, S. Maegawa, T. Inami, Y. Oka, Can. J. Phys. 79 (11/12) (2001) 1511.
[219] J. Robert, V. Simonet, B. Canals, R. Ballou, P. Bordet, P. Lejay, A. Stunault, Phys. Rev. Lett. 96 (2006) 197205.
[220] K. Marty, V. Simonet, E. Ressouche, R. Ballou, P. Lejay, P. Bordet, Phys. Rev. Lett. 101 (2008) 247201.
[221] H.D. Zhou, B.W. Vogt, J.A. Janik, Y.-J. Jo, L. Balicas, Y. Qiu, J.R.D. Copley, J.S. Gardner, C.R. Wiebe, Phys. Rev. Lett. 99 (2007) 236401.
[222] N. Rogado, M.K. Haas, G. Lawes, D.A. Huse, A.P. Ramirez, R.J. Cava, J. Phys. Condens. Matter. 15 (2003) 907.
[223] G. Lawes, M. Kenzelmann, N. Rogado, K.H. Kim, G.A. Jorge, R.J. Cava, A. Aharony, O. Entin-Wohlman, A.B. Harris, T. Yildirim, Q.Z. Huang, S. Park, C. Broholm, A.P. Ramirez, Phys. Rev. Lett. 93 (2004) 247201.
[224] G. Ehlers, A.A. Podlesnyak, S.E. Hahn, R.S. Fishman, O. Zaharko, M. Frontzek, M. Kenzelmann, A.V. Pushkarev, S.V. Shiryaev, S. Barilo, Phys. Rev. B 87 (2013) 214418.
[225] L. Regnault, P. Burlet, J. Rossat-Mignod, Phys. B 86−88 (1977) 660.

[226] L.P. Regnault, J.Y. Henry, J. Rossat-Mignod, A. De Combarieu, J. Magn. Magn. Mater. 15–18 (1980) 1021.
[227] L.P. Regnault, J. Rossat-Mignod, in: L.J. de Jongh (Ed.), Magnetic Properties of Layered Transition Metal Compounds, Kluwer, Dordrecht, 1990.
[228] N. Rogado, Q. Huang, J.W. Lynn, A.P. Ramirez, D. Huse, R.J. Cava, Phys. Rev. B 65 (2002) 144443.
[229] J.H. Roudebush, N.H. Andersen, R. Ramlau, V.O. Garlea, R. Toft-Petersen, P. Norby, R. Schneider, J.N. Hay, R.J. Cava, Inorg. Chem. 52 (2003) 6083.
[230] L. Viciu, Q. Huang, E. Morosan, H.W. Zandbergen, N.I. Greenbaum, T. McQueen, R.J. Cava, J. Solid State Chem. 180 (2007) 1060.
[231] M.D. Lumsden, G.E. Granroth, D. Mandrus, S.E. Nagler, J.R. Thompson, J.P. Castellan, B. D. Gaulin, Phys. Rev. B 62 (2000) R9244.
[232] A.N. Vasiliev, O.L. Ignatchik, M. Isobe, Y. Ueda, Phys. Rev. B 70 (2004) 132415.
[233] G. Nenert, M. Isobe, I. Kim, C. Ritter, C. Colin, A.N. Vasiliev, K.H. Kim, Y. Ueda, Phys. Rev. B 82 (2010) 024429.
[234] T. Drokina, G. Petrakovskiĭ, L. Keller, J. Schefer, A.D. Balaeva, A.V. Kartasheva, D.A. Ivanova, J. Extheor. Phy. 112 (1) (2011) 121.
[235] J. Cheng, W. Tian, J. Zhou, V.M. Lynch, H. Steinfink, A. Manthiram, A.F. May, V.O. Garlea, J.C. Neuefeind, J. Yan, J. Am. Chem. Soc. 135 (7) (2013) 2776.
[236] M. Kurmoo, Chem. Soc. Rev. 38 (2009) 1353.
[237] T. Yankova, D. Hüvonen, S. Mühlbauer, D. Schmidiger, E. Wulf, S. Zhao, A. Zheludev, T. Hong, V. Garlea, R. Custelcean, G. Ehlers, Philos. Mag. 92 (2012) 2629.
[238] R. Sibille, E. Lhotel, T. Mazet, B. Malaman, C. Ritter, V. Ban, M. Francois, Phys. Rev. B 89 (2014) 104413.
[239] V. Garlea, C. Darie, O. Isnard, P. Bordet, Solid State Sci. 8 (2005) 457.
[240] R.H. Colman, T. Fennell, C. Ritter, G. Lau, R.J. Cava, A.S. Wills, J. Phys. Conf. Ser. 145 (2009) 012037.
[241] R.P. Santoro, R.E. Newnham, S. Nomura, J. Phys. Chem. Solids 27 (1966) 655.
[242] G.C. Lau, B.G. Ueland, R.S. Freitas, M.L. Dahlberg, P. Schiffer, R.J. Cava, Phys. Rev. B 73 (2006) 012413.
[243] J.S. White, T. Honda, K. Kimura, T. Kimura, C. Niedermayer, O. Zaharko, A. Poole, B. Roessli, M. Kenzelmann, Phys. Rev. Lett. 108 (2012) 077204.
[244] V.O. Garlea, L.D. Sanjeewa, M.A. McGuire, P. Kumar, D. Sulejmanovic, J. He, S.-J. Hwu, Phys. Rev. B 89 (2014) 014426.
[245] O.A. Petrenko, Low Temp. Phys. 40 (2) (2014) 106.
[246] H.-F. Li, C. Zhang, A. Senyshyn, A. Wildes, K. Schmalzl, W. Schmidt, M. Boehm, E. Ressouche, B. Hou, P. Meuffels, G. Roth, T. Brückel, Front. Phys. 2 (2014) 42.
[247] A. Aczel, L. Li, V. Garlea, J.-Q. Yan, F. Weickert, M. Jaime, B. Maiorov, R. Movshovich, L. Civale, V. Keppens, D. Mandrus, Phys. Rev. B 90 (2014) 134403.
[248] J.-J. Wen, W. Tian, V.O. Garlea, S.M. Koohpayeh, T.M. McQueen, H.-F. Li, J.-Q. Yan, J. A. Rodriguez-Rivera, D. Vaknin, C.L. Broholm, Phys. Rev. B 91 (2015) 054424.
[249] F. Heidrich-Meisner, I. Sergienko, A. Feiguin, E.R. Dagotto, Phys. Rev. B 75 (2007) 064413.
[250] A. Zheludev, Z. Honda, K. Katsumata, R. Feyerherm, K. Prokes, Europhys. Lett. 55 (6) (2001) 868.
[251] A. Zheludev, S.M. Shapiro, Z. Honda, K. Katsumata, B. Grenier, E. Ressouche, L.-P. Regnault, Y. Chen, P. Vorderwisch, H.-J. Mikeska, A.K. Kolezhuk, Phys. Rev. B 69 (2004) 054414.

[252] A. Zheludev, B. Grenier, E. Ressouche, L.-P. Regnault, Z. Honda, K. Katsumata, Phys. Rev. B 71 (2005) 104418.
[253] M. Hagiwara, L.-P. Regnault, A. Zheludev, A. Stunault, N. Metoki, T. Suzuki, S. Suga, K. Kakurai, Y. Koike, P. Vorderwisch, J.-H. Chung, Phys. Rev. Lett. 94 (2005) 177202.
[254] V.O. Garlea, A. Zheludev, T. Masuda, H. Manaka, L.-P. Regnault, E. Ressouche, B. Grenier, J.-H. Chung, Y. Qiu, K. Habicht, K. Kiefer, M. Boehm, Phys. Rev. Lett. 98 (2007) 167202.
[255] V.O. Garlea, A. Zheludev, K. Habicht, M. Meissner, B. Grenier, L.-P. Regnault, E. Ressouche, Phys. Rev. B 79 (2009) 060404(R).
[256] S. Rosenkranz, R. Osborn, Pramana J. Phys. 71 (2008) 705.
[257] W. Marshall, J. Phys. C (Proc. Phys. Soc.) 1 (1968) 88.
[258] O. Young, A.R. Wildes, P. Manuel, B. Ouladdiaf, D.D. Khalyavin, G. Balakrishnan, O.A. Petrenko, Phys. Rev. B 88 (2013) 024411.
[259] J.R. Stewart, P.P. Deen, K.H. Andersen, H. Schober, J.-F. Barthélémy, J.M. Hillier, A.P. Murani, T. Hayes, B. Lindenau, J. Appl. Cryst. 42 (2009) 69.
[260] [Online]. Available: http://www.ill.eu/instruments-support/instruments-groups/instruments/d10/ (accessed 27.07.15).
[261] L.C. Chapon, P. Manuel, P.G. Radaelli, C. Benson, L. Perrott, S. Ansell, N.J. Rhodes, D. Raspino, D. Duxbury, E. Spill, J. Norris, Neutron News 22 (2011) 22.
[262] J.A.M. Paddison, J.R. Stewart, P. Manuel, P. Courtois, G.J. McIntyre, B.D. Rainford, A.L. Goodwin, Phys. Rev. Lett. 110 (2013) 267207.
[263] K. Kamazawa, Y. Tsunoda, H. Kadowaki, K. Kohn, Phys. Rev. B 68 (2003) 024412.

Chapter 5

Multiferroics

William D. Ratcliff II and Jeffrey W. Lynn*

NIST Center for Neutron Research, National Institute of Standards and Technology, Gaithersburg, Maryland, USA
Corresponding author: E-mail: Jeffrey.Lynn@nist.gov

Chapter Outline

5.1	Introduction	291	5.5.2	HoMnO$_3$	315
5.2	Symmetry Considerations for Ferroelectrics	293	5.6	Thin Films and Multilayers	316
			5.6.1	BiFeO$_3$	316
5.3	Type-I Proper Multiferroics	294	5.6.2	TbMnO$_3$	319
	5.3.1 HoMnO$_3$	294	5.7	Spin Dynamics	322
	5.3.2 BiFeO$_3$	297		5.7.1 HoMnO$_3$	322
	5.3.3 (Sr-Ba)MnO$_3$	300		5.7.2 BiFeO$_3$	323
5.4	Type-II Improper Multiferroics	301		5.7.3 (Sr-Ba)MnO$_3$	323
	5.4.1 Spin-Spiral Systems: TbMnO$_3$, MnWO$_4$, and RbFe(MoO$_4$)$_2$	302		5.7.4 Electromagnons	325
				5.7.5 MnWO$_4$	328
				5.7.6 YMn$_2$O$_5$	329
	5.4.2 Exchange-Striction Systems: Ca$_3$CoMnO$_6$ and YMn$_2$O$_5$	307		5.7.7 RbFe(MoO$_4$)$_2$	330
			5.8	Future Directions	332
				Acknowledgments	333
5.5	Domains	315		References	333
	5.5.1 BiFeO$_3$	315			

5.1 INTRODUCTION

Multiferroics (MFs) are materials that possess both ferroelectric (FE) and magnetic order, and these two types of order have symmetry requirements that a MF must simultaneously satisfy. FEs must have their space inversion symmetry broken, requiring that only crystal structures with noncentrosymmetric space groups can exhibit ferroelectricity. Often the ferroelectricity develops from a high-temperature symmetric phase via a structural phase transition involving a soft phonon, and neutron scattering has played a pivotal role in illuminating the relation between the soft phonon, structural phase transition, and FE order [1]. For magnetic materials, on the other hand, it is time reversal symmetry that has to be broken to allow the ionic magnetic moments to align. Hence, both symmetries must be violated for a material to be a candidate MF, making them a rather rare but important commodity. The recent surge in research interest in MFs stems both from a desire to understand the nature of the order and coupling of these two disparate order parameters from a fundamental viewpoint, as well as from the practical perspective of realizing the intriguing possibility of controlling the magnetic properties electronically and vice versa, thereby enabling completely new technological capabilities [2,3].

Classic FEs such as $BaTiO_3$ lower their symmetry through bonding of the empty 3d orbitals on the Ti ions with the oxygen 2p orbitals, causing the oxygen octahedron to tilt and the Ti ions to displace off their center of symmetry [4–6]. But, many systems hybridize; what makes some FE and others not? Fillipetti and Hill (now Spaldin) [7] tell us that the answer lies in the nature of which orbitals participate in the hybridization. If the orbitals hybridize in a direction which is susceptible to a FE instability, then this hybridization will lead to ferroelectricity. In the case of $BaTiO_3$, the hybridization occurs between nominally empty d orbitals and the oxygen p orbitals (because of the hybridization, the bonds cannot be thought of as purely ionic and some of the electrons from the oxygen p orbitals reside on the Ti site). In particular, if they allow a tetragonal distortion of the lattice, they find that the d_{xz}, d_{yz}, and d_z^2 orbitals hybridize with the oxygen p orbitals and show a dramatic change in occupation. This results in a polarization, and thus the ferroelectricity depends on (relatively) unoccupied d orbitals. In fact, for $BaTiO_3$ all of the d orbitals are unoccupied, so it should be sensitive to displacements along any and all directions. For more delocalized d orbitals such as 4d and 5d systems these destabilizations are harder to generate, so we are often limited to 3d electrons. Because this mechanism of ferroelectricity requires *empty* d *orbitals*, it excludes any possibility of magnetism in the system, disqualifying them as MFs. Consequently, MFs are rare in nature and typically an alternative avenue must be found to have both properties. Hill (Spaldin) has addressed this issue in an excellent review article [8], where some additional difficulties are noted for materials with occupied d orbitals, which are often resistant to hybridization-induced distortions from the symmetrical state, or alternatively are subject to Jahn Teller distortions which are symmetric.

One alternate pathway is for a material to have a sublattice of magnetic ions that is separate from the FE sublattice. Generally these are designated as type-I or "proper" MFs as the FE mechanism is the standard one. Commonly they have robust FE order but also often have quite different magnetic and FE ordering temperatures. More importantly, since the magnetism and ferroelectricity occur on different sublattices, these two disparate order parameters tend to be weakly coupled. A second possibility is to have the magnetism itself break the inversion symmetry, thereby enabling FE order to develop. In this case, the ferroelectricity is a secondary order parameter, only permitted to exist by the magnetism and completely controlled by it. These are designated type-II or "improper" MFs.

Typically the magnetic order develops at low temperatures, and with a FE order parameter that is two to three orders-of-magnitude smaller than robust FEs such as the prototypical $BaTiO_3$ currently used in many contemporary applications.

Neutron scattering plays a number of very important roles in determining the properties of MFs. Neutron diffraction on both powders and single crystals can be employed to reveal the detailed crystal structures and magnetic structures as a function of temperature, pressure, magnetic field, and electric field, while inelastic scattering plays a central role in elucidating the lattice dynamics and spin dynamics, and in understanding the origin of how the magnetic and FE order parameters are coupled. Small-angle neutron scattering (SANS) is used to explore the nature of large-scale structures such as domains and domain walls. Reflectometry is the technique of choice to investigate the properties of thin films and multilayers. The neutron techniques themselves are reviewed in detail elsewhere [9].

Here we focus on the materials themselves and what has been learned using neutron scattering. We first discuss three systems which are type-I (proper) materials: hexagonal $HoMnO_3$, which is a prototype system with separate magnetic and FE sublattices; $BiFeO_3$, which develops both types of order well above room temperature and currently is a leading candidate for applications; and orthorhombic $(Sr-Ba)MnO_3$, which breaks the "type-I" rules in that the Mn ion performs both FE and magnetic functions. These "proper" FEs typically have high FE transition temperatures—well above room temperature—but low magnetic ordering temperatures, and weakly coupled order parameters. $(Sr-Ba)MnO_3$ is an exception to this type-I rule. We then discuss a few prototypical type-II (improper) FEs, such as $TbMnO_3$ and $(Mn-Fe)WO_4$. These are materials with low magnetic ordering temperatures where the magnetic structure itself breaks the inversion symmetry and permits typically very weak FE order to develop. We then discuss the results from a SANS study to explore the coupling between magnetic and FE domains, and reflectometry and diffraction studies of thin films and multilayers. While both types of MFs are interesting from a fundamental point of view, so far only some of the "proper" materials exhibit both types of order near to or above room temperature and thus provide the most promise for applications such as for multifunctional memory, spintronics, and sensors. One such proper MF is $BiFeO_3$ and we discuss the results for thin films. We then discuss the spin and lattice dynamics of these, and their coupling to produce electromagnons. We conclude with prospects for further work.

5.2 SYMMETRY CONSIDERATIONS FOR FERROELECTRICS

To understand FEs, let us start with electric dipoles on atoms or molecules. We define the polarization of a material as the average electric dipole moment, $P = \sum q R$. A FE material is one in which a macroscopic spontaneous polarization develops below some temperature T_C. Furthermore, this polarization

TABLE 1 Division of Crystallographic Point Groups into Centrosymmetric and Non-centrosymmetric Groups

Crystal System	Centrosymmetric	Non-centrosymmetric
Triclinic	−1	1
Monoclinic	2/m	2, m
Orthorhombic	mmm	222, mm2
Tetragonal	4/m, 4/mmm	4, −4, 422, 4mm, −42m
Trigonal	−3, −3m	3, 32, 3m
Hexagonal	−6/m, 6/mmm	6, −6, 622, 6mm, −6m2
Cubic	m3, m3m	23, −43m, 432

Based on Ref. [10].

must be switchable by an electric field. If we restrict ourselves to ordered solids, what can symmetry tell us about such materials? The first thing is that if we want to have such a material, we cannot have a center of inversion, or the polarization will vanish. The absence of a center of inversion will give us crystals that are at least piezoelectric—polarization that develops with stress. For a FE we require in addition that the polar axis be unique. What do we mean by this? According to Von Neumann [11], the symmetry property of a crystal is a superset of the symmetry elements of the point group of the crystal. Then if we have a unique axis for a property like the polarization, the symmetry elements of the point group of the crystal must leave that unique axis invariant. This is not the same as a polar axis in which the symmetry elements of the group are not allowed to invert the order of points along the axis. This consideration greatly limits the number of point groups which can allow ferroelectricity. They are summarized in Table 1.

5.3 TYPE-I PROPER MULTIFERROICS

Type-I MFs are ones with a large, robust FE order parameter which typically develops well above the magnetic ordering temperature and where the coupling between the two types of orders is relatively weak. Below are several examples of such MFs, including one where the coupling is actually very strong.

5.3.1 HoMnO$_3$

The hexagonal rare earth manganese oxides (RMnO$_3$ with R = Ho, Er, Tm, Yb, Lu, and Y) are a family of MF materials which exhibit very high FE transition temperatures, and HoMnO$_3$ is a prototype system that has been

(a)

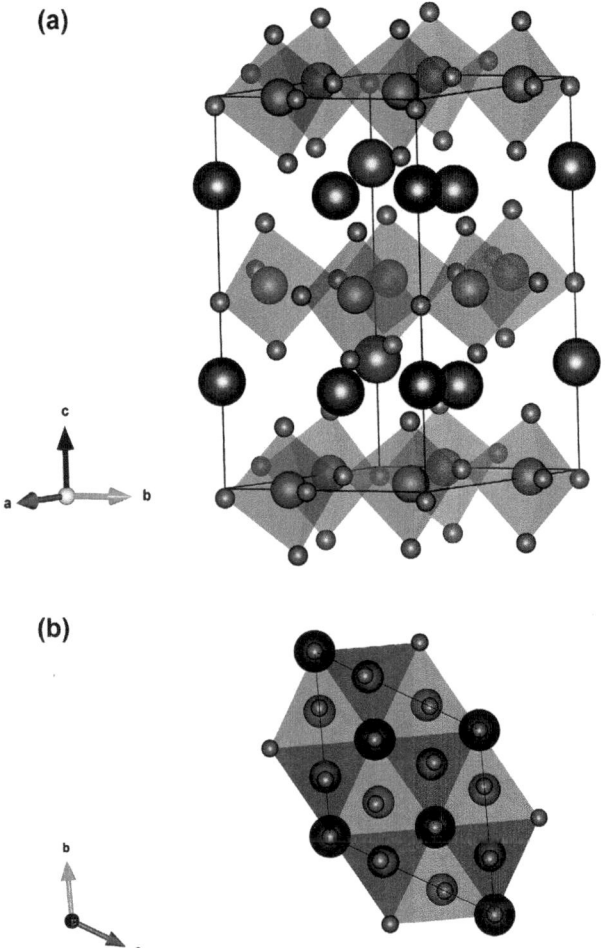

(b)

FIGURE 1 Crystal structure for hexagonal HoMnO$_3$. (a) View from the side (c-axis is up). Large blue (dark gray in print versions) balls are Ho, red (light gray in print versions) balls are oxygen. The Mn ions sit at the center of each tetrahedron of oxygen ions. (b) View along the c-axis. Note the triangular configuration of the Mn (oxygen tetrahedra) as well as the Ho. The oxygen that makes up the sides of each tetrahedron also exhibits triangular coordination.

studied in considerable detail. The FE behavior arises from the holmium− oxygen ions which undergo a displacement at very high temperature ($T_C = 1375$ K) [12] and gives rise to a FE moment along the c axis. Long-range magnetic order develops for the separate Mn sublattice at $T_N = 72$ K [13−16], with such disparate ordering temperatures indicating that the order parameters are weakly coupled. The Mn moments occupy a classic frustrated triangular lattice as shown in Figure 1, and order in a noncollinear 120° spin structure. At 40 K a spin reorientation ($T_{SR} = 40$ K) takes place in which the

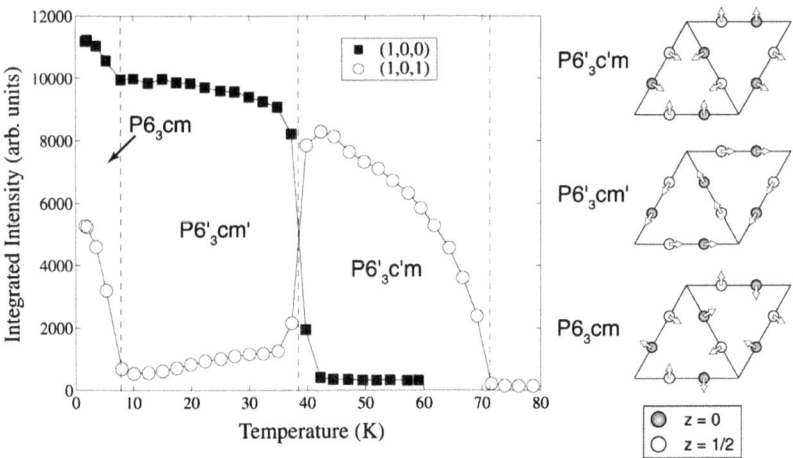

FIGURE 2 Neutron diffraction measurements of the integrated intensities of the (1,0,0) and (1,0,1) magnetic Bragg reflections of HoMnO$_3$. Two spin-reorientation transitions (indicated by dashed lines) lead to changes in the intensities of both peaks. The Mn^{3+} spin configurations for these phases are shown on the right. Ho^{3+} moments (not shown) order in the low-temperature phase. *Reproduced from Ref. [16].*

Mn moments rotate in the plane, changing the magnetic symmetry and adopting a different 120° spin structure. At low temperatures, the Ho moments order antiferromagnetically ($T_{Ho} = 8$ K) with moments aligned along the c axis, accompanied by a second spin reorientation transition of the Mn moments into the $P6_3cm$ 120° spin structure phase [17–20]. The spin structures in these phases are shown in Figure 2. The sharp magnetic transitions shown in the figure are accompanied by sharp anomalies in the dielectric constant, which increase in amplitude with decreasing temperature as the (FE) holmium moments become more involved in the magnetism [21]. Thus the magnetic and FE order parameters do exhibit some weak coupling. Second harmonic generation (SHG) measurements as a function of magnetic field also show the change in magnetic symmetry and the reentrant phase as a function of magnetic field suggested by dielectric susceptibility [22].

A magnetic field shifts T_{SR} to lower T and broadens the transition, with the transition becoming reentrant as shown in the magnetic field (H) versus T phase diagram of Figure 3 [16,20,23,24]. In the intermediate-temperature phase, a sufficiently strong applied magnetic field along the c axis pushes HoMnO$_3$ into the higher-temperature phase. The phase diagram becomes considerably more complex as the holmium moments order, including competing interactions that result in a critical end point around 2 T and 2 K [23]. The FE and magnetic order parameters are coupled via the Ho ions which participate in both types of order. SANS measurements indicate that the FE and magnetic domains are coupled together with a net magnetization in the domain walls [25].

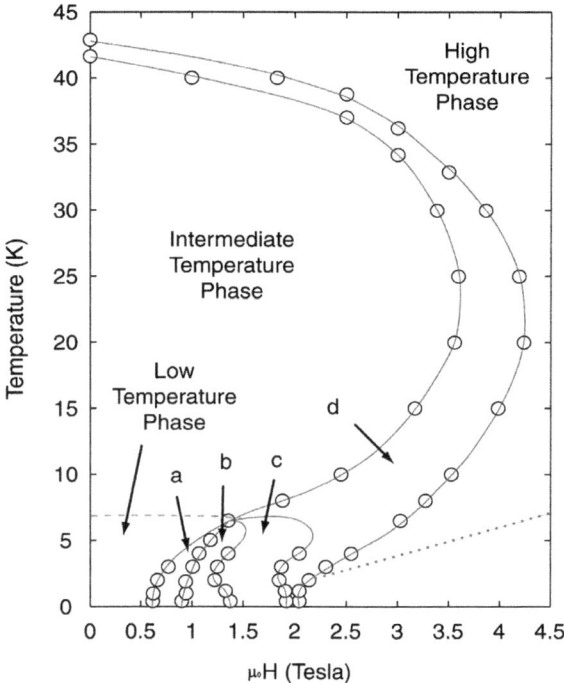

FIGURE 3 Temperature versus magnetic field phase diagram for $HoMnO_3$ obtained from neutron diffraction measurements. Curves are guides-to-the-eye. The dashed line and dotted lines indicate approximate phase boundaries for previously reported transitions not observable from the diffraction data. The designated regions correspond to intermediate phases and hysteretic overlap regions that occur at low T. *Reproduced from Ref. [24].*

5.3.2 $BiFeO_3$

$BiFeO_3$ is perhaps the most studied MF material, primarily because it has compelling properties for possible device applications. The FE transition at ~1100 K is driven by the Bi lone-pair mechanism. While the exact sequence of distortions is debated, we know that at high temperatures it is cubic and at room temperature it is rhombohedral. $BiFeO_3$ is a proper MF with a polarization comparable to $BaTiO_3$. The material also orders magnetically above room temperature, making $BiFeO_3$ suitable for device applications.

$BiFeO_3$ was first studied with neutrons by Sosnowska et al. [26]. Initial measurements were performed on powders and it is rather amazing, given the data shown in Figure 4, how accurate the incommensurate magnetic structure was determined. They found that the magnetic structure can basically be described as G-type [27], but with a long wavelength modulation of around 600 Å. Powder averaging often results in the loss of phase information and subsequent work showed that it is not possible to distinguish between

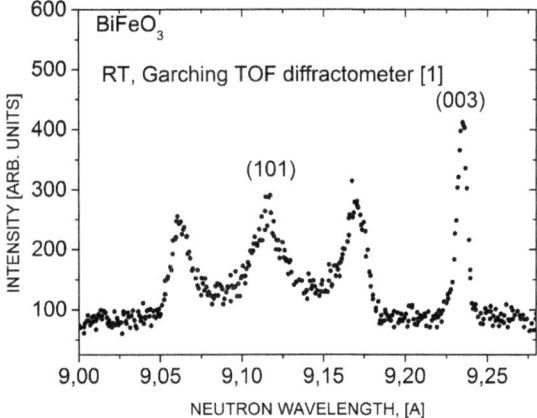

FIGURE 4 Normalized time-of-flight powder diffraction pattern from $BiFeO_3$ (full circles). Reproduced from Ref. [26].

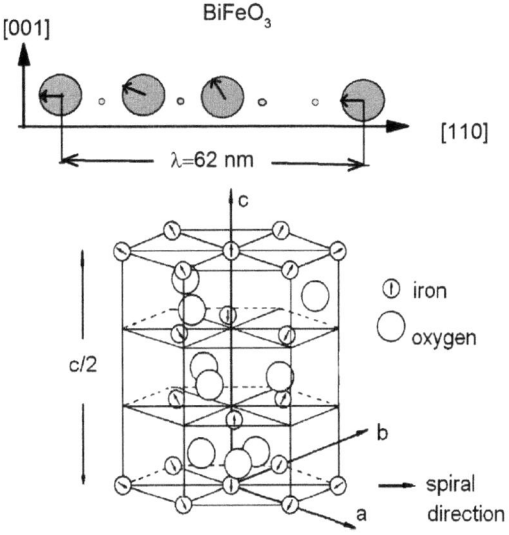

FIGURE 5 Portion of the $BiFeO_3$ lattice with only iron and oxygen ions shown. The arrows indicate the Fe^{3+} moment direction of our model. The spiral period is reduced for illustration purpose. Reproduced from Ref. [26].

amplitude-modulated and spiral magnetic structures [28]. However, they chose a cycloidal magnetic structure, Figure 5, which agrees with later studies performed on single crystals. Lebeugle and coworkers [29] measured a large number of magnetic reflections using a four-circle diffractometer and were able to fit the magnetic structure to determine that it was a cycloid with the magnetic moment in the plane containing the FE polarization and the magnetic propagation vector. This is shown in Figure 6.

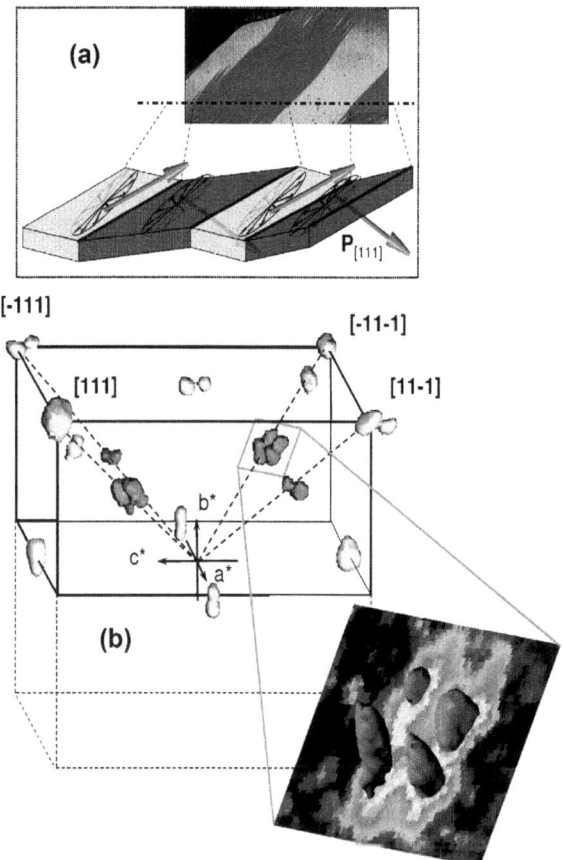

FIGURE 6 (a) Polarized light microscope images of the crystal showing stripe domains with two different polarization directions. The lower portion shows a cartoon of the buckling of the crystal and the different polarizations of the stripe domains. It also shows the plane of the cycloid with respect to the polarization in a given domain. (b) Mapping of the neutron intensity of $BiFeO_3$ in reciprocal space. Two sets of splittings appear for the nuclear intensity (yellow (white in print versions) or light gray spots) due to the presence of two ferroelastic domains (see (a)): one because of the presence of two rhombohedral distortions along [1,1,1] and [1,−1,1] and the second because of a physical buckling of the crystal induced by the twinning. Magnetic peaks (shown in red (dark gray in print versions)) are further split because of the cycloids. Note that because the splitting is small, the scale has been magnified by a factor of 10 around each peak position. *Reproduced from Ref. [29].*

Lee et al. [30] used polarized neutron diffraction to investigate the magnetic structure and basically found the same result. Polarized neutrons are sensitive to the "chirality" of a magnetic structure. (There is a subtle point in that they are sensitive to the chirality with respect to a coordinate system defined by the neutron polarization. A cycloid is of course not inherently chiral.) They were able to show that their as-grown crystal consisted of a single FE domain, and were able to determine that the system ordered as a single magnetic domain and

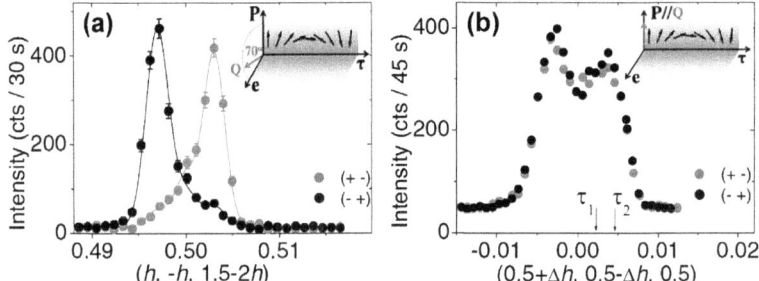

FIGURE 7 Polarized neutron scans of BiFeO$_3$ in the (+/−) and (−/+) spin-flip channels. (a) ($h,-h,1.5-2h$) scan through domain τ_1 near the (0.5,−0.5,0.5) position in the ($h,-h,l$) plane. Error bars are from counting statistics and represent one standard deviation throughout this chapter. (b) (0.5 + δh, 0.5−δh, 0.5) scan through domains τ_1 and τ_2 near the (0.5,0.5,0.5) position in the ($h,2l-h,l$) plane. Arrows show the positions of the peaks due to these domains. Insets show the orientation of the scattering vector **Q** with respect to the plane of the magnetic spiral, as well as vectors, **e**, **P**, and **τ**, where **e** is a vector describing the spiral axis. *Reprinted with permission from Ref. [30].*

even as a single chirality domain. This is shown in Figure 7 and will be discussed in more detail later. Subsequent measurements revealed that there was a slight anharmonicity to the cycloid [31], manifested as a weak third harmonic. It was also found that there were no significant low temperature anomalies in the magnetic order. Interestingly, SANS measurements revealed that in addition to the cycloidal order, there is also a weak spin-density wave (SDW) present that serves to slightly tilt the spins out of the plane of the cycloid (see Figure 8) [32]. This SDW collapses to a weak ferromagnet (FM) moment in thin films, which is important for devices that depend on multilayer coupling between BiFeO$_3$ and another magnetic compound such as CoFeB.

5.3.3 (Sr-Ba)MnO$_3$

For MFs to be useful, an alternative mechanism for ferroelectricity is highly desirable, where higher ordering temperatures and strong coupling might be realized. One such alternative mechanism for magnetic−FE coupling was pointed out for a class of perovskite manganites such as (R,Ca)MnO$_3$ (R = lanthanide) via the charge and/or orbital ordering that can occur in these systems [33,34], which is a class of materials perhaps more familiar as colossal magnetoresistive materials at lower doping [35,36]. The advantage of the perovskite manganites is that these are type-I MFs that can exhibit a large electric polarization, with an electric polarization comparable to BaTiO$_3$.

(Sr$_{1-x}$-Ba$_x$)MnO$_3$ crystallizes in the cubic space group symmetry of $Pm\overline{3}m$, but sufficient substitution of the larger Ba ions for Sr produces a robust FE phase, with a tetragonal distortion that produces the polarization and ordering temperature comparable to BaTiO$_3$ [37]. T_C can exceed 400 K while the magnetic ordering develops around 200 K, as shown in Figure 9 [38]. Since it

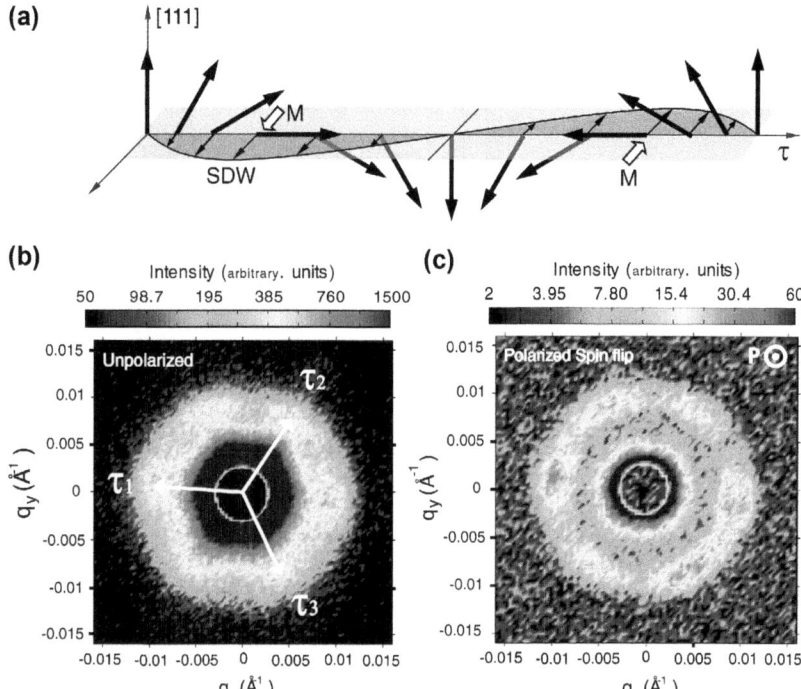

FIGURE 8 (a) The spin-density wave (SDW) in $BiFeO_3$ (thin arrows normal to both [1,1,1] and τ) produced by the tilt of the Fe spins out of the plane of the magnetic cycloid (thick arrows in the plane defined by [1,1,1] and τ). The amplitude of the SDW is exaggerated and only a few representative spins are shown. Open arrows show local magnetization due to the SDW. (b) and (c) show the diffraction pattern in the (1,1,1) scattering plane with (b) unpolarized neutrons, (c) neutrons polarized with $P\|k_i$ that scatter with spin-flip. The images were prepared by summing several detector measurements over a range of rocking angles β, allowing all Bragg reflections from the long-wavelength modulation to be presented in a single image. A beam stop blocks the direct beam in the center of each image. *Reproduced from Ref. [32].*

is the magnetic Mn ions that displace to produce the ferroelectricity, the two orders parameters are naturally strongly coupled together. As the magnetic order develops as shown in the inset, the FE order parameter is greatly reduced, demonstrating their strong interaction. Calculations show that the ferroelectricity is stabilized by an enhanced in-plane Mn-O hybridization, while the exchange interaction is very sensitive to the bond-bending distortion, explaining this very strong coupling [39].

5.4 TYPE-II IMPROPER MULTIFERROICS

Type-II MFs are materials where the magnetic order breaks the inversion symmetry and permits ferroelectricity to develop. In these cases the ferroelectricity is a secondary order parameter that is controlled by the magnetism.

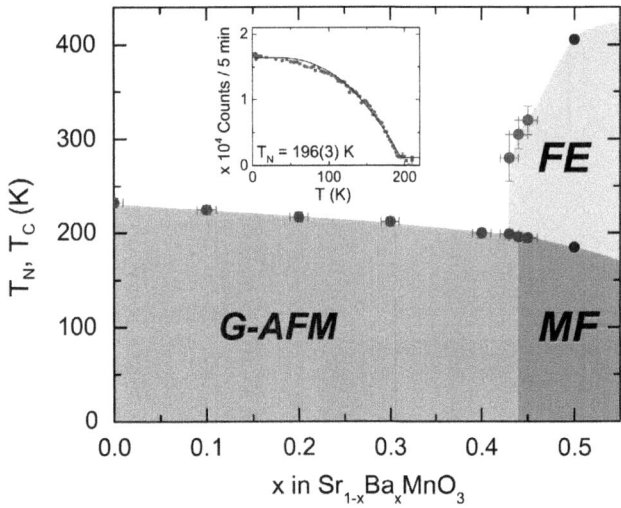

FIGURE 9 Phase diagram for $Sr_{1-x}Ba_xMnO_3$ showing the concentration region where ferroelectric and multiferroic behaviors develop. The point for $x = 1/2$ is from Sakai et al. [37]. The inset shows the smooth development of the integrated intensity of the $\{1/2,1/2,1/2\}$ antiferromagnetic peak of this G-type magnetic structure. The curve is a simple mean-field fit of the square of the magnetic order parameter to estimate the antiferromagnetic transition T_N of 196(3) K. *Reproduced from Ref. [38].*

In the following, we discuss two types of materials, one where a spiral magnetic phase develops, and the other where exchange striction is the driving force.

5.4.1 Spin-Spiral Systems: TbMnO$_3$, MnWO$_4$, and RbFe(MoO$_4$)$_2$

TbMnO$_3$ crystallizes in the orthorhombic *Pbnm* structure (isostructural to LaMnO$_3$) and orders magnetically at 41 K. As shown in Figure 10(a), the initial order is an amplitude-modulated SDW, meaning that the spin direction is parallel to the propagation vector of $(0,k,0)$ [40]. Just below T_N, $k \approx 0.27$ and is only weakly temperature dependent with no tendency to lock-in to a commensurate value (see Figure 10(c)). No higher-order harmonics are observed in this temperature range so that the magnetic structure is a pure sine wave and belongs to a single representation. At 27 K a second representation develops with an elliptical (spiral) component having components along the *c*-axis (see Figure 10(b)), which breaks the spatial-inversion symmetry and allows FE order to develop. In this phase, it was found that the application of a magnetic field (see Figures 10 and 11) along the *b*-axis suppressed the electric polarization along the *c*-axis and instead a polarization developed along the *a*-axis, which was accompanied by a change in the magnetic structure where

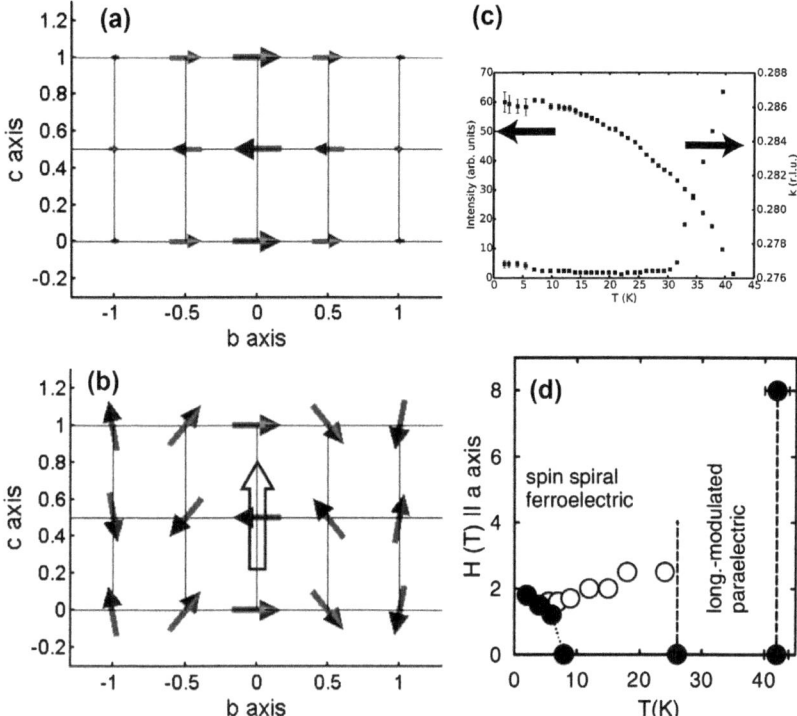

FIGURE 10 Schematic of the magnetic structure of TbMnO$_3$ at (a) $T = 35$ K and (b) $T = 15$ K, projected onto the $b-c$ plane. Filled arrows indicate the direction and magnitude of Mn moments. The longitudinally modulated phase (a) respects inversion symmetry along the c-axis, but the spiral phase (b) violates it allowing an electric polarization (unfilled arrow). (c) Temperature dependence of the intensity (order parameter squared) and incommensurate wave vector. Arrows indicate with which axis symbols are associated. (d) Phase diagram as a function of temperature and field applied along the a-axis. Solid circles indicate second-order phase transitions. Open circles indicate the characteristic field for the reduction of magnetic $(0,1-q,1)$ Bragg scattering from Tb moments by 50% from its zero-field intensity. *Reproduced from Ref. [40].*

the plane of the spiral "flops" [40]. The trilinear coupling term in the free energy for this new structure gives rise to the electric polarization observed in the spiral phase, which must vanish by symmetry in the simple SDW phase. There have been a number of phenomenological and microscopic models to explain the magnetic order-driven ferroelectricity in this system [41–44]. Further refinements to these magnetic structures, including contributions from the (induced) Tb moments and low temperature ordering, were investigated by magnetic X-ray scattering [45].

The mineral hubnerite (MnWO$_4$) is another example of a spiral-induced FE. In this structure, MnO$_6$ octahedra form chains along the c-axis and the system undergoes a series of magnetic phase transitions (see Figure 12). It orders into

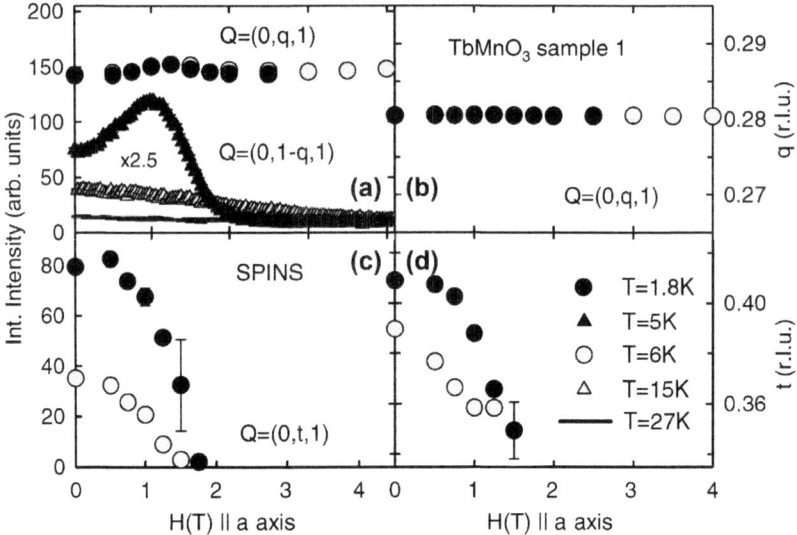

FIGURE 11 Field dependence of the magnetic Bragg scattering from TbMnO$_3$. (a and b) show data for the incommensurate peaks that occur for $T < 41$ K. (a) The $(0,q,1)$ peak that is mostly sensitive to the staggered magnetization on Mn sites and the $(0,1-q,1)$ peak that is sensitive to staggered magnetization on Tb sites. (c) and (d) show data for the incommensurate peaks that develop for $T < 7$ K. *Reproduced from Ref. [40].*

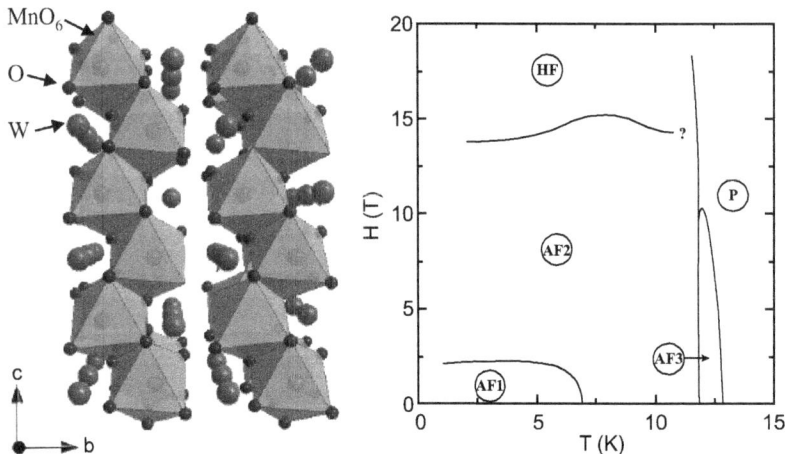

FIGURE 12 Crystal structure of MnWO$_4$ and schematic H–T phase diagram for a magnetic field applied along the easy axis according to Refs [46,47], respectively. *Reproduced from Ref. [48].*

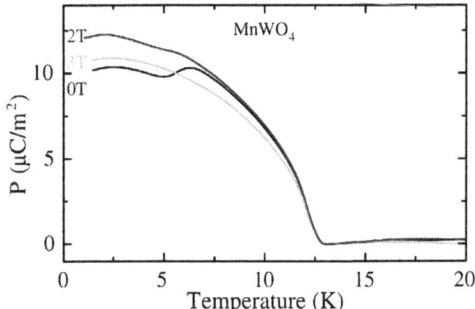

FIGURE 13 Ferroelectric polarization along the a-axis of MnWO$_4$ for different magnetic fields applied along the c-axis. *Reproduced from Ref. [48].*

an amplitude-modulated incommensurate structure called the AF3 phase at $T_N \sim 13.5$ K, with $k = (-0.214, 0.5, 0.457)$. At 12.7 K, this sinusoidally modulated phase develops into a helical structure (the AF2 phase) with almost the same wave vector. Finally, at 7.6 K the system locks-in to a commensurate helical phase with $k = (-0.25, 0.5, 0.5)$. Ferroelectricity is only present in the helical phases, as shown in Figure 13.

In an elegant polarized beam experiment [49], Sagayama and coauthors measured the $(-1,0,2) +/-k$ reflections as a function of temperature. They showed that in the helical phase the polarization along the b-axis, $P_b > 0$ was associated with a helix with one handedness and that $P_b < 0$ was associated with the opposite handedness (see Figure 14). The asymmetry between the left and right satellites (indicative of chirality) vanished in the nonhelical magnetic phases, along with the electric polarization. They were also able to use the simple spin current model to predict the FE polarization as a function of temperature. This is shown in Figure 14. A similar polarized beam experiment by Cabrera et al. demonstrated this handedness in the FE phase of MF Ni$_3$V$_2$O$_8$ [50].

Finally, we discuss the RbFe(MoO$_4$)$_2$ system, which is an example of antiferromagnetically coupled xy spins placed on the 2D triangular lattice [51]. At room temperature this system orders in the $P\bar{3}m1$ structure, and below 180 K the system undergoes a structural ferroaxial phase transition to the trigonal $P\bar{3}$ structure. In this structure, there are two types of triangles in the plane, one in which the Rb-O tetrahedra are above the plane and point "up" along the c-axis (up triangles, Figure 15), and the other where these tetrahedra point down and are below the plane (down triangles). For a single triangle in the 120° magnetic structure, if we traverse the vertices clockwise the spins can either rotate by 120° in the clockwise direction or in a counterclockwise fashion. If one considers only the "up" triangles we can have two order parameters: one associated with a "chiral" rotation in the clockwise direction and another associated with a chiral rotation in the counterclockwise direction. In this material, the magnetic propagation vector is given by $q = (1/3, 1/3, q_z)$,

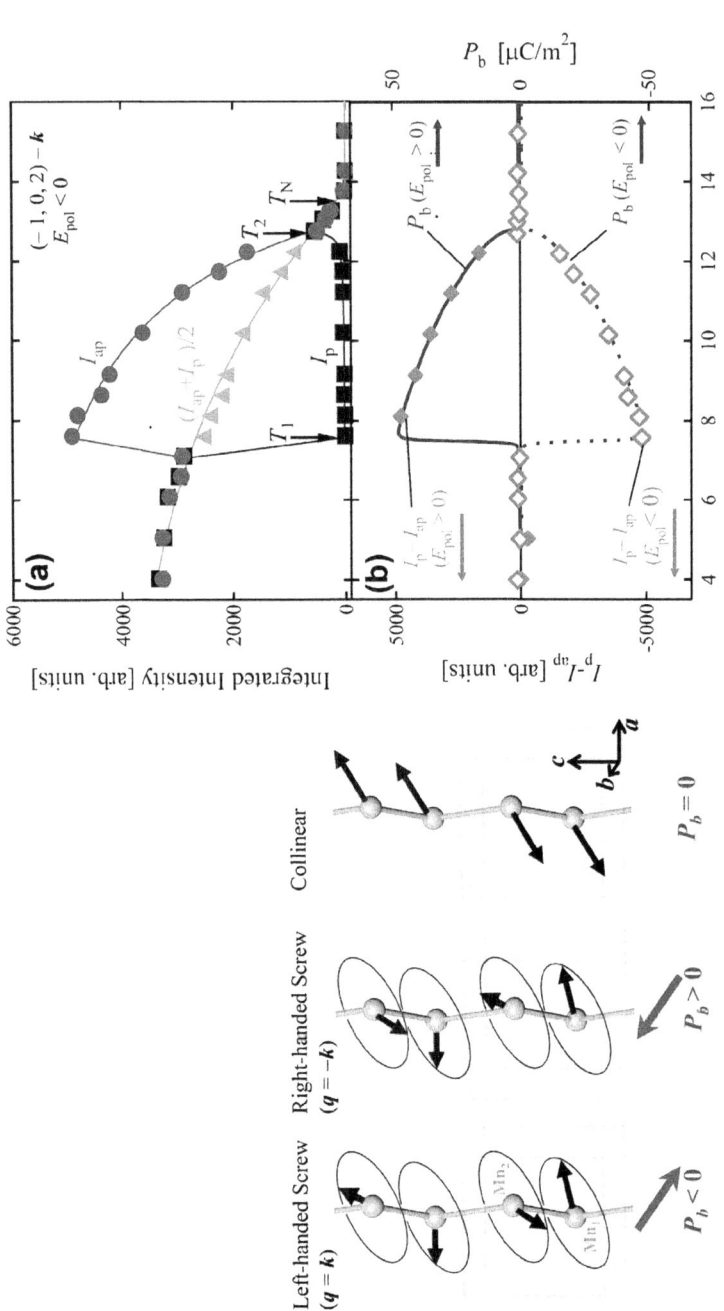

FIGURE 14 The relation between the electric polarization and spin structure in MnWO$_4$. The left- and right-handed screw configurations on a zigzag chain correspond to the $P_b > 0$ and $P_b < 0$ states, respectively. (a) The temperature dependence of the integrated intensities of the magnetic reflection at $(-1,0,2-k)$ in the $E_{pol} < 0$ case. (b) Temperature dependence of the electric polarization and the difference between the integrated intensities I_{up} and I_p at $Q = (-1,0,2-k)$. Adapted from Ref. [49].

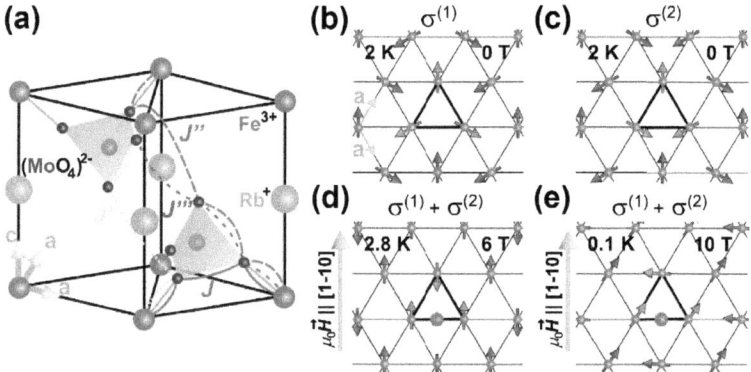

FIGURE 15 (a) The low-temperature $P\bar{3}$ structure of RbFe(MoO$_4$)$_2$. O$_2$-mediated superexchange interaction paths are indicated, and the exchange hierarchy is $J \gg J' > J'' \sim J'''$. (b–e) Magnetic structures within a single triangular lattice layer for fields up to 10 T [51]. (b) and (c) show degenerate zero-field 120° spin structures of antiphase chirality. Each can be described by a single phenomenological order parameter $\sigma^{(1)}$ or $\sigma^{(2)}$ [51,52]. The magnetic structure under $\mu_0 H \| [1, -1, 0]$ of (d) 6 R and (e) 10 T are each described by a combination of $\sigma^{(1)}$ and $\sigma^{(2)}$. Green (dark gray in print versions) circles show inversion centers, and red spin triangles are representative for each magnetic structure. *Reproduced from Ref. [53].*

where $q_z \sim 0.44$. Representational analysis allows us to write the components of the spin in terms of the complex value as:

$$S_x(r) = \left[\sigma^{(1)}(q_z) + \sigma^{(2)}(q_z)\right] e^{i q \cdot r} + cc \qquad (1)$$

$$S_y(r) = i\left[\sigma^{(1)}(q_z) - \sigma^{(2)}(q_z)\right] e^{i q \cdot r} + cc \qquad (2)$$

Thus, if the 120° structure orders with either $\sigma^{(1)}$ or $\sigma^{(2)}$, inversion symmetry along the c-axis will be lost. Thus a FE polarization develops along the c-axis and tracks the intensity of the (1/3,1/3,0.44) reflection (Figure 16). Actually, according to Landau theory the polarization depends on the difference between the two order parameters, $\sigma^{(1)}$ and $\sigma^{(2)}$. With the application of an in-plane magnetic field, the magnetic structure changes and instead of being solely determined by $\sigma^{(1)}$, is described by $\sigma^{(1)}$ and $\sigma^{(2)}$, and the FE polarization vanishes. The phase diagram in field and temperature is shown in Figure 17.

Further spherical polarimetry experiments performed under electric field revealed that there are both chiral and helical domains present in this material and that their populations can be changed by electric field. The off-diagonal elements of the polarization tensor are particularly sensitive to these populations. This is shown in Figure 18.

5.4.2 Exchange-Striction Systems: Ca$_3$CoMnO$_6$ and YMn$_2$O$_5$

Another example of a type-II MF is Ca$_3$CoMnO$_6$, in which exchange striction breaks inversion symmetry [54]. In Figure 19, we show the crystal structure.

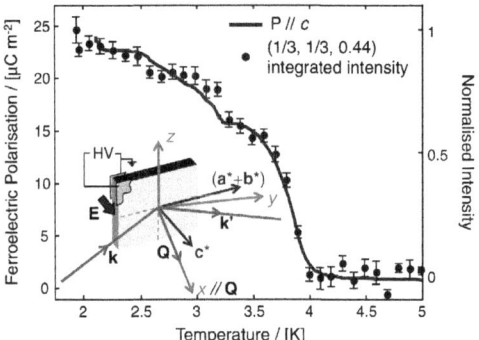

FIGURE 16 c-axis polarization (obtained by integrating the pyroelectric current) of Rb(FeMoO$_4$)$_2$ and intensity of the (1/3,1/3,q_z) magnetic reflection (measured without polarization analysis) as a function of temperature. Inset: Experimental geometry showing the crystal orientation, direction of the applied **E** field, and scattering vector $\mathbf{Q} = \mathbf{k}' - \mathbf{k}$. *Reproduced from Ref. [55].*

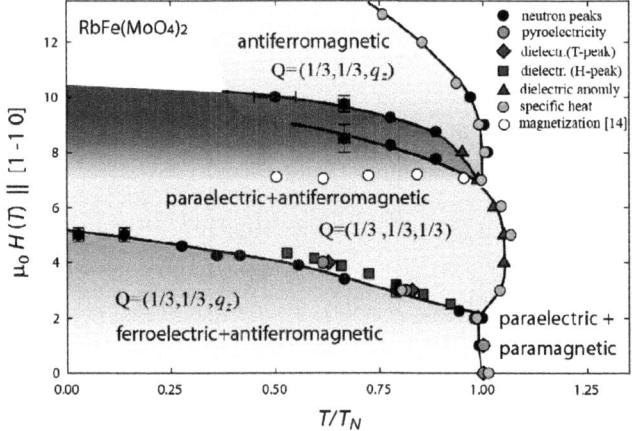

FIGURE 17 H–T phase diagram of RbFe(MoO$_4$)$_2$ for fields applied along the [1,−1,0] direction with $T_N \sim 3.8$ K, obtained through various techniques. Open circles represent the end of the magnetization plateau as a function of field [56]. The color gradient indicates the estimated increasing magnetization as a function of magnetic field. The boundary of the high-field incommensurate phase has not been determined for $T < 2$ K–0.5 T_N. Subtle phase modifications which may occur in the regions adjacent to the $Q = (1/3,1/3,1/3)$ phase [56] are not shown. The commensurate and high-field incommensurate magnetic orders coexist in a narrow field region for $\mu_0 H \approx 9$ T. The two solid lines indicate where each type of order was observed in diffraction data. *Reproduced from Ref. [51].*

Co^{2+}/Mn^{4+} ions form chains along the c-axis, with Co^{2+} and Mn^{4+} ions charge ordering in an alternating fashion. From the neutron powder diffraction measurements shown in Figure 20, it was found that the moments order in the up-up-down-down pattern shown in Figure 19(a), where exchange striction leads to a net polarization due to the difference in valence between Co^{2+} and Mn^{4+}. With the application of a magnetic field, the up-up-down-down magnetic structure is transformed to an up-up-up-down magnetic structure. In this

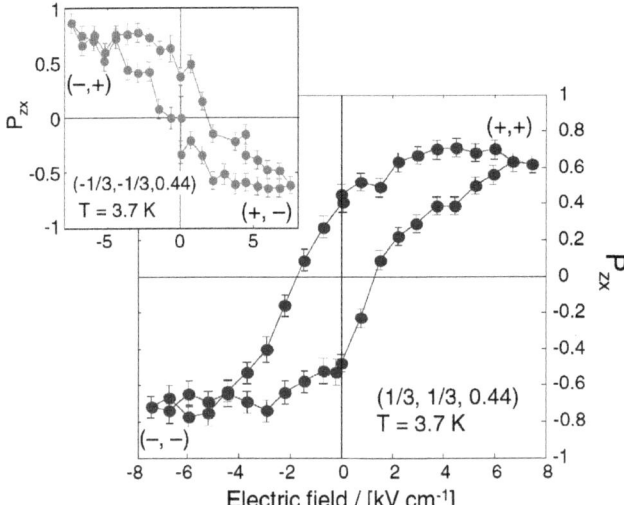

FIGURE 18 Hysteresis loops in P_{zx} as a function of the applied electric field for two magnetic reflections of RbFe(MoO$_4$)$_2$. The chiralities $(\sigma_t, \sigma_h) = (\pm, \pm)$ are indicated, where $\sigma_t = (S_1 \times S_2 + S_2 \times S_3 + S_3 \times S_1) \cdot v/S^2$ is the staggered triangular chirality (v is a unit vector describing the direction of the MoO$_4$ tetrahedra), where S_i is the spin on site i. The helicity is $\sigma_h = (S_{z=0} \times S_{z=1}) \cdot \widehat{r_{01}}/S^2 = \pm 1$ for right- and left-handed helices, where $\widehat{r_{01}}$ is the position vector between the spins at $z = 0$ and $z = 1$. Reproduced from Ref. [55].

structure (shown in Figure 21), the inversion center is restored, so the polarization should vanish, as is observed [57].

The REMn$_2$O$_5$ (RE = Ho, Tb, Tm, Gd, Er, Dy, Y) materials are also examples of exchange-striction based MFs. The crystal and magnetic structures are shown in Figure 22, where for simplicity we focus on YMn$_2$O$_5$ to avoid the complications of the rare earth moments. YMn$_2$O$_5$ crystallizes in the $Pbam$ structure, which consists of Mn^{4+} ions in an octahedral oxygen environment and Mn^{3+} ions within a trigonal pyramid oxygen environment, with a total of 8 Mn atoms in the unit cell. The Mn^{4+} octahedra are edge-sharing and form chains along the c-axis (Figure 22(b)). Within the $a-b$ plane, the Mn^{3+} and Mn^{4+} polyhedra are corner sharing and form zigzag antiferromagnetic chains. There are also five-sided rings within this plane which lead to frustrated interactions. Measurements of the polarization show that ferroelectricity develops in this material when it orders magnetically as shown in Figure 23, then drops to a much lower value at lower temperatures, which is accompanied by an additional magnetic transition.

Given the complexity of the lattice, it is not surprising that solving the magnetic structure was complicated. A naive application of representational analysis leaves 48 independent parameters to be determined by the data. While in the commensurate phase there was some success using powder diffraction measurements, single crystal measurements offered far more constraints on the structure. Also, as pointed out by Harris [58], there is an important constraint due to

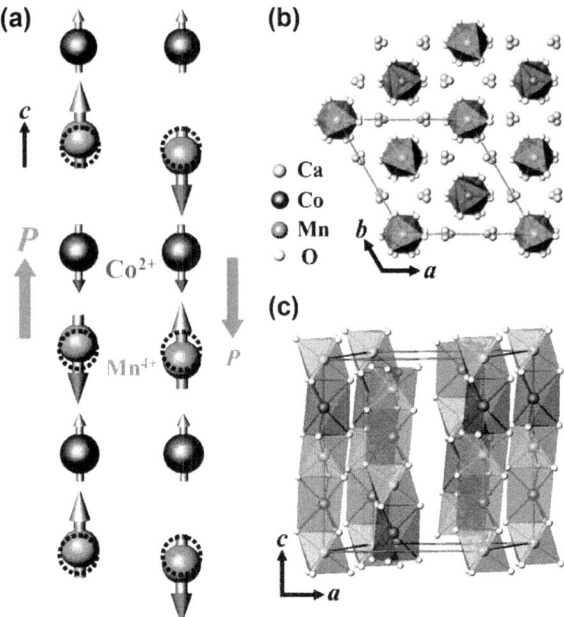

FIGURE 19 (a) Ising chains with the up-up-down-down spin order and alternating ionic order, in which electric polarization is induced through symmetric exchange striction. The two possible magnetic configurations leading to the opposite polarizations are shown. (b) The atomic positions in the undistorted chains are shown with dashed circles. (c) Crystal structure of the Ca_3CoMnO_6 chain magnet. The green (gray in print versions) boxes represent the crystallographic unit cell. *Reproduced from Ref. [59]*.

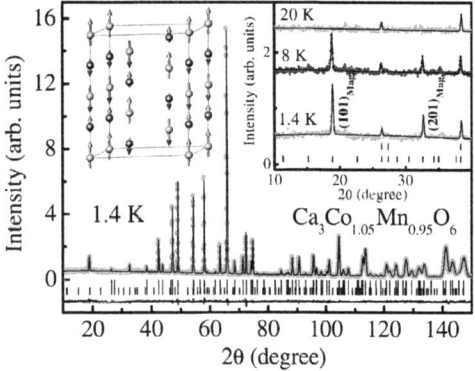

FIGURE 20 Observed (symbols) and calculated (line) powder neutron diffraction patterns for polycrystalline $Ca_3Co_{1-x}Mn_xO_6$ ($x = 0.95$) at $T = 1.4$ K. The first row of bars below the diffraction pattern indicates the positions of the nuclear Bragg peaks, and the second row depicts the magnetic Bragg peaks. The blue (gray in print versions) line shows the difference between the observed and calculated diffraction patterns. The insets show the low-angle patterns for $T = 1.4$, 8, and 20 K and the refined spin structure. *Reproduced from Ref. [59]*.

Multiferroics Chapter | 5 **311**

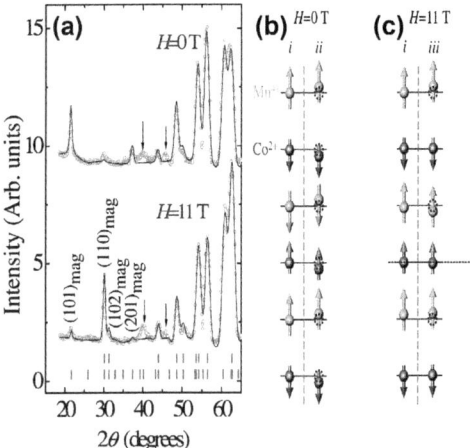

FIGURE 21 (a) Observed (symbols) and calculated (line) powder neutron diffraction patterns of Ca_3CoMnO_6 for $B = 0$ and 11 T, both at $T = 1.6$ K. The bars below the patterns indicate positions for nuclear (first row) and magnetic (second row) Bragg peaks. Arrows show an impurity phase. (b,c) Sketch representing the Ising chains with zero-field up-up-down-down spin order and alternating ionic charge order, in which electric polarization is induced through symmetric exchange striction. The corresponding displacements of the Mn (longer arrow) and Co (shorter arrow) ions along the spin chain are depicted in (ii) relative to their positions (i) in a uniform chain. The up-up-up-down plateau state without (i) and with (ii) ionic displacements due to the exchange striction. This magnetic and charge order posses an inversion symmetry center (horizontal line). *Adapted from Ref. [57].*

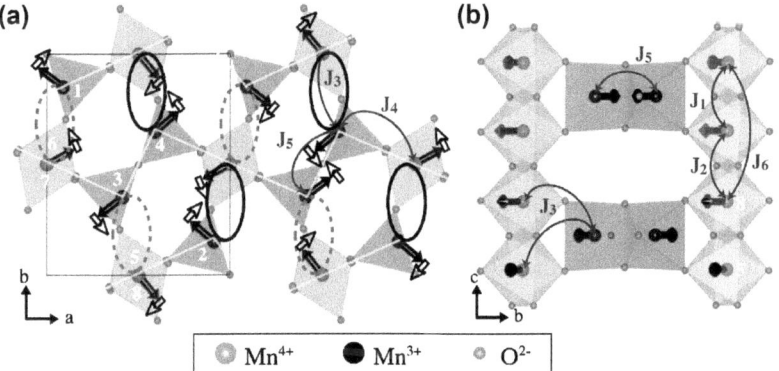

FIGURE 22 Crystal and magnetic structures in the LTI phase of YMn_2O_5 projected onto the (a) $a-b$ and (b) $b-c$ planes. Thin white lines indicate the antiferromagnetic chains along the a-axis. Also shown are the exchange coupling constants J_1-J_6 used to calculate the spin-wave spectrum. Black-filled arrows indicate the spin directions for the numerically obtained ground state of the spin Hamiltonian—see Eqn (1). Unfilled arrows in (a) represent the directions of the oscillating Mn spins in the optical phason mode. Ellipses with solid (dashed) lines show that the angle between neighboring spins is increased (decreased) by this mode. *Reproduced from Ref. [60].*

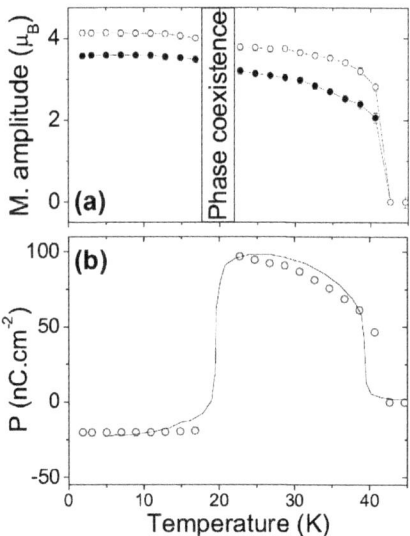

FIGURE 23 Refined values of the magnetic wave amplitudes on Mn^{3+} (open symbols) and Mn^{4+} (solid symbols) as a function of temperature. The average moment on each site is $2^{-1/2}$ of the wave amplitude. (b) Symbols: Calculated electrical polarization of YMn_2O_5. Solid line: Experimental values of the ferroelectric polarization from Kagomiya et al. [61]. The calculated polarization has been scaled by a single constant to account for the unknown magnetoelastic coupling parameter. Reproduced from Ref. [62].

inversion symmetry that dramatically reduces the number of free parameters to 15 (including a scale factor). Kim and coworkers [63] fit a structure using these constraints to both single crystal unpolarized neutron data as well as to spherical polarimetry data, which offered even tighter constraints on models. At this point, there is consensus on the magnetic structure in the FE, commensurate phase. This structure has nearly collinear spins for both the zigzag spin chains and for the ab components of the Mn^{4+} moments along the c-axis (see Figure 22), while the ac and bc components of the Mn^{4+} moments form spirals along the c-axis.

Landau theory predicts that such a magnetic structure will give rise to a polarization along the b-axis. However, microscopically, it is not clear whether the polarization arises from a spin current model or from exchange striction. To understand that, we must look to the low-temperature magnetic structure in which the FE polarization is greatly reduced, a rather difficult task. As mentioned earlier, there are naively 49 (including a scale factor) parameters required to determine the magnetic structure. Kim and coworkers believed that there was a first-order transition between the commensurate and incommensurate phase, and so relaxed the restraints of Landau theory and

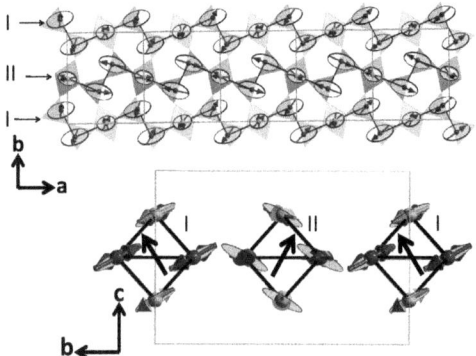

FIGURE 24 Low-temperature incommensurate magnetic structure of YMn_2O_5. Top: projection of the magnetic structure onto the $a-b$ plane. Mn^{3+} and Mn^{4+} are shown with red (light gray in print versions) and green (dark gray in print versions) colors, respectively. The ellipsoidal contours, more elongated on Mn^{3+}, show the envelope of the magnetic modulations. Adjacent antiferromagnetic zigzag chains, labeled I and II, are shown by thick black lines. Bottom: projection of the magnetic structure onto the $b-c$ plane. The black arrows represent the normal direction to the ellipsoidal modulates in chains I and II. *Reproduced from Ref. [64].*

obtained a rather complicated magnetic structure that fit their data. However, Radaelli and coworkers [64] incorporated the constraints of inversion symmetry and the requirement of a polarization along the b-axis. They expressed this in the language of corepresentations rather than the free energy approach taken by Harris, and found that they could achieve an excellent fit to the low-temperature incommensurate structure through the use of a mere 15 parameters (including a scale factor). The magnetic structure also should be invariant up to a global shift in the phases of the Fourier components, so that reduced their total number of parameters to 14.

In their magnetic structure (Figure 24), neighboring zigzag chains along the a-axis are out of phase and thus neighboring moments on adjacent chains are almost orthogonal to each other. As the polarization induced by the exchange striction depends on the overlap of these moments, it vanishes. The authors calculated the spin-orbit coupling (SOC) and found that it increases in the low-temperature phase, and that the exchange striction decreases. Hence it seems likely that exchange striction (EC) is the driving mechanism for the induced polarization in these compounds, though the spin-orbit-coupling-induced spin currents may play a role in the small residual ferroelectricity which remains in the low-temperature incommensurate phase.

A summary of both type-I and type-II multiferroics is provided in Table 2, including ordering temperatures, space group symmetry, ferroelectric polarization, and coupling mechanism.

TABLE 2 Summary of Multiferroic Properties of Compounds

	T_N (K)	T_C (K)	P ($\mu C/m^2$)	Mechanism	SG	References
Ba_2CoGeO_7	6.7	6.7	125	On-site SOC	$P\bar{4}2_1m$	[65,66]
$MnWO_4$	12.3	12.3	5	IDM	$P2/c$	[48]
Ca_3CoMnO_6	16.5	16.5	90	ES	$R\bar{3}c$	[59]
$h\text{-}HoMnO_3$	72	1375	56,000	Type I	$P63cm$	[12,13–16,67]
YMn_2O_5	45, 17	$17 < T < 45$	1000	ES	$Pbam$	[62]
CuO	213, 230	$213 < T < 230$	~75	IDM	$C2/c$	[68]
$BiFeO_3$	643	1093	100,000	Type I	$R3c$	[29]
$TbMnO_3$	42	27	800	IDM	$Pbnm$	[69]
$CuFeO_2$	10, 15	$10 < T < 15$	400 (under field)	IDM	$R\bar{3}m$	[69]
$CoCr_2O_4$	95, 27	27	6	IDM	$Fd\bar{3}m$	[70]
$(Sr,Ba)MnO_3$	200	400	130,000	Type I	$Pm\bar{3}m$	[37]
$RbFe(MoO_4)_2$	3.7	3.7	5	IDM	$P\bar{3}$	[51]
$BaTiO_3$		393	260,000		$P4mm$	[71]

5.5 DOMAINS

5.5.1 BiFeO$_3$

At room temperature, BiFeO$_3$ is rhombohedral and orders in the R3c structure, but here we will use the pseudocubic notation for simplicity, where the magnetic propagation vector is $(\delta,\bar{\delta},0)$. Due to the threefold axis along the (1,1,1) direction, we expect three magnetic domains, with analogous propagation vectors of $(\delta,0,\bar{\delta})$ and $(0,\delta,\bar{\delta})$. Lee and coauthors showed [30] that it is possible to prepare a single crystal such that it crystallizes in a single FE domain. Moreover, this single FE domain consisted of a single magnetic domain and a single "chiral" domain (that is a cycloid with a single handedness with respect to the coordinate system defined by the propagation vector of the cycloid and the FE polarization). This was determined by a polarized beam experiment, where if there is a cycloid of a single helicity the $+/-$ and $-/+$ spin-flip channels will split. This splitting is maximal when the neutron polarization is parallel to $\mathbf{Q}$ and the helicity axis of the spiral is parallel to the neutron polarization; it is minimal when the helicity axis is perpendicular to the neutron polarization. This is exactly what is observed in Figure 7 and determines that the cycloid forms in the plane defined by the propagation vector and the FE polarization. This result is confirmed by the unpolarized beam measurements of Lebeugle (see Figure 6) and coworkers [29]. In both works, it was found that the application of a strong electric field along the [001] direction switches the FE polarization. As the cycloid is induced by an inverse Dzyaloshinskii-Moriya (IDM) interaction, it is intimately coupled to the FE polarization and rigidly switches with it. Lee et al. also noted that even when the electric field did not switch the FE polarization, magnetostriction would cause one magnetic domain to be favored over the others (see Figure 25). A similar effect can be achieved with pressure [72].

5.5.2 HoMnO$_3$

Understanding the behavior of domains is essential when it comes to applications, and for MFs, this naturally involves both FE and magnetic domains, and their coupling. For HoMnO$_3$, one of the striking results from optical experiments was the claim that bulk ferromagnetic order of the Ho moment can be induced and reversibly switched through application of an electric field. However, X-ray and neutron scattering experiments found no evidence for any electric field-induced ferromagnetism [24,73], indicating a different origin for the magnetic behavior, such as coupling between magnetic and FE domains and domain walls. Figure 26 shows SANS measurements while applying an electric (and magnetic) field, which reveal a net magnetization occurring within antiferromagnetic domain walls [25]. Figure 26(a) shows the temperature dependence of the total counts, which increases quickly below the spin reorientation transition $T_{SR} = 40$ K (see Figure 2). This scattering follows a power law for small Q with an exponent of -4, as expected from the Porod

316 Neutron Scattering - Magnetic and Quantum Phenomena

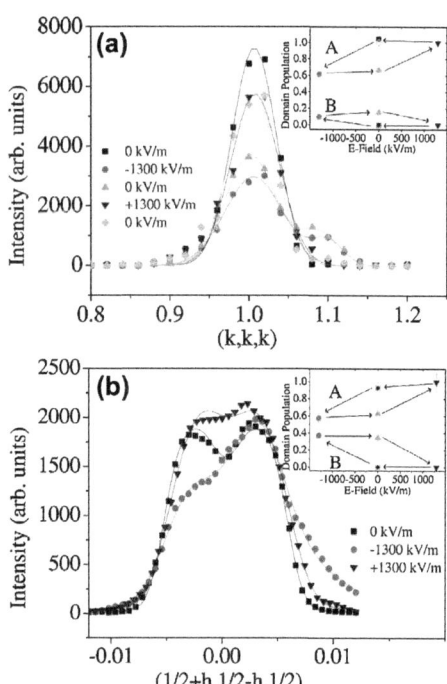

FIGURE 25 (a) Scans through the (1,1,1) peak (domain A) and (1,−1,1) or equivalent peak (domain B) in BiFeO$_3$. (b) Scans through the (0.5,0.5,0.5) +/− τ (domain A) and (0.5,−0.5,0.5) +/− τ or equivalent (domain B) magnetic peaks. In (b), zero-field data are shown only for the as-prepared sample for clarity. Solid lines are fits. The insets show ferroelastic domain populations for various applied electric fields. Electric fields were applied in the sequence shown in the legend. *Reproduced from Ref. [74].*

law for scattering from large objects. The Porod amplitude versus temperature (Figure 26(d)) follows the same behavior as a function of temperature, abruptly increasing with decreasing temperature below T_{SR}. Polarized SANS measurements of the spin-flip cross section (inset to Figure 26(a)) show a rise in scattering below T_{SR}, similar to the rises observed in Figure 26(a), (b), and (d), demonstrating that the abrupt increase in the scattering intensity below T_{SR} indeed has a magnetic component. These data demonstrate that the observed increase in the Porod scattering below T_{SR} is associated with a net magnetization developing within antiferromagnetic domain walls.

5.6 THIN FILMS AND MULTILAYERS

5.6.1 BiFeO$_3$

Some of the earliest studies of MF thin films focused on the BiFeO$_3$ system because of its potential for applications. There are several technical problems that come with attempting to do diffraction on thin films, mostly from the weakness of the signal. The earliest studies of BiFeO$_3$ films were performed

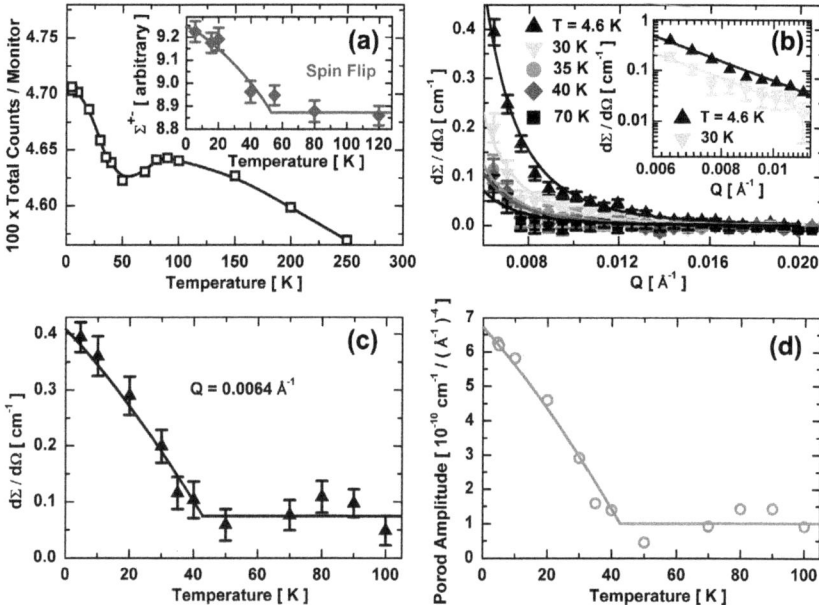

FIGURE 26 (a) Temperature dependence of the normalized total counts for HoMnO$_3$. The inset shows the spin-flip scattering cross section determined from the sum of the radially averaged polarized small-angle neutron scattering data. (b) Temperature dependence of the average scattering cross section for $Q = 0.0064$ Å^{-1}. (c) Semilog plot of the average scattering cross section at two temperatures. The lines are fits to the Porod law which has an $\sim Q^{-4}$ dependence. (d) The Porod amplitude versus temperature relation as determined from the fits is shown in (c). Lines are guides to the eye unless otherwise noted. *Reproduced from Ref. [25].*

by Bea et al. [75] on films grown on the 001 SrTiO$_3$ substrate. For films of thicknesses less than 70 nm, they found that the structure was tetragonal, but as the thickness increased to 240 nm the structure changed to monoclinic (consistent with X-ray results). Furthermore, they found that for the 240-nm-thick films, the cycloid collapsed to simple commensurate G-type order.

Later work was done by Ke and coworkers for films of 800 and 200 nm on SrTiO$_3$ substrates [76]. These substrates either were flat, or had a miscut along the [1,1,0] or [1,0,0] directions to change the FE domain populations. They found that for the sample with a miscut along the [110] direction a single FE domain was found, with the propagation vector along the [1,$\bar{1}$,0] direction. They concluded that this was a spiral in the plane containing the propagation vector and the FE polarization as in bulk (see Figure 27), but based only on a single reflection. The splitting of the reflection only tells us that the ordering is incommensurate, but the magnetic structure may either be amplitude modulated or a spiral. For the miscut along the [1,0,0] direction, they found that there are two FE domains present, one with a propagation vector along the [1,−1,0] direction and one along the [1,1,0] direction. For the film with no miscut, all FE domains were present. Here, the combinatorial possibilities were too large for them to simulate the neutron diffraction pattern. Finally, they suggested that for the 200-nm-thick film the magnetic structure remained G-type.

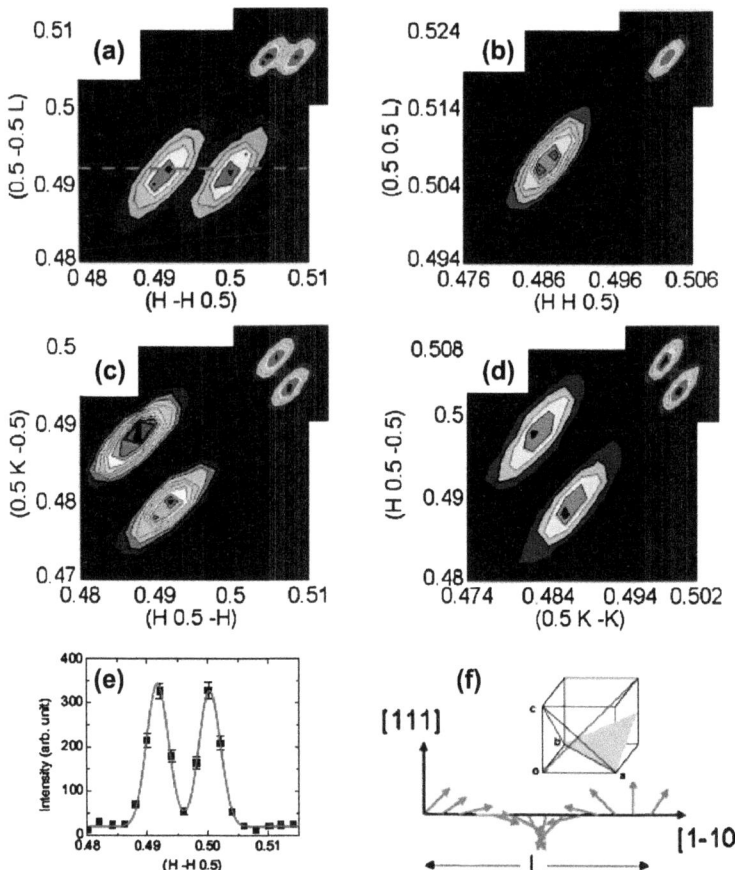

FIGURE 27 Results of neutron diffraction measurements performed on an 800-nm film of BiFeO$_3$ grown on SrTiO$_3$ 001, with a 4° miscut along [1,1,0]. (a)–(d) Contour maps showing the neutron intensity distribution of S1 in different scattering planes. The insets represent simulations of the magnetic structure for a monodomain sample with a propagation vector along the [1,−1,0] direction. (e) Line cut cross the center of two ellipsoids parallel to the [1,−1,0] direction, with the solid curve a fit to two Gaussians. (f) Schematics showing the cycloidal spiral spin structure of S1. The shaded triangle represents the spin rotation plane defined by the polarization direction [1,1,1] and the propagation wave vector [1,−1,0]. Spins rotate along the [1,−1,0] direction. *Reproduced from Ref. [76].*

Subsequent work by Ratcliff and coworkers [77] used polarized beam measurements on 1-micron-thick films grown on the 001 surface of SrTiO$_3$ with a 4° miscut along 001 (analogous to the S1 sample of Ke and coworkers). They found that the plane of the spiral was different from the bulk (see Figure 28), with the moments in the plane normal to the FE polarization. This is consistent with theoretical predictions of a 111 easy plane anisotropy [78]. Shown in Figure 27(a) are unpolarized beam measurements of the (0.5,0.5,0.5) reflection in the $(H,K,H + K/2)$ scattering plane. Other work [79] showed that

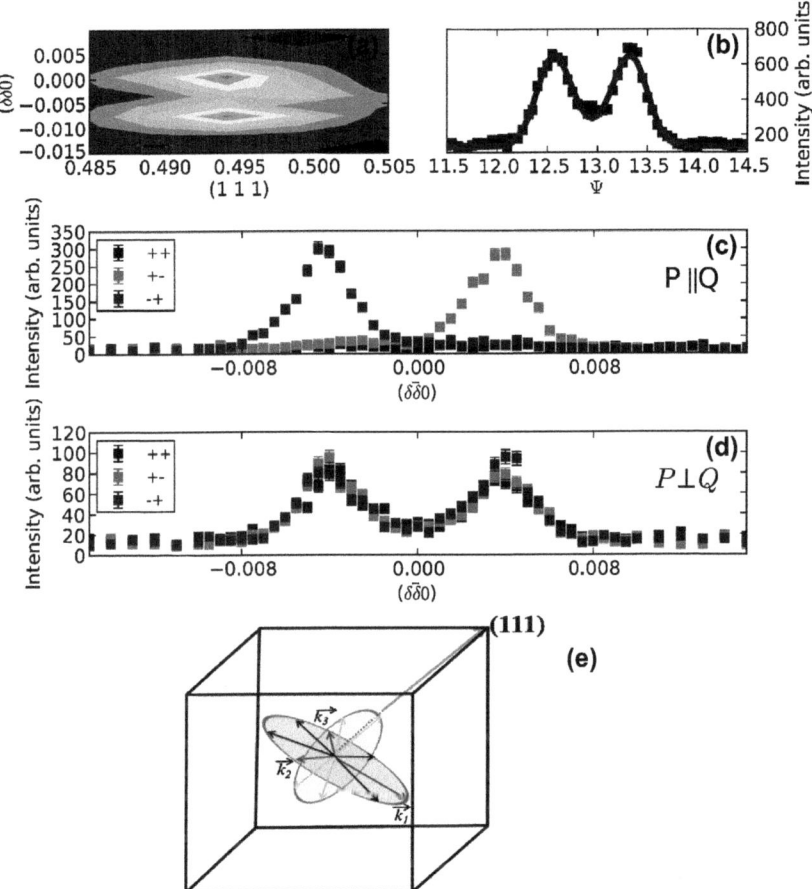

FIGURE 28 (a) Contour map of the neutron diffraction in the [h,k,h+k/2] zone for the S2 sample of BiFeO$_3$ about the (0.5,0.5,0.5) position in reciprocal space, after application of an electric field of −200 kV/cm. (b) ψ cut through the two peaks. Spin-flip and Non-spin-flip cross sections for (c) **P**||**Q** and (d) **P** ⊥ **Q** about the (0.5,0.5,0.5) position. (e) Magnetic structure. Red (gray in print versions) arrows show propagation wave vectors for the three possible magnetic domains; the plane of the spiral is in red (gray in print versions); blue (dark gray in print versions) arrows show moments in this plane; the plane of the spiral is purple for single crystals (for domain defined by k_3 propagation vector); and green (light gray in print versions) arrows show moments in this plane. *Reproduced from Ref. [77].*

the actual nature of the magnetic order in these materials depends upon the orientation of the substrate.

5.6.2 TbMnO$_3$

TbMnO$_3$ films were studied using polarized neutron reflectivity (PNR) measurements by Kirby and coworkers [80]. They looked at films grown by pulsed laser deposition on the 001 oriented SrTiO$_3$ (STO) substrates. Due to the lattice

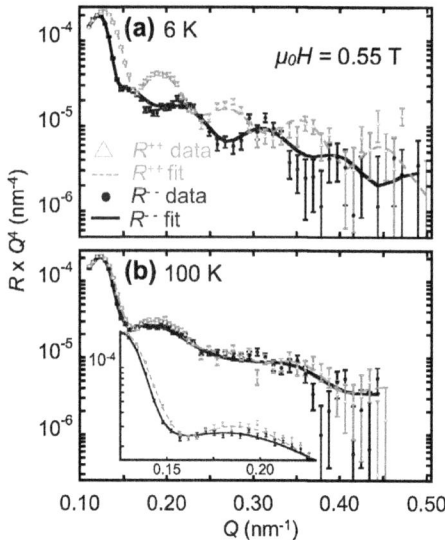

FIGURE 29 Fitted polarized neutron reflectivity of a TbMnO$_3$ thin film. Spin-dependent oscillations are present at 6 K (a) and at 100 K (b), where the inset shows an expanded view. *Reproduced from Ref. [80].*

mismatch between the film and the substrate, the film is compressively strained. They performed their measurements after field-cooling the sample from 290 to 6 K in 0.55 mT. The results (Figure 29) show that the magnetization in the film is parallel to the applied field. They also showed that the whole film is magnetized. Oddly, this magnetization decreases but persists up to 100 K, well above the bulk FE and magnetic ordering temperatures. The power of this measurement is that it shows that strain gives rise to ferromagnetism throughout the film, rather than just at surface. It would be interesting to see whether this ferromagnetism interacts with the low-temperature FE properties of the film.

Besides TbMnO$_3$, orthorhombic LuMnO$_3$ MF films grown on the 110 surface of YAl$_2$O$_3$ were measured by White et al. using PNR [81]. They found the development of ferromagnetism in a 56-nm-thick film, with a rather large $\sim 1\ \mu_B$ moment at the interface (the average moment is $\sim 0.49\ \mu_B$) that develops below ~ 100 K (Figure 30). They also studied the antiferromagnetic order using wide-angle neutron diffraction (on a slightly thicker, 80-nm-thick film) and observed that antiferromagnetic order sets in at ~ 40 K. The peaks have a propagation vector of $(0,k,1)$ and are incommensurate with $k = 0.482(3)$. This is interesting because the dielectric anomaly [82] is also observed at this temperature (see Figure 31), similar to bulk TbMnO$_3$. The current thought is that there is a strain gradient in these films, which results in part of the film being ordered ferromagnetically near the interface with the substrate and gradually transitioning to an incommensurate order farther away.

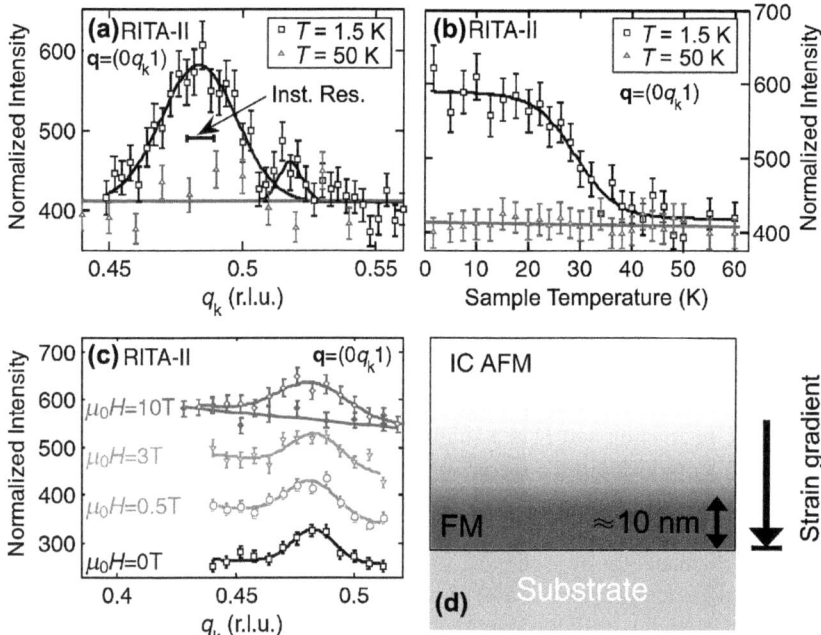

FIGURE 30 (a) The $(0,q_k,1)$ antiferromagnetic (AFM) peak in a LuMnO$_3$ thin film. A Gaussian fit of the main peak gives the peak center at $q_k = 0.482(3)$. The solid bar indicates the instrumental resolution. (b) T dependence of the $(0,q_k,1)$ for $B||a$ up to 10 T. Empty (filled) symbols denote data taken at $T = 2$ K ($T = 50$ K). Curves are displaced vertically for clarity and peaks are fit with Gaussian line shapes. (d) Sketch to illustrate the magnetic situation in the film. The FM layer located near the strained film substrate interface evolves toward a likely cycloidal incommensurate (IC) AFM order when moving down the strain gradient toward the film surface. *Reproduced from Ref. [81].*

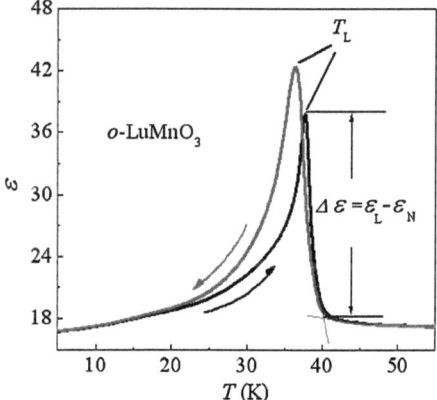

FIGURE 31 Temperature dependence of the dielectric constant ε at 1 kHz for an o-LuMnO$_3$ sample, in both cooling and warming cycles. $\Delta\varepsilon$ is the difference between the dielectric constant at T_L (ε_L) and at T_N (ε_N). L denotes the lock-in temperature ~ 37 K, while $T_N \sim 40$ K. *Reproduced from Ref. [82].*

5.7 SPIN DYNAMICS

5.7.1 HoMnO$_3$

The spin dynamics for the hexagonal MFs has been investigated in some detail. Within the plane, the triangular magnetic lattice is frustrated and yields noncollinear magnetic structures, with each spin oriented at 120° with respect to its neighbors as discussed previously. One simplification is that the coupling along the c-axis is very weak, so the spin dynamics is basically two-dimensional in nature. A Hamiltonian that captures the essential physics can be written as

$$H = -\sum_{i,j} J_{ij} \mathbf{S}_i \cdot \mathbf{S}_j - D \sum_i \left(S_i^z\right)^2 \qquad (3)$$

where $J_{ij} = J$ is the strength of the nearest-neighbor in-plane antiferromagnetic exchange interaction and D is a single-ion anisotropy. The Hamiltonian can be solved by introducing three flavors of bosons, and linear spin-wave theory then consists of three separate modes that propagate within the hexagonal plane, which are equivalent in energy but their origins are offset in wave vector as shown schematically in Figure 32 [16]. Fitting the model to the data at 20 K (intermediate temperature phase) along the high-symmetry directions in reciprocal space results in fitting parameters $J = 2.44$ meV and $D = 0.38$ meV. The authors found that the anisotropy has a substantial temperature dependence, and this is what drives the spin reorientation transitions observed in this system.

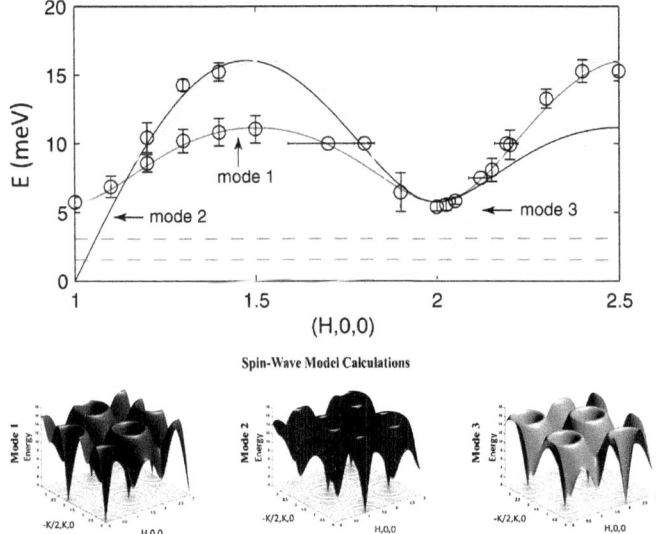

FIGURE 32 In-plane spin-wave dispersion for HoMnO$_3$ at 20 K for the three modes. Dashed lines indicate two (dispersionless) crystal field levels of Ho at 1.5(1) and 3.1(1) meV. The bottom part of the figure shows the calculated spin wave dispersion relations for the three modes. The dispersion curves for the three modes are identical, but the origins are offset for the three. *Adapted from Ref. [16]*.

We note that such spin reorientations do not occur in YMnO$_3$ [83], so it is clear that they originate from the holmium, as has been observed in other rare earth systems such as Nd$_2$CuO$_4$ [84]. We note that the full magnetic Hamiltonian for HoMnO$_3$, including the Ho moments and FE coupling, must be considerably more complex than Eqn (3), but this very simple magnetic model establishes the dominant magnetic interactions for the Mn ions in this system.

5.7.2 BiFeO$_3$

For BiFeO$_3$, some of the earliest measurements of the spin dynamics in single crystals were performed by Jeong et al. [85]. They fit their observations to a relatively simple Hamiltonian where

$$H = J \sum_{NN} S_i \cdot S_j + J' \sum_{NNN} S_i \cdot S_j - D \cdot \sum_i S_i \times S_{i+\hat{\delta}}. \quad (4)$$

In this model, there is an exchange interaction between nearest neighbors, J, and another exchange interaction between next-nearest neighbors, J'. They also included a Dzyaloshinskii—Moriya (DM) antisymmetric exchange interaction D, where D is parallel to the propagation vector, $[1,\bar{1},0]$. $S_{i+\hat{\delta}}$ is the next-nearest neighbor spin along the [1,1,0] direction. This is an effective DM interaction, which is the one that is induced from the FE polarization as predicted by Landau theory [86]. In Figure 33, we show their measurements and fits to the data. They found that their values of J were consistent with the ordering temperature. Furthermore, their determination of D was consistent with the period of the long-wavelength spiral observed in BiFeO$_3$. They found $J = 4.38$ meV, $J' = 0.15$ meV, and $D = 0.107$ meV.

Higher-resolution measurements were performed by Matsuda and co-workers [87], where they found interaction parameters of $J = 6.48$ meV, $J' = 0.29$ meV, and $D = 0.1623$ meV. They also included a single-ion anisotropy term in their Hamiltonian, K, and found $K = 0.0068$ meV, which directs the moments normal to the [1,1,1] polarization direction. In contrast to the work of Jeong et al., they placed D along $[1,1,-2]$ which is orthogonal to $[1,-1,0]$. Furthermore, the nearest neighbors in their sum for the DM term are now along $[1,-1,0]$ instead of [1,1,0]. The additional K term induces an anharmonicity in the cycloid. For their value of K, the calculated ratio of the intensity of the primary to the third-order harmonic, I1/I3, is 500 while from the neutron diffraction measurements described earlier found 120. From NMR measurements, it was 25. We should note that later Raman measurements [88] were able to refine these values of D and K.

5.7.3 (Sr-Ba)MnO$_3$

To determine the exchange interactions in materials, inelastic neutron scattering information is needed, which can best be obtained with measurements on single crystals as discussed above for HoMnO$_3$ and BiFeO$_3$. Generally,

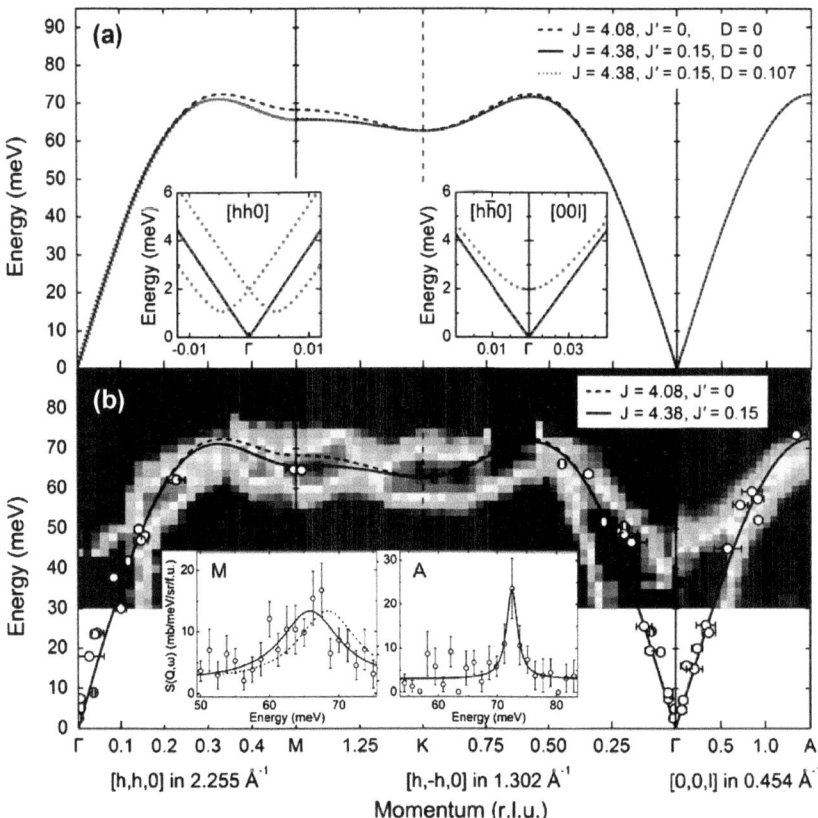

FIGURE 33 Spin-wave dispersion of BiFeO$_3$. (a) Theoretical spin waves calculated with three different Hamiltonians. The two inserts are for the blown-up figures of the low-energy regions to illustrate the effects of the Dzyaloshinskii−Moriya-like term (D) on the spin waves along the $\Gamma - M$ and $\Gamma - A$ directions, respectively. (b) Experimental spin waves measured on the AMATERAS (circles) and MERLIN (contour plot) spectrometers together with the theoretical spin waves (solid line). The dashed line is for the theoretical spin waves calculated with the Hamiltonian having the nearest-neighbor interaction alone. Inserts are for the momentum cut at the M and A points. *Reproduced from Ref. [85].*

crystals of sufficient (gram) size samples are required for these measurements, which are not always available. However, important information can still be extracted with inelastic measurements on polycrystalline samples. Figure 34 exhibits maps of the ground state magnon density-of-states emanating from the {1/2,1/2,1/2} magnetic Bragg peak position located at $|\mathbf{Q}| = |\mathbf{Q}_{AFM}| = 1.42$ Å^{-1} for the same (Sr$_{0.56}$Ba$_{0.44}$)MnO$_3$ sample discussed in Section 5.3.3 [38]. There is a gap in the excitation spectrum of 4.6 meV, with a conical magnon density-of-states emerging for energies above that. These data can be modeled by a simple nearest-neighbor exchange model to

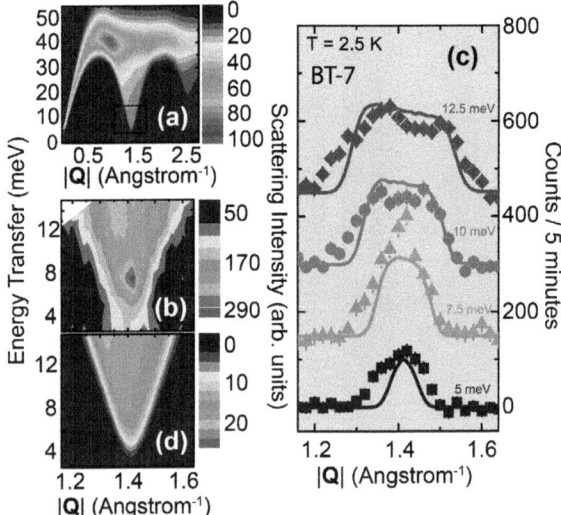

FIGURE 34 (a) Contour plot of the calculated magnon density-of-states from $(Sr_{0.56},Ba_{0.44})MnO_3$ using $J = 4.8$ meV and a spin-wave gap $\Delta_{AFM} = 4.6$ meV extracted from fitting the experimental data for the ground state spin dynamics at low energies shown in (b). The wave vector and energy ranges shown in (b) and (d) are indicated by the bounded box in (a), and (d) shows the calculated magnon density-of-states over the same range to compare directly with the data in (b). (c) Constant-energy cuts through the low-temperature data (points). Data have been offset by 150 counts for each energy for clarity. The solid curves through the data are the calculated results from the model. *Adapted and reproduced from Ref. [38].*

determine an average, isotropic exchange interaction of $J = 4.8$ meV. The top of the magnon band of excitations then is predicted to occur at ≈ 40 meV, which was confirmed by subsequent measurements. Thus straightforward spectroscopic measurements on polycrystalline samples can determine the basic interactions of the system, but not in the exquisite detail that single crystal measurements allow.

5.7.4 Electromagnons

Electromagnons are excitations that involve both the magnetism and the ferroelectricity simultaneously, and we now discuss the nature of such combined excitations. We start by discussing incommensurate magnetic materials such as $TbMnO_3$ where the excitations are complicated [89]. In the cycloid phase, for example, there is a "phason" or "sliding mode" excitation, which corresponds to a change in the wavelength of the excitation. In $TbMnO_3$, the wave vector of this cycloid is along the a-axis. If one coherently rotated each moment in the system by a fixed amount, each moment's relative orientation would remain unchanged, so the wavelength is unchanged and such an excitation would not cost any energy to make this rearrangement. Hence, this is the Goldstone mode

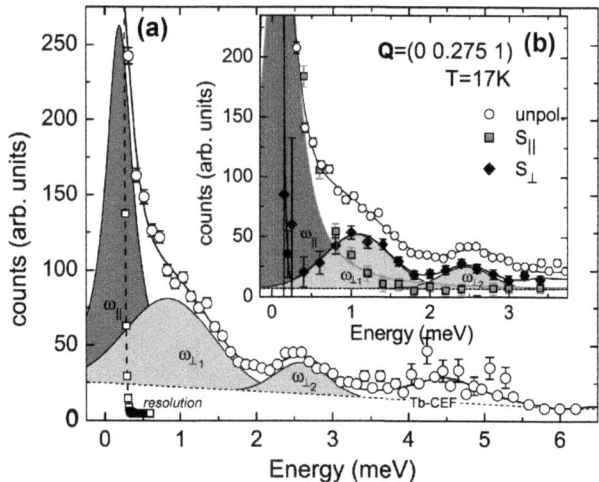

FIGURE 35 (a) Magnetic excitation spectrum in TbMnO$_3$ at the antiferromagnetic zone center $Q = (0,0.275,1)$ in the ferroelectric spiral phase at $T = 17$ K using unpolarized neutrons. Solid lines denote fits to the data, and shaded areas mark the different magnon signals. Open squares and dashed lines correspond to the experimental resolution determined using a vanadium standard sample. (Inset, b) polarization analysis of the excitation spectrum with the decomposition of the unpolarized spectrum into the components of the dynamic magnetization parallel and perpendicular to the ab spiral plane, $S_{\parallel}$ and $S_{\perp}$, respectively. *Reproduced from Ref. [90].*

of the system. If instead we change the relative phase of each spin by a small amount, in other words make a small change in the wavelength of the cycloid, this would cost energy and would correspond to a low-energy phason magnetic excitation. This is the $w_{\parallel}$ mode that was measured by Senff et al. [90] and is shown in Figure 35. However, this mode does not change the induced electric polarization, and so is a pure magnetic excitation which does not involve the ferroelectricity. This is an important type of excitation, but is not an electromagnon.

However, there are two other interesting excitations, one which effectively rotates the plane of the cycloid around the b-axis (the propagation vector direction) and the other around the c-axis (see Figure 36). The rotation about the b direction affects the electric polarization through the spin-current model [42], and will change the electric polarization along the a direction (Figure 37).

Senff et al. [90] performed both polarized and unpolarized neutron scattering measurements to observe these modes. They first performed unpolarized inelastic measurements at the antiferromagnetic zone center and fit their data to three modes, along with a low-lying Tb crystalline electric field excitation. The nature of these modes is revealed by polarized inelastic neutron measurements. The low-energy $w_{\parallel}$ mode consists of a fluctuating moment parallel to the scattering vector, whereas the two higher-energy modes correspond to fluctuations perpendicular to the scattering vector. Given that they measured in a geometry with the a-axis perpendicular to the scattering plane, these

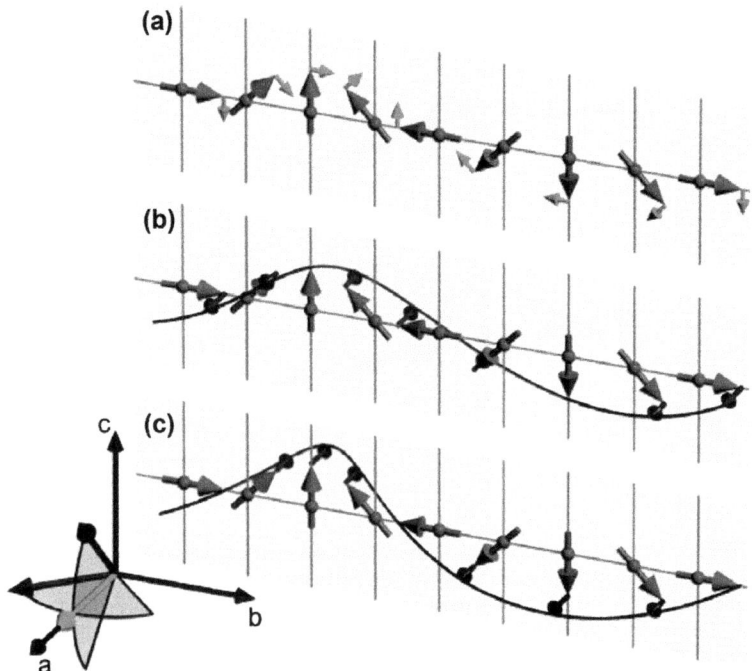

FIGURE 36 Sketch of the polarization schemes of the different magnon excitations at the magnetic zone center of the cycloid structure TbMnO$_3$. The static spin structure is marked by thick gray arrows, and the local fluctuations by smaller colored arrows. In addition, the spin chirality $C = S_i \times S_{i+1}$ is shown for the sketched instantaneous spin arrangement with the different modes marked by the color scheme. The bc-polarized sliding mode can be regarded as a rotation of the spin plane around the a-axis (a). For the a-polarized modes the fluctuations can either be in-phase with the static b-component (b) or with the static c-component (c), resulting in a rotation of the spiral plane around c and b, respectively. In the lower-left edge we show the vector diagram of the ferroelectric oscillating polarization associated with the three magnons. *Reproduced from Ref. [90].*

fluctuations are then either parallel or perpendicular to the plane of the spiral. Hence the lower energy phason mode is within the plane of the spiral, and the higher energy modes are fluctuations along the chirality axis of the spiral, the a-axis. We also note that the separation is much more distinct when measured with polarized neutrons. Both modes are shown in Figure 35. The higher energy mode matches the energy that Pimenov and coauthors reported in Ref. [91] based on optical measurements.

It is also worthwhile to perform inelastic measurements in the paraelectric longitudinal SDW phase. Analogously to the cycloid phase, there is a phason Goldstone mode (see Figure 37). There is another mode that corresponds to a change in the amplitude of the wave without changing the phase or wavelength (amplitudon). This mode has a gap and is expensive energetically and was not observed. There are also four transverse modes. Two of these will be in-phase

328 Neutron Scattering - Magnetic and Quantum Phenomena

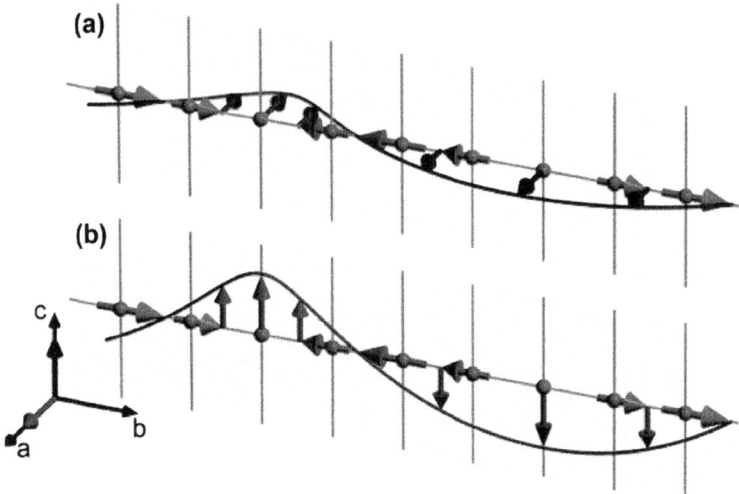

FIGURE 37 Polarization schemes of the two $\pi/2$ modes in a spin-density wave (SDW) structure of TbMnO$_3$. The a-polarized mode transforms the SDW into a magnetic cycloid constrained to the $a-b$ plane (a), while the c-polarized mode results in a bc cycloidal structure (b). In the lower-left corner the vector product $S_i \times S_{i+1}$ is shown for the frozen-in states with the different modes marked by the color code. *Reproduced from Ref. [90].*

and two out-of-phase with the underlying static structure. Both of these will give rise to a "dynamic" cycloid either in the $a-c$ or in the $b-c$ plane. Based on the spin-current model, one should give rise to a dynamic polarization along the a-axis and the other along the c-axis. The one giving rise to the cycloid in the $b-c$ plane should be our magnetic soft mode. The $a-b$ cycloid mode should be infrared active and was observed by Pimenov and coworkers [91]. Again, neutron polarized beam measurements were critical in observing these modes directly (see Figure 38). Finally, it was possible to measure the spin wave dispersion and to determine the interaction parameters and the magnitude of the single-ion anisotropy.

5.7.5 MnWO$_4$

As discussed earlier, the magnetic phase diagram of the MnWO$_4$ system is rather complex, and a determination of the spin wave dispersion relations and concomitant interaction parameters is essential to understand the nature of the ground state. The first study [46] required nine exchange parameters to describe the spin wave excitations observed, along with a single-ion anisotropy term. These fits provided an adequate description of the data and correctly predicted the ordering wave vector in the AF3 phase. A later higher resolution study [92] required 11 exchange parameters and (see Figure 39) a single-ion anisotropy term to adequately describe the data taken

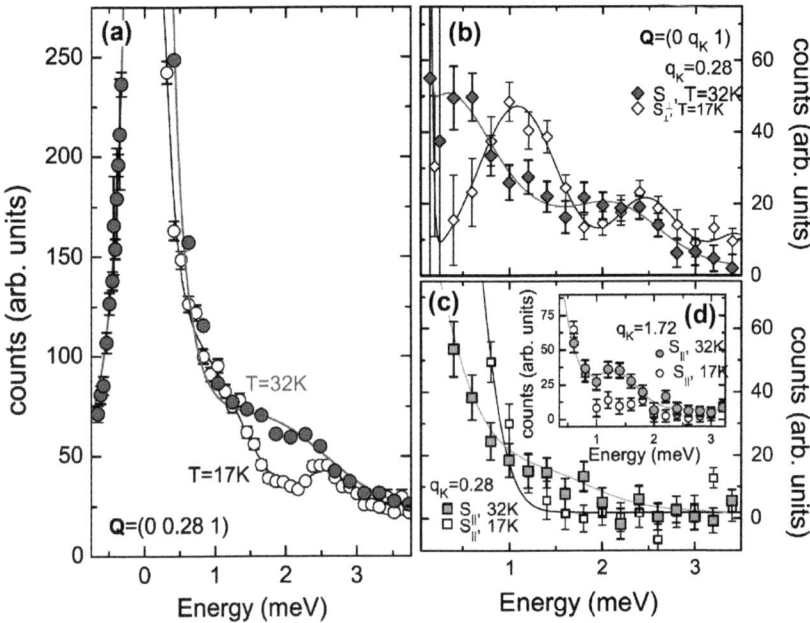

FIGURE 38 Comparison of the energy scan in TbMnO$_3$ at the magnetic zone center $Q = (0,0.28,1)$ in the ferroelectric cycloid phase at $T = 17$ K, and in the paraelectric SDW phase at $T = 32$ K: (a) superposition of both components, (b) the components perpendicular to the b–c plane, and (c) the components parallel to the b–c plane. The inset in (c) shows the parallel component at $Q = (0,1.72,1)$. *Reproduced from Ref. [90].*

along several different cuts in reciprocal space (see Figure 40). In both cases, the interactions are primarily antiferromagnetic in nature. However, many of the interactions are competing and comparable in magnitude. This makes the system rather sensitive to doping [93–96], with rather dramatic effects for the magnetic structure and thus for the electric polarization.

5.7.6 YMn$_2$O$_5$

To understand the interactions in YMn$_2$O$_5$, Kim et al. [60] studied the excitations in the incommensurate low-temperature phase. They started with a simple spiral model in the a–b plane, with a standard Heisenberg exchange Hamiltonian. They also allowed for different single-ion anisotropy terms for the Mn^{3+} and Mn^{4+} ions. Based on inelastic data (see Figure 41), they found $J_1 = -0.2$, $J_2 = -1.4$, $J_3 = 0.15$, $J_4 = 3.75$, $J_5 = 2.75$, $J_6 = 0.72$, $D_{Mn}^{3+} = 0.13$, and $D_{Mn}^{4+} = 0.11$ meV.

Besides spin wave excitations, they found the usual phason excitations associated with spin spirals. They also found excitations associated with rotations of the spiral plane about either the a or b axis. There are optical counterparts to these acoustic modes, which are electromagnons [60].

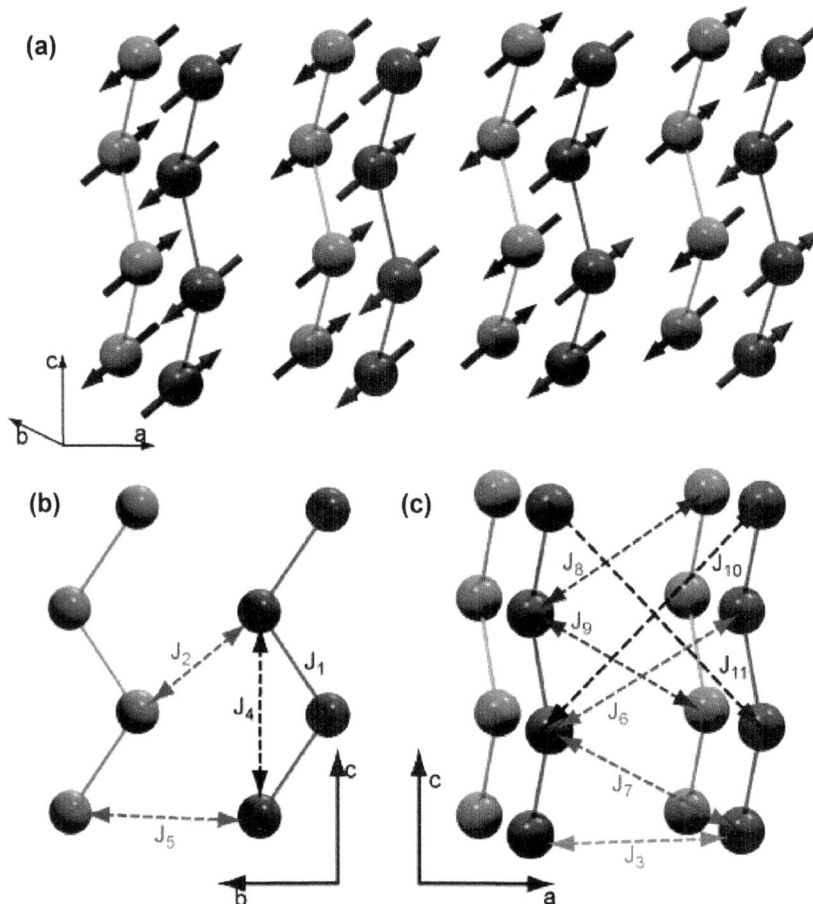

FIGURE 39 (a) Magnetic structure of MnWO$_4$ in the collinear, commensurate phase at low temperature. The magnetic spins lie in the $a-c$ plane with the moment canted to the a-axis about 35°. The magnetic spins form zigzag up-up-down-down chains along the c-axis and are coupled antiferromagnetically along the b-axis. (b) The magnetic interactions along and between spin chains in the $b-c$ plane. (c) Higher-order magnetic interactions along the a-axis. The magnetic couplings are labeled with increasing bond distance. Note the monoclinic crystal structure ($\beta = 91.4°$) makes J_6/J_7, J_8/J_9, J_{10}/J_{11} pairs different. *Reproduced from Ref. [92].*

5.7.7 RbFe(MoO$_4$)$_2$

Spin wave measurements were performed to determine the exchange interactions in this material. The interactions include an in-plane interaction, J; an effective interplane interaction, J_p; and a single ion anisotropy, D. The fits are good (see Figure 42) and show that this material is well described as a triangular lattice antiferromagnet with xy anisotropy.

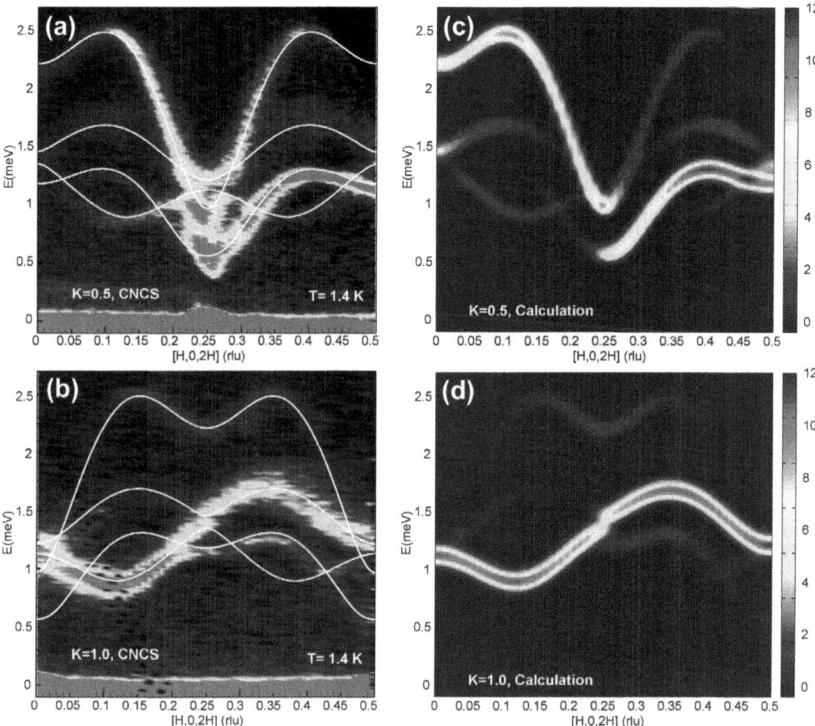

FIGURE 40 (a) Spin-wave dispersion along the [1,0,2] direction through the magnetic peak ($\frac{1}{4}$, $\frac{1}{2}$, $\frac{1}{2}$) of MnWO$_4$. (b) Magnetic excitation spectra along the same direction with $K = 1.0$. (c and d) Calculated spectra along the [1,0,2] direction with $K = 0.5$ and $K = 1.0$ using magnetic exchange interactions up to J_{11} with instrument resolution convoluted. The solid lines overlapped with experimental data in (a) and (b) are the predicted dispersion curves. *Reproduced from Ref. [92].*

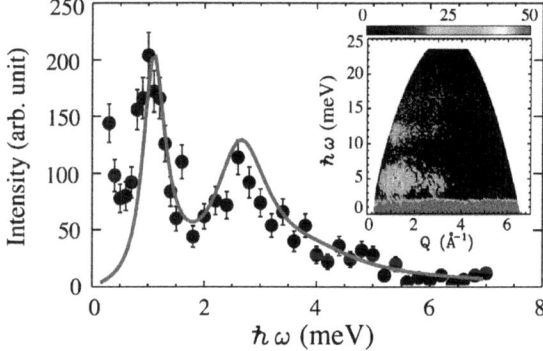

FIGURE 41 Constant-Q scan on single crystals of YMn$_2$O$_5$ at $Q = (1,0,0) + q_{IC}$ at 4 K. Blue (dark gray in print versions) symbols are the neutron scattering data, while the red (light gray in print versions) line is the result of spin-wave calculations. The inset is a contour map obtained from a powder sample of YMn$_2$O$_5$ at 1 K. *Reproduced from Ref. [60].*

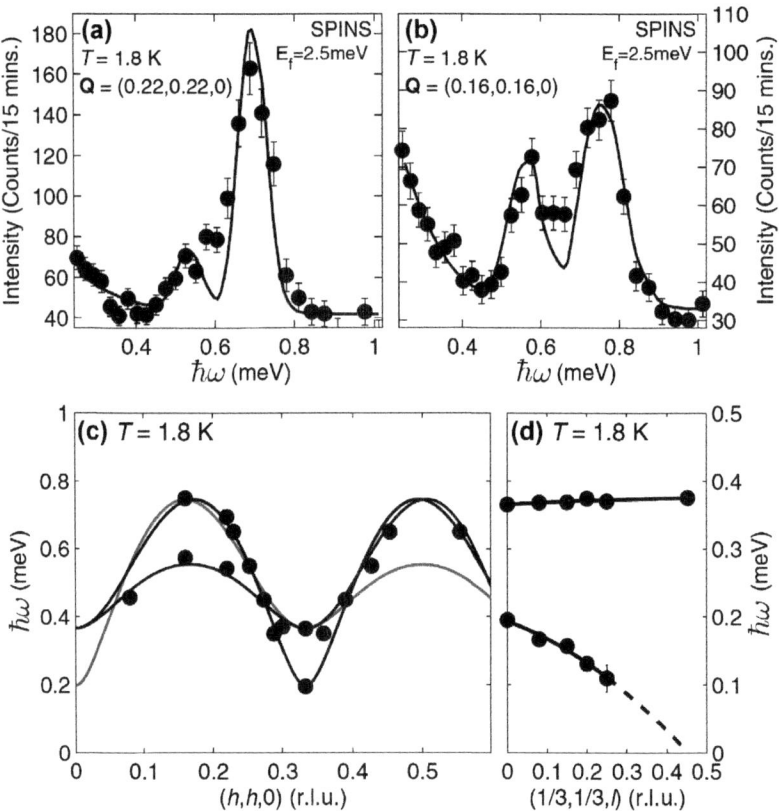

FIGURE 42 RbFe(MoO$_4$)$_2$ dispersion measurements. Constant wave-vector scans of the spin-wave excitations at zero field, and $T = 1.8$ K at (a) $Q = (0.22, 0.22, 0)$ and (b) $Q = (0.16, 0.16, 0)$. The solid curves result from a numerical convolution of Lorentzian energy profiles with the spectrometer resolution function. (c) The in-plane dispersion constructed from the analysis of the constant $(h,h,0)$ wave-vector scans. The lines in (c) show the dispersion of the three modes described by the spin Hamiltonian given in the text. (d) The interplane dispersion determined along $(1/3,1/3,l)$ with lines as guides for the eye. *Reproduced from Ref. [53].*

5.8 FUTURE DIRECTIONS

We have provided an overview of the types of materials and mechanisms that MFs enjoy. To a large extent, the field of MFs has been driven by the discovery of new materials, and no doubt this will be the case for the foreseeable future. It will be especially interesting if new classes of MFs with alternative mechanisms of magnetic-FE coupling are discovered. From an applications standpoint, on the other hand, the strongest challenge is to find new materials with higher transition temperatures and strong coupling between the magnetic order and FE polarization. Neutron diffraction measurements will no doubt be vital in determining the magnetic structures of these materials, while inelastic measurements will be essential in developing an understanding of the nature of the magnetic phases and how these might be modified to tailor their properties. This will be particularly important in investigations of thin films and multilayers, where the properties are sensitive to thickness and strain. Often these sensitivities can be employed to engineer the behavior for specific applications.

Recently, density functional theory has been successful in predicting new MF materials. Some of these new materials can only be stabilized in thin-film form, but may still be quite suitable for devices developed in thin films. Already neutron diffraction measurements have proved useful in determining the magnetic structure of films and how they differ from the bulk. Moving beyond thin films, heterostructures have shown great promise for device applications, but more work needs to be done to understand the interactions between the various layers. Neutron diffraction and reflectivity experiments will be critical in developing this understanding. There is no doubt that MF research will continue to be very active in both the fundamental and applications regimes.

ACKNOWLEDGMENTS

We thank our many coauthors with whom we have worked on these fascinating materials. We particularly would like to thank Brooks Harris, Michel Kenzelmann, Paolo Radaelli, Izabela Sosnowska Owen Vajk and Feng Ye for their help in preparing this review. We note that there have been a great many neutron studies of multiferroics, and in this brief overview only prototypical examples can be discussed, and with apologies to all the many other authors in this field, a number of these are taken from our own work for our convenience.

REFERENCES

[1] G. Shirane, Neutron scattering studies of structural phase transitions at Brookhaven, Rev. Mod. Phys. 46 (3) (1974) 437–449.
[2] S.-W. Cheong, M. Mostovoy, Multiferroics: a magnetic twist for ferroelectricity, Nat. Mater. 6 (2007) 13–20.
[3] R. Ramesh, N.A. Spaldin, Multiferroics: progress and prospects in thin films, Nat. Mater. 6 (1) (2007) 21–29.
[4] R.E. Cohen, Origin of ferroelectricity in perovskite oxides, Nature 358 (6382) (July 1992) 136–138.
[5] R. Cohen, H. Krakauer, Lattice dynamics and origin of ferroelectricity in $BaTiO_3$: linearized-augmented-plane-wave total-energy calculations, Phys. Rev. B 42 (10) (October 1990) 6416–6423.
[6] P. Ghosez, J.P. Michenaud, X. Gonze, The physics of dynamical atomic charges: the case of ABO_3 compounds, Phys. Rev. B 58 (10) (September 1998) 6224.
[7] A. Filippetti, N. Hill, Coexistence of magnetism and ferroelectricity in perovskites, Phys. Rev. B 65 (19) (May 2002) 1–11.
[8] N.A. Hill, Density functional studies of multiferroic magnetoelectrics, Annu. Rev. Mater. Res. 32 (3) (2002) 1–37.
[9] D.L. Price, F. Fernandez-Alonso (Eds.), Neutron Scattering–Fundamentals, Academic Press, 2013.

[10] A.L. Kholkin, N.A. Pertsev, A.V. Goltsev, Piezoelectricity and crystal symmetry, in: Piezoelectric and Acoustic Materials for Transducer Applications, Springer, US, 2008, pp. 17–38.
[11] J.F. Nye, Physical Properties of Crystals: Their Representation by Tensors and Matrices, Oxford University Press, 1957.
[12] S.C. Chae, N. Lee, Y. Horibe, M. Tanimura, S. Mori, B. Gao, S. Carr, S.-W. Cheong, Direct observation of the proliferation of ferroelectric loop domains and vortex-antivortex pairs, Phys. Rev. Lett. 108 (16) (2012) 167603.
[13] W.C. Koehler, H.L. Yakel, E.O. Wollan, J.W. Cable, A note on the magnetic structures of rare earth manganese oxides, Phys. Lett. 9 (2) (1964) 93–95.
[14] A. Munoz, J.A. Alonso, M.J. Martinez-Lope, M.T. Casais, J.L. Martinez, M.T. Fernandez-Diaz, A. Muñoz, J.A. Alonso, M.J. Martínez-Lope, M.T. Casáis, J.L. Martínez, M.T. Fernández-Díaz, Evolution of the magnetic structure of hexagonal HoMnO$_3$ from neutron powder diffraction data, Chem. Mater. 13 (5) (2001) 1497–1505.
[15] T. Lonkai, D. Hohlwein, J. Ihringer, W. Prandl, The magnetic structures of YMnO$_{3-\delta}$ and HoMnO$_3$, Appl. Phys. A 74 (2002) s843–s845.
[16] O.P. Vajk, M. Kenzelmann, J.W. Lynn, S.B. Kim, S.-W. Cheong, Magnetic order and spin dynamics in ferroelectric HoMnO$_3$, Phys. Rev. Lett. 94 (8) (2005) 087601.
[17] P. Brown, T. Chatterji, Determination of the magnetic structures of the field-induced phases of HoMnO$_3$, Phys. Rev. B 77 (10) (2008) 104407.
[18] R.C. Rai, J. Cao, L.I. Vergara, S. Brown, J.L. Musfeldt, D.J. Singh, G. Lawes, N. Rogado, R.J. Cava, X. Wei, High-energy magnetodielectric effect in kagome staircase materials, Phys. Rev. B 76 (17) (2007) 174414.
[19] P.J. Brown, T. Chatterji, Neutron diffraction and polarimetric study of the magnetic and crystal structures of HoMnO$_3$ and YMnO$_3$, J. Phys. Condens. Matter. 18 (2006) 10085–10096.
[20] B. Lorenz, F. Yen, M.M. Gospodinov, C.W. Chu, Field-induced phases in HoMnO$_3$ at low temperatures, Phys. Rev. B 71 (1) (2005) 014438.
[21] B. Lorenz, A.P. Litvinchuk, M.M. Gospodinov, C.W. Chu, Field-induced reentrant novel phase and a ferroelectric-magnetic order coupling in HoMnO$_3$, Phys. Rev. Lett. 92 (8) (2004) 087204.
[22] T. Lottermoser, T. Lonkai, U. Amann, D. Hohlwein, J. Ihringer, M. Fiebig, Magnetic phase control by an electric field, Nature 430 (May 2004) 541–544.
[23] Y.J. Choi, N. Lee, P.A. Sharma, S.B. Kim, O.P. Vajk, J.W. Lynn, Y.S. Oh, S.-W. Cheong, Giant magnetic fluctuations at the critical endpoint in insulating HoMnO$_3$, Phys. Rev. Lett. 110 (April 2013) 157202.
[24] O.P. Vajk, M. Kenzelmann, J.W. Lynn, S.B. Kim, S.-W. Cheong, Neutron-scattering studies of magnetism in multiferroic HoMnO$_3$ (invited), J. Appl. Phys. 99 (2006) 08E301.
[25] B.G. Ueland, J.W. Lynn, M. Laver, Y.J. Choi, S.-W. Cheong, Origin of electric-field-induced magnetization in multiferroic HoMnO$_3$, Phys. Rev. Lett. 104 (April 2010) 147204.
[26] I. Sosnowska, T.P. Neumaier, E. Steichele, Spiral magnetic ordering in bismuth ferrite, J. Phys. C Solid State Phys. 15 (23) (August 1982) 4835–4846.
[27] E.O. Wollan, W.C. Koehler, Neutron diffraction study of the magnetic properties of the series of perovskite-type compounds [(1−x)La, xCa]MnO{3}, Phys. Rev. 100 (2) (1955) 545–563.
[28] R. Przeniosło, M. Regulski, I. Sosnowska, Modulation in multiferroic BiFeO$_3$: cycloidal, elliptical or SDW? J. Phys. Soc. Jpn. 75 (8) (2006) 44–46.
[29] D. Lebeugle, D. Colson, A. Forget, M. Viret, A.M. Bataille, A. Gukasov, Electric-field-induced spin flop in BiFeO$_3$ single crystals at room temperature, Phys. Rev. Lett. 100 (June 2008) 227602.

[30] S. Lee, T. Choi, W. Ratcliff, R. Erwin, S.-W. Cheong, V. Kiryukhin, Single ferroelectric and chiral magnetic domain of single-crystalline BiFeO$_3$ in an electric field, Phys. Rev. B 78 (10) (2008) 100101.
[31] M. Ramazanoglu, W. Ratcliff, Y.J. Choi, S. Lee, S.-W. Cheong, V. Kiryukhin, Temperature-dependent properties of the magnetic order in single-crystal BiFeO$_3$, Phys. Rev. B 83 (17) (2011) 174434.
[32] M. Ramazanoglu, M. Laver, W. Ratcliff, S.M. Watson, W.C. Chen, A. Jackson, K. Kothapalli, S. Lee, S.-W. Cheong, V. Kiryukhin, Local weak ferromagnetism in single-crystalline ferroelectric BiFeO$_3$, Phys. Rev. Lett. 107 (20) (November 2011) 207206.
[33] D.V. Efremov, J. van den Brink, D.I. Khomskii, Bond- versus site-centred ordering and possible ferroelectricity in manganites, Nat. Mater. 3 (12) (2004) 853−856.
[34] D. Khomskii, Classifying multiferroics: mechanisms and effects, Physics (College. Park. Md) 2 (2009) 20.
[35] C.N.R. Rao, B. Raveau, Colossal Magnetoresistance Charge Ordering and Related Properties of Manganese Oxides, World Scientific, Singapore, 1998.
[36] E. Dagotto, Nanoscale Phase Separation and Colossal Magnetoresistance, Springer, Berlin, 2003.
[37] H. Sakai, J. Fujioka, T. Fukuda, D. Okuyama, D. Hashizume, F. Kagawa, H. Nakao, Y. Murakami, T. Arima, A.Q.R. Baron, Y. Taguchi, Y. Tokura, Displacement-type ferroelectricity with off-center magnetic ions in perovskite Sr$_{1-x}$Ba$_x$MnO$_3$, Phys. Rev. Lett. 107 (13) (2011) 137601.
[38] D.K. Pratt, J.W. Lynn, J. Mais, O. Chmaissem, D.E. Brown, S. Kolesnik, B. Dabrowski, Neutron scattering studies of spin dynamics in the Type-I multiferroic Sr$_{0.56}$Ba$_{0.44}$MnO$_3$, Phys. Rev. B 90 (2014) 140401(R).
[39] G. Giovannetti, S. Kumar, C. Ortix, M. Capone, J. van den Brink, Microscopic origin of large negative magnetoelectric coupling in Sr$_{1/2}$Ba$_{1/2}$MnO$_3$, Phys. Rev. Lett. 109 (September 2012) 107601.
[40] M. Kenzelmann, A.B. Harris, S. Jonas, C. Broholm, J. Schefer, S.B. Kim, C.L. Zhang, S.-W. Cheong, O.P. Vajk, J.W. Lynn, Magnetic inversion symmetry breaking and ferroelectricity in TbMnO$_3$, Phys. Rev. Lett. 95 (8) (2005) 087206.
[41] M. Mostovoy, Ferroelectricity in spiral magnets, Phys. Rev. Lett. 96 (February 2006) 067601.
[42] H. Katsura, N. Nagaosa, A.V. Balatsky, Spin current and magnetoelectric effect in noncollinear magnets, Phys. Rev. Lett. 95 (July 2005) 057205.
[43] I.A. Sergienko, E. Dagotto, Role of the Dzyaloshinskii-Moriya interaction in multiferroic perovskites, Phys. Rev. B 73 (2006) 094434.
[44] J. Hu, Microscopic origin of magnetoelectric coupling in noncollinear multiferroics, Phys. Rev. Lett. 100 (February 2008) 077202.
[45] S.B. Wilkins, T.R. Forrest, T.A.W. Beale, S.R. Bland, H.C. Walker, D. Mannix, F. Yakhou, D. Prabhakaran, A.T. Boothroyd, J.P. Hill, P.D. Hatton, D.F. McMorrow, Nature of the magnetic order and origin of induced ferroelectricity in TbMnO$_3$, Phys. Rev. Lett. 103 (November 2009) 207602.
[46] H. Ehrenberg, H. Weitzel, H. Fuess, Magnon dispersion and magnetic phase diagram of MnWO$_4$, Phys. B 234−236 (1997) 560−563.
[47] G. Lautenschläger, H. Weitzel, T. Vogt, R. Hock, A. Böhm, M. Bonnet, H. Fuess, Magnetic phase transitions of MnWO$_4$ studied by the use of neutron diffraction, Phys. Rev. B 48 (1993) 6087−6098.
[48] O. Heyer, N. Hollmann, I. Klassen, S. Jodlauk, L. Bohaty, P. Becker, J.A. Mydosh, T. Lorenz, D. Khomskii, A new multiferroic material: MnWO$_4$, J. Phys. C Solid State Phys. 18 (2006) L471−L475.

[49] H. Sagayama, K. Taniguchi, N. Abe, T.H. Arima, M. Soda, M. Matsuura, K. Hirota, Correlation between ferroelectric polarization and sense of helical spin order in multiferroic $MnWO_4$, Phys. Rev. B 77 (2008) 2−5.

[50] I. Cabrera, M. Kenzelmann, G. Lawes, Y. Chen, W.C. Chen, R. Erwin, T.R. Gentile, J.B. Leão, J.W. Lynn, N. Rogado, R.J. Cava, C. Broholm, Coupled magnetic and ferroelectric domains in multiferroic $Ni_3V_2O_8$, Phys. Rev. Lett. 103 (8) (2009) 087201.

[51] M. Kenzelmann, G. Lawes, A.B. Harris, G. Gasparovic, C. Broholm, A.P. Ramirez, G.A. Jorge, M. Jaime, S. Park, Q. Huang, A.Y. Shapiro, L.A. Demianets, Direct transition from a disordered to a multiferroic phase on a triangular lattice, Phys. Rev. Lett. 98 (26) (June 2007) 267205.

[52] A.B. Harris, Landau analysis of the symmetry of the magnetic structure and magnetoelectric interaction in multiferroics, Phys. Rev. B 76 (5) (2007) 74706.

[53] J.S. White, C. Niedermayer, G. Gasparovic, C. Broholm, J.M.S. Park, A.Y. Shapiro, L.A. Demianets, M. Kenzelmann, Multiferroicity in the generic easy-plane triangular lattice antiferromagnet $RbFe(MoO_4)_2$, Phys. Rev. B 88 (2013) 1−5.

[54] V. Kiryukhin, S. Lee, W. Ratcliff, Q. Huang, H.T. Yi, Y.J. Choi, S.-W. Cheong, Order by static disorder in the ising chain magnet $Ca_3Co_{2-x}Mn_xO_6$, Phys. Rev. Lett. 102 (18) (2009) 187202.

[55] A.J. Hearmon, F. Fabrizi, L.C. Chapon, R.D. Johnson, D. Prabhakaran, S.V. Streltsov, P.J. Brown, P.G. Radaelli, Electric field control of the magnetic chiralities in ferroaxial multiferroic $RbFe(MoO_4)_2$, Phys. Rev. Lett. 108 (23) (2012) 237201.

[56] L.E. Svistov, A.I. Smirnov, L.A. Prozorova, Quasi-two-dimensional antiferromagnet on a triangular lattice $RbFe(MoO_4)(2)$, Phys. Rev. B (2003) 1−9.

[57] Y.J. Jo, S. Lee, E.S. Choi, H.T. Yi, W. Ratcliff, Y.J. Choi, V. Kiryukhin, S.-W. Cheong, L. Balicas, 3:1 magnetization plateau and suppression of ferroelectric polarization in an Ising chain multiferroic, Phys. Rev. B 79 (1) (2009) 012407.

[58] A.B. Harris, Landau analysis of the symmetry of the magnetic structure and magnetoelectric interaction in multiferroics, Phys. Rev. B 76 (5) (2007) 054447.

[59] Y.J. Choi, H.T. Yi, S. Lee, Q. Huang, V. Kiryukhin, S.-W. Cheong, Ferroelectricity in an ising chain magnet, Phys. Rev. Lett. 100 (February 2008) 047601.

[60] J.H. Kim, M.A. van der Vegte, A. Scaramucci, S. Artyukhin, J.H. Chung, S. Park, S.-W. Cheong, M. Mostovoy, S.H. Lee, Magnetic excitations in the low-temperature ferroelectric phase of multiferroic YMn_2O_5 using inelastic neutron scattering, Phys. Rev. Lett. 107 (August 2011) 097401.

[61] Y.O.I. Kagomiya, S. Matsumoto, K. Kohn, Y. Fukuda, T. Shoubu, H. Kimura, Y. Noda, N. Ikeda, Lattice distortion at ferroelectric transition of YMn_2O_5, Ferroelectrics 286 (1) (January 2003) 167−174.

[62] L.C. Chapon, P.G. Radaelli, G.R. Blake, S. Park, S.-W. Cheong, Ferroelectricity induced by acentric spin-density waves in YMn_2O_5, Phys. Rev. Lett. 96 (9) (2006) 097601.

[63] J.H. Kim, S.H. Lee, S.I. Park, M. Kenzelmann, A.B. Harris, J. Schefer, J.H. Chung, C.F. Majkrzak, M. Takeda, S. Wakimoto, S.Y. Park, S.-W. Cheong, M. Matsuda, H. Kimura, Y. Noda, K. Kakurai, Spiral spin structures and origin of the magnetoelectric coupling in YMn_2O_5, Phys. Rev. B 78 (2008) 1−10.

[64] P.G. Radaelli, C. Vecchini, L.C. Chapon, P.J. Brown, S. Park, S.-W. Cheong, Incommensurate magnetic structure of YMn_2O_5: a stringent test of the multiferroic mechanism, Phys. Rev. B 79 (2) (2009) 020404.

[65] H. Murakawa, Y. Onose, S. Miyahara, N. Furukawa, Y. Tokura, Ferroelectricity induced by spin-dependent metal-ligand hybridization in $Ba_2CoGe_2O_7$, Phys. Rev. Lett. 105 (13) (2010) 137202.

[66] K. Yamauchi, P. Barone, S. Picozzi, Theoretical investigation of magnetoelectric effects in $Ba_2CoGe_2O_7$, Phys. Rev. B 84 (16) (2011) 1–6.
[67] P. Coeuré, F. Guinet, J.C. Peuzin, G. Buisson, E.F. Bertaut, Proc. Int. Meeting on Ferroelectricity, in: V. Dvorák (Ed.), Institute of Physics of the Czechoslovak Academy of Sciences, Prague, 1996, pp. 332–340.
[68] P. Babkevich, A. Poole, R.D. Johnson, B. Roessli, D. Prabhakaran, A.T. Boothroyd, Electric field control of chiral magnetic domains in the high-temperature multiferroic CuO, Phys. Rev. B 85 (2012) 1–6.
[69] T. Kimura, J.C. Lashley, A.P. Ramirez, Inversion-symmetry breaking in the noncollinear magnetic phase of the triangular-lattice antiferromagnet $CuFeO_2$, Phys. Rev. B 73 (22) (2006) 220401.
[70] Y.J. Choi, J. Okamoto, D.J. Huang, K.S. Chao, H.J. Lin, C.T. Chen, M. van Veenendaal, T. A. Kaplan, S.-W. Cheong, Thermally or magnetically induced polarization reversal in the multiferroic $CoCr_2O_4$, Phys. Rev. Lett. 102 (6) (2009) 067601.
[71] H. Wieder, Electrical behavior of barium titanate single crystals at low temperatures, Phys. Rev. 99 (4) (1955) 1161.
[72] M. Ramazanoglu, W. Ratcliff, H.T. Yi, A.A. Sirenko, S.-W. Cheong, V. Kiryukhin, Giant effect of uniaxial pressure on magnetic domain populations in multiferroic bismuth ferrite, Phys. Rev. Lett. 107 (6) (August 2011) 067203.
[73] S. Nandi, A. Kreyssig, L. Tan, J.W. Kim, J.Q. Yan, J.C. Lang, D. Haskel, R.J. McQueeney, A.I. Goldman, Nature of Ho magnetism in multiferroic $HoMnO_3$, Phys. Rev. Lett. 100 (May 2008) 1–4.
[74] S. Lee, W. Ratcliff, S. Cheong, V. Kiryukhin, Electric field control of the magnetic state in $BiFeO_3$ single crystals, Appl. Phys. Lett. 92 (19) (2008) 192906.
[75] H. Béa, M. Bibes, S. Petit, J. Kreisel, A. Barthélémy, Structural distortion and magnetism of $BiFeO_3$ epitaxial thin films: a Raman spectroscopy and neutron diffraction study, Philos. Mag. Lett. 87 (2007) 165–174.
[76] X Ke, P.P. Zhang, S.H. Baek, J. Zarestky, W. Tian, C.B. Eom, Magnetic structure of epitaxial multiferroic $BiFeO_3$ films with engineered ferroelectric domains, Phys. Rev. B 82 (13) (October 2010) 134448.
[77] W. Ratcliff, Z. Yamani, V. Anbusathaiah, T.R. Gao, P.A. Kienzle, H. Cao, I. Takeuchi, Electric-field-controlled antiferromagnetic domains in epitaxial $BiFeO_3$ thin films probed by neutron diffraction, Phys. Rev. B 87 (14) (April 2013) 140405.
[78] C. Ederer, N.A. Spaldin, Weak ferromagnetism and magnetoelectric coupling in bismuth ferrite, Phys. Rev. B 71 (6) (February 2005) 060401.
[79] W. Ratcliff, D. Kan, W. Chen, S. Watson, S. Chi, R. Erwin, G.J. McIntyre, S.C. Capelli, I. Takeuchi, Neutron diffraction investigations of magnetism in $BiFeO_3$ epitaxial films, Adv. Funct. Mater. 21 (May 2011) 1567–1574.
[80] B.J. Kirby, D. Kan, A. Luykx, M. Murakami, D. Kundaliya, I. Takeuchi, Anomalous ferromagnetism in $TbMnO_3$ thin films, J. Appl. Phys. 105 (7) (2009) 07D917.
[81] J.S. White, M. Bator, Y. Hu, H. Luetkens, J. Stahn, S. Capelli, S. Das, M. Döbeli, T. Lippert, V.K. Malik, J. Martynczuk, A. Wokaun, M. Kenzelmann, C. Niedermayer, C.W. Schneider, Strain-induced ferromagnetism in antiferromagnetic $LuMnO_3$ thin films, Phys. Rev. Lett. 111 (July 2013) 037201.
[82] L.J. Wang, Y.S. Chai, S.M. Feng, J.L. Zhu, N. Manivannan, C.Q. Jin, Z.Z. Gong, X.H. Wang, L.T. Li, Large magneto (thermo) dielectric effect in multiferroic orthorhombic $LuMnO_3$, J. Appl. Phys. 111 (11) (2012) 114103.
[83] T.J. Sato, S.-H. Lee, T. Katsufuji, M. Masaki, S. Park, J.R.D. Copley, H. Takagi, Unconventional spin fluctuations in the hexagonal antiferromagnet $YMnO_3$, Phys. Rev. B 68 (2003) 014432.

[84] S. Skanthakumar, H. Zhang, T.W. Clinton, W.-H. Li, J.W. Lynn, Z. Fisk, S.-W. Cheong, Magnetic phase transitions and structural distortion in Nd_2CuO_4, Phys. C 160 (2) (1989) 124−128.
[85] J. Jeong, E.A. Goremychkin, T. Guidi, K. Nakajima, G.S. Jeon, S.-A.A. Kim, S. Furukawa, Y.B. Kim, S. Lee, V. Kiryukhin, S.-W. Cheong, J.-G.G. Park, Spin wave measurements over the full Brillouin zone of multiferroic $BiFeO_3$, Phys. Rev. Lett. 108 (7) (February 2012) 077202.
[86] A.M. Kadomtseva, A.K. Zvezdin, Y.F. Popov, A.P. Pyatakov, G.P. Vorob'ev, Space-time parity violation and magnetoelectric interactions in antiferromagnets, JETP Lett. 79 (11) (June 2004) 571−581.
[87] M. Matsuda, R.S. Fishman, T. Hong, C.H. Lee, T. Ushiyama, Y. Yanagisawa, Y. Tomioka, T. Ito, Magnetic dispersion and anisotropy in multiferroic $BiFeO_3$, Phys. Rev. Lett. 109 (August 2012) 1−5.
[88] R.S. Fishman, N. Furukawa, J.T. Haraldsen, M. Matsuda, S. Miyahara, Identifying the spectroscopic modes of multiferroic $BiFeO_3$, Phys. Rev. B 86 (22) (December 2012) 220402.
[89] A.I. Milstein, O.P. Sushkov, Magnetic excitations in the spin-spiral state of $TbMnO_3$ and $DyMnO_3$, Phys. Rev. B 91 (2015), 094417.
[90] D. Senff, N. Aliouane, D.N. Argyriou, A. Hiess, L.P. Regnault, P. Link, K. Hradil, Y. Sidis, M. Braden, Magnetic excitations in a cycloidal magnet: the magnon spectrum of multiferroic $TbMnO_3$, J. Phys. Condens. Matter. 20 (2008) 434212.
[91] A. Pimenov, A.A. Mukhin, V.Y. Ivanov, V.D. Travkin, A.M. Balbashov, A. Loidl, Possible evidence for electromagnons in multiferroic manganites, Nat. Phys. 2 (February 2006) 97−100.
[92] F. Ye, R.S. Fishman, J.A. Fernandez-Baca, A.A. Podlesnyak, G. Ehlers, H.A. Mook, Y. Wang, B. Lorenz, C.W. Chu, Long-range magnetic interactions in the multiferroic antiferromagnet $MnWO_4$, Phys. Rev. B 83 (2011) 140401.
[93] F. Ye, S. Chi, J.A. Fernandez-Baca, H. Cao, K.C. Liang, Y. Wang, B. Lorenz, C.W. Chu, Magnetic order and spin-flop transitions in the cobalt-doped multiferroic $Mn_{1-x}Co_xWO_4$, Phys. Rev. B 86 (9) (2012) 094429.
[94] Y.S. Song, J.H. Chung, J.M.S. Park, Y.N. Choi, Stabilization of the elliptical spiral phase and the spin-flop transition in multiferroic $Mn_{1-x}Co_xWO_4$, Phys. Rev. B 79 (22) (2009) 224415.
[95] R.P. Chaudhury, F. Ye, J.A. Fernandez-Baca, B. Lorenz, Y.Q. Wang, Y.Y. Sun, H.A. Mook, C.W. Chu, Robust ferroelectric state in multiferroic $Mn_{1-x}Zn_xWO_4$, Phys. Rev. B 83 (2011).
[96] I. Urcelay-Olabarria, E. Ressouche, A.A. Mukhin, V.Y.Y. Ivanov, A.M. Balbashov, G.P. Vorobev, Y.F. Popov, A.M. Kadomtseva, J.L. García-Muñoz, V. Skumryev, J.L. Garc, Neutron diffraction, magnetic, and magnetoelectric studies of phase transitions in multiferroic $Mn_{0.90}Co_{0.10}WO_4$, Phys. Rev. B 094436 (2012) 1−10.

Chapter 6

Neutron Scattering in Nanomagnetism

Boris P. Toperverg[1,2] and Hartmut Zabel[1,3,*]
[1]*Institute for Experimental Condensed Matter Physics, Ruhr-University Bochum, Bochum, Germany;* [2]*Petersburg Nuclear Physics Institute, St Petersburg, Russia;* [3]*Johannes Gutenberg-Universität Mainz, Mainz, Germany*
Corresponding author: E-mail: hartmut.zabel@rub.de

Chapter Outline

6.1 Introduction	340	
6.1.1 Topics in Nanomagnetism	340	
6.1.2 Magnetic Neutron Scattering	341	
6.2 Theoretical Background	344	
6.2.1 Basic Interactions and Scattering Amplitudes	344	
6.2.2 Scattering Cross Section of Polarized Neutrons	347	
6.2.3 Grazing Incidence Kinematics	352	
6.2.4 Specular Polarized Neutron Reflectivity for 1D Potential	359	
6.2.5 Off-Specular Neutron Scattering	363	
6.3 Instrumental Considerations	366	
6.3.1 Design of Neutron Reflectometers	366	
6.3.2 Neutron Optics	370	
6.3.3 Detection and Acquisition of Data	373	
6.3.4 Modeling and Fitting of Data	373	
6.4 Case Studies: Static Experiments	374	
6.4.1 Thin Films	374	
6.4.1.1 The Saturated State	374	
6.4.1.2 Magnetization Rotation	380	
6.4.1.3 Magnetic Domain State	384	
6.4.1.4 Magnetization Reversal	390	
6.4.1.5 Domain Walls	394	
6.4.2 Multilayers	396	
6.4.2.1 Magnetic Domains	397	
6.4.2.2 Enhancing Interface Sensitivity	401	
6.4.3 Stripes, Islands, and Nanoparticles	403	
6.4.3.1 Laterally Patterned Nanostructures	404	
6.4.3.2 Magnetic Nanoparticles	407	
6.4.3.3 Grazing Incidence Small Angle Scattering	413	
6.5 Time Dependent Polarized Neutron Scattering	416	
6.6 Summary, Conclusion, and Outlook	421	
Acknowledgments	422	
References	423	

6.1 INTRODUCTION
6.1.1 Topics in Nanomagnetism

aaaNanomagnetism is the magnetism on the nanoscale. More specifically we understand by nanomagnetism ordered ferro-, ferri-, or antiferromagnetic states that are artificially confined to lower dimensions like thin films, multilayers, laterally patterned magnetic islands, spherical nanoclusters, or nanopillars. Important concepts of nanomagnetism are the shape anisotropy versus crystal anisotropy, proximity effects via common interfaces with neighboring films, interdiffusion at the interface, and scaling of spin ordering temperatures. Up-to-date reviews on the field of nanomagnetism can be found in Refs [1−3]. Among the many analytical tools that are available for the analysis of nanomagnetic systems, polarized neutron scattering is particularly powerful for the analysis of magnetization profiles across interfaces, for the characterization of magnetization reversal processes, and for determining vector fluctuations of the magnetization. It should be emphasized that all results are statistical averages across the illuminated surface and that numbers provided are quantitative and in absolute units of Bohr magnetons. This is particular important as many other methods either average over the total sample volume superconducting quantum interference device (SQUID), vibrating sample magnetometer (VSM), magneto-optic Kerr effect (MOKE) or are surface sensitive and cannot provide information on layers laying deeper below the surface magnetic force microscopy (MFM), photoemission electron microscopy (PEEM), etc. The great power of polarized neutron scattering is only hampered by the low flux of neutron sources compared to photon sources and the access to state-of-the-art polarized neutron reflectivity instruments at cold neutron sources.

The research field of nanomagnetism is closely related to the topic of spintronics. Spintronics utilizes the spin in addition to the charge of electrons for novel types of electronics based on spin transport. This includes the movement of domain walls (DWs) via spin transfer torque [4], the separation of up and down spins in the spin Hall effect [5], the pure spin transport without charge transport via spin pumping [6], and spin caloric effects, such as the spin Seebeck effect [7,8]. Overviews on the field of spintronics can be found in Refs [9−13]. All these spintronic effects require the manipulation and control of magnetic domains, magnetic DWs, magnetic anisotropy, spin polarization of materials, and their ordering temperatures. The control of these properties is achieved by combining different magnetic and nonmagnetic materials in heterostructures with common interfaces, by shaping magnetic materials into nanoislands, nanorods, nanoparticles (NPs), or other shapes, by selecting specific alloys and compounds with high spin polarization at the Fermi level, and by manipulating the interface for specific anisotropies. Some examples are sketched in Figure 1.

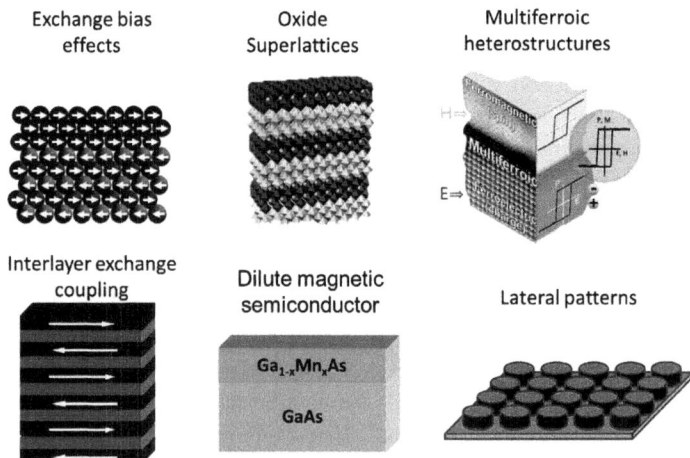

FIGURE 1 Schematically shown are some topics in nanomagnetism which are investigated by polarized neutron reflectivity. Those include among others heterostructures of dissimilar materials like ferro- and antiferromagnetic layers, soft and hard magnetic layers, multiferroics, metal and oxide multilayers, dilute magnetic semiconductors, spin valves, self-assembled nanoclusters, and patterned arrays of magnetic islands.

6.1.2 Magnetic Neutron Scattering

Magnetic neutron scattering is generally known as a bulk probe for the investigation of spin structures, spin fluctuations, and excitations. Antiferromagnetic structure analysis, crystal field excitations, and magnon dispersions are the hallmarks of very successful bulk magnetic elastic and inelastic neutron scattering work. Much of what we know today about antiferromagnets and which has been essential for the design of spintronic devices is based on neutron scattering work in the 1960s and 1970s of the last century. The basic concepts of magnetic neutron scattering are reviewed in the first volume of this book series by Price and Fernandez-Alonso [14].

In case of nanomagnetism, it is the interface sensitivity which is essential for studies of magnetic thin films, multilayers, and lateral arrays of magnetic islands or clusters. Since the introduction of neutron reflectivity by Felcher in the early 1980s [15], neutron reflectometers have flourished all over the world at steady state and pulsed neutron sources, an overview of presently existing neutron reflectometers is provided by Rennie in the Internet [16]. Next to neutron powder diffraction and neutron small-angle scattering, neutron reflectometry (NR) has matured to one of the most important and fruitful scattering methods for materials scientists and alike. Any steady state or pulsed neutron source features nowadays at least a few neutron reflectometers, specialized for different topics in materials science, such as fluids, polymers, biomaterials, and condensed matter. For the analysis of nanomagnetic materials, it is essential that the neutron reflectometer instrument is equipped with a neutron polarizer, and it is highly desirable if the instrument also offers neutron spin analysis after scattering from the sample. Each year more than 100 papers are being published on work that utilizes NR, roughly 10% of those use the polarized version for the analysis of magnetic materials. But also in neutron reflectivity studies of soft matter often magnetic reference layers are inserted either to increase the contrast or to pinpoint the layer sequence in a stack of layers [17−19].

The sensitivity of neutron reflectivity to interfaces is due to the fact that their wavelength projection onto the surface normal matches the thickness of thin films, and that the neutron wave field becomes strongly distorted near surfaces when nuclear potential steps are encountered. This interface sensitivity is exploited in specular NR and off-specular neutron scattering, independent of the state of the matter, amorphous, polymeric, micellar, liquid, epitaxial, or other. Neutron reflectivity probes the nuclear density profile perpendicular to surfaces and interfaces in contrast to the electron density contrast probed by X-ray reflectivity (XRR), interface roughness, and interface correlations. In case of magnetic films and multilayers, magnetization profiles are probed in addition to nuclear profiles. The most significant information is gained if the incident neutron polarization is fixed and the polarization state of the scattered beam is analyzed before the detector. This variant of NR is referred to as polarized neutron reflectometry (PNR) and it is used mainly for the investigation of magnetic thin films, magnetic superlattices, or any other kinds of magnetic heterostructure. Most of the PNR work published so far assumes a static or quasistatic sample environment. Time-resolved experiments are not practical when a single scan takes a few hours. For probing specific features, the scanning time can be reduced to a few minutes or even seconds. This allows in situ investigations of growth kinetics, interdiffusion at interfaces, and slow relaxation processes. But milli- to microseconds time resolution was so far out of reach. This is no longer the case as recent frequency-dependent PNR experiments have shown conducted on the magnetization reversal process up into the megahertz regime [20].

In the field of nanomagnetism not only ferromagnets but also antiferromagnets play an essential role, for instance, as pinning layer in spin valve. In fact new developments emphasize all antiferromagnetic spintronics, where not only the pinning layers are antiferromagnets but also the ferromagnetic electrodes are replaced by antiferromagnets [21]. The spin structure of antiferromagnets cannot be studied via PNR but requires high angle scattering experiments either with a powder diffractometer, if the material is polycrystalline, or otherwise with a single-crystal diffractometer. Those neutron techniques are described in the chapter on "Magnetic Structures" by Garlea and Chakoumakos in this book. Often antiferromagnets have a fine grain domain structure after nucleation below the Néel temperature with uncompensated spins in the DWs. PNR would be sensitive to DWs in antiferromagnets, but this has not been studied yet.

Another important area in nanomagnetism is magnetic NPs (also called nanoclusters). If the NPs are deposited on a flat substrate, PNR is again useful for their analysis. However, if the NPs are arranged more randomly in space, then it may be advantageous to use small-angle neutron scattering (SANS) instead, or better the polarized version of SANS, which is known as SANS-POL. For the internal crystal and spin structure of the NPs, again neutron powder diffraction is the best choice. SANS techniques are described in the first volume of Neutron Scattering, Chapter 5 on *Large-scale Structures*, by Penfold and Tucker [22]. In this review, we will mainly concentrate on PNR methods with special emphasis of nanomagnetic systems.

Serious competition to PNR has in recent years evolved from X-ray resonant magnetic scattering (XRMS) with soft X-rays. The usually weak cross section for magnetic X-ray scattering is enhanced by resonant excitation of core levels with circularly polarized X-ray photons, making the method, in addition, element specific. Using XRMS methods with photon wavelengths comparable to cold neutrons, the magnetization profile can be probed, element-specific magnetic hysteresis can be measured, ferro- and antiferromagnetic orderings in thin films and multilayers can be analyzed, and charge as well as orbital ordering effects can be detected. Recent excellent reviews on this topic have appeared in Refs [23–25]. PNR does not feature element selectivity, but often magnetic layers can be identified by their scattering length density (SLD), and if required, special isotopes can be used such as ^{57}Fe to either reduce the nuclear contribution to the scattering length [26] or to use it, in addition, for nuclear resonant scattering experiments [27]. Nevertheless, there are distinct differences in PNR and XRMS methods that concern the selection rule for magnetic scattering and the spin-charge cross terms. In general, one can state that XRMS is an extremely powerful method, in particular for the investigation of nanomagnetic systems, but more difficult to analyze for retrieving absolute quantitative data as compared to PNR. In any case, it has become common practice in recent years to use both methods for a complete analysis of nanomagnetic systems. This practice is often facilitated by the availability of neutron and synchrotron radiation sources on the same campus in close proximity and in some cases by combined beam time proposal schemes.

Earlier reviews on the theoretical background of PNR, experimental techniques, and applications up to about 2005 can be found in Refs [28–39]. More recent reviews published since about 2005 appeared in Refs [40–51]. A review with many practical hints was published by Fitzsimmons and Majkrzak [41], and a review with special emphasis on off-specular diffuse scattering in magnetic multilayers is available from Lauter-Pasyuk [46]. Articles with focus on the theory of PNR are published in Refs [42,44,52−56]. In particular, we would like to refer to the reviews in Refs [42,44], which provide a comprehensive overview of the theory of PNR. The main aim of the present review is not to duplicate the earlier versions, but to provide a simplified approach to the theoretical background with practical applications, whenever appropriate, exemplified by recent experimental work.

6.2 THEORETICAL BACKGROUND

One of the main advantages of neutron scattering as a nondestructive tool for condensed matter investigations is usually ascribed to the deep penetration of neutrons into materials. This is mostly due to the weak absorption of slow neutrons by a vast majority of isotopes, as well as to a relatively small scattering cross section. As a result, scattering can be considered as a small perturbation of the free propagating neutron wave impinging onto the sample and is usually well described within the framework of the Born approximation (BA). The relative straightforward interpretation of experimental results related to weak scattering is at the same time one of the main disadvantages of the method which hence requires a bulk of scattering material, typically of a few cubic millimeters, even at the most intense neutron sources. This disadvantage can be partially compromised via the choice of the scattering kinematics when a well-collimated neutron beam is incident onto a flat surface at shallow glancing angles. Then, as we shall see, scattering can be detected even from a few atomic layers deposited over the surface of about a centimeter squared. However, at shallow angles of incidence BA may fail to describe scattering from thicker layers, or thin layers deposited onto a bulk substrate. Then more sophisticated theoretical methods discussed in this section should be applied to quantitatively describe experimental data.

6.2.1 Basic Interactions and Scattering Amplitudes

The applicability range of the BA depends on two dimensionless parameters. The first one is the ratio between the effective energy V of interaction of the neutron with matter and its kinetic energy $E = \hbar^2 k_n^2/2m_n$, where $k_n = |\mathbf{k}_n|$, is the absolute value of the neutron wave vector $\mathbf{k}_n$. The second parameter is the ratio of the neutron wavelength $\lambda_n = 2\pi/k_n$ to the range of the interaction potential $V = V(\mathbf{r})$.

The interaction, or scattering potential,

$$\widehat{V} = \widehat{V}_{\text{nucl}} + \widehat{V}_{\text{magn}}, \tag{1}$$

is generally represented as a sum of nuclear, $\widehat{V}_{nucl} = \widehat{V}_{nucl}(\mathbf{r})$, and magnetic, $\widehat{V}_{magn} = \widehat{V}_{magn}(\mathbf{r})$, terms which due to the neutron spin $\widehat{\mathbf{s}}_n$ are proportional to 2×2 matrices in the spin space.
The first term,

$$\widehat{V}_{nucl}(\mathbf{r}) = \sum_l \widehat{V}_l^N (\mathbf{r} - \mathbf{r}_l) \qquad (2)$$

in Eqn (1) refers to the interaction of a neutron with nuclei located at positions $\mathbf{r}_l$ numerated in the sample by l and, in general, depends on the nuclear spin state. Usually, nuclear spins at finite temperatures are disordered and this dependence, as we shall see, for further consideration is of minor importance with respect to the spin-independent part of the nuclear interaction and therefore the spin matrix $\widehat{V}_{nucl}$ is diagonal with respect to spin variables. Therefore the matrix $\widehat{V}_{nucl}$ is set proportional to the unit 2×2 matrix $\widehat{1}$ which is not indicated explicitly in the following.

The major part of neutron–nuclei interaction is classified as a type of strong fundamental interaction acting between nucleons over the short range in the order of $r_n \sim 10^{-13}$ cm, i.e., much smaller than the typical wavelength $\lambda_n \sim 0.1 - 3$ nm of thermal or cold neutrons and interatomic distances in solids. Therefore, the nuclear scattering potential is approximated by a pointlike Fermi pseudopotential,

$$\widehat{V}_l^N(\mathbf{r} - \mathbf{r}_l) = \frac{2\pi\hbar^2}{m_n} b_l^N \delta(\mathbf{r} - \mathbf{r}_l). \qquad (3)$$

It totally ignores the range of nuclear–neutron interaction and instead introduces the nuclear scattering length b_l^N, which—in general—is a complex quantity. Its real part for different elements and isotopes may be positive, as well as negative, and typically ranges from -0.373×10^{-12} cm for ^{55}Mn up to 1.44×10^{-12} cm for ^{58}Ni. The imaginary part of the scattering length takes into account absorption of neutrons by nuclei. For most of the isotopes, it is much below 1% of the real part and only for a few of them, such as Gd, may reach up to 30% of the latter. Note that neutron scattering lengths are nowadays experimentally measured and well tabulated for the vast majority of isotopes hence providing a solid background for quantitative evaluation of data on, e.g., NR. A comprehensive table of up-to-date scattering lengths can be found in the first volume on "Neutron Scattering" by Dawidowski et al. [57].

As was just mentioned, if nuclei have spin then due to its strong exchange interaction with the spin of neutrons, as well as dipole–dipole interaction between magnetic moments of neutrons and nuclei, nuclear scattering also depends on neutron and nuclear spins. However, more important is the other type of spin-dependent interaction constituting the magnetic part of the scattering potential,

$$\widehat{V}_{magn}(\mathbf{r}) = -\widehat{\boldsymbol{\mu}}_n \cdot \sum_l \mathbf{B}_l(\mathbf{r} - \mathbf{r}_l), \qquad (4)$$

describing interaction between the vector of neutron magnetic moment $\hat{\boldsymbol{\mu}}_n$ and the vector of magnetic induction,

$$\mathbf{B}(\mathbf{r}) = \sum_l \mathbf{B}_l(\mathbf{r} - \mathbf{r}_l), \quad (5)$$

created by spin and orbital currents of moving electrons of atoms. The vector operator of neutron magnetic moment $\hat{\boldsymbol{\mu}}_n = 2\mu_n \hat{\mathbf{s}}_n$ is proportional to its spin operator $\hat{\mathbf{s}}_n$, with the coefficient $\mu_n = \gamma \mu_N$ being the neutron magnetic moment expressed in units of the nuclear magneton $\mu_N = e\hbar/2m_p c$ and $\gamma \approx -1.913$. The spin-1/2 operator $\hat{\mathbf{s}}_n = \hat{\boldsymbol{\sigma}}/2$ is represented by the set of three mutually orthogonal Pauli matrices: $\hat{\sigma}_x, \hat{\sigma}_y,$ and $\hat{\sigma}_z$, completing a set of Cartesian projections of the vector $\hat{\boldsymbol{\sigma}}$.

Therefore, magnetic interaction is represented by 2×2 matrix in the neutron spin space. In contrast to $\hat{V}_{\text{nucl}}$ magnetic interaction in nondiagonal and has, in general, all four nonzero elements. This interaction is essentially nonlocal and anisotropic even in the simplest, but most common case when this field is associated with only unpaired spins of electronic shells of atoms. Then the field,

$$\mathbf{B}_l(\mathbf{r} - \mathbf{r}_l) = \int \frac{d\mathbf{q}}{(2\pi)^3} e^{-i\mathbf{q}(\mathbf{r}-\mathbf{r}_l)} \mathbf{B}_l(\mathbf{q}), \quad (6)$$

is expressed via its Fourier transform,

$$\mathbf{B}_l(\mathbf{q}) = 4\pi\mu_l \mathbf{m}_l^\perp f_l(\mathbf{q}), \quad (7)$$

where $f_l(\mathbf{q})$ is the magnetic form factor of the lth atom, $\mu_l = |\boldsymbol{\mu}_l|$ is the absolute value of the atomic magnetic moment $\boldsymbol{\mu}_l$ directed along the unit vector $\mathbf{m}_l = \boldsymbol{\mu}_l/\mu_l$, and due to the classical electrodynamics [58] the vector

$$\mathbf{m}_l^\perp = \mathbf{m}_l - \mathbf{q}(\mathbf{q} \cdot \mathbf{m}_l)/q^2 \quad (8)$$

is orthogonal to the vector $\mathbf{q}$.

In accordance with Eqns (6–9) the magnetic field and hence the interaction potential is basically concentrated within the range of the atomic electron spin density function,

$$\rho_l^e(\mathbf{r}) = \int \frac{d\mathbf{q}}{(2\pi)^3} e^{-i\mathbf{q}\cdot\mathbf{r}} f_l(\mathbf{q}), \quad (9)$$

instead of the pointlike δ-function in Eqn (2). Therefore, in contrast to nuclear interaction, the spacial distribution of the magnetic potential is extended over the atomic size on the order of 10^{-8} cm, while decaying as r^{-3} at larger distances.

Although, the magnetic potential cannot be written in the form of Eqn (1), one can formally introduce the magnetic scattering length,

$$b_l^M = \frac{m_n}{2\pi\hbar^2} \mu_n \overline{B}_l v_l, \quad (10)$$

where $\overline{B}_l = 4\pi\mu_l$ and v_l is the volume per lth magnetic atom. It is important to note that magnetic scattering length b^M for most of magnetic atoms is comparable with the nuclear scattering length b^N in spite of their different basic mechanisms and ranges of interaction. This coincidence gives rise to the other advantage of neutron scattering as a tool for studying magnetic materials.

6.2.2 Scattering Cross Section of Polarized Neutrons

In order to examine neutron scattering from nuclear and magnetic potentials in Eqn (1) in further detail, let us consider a large distance solution of the stationary Schrödinger equation (SE),

$$\left\{-\frac{\hbar^2}{2m_n}\nabla^2 + \widehat{V}(\mathbf{r}) - E\right\}|\Psi(\mathbf{r})\rangle = 0, \quad (11)$$

for the wave function $|\Psi(\mathbf{r})\rangle$. Due to the neutron spin-1/2 the 2D vector describing the spin states is represented by a column vector

$$|\Psi(\mathbf{r})\rangle = \begin{pmatrix} \Psi_+(\mathbf{r}) \\ \Psi_-(\mathbf{r}) \end{pmatrix}. \quad (12)$$

Its elements $\Psi_\pm(\mathbf{r})$ denote probability amplitudes to find a neutron with positive or negative spin projection onto the quantization axis. The vector $|\Psi\rangle$ evolves under the action of the 2×2 matrix of interaction potential:

$$\widehat{V}(\mathbf{r}) = \begin{pmatrix} V_{++} & V_{+-} \\ V_{-+} & V_{--} \end{pmatrix}. \quad (13)$$

The solution $|\Psi(\mathbf{r})\rangle$ of Eqn (11) is usually written as a superposition of two waves:

$$|\Psi(\mathbf{r})\rangle = |\Psi_i(\mathbf{r})\rangle + |\Psi_s(\mathbf{r})\rangle \quad (14)$$

propagating after scattering of the incident plane wave,

$$|\Psi_i(\mathbf{r})\rangle = e^{i\mathbf{k}_i \cdot \mathbf{r}}|\psi_i\rangle. \quad (15)$$

As the solution of Eqn (11) with $V(\mathbf{r}) = 0$ it is obviously represented by a plane wave describing propagation of the neutron with initial wave vector $\mathbf{k}_i$, whose length $|\mathbf{k}_i| = k$ is determined by the neutron energy $E = \hbar^2 k^2/(2m_n)$, and $|\psi_i\rangle$ is the 2D vector of initial spin states.

The second term in Eqn (14) describes propagation of scattered neutrons represented at large distances as a divergent spherical wave:

$$|\Psi_s(\mathbf{r})\rangle = \frac{e^{ikr}}{r}\widehat{f}_n|\psi_i\rangle. \quad (16)$$

As a side remark, one should note that the incident wave defined at times $t \to -\infty$ is described solely by the first term, while after scattering, at

$t \to +\infty$, both terms are present. This looks like a paradox assuming a creation of extra neutrons by the target. This, however, is not the case and the paradox is resolved by considering neutron flux instead of the wave function. Then one can readily recognize that contributions of two terms experience destructive interference reducing flux in the forward direction by exactly the same amount as that scattered into the solid angle. This constitutes the famous optical theorem [59], which is further generalized for PNR in Refs [60,61]. The spin components of $|\Psi_s(\mathbf{r})\rangle$ are determined by four elements of the scattering amplitude matrix $\widehat{f}_n = \widehat{f}_n(\mathbf{k}_i, \mathbf{k}_f)$ generally depending on the incident, $\mathbf{k}_i$, and the scattered, $\mathbf{k}_f = k\mathbf{r}/r$, neutron wave vectors. The scattering matrix is a function of the neutron spin $\widehat{\boldsymbol{\sigma}}/2$ and, as any function of 2×2 matrix, is a matrix 2×2. Diagonal elements of $\widehat{f}$ conserve spin states in the scattering process, while nondiagonal ones describe transitions between these states. The probability amplitude of the transition between the initial, $|\psi_i\rangle$, and the scattered, $|\psi_f\rangle$, spin states is determined by the matrix element,

$$f_n(\mathbf{k}_i, \mathbf{k}_f) = \langle \psi_f | \widehat{f}(\mathbf{k}_i, \mathbf{k}_f) | \psi_i \rangle, \tag{17}$$

called *the scattering amplitude*.

The scattering amplitude matrix $\widehat{f}_n$ as any 2×2 matrix can be formally decomposed over the set of three mutually orthogonal Pauli matrices $\widehat{\sigma}_x, \widehat{\sigma}_y, \widehat{\sigma}_z$ and the unit matrix:

$$\widehat{f}_n = f_0 + (\mathbf{f} \cdot \widehat{\boldsymbol{\sigma}}), \tag{18}$$

where three Pauli matrices are combined into the 3D vector $\widehat{\boldsymbol{\sigma}}$ so that in a particular coordinate system the scalar product $(\mathbf{f} \cdot \widehat{\boldsymbol{\sigma}}) = f_x\widehat{\sigma}_x + f_y\widehat{\sigma}_y + f_z\widehat{\sigma}_z$.

Here the scalar component $f_0 = f_0(\mathbf{k}_i, \mathbf{k}_f)$ is the average of diagonal elements: $f_0 = \frac{1}{2}\text{Tr}\{\widehat{f}\}$, while the 3D vectorial function $\mathbf{f} = \mathbf{f}(\mathbf{k}_i, \mathbf{k}_f)$ is determined as $\mathbf{f} = \frac{1}{2}\text{Tr}\{\widehat{f}\widehat{\boldsymbol{\sigma}}\}$. One of the main advantages of the invariant form of Eqn (18) is that it is independent of a laboratory coordinate system, which can be set, when necessary, i.e., after all formal calculations are accomplished.

The parametrization of the scattering matrix in Eqn (18) is exact and exploits only basic properties of the spin-1/2 algebra, but further calculations of parameters f_0 and $\mathbf{f}$ can be carried out via the use of relevant approximate solutions of SE. The solution given by Eqns (14–16) is asymptotically exact in the limit of distances much greater than the range of the interaction potential and hence the size of the scattering target. However, explicit exact expressions for the scattering amplitude matrix can be found for only a few particular examples of $\widehat{V}(\mathbf{r})$, and usually approximate solutions for $\widehat{f}(\mathbf{k}_i, \mathbf{k}_f)$ are used in practice. One of them is based on the perturbation theory approach, which assumes that the total probability of scattering, i.e., the flux scattered into the solid angle is small with respect to the initial flux far away of the scattering potential. Then the scattering amplitude describing the transition in the first

BA between the incident and final plane waves is linear with respect to the scattering potential and proportional to the matrix element

$$f_n(\mathbf{k}_i, \mathbf{k}_f) = -\frac{m_n}{2\pi\hbar^2} \int d\mathbf{r} \langle \Psi_f(\mathbf{r}) | \widehat{V}(\mathbf{r}) | \Psi_i(\mathbf{r}) \rangle. \quad (19)$$

Here the initial plane wave $|\Psi_i(\mathbf{r})\rangle$ is determined by Eqn (15), while

$$\langle \Psi_f(\mathbf{r}) | = \langle \psi_f | e^{-i\mathbf{k}_f \cdot \mathbf{r}}, \quad (20)$$

corresponds to the unperturbed plane wave freely propagating in the direction opposite to that of observation $\mathbf{r}/r$, and $\langle \psi_f |$ corresponds to the final spin states.

Substitution of Eqns (15) and (20) into Eqn (19) immediately yields the explicit equation for the Born scattering amplitude,

$$f_n(\mathbf{Q}) = \langle \psi_f | \widehat{f}_n(\mathbf{Q}) | \psi_i \rangle. \quad (21)$$

Hence, the Born scattering amplitude matrix $\widehat{f}_n(\mathbf{Q})$ is simply proportional to the Fourier transform,

$$\widehat{f}_n(\mathbf{Q}) = -\frac{m_n}{2\pi\hbar^2} \int d\mathbf{r} e^{-i\mathbf{Q}\cdot\mathbf{r}} \widehat{V}(\mathbf{r}), \quad (22)$$

of the interaction potential $\widehat{V}(\mathbf{r})$ matrix. It depends on the wave vector transfer $\mathbf{Q} = \mathbf{k}_f - \mathbf{k}_i$, but not on $\mathbf{k}_i$ and $\mathbf{k}_f$ separately, as it would be in the general case.

Consequently, in the BA the scalar parameter in Eqn (18) $f_0 = f_N$ and the vector parameter $\mathbf{f} = \mathbf{f}_M$ are determined in accordance with Eqns (1), (2), and (22) as:

$$f_N = -\sum_l b_l^N e^{-i\mathbf{Q}\cdot\mathbf{r}_l}, \quad (23)$$

$$\mathbf{f}_M = -\sum_l b_l^M f_l(\mathbf{Q}) e^{-i\mathbf{Q}\cdot\mathbf{r}_l} \mathbf{m}_l^\perp. \quad (24)$$

The probability of scattering into the solid angle element $d\Omega$ is given by the differential scattering cross section equal to the scattering amplitude modulus squared:

$$\frac{d\sigma}{d\Omega} = \overline{|\langle \psi_f | \widehat{f}_n | \psi_i \rangle|^2} = \text{Tr}\left\{ \widehat{\rho}_f \widehat{f}_n \widehat{\rho}_i \widehat{f}_n^+ \right\} \quad (25)$$

averaged over spin states in the incoming and scattered neutron beams. Such averaging can readily be accomplished taking into account that the polarizing ability of the spin polarizer is independent of the spin efficiency of the spin analyzer. Therefore, after the cycling of the spinor $|\psi_f\rangle$ under the sign of averaging in Eqn (25) from the last place to the first, one can introduce two spin density matrices $\widehat{\rho}_i = \overline{|\psi_i\rangle\langle\psi_i|}$ and $\widehat{\rho}_f = \overline{|\psi_f\rangle\langle\psi_f|}$. These

2 × 2 matrices can be commonly decomposed over a set of Pauli matrices, so that

$$\hat{\rho}_i = \frac{1}{2}\left(1 + \mathbf{P}_i \cdot \hat{\boldsymbol{\sigma}}\right), \qquad (26)$$

$$\hat{\rho}_f = \frac{1}{2}\left(1 + \mathbf{P}_f \cdot \hat{\boldsymbol{\sigma}}\right), \qquad (27)$$

where the vectorial parameter $\mathbf{P}_i = 2\langle \mathbf{s}_n \rangle$ is the polarization vector customary defined as a doubled mean value of the spin operator $\mathbf{s}_n$ averaged over spin states in accordance with the spin density matrix. The length $|\mathbf{P}_i| \leq 1$ of the vector $\mathbf{P}_i$ is then the degree of the polarization, and the vector $\mathbf{P}_f$ determines the direction of the polarization analysis with the efficiency $|\mathbf{P}_f| \leq 1$.

Substitution of density matrices from Eqns (26) and (27) into Eqn (25) yields the equation for scattering cross section consisting of four terms:

$$\frac{d\sigma}{d\Omega} = \left(\frac{d\sigma}{d\Omega}\right)_0 + \left(\frac{d\sigma}{d\Omega}\right)_i + \left(\frac{d\sigma}{d\Omega}\right)_f + \left(\frac{d\sigma}{d\Omega}\right)_{if}. \qquad (28)$$

Here the first term is apparently independent of the initial polarization and polarization analysis and equal to:

$$\left(\frac{d\sigma}{d\Omega}\right)_0 = \frac{1}{2}\left\{|f_0|^2 + |\mathbf{f}|^2\right\}, \qquad (29)$$

where $|\mathbf{f}|^2 = (\mathbf{f} \cdot \mathbf{f}^*)$. This term in Eqn (28) describes scattering of unpolarized neutrons when $\mathbf{P}_i = \mathbf{P}_f = 0$ and all three last terms vanish. Then one can measure the sum of two scattering cross sections $|f_0|^2$ and $|\mathbf{f}|^2$, corresponding in BA to nuclear and magnetic scattering, respectively, but one cannot discriminate between these two contributions.

The second and the third terms in Eqn (28) are linear with respect to either the vector $\mathbf{P}_i$ or $\mathbf{P}_f$:

$$\left(\frac{d\sigma}{d\Omega}\right)_i = \left\{\Re(f_0^* \mathbf{f}) + \frac{1}{2}\Im[\mathbf{f}^* \times \mathbf{f}]\right\} \cdot \mathbf{P}_i, \qquad (30)$$

$$\left(\frac{d\sigma}{d\Omega}\right)_f = \left\{\Re(f_0^* \mathbf{f}) - \frac{1}{2}\Im[\mathbf{f}^* \times \mathbf{f}]\right\} \cdot \mathbf{P}_f. \qquad (31)$$

These terms together with that in Eqn (29) describe scattering of either polarized neutrons, but without polarization analysis, or of neutrons initially unpolarized, but spin analyzed after scattering. In the first case $\mathbf{P}_f = 0$ and two last terms in Eqn (28) are equal to zero, while in the second case $\mathbf{P}_i = 0$ and only the first and the third terms contribute to scattering cross section. The first term in Eqns (30) and (31) gives access to the additional information, e.g., about the direction of the vector $\mathbf{f}$. In BA these terms describe interference between nuclear and magnetic scattering, and according to Eqn (24) the vector $\mathbf{f} = \mathbf{m}_\perp f_M$ is directed along the vector of magnetic induction $\mathbf{B}$ defined in Eqn (5). In BA scattering amplitudes are purely real and the second terms in Eqns (30) and (31) vanish. However, as we shall see, they play an essential role beyond BA. The last term in Eqn (28), bilinear with respect to $\mathbf{P}_i$ and $\mathbf{P}_f$, reads:

$$\left(\frac{d\sigma}{d\Omega}\right)_{if} = \frac{1}{2}\left(|f_0^2| - |\mathbf{f}|^2\right)(\mathbf{P}_i \cdot \mathbf{P}_f) + \Re\{(\mathbf{f}^* \cdot \mathbf{P}_i)(\mathbf{f} \cdot \mathbf{P}_f)\} - \frac{1}{2}\Im\{f_0^*(\mathbf{f} \cdot [\mathbf{P}_i \times \mathbf{P}_f])\}.$$

(32)

This term is only present when incident neutrons are polarized and scattered spin states are analyzed. Equation (32) together with Eqns (29–31) describes the general case of scattering of initially polarized neutrons with the polarization analysis of the scattered beam. The first term in Eqn (32) is independent of the direction of the vector $\mathbf{f}$ and turns to zero when $\mathbf{P}_i \perp \mathbf{P}_f$. The second term is determined by the projections of $\mathbf{P}_i$ and $\mathbf{P}_f$ onto the vector $\mathbf{f}$. The last term may contribute to cross section only if all three vectors $\mathbf{P}_i$, $\mathbf{P}_f$, and $\mathbf{f}$ are not collinear.

In BA scattering amplitudes, Eqns (29–32) are substantially simplified:

$$\left(\frac{d\sigma}{d\Omega}\right) = \frac{1}{2}f_N^2(1 + \mathbf{P}_i \cdot \mathbf{P}_f) + (f_N f_M)\mathbf{m}_\perp \cdot (\mathbf{P}_i + \mathbf{P}_f)$$
$$+ \frac{1}{2}f_M^2[1 - \mathbf{P}_i \cdot \mathbf{P}_f + 2(\mathbf{m}_\perp \cdot \mathbf{P}_i)(\mathbf{m}_\perp \cdot \mathbf{P}_f)],$$

(33)

and manipulating with directions and length of vectors $\mathbf{P}_i$ and $\mathbf{P}_f$ one can separate contributions of nuclear scattering, magnetic scattering, and nuclear–magnetic scattering interference.

Equations (29–32) written in the invariant form are exact and valid for arbitrary orientation between directions of the vector $\mathbf{P}_i$ of the incident polarization and of the vector $\mathbf{P}_f$ of polarization analysis. Each of these vectors is defined by three Cartesian projections P_x, P_y, and P_z, and at least, nine scattering cross sections can be measured via directing $\mathbf{P}_i$ and $\mathbf{P}_f$ along orthogonal axes, although not all of those cross sections are independent. In particular, cross section in Eqn (28) is invariant with respect to the simultaneous interchanging between the pair of initial vectors $\mathbf{P}_i$, $\mathbf{k}_i$ and the pair of final ones $\mathbf{P}_f$, $\mathbf{k}_f$, as required by the reciprocity principle [59].

On the other hand, in accordance with Eqn (18) the scattering amplitude matrix is totally determined by four complex elements: the scalar f_0 and three complex Cartesian projections of the 3D vector $\mathbf{f}$. They altogether amount eight real quantities including four absolute values of the scalar f_0 and three projections of the vector $\mathbf{f}$, as well as four phases. However, the global phase remains unknown and only three phase shifts may experimentally be determined. Therefore, the scattering amplitude matrix is generally determined by seven real functions of the incident and final wave vectors.

The so-called 3D polarization vector analysis (3DPVA) proposed in the early 1970s [62,63] is nowadays developed for SANS [64] and neutron inelastic scattering spectrometry [65]. Although the method of 3DPVA can provide apparent benefits for PNR [66], it has not yet been implemented in practice. One of the benefits is that (3DPVA) permits to determine a spatial orientation of the vector **f**. In the simplest case this is the vector of magnetic induction **B** in Eqns (7) and (24).

However, in the commonly used version of PNR polarization vectors P_i and P_f are directed either parallel or antiparallel to each other. Then a maximum of four essentially different scattering cross sections: two non-spin flip (NSF) and two spin flip (SF) are measured and are available for the theoretical analysis. NSF configurations with $P_i \uparrow \uparrow P_f$, or $P_i \downarrow \downarrow P_f$, select neutrons maintaining sign of their spin projections onto external magnetic field guiding neutron spins from the polarizer to the sample and further to the spin analyzer. Setting $P_i \uparrow \downarrow P_f$, or $P_i \downarrow \uparrow P_f$, SF scattering processes altering the sign of the spin projection are probed. These processes are certainly due to the components of the vector **f**, e.g., those determined by the vector of magnetic induction, perpendicular to the spin quantization axis chosen collinear with the guide field.

6.2.3 Grazing Incidence Kinematics

As indicated above, the set of Eqns (28—32) is substantially simplified if the scattering effect is small so that BA is valid. This, however, is not always the case in spite of the fact that the scattering amplitudes in Eqns (23) and (24) are small. Indeed, the sum in this equation runs over all scatterers in a macroscopic sample and may contain a very large number of terms. Due to oscillating phase factors, most of the terms at arbitrary wave vector transfer **Q** cancel each other, but for some definite values of **Q** oscillations may be in-phase providing constructive interference. Such interference takes place, for instance, for Bragg diffraction from perfect single crystals at $\mathbf{Q} = \tau$, where τ is one of the lattice reciprocal vectors. Then, taking into account that the scattering length $|b| \sim 10^{-12}$ cm and that the number density n of condensed matter is typically $\sim 10^{22}$ cm^{-3}, one can estimate the SLD $nb \sim 10^{10}$ cm^{-2}. Therefore, at $\mathbf{Q} = \tau$ the scattering amplitude $|f(\tau)| = nbv \sim 10^{-5}$ cm for the crystal volume $v \sim 10^{-15}$ cm^3. This means that for such crystals the scattering length may be compensated by the large number $N_0 \sim 10^7$ of scattering atoms. As a result, the scattering cross section of $\sim 10^{-10}$ cm^2 becomes comparable with the value of classical geometrical cross section of the target. In this case all neutrons experience at least one collision with the target atoms. On the other hand, BA is based on the assumption that only a small fraction of neutrons is scattered. Hence, even for relatively small targets BA may be inapplicable, and then more complicated approaches, such as the theory of dynamical diffraction, should be employed.

Estimations above also hold for the scattering wave vectors slightly deviating from exact Bragg conditions within the range of $\mathbf{Q}$, where $|\mathbf{Q} - \tau| \sim \pi v^{-1/3}$. At larger deviations dephasing start to suppress coherent contribution of atoms to the sums in Eqns (28—32), and the scattering amplitudes decay.

The above consideration is certainly valid at $|\mathbf{Q}| \ll n^{1/3}$, i.e., in the vicinity of $\tau = 0$ corresponding to the Bragg diffraction of the zero order. For elastic scattering the wave vector transfer $|\mathbf{Q}| = (4\pi/\lambda)|\sin(\theta/2)|$ is small if either the wavelength $\lambda \gg n^{-1/3}$ is large or/and scattering angles θ between vectors $\mathbf{k}_i$ and $\mathbf{k}_f$ are small. At low values of $|\mathbf{Q}|$ constructive interference does not actually require a periodicity of scattering potentials, and one can use the continuous media approximation introducing SLD $\rho(\mathbf{r}) = nb(\mathbf{r})$ as a continuous function of $\mathbf{r}$. As a result, the sums in Eqns (23) and (24) can be substituted by integrals so that:

$$f_N = -\int d\mathbf{r} \rho^N(\mathbf{r}) e^{-i\mathbf{Q} \cdot \mathbf{r}}, \tag{34}$$

$$\mathbf{f}_M = -\int d\mathbf{r} \rho^M(\mathbf{r}) e^{-i\mathbf{Q} \cdot \mathbf{r}} \mathbf{m}_\perp, \tag{35}$$

where at low $|\mathbf{Q}|$ the atomic magnetic form factor $f(\mathbf{Q}) \approx 1$. Integrations in these equations are performed over the volume of the target, and the scattering amplitude crucially depends on its shape and orientation with respect to the incident neutron wave vector $\mathbf{k}_i$.

This can be illustrated for the simplest, but the most common case, when the neutron plane wave is incident onto a thin slab at a shallow glancing angle $\alpha_i \ll 1$ with respect to the largest face of the slab, as sketched in Figure 2. Then the vector $\mathbf{k}_i$ has the projection $p_i = k\sin\alpha_i \approx k\alpha_i$ onto the normal to the slab face and the lateral component κ_i with the length

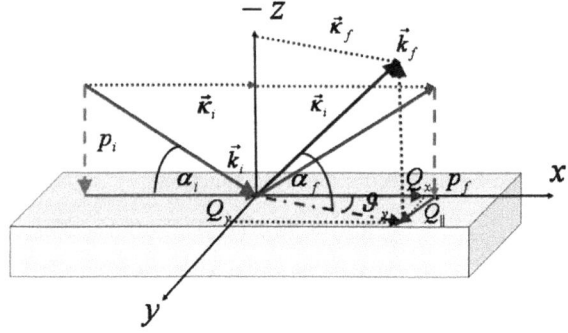

FIGURE 2 Schematically shown is the layout of the scattering kinematics from a slab-shaped target.

$\kappa_i = k \cos \alpha_i \approx k(1 - \alpha_i^2/2)$. At the same time the wave vector $\mathbf{k}_f$ scattered at the shallow glancing angle α_f has the normal projection $p_f = k \sin \alpha_f \approx k\alpha_f$ and the lateral component κ_f with projections $k_f^x = k \cos \alpha_f \cos \theta_y$ and $k_f^y = k \cos \alpha_f \sin \theta_y$. Correspondingly, the normal projection of the scattering vector $\mathbf{Q}$ is represented as

$$Q_z = k\{\sin \alpha_f + \sin \alpha_i\} \approx k(\alpha_f + \alpha_i), \tag{36}$$

while the lateral component $\mathbf{Q}_\| = \kappa_f - \kappa_i$ has projections

$$Q_x = \{\cos \alpha_f \cos \theta_y - \cos \alpha_i\} \approx k\left(\alpha_i^2 - \alpha_f^2 - \theta_y^2\right)/2 \tag{37}$$

$$Q_y = k \cos \alpha_f \sin \theta_y \approx k\theta_y. \tag{38}$$

The integration procedure in Eqns (34) and (35) can be further simplified assuming a homogeneous distribution of SLD's $\rho^{(N,M)}$ over the thin slab volume and that at shallow incidence the scattering vector is almost normal to the slab face, and hence $\mathbf{m}_\perp = \mathbf{m}$ is mostly displayed within the x, y plane. (As a side remark, a homogeneous magnetization is, strictly speaking, only possible for samples having an ellipsoidal shape, and if any external magnetic field is applied along one of its principle axes. Then magnetic induction components normal to the ellipsoid surface are continues creating inhomogeneous magnetic stray fields depending also on the magnetic surrounding. In some cases [67] these fields may cause interesting scattering effects which are beyond the scope of present consideration. Under these circumstances $\mathbf{f}_M = \mathbf{m}f_M$ and Eqns (34) and (35) factorize into products:

$$f_{N,M} = \rho^{N,M} \Lambda_x(Q_x) \Lambda_y(Q_y) \Lambda_z(Q_z). \tag{39}$$

Here the shape functions Λ_ν with $\nu = x, y, z$ can be written in a unique form:

$$\Lambda_\nu(Q_\nu) = 2\frac{\sin(Q_\nu L_\nu/2)}{Q_\nu}, \tag{40}$$

where L_x, L_y, and $L_z = d$ are the dimensions of the slab along corresponding Cartesian axes. At $Q_\nu \gg L_\nu^{-1}$ these functions decay as Q_ν^{-1} oscillating with the period $Q_\nu = 2\pi/L_\nu$ and revealing main maxima at $Q_\nu = 0$ with $\Lambda_\nu(0) = L_\nu$ and the width $\sim L_\nu^{-1}$, while each of the integrals of $\Lambda_\nu(Q_\nu)$ over Q_ν is equal to 2π. For the thin slab with $L_x, L_y \gg d$ maxima of $\Lambda_{x,y}$ are much sharper than those of Λ_z and $\Lambda_{x,y}$ can be approximated by δ-functions,

$$\Lambda_{x,y} \approx 2\pi\delta(Q_{x,y}). \tag{41}$$

Due to this equation, Eqn (33) can be written as follows:

$$\frac{d\sigma}{d\Omega} \approx S_b\{R\delta(\alpha_i - \alpha_f) + T\delta(\alpha_i + \alpha_f)\}\delta(\theta_y), \tag{42}$$

where $S_b = S \sin \alpha_i$ is the visible cross section of the target, i.e., its footprint onto the plane normal to the incident wave vector, δ-functions emphasize that scattering is concentrated in either the direction of *specular reflection* when $\alpha_f = \alpha_i$ or transmission when $\alpha_f = -\alpha_i$.

The dimensionless reflectivity, R, and transmissivity, T, obey the same Eqn (33) in which the scattering cross section is substituted for R, or T, and scattering amplitudes are substituted for Born amplitudes of reflection, $r_{M,N}$, or correspondingly, transmission, $t_{N,M} = r_{N,M}$, coinciding in this approximation and equal to:

$$r_{N,M} = \frac{1}{2}\left(\frac{Q_c^{N,M}}{Q}\right)^2 \sin(Q_z d/2), \tag{43}$$

with $Q_c^{N,M} = 4\sqrt{\pi \rho^{N,M}}$. Hence, in BA the scattering cross section for specular reflection is proportional to $S\alpha_i (Q_c^{N,M}/Q_z)^4$, i.e., increases with the neutron wavelength as λ^4 and diverges as α_i^{-3} when the incident angle approaches 0. Formally, BA cross sections in Eqn (42) may exceed the target area $S = L_x L_y$, but the divergence at $\alpha_i \to 0$ does not actually take place as the validity of the approximation in Eqn (41) fails at very small angles $\alpha_i \leq (d/L_x) \ll 1$ when the incident wave mostly illuminates the side edge of the target.

However, the scattering cross section may become large so that BA approximation fails even at angles α_i where the Eqns (41) and (42) are still valid. Due to the fact that Eqn (42) allows for only specular reflection and transmission, one can readily go beyond BA and obtain an asymptotically exact solution of SE for a scattering potential which is independent of the lateral coordinates due to, e.g., ignoring side edge effects.

In the representation with the quantization axis chosen along the vector $\mathbf{m}$, the scattering potential matrix $\hat{V}$ in Eqn (13), which has eigenvalues $V_\pm = V_{\text{nucl}} \pm V_{\text{magn}}$, is diagonal. Therefore, the matrix equation Eqn (11) for the spinor $|\Psi(\mathbf{r})\rangle$ in Eqn (12) is decoupled into two separate equations: the one for the wave $\Psi_+(\mathbf{r})$ describing propagation of the positive spin projection and the other for the wave $\Psi_-(\mathbf{r})$ referring to the negative spin projection onto the quantization axis.

The general solutions for each of those spin components inside the slab, i.e., within the range $0 \leq z \leq d$ of the constant scattering potential, can be found in the following factorized form:

$$\Psi_\pm(\mathbf{r}) = e^{i\boldsymbol{\kappa}\cdot\mathbf{r}_\parallel} \psi_\pm(z), \tag{44}$$

in which the neutron wave propagation is represented as a combination of two independent types of motion. The first oscillating exponential factor in this equation describes the free propagation of the neutron plane wave with the wave vector $\boldsymbol{\kappa}$ in the lateral direction. The second factor refers to the motion in the orthogonal direction and is described by the superposition of plane waves,

$$\psi_\pm(z) = e^{ip_{1\pm}z} t_{1\pm} + e^{-ip_{1\pm}z} r_{1\pm}, \tag{45}$$

propagating with transmission, $t_{1\pm}$, and reflection, $r_{1\pm}$, amplitudes of waves along the z-axis in the positive and, correspondingly, negative directions. Their wave numbers $p_{1\pm} = \sqrt{p_0^2 - p_{c\pm}^2}$ inside the sample are determined by the wave vector projection $p_0 = \sqrt{k^2 - \kappa^2}$ onto the normal to the surface, and parameters $p_{c\pm} = \sqrt{4\pi(\rho^N \pm \rho^M)}$ correspond to critical wave numbers of total reflection for positive or negative spin components of the neutron wave. Hence, neutron waves refracted into the sample with $p_0 \geq p_{c\pm}$ are split into two revealing spin-dependent birefringence effects [68,69]. If $p_0 \leq p_{c+}$, or $p_0 \leq p_{c-}$ then corresponding spin components of the neutron wave become evanescent, exponentially decaying into depth of the sample and may be totally reflected, if $p_\pm d \gg 1$.

Transmission amplitudes, $t_{1\pm}$, and reflection amplitudes, $r_{1\pm}$, are determined from boundary conditions at the front and the back faces of the slab. In the first case, just above the front surfaces, the wave function is a superposition of two waves: the one incident onto the surface with the unity amplitude $t_\pm = 1$ and the other one reflected with the amplitude $r_\pm$, so that at $z \leq 0$:

$$\psi_\pm(z) = e^{ip_0 z} + e^{-ip_0 z} r_\pm. \quad (46)$$

Below the back face at $z \geq d$ there exists only transmitted waves: $\psi_\pm(z) = t_{s\pm} e^{ip_0 z}$ propagating in free space. Matching "+" and "−" wave functions and their first derivatives at both surfaces, one obtains a system of four linear equations with respect to four unknown amplitudes: $r_\pm$, $t_{1\pm}$, $r_{1\pm}$, and $t_{s\pm}$. Solving this system one readily finds all amplitudes:

$$r_\pm = p_{c\pm}^2 \left(1 - e^{2i\varphi_\pm}\right) \Delta_\pm^{-1}, \quad (47)$$

$$r_{1\pm} = 2p_0 (p_\pm - p_0) e^{2i\varphi_\pm} \Delta_\pm^{-1}, \quad (48)$$

$$t_{1\pm} = 2p_0 (p_\pm + p_0) \Delta_\pm^{-1}, \quad (49)$$

$$t_{s\pm} = 4p_0 p_\pm e^{i\varphi_\pm} \Delta_\pm^{-1}, \quad (50)$$

$$\Delta_\pm = (p_\pm + p_0)^2 - (p_\pm - p_0)^2 e^{2i\varphi_\pm}, \quad (51)$$

where $\varphi_\pm = p_\pm d$ is the phase shift gained over the thickness d.

If this thickness is so large that due to the finite absorption $e^{2i\varphi_\pm} \approx 0$, then Eqns (47) and (49) boil down to the equations for Fresnel amplitudes of reflection, $r_\pm = r_\pm^F$, and transmission, $t_{s\pm} = t_\pm^F$:

$$r_\pm^F = \frac{p_0 - p_\pm}{p_0 + p_\pm}, \quad (52)$$

$$t_\pm^F = \frac{2p_0}{p_0 + p_\pm}, \quad (53)$$

while $r_{1\pm} \approx 0$ and $t_{s\pm} \approx 0$. Note that usually neutron absorption is small and at low values of p_0 such that $p_{c-} < p_0 < p_{c+}$ only the negative spin components penetrating into depth of material is absorbed, while the positive spin component is totally reflected. This occurs if the length of extinction for the positive spin component is much smaller than the absorption length.

In all cases when optical effects, i.e., refraction and specular reflection are strong, BA (Eqn (33)) is not applicable. However, Eqn (42) is valid also in this general case, if the reflection coefficient R is calculated in accordance with Eqns (28−32) in which scattering cross section in Eqn (28) is substituted by R, and the parameters f_0 and $\mathbf{f}$ in the scattering amplitude matrix $\widehat{f}_n$ in Eqn (18) is substituted by corresponding parameters r_0 and $\mathbf{r}_0$ of the exact 2 × 2 reflection amplitude matrix,

$$\widehat{r} = r_0 + (\mathbf{r}_0 \cdot \boldsymbol{\sigma}). \tag{54}$$

These parameters are determined via eigenvalues $r_\pm$ of the reflection amplitude matrix as follows:

$$r_0 = \frac{1}{2}(r_+ + r_-), \tag{55}$$

$$\mathbf{r}_0 = \frac{\mathbf{m}}{2}(r_+ - r_-). \tag{56}$$

In the limit $p_0 \gg p_{c\pm}$, one can neglect refraction effects assuming that in the denominator of Eqn (47) $p_\pm \approx p_0$. Then, taking into account that $Q_z = 2p_0$ and that $p_{c+}^2 \pm p_{c-}^2 = 8\pi\rho^{N,M}$, from Eqn (47) one obtains the BA result in Eqn (43). However, in accordance with the exact Eqn (47) at $Q_z \leq Q_c^{N,M}$ the divergence of scattering cross section of specular reflection cuts off at $Q_z \approx Q_{c\pm} = 2p_{c\pm}$.

This can be illustrated via calculations of NSF, $R^{\pm\pm} = R(\pm\mathbf{P}_i, \pm\mathbf{P}_f)$, and SF, $R^{\pm\mp} = R(\pm\mathbf{P}_i, \mp\mathbf{P}_f)$, reflectivities, in accordance with Eqns (28−32) in which parameters of the scattering amplitude matrix are substituted with corresponding parameters of the reflection amplitude matrix finally expressed via amplitudes $r_\pm$. Then, assuming that incident and final polarization vectors are directed along with, or opposite to the Y-axis, as sketched in Figure 3, and that $|P_i^y| = |P_f^y| = 1$, one can write down the following set of equations:

$$R^{++} = \frac{1}{4}|r_+(1 + \cos\gamma) + r_-(1 - \cos\gamma)|^2, \tag{57}$$

$$R^{--} = \frac{1}{4}|r_+(1 - \cos\gamma) + r_-(1 + \cos\gamma)|^2, \tag{58}$$

$$R^{+-} = R^{-+} = \frac{1}{4}|r_+ - r_-|^2 \sin^2\gamma, \tag{59}$$

where γ is the angle between the magnetic induction and the Y-axis.

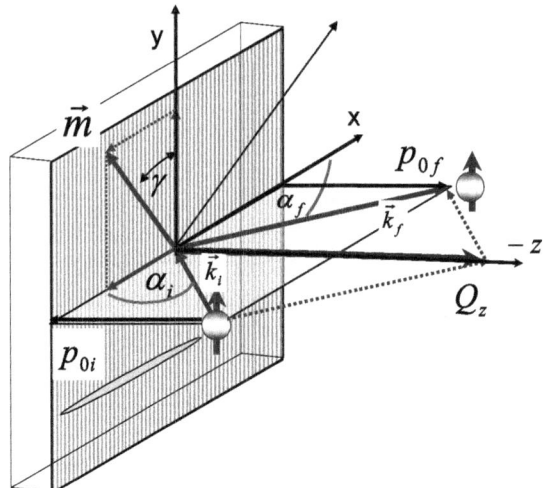

FIGURE 3 Schematically sketched is the geometry of polarized neutron reflectometry experiments with one-dimensional polarization analysis along Y-axis. The elongated ellipse indicates the coherence volume of neutrons in PNR experiments.

In the reduced version of PNR, no spin analysis of the reflected neutrons initially polarized parallel to the Y-axis is assumed. Then only two reflectivities $R^+ = R^{++} + R^{+-}$ and $R^- = R^{--} + R^{-+}$ are available:

$$R^{\pm} = \frac{1}{2}\left\{|r_+|^2(1 \pm \cos \gamma) + |r_-|^2(1 \mp \cos \gamma)\right\}. \tag{60}$$

For a homogeneously magnetized sample, both versions of PNR contain the same set of parameters, i.e., nuclear SLD (nSLD), magnetic SLD (mSLD), and the angle γ which can be experimentally determined. The sign of the latter, however, in neither of cases can be fixed from 1D PNR, and 3D vector analysis is required. On the other hand, in case of inhomogeneous lateral magnetization full 1D PNR analysis, as will be demonstrated below, has certain advantage over the reduced version.

In Figure 4(a) and (b), NSF R^{++} and R^{--} as well as SF R^{+-} and R^{-+} are plotted for a "free-standing" Fe film, i.e., for a film that has vacuum on both sides, of thickness $d = 100$ nm: Figure 4(a) with the vector **m** collinear and Figure 4(b) with **m** orthogonal to the polarization axis. In the first case $R^{+-} = R^{-+} = 0$, while R^{++} and R^{--} are split revealing critical edges of total reflection $Q_{c\pm}$ below which both curves merge at the level 1, and at $Q \geq Q_{c\pm}$ show thickness oscillations decaying as Q^{-4} at $Q \gg Q_{c\pm}$. Similar oscillations can also be recognize for the orthogonal case in Figure 4(b), but here $R^{++} = R^{--}$ showing both critical edges, saturating only at $Q \ll Q_{c\pm}$ where $R^{+-} = R^{-+}$ tend to be zero. Note that oscillations persist, albeit with lower

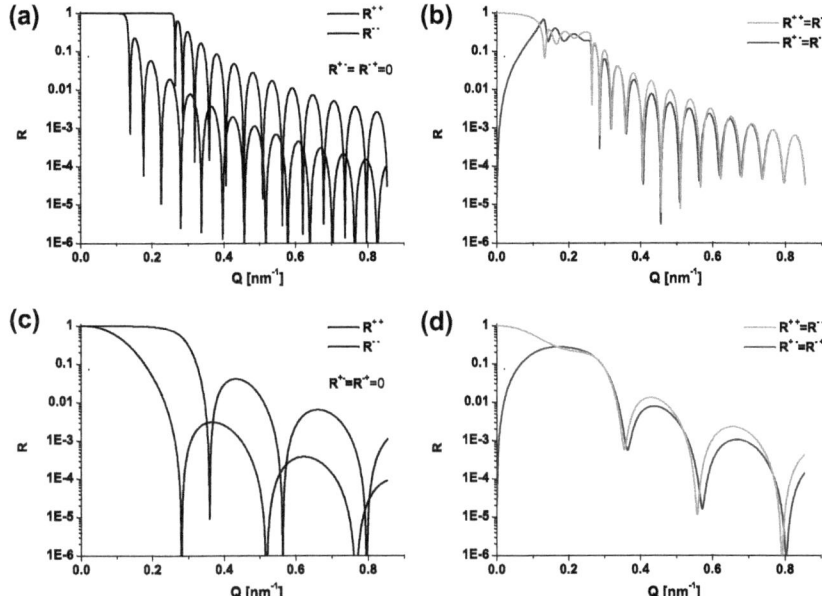

FIGURE 4 Non-spin-flip and spin-flip polarized neutron reflectometry from a free-standing Fe film (vacuum on both sides): (a) with thickness $d = 100$ nm and with the angle $\gamma = 0$, (b) with $\gamma = 90°$; (c) and (d) are the same as in (a) and (b) but for a film thickness of $d = 25$ nm. Here γ is the angle between the magnetization and the y-axis.

amplitude even below Q_{c+} and vanish only when $Q \leq Q_{c-}$, where both spin components no longer reach the back face of the film.

Similar behavior is seen also in Figure 4(c) and (d) calculated for $d = 25$ nm. However, in this case the critical edges especially the one for R^{--} are smeared out as the thickness is lower than the extinction length and neutron waves are partially transmitted through the backside of the sample except for very low $Q \approx 0$.

6.2.4 Specular Polarized Neutron Reflectivity for 1D Potential

Exact solutions of the 1D SE can also be obtained for SLD profiles arbitrarily varying across the slab thickness, but independent of lateral coordinates. If the scattering potential $\widehat{V}(\mathbf{r}) = \widehat{V}(z)$ in Eqn (11) depends only on the coordinate z, then dividing the slab into a set of N thin flat layers, its depth profile can be approximated by a sequence of step functions with required accuracy:

$$\widehat{V}(z) = \sum_{j=1}^{N} \widehat{V}_j(z - z_{j-1}). \tag{61}$$

where $z_0 = 0$ and $\widehat{V}_j(z - z_{j-1}) = \widehat{V}_j$ is a constant at $z_{j-1} \leq z \leq z_j$ and equal to zero elsewhere.

The general solution of Eqn (11) for such 1D potential can be found in the following factorized form,

$$|\Psi(\mathbf{r})\rangle = e^{i\boldsymbol{\kappa}\cdot\mathbf{r}_\parallel} \sum_{j=1}^{N+1} \widehat{S}_j(z)|\psi_0\rangle, \qquad (62)$$

in which the neutron wave propagation is again represented as a combination of two types of motion. The first of them is described by the oscillating exponential factor corresponding to the free propagation of the neutron plane wave with the wave vector $\boldsymbol{\kappa}$ in the lateral direction. The second one refers to the motion in the orthogonal direction and is described by the propagation matrix S represented as a sum:

$$\widehat{S}_j(z) = e^{i\widehat{p}_j(z-z_{j-1})}\widehat{t}_j + e^{-i\widehat{p}_j(z-z_{j-1})}\widehat{r}_j, \qquad (63)$$

in which transmission, $\widehat{t}_j$, and reflection, $\widehat{r}_j$, amplitude matrices are independent of z and determined within the layer j, i.e., at $z_{j-1} \leq z \leq z_j$ turning to zero elsewhere. Taking into account that in the vacuum above the surface at $z \leq 0$ there simultaneously coexist incident and reflected waves, one can add to Eqn (63) one more equation:

$$\widehat{S}_0(z) = e^{i\widehat{p}_0 z}\widehat{t}_0 + e^{-i\widehat{p}_0 z}\widehat{r}_0, \qquad (64)$$

where $\widehat{t}_0 = 1$ and $\widehat{r}_0 = \widehat{r}$ are the amplitude matrices of reflection to be finally determined. Additionally, one should note that the wave propagating in the last, usually thick, layer $N+1$ may leave it through its side edge, or absorbed before reaching the back face. Then $\widehat{r}_{N+1} = 0$ and $\widehat{t}_{N+1} = \widehat{t}_s$ is the transmission amplitude matrix.

Substitution of Eqns (62) and (63) into Eqn (11) immediately yields the relationship $\hbar^2(\widehat{p}_j^2 + \kappa^2) = 2m_n(E - \widehat{V}_j)$ which determines the matrix $\widehat{p}_j = \sqrt{p_0^2 - \widehat{p}_{cj}^2}$, where $\widehat{p}_{cj}^2 = 4\pi\widehat{\rho}_j$. Eigenvalues $p_{j\pm} = \sqrt{p_0^2 - p_{cj\pm}^2}$ of the matrix $\widehat{p}_j$ correspond to wave numbers of neutrons with positive or negative spin projection transmitted into the layer j, or reflected from its $j+1$ interface. Eigenvalues $p_{cj\pm}$ of the matrix $\widehat{p}_{cj}$ have the sense of two critical wave numbers of the total reflection for that or for the other spin components of the neutron wave. Hence, at $p_0 \geq p_{cj\pm}$ refracted into the layer j the neutron waves are split revealing spin-dependent birefringence effects [68,69]. If $p_0 \leq p_{cj+}$, or $p_0 \leq p_{cj-}$, then the corresponding spin component of the neutron wave becomes evanescent exponentially decaying into the depth of the layer j.

Eigenvalues $t_{j\pm}$ and $r_{j\pm}$ of transmission matrices $\widehat{t}_j$ and reflection matrices $\widehat{r}_j$, respectively, determine amplitudes of spin components of transmitted and reflected neutron waves. These matrices, as well as all other matrices: $\widehat{V}_j$, $\widehat{p}_{cj}$, $\widehat{p}_j$, and $\widehat{S}_j(z)$ are simultaneously diagonal in the representation with the quantization axis directed along induction vector $\mathbf{B}_j$ in the layer j. If these vectors are collinear over the whole set of layers then matching wave fields at interfaces between neighboring layers, one can formulate the conventional Parratt [70] recursion algorithm for calculations of amplitudes with positive or negative spin projections onto the unique quantization axis.

In general, nondiagonal elements of reflection and transmission amplitude matrices are nonzero and mix up neutron spin states. Then conditions for continuity of each spin component of the wave function and its first derivative can be formulated in the form of the following pair of matrix equations:

$$\widehat{t}_{j+1} + \widehat{r}_{j+1} = e^{i\widehat{\varphi}_j}\widehat{t}_j + e^{-i\widehat{\varphi}_j}\widehat{r}_j, \tag{65}$$

$$\widehat{p}_{j+1}\left\{\widehat{t}_{j+1} - \widehat{r}_{j+1}\right\} = \widehat{p}_j\left\{e^{i\widehat{\varphi}_j}\widehat{t}_j - e^{-i\widehat{\varphi}_j}\widehat{r}_j\right\}, \tag{66}$$

where $\widehat{\varphi}_j = \widehat{p}_j d_j$ is the matrix of the phase shift gained over the layer thickness d_j.

This system actually consists of eight equations connecting eight unknown matrix elements: four elements of the reflection and four elements of the transmission amplitude matrices, which needs to be determined.

In principle, closed form solutions for the reflection, $\widehat{r}$, and transmission, $\widehat{t}$, amplitude matrices can be found when applying to Eqns (65–66) a customarily used matrix formalism [71] generalized [68] for the case of the spin-1/2 particles via combining two pairs of matrix equations in Eqns (65) and (66) into a single supermatrix (SM) equation [54]. Then multiplication of supermatrices determined for each layer results in a global supermatrix, linking together $\widehat{r}$ and $\widehat{t}_{N+1}$ matrices. This allows to express the elements of $\widehat{r}$ and $\widehat{t}_{N+1}$ matrices via the elements of the global supermatrix for the whole multilayer stack. The SM approach may, however, cause some numerical problems for a large number and/or large thicknesses of the layers, as is indicated in Ref. [55] where instead of SM formalism a generalization of the Parratt formalism [70] was proposed.

A recurrent solution of Eqns (65) and (66) can readily be accomplished, in general form, in two steps. First, one can introduce the auxiliary matrices $\widehat{R}_j = (\widehat{r}_j \widehat{t}_j^{-1})$, which in accordance with Eqns (65) and (66) obey the recursion matrix equation,

$$\begin{aligned}\widehat{R}_j = e^{i\widehat{\varphi}_j}&\left\{\left(1 - \widehat{p}_j^{-1}\widehat{p}_{j+1}\right) + \left(1 + \widehat{p}_j^{-1}\widehat{p}_{j+1}\right)\widehat{R}_{j+1}\right\} \\ \times &\left\{\left(1 + \widehat{p}_j^{-1}\widehat{p}_{j+1}\right) + \left(1 - \widehat{p}_j^{-1}\widehat{p}_{j+1}\right)\widehat{R}_{j+1}\right\}^{-1} e^{i\widehat{\varphi}_j}.\end{aligned} \tag{67}$$

This equation explicitly expresses the matrix $\widehat{R}_j$ via the matrix $\widehat{R}_{j+1}$ in the next layer. The recursion solution starting from $\widehat{R}_{N+1} = 0$ provides all intermediate matrices $\widehat{R}_j$ till the final reflection amplitude matrix $\widehat{r}$.

In additional one can calculate, if necessary, the transmittance matrices via the recursive solution of the following set of equations

$$\widehat{t}_{j+1} = \left\{\widehat{1} + \widehat{R}_{j+1}\right\}^{-1}\left\{\widehat{1} + e^{-i\widehat{\varphi}_j}\widehat{R}_j e^{-i\widehat{\varphi}_j}\right\}e^{i\widehat{\varphi}_j}\widehat{t}_j. \tag{68}$$

In this case the recursion begins with $\widehat{t}_0 = \widehat{1}$ and ends with $\widehat{t}_{N+1} = \widehat{t}_s$. Finally, all reflection $\widehat{r}_j = \widehat{R}_j\widehat{t}_j$ and all transmission amplitude matrices can be determined for each layer hence determining the neutron wave field inside the multilayer.

In analogy to the SM approach, this method may be called super recursion (SR) because it uses recursion routine for the solution of coupled matrix equations. The SR formalism allows to compute spin components of the reflection amplitude matrix for arbitrary orientations between magnetization directions in subsequent layers.

After the reflection amplitude matrix $\widehat{r}$ is computed, it can be parametrized in accordance with Eqns (54–56), where the vector **m** determines the coordinate system in which this matrix is diagonal. Finally, the parameters r_0 and $\mathbf{r}_0$ expressed via eigenvalues $r_\pm$ should be substituted into Eqns (28–32) instead of f_0 and **f**. This would yield the final equation determining the reflectivity $R = R(\mathbf{P}_i, \mathbf{P}_f)$, or reflectivity for arbitrary distribution of magnetization vector directions over a sequence of layers and at arbitrary orientation between vectors $\mathbf{P}_i$ and $\mathbf{P}_f$ in the case of 3D polarization analysis [66].

The result of the procedure for 1D polarization analysis is illustrated in Figure 5 where NSF, $R^{\pm\pm}$ and SF, $R^{\pm\mp}$ are computed for a [Fe(8 nm)/Cr(1 nm)] × 12 multilayer in different magnetic states. 12 is the number of repeats of the Fe/Cr bilayer, where Fe is assumed to be in a homogeneous ferromagnetic (FM) state, whereas Cr is assumed to be nonmagnetic. In saturation (Figure 5(a)), when the magnetization in all Fe layers is aligned parallel to the external field, no SF is seen, while NSF reflectivities are split revealing two critical edges of total reflection. R^{++} manifests strong Bragg reflection corresponding to the periodic stack of Fe/Cr bilayers. At the same time no Bragg reflection is seen in R^{--} curve as for the negative spin component of the scattering potential there is almost no contrast between iron and chromium: $\rho_{Fe}^N - \rho_{Fe}^M \approx \rho_{Cr}^N$. Figure 5(b) refers to the antiferromagnetic (AFM) state when magnetization directions alter in neighboring layers at low external fields, remaining collinear with the polarization axis. In this case the two NSF reflectivities almost perfectly merge into one curve, showing doubling of the unit cell via magnetic superstructure Bragg peaks, also known as half-order peaks. These half-order peaks are suppressed in NSF reflection if the magnetization vectors are turned perpendicular to the guiding field, known as the spin-flop phase and shown in Figure 5(c). Simultaneously, a set of only half-order superstructure Bragg reflections appears in the SF channel so that a combination of both NSF and SF channels allows to detect a doubling of the

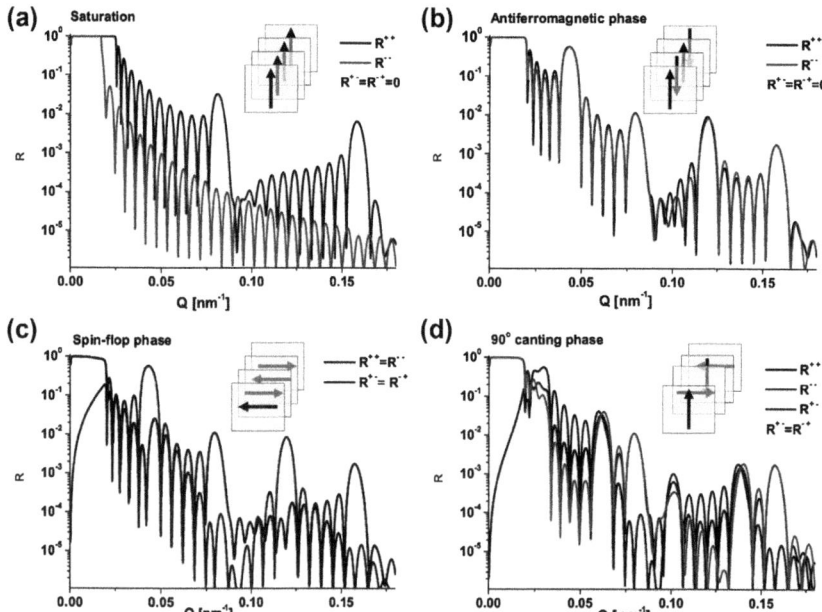

FIGURE 5 Non-spin-flip and spin-flip reflectivities simulated for [Fe(8 nm)/Cr(1 nm)] × 12 in four different magnetic states: (a) in saturation when the angle $\gamma = 0$ for all layers; (b) in antiferromagnetic configuration when $\gamma = 0$ and $\gamma = 180°$ for neighboring layers; (c) spin-flop phase when in neighboring layers $\gamma = +90°$ and $\gamma = -90°$; (d) for the 90° canted state when magnetization in neighboring layers is subsequently tilted by 90° hence making angles $\gamma = 0$, $\gamma = 90°$, $\gamma = 180°$, and $\gamma = 270°$ against Y-axis.

magnetic unit cell. In fact, all these three examples do not necessarily require a super-iterative routine as long as the magnetization directions in neighboring layers are collinear. However, it is mandatory in a noncollinear arrangement, as it take place, for instance, in the case of biquadratic interlayer coupling. Then, for the case of a spiral phase with 90° rotation of the magnetization direction from one layer to the next as shown in the inset of Figure 5(d), a set of new Bragg reflections emerge in both NSF and SF channels manifesting the fact that now the magnetic unit cell contains already eight sublayers. Interestingly, the strong AFM peaks seen as NSF scattering in Figure 5(b) plot, or as SF scattering in Figure 5(c) are totally canceled out in the plot in Figure 5(d).

6.2.5 Off-Specular Neutron Scattering

Equations derived above for specular reflectivity are essentially based on Eqn (41), which not only ignores finite sample size in lateral directions, but also assumes that it has ideal surfaces and interfaces so that scattering potential in the SE depends only on one coordinate z in the direction normal to the sample face. In reality this is often not the case due to a number of reasons. One of them is simply an atomic structure which may cause in-plane grazing incidence Bragg diffraction [68,69,72] and diffuse scattering into large azimuthal angles θ_y in Figure 2, but into small angles α_f. The other example is unavoidable surface and interfacial structural and magnetic roughness characterized by mean square deviations of a real surface from its average position. Lateral dimensions of roughness determined by the lateral correlations between these deviations are scaled with the so-called height–height correlation length [73]. Usually this length is quite large and therefore roughness correlations contribute into the range of small azimuthal angles θ_y. The latter are often not resolved experimentally and off-specular scattering (OSS) is observed as diffuse scattering at $\alpha_f \neq \alpha_i$, i.e., via the lateral wave vector transfer projection Q_x. Similar OSS effect caused by magnetic domains also lead to a violation of the translational invariance in the lateral direction and having comparable dimensions. Last but not least, one should mention various laterally patterned films into stripes or islands as well as nanocomposite materials. In the latter case, as thoroughly discussed below, off-specular grazing incidence SANS (GISANS) can be resolved in both lateral directions, i.e., over Q_x and Q_y.

Despite of the variety of subjects, scattering at grazing incidence can be treated within a common approach called the distorted-wave Born approximation (DWBA) if ever specular reflection, or transmission, is sufficiently strong and well distinguishable from OSS. This means that the width and the shape of the specular peak in Q_x, Q_y correspond to that expected for a laterally flat target of the same dimensions, although this peak is superimposed onto a broader scattering component with lower intensity attributed to OSS. Then this scattering can be treated within the framework of the perturbation theory formally decomposing the interaction potential matrix $\widehat{V}(\mathbf{r})$ in SE (11) into two parts:

$$\widehat{V}(\mathbf{r}) = \sum_j \left\{ \widehat{V}_{j0}(z) + \Delta \widehat{V}_j(\mathbf{r}) \right\}. \tag{69}$$

Here $\widehat{V}_0(z) = \langle \widehat{V}_j(\mathbf{r}) \rangle_{\text{lateral}}$ is the mean potential matrix averaged over lateral coordinates and providing specular reflection. Deviations $\Delta \widehat{V}(\mathbf{r}) = \widehat{V}(\mathbf{r}) - \widehat{V}_0(z)$ from this mean potential cause OSS. The DWBA scattering amplitude matrix for OSS can be calculated employing the same Eqn (19) in which $\widehat{V}(\mathbf{r})$ is substituted by $\Delta \widehat{V}(\mathbf{r})$ and exact solutions in the form of Eqn (62) for 1D potential matrix $\widehat{V}_0(z)$ are used instead of the plane waves in Eqns (15) and (20).

Substitution in the matrix element in Eqn (19) of corresponding plane waves by those distorted in the mean potential yields the equation,

$$\widehat{f}_n\left(\mathbf{Q}_\|; p_{0i}, p_{0f}\right) = -\frac{m_n}{2\pi\hbar^2} \sum_j \int d\mathbf{r} e^{-i\mathbf{Q}_\| \cdot \mathbf{r}_\|} \widehat{S}_j^{f+}(z) \Delta \widehat{V}_j(\mathbf{r}_\|, z) \widehat{S}^i(z), \tag{70}$$

for the DWBA scattering matrix amplitude. Now it depends not on the 3D wave vector transfer $\mathbf{Q}$, but on its 2D lateral projections $\mathbf{Q}_\| = \boldsymbol{\kappa}_f - \boldsymbol{\kappa}_i$ and normal to the surface projections $p_{i0} = k \sin \alpha_i$ and $p_{f0} = k \sin \alpha_f$ of the incoming, $\mathbf{k}_i$, and outgoing, $\mathbf{k}_f$, wave vectors. In Eqn (70) S-matrices are determined in accordance with the same (63) in which matrices $\widehat{p}_j$ are substituted by $\widehat{p}_j^i = \sqrt{p_{i0}^2 - \widehat{p}_{cj}^2}$ and, respectively, $\widehat{p}_j^f = \sqrt{p_{f0}^2 - \widehat{p}_{cj}^2}$.

Further calculation of scattering cross sections in DWBA is quite straightforward and results in the set of Eqns (28−32) in which the scalar, f_0, and the vector, $\mathbf{f}$, parameters in Eqn (18) are substituted by the scalar and vector components of the matrix in Eqn (70). Examples illustrating DWBA applications to various systems will be given in the following sections. These examples show that DWBA correctly describes most of the prominent features arising from the fact that neutron waves incident onto a sample and scattered at shallow glancing angles are heavily distorted in the mean optical potential, so that OSS experiences not only waves refracted through, but also those reflected from interfaces. These waves may experience constructive and destructive interference providing patterns rich in details of the scattered intensity distribution over the 2D plane of incident and scattered wave vectors.

Concluding the theoretical section, it is worth to note that above consideration is limited to the analysis of matrix elements between ideally monochromatic incoming and outgoing plane waves propagating with a unique wave vector $\mathbf{k}$ and with the wave front extended over a distance much larger than the size of the target. In reality, the initial neutron beam produced by a finite size thermal source contains an incoherent sum of divergent neutron waves erupted from different points at different moments of time. The beam is further subjected to monochromatization and collimation, which select neutrons with desired velocities and directions as described in Section 6.3.1. Such selection is accomplished with a finite accuracy so that the incident wave vector $\mathbf{k}_i$ has an uncertainty $\Delta \mathbf{k}_i$. The latter is quite different along the vector $\mathbf{k}$ and within the plane normal to its mean direction. Similarly, the outgoing wave vector $\mathbf{k}_f$ is also determined with a finite accuracy $\Delta \mathbf{k}_f$. As a result, the wave vector transfer $\mathbf{Q}$ is known within the uncertainty range $\Delta \mathbf{Q}$ related with the resolution ellipsoid. There the integration in definitions of the scattering amplitude in Eqns (19) and (22) is effectively restricted by either the range of the interaction potential where $\widehat{V}(\mathbf{r}) \neq 0$, or by parameters $l_x \sim \Delta Q_x^{-1}$, $l_y \sim \Delta Q_y^{-1}$, and $l_z \sim \Delta Q_z^{-1}$ referred to lengths of principle axes of the coherence ellipsoid, or coherence lengths (see, Refs [74,75] for details).

Usually, at grazing incidence and scattering the in-plane projection of this ellipsoid is much smaller than the sample area, and measured scattering cross section actually is just an incoherent sum of cross sections calculated above for individual ellipsoids. For laterally homogeneous and flat samples, this summation is trivial and providing measured cross section of specular reflection with a factor equal to the total sample area illuminated by the beam. However, in case of laterally inhomogeneous samples a particular line shape of the scattered intensity distribution depends on the ratio between the size of inhomogeneities determined by the condition $\Delta \widehat{V}_j(\mathbf{r}_\|, z) \neq 0$ in Eqn (70) and coherence lengths l_x and l_z. If this ratio is small, then a number of inhomogeneities comprised within each of ellipsoids is large and the mean value $\langle \Delta \widehat{V}_j(\mathbf{r}_\|, z) \rangle_{l_x, l_y} \approx 0$. However, the bilinear combination, $\langle \Delta \widehat{V}_j(\mathbf{r}_\|, z) \Delta \widehat{V}_j(\mathbf{r}'_\|, z') \rangle_{l_x, l_y} \neq 0$, determining OSS cross section, which is present in addition to specular reflection from the mean scattering potential $\widehat{V}_0$. If, alternatively, no OSS can be seen as will be thoroughly discussed and illustrated in the following sections, then the lateral potential has only contributions to the average potential.

6.3 INSTRUMENTAL CONSIDERATIONS

Neutron reflectometers are comparatively easy to build and fairly economical to run. The main challenge is the delivery of a sufficiently high flux and maintaining a low background even close to the primary beam direction. Therefore, well-thought-out neutron optics and radiation shielding are essential for the success of a neutron reflectometer. Two main types are available: angle dispersive—fixed neutron wavelength instruments and wavelength dispersive—fixed angle instruments. The latter one requires a pulsed neutron beam for time-of-flight (ToF) detection. This can be achieved either by using a pulsed neutron source or in case of a continuous beam source by using a chopper system. For use of polarized neutrons additional polarizing optics and spin flippers have to be installed. Some aspects of polarized neutron reflectometers are discussed in the following.

6.3.1 Design of Neutron Reflectometers

Figure 6 shows a classical instrument design for PNR using a monochromatic incident beam from a cold neutron source and an angle dispersive diffractometer. The schematics are taken from the NGD reflectometer at the NIST

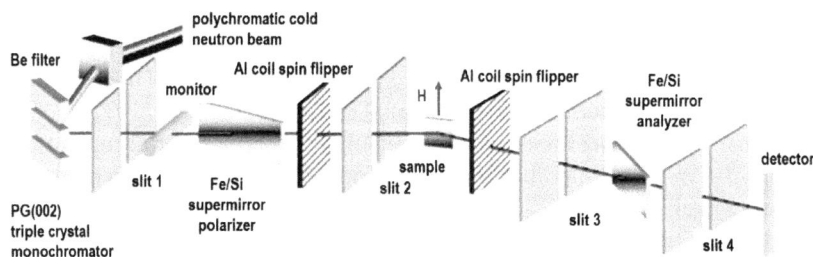

FIGURE 6 Schematic instrument layout of the NG1 reflectometer at the NIST Center for Neutron Research.

NCNR [33]. It follows the principles of polarized neutron scattering for triple-axis spectrometers, first described by Moon, Riste, and Koehler in 1969 [76]. The main difference is the much more collimated beam that has to be used when scattering is performed close to the incident beam and the use of supermirrors instead of Cu_2MnGe Heusler crystals for polarizing and analyzing the neutron spin state. In the layout of Figure 6 (side view) the polychromatic beam in the neutron guide first passes a Be filter to cut out short wavelengths. Then the beam strikes a monochromator, here made of highly oriented pyrolytic graphite (HOPG) that selects a wavelength band $\Delta\lambda/\lambda$ of typically 1.0−1.5% from the polychromatic beam. The HOPG lamella is tilt as to focus in the Y-direction the neutron beam onto the sample. Variable slits 1−4 define and confine the neutron beam path, which is also monitored after the first slit. The neutron beam then hits a fully magnetized and saturated Fe/Si supermirror, which reflects one spin direction and transmits the other. The magnetization of the supermirror points in the direction perpendicular to the scattering plane, which is defined as the Y-direction. Thus neutrons that have passed the supermirror in transmission mode have a polarization parallel to the Y-direction. After the supermirror follows a crossed coil, also known as Mezei spin flipper [77], which rotates the neutron spin adiabatically to a defined orientation when activated. It is customary to flip the neutron spin from up (parallel to the Y-direction) by 180° to down. In the schematics of Figure 6 the sample is magnetized in an external H-field parallel to the Y-direction. The exit beam after scattering from the magnetized sample passes again a spin-flipping coil and a Fe/Si supermirror with the same magnetization direction as the one on the front side. The supermirror on the backside serves for the analysis of the neutron spin after scattering from the sample. Once the neutrons passed the analyzer supermirror, they are detected in a ^{3}He pencil detector. Not seen in the schematics but always present is a neutron guiding field of a few oersted parallel to the Y-direction applied from the first supermirror up to the second. Also not shown in the schematics are evacuated flight tubes along the beam path in order to reduce any scattering from air. Assuming that both spin flippers are not activated (off), then the reflectivity R^{++} is measured. If both flippers are activated, the reflectivity R^{--} is measured. Those are the NSF channels of neutron detection. If the front flipper is activated but the back flipper is off, the SF reflectivity R^{-+} is detected, vice versa R^{+-}.

A number of polarized neutron reflectometers are designed according to the scheme shown in Figure 6, such as the VG-6 instrument at the Helmholtz Zentrum Berlin [78], the ADAM reflectometer at the ILL [79], NREX and MIRA at the FRMII in Munich [80], and SUIREN at JRR-3 [81]. Simplified versions of a polarized neutron reflectometer may dispense the back flipper and supermirror analyzer. Then a distinction between NSF and SF reflectivity is no longer possible. However, for depth profiling of samples which are magnetically homogeneous, the nSLD and mSLD can still be determined from measuring the reflectivities R^+ and R^- according to Eqn (60). The neutron

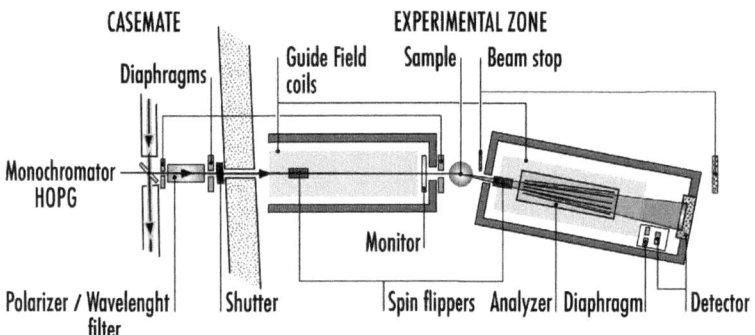

FIGURE 7 Schematic instrument layout of the reflectometer Super ADAM at the Institute Laue-Langevin, Grenoble. HOPG, highly oriented pyrolytic graphite. *Reprinted with permission from Ref. [83].*

reflectometer at Dhruva, India, works according to this scheme [82], which is sufficient for many purposes.

For the upgrade and redesign of the reflectometer ADAM to Super ADAM, many optical components were replaced. A schematic outline of the new Super ADAM instrument is shown in Figure 7 (top view) and described in Ref. [83]. The Be filter was removed and instead a sequence of supermirrors or Bragg mirrors are combined into an optical wavelength filter consisting either of polarizing or nonpolarizing mirrors with a double bounce that cuts off higher harmonics and reduces strongly background radiation. The cross coil Mezei-type spin flippers [77] are replaced by contact free radio frequency Drabkin flippers [84], and the single blade supermirror analyzer is replaced by a wide angle solid state supermirror analyzer array that allows spin analysis not only of the specularly reflected beam but also of off-specular diffuse intensity [85,86]. Alternatively, this supermirror array can be replaced by a ^{3}He spin filter [87], which also allows spin analysis over a large solid angle. Finally the ^{3}He pencil detector is replaced by a 300 × 300 mm^2 area detector, allowing to record specular as well as OSS intensity over a wide angular range. All these components are surrounded by guide field coils and shielded against background radiation via B_4C containing mats covering up boxes before and after the sample table (indicated by dark gray lines).

A different design is required for ToF NR. ToF instruments require a polychromatic pulsed beam structure. This can be achieved either by inserting a chopper into the primary beam of the neutron guide or by using the intrinsic time structure of a pulsed neutron source. A typical outline of a ToF instrument at a steady state source is shown from the Platypus reflectometer at the OPAL reactor in Figure 8 [88,89]. The main difference to monochromatic instruments is a set of choppers in the incident beam, which define the pulse structure. The subsequent neutron optics is similar to the one at angle dispersive instruments. The main advantage of ToF instruments is the adjustable wavelength

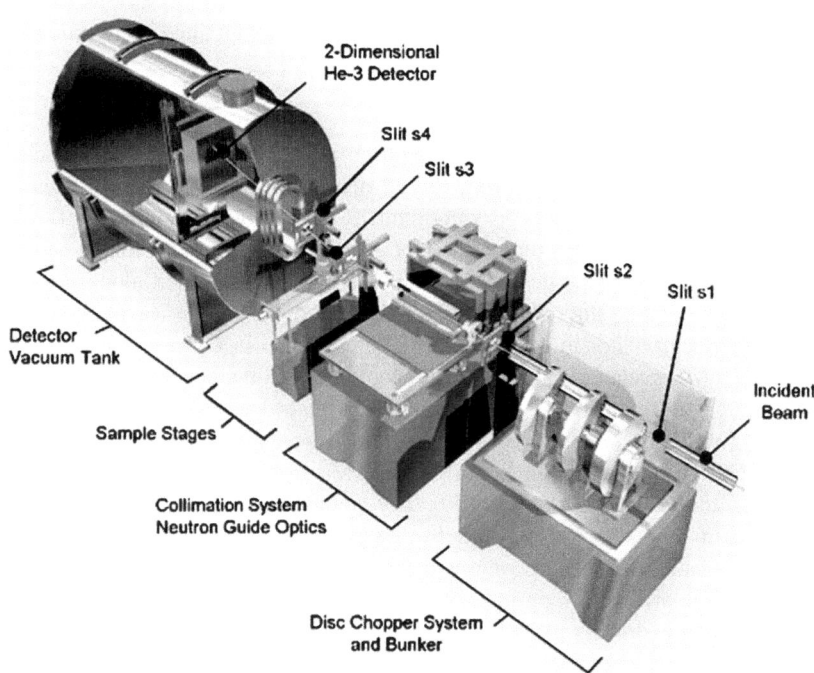

FIGURE 8 Schematic instrument overview of the time-of-flight instrument Platypus at the OPAL research reactor, Sydney. *Reprinted with permission from Ref. [88].*

resolution via setting chopper speeds and angles and the fast and simultaneous collection of data for a certain Q-range corresponding to the wavelength band that is transmitted through the choppers at a fixed incident angle. In most cases this Q-range is not sufficient to cover the area of interest, and therefore the incident angle has to be reset a few times to complete a scan. In case of monochromatic instruments all optical components, including polarizer, analyzer, and spin flipper, are optimized for one specific operating wavelength. For ToF instruments the optimization has to be achieved for a wide wavelength band, which is definitely a challenge. For more information on ToF techniques, we refer to the chapter of Arai in Chapter 3 of the first volume on Neutron Scattering [14] and the chapter of Major, Farago, and Mezei in this volume. A new concept is introduced for the design of a reflectometer to be build at the European Spallation Source. The reflectometer ESTIA will have a horizontal scattering plane and it is optimized for a small sample area by using focusing neutron guides. The focusing optics of ESTIA is a novel concept among neutron instruments, allowing users to control both the footprint of the beam on the sample and the divergence of the beam. In addition, polarizers and spin flippers will be available for polarized measurements. A list of all polarized and unpolarized neutron reflectometers existing presently at various neutron sources around the world is posted on the web page maintained by Rennie [16].

6.3.2 Neutron Optics

The discussion of neutron optics for neutron reflectometers starts with neutron guides that deliver neutrons from a cold neutron source into a neutron guide hall. A cold neutron source is important as neutron wavelengths $\lambda \geq 0.3$ nm work best for reflectivity studies. The location of the instrument along a neutron guide is decisive for its design. If the instrument is positioned alongside of a neutron guide, then the only option is to deflect out a narrow wavelength band via a monochromator in the neutron guide as shown in Figures 6 and 7. With this narrow wavelength band the only option is to build an angle-dispersive reflectometer. However, if the instrument is placed at the end of a neutron guide, further options are open. The wide wavelength distribution available combined with a divergent beam exit allow a range of different instrument designs: monochromatic, prism [90], ToF, velocity selector-based instruments, multibeam monochromators, and further options for using a divergent and polychromatic beam. End positions definitely offer more flexibility and usually also more floor space to optimize such an instrument.

We cannot discuss all possible design options, but retreat to more conventional optics used for angle-dispersive instruments. Usually HOPG is used as monochromator material, providing wavelengths in the range of 0.3−0.5 nm, depending on the takeoff angle. Sometimes one would like to use longer wavelengths. This requires monochromators with larger lattice parameters. Monochromators on the basis of graphite intercalation compounds have recently been developed for this purpose. In particular the stage 1 compound KC_8 [91,92], featuring a lattice parameter of 0.53 nm, turns out to be very useful. Larger lattice parameters could in principle be achieved with higher stage compounds, but stage mixing hampers their applicability.

The resolution of angle dispersive reflectometers is given by:

$$\frac{\Delta Q}{Q} = \sqrt{\left(\frac{\Delta \theta}{\theta}\right)^2 + \left(\frac{\Delta \lambda}{\lambda}\right)^2} \tag{71}$$

For many routine experiments it is a sufficient to aim for a 5% scattering vector resolution $\Delta Q/Q$. The angular resolution of an instrument $\Delta \theta/\theta$, depending on the accuracy the diffractometer and the beam divergence defined by the collimating slits, is usually better than 1−2%. Therefore one could relax the wavelength resolution $\Delta \lambda/\lambda$ to about 5%. However, HOPG monochromators deliver a wavelength resolution of about 1% on the expense of intensity. The use of lower grade graphite with a larger mosaicity does not remedy the problem as the reflectivity dramatically decreases. Also other concepts such as the use of double or triple stacks of HOPG showed limited success. Only with Fermi velocity selectors a variable wavelength resolution up to $\Delta\lambda/\lambda \approx 10\%$ can be reached (see further below). Another problem of HOPG monochromators is a severe higher harmonic wavelength contamination. Therefore an LN_2-cooled polycrystalline Be filter before or after the monochromator is mandatory, providing a short wavelength cutoff. As the Be filter produces Debye–Scherrer like powder diffraction cones, for safety reasons it must be kept inside the radiation shield. Alternatively an optical filter composed of deflecting mirrors can also remove the higher harmonics. If the mirror is at the same time magnetic, beam polarization and higher harmonic removal can be combined. This concept is followed at the Super ADAM instrument (ILL) and at the SUIREN reflectometer (JRR-3).

For polarizers usually supermirrors (SMi) of graded Fe/Si, Fe/Ge, or other types of multilayers are used. Grading the multilayer periodicity parallel to the surface normal provides a stretched plateau for the total reflection for one spin state, whereas the other spin state is transmitted [93,94]. Supermirrors are characterized by a multiplicity factor m, indicating how much Q_c^{SMi} of the supermirror is larger than Q_c of a Ni film: $Q_c^{SMi} = m \times Q_c^{Ni}$. Supermirrors with m values up to 4 are commercially available. Supermirrors can be used as polarizers and analyzers in transmission or in reflection mode. In transmission mode the instrument is easier to align being straight from monochromator to sample and from sample to the detector. On the other hand, transmission causes small-angle scattering, which is to be avoided. Furthermore, if the polarizer is also used as higher harmonic wavelength filter, then the polarizer has to be used in reflection mode as the higher harmonics will be transmitted. Similarly, if not a single analyzer is used but a fan of blades for analyzing the spin state of the scattered beam, then one has to revert to the reflection mode.

Supermirrors are also characterized with respect to their ability to polarize a neutron beam, i.e., to select one spin state in favor over the other. The polarization is defined:

$$P = \frac{I^+ - I^-}{I^+ + I^-}, \quad (72)$$

where I^+ and I^- are the intensities reflected from the supermirror with spin up and spin down, respectively. This equation can be recast into:

$$P = \frac{F - 1}{F + 1}, \quad (73)$$

where $F = I^+/I^-$ is called the flipping ratio. As an example, let us assume that we want to select the "+" state. However, as the polarizer is not perfect, with some probability the "−" polarized neutrons can pass. If out of 100 neutrons 90 of them have the correct polarization and 10 the wrong one, the flipping ratio is $F = 9$ and the polarization is 80%. Nowadays, supermirrors provide neutron beam polarization routinely of more than 95%, some reach even 99%. There are two problems which need to be kept in mind when working with supermirrors.

First, the soft magnetic coating has to be placed in a strong and homogeneous magnetic field to warrant a saturated single domain state. Second, SANS from the multilayer should be kept as low as possible. In reflection mode this means that the interfaces in the multilayer should be very smooth, and in transmission mode the absorption in the supporting substrate should be kept small.

Most neutron reflectometers with polarization analysis offer ^{3}He transmission spin filters as one of the options for spin analysis. When the ^{3}He gas in a flask is highly polarized, neutrons with spin polarization parallel to the polarization of ^{3}He have a very low absorption cross section of 5 barn, neutrons with opposite polarization have a very high absorption cross section of 5333 barn [95]. The flasks containing ^{3}He gas is usually polarized in an outside facility and transported within a "magic box" to the instrument [96]. The advantage of such filters is a wide acceptance angle and very low background. Of disadvantage is the fact that high transmissivity is sacrificed for high analyzing efficiency. Furthermore, the polarization and analyzing power of the filter decays with time, and the operation of the filter is not maintenance free [87,97]. An alternative to spin filters for wide angle spin analysis are solid state neutron spin analyzer, which reach efficiencies of more than 98% at a transmission of 65% for neutrons of 0.5 nm wavelength [86].

In the past Mezei spin flippers were used because of their ease of operation. However, the transmission of neutrons through polycrystalline alumina wires causes some absorption but more importantly some small-angle scattering. Therefore modern designs use contact free RF flippers, which are easy to construct [41,84,89]. Most of the feature discussed so far have been implemented at the Super ADAM instrument, the optical components are schematically reproduced in Figure 9.

As alluded to already, if a reflectometer is situated at the end of a neutron guide, many more options in the design are open. This has been utilized in case of the MARIA reflectometer at the FRMII in Munich [98]. Instead of using a monochromater to filter out a narrow wavelength band, a velocity selector is installed that provides a continuous beam with variable wavelength and bandwidth. Alternatively the velocity selector can be replaced by a Fermi

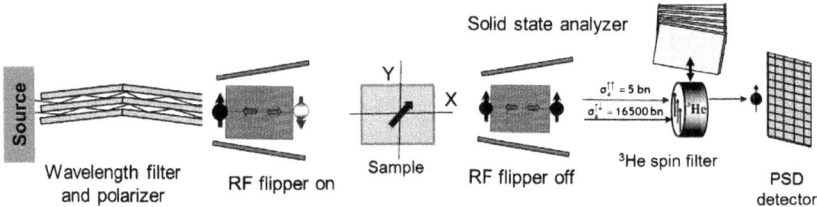

FIGURE 9 Optical components of the Super ADAM reflectometer at the ILL. The Be filter and the polarizer are combined in one wavelength filter and polarizer, the usual Mezei spin flipper are replaced by RF flippers, and the neutron spin analysis can be either performed by a solid state fan analyzer or by a ^{3}He spin filter. PSD, position sensitive detector.

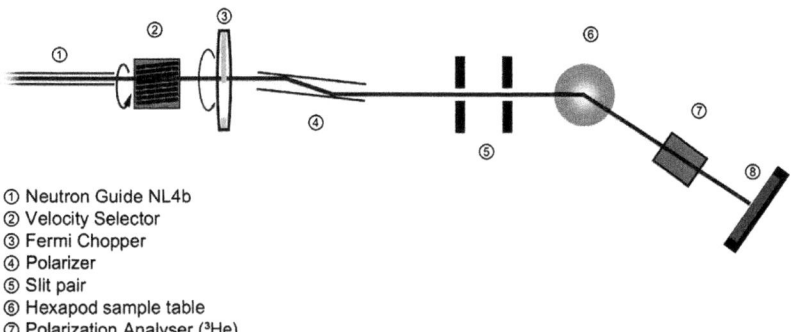

① Neutron Guide NL4b
② Velocity Selector
③ Fermi Chopper
④ Polarizer
⑤ Slit pair
⑥ Hexapod sample table
⑦ Polarization Analyser (^{3}He)
⑧ Detector

FIGURE 10 Schematic layout of the MARIA reflectometer at the FRMII reactor, which is situated at the end of a neutron guide. The reflectometer features two experimental options: a velocity selector for monochromatic angle-dispersive experiments and a Fermi chopper for time-of-flight experiments. The neutrons are polarized by a polarizing guide. Spin flipping is performed by RF spin flippers, which are omitted in the layout.

chopper to generate a pulsed beam of an adjustable wavelength band, which is then combined with ToF detection. The main components of the MARIA instrument are shown in Figure 10.

6.3.3 Detection and Acquisition of Data

For the detection of neutrons there are only a few options available. Single well-shielded tubes filled with ^{3}He gas, so-called pencil detectors, are often used for specular reflectivity measurements. For detecting off-specular diffuse intensity, it is more time effective to use an area detector or position sensitive detector (PSD). Position sensitivity is usually provided by an $x-y$ multiwire proportional counter. The use of a PSD goes on the account of background protection. Therefore a comprise has to be found between well-shielding and simultaneous detection over a certain solid angle. The resolution of the PSD is a function of the wire separation versus distance of the detector from the sample.

6.3.4 Modeling and Fitting of Data

There are a number of publicly available fitting packages for polarized neutron reflectivity. The package developed and used at the NIST center for neutron research [42] is the program *Refl1D*. Supporting programs and online calculators are also offered [99]. Users of Super ADAM usually employ an in-house developed software package for online automatic fitting of specular PNR and modeling of OSS. The routine is based on the super-iterative algorithm [43,55,56,100], allowing to model and evaluate PNR data from noncollinear magnetic structures. The least square routine is applied to the theoretical curves which are convoluted with the instrumental resolution function and take into account polarization efficiencies. Superfit is a program for LINUX computers for simulating and fitting of specular and diffuse neutron reflectivity data obtained with polarized or nonpolarized neutrons on magnetic or nonmagnetic samples. This program is based on the combination of the supermatrix method and the DWBA [54,68].

It is recommended to take first XRR data from the sample in question and to fit the chemical structure as well as possible. The fit parameters, including layer thickness, interface roughness, and mass density, can then be used as input parameters for the nuclear profile. Often it is sufficient to fix those according to the XRR results and to fit only the magnetic profile. Concerning fitting of PNR data, users usually follow different philosophies. One may either correct the data first with respect to background, illumination factor, and polarization efficiency and then fit the corrected data, or, which is the preferable procedure, not to apply any corrections or manipulations of the raw data. Instead to include in the fitting routine all correction factors convoluted with the resolution function, and only then fit the data with a model reflectivity curve.

6.4 CASE STUDIES: STATIC EXPERIMENTS

6.4.1 Thin Films

From the discussion so far it is clear that there are a number of important issues of magnetic thin films that can be addressed successfully and uniquely by polarized neutron reflectivity. Those include:

1. vector magnetization depth profiling;
2. magnetic domain state as a function of magnetic field;
3. collinear and noncollinear magnetization vectors in multilayers;
4. interfacial magnetism in heterostructures;
5. interdiffusion at interfaces.

In the following subsections, we will discuss these issues in more detail.

6.4.1.1 The Saturated State

We start with a most straightforward case, the magnetically saturated state of thin films, which nevertheless is instructive and has model character for all later discussions. Furthermore the saturated state is the reference state of any sample, which needs to be analyzed in all details with respect to structural and magnetic parameters, before any other magnetic state of the sample should be investigated. Thus, we consider first an idealized single magnetic film (Fe) deposited on a silicon substrate with an ideally smooth surface and interface to the substrate. We also assume a homogeneous magnetization distribution within the film plane and perpendicular to it, i.e., a magnetic state referred to as a magnetically saturated single domain state. Saturation can be achieved either by applying a field larger than the saturation field or assuring that the film is oriented along the easy magnetization axis exhibiting 100% remanence after taking off the magnetic field (aside from a small neutron guide field). Now we orient the magnetization of the film such that it is in the film plane and parallel to the neutron spin quantization axis, i.e., parallel to the polarization axis of the incident neutron beam, which in our coordinate system is the Y-axis. PNR on such a system yields two distinct reflectivities for R^{++} and R^{--}. As there is no spin-flip scattering under the assumptions made, it would be sufficient to consider R^+ and R^- reflectivities only without applying a spin analysis. These two reflectivities can be measured either by changing the incident neutron spin polarization or by taking measurements at positive and negative saturation of the film magnetization. These two reflectivities plotted in Figure 11(b) are distinguished by different critical scattering vectors for total reflection Q_{c+} and Q_{c-}, which shifts the thickness oscillations, also named Kiessig fringes, against each other, while the periodicity of the oscillations $\Delta Q = 2\pi/d$ is reciprocal to the film thickness d at sufficiently high Q

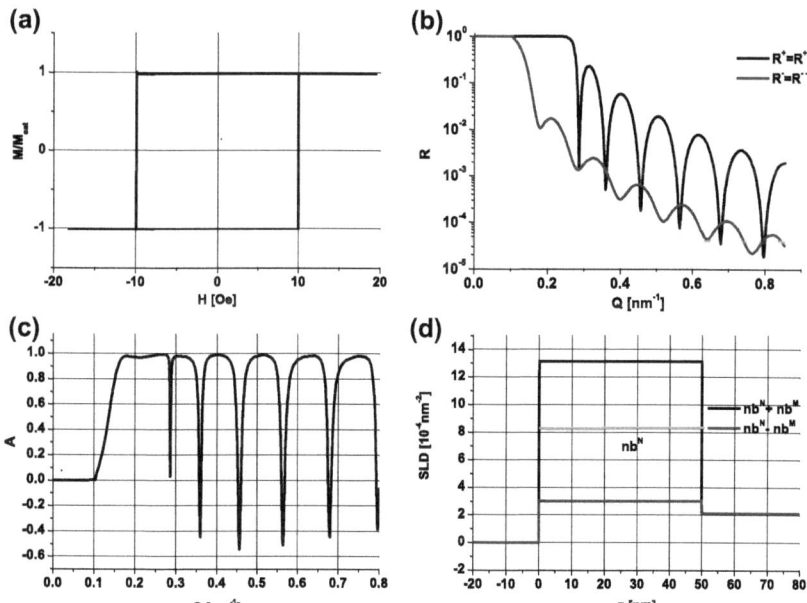

FIGURE 11 Model 50-nm-thick Fe film on a silicon substrate with perfectly smooth interfaces. (a) Assumed square form of the magnetic hysteresis. The magnetization vector is aligned parallel to the neutron polarization axis; (b) calculated neutron reflectivity for neutron polarization parallel to the magnetization (R^+) and antiparallel to it (R^-); (c) asymmetry calculated from the reflectivities plotted in panel (b); (d) magnetic and nuclear scattering length densities for the Fe film on Si substrate.

values. Due to the difference in amplitudes r_+ and r_- in Eqn (60), the asymmetry

$$A = \frac{R^+ - R^-}{R^+ + R^-} \qquad (74)$$

shown in Figure 11(c) features pronounced oscillations, which are characteristic for a ferromagnetic single domain state of the sample. Another way of plotting the reflectivity, which is often used in the literature to make faint effects at high Q values better visible, is the normalization of the measured reflectivity by the unpolarized Fresnel reflectivity $R^F = (4\pi)^2/Q^4$ in BA. From fits to the measured reflectivities the nSLD and mSLD are obtained according to procedures described and discussed in Section 6.2, for instance, via a modified Parratt formalism or super-iterative algorithm [55], discussed in Subsection 6.2.4. For such an artificial sample the SLD profile can simply be written as: $n\rho(z) = nb^N(z) \pm nb^M(z)$, where n is the number density of nuclei and of magnetic ions in the sample. In a more general case these densities do not need to be the same, but for now we take them to be so. Then nb^N is the nSLD and nb^M is the mSLD. The z dependence considers any possible gradient in the sample perpendicular to the film plane. This equation is also often written as $\rho(z) = \rho_n(z) \pm CM_y(z)$, where $C = 2.853 \times 10^{-3}\,\mathrm{nm}^{-2}\,\mathrm{T}^{-1}$ and $M(z)$ (in T) is the magnetization in the film plane parallel to the Y-axis [41]. The density profiles for model calculations are reproduced in Figure 11(e). They are, as expected, split into the SLD for neutrons polarized parallel to the magnetization ($nb^N + nb^M$) and for neutrons polarized antiparallel to the magnetization ($nb^N - nb^M$), while in the region of the nonmagnetic substrate the SLD is only determined by the nuclear SLD of the substrate material.

There are numerous PNR studies published in the literature from magnetic thin films, heterostructures, and multilayers in magnetic saturation for the determination of depth-resolved magnetization profiles on an absolute scale. This is in fact one of the most frequent and fruitful applications of PNR to nanomagnetism. Magnetization measurements with a VSM or with a SQUID do not provide depth-resolved information and therefore hybridization or interdiffusion effects at interfaces can lead to wrong conclusions. On the other hand, if R^+ and R^- reflectivity measurements do not show any asymmetry over the entire Q-range scanned by applying a saturation magnetic field parallel to the Y-axis, we can safely conclude that the sample is nonmagnetic. Vice versa the sensitivity of PNR to any residual magnetization is very high. This is particularly useful for studying any emergent magnetization at interfaces between two materials due to doping or proximity effects. Therefore, even weak magnetic signals can be detected and localized with PNR, as we will see later. In the following, we discuss a few illustrative examples from recent publications on magnetization depth profiling of thin films, heterostructures, and multilayers in saturation.

PNR measurements and the resulting magnetization profile of a 100-nm-thick Fe film in saturation was recently published by Devishvili et al. in order to demonstrate the capabilities of the Super ADAM reflectometer installed at the Institut Laue-Langevin [83]. The data together with a fit are shown in Figure 12. After fitting a footprint correction was applied. One can easily recognize two critical scattering vectors for the up (parallel) and down (antiparallel) polarized neutrons with respect to the sample magnetization and the plateau region for $Q < Q_{c\pm}$. For $Q > Q_{c\pm}$ the splitting of the reflectivities R^{++} and R^{--} and the thickness oscillations confirm the ferromagnetic state of the sample in a confined thickness of 100 nm. In addition spin-flip scattering can be recognized, which in this case is only due to the finite efficiencies of the spin polarizer and analyzer. Otherwise a completely saturated ferromagnetic film parallel to the Y-axis should not exhibit any spin-flip scattering. At a scattering vector of about 0.1 $Å^{-1}$ the intensity goes into the background. Thus with good background shielding, reflectivities over five to six orders of magnitude can be measured. The inset shows the depth profile for the nSLD and mSLD. At the surface there is a 12-nm-thick nonmagnetic protecting Pd layer, in the region of the Fe film the SLDs are split for the up and down state of the neutrons, and the MgO substrate is again nonmagnetic, as expected. The mSLD was determined to be $4.962 \times 10^{-6} Å^{-2}$, corresponding to a magnetization of 2.145 T, which is close to the nominal iron saturation value at room temperature of 2.2 T and consistent with magnetometry measurements.

Upon doping of Fe with nitrogen it is theoretically expected that Fe attains a giant magnetic moment for the stoichiometry of $Fe_{16}N_2$. As the fabrication of such compounds is nontrivial and the magnetization may develop a

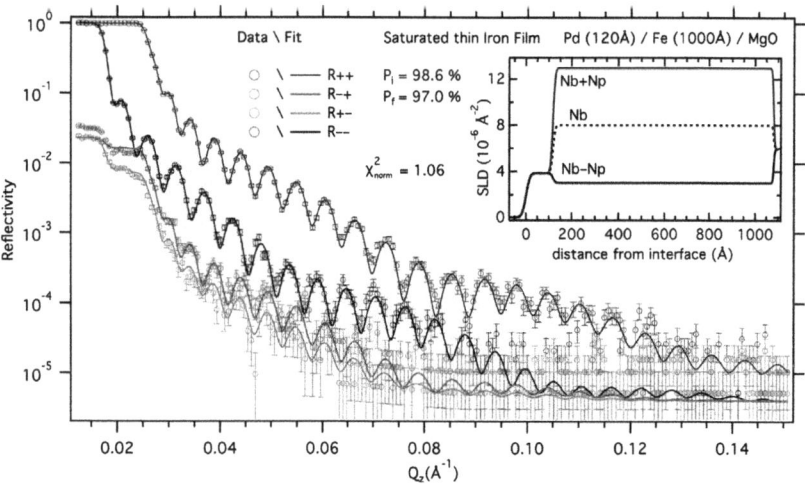

FIGURE 12 Polarized neutron reflectivity from 100-nm-thick Fe film in saturation. The inset shows the nuclear and magnetic scattering length profile. *Reprinted with permission from Ref. [83].*

gradient, it is advantageous to use a magnetization depth profiling method instead of determining the magnetization by VSM or SQUID magnetometer, which average over the total thickness. Indeed by PNR measurements a giant magnetization close to the substrate interface could be discerned reaching 3 T, the rest of the film has lower moment due to off-stoichiometry [101,102]. The data were taken with the Magnetism Reflectometer at the Spallation Neutron Source, Oak Ridge National Laboratory [103] (Figure 13).

PNR is also helpful in disclaiming conjectures about giant magnetic moments in Fe_3O_4 thin films epitaxially grown on MgO substrates and measured by VSM and SQUID [104]. With PNR the authors could show that the presence of Fe impurities embedded in the substrate close to the interface can explain this effect and that the intrinsic magnetic moment of Fe_3O_4 is in accordance with literature values. These results nicely demonstrate how PNR can discriminate between intrinsic and extrinsic effect.

Chemical uniformity versus nonuniformity plays a decisive role in particular in perovskites, where minute changes in the composition can dramatically change local magnetic moments and phase transitions. In thin films it is difficult to maintain a homogeneous magnetization because of substrate and surface effects. At the same time, macroscopic measurement techniques with VSM or SQUID cannot resolve any magnetization inhomogeneities. Using PNR to probe the nuclear and magnetic depth profile in $(La_{1-x}Pr_x)_{1-y}Ca_yMnO_3$ thin films epitaxially grown on (110) $NdGaO_3$ substrates, Singh et al. [105] could demonstrate that the magnetization is highest in the middle of the layer and drops toward the substrate and surface. This magnetization profile goes along with a change of the stoichiometry. All this occurs below the Curie temperature of the film, which is about 100 K.

Similar studies on the location of magnetism in $Fe_{1-x}Rh_x$ compounds using PNR methods were reported in a number of articles [106–108]. FeRh is

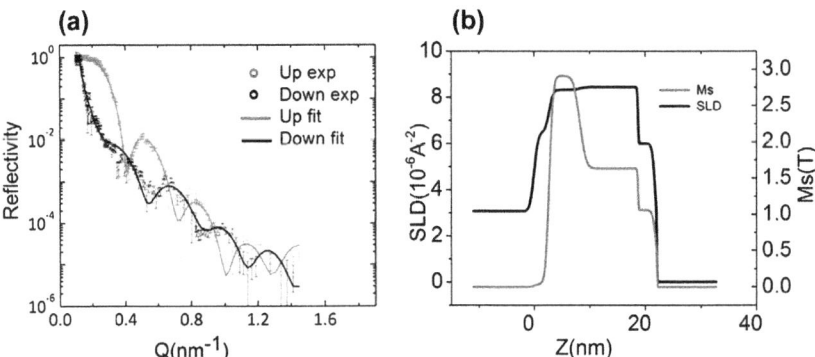

FIGURE 13 Panel (a) shows polarized neutron reflectivity of FeN on GaAs substrate. Panel (b) shows the nuclear and magnetic scattering length density profiles. An enhancement of the magnetization is visible only close to the GaAs substrate, which is to the left. *Reprinted with permission from Ref. [101].*

antiferromagnetic at room temperature and therefore useful as pinning layer in spin valves. However, minute changes of stoichiometry or capping the surface can dramatically change the magnetic state from AFM to FM. PNR methods were used to scrutinize ferromagnetic components at the surface of interface of these compound films.

As a last example, we discuss an experiment where PNR is the only method that can verify or falsify conjectures about interfacial magnetism. It has been debated in the past using various bulk measurement techniques whether there is a magnetic moment at the interface of $LaAlO_3/SrTiO_3$ heterostructures in high magnetic fields and at low temperatures. Fitzsimmons and coworkers have used PNR for pinpointing the existence of interfacial magnetism by plotting the asymmetry (see Eqn (74)) from their R^+ and R^- reflectivity results on $LaAlO_3/SrTiO_3$ superlattices measured in fields up to 11 T and temperatures down to 1.7 K [109]. The asymmetry inferred from data taken at the ASTERIX reflectometer of LANSCE is shown in Figure 14. Various assumptions made for the magnitude of the interface magnetism are not in accordance with the data, as can be seen by the model calculations. With these experiments the authors confirmed that no magnetization exists for any of the measured samples with an upper limit of ≈ 2 G. Not only the sensitivity of the PNR method to weak interfacial magnetism is impressive, it is also the choice of the proper method. As the origin of the magnetism at $LaAlO_3/SrTiO_3$ interfaces, if existing, is not clear and may arise from intrinsic as well as extrinsic effects, XRMS would not yield unambiguous results. PNR is free of any assumptions and magnetization can be detected even if there were no spins present, since PNR is sensitive to the vector of magnetic induction **B**.

In general, we can state that taking PNR data on films, heterostructures, and multilayers in a saturating magnetic field is extremely useful for locating

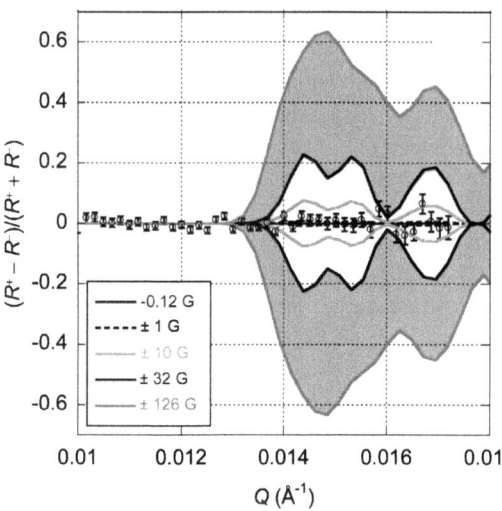

FIGURE 14 Asymmetry evaluated from polarized neutron reflectometry measurements on $LaAlO_3/SrTiO_3$ supelattice for detecting any possible magnetism at the interfaces. *Reprinted with permission from Ref. [109].*

magnetic layers even deep below the surface, for determining magnetization profiles, and for obtaining an accurate magnetization value on an absolute scale. The experiment is straightforward. The only requirement is a well-polarized and collimated incident neutron beam and a low radiation background to reach high Q_z values for extracting reliable depth profiles from the data analysis. Analysis of the off-specular diffuse intensity is not required, as the system by definition is in a single domain state at saturation. These experimental methods can be combined with temperature-dependent studies [105,110], phase transitions [111,112], interfacial magnetism of superconducting/ferromagnetic and multiferroic heterostructures [113—116], interdiffusion at interfaces [117—119], annealing studies [120,121], kinetics of alloy formation [122], in situ growth investigations [123], for detecting ferromagnetic phases in dilute magnetic semiconductors [124,125], hydrogen/deuterium absorption in magnetic multilayers [126], oxygen deficiency in magnetic oxides [127], etc. About 80—90% of all PNR studies are of this kind. Any deviation of the magnetization vector from collinearity with the neutron polarization axis requires additional analysis, which we will discuss in the following sections.

6.4.1.2 Magnetization Rotation

So far we have discussed the saturation state with the mean magnetization of the sample parallel to the polarization axis. Under these conditions spin-flip scattering is not expected and measurements of R^+ and R^- are rather straightforward. As soon as we turn the magnetization vector by an angle γ with respect to the polarization axis Y (see Figure 3) the reflectivity curves $R^+ = R^{++} + R^{+-}$ and $R^- = R^{--} + R^{-+}$ indicated by solid lines in Figure 4 remain the same for $Q < Q_c$, but for $Q > Q_c$ characteristic changes occur that can be best recognized on a linear scale of reflected intensities calculated for the same model as in Figure 11. The top left panel in Figure 15 shows the reflectivity with angle of the magnetization rotation $\gamma = 0$. With increasing γ, both R^+ and R^- reflectivities exhibit two critical edges at the respective scattering vectors $Q_{c\pm}$ and an intensity plateau in between. The plateau is high for R^+ and low for R^-. Note that between Q_{c-} and Q_{c+} oscillations of R^{++} and R^{--} become visible. With increasing angle γ the plateaus approach each other (see top right and bottom left plots in Figure 15 and saturate at half height once γ reaches 90°. As each reflectivity R^+ and R^- expresses both critical edges at positions independent of the tilt angle, from measuring only one of the reflectivities and determining the difference

$$\Delta Q_c^2 = Q_{c+}^2 - Q_{c-}^2 = 32\pi nb^M = 32\pi CM \qquad (75)$$

the total average magnetization M of the film can be derived. This indeed provides an easy estimate of the layer magnetization, which does, however, not contain any depth resolution.

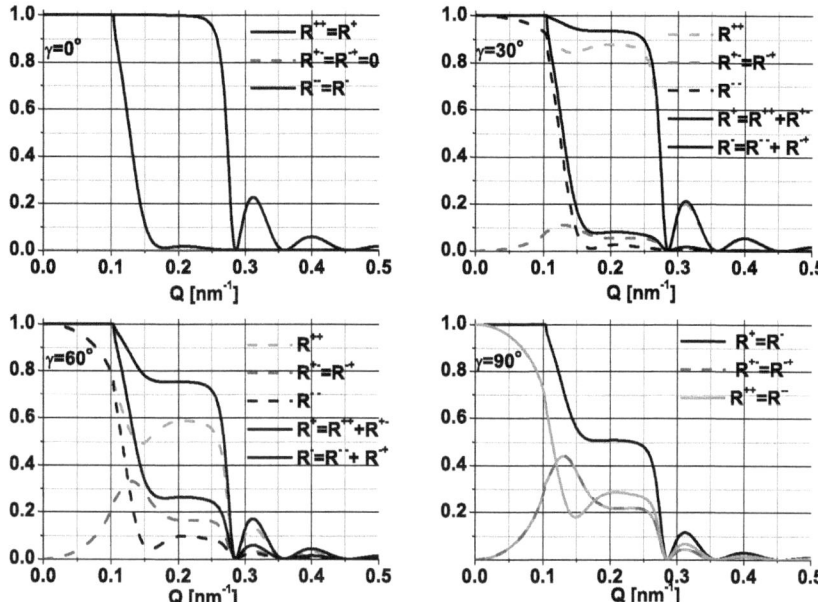

FIGURE 15 Model calculations for non-spin-flip and spin-flip reflectivity curves from 50-nm Fe film on a silicon substrate with perfectly smooth interfaces and magnetization tilted with respect to Y-axis at angles: $\gamma = 0°$; $\gamma = 30°$; $\gamma = 60°$; $\gamma = 90°$.

Further details become visible when measuring all NSF and SF channels. In fact the most dramatic effect that occurs by tilting the magnetization is the SF scattering. The SF intensity immediately signifies a magnetization vector with a projection M_x onto the X-axis. From the SF reflectivity, also shown in Figure 15, it is possible to determine the modulus of the angle γ but not the sign, since the SF reflectivity R^{+-} and R^{-+} in Eqn (59) are proportional to $M_x^2 \propto \sin^2 \gamma$, and:

$$|\gamma| = \arcsin \frac{2\sqrt{R^{+-}}}{|r^+ - r^-|} \qquad (76)$$

In case of reflection from large domains with magnetization tilted to the left and to the right with respect to the vector of mean magnetization parallel to Y-axis the mean value $\langle M_x \rangle \propto \langle \sin \gamma \rangle = 0$. However, the mean value of magnetization projection M_x squared and averaged over domains, $\langle M_x^2 \rangle = M^{2,\text{sat}} \langle \sin^2 \gamma \rangle \neq 0$. This mean value can be found if the saturation magnetization $M^{2,\text{sat}}$ is determined from Eqn (75). The average is taken over the sample surface and thickness. In real experiments, one should keep in mind that some SF intensity is always present, even for $M_x = 0$, because of nonideal polarization of the incident neutron beam and finite flipping efficiency. Only after a proper fitting, one can decide whether the observed SF intensity is due to a M_x component or whether it is an artifact due to finite polarization and flipping ratio.

When rotating the magnetization by $\gamma = 90°$, then the NSF intensities merge together, so that $R^{++} = R^{--}$, and the asymmetry A in Eqn (74) vanishes, as already discussed in Section 6.2.3 and in connection with Figure 4. This fact, however, does not mean that the NSF reflectivities are solely due to nuclear scattering. The fact that NSF reflectivities contain information on magnetic interaction is unambiguously demonstrated in the last plot in Figure 15 where both critical edges $Q_{c\pm}$ are clearly manifested in the NSF reflectivities. Moreover, in accordance with Eqn (59) the SF reflectivity depends on the exact reflection amplitudes $r_\pm$ which, in turn, depend nonlinearly on the mSLD. This is demonstrated by the fact that the SF reflectivity curve has a spike at $Q = Q_{c-}$ and an inflection point at $Q = Q_{c+}$ which depend nonlinearly on both nSLD and mSLD.

The SF reflectivity is clearly due to magnetic interaction as it vanishes for nonmagnetic samples. This fact may render an easy experimental possibility to separate completely nuclear and magnetic reflection for $\gamma = 90°$. In spite of the foregoing discussion, this is indeed possible if the BA is valid, i.e., for scattering vectors far away from the total reflection edges.

Saturation of the magnetization in X-direction is not possible as this would change the polarization of the neutron beam from the Y-axis to the X-axis. Therefore the only option is to rotate the sample by 90° and this should only be done if the magnetic hysteresis has a square shape combined with a coercive field that is higher than the neutron guide field.

The orientation of sample magnetization parallel to the X-axis and hence perpendicular to the incident polarization vector has not often been exploited in experiments. One example of a PNR measurement is shown in Figure 16

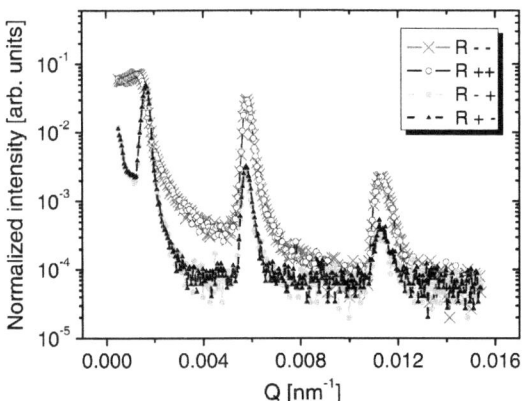

FIGURE 16 Polarized neutron reflectometry from a Heusler multilayer Co_2MnGe/Al_2O_3 with magnetization vector pointing along the X-axis. *Reprinted with permission from Ref. [128]*.

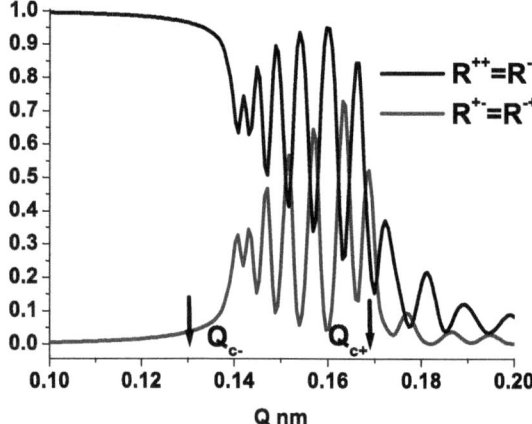

FIGURE 17 Non-spin-flip and spin-flip reflectivities calculated with high resolution for the same multilayer as in Figure 16 with magnetization vector turned by 90° parallel to the X-axis. Note the linear scale of the y-axis.

from a Heusler multilayer: $[Co_2MnGe(3\ nm)/Al_2O_3(9\ nm)]_{50}$ [128]. The NSF reflectivity is not split, while the SF reflectivity is very strong. Note that the first peak in the SF channel should not be mistaken as a multilayer Bragg peak. This peak is situated at $Q_{c-} \leq Q_z \leq Q_{c+} \approx Q_c$ in the SF channel and related to a totally different phenomenon, namely to pseudoprecession [129] of the neutron polarization vector about the magnetic induction in the film. Corresponding high-frequency antiphase oscillations [130] in SF and NSF reflection channels should be seen below Q_{c+}, but above Q_{c-}. In Figure 16, such oscillations are completely smeared out by a relatively crude resolution, but they can readily be seen in model PNR curves calculated for the same multilayer with high resolution and presented in Figure 17.

For another experimental realization and analysis of neutron reflectivity upon rotation of an Fe film, we refer to Ref. [131] and extensive discussions in our previous review [44].

Before closing the discussion on 90° magnetization orientation, it should be pointed out that for specular reflection both SF reflectivities R^{+-} and R^{-+} need to be always identical independent of any complexities with respect to magnetization profiles, collinearity or noncollinearity across the film, presence or absence of magnetic roughness, domains, etc. This is just a consequence of the reciprocity principle and can be violated only in case of the difference between efficiencies (and/or directions) of initial polarization and polarization analysis, i.e., instrumental parameters.

Surface and interface roughness will change the picture drawn so far only slightly. The measured reflectivities will drop faster than predicted by the Fresnel reflectivity and the nSLD and mSLD will be smoother at the surface and interface. In fact, we have to distinguish between structural and magnetic roughness. The structural roughness in PNR experiments has the same effect as for XRR measurement and can be described by the same effective Debye—Waller or Névot—Groce factor [132]. For any meaningful experiment the interfacial roughness should not exceed 4—5 nm, and the mean square roughness should always be smaller than the film thickness. Magnetic roughness is more difficult to treat. Structural and magnetic roughness, as well as magnetic domains, causes off-specular diffuse scattering to be discussed next.

6.4.1.3 Magnetic Domain State

Up to now we have assumed that the magnetization vector has a unique orientation over the total surface without fluctuations about its mean value. The world of nanomagnetism is more complex and PNR provides the proper answer. From the asymmetry defined in Eqn (74), we determine the Y-component of the magnetization. In case of a domain state this is a measure of the average projection $\langle \cos \gamma \rangle$. From a measurement of the SF reflectivities, we have the squared average of the X-component of the magnetization: $\langle \sin \gamma \rangle^2$. From these two measurements, we find the dispersion [133] defined as $\Delta = \langle \cos^2 \gamma \rangle - \langle \cos \gamma \rangle^2$, which is a fingerprint for magnetic fluctuations and domain formation. The brackets indicate a configurational average over the total illuminated sample. For a homogeneous magnetization independent of the tilt angle γ, the dispersion Δ is always zero. However, if there are some fluctuations about the mean magnetization axis, then $\langle \cos \gamma \rangle^2$ is smaller than $\langle \cos^2 \gamma \rangle$ and Δ becomes larger than zero. It reaches a maximum of 1 for the completely demagnetized state, which is achieved by $\langle \cos^2 \gamma \rangle = 1$, while $\langle \cos \gamma \rangle^2 = 0$. In contrast, a single domain state always has a $\Delta = 0$, irrespective of the rotation angle γ of the in-plane magnetization, because coherent rotation implies $\langle \cos^2 \gamma \rangle = \langle \cos \gamma \rangle^2$. Thus for characterizing the magnetic domain state, it is essential to determine Δ. Some exemplary cases are shown schematically in Figure 18.

From these examples it is seen that the dispersion Δ is a very useful quantity, although it does not always unambiguously discriminate between a coherent rotation of the magnetization vector and domains in a particular experimental configuration. For instance, 1D PNR cannot distinguish between a single domain, whose magnetization vector is homogeneously rotated by an angle γ, and two large domains with the magnetization tilted by an angle $+\gamma$ in one domain and $-\gamma$ in the other domain. These two cases can, however, be reliably discriminated via additional PNR measurements [133] in which the whole sample is turned at certain angles with respect to the polarization axis.

Moreover, with specular PNR preliminary information on the presence of magnetic domains and their lateral size can be gained by inspecting the positions of the critical edges for total reflection Q_{c+} and Q_{c-}. As was discussed above and illustrated in Figures 4, 16, and 17, these critical edges are independent of the tilt angle γ. Assuming that the magnetization in individual

Single domain state:

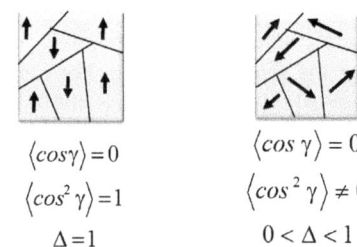

$\langle cos^2 \gamma \rangle = \langle cos\gamma \rangle^2 = 1$ $\langle cos^2 \gamma \rangle = \langle cos\gamma \rangle^2 = 0$
$\Delta = 0$ $\Delta = 0$

FIGURE 18 Examples for different domain states of ferromagnetic samples and the respective value for the dispersion defined as $\Delta = \langle \cos^2 \gamma \rangle - \langle \cos \gamma \rangle^2$.

Multi domain state:

$\langle cos\gamma \rangle = 0$ $\langle cos\,\gamma \rangle = 0$
$\langle cos^2 \gamma \rangle = 1$ $\langle cos^2 \gamma \rangle \neq 0$
$\Delta = 1$ $0 < \Delta < 1$

domains is close to its saturation value, one should expect that $Q_{c\pm}$ are the same as in a single domain state. This is, however, not always the case and depends on the domain size. In fact, the consideration above implicitly assumed that each neutron probes only one of the domains, while the statistical average over all neutrons in the beam impinging onto the sample surface is substituted by the configurational average over all domains. This, in turn, implies that the domain size is sufficiently large as to be illuminated by the neutron *coherently*. Then the reflectivity of the neutron from such a domain can be described neglecting its finite lateral domain size, i.e., with reflectivity R_i from ith domain, while the reflectivity R of the whole sample is given by the weighted sum:

$$R = \sum_i w_i R(\gamma_i), \qquad (77)$$

where $w_i = w(\gamma_i)$ is the surface fraction of domains with magnetization axis tilted at the angle γ_i with respect to the polarization axis.

A reason for such incoherent averaging is that, unlike laser beams, neutron sources are neither monochromatic nor have the neutrons, impinging on a sample, any fixed phase relationship. After being reflected from a monochromator the neutrons have a Gaussian distribution with a mean wavelength λ and a full width at half maximum $\Delta\lambda$, which depends on the mosaicity of the monochromator and the divergence of the beam defined by slits in the scattering plane (x-direction). The incident beam is further collimated after the monochromator, and the wave vector $\mathbf{k}_i$ is then determined with an accuracy $\Delta\mathbf{k}_i$ generally depending on both factors: the monochromatization degree $\Delta\lambda/\lambda$ and the beam divergency $\Delta\alpha_i$. Similarly, the accuracy in the wave vector $\mathbf{k}_f$ of elastic scattering is composed by the same parameter $\Delta\lambda/\lambda$ and the accuracy of the scattering angle determination $\Delta\alpha_f$. As a result, the length of the wave vector transfer, or its projection onto normal to the surface in Figures 2 and 3, is determined with the accuracy given in Eqn (71). If the collimation is perfect and the surface is flat, then the scattering vector is perfectly orthogonal to the surface while the accuracy in its determination is given as $\Delta Q_z = 2\pi \sin\theta \Delta\lambda/\lambda^2$, where $\theta = \alpha_i + \alpha_f = 2\alpha_i$. Correspondingly, by Heisenberg's uncertainty principle $\Delta z \Delta Q_z = \pi$, we find for $\Delta z = \lambda^2/(2\sin\theta\Delta\lambda)$. We want to call Δz the *longitudinal coherence length* l_z^{coh} as it is defined along the scattering vector $\mathbf{Q}_z$. One can consider this coherence length of the neutrons as the length of a wave train over which constructive or destructive interference is possible. It is manifested, e.g., in interference fringes on the reflectivity curve. For neutron scattering with a resolution $\Delta\lambda/\lambda \sim 1\%$, and an angle of incidence of 10 mrad, the coherence length $l_z^{coh} \sim 2000$ nm. In practice, however, such longitudinal coherence length is hardly achievable as the collimation is rarely better than $0.2 \sim 0.3$ mrad. Then the collimation and angular resolution of detection are dominating over the contribution of the monochromatization, yielding coherence lengths on the order of 50–300 nm, depending on the instrumental settings.

There is also a transverse coherence length l_y^{coh} of the neutron beam, defined by the opening of the slit in the direction normal to the reflection plane, which is the Y-direction. Because of intensity reasons, the slit is usually wide open and often the beam is focused onto the sample. This condition translates into a rather small transverse coherence length of not more than a few nanometers for most of the angle-dispersive instruments. In contrast, the *lateral* component Q_x within the reflection plane is, in accordance with Eqns (37) and (38), much smaller than Q_z and Q_y even with the same monochromatization and collimation. Therefore, the corresponding coherence length $l_x^{coh} \gg l_z^{coh}$, l_y^{coh} is much longer and at incident angles of a few milliradians may readily cover the range up to submillimeters! Clearly, this coherence length shrinks with increasing incident angle. The longitudinal and two lateral coherence lengths define a *coherence volume*, which has an elliptical shape with the longest axis in the X-direction and the shortest axis in the Y-direction, also shown schematically in Figure 3.

The importance of the coherence volume is as follows. Scattering amplitudes from objects that lie in the XY sample plane within the coherence volume of the neutron beam add up coherently, i.e., the scattered intensity from all domains in the ith ellipsoid is a square of the sum of the amplitudes:

$$\left(\frac{d\sigma}{d\Omega}\right)_i^{coh} = \left(\sum_j f_j\right)^2 \quad (78)$$

where f_j is the scattering amplitude for jth domain. Vice versa, scattering amplitudes from objects which are separated by a distance larger than the coherence volume add up incoherently, i.e., the total intensity is the sum of all scattering cross sections indexed by the subscript i:

$$\frac{d\sigma}{d\Omega} = \sum_i \left(\frac{d\sigma}{d\Omega}\right)_i^{coh}, \qquad (79)$$

similar to Eqn (77).

The footprint of the incident neutron beam at small incident angles usually completely covers up the sample surface. Within illuminated surface of the sample, we imagine small overlapping patches of coherence volumes which take a statistical average over all structural and magnetic configurations in the sample plane. If magnetic domains are smaller than the coherence volume, the scattering amplitudes within the coherence ellipsoid add up either constructively or destructively. In case of constructive interference the intensity distribution from these small domains is centered at the specular condition $\alpha_i = \alpha_f$ and spreads out into the off-specular regime Q_x on either side of the specular ridge, where some dephasing in the sum of Eqn (78) starts to play a role. The smaller the domains, the more important is dephasing, hence leading to further reduction of the intensity and increasing broadening of the diffuse peak.

Vice versa, if the domains are much larger than the coherence volume, the intensity has solely a sharp Gaussian maximum at the specular ridge with a width that corresponds to the instrumental resolution. Coherence volumes in relation to domain sizes and the resulting intensity distribution are shown schematically in Figure 19. In the lower panel the specular peak is superimposed with a broader diffuse scattering component unresolved from

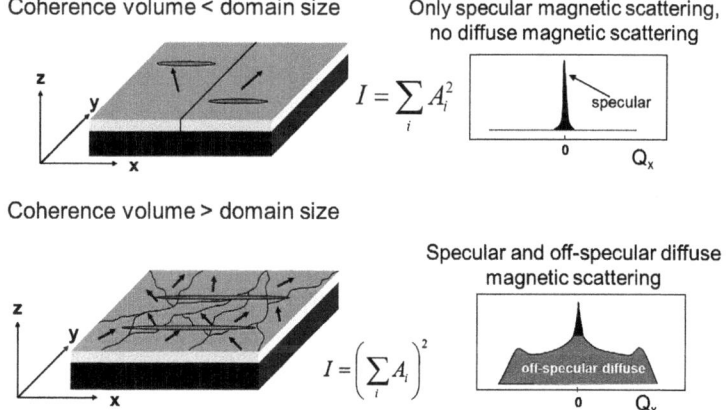

FIGURE 19 Coherence volume of the neutron beam in relation to magnetic domain sizes. In the top panel the coherence volume is assumed to be smaller than the average domain size, resulting in specular intensity, which is sharply peaked at $Q_x = 0$. The bottom panel shows the case when the coherence volume covers several domains causing off-specular diffuse scattering.

specular peak at $Q_x \approx 0$ and spreads out on either side with small maxima at the end, called "Yoneda" wings, which are due to the fact that either the incident or the exit beam fulfills the condition of total reflection [73,134].

For further discussions of the coherence volume and the resulting intensity modeled within DWBA, we refer to Section 6.2.5. Here we show some simulation results of how the domain size affects the intensity distribution for a fixed longitudinal coherence length l_x^{coh}, fixed to 50 µm. First we assume a magnetic domain state with a mean domain size of 200 µm in the film plane, i.e., much larger than l_x^{coh}, with fluctuating magnetization directions tilted by angles $\pm \gamma = 45°$ with respect to the polarization axis. The corresponding intensity maps for NSF and SF scattering is shown in Figure 20(a), where the scattered intensity is plotted as a function of the normal projections of the incident and scattered wave vectors p_i and p_f, respectively. One can recognize ridges of specular reflection in all NSF and SF maps running along the main diagonal $p_i = p_f$ corresponding to the condition of specular reflection, but no diffuse scattering. The intensity along the ridges is modulated by the film thickness oscillations (Kiessig fringes). In the NSF maps, different total reflectivity edges for R^{++} and R^{--} can be spotted.

Now we assume that the domain size is 10 µm and hence much smaller than $l_x^{coh} = 50$ µm. We also assume that magnetization of small domains deviate by angles $\Delta \gamma$ from the mean magnetization, while the latter is turned by 45° with respect to the polarization axis. In the simulated maps reproduced in Figure 20(b) we notice a dramatic change of the intensity distribution. All four maps exhibit off-specular diffuse intensity, which is most pronounced at small angles. In the NSF maps two diffuse intensity island occur on either side of the specular ridge, which are due to the Yoneda enhancement of intensity. In the SF maps the off-specular diffuse scattering is characteristically asymmetric. While the specular SF reflectivities are identical $R^{+-} = R^{-+}$, the off-specular SF intensities are not. Yoneda enhancements are recognizable again on either side. At higher angles the diffuse intensity fades away. For one this is due to the fact that the specular and off-specular intensities drop off rapidly with the inverse scattering vector to the forth power Q_z^{-4}. For another reason this is also due to the shrinking coherence length of the neutron beam. With increasing scattering angle the effective coherence length may become smaller than the domain size, which implies a crossover from coherent addition of amplitudes to incoherent addition of intensities.

For the sake of completeness, we briefly discuss the case when the mean magnetization is parallel to the Y-axis (polarization axis) so that $\langle \cos \gamma \rangle^2 = \langle \cos^2 \gamma \rangle = 0$. Furthermore we assume that domain fluctuations $\Delta \gamma$ with average size smaller than the coherence volume occur symmetrically about the Y-axis. As shown in Figure 20(c) the simulated NSF maps contain only specular intensity but no off-specular diffuse intensity. Contrary, the SF maps contain no specular intensity but only off-specular diffuse intensity. The diffuse intensity in the SF maps I^{-+} and I^{-+} exhibit a maximum asymmetry.

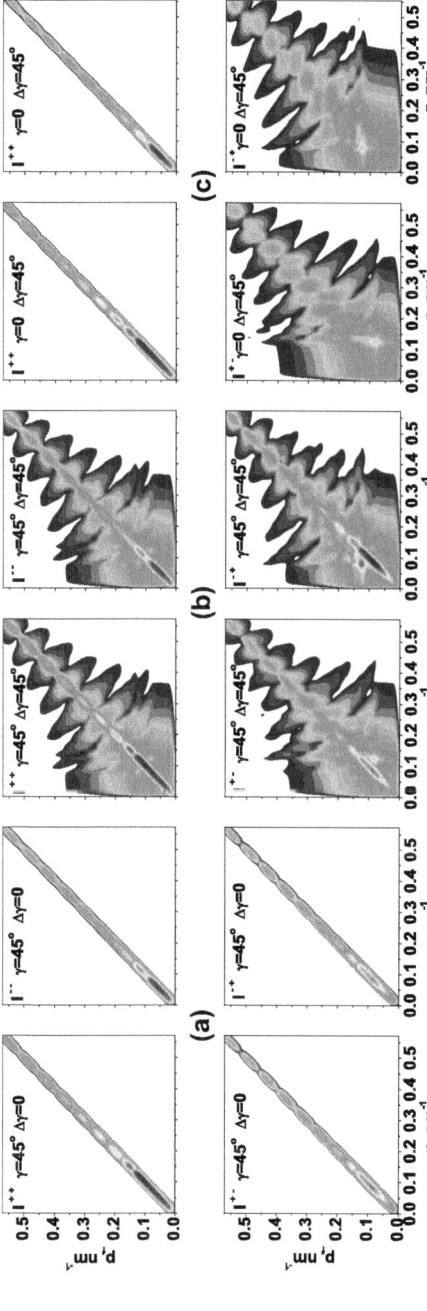

FIGURE 20 Simulated intensity maps for non-spin-flip and spin-flip scattering of a magnetic domain state. (a) Simulated maps under the assumption that the average magnetization makes an angle of $\langle \gamma \rangle = 45°$ with respect to the polarization axis Y, and for the case that the average magnetic domain size is much larger than the coherence length of the neutrons. (b) Simulated maps for the case that the average magnetic domain size is much smaller than the coherence length of the neutrons. (c) Same as in (b) but for an average tilt angle of $\langle \gamma \rangle = 0°$ with respect to the polarization axis, implying that fluctuations $\Delta \gamma$ are also symmetric about the Y-axis.

In this special case specular and diffuse scattering are completely separated in different maps.

Although most instruments for PNR measurements are equipped with PSD detectors, it is quite rare that off-specular diffuse intensity maps are analyzed. Clearly, simulation of the diffuse intensity within the DWBA is nontrivial. But even visual inspection of the diffuse maps can already tell much about the domain state of the sample. In particular the domain size can be estimated knowing the coherence length of the neutrons, and the mean magnetization direction can be gleaned. The maps shown here are intended as a field guide for the interpretation of off-specular diffuse intensity. With this information in mind, we now return to our discussion of the magnetization reversal, which often is marked by domain formation.

6.4.1.4 Magnetization Reversal

We continue with our simple in-plane magnetized film and discuss the magnetization reversal from one saturation state to the other. Magnetization reversal can proceed via nucleation and growth of domains, by coherent rotation of the magnetization vector from one saturation state to the opposite, or by domain formation and growth. These three cases are schematically sketched in Figure 21. There are a number of methods to characterize the magnetization reversal either by magnetometry, by MOKE, or by real space imaging of magnetic domains. Similar to vector MOKE, PNR methods allow to determine the orientation and the modulus of the magnetization, which is required to distinguish the three reversal cases sketched in Figure 21. Additional information that PNR provides in comparison to MOKE is the depth sensitivity. This is important for thicker films of 20 nm and more beyond the penetration depth of laser light into metal films, and whenever the magnetization reversal is not homogeneous across film thicknesses or multilayer periods.

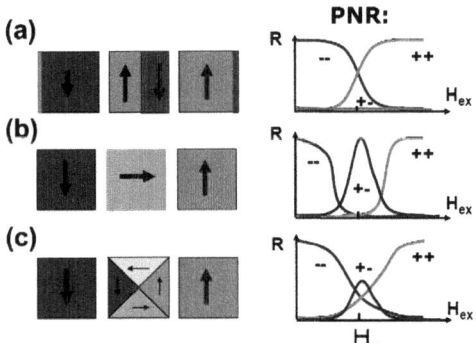

FIGURE 21 Three possibilities for the magnetization reversal are sketched: (a) nucleation and domain wall propagation; (b) coherent magnetization rotation; (c) domain formation. In the right-hand panels the respective specular non-spin-flip (++,−−) and spin-flip reflectivities (+−,−+) are shown for magnetic fields at the coercivity H_c of ferromagnetic films. PNR, polarized neutron reflectometry.

The first case shown in Figure 21(a) is governed by domain formation and growth, while the magnetization vectors always stay parallel to the external field and parallel to the neutron polarization axis. Therefore the intensity of the NSF reflectivities will change and cross over at the coercive field H_c of the magnetization reversal, but there will be no SF intensity as indicated in the right-hand panel. The second case of coherent magnetization rotation (Figure 21(b)) has already been discussed in Section 6.4.1.2. As soon as the magnetization vector makes an angle γ with respect to the polarization axis, there is a magnetization component M_x which causes SF scattering. This SF scattering has a maximum for the angle $\gamma = 90°$ and diminishes again upon reaching opposite saturation, as sketched in the corresponding right-hand panel.

The third case is a mixture of the first two cases. However, the discussion of the scattering of this situation is rather delicate, as we have to consider the coherence volume of the incident neutron beam and compare it with the magnetic domain size. At coercivity the sum of the magnetization vectors yields zero. Macroscopically the sample is completely demagnetized by domain formation. If the domains are larger than the coherence volume of the neutron beam, each domain will contribute to the respective reflectivity. The horizontal domains will both contribute to specular SF intensities $R^{+-} = R^{-+}$, and the vertical domains will contribute to specular NSF reflectivities $R^{++} \ne R^{--}$. Therefore, at the coercive field, specular intensity of all four reflectivities will be detected, as indicated in the respective panel on the right-hand side.

For domains which are smaller than the coherence volume of the incident neutron beam, the reflectivities from all domains will destructively interfere, resulting in equal NSF reflectivities $R^{++} = R^{--}$ and $\Delta Q_c^2 = Q_{c+}^2 - Q_{c-}^2 = 0$, which are due to the nuclear potential, and zero specular SF reflectivities $R^{+-} = R^{-+} = 0$. However, this domain state will produce off-specular diffuse scattering, as already discussed in the previous Section 6.4.1.3. The smaller the domain sizes, the larger will be the width of the off-specular diffuse scattering and the lower its magnitude. Thus, we need to record and analyze off-specular polarized neutron scattering in order to reach conclusions about the average size of domains in the domain state of the sample.

Specular reflectivity and OSS indeed provide a wealth of information on the magnetic state of thin films and their behavior in a magnetic field upon magnetic field reversal, which would not be available otherwise. PNR can distinguish between three main types of reversal processes and in addition yields information on the domain size in the domain state at coercivity. The static case discussed here can even be supplemented by dynamic magnetization reversal investigations to be discussed later in Section 6.5.

Investigations of the magnetization reversal process are particularly important for exchange-biased (EB) magnetic heterostructures, composed of a pinning antiferromagnetic (AFM) layer and a ferromagnetic (FM) layer, which share a common interface. The EB effect is manifested after saturating the FM layer in an external magnetic field while cooling the system through the

Néel temperature of the AFM. The fingerprint for the existence of an EB effect is a shift of the magnetic hysteresis in the direction opposite to the cooling field. The strength of the EB effect is determined by an offset of the center of the magnetic hysteresis from zero field, which is referred to as the EB field H_{EB}. Simultaneously in many cases a broadening of the magnetic hysteresis is observed. Usually the EB effect is not observed just below T_N but below a blocking temperature $T_B < T_N$ that signifies a blocking of DW movement and/ or a sufficient spin stiffness in the AFM. As the EB shift is usually smaller than theoretically expected, particular attention has to be paid to the spins at the interface. Another important aspect is the role of the magnetocrystalline anisotropies in the FM and AFM layers and their mutual interaction. In fact, in order for the EB effect to occur, the anisotropy energy of the AFM needs to be larger than the anisotropy energy of the FM layer and also larger than the FM-AFM exchange energy at the interface. For a general background and discussion of the EB effect, we refer to pertinent review articles [135−139].

Two illustrative examples from recent experiments on EB systems are discussed in a bit more detail. The "fruit fly" for studying the EB effect is the heterostructure Co/CoO or variants of it, such as Fe/CoO or Py/CoO. Py stands for the permalloy composition $Fe_{0.20}Ni_{0.80}$. CoO is an antiferromagnet with a Néel temperature of $T_N = 291$ K. In a recent study by Chen et al. [140], PNR was used for discriminating between spins at the interface that reverse in an applied magnetic field and those which do not switch in Py/CoO. The nonswitchable spins appear to exist already above the Néel temperature of the AFM layer and they are seemingly attached to interfacial grains with $CoFe_2O_4$ stoichiometry generated during the growth process. The density distribution of the nonswitchable spins is more extended in polycrystalline Py/CoO bilayers than in epitaxially grown ones in accordance with the interfacial chemical roughness. For quantitatively accounting the amount of nonswitchable spins, the authors have determined the magnetization profile after saturating in a positive field and subsequently in a negative field. The difference of both profiles shows the partial magnetization that did not reverse upon field reversal, which amounts to about 10−15% of the total magnetization. The magnetization profile of the nonswitchable spins is shown for both samples in Figure 22. It is more spread out for the polycrystalline sample than for the epitaxial bilayer.

Magnetization reversal in epitaxial Co/CoO was recently also investigated by Demeter and coworkers using vector MOKE and vector PNR [141]. The bilayer was field cooled along two different orientations: the easy Co[001] axis and the hard Co[011] axis. The results are plotted in Figure 23. Upon first reversal parallel to the easy axis and along the descending branch of the hysteresis, the NSF and SF reflectivities behave as shown in the schematics of Figure 21(a), i.e., at the coercive field the reflectivities R^{++} and R^{--} cross over and there is no discernible SF reflectivity, meaning that the reversal is characterized by domain nucleation and DW motion. This behavior can be rationalized by recognizing that coherent rotation of the magnetization

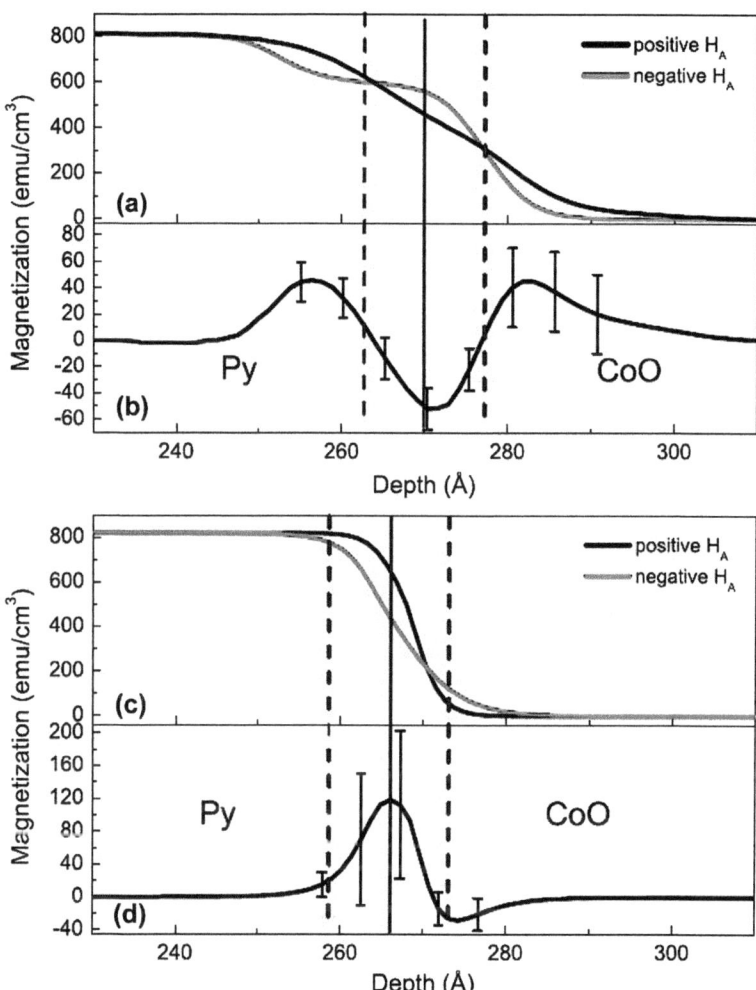

FIGURE 22 Magnetization profile of the nonswitchable spins at the interface between Py and CoO. (a) Magnetization profiles after saturating in a positive and negative saturation field for a polycrystalline Py/Co bilayer; (b) Magnetization difference curve, showing the location of nonswitchable spins at the Py/CoO interface. (c,d) same as for (a,b) but for an epitaxial Py/CoO bilayer. The vertical line indicates the position of the chemical interface, the dashed lines mark the interfacial width. *Reproduced with permission from Ref. [140].*

would imply passing over the hard axis Co[011] direction, which is energetically unfavorable. After several reversals the EB field decreases, which is known as the training effect, and simultaneously the SF intensity increases at the coercive field in the descending as well as in the ascending branch of the hysteresis. This is interpreted by the authors as a clear sign for a change of anisotropy at the interface caused by changes in the AFM domain structure

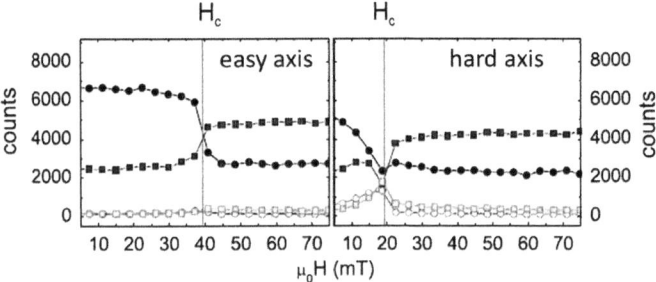

FIGURE 23 Magnetization reversal of Co/CoO bilayer after field cooling to 10 K parallel to the easy axis of the ferromagnetic Co layer. In the left panel the first reversal of the virgin sample is shown and in the right panel is shown the reversal after training. *Adapted with permission from Ref. [141].*

favoring coherent rotation of the magnetization. Overall the intensity of the NSF reflectivities decreases after several reversals, which is likely due to the formation of small magnetic domains that channel intensity from the specular ridge to the off-specular regime. Along the hard axis a different reversal mechanism can be observed, which is a mixture of DW nucleation and motion and of coherent rotation. After training the magnetization reversal processes parallel to the easy axis and the hard axis are rather similar and in good agreement with earlier PNR experiments on textured Co/CoO experiments by Radu et al. [142].

6.4.1.5 Domain Walls

So far we have discussed various domain structures. Domains are separated by DWs, characterized by their type and width. In thin films with in-plane magnetization the preferred DW is of Néel type instead of Bloch type present in bulk materials. The DW width depends on the competition between exchange coupling, tending to increase the DW width, and magnetocrystalline anisotropy, preferring narrow DWs. In Fe thin films the DW width is on the order of 40 nm, while in Co thin films the DW is much more narrow on the order of 10 nm. Usually in the domain state of thin films the magnetic domains are much larger in area than the DWs. In PNR experiments scattering from magnetic domains dominates the intensity and there is little chance to obtain information on DWs. This is different for magnetic films with perpendicular anisotropy. By the selection rules of magnetic neutron scattering, any magnetization component that is perpendicular to the scattering vector is blinded out. Then PNR can focus on in-plane components of DWs and discriminate between Bloch and Néel walls.

Navas et al. have investigated the domain type in CoCrPt thin films with easy axis perpendicular to the film plane [143]. Although MFM maps clearly show the meander type of domain structure typical for perpendicular anisotropy [144,145], from these images the type of DW cannot be discerned.

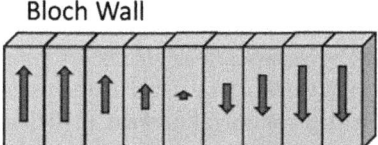 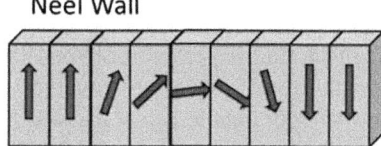

FIGURE 24 Bloch and Néel domain walls are schematically sketched for magnetic films with perpendicular anisotropy. In a slice cutout of the film, spins in Bloch walls rotate out of the front plane, whereas spins in Néel walls remain in the front plane.

Micromagnetic simulations of domains with perpendicular anisotropy and magnetization components of DWs are shown in Figure 25(a). For Néel DWs the magnetization inside the walls continuously rotates from up to down within the same plane, whereas in Bloch DWs the magnetization rotates in the third direction as sketched in Figure 24. Applied to perpendicular anisotropy, both types of DWs have in-plane components with one important difference. Néel walls between neighboring domains have in-plane components with alternating directions canceling out when averaged over a coherence volume. In contrast, Bloch walls between opposite domains always rotate in the same direction when a small field is applied. Therefore their in-plane component remains. PNR results from a 20-nm-thick CoCrPt film measured in low fields of 200 Oe, well below saturation, suggest that the magnetic domains are separated via Bloch walls, capped at the surface with Néel closure domains. The PNR results are reproduced in Figure 25(b). This is so far the first and only example of PNR experiments contributing to the characterization of DWs.

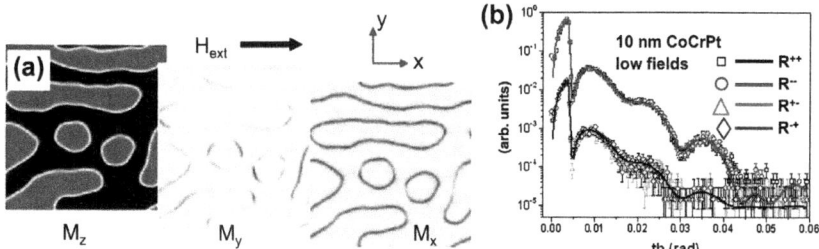

FIGURE 25 (a) Domain walls of a film with perpendicular anisotropy according to micromagnetic simulations. The three maps show the z-, y-, and x-component of the magnetization, the x-component being the most relevant for polarized neutron reflectometry (PNR) measurements. (b) PNR measurements in low fields. Magnetic contrast is due to magnetization components from Bloch domain walls in the film plane. Non-spin-flip reflectivity data are plotted with black and red (gray in print versions) symbols, spin-flip reflectivity data are plotted with blue (dark gray in print versions) and green (light gray in print versions) symbols. Solid lines are fits to the data points. *Reproduced with permission from Ref. [143].*

6.4.2 Multilayers

Magnetic multilayers of the type [FM/NM] $\times$ n, where FM stands for a magnetic layer and NM for a nonmagnetic layer, are defined by a repeat unit usually consisting of double layers FM and NM with a common interface, a thickness $\Lambda = d_1 + d_2$, where $d_{1,2}$ are the layer thicknesses within the repeat unit and the number of repeats n. The artificial periodicity generates Bragg reflections of mth order at positions in reciprocal space defined by the following equation [146]:

$$Q_z = \sqrt{(Q_c)^2 + \left(m\frac{2\pi}{\Lambda}\right)^2}, \tag{80}$$

where Q_z is the reciprocal lattice vector perpendicular to the multilayer stack. Multilayers have many advantages over bilayers. The number of interfaces can be multiplied by the number of repeats, which may be important for weak interfacial effects to make them visible at all. The multilayer Bragg reflections are the Fourier coefficients of the multilayer structure and are highly sensitive to any structural or magnetic changes. The sensitivity can even be enhanced by designing the layer thicknesses to be $d_1 = d_2$. This removes all even-order Bragg reflections, an effect which is well known in the optics of gratings for the case that the slit opening is half the size of the slit separation and which is well sought after in the design of multilayer monochromators to eliminate higher-order harmonics. Any intensity change at the position of the even-order peaks indicates deviations from the equality $d_1 = d_2$, which may be due to interdiffusion, strain, or proximity effects.

The main disadvantage of multilayers is the difficulty to fabricate them with precision and perfection over many periods. One layer may grow well on top of the other layer because of a good match of surface energies, resulting in layer-by-layer growth. However, the other way around the growth may not be as perfect and islands may form. There is an intrinsic asymmetry between top and bottom interfaces in multilayers, which has often been observed and which is hard to overcome, but needs to be taken into account in fitting procedures.

Various types of magnetic multilayers with collinear and noncollinear magnetization vectors are discussed in Section 6.2.4, and simulated specular reflectivity curves are shown in Figure 5. Magnetic multilayers with noncollinear magnetic structures due to various competing interlayer coupling effects have been determined by PNR and analyzed in detail by Schreyer et al. [147,148] and later by Lauter-Pasyuk et al. [26], te Velthuis et al. [149], Saerbeck et al. [150], and Brüssing et al. [151].

For the following discussion, we neglect possible structural imperfections of multilayer interfaces and concentrate on magnetic domain structures. We assume that the ferromagnetic layers in an AFM-coupled multilayer decompose in a number of lateral domains, which are correlated in the perpendicular direction. The latter assumption is necessary in order to preserve the

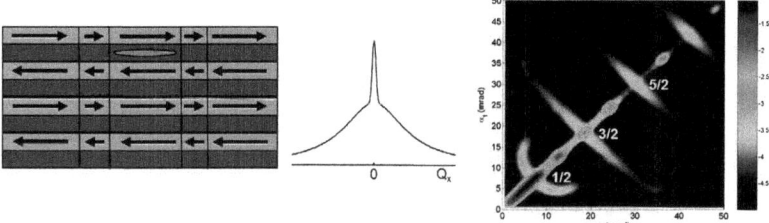

FIGURE 26 Left panel: Antiferromagnetically ordered multilayer with the magnetization vectors in the film plane broken up into domains. The domains are perfectly correlated in the perpendicular direction and exhibit a bimodal size distribution in the plane. The green (gray in print versions) ellipse indicates the neutron coherence volume. The schematic drawing is not to scale. In reality the total multilayer thickness may be a few hundred nanometers, whereas the lateral size is on the order of tens of millimeter. Middle panel: Transverse scan across the half-order antiferromagnetic Bragg reflection reveals a sharp peak and a broad diffuse peak due to the bimodal domain size distribution. Right panel: Intensity map plotted as a function of glancing incident angle α_i and exit angle α_f for the sketched antiferromagnetically coupled multilayer decomposed into lateral domains.

antiferromagnetic half-order Bragg reflections. In real experiments the longitudinal width of the half-order peaks ΔQ_z provides information on the actual correlation length $\xi_z = 2\pi/\Delta Q_z$ in the z-direction, which often is observed to be shorter than the multilayer stack height. Next we assume that the magnetic domains have a bimodal in-plane distribution, composed of domains much smaller and much larger than the coherence length l_x^{coh} of the incident neutrons projected onto the sample surface. Those domains which are much larger than l_x^{coh} give rise to a sharp peak in the transverse Q_x direction across the specular ridge, whereas domains smaller than the neutron coherence length generate a much wider diffuse peak in the transverse direction. This diffuse intensity perpendicular to the specular intensity ridge is often referred to as *Bragg sheets*. Thus the bimodal domain distribution is expressed in a bimodal shape of the transverse peak when scanned across the half-order position, as shown in the middle panel of Figure 26. The intensity map summed up from NSF and SF scattering is plotted in the right panel of Figure 26. We clearly recognize the structural Bragg peaks, which carry no streaks in the transverse direction because of the assumed ideal interfaces, and the half-order antiferromagnetic peaks with extended streaks from small domains superimposed on a sharp half-order Bragg peak due to large domains. In the actual simulation the domain sizes are 300 nm and 30 μm for the small and large domains, respectively.

6.4.2.1 Magnetic Domains

As an example for an antiferromagnetically coupled multilayer, we show in Figure 27 neutron scattering results and simulations of intensity maps taken from [Co_2MnGe(3 nm)/V(3 nm)]$_{50}$ Heusler multilayers [152]. From inspection

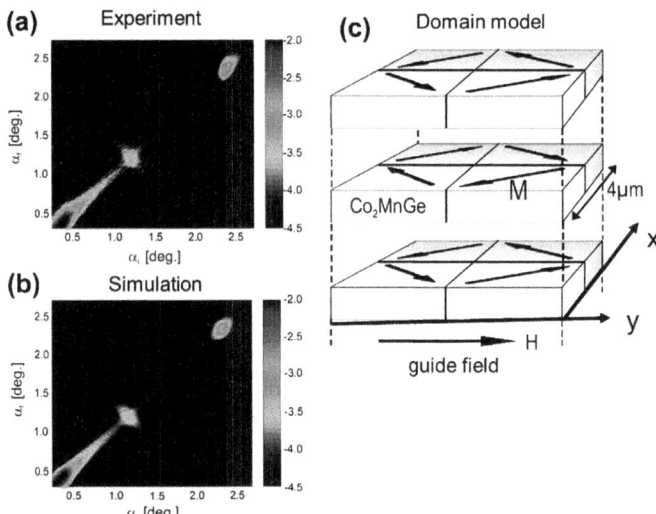

FIGURE 27 (a) Two-dimensional intensity map from neutron intensity of the antiferromagnetically coupled [Co$_2$MnGe(3 nm)/V(3 nm)]$_{50}$ Heusler multilayer. The data were taken in remanence at 15 K. The Bragg sheet at the half-order position is due AFM coupling and the width is determined by the in-plane domain structure and out-of-plane correlation length. (b) Simulation of the intensity map within a super-iterative based version of the distorted-wave Born approximation. (c) Model of the in-plane domain structure. *Reproduced with permission from Ref. [152].*

of the measured intensity map taken with unpolarized incident neutrons, we notice that the half-order AFM Bragg peak has about the same width ΔQ_z along the specular ridge as the first structural Bragg peak. This indicates that the AFM domain correlation perpendicular to the layer stack is on the same order of magnitude as the structural correlation.

Furthermore we notice that the transverse width of the half-order peak is broadened as compared to the structural first-order Bragg, revealing that the lateral magnetic domains are smaller than the extent of the structural correlation and smaller than the neutron coherence length. Additional PNR measurements show that the half-order peak is present in the NSF and in the SF channels. Thus the magnetization orientation in the AFM domains has projections in the Y- and X-axis. From fitting the data it was found that a Landau domain pattern perfectly stacked in the perpendicular directions with an in-plane domain size of 2 μm, as sketched in the right-hand panel of Figure 27, can very well describe the data and yields a good match of the simulated maps with the measured ones.

Another illustrative example is shown in Figure 28 from the work of Saerbeck et al. [89,150]. The well-known phenomenon of interlayer exchange coupling (IEC) in magnetic multilayers, reviewed by Grünberg [153], is here revisited with a new twist. In the classical version the IEC in Co/Cu multilayers is mediated by the Cu 4s electrons yielding either a parallel or antiparallel

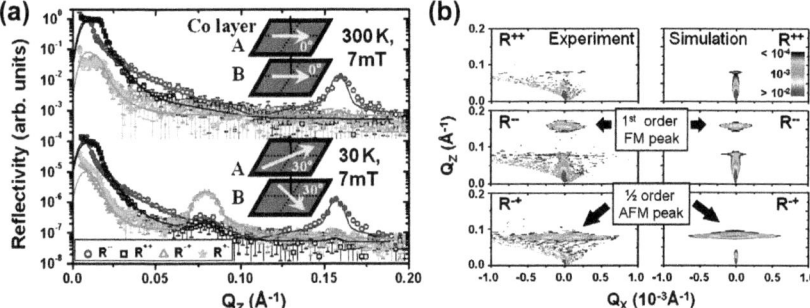

FIGURE 28 (a) Non-spin-flip reflectivities (red (dark gray in print versions) circles, blue (black in print versions) squares) and spin-flip reflectivities (green (gray in print versions) triangles, cyan (light gray in print versions) stars) of a Co/Cu multilayer with a 6% Mn doping in the Cu spacer layer. The polarized neutron reflectometry data are fitted to the model sketched in the inset. (b) Intensity maps of off-specular scattering at 30 K and in a field of 7 mT data (left) and corresponding simulations (right). *Reproduced with permission from Ref. [150].*

coupling of the adjacent Co layers as a function of the Cu spacer thickness, known as bilinear coupling and described by a two-dimensional version of the Rudermann−Kittel−Kasuya−Yosida interaction [154]. When doping Cu with Mn ions, a new impurity-mediated exchange interaction is introduced causing a noncollinear coupling at low temperatures. This is similar to the loose spin model of Slonczewski [155], generating a biquadratic coupling in addition to the bilinear coupling, with the difference that in the original model the loose spins are due to disorder at the interfaces, whereas here they are introduced by doping paramagnetic Mn ions into the nonmagnetic Cu spacer layer. In Figure 28, we recognize a ferromagnetic coupling of the Co magnetization vectors at 300 K by the strong splitting of the R^{++} and R^{--} reflectivities at the first-order Bragg peak at $Q_z = 0.158$ Å^{-1} and no spin-flip intensity other than due to finite flipping ratio. At low temperatures of 30 K the situation is quite different. Here we observe a half-order peak in the spin-flip channels R^{+-} and R^{-+}, and to a lesser extend also in the nonflip reflectivities, while the first-order Bragg peak still exhibits a ferromagnetic component. On the one hand the half-order peak signals an antiferromagnetic doubling of the magnetic periodicity, on the other hand there is still a strong ferromagnetic component seen at the first-order Bragg peak. The superposition of both yields a canted noncollinear magnetization with projections onto the non-spin-flip Y-axis and spin-flip X-axis. A fit reveals a canting angle of ±30° with respect to the Y-axis. Furthermore, in the $Q_z - Q_x$ maps reproduced in Figure 28(b) it can be seen that the half-order peak has almost no specular component but is mainly spread out in the X-direction. Simulations of the intensity maps indicate an average lateral domain size of 0.43 μm. The main new insight of this work is the fact that doping paramagnetic impurity ions in a nonmagnetic matrix enhances the IEC and contributes to biquadratic coupling.

In AFM-coupled asymmetric superlattices of the type $FM_1/NM/FM_2/NM$ the half-order Bragg peak vanishes since the magnetic period is identical with the chemical period. In fact this is always the case independent of the orientation of the magnetization vectors in FM_1 and FM_2, being parallel, antiparallel, or noncollinear. Such superlattices were studied by Brüssing et al. [151] with the layer sequence Fe/Cr/Co/Cr. As Fe and Co exhibit very different coercive fields, at remanence with negligible exchange coupling mediated by a 3-nm-thick Cr layer, the superlattice can be considered as a pseudospin valve system. Fitting of the data of all four reflectivities confirms the antiparallel magnetization state of the Fe and Co layers at remanence, as shown in Figure 29(b). A completely different picture emerges for the sample magnetization in the

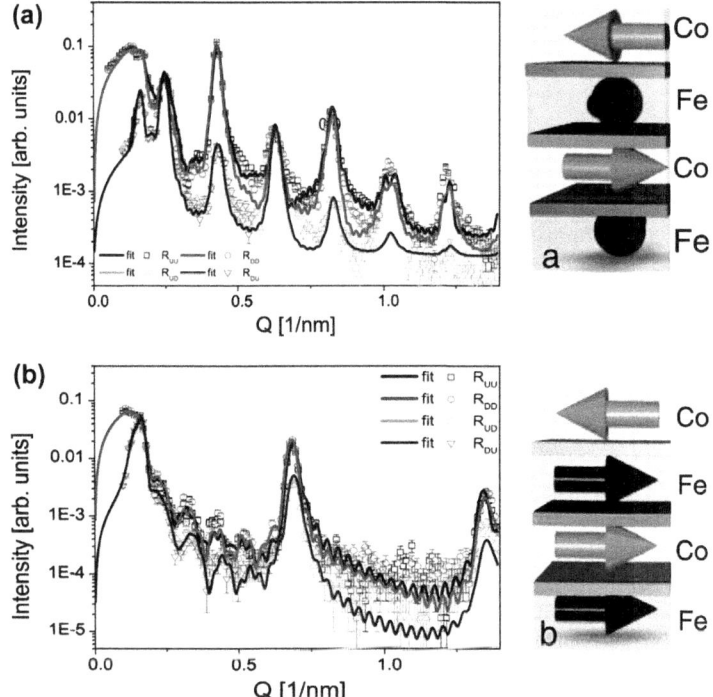

FIGURE 29 Polarized neutron reflectometry (PNR) scans from Co/Cr/Fe/Cr superlattices with a total periodicity of 16 nm and 20 periods. (a): as-grown state, before applying any magnetic field. New half-order Bragg peaks signal a doubling of the magnetic unit cell, comprising four magnetic layers forming a spiral, sketched in the upper right-hand panel. (b) PNR from the same superlattice as in panel (a) but after saturation and returning to remanence. The Co and Fe layers at remanence are in an antiparallel state as sketched in the lower right-hand panel. The outlines to the right indicate the magnetization vectors in the individual layers, red (gray in print versions) arrows for Co and green (dark gray in print versions) arrows for Fe. The magnetic layers are separated by Cr spacer layers shown in gray. *Adapted and reproduced with permission from Ref. [151].*

as-grown state. As can be recognized by the top panel of Figure 29(a) new half-order Bragg peaks appear, indicating a doubling of the periodicity. This doubling vanishes with increasing magnetic field and cannot be reproduced by reversing the field, only by preparing a new sample. Obviously during growth a spiral magnetic structure is imprinted into the superlattice similar to the case of Figure 5(d) and also sketched in the upper right-hand panel of Figure 29 that is confirmed by model fits, providing a very weak anisotropy easily overcome by an external field.

6.4.2.2 Enhancing Interface Sensitivity

The concept of enhancing interface sensitivity by using equal thicknesses of bilayers in superlattices has been used for studying proximity effects between cuprate high T_c films ($YBa_2Cu_3O_7$ or (YBCO)) in combination with manganite layers. Because of the antagonistic nature of superconductivity and ferromagnetism the combination of both layers in superlattices with common interfaces leads to new and fascinating properties [156−158]. Either the ferromagnetic order parameter close to the interface is reduced via a cryptomagnetic state, or an inverse proximity effect enhances the magnetization in the superconducting layer close to the interface. Using two different YBCO/manganite superlattices, the authors showed that the magnetic proximity effect is present in case of cuprate superlattices together with the metallic ferromagnetic manganite $La_{2/3}Ca_{1/3}MnO_3$ (YBCO/LCMO), but absent in case of the insulating ferromagnetic manganite $LaMnO_{3+\delta}$ (YBCO/LMO).

Two sets of superlattices were deposited by pulsed laser deposition on $La_{0.3}Sr_{0.7}Al_{0.65}Ta_{0.35}O_3$ (LSAT) substrates: [YBCO(10 nm)/LCMO(10 nm)]$_{10}$ with a superconducting T_c^{SC} of 88 K and a ferromagnetic T_c^{FM} of 200 K and [YBCO(10 nm)/LMO(10 nm)]$_{10}$ with $T_c^{SC} \approx 77$ K and $T_c^{FM} \approx 140$ K. The choice of the substrate is essential, because LSAT does not undergo a structural phase transition at low temperatures, which otherwise would corrugate the surface and hamper high-quality PNR data up to high Q_z values [159]. In the past similar superlattices were grown on $SrTiO_3$ substrates. But because of low-temperature structural phase transitions, scans only up to the second-order superlattice Bragg peak could be recorded so far. However, this is not sufficient, if fine details at the interfaces are to be analyzed. With the LSAT substrate indeed much better data could be gained up to the fourth-order Bragg peak, as can be seen in Figure 30. The PNR data were recorded with the polarized neutron reflectometers, Super ADAM at the ILL and NREX at the FRMII. Unpolarized reflectivity scans were taken at 300 K and polarized reflectivity scans were measured below the respective temperatures T_c^{SC} and T_c^{FM}. At 300 K the second-order Bragg peak is essentially distinct, confirming that the YBCO and manganite layers have equal thickness from a structural/chemical point of view. At low temperatures an absolutely remarkable effect occurs, which can already be well interpreted just by visual inspection. For

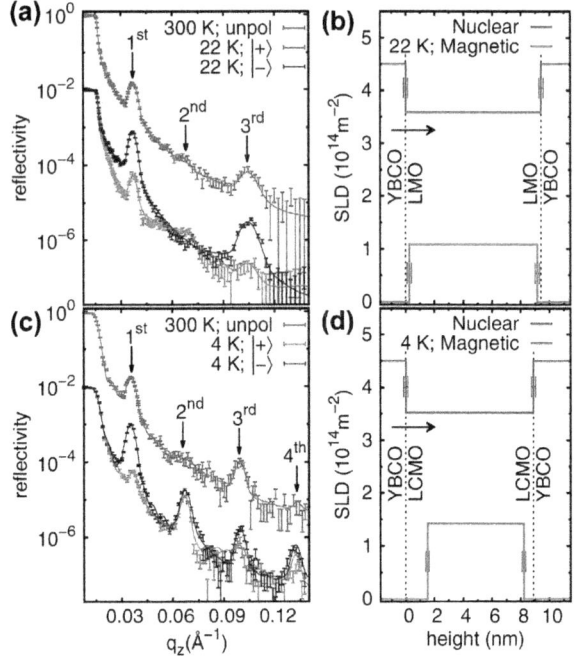

FIGURE 30 (a) Unpolarized reflectivity scan at 300 K and R^+ and R^- reflectivity scans at 22 K of YBCO/LMO below the superconducting and ferromagnetic transition temperatures of YBCO and LCMO, respectively. Note that at both temperatures the even-order Bragg peaks are extinct due to the same thicknesses of the YBCO and LMO layers. (b) Fit results for the nuclear and magnetic SLD profiles for the low temperature scan. (c) Similar measurements as in (a) but with LCMO replacing LMO. At low temperatures the even-order Bragg peaks become visible, indicating that the effective magnetic thickness of the LCMO layer has changed in the superconducting state of the sample. (d) Fit results of the SLC profile as in (b) but with a dramatically changed profile for the magnetic layer. *Reprinted with permission from Ref. [114].*

[YBCO(10 nm)/LMO(10 nm)]$_{10}$ the R^+ and R^- reflectivities, shown in Figure 30(a), are split at the first and third order, indicative for the ferromagnetic state of the LMO layers. The second-order peak remains essentially extinct. From this we infer that the nSLD and the mSLD for YBCO and LMO remain to have equal thicknesses, as shown in Figure 30(b). In contrast, for the [YBCO(10 nm)/LCMO(10 nm)]$_{10}$ superlattice at low temperatures, we recognize not only a splitting of the R^+ and R^- reflectivities, but also pronounced second- and fourth-order Bragg peaks. This immediately signifies that the mSLD must have different widths (not amplitude) in both layers, being either more or less extended. It is unlikely that the nSLD profile changes at low temperature and therefore does not need to be further discussed. We also notice that the second- and fourth-order Bragg peak show almost no splitting, in contrast to the first and third order. From this we may infer that the changes that took place at the interface are nonmagnetic. This notion is confirmed by a more detailed fitting procedure, providing the nSLD and mSLD shown in Figure 30(d). Thus in [YBCO(10 nm)/LCMO(10 nm)]$_{10}$ a pronounced proximity effect can be observed at low temperatures, which suppresses the ferromagnetic state in LCMO close to the superconducting interface. It is clear that due to the missing phase information the exact position of the SLD depth profiles cannot be determined. However, by repeating the reflectivity scans to even higher Q_z values and by testing many other competing models, the one proposed has a high confidential level [115].

6.4.3 Stripes, Islands, and Nanoparticles

Confining magnetic films in the lateral direction adds another degree of freedom to the design of magnetic nanostructures as compared to thin films and multilayers. When reducing the lateral extension of a film, the shape anisotropy starts to play a major role competing with the exchange interaction and magnetocrystalline anisotropy. By the island size, shape, and aspect ratio the domain state can be selected as well as the type of DWs separating them. Furthermore, in patterns of islands the interisland interaction via magnetic dipole or higher pole interaction gives rise to new magnetic ordering phenomena [3,160]. Some patterns are shown schematically in Figure 31.

Magnetic islands and lateral patterns are usually investigated by a number of different imaging techniques, including MFM, Kerr microscopy, X-ray

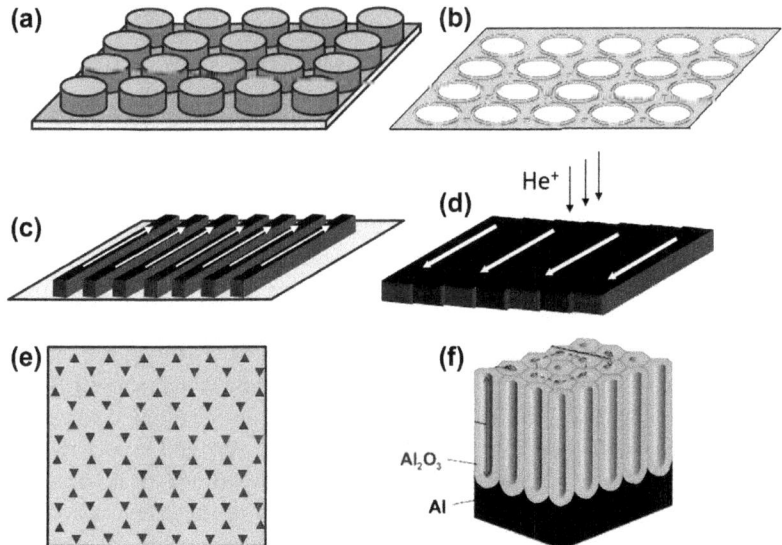

FIGURE 31 Different lateral patterns in the form of dots (a), antidots (b), and stripes (c). Homogeneous magnetic films can also be structured by ion-beam modification of interfaces (d); self-assembled arrays of nanospheres can be taken as mask for evaporation (e); nanopillars can be grown inside of anodized aluminum oxide cavities (f).

PEEM, scanning transmission X-ray microscopy, spectro-holography, scanning electron microscopy with polarization analysis, and Lorentz microscopy. Investigating lateral magnetic patterns with PNR is a big challenge because of the small scattering volume and the need to explore Bragg reflections in the off-specular regime, where the intensity is naturally much lower. Aggravating the problem is the nuclear scattering from the uncovered part of the substrate, whereas the main interest is in magnetic scattering only from the islands. Therefore it is mandatory to study magnetic patterns via off-specular Bragg reflections to filter out the substrate contribution. Because of these difficulties only a few examples exist in the literature of successful PNR studies from lateral magnetic patterns.

With PNR, one may study magnetic contributions to Bragg peaks in Q_x direction but not in Q_y direction, where the resolution would be too low. The off-specular Bragg peaks represent Fourier coefficients of the magnetization distribution within islands and correlation effects between them. Depending on the lateral periodicity quite a limited number of Bragg peaks may be reached in the Q_x direction, often hardly sufficient to get a complete picture of the in-plane magnetization distribution. Similar studies can also and more easily be performed with vector MOKE in combination with diffraction MOKE [161,162] or with XRMS [163,164]. Therefore any PNR experiment on laterally patterned arrays should be carefully planned, including considerations of the proper orientation of the neutron coherence volume with respect to the orientation of the pattern. The effort to study lateral magnetic patterns via PNR is firmly justified when the high sensitivity to vector magnetization in buried layers is combined with an analysis of the correlation effects between magnetic island during the magnetization reversal process.

6.4.3.1 Laterally Patterned Nanostructures

A proof of principle for OSS from arrays of magnetic islands was provided for the first time by Toperverg et al. [166]. Shortly after Temst and coworkers demonstrated using a Co dot array that the magnetization reversal process can be characterized by studying all four cross sections at Bragg peaks in the off-specular regime. They also pointed out the power of PNR for measuring the vector magnetization simultaneously with specular and OSS [167]. The same group has also studied the reversal in rectangular Co islands [168] and the magnetization reversal of Co/CoO rectangular islands, which show EB effect after field cooling (FC) [165,169–171]. The aim was to investigate whether the reversal mechanism changes as a function of island size and aspect ratio. All four cross sections were measured at the first-order Bragg peak in the off-specular regime. They find that the spin-flip scattering intensity at coercivity is lower for islands than for continuous films, indicating that in islands the portion of magnetization that undergoes coherent rotation is smaller. The results shown in Figure 32 are representative of independent island switching. The interplay between uniaxial shape anisotropy and unidirectional EB in patterned

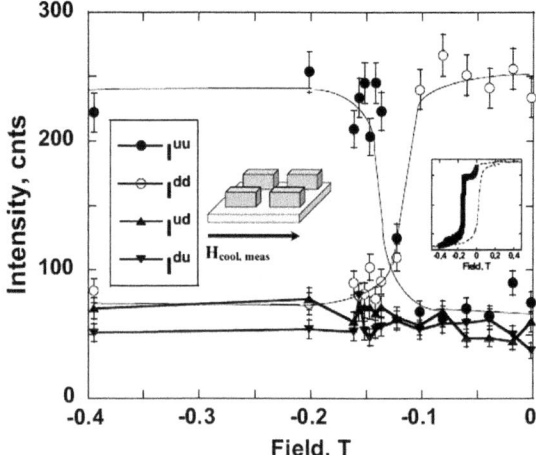

FIGURE 32 The magnetization reversal of rectangularly shaped Co/CoO islands determined by all four cross sections at the first lateral Bragg peak of the pattern. The inset shows the schematics of the pattern with the direction of the cooling field and a vibrating sample magnetometer magneto-optic Kerr effect hysteresis. *Reprinted with permission from Ref. [165].*

stripe arrays was studied by Mohanty et al. [172] for the EB system NiFe/FeMn, and the results were compared to the magnetization reversal of a continuous film with otherwise same parameters. The authors find that for both, continuous and patterned films and for fields applied parallel and perpendicular to the stripes, the reversal mechanism is dominated by nucleation and DW motion, rather than by coherent rotation. Similarly Theis-Brühl et al. [173,174] investigated the effect of uniaxial shape anisotropy on the magnetization reversal process of FeCo stripes, but without EB. These studies showed that while the magnetic hysteresis taken with MOKE and with PNR agrees, PNR—at the same time—offers information on the direction of the mean domain magnetization vectors as well as longitudinal and transverse fluctuations about them. In particular the data unambiguously show that the domain magnetization vectors are strongly correlated across many stripes during magnetization reversal, most likely due to dipolar coupling via stray magnetic fields produced by magnetic charges that occur at the edges of the stripes during reversal.

Patterning of films can also be achieved by ion-beam modification of interfaces, sketched in Figure 31(d). For instance, FM/AFM bilayers exhibiting EB at the interface can be modified by ion-beam bombardment to locally and periodically change the EB field H_{EB}. Similar to perpendicular spin valves this allows to construct lateral spin valves with antiparallel magnetization in neighboring stripes [175,176]. From a neutron point of view, the advantage of periodically modifying the interface is a much bigger scattering volume. However, as the stripes are not spatially separated, there will always be a strong ferromagnetic exchange coupling acting at the interface between the stripes in addition to the EB with the AFM substrate. The competition leads to a canting of the magnetization vectors in neighboring stripes with a resulting ferromagnetic contribution oriented perpendicular to the stripe axis [176,177]. One can estimate the canting angle from the AFM/FM interface area on the one hand, and the area of DWs separating the ferromagnetic stripes on the other hand. This yields a ratio roughly proportional to t_F/D_{st}, t_F is the film thickness of the stripe with width D_{st}. Thus, an antiparallel magnetization configuration in neighboring stripes can only be expected for sufficiently thin FM layers, whereas stripes in thick layers tend to align parallel. In the work of Hamann et al. [175], this conjecture was investigated in more detail by PNR and OSS using samples with three different film thicknesses. A sketch of the sample used is shown in Figure 33(a). It consists of a homogeneous NiFe layer covered by an antiferromagnetic IrMn top layer. The stripelike character of the sample was generated by allowing the IrMn layer to oxidize in regions which were not protected by top Ta stripes. This alleviates the AFM structure of IrMn and quenches the EB field in a periodic fashion. Indeed the artificial periodicity could be verified by Bragg reflections in the Q_x direction, as shown in Figure 33. Furthermore, PNR measurements confirmed that the canting angle between neighboring stripe

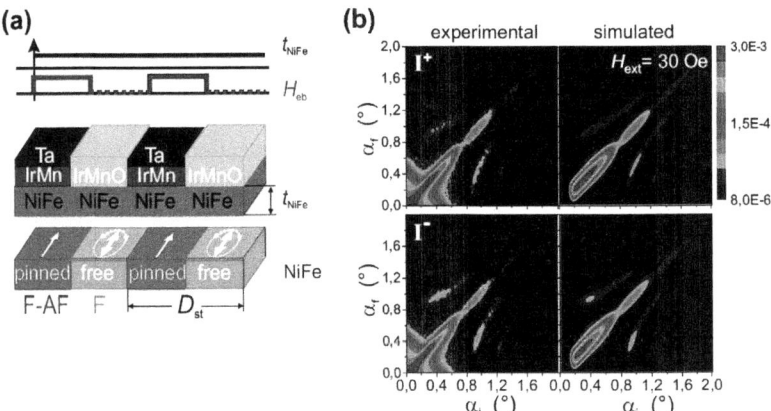

FIGURE 33 (a) Sketch of the geometry of a stripe pattern produced by modifying the interface between a NiFe layer covered by an antiferromagnetic IrMn layer with a stripe period of D_{st}. The top panel indicates a homogeneous thickness of the sample and a periodic variation of the exchange bias field H_{EB}, the middle panel shows the lateral arrangement of the stripe pattern, and the bottom pattern designates the free and pinned stripes. (b) Experimental and simulated polarized neutron scattering intensity maps of NiFe(20 nm)/IrMn-IrMnO/Ta-IrMnO stripe array with a stripe periodicity D_{st} of 4 μm and thickness of 20 nm. The intensities on a logarithmic scale of R^+ and R^- are plotted as the function of the incident and scattering angle α_i and α_f, respectively. *Reprinted with permission from Ref. [175].*

magnetization vectors decreases with decreasing ferromagnetic layer thickness and increasing stripe period, as expected.

6.4.3.2 Magnetic Nanoparticles

This review would not be complete without mentioning magnetic NPs. Although this research field is not the typical application of PNR, in a number of recent papers PNR and other magnetic neutron scattering methods have been applied for their analysis, such as neutron powder diffraction, SANS, or the polarized version of SANS termed SANSPOL, and SANS at grazing incidence termed GISANS. As this field is expanding rapidly, a short overview is warranted. For a more details references to pertinent original work and comprehensive reviews including suitable neutron scattering methods is provided.

Magnetic NPs are small clusters of magnetic ions. Each NP is assumed to be in a single domain ferromagnetic state, according to Brown's fundamental theorem: "As a magnet is reduced in size, there should be a point where exchange dominates over demagnetization and where the magnet must, hence, adopt the single-domain state" [179]. The single-domain state is usually reached for particle sizes in the range of 30—100 nm. Thus, each NP can be considered as a macrospin containing thousands of ions with a total spin on the order of 1000 μ_B. Magnetic NPs are promising for applications in numerous fields, such as spintronics [1,180,181], photonics [182], plasmonics [183], and biomedicine [184].

In an ensemble of many noninteracting NPs the average magnetization follows the classical Langevin equation as a function of magnetic field and temperature:

$$M(T,H) = \frac{N}{V} m_z L\left(\frac{m_z H_z}{k_B T}\right), \quad (81)$$

where $L(x)$ is the Langevin function:

$$L(x) = \cosh(x) - \frac{1}{x}; \quad x = \left(\frac{m_z H_z}{k_B T}\right). \quad (82)$$

As long as the Langevin equation holds for the magnetization of NPs, the state is referred to as *superparamagnetic*. Because of scaling effects, the Curie temperature determined by the local exchange interaction is usually lower than in the corresponding bulk system. For temperatures below the Curie temperature of individual NPs, another temperature, the so-called *blocking temperature* T_B is most characteristic and relevant for NPs. For temperatures $T > T_B$ the macrospins will strongly fluctuate, whereas for $T < T_B$ the magnetic moments appear blocked within the timescale of the measurement. T_B is also the temperature where the average magnetization of NPs after zero field cooling (ZFC) and FC split (see Figure 34). The ZFC measurement protocol requires the NPs first to be cooled in zero magnetic field from $T > T_B$ to the lowest

408 Neutron Scattering - Magnetic and Quantum Phenomena

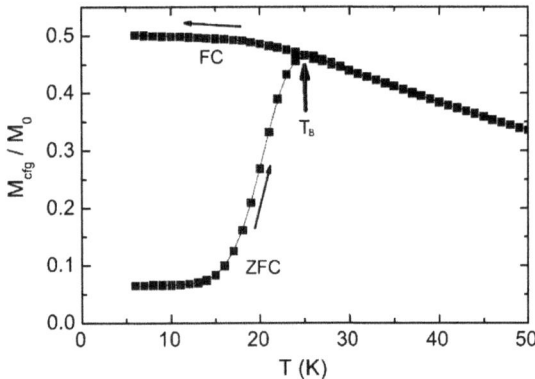

FIGURE 34 Zero field cooling (ZFC) points: The sample is first cooled in zero magnetic field to low temperatures. Data points are then taken during warming up in a finite constant field of 50−100 Oe. Field cooling (FC) points: Data are taken in the same constant magnetic field during cooling of the sample to low temperatures. The points plotted are actually from a Monte-Carlo simulations of an ensemble of noninteracting superparamagnetic particles with $KV/k_B = 315$ K and $M_s V/k_B = 1475$ K T^{-1} at $\mu_0 H = 0.04$ T. The blocking temperature, T_B, is marked by an arrow. Reprinted with permission from Ref. [178].

temperature followed by warming up in a specified field. The FC magnetization curve is then measured by applying the same field during cooling of the sample. The splitting temperature and the blocking temperature are usually identical and characterize the temperature, where the thermal energy $k_B T$ equals the anisotropy energy KV, where K is an anisotropy constant and V is the particle volume. KV is the energy barrier, which NPs have to overcome for a magnetic reorientation. The anisotropy constant K collects all contributions from shape anisotropy, magnetocrystalline anisotropy, and surface anisotropy. For constant K, T_B scales with the volume. For instance, Co NPs with a diameter of 1 nm have a T_B of about 15 K [185]. As soon as interaction between NPs is turned on, for instance, by packing them closely, all properties are changed. Depending on the type and strength of the interaction, new collective states can be observed, such as modified superparamagnetic state, glasslike spin freezing, superspin glass, and superferromagnetic states. Interaction between NPs may have different origins. These states may be due to direct or indirect exchange, to magnetic dipole interaction from dipolar fields of NPs, or due to Van der Waals interaction mediated by the solvent in which NPs often are embedded. In addition to NPs with ferromagnetic spin structure, also NPs with ferri- or antiferromagnetic spin structures have been intensively investigated. Another important aspect of magnetic NPs is the distinction between core and shell. While the core is in an ordered magnetic state, the shell may be magnetically less ordered with spin glasslike features [186−188] or canted with respect to the core magnetization [189]. For a detailed overview on the physics of magnetic Nps, we refer to some recent reviews in Refs [178,190–193].

Magnetic NPs can be investigated in various forms: in solution of a solvent [194,195] (ferrofluids), in form of a dry powder [196], deposited and self-organized into close packed structures on flat substrates [197], on pre-structured substrates [198,199], or embedded layer wise in ceramic [200] or polymer matrices [201]. The main analytical tools for analyzing their magnetic properties are magnetization measurements with SQUID or VSM. In the past, neutron scattering has played a minor role for the analysis of magnetic NPs, but this appears to be changing. In fact, a whole array of neutron scattering methods is available for a full and detailed analysis, unlike the analysis of thin magnetic films and multilayers, which relies mainly on polarized neutron reflectivity. For magnetic NPs neutron powder diffraction is essential for unraveling spin structures on the atomic scale. For AFM NPs this is the only way to determine the magnetic space group. Furthermore, from the intensity of the ferro- or antiferromagnetic Bragg peaks the temperature dependence of the order parameter and therefore also the Curie or Néel temperature, respectively, can be determined. Excellent recent examples are the determination of the antiferromagnetic space group of Mn_2Au [202] particles and a reinvestigation of the antiferromagnetic order in NiO NPs [203] by neutron powder diffraction.

On a larger length scale of 1−100 nm with SANS the particle form factor $P(Q)$ and the structure factor $S(Q)$ can be determined. $P(Q)$ describes the morphology or shape of individual particles fulfilling the boundary condition $P(0) = V$, where V is the particle volume. This also allows a determination of the NP size distribution and a distinction between core and shell. $S(Q)$ yields information on correlations between NPs and resulting interference patterns due to scattering from different particles provide information on particle density and order. Adding the option of a polarized incident beam, referred to as SANSPOL, enhances the information on magnetic cluster sizes and correlations. Similar to magnetization depth profiling of thin films and multilayers, SANSPOL is also executed in a saturation field applied in the direction perpendicular to the incoming beam. For obtaining magnetic contrast either the field or the incident polarization is switched. Even more useful is a complete SANS polarization analysis, which is possible nowadays due to the availability of ^{3}He spin filters for the scattered beam.

SANS experiments are very versatile. No special requirement for the sample is necessary. Therefore in case of SANS the BA is always applicable, if NPs are small enough, in contrast to PNR, where DWBA has to be applied at small glancing angles close to total reflection. In SANS experiments the scattering vector $Q_{x,y}$ is essentially perpendicular to the incident beam with cylindrical symmetry about it, thus probing correlations in the x, y plane perpendicular to the incoming beam.

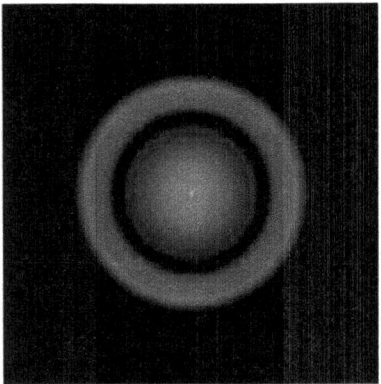 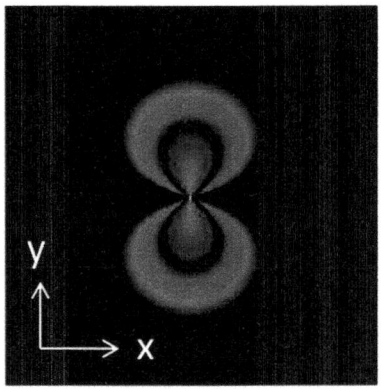

FIGURE 35 Small-angle neutron scattering intensity patterns from magnetic nanoparticles (NPs). Left panel: intensity distribution for randomly oriented magnetic NPs at zero fields; right panel: intensity distribution for magnetization of NPs partially oriented parallel to the X-axis. *Adapted and reproduced with permission from Ref. [205].*

For an isotropic sample of NPs, $S(Q)$ has a ringlike shape. Now we assume that the incoming beam is unpolarized and that we apply a saturation field parallel to the X-axis such that the magnetization $\mathbf{M} \parallel \mathbf{H}_x$. In this case, the scattering in the Q_x direction contains only nuclear scattering, whereas scattering along Q_y has nuclear and magnetic components. Therefore the magnetic scattering is anisotropic showing a typical $\sin^2\psi$ distribution, where ψ is the angle between the vectors $\mathbf{M}$ and $\mathbf{H}$, while the nuclear scattering is isotropic. In Figure 35, simulated magnetic SANS intensity maps for unpolarized incoming neutrons are shown for randomly oriented magnetic NPs at zero fields (left panel), and for NPs partially magnetized parallel to the X-axis (right panel), yielding a characteristic dumbbell-like intensity distribution. In case that the incoming neutron beam is polarized in the X-direction, a similar intensity distribution occurs, however, with additional contrast between the intensity for "up" polarization $S^+(Q)$ and "down" polarization $S^-(Q)$, similar to the case of PNR without polarization analysis. SANS is discussed in more detail by Penfold and Tucker in Chapter 5 of the first volume on Neutron Scattering [14] and SANSPOL on magnetic NPs has been reviewed by Wiedemann [204].

Using the SANSPOL techniques, Disch et al. [206] have analyzed the SLD of the ferrimagnet maghemite (γ-Fe_2O_3) NPs with spherical shape of 9.2 nm diameter and cubical shape with an edge length of 8.6 nm. Applying a field of 1.5 T at room temperature, they found a reduced magnetization in the core of the NPs and a further reduction of the magnetization toward the surface. The reduction in the core was unexpected, but the further decrease in the shell is attributed to spin canting, as has been observed already in many other cases. From measuring the field dependence of the mSLD in the particle core and fitting to the Langevin equation for noninteracting NPs, an average magnetic moment per Fe^{3+} ion in Fe_2O_3 of 0.77 μ_B was determined, as compared to 1 μ_B expected at 300 K. The difference is ascribed to defects, off-stoichiometry, and strain in the core of the NPs.

SANSPOL has also been used by Lister and coworkers for the analysis of size-dependent switching fields of CoCrPt granular NPs sputter deposited on SiO_x substrate [207]. As the magnetic field was applied parallel to the incident beam and perpendicular to the sample surface, the ringlike radial scattering pattern is isotropic. By taking the difference of up and down polarization, the radial Q-dependence of $I^D(Q) = I^+(Q) - I^-(Q)$ is proportional to the nuclear–magnetic cross term and depends on the fraction of particles switched upon reversing the applied magnetic field from positive to negative saturation, allowing a determination of the size dependence of the reversal process [207,208].

SANS experiments on iron oxide NPs with the stoichiometry Fe_3O_4 and an average size of 9 nm were analyzed by SANS with a full polarization analysis by Krycka et al. [189,209]. A schematic of the setup with supermirror polarizer and ^{3}He spin filter analyzer is shown in Figure 36 together with two characteristic SANS patterns recorded with a PSD. A field of 1.2 T was applied in the X-direction at 200 K and all four scattering cross sections I^{++}, I^{+-}, I^{-+}, I^{--} were analyzed for scattering vectors Q_x, Q_y and directions in between.

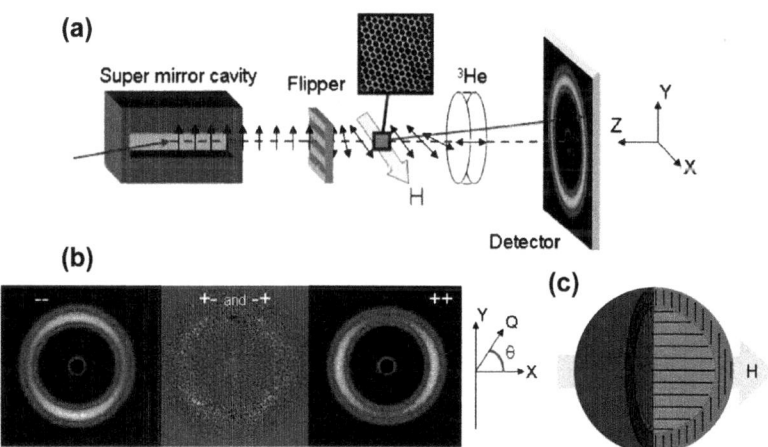

FIGURE 36 SANSPOL with polarization analysis. (a) The experimental setup includes a polarizing supermirror, Mezei flippers, sample holder with cryostat and variable magnetic field, ^{3}He filter analyzer, and a position sensitive gas detector. Arrows indicate the neutron polarization direction. Note that the neutron polarization is rotated by 90° in the X-direction at the sample position; (b) 2D small-angle neutron scattering intensity maps for (−−), (+−) + (−+), (++) polarization and for 1.2 T at 200 K corrected for polarization efficiency; (c) nanoparticle model containing a ferrimagnetic core (diameter 7.4 nm) and a magnetic shell (thickness of 0.8 and 1.2 nm) with a 90° canted magnetization. *Adapted and reproduced with permission from Ref. [189].*

Keeping the neutron polarization parallel or antiparallel to the X-axis, i.e., parallel to the applied field axis, the sum of NSF intensity for scattering vectors in the x-direction is proportional to the nuclear scattering $I^{++}(Q_x) + I^{--}(Q_x) \sim N^2(Q_x)$, whereas the NSF intensity difference for scattering vectors in the perpendicular y-direction is proportional to the magnetization component parallel to the X-axis: $(I^{++}(Q_y) - I^{--}(Q_y))^2 \sim M^2(Q_x)$. The spin-flip scattering for scattering vectors in the X- or Y-direction is sensitive to the magnetization perpendicular to the X-direction according to $I^{+-}(Q_{x,y}) + I^{-+}(Q_{x,y}) \sim M^2(Q_y)$. Thus both magnetization components M_x and M_y can be determined from a full polarization analysis. The results obtained show that while the magnetization in the core of the NPs is aligned parallel to the applied field, in the shell with a thickness of about 1 nm the magnetization is canted by about 90°, as depicted in Figure 36.

Magnetic NPs may also be studied by the use of PNR geometry. When spin coated on a flat substrate, the NPs self-assemble into a close packed structure and stack nearly layer by layer [197]. Mishra and coworkers have investigated monolayers of magnetite NPs [210] as well as multilayers [211]. PNR results of NP multilayers are shown in Figure 37. The ordering of the multilayer stack composed of five monolayers of NPs is so perfect that one can recognize Kiessig fringes from the total stack height and a Bragg peak at 0.45 nm^{-1} from the intrinsic periodicity. Because of close packing the interaction between the

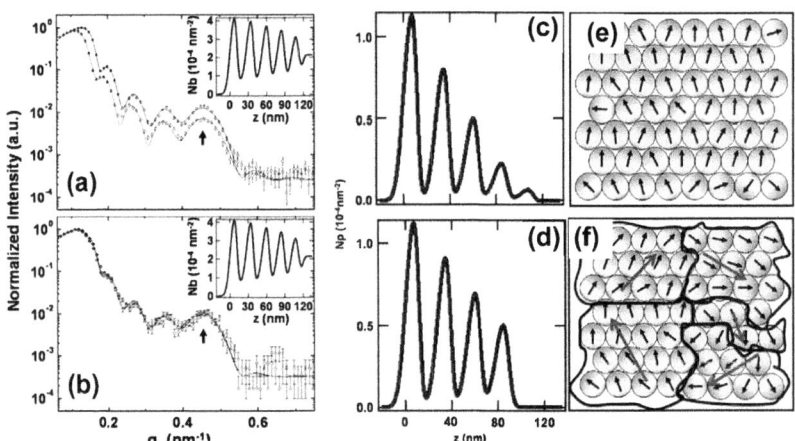

FIGURE 37 (a) Polarized neutron reflectometry (PNR) of self-assembled nanoparticle (NP) multilayer containing five monolayers exposed to a saturation field of 5 kOe at 300 K. The arrow indicates the first-order Bragg peak of the multilayer. The inset shows the nuclear scattering length density (SLD) (Nb) from the fit result. (b) PNR of the same NP multilayer at remanence. The nuclear SLD shown in the inset has not changed. (c) Magnetic SLD from the fit results (solid lines) in saturation and (d) at remanence. (e) Corresponding sketch of magnetic moment distribution in saturation and (f) at remanence. Solid lines indicate domain walls separating quasidomains at remanence. *Adapted and reproduced with permission from Ref. [211].*

NPs is rather strong. The interaction between the NPs is mainly due to magnetic dipole interaction, Van der Waals type interaction can be neglected in comparison. In a high magnetic field the magnetization of the particles aligns in the field direction. At remanence the magnetization is macroscopically zero. However, PNR analysis has shown that the magnetization of the individual particles is not randomly orientated. Rather the data can be interpreted by assuming a quasiferromagnetic domain state. In remanence the layer magnetization persists within some areas larger than the neutron coherence length, but becomes misdirected over macroscopic length scales. Figure 37(e) and (f) shows sketches of the "superspin" distributions at 5 kOe and at remanence, respectively.

Recently PNR has also been used for the analysis of magnetite NP assemblies in water-based ferrofluids close to a silicon surface under three different conditions: static, under application of a shear force, and in a magnetic field [212].

Alternatively magnetic NPs may be embedded periodically in polymers [201,213] or ceramic matrices [200,214,215], yielding a multilayer-like structure. Depending on the density of the magnetic NPs, their physical properties may range from superparamagnetic to superferromagnetic behavior. This has been investigated in much detail by Bedanta et al. [200] via multilayers of $[Co_{80}Fe_{20}(1.3\ nm)/Al_2O_3(3\ nm)]_{10}$. The $Co_{80}Fe_{20}$ layer is granular and non-percolating, i.e., without direct contact between the NPs. However, due to strong interparticle interactions through dipolar stray fields a collective superferromagnetic state was discerned by DC and AC magnetization measurements and confirmed by PNR. The magnetization reversal is characterized by nucleation and switching of pinned DWs without appreciable magnetization rotation. The only difference to a continuous ferromagnetic layer is the observation of a very slow switching behavior with exponential relaxation at fields close to the coercive fields. This is shown in Figure 38. When inspecting the PNR data at the coercive field more closely, Bedanta et al. noticed that the magnetization is not compensated by an appropriate domain structure in each layer, but by a correlated and periodic magnetization reversal comprising a total of five double layers [215]. This unique magnetization modulation perpendicular to the multilayer stack is controlled on the one hand by interparticle dipolar interaction, and on the other hand by Néel coupling between the particles in multilayers with interface roughness.

6.4.3.3 Grazing Incidence Small Angle Scattering

Finally we want to discuss a method for the investigation of magnetic NPs that combines techniques and geometries of SANSPOL and PNR. This technique is known as grazing incidence SANS also called GISANS (see recent review [213]), the polarized option accordingly P-GISANS. P-GISANS is similar to PNR with the important difference that the incident beam is well collimated not only in the X-direction but also in the Y-direction. Usually in PNR the beam

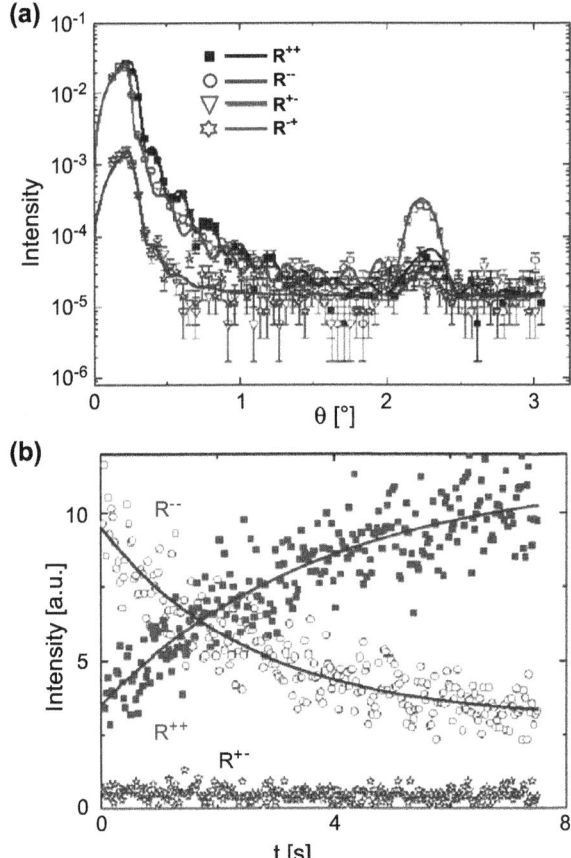

FIGURE 38 (a) Polarized neutron reflectometry measurements of a multilayer [$Co_{80}Fe_{20}$ (1.3 nm)/Al_2O_3(3 nm)]$_{10}$. All four non-spin-flip (NSF) and spin-flip (SF) reflectivities were taken in equilibrium at $T = 150$ K and at a field of 1.85 mT close to the coercive field applied parallel to the polarization axis in the Y-direction. The SF reflectivity is solely due to finite polarization and flipping efficiency. (b) Relaxation of the neutron reflectivities R^{++}, R^{--}, and R^{+-} versus time at 150 K and at the position of the Bragg peak at $2\theta = 0.7°$ in a constant field of 1.7 mT. NSF reflectivities show a reversal within a few seconds, whereas SF reflectivity remains negligibly small, indicating that any transverse magnetization component does not play any role during reversal. *Adapted and reproduced with permission from Ref. [216].*

height in the Y-direction is divergent (convergent) and the intensity is integrated over the Y-direction. With fine collimation in the Y-direction Q_y scans are possible. Thus in GISANS geometry all three directions of the scattering vector are accessible according to Eqns (36) and (37).

In SANS geometry and without applying a magnetic field, Q_x and Q_y are completely equivalent, while Q_z is not accessible. In contrast, in PNR experiments Q_z and Q_x are very different not only with respect to the probing correlation perpendicular versus parallel to the plane, respectively, but also with respect to the probing length scale. In Q_z direction a spatial sensitivity of subnanometers is achieved by default, whereas the spatial sensitivity in Q_x direction is on the order of 100 nm to 100 μm. In Q_y direction there is no sensitivity at all due to the integration over the beam divergence in the Y-direction. This shortcoming of PNR is remedied by collimating the Y-direction of the incoming beam, allowing Q_y scans with the same resolution as Q_z scans. Furthermore, at low glancing incident angles the sensitivity to the surface is much enhanced, which is the major advantage as compared to SANS or SANSPOL. The spatial sensitivity in Q_y scans is similar to Q_z scans. This can be recognized by inspection of Eqns (36) and (37).

GISANS patterns recorded by an area detector have features similar to SANS. The primary beam penetrating the substrate is usually visible below the horizon of the sample surface (point D in Figure 39). The specular reflected beam is seen above the sample horizon H in point R of Figure 39. The SANS scattering ring is still centered about D but becomes distorted and interrupted by the sample horizon. Below the surface the ring pattern is completed. If the surface is well ordered than truncation rods from the ordered structure become visible.

While grazing incidence small angle X-ray scattering (GISAXS) is widely applied for the analysis of surface structures [217], GISANS has not often been used until recently and even less so in the polarized version. A proof of principle was given by Pannetier et al. [218]. With the development of powerful advanced SANS machines or modified neutron reflectometers, this method will become more important in the future. One example should be mentioned here. Theis-Bröhl and coworkers have employed P-GISANS to analyze the structural and magnetic parameters of cobalt−oleyl amine nanocomplexes in thin films [219]. In this work they have demonstrated that inside the film the NPs are self-organized into a three-dimensional paracrystalline-like lattice with the positional order well defined over a few interparticle

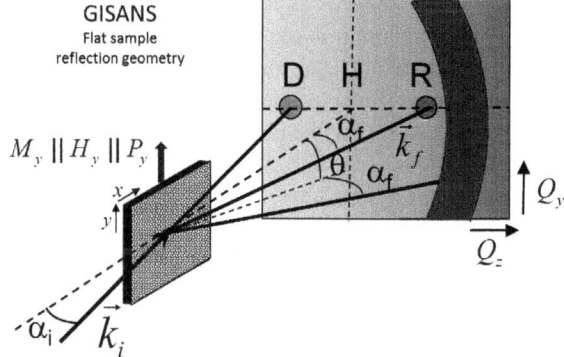

FIGURE 39 Scattering geometry for grazing incidence small-angle scattering (GISANS) experiments. D, direct transmitted beam; H, sample horizon; R, specular reflected beam.

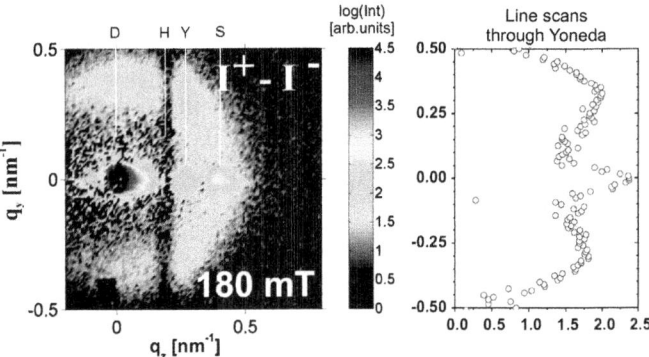

FIGURE 40 Polarized GISANS data taken from self-assembled Co/CoO nanoparticles deposited on a Si/SiO$_2$ substrate. Left panel shows a $Q_y - Q_z$ map of the intensity difference $I^+ - I^-$ taken at a field of 180 mT. The labels "D," "H," "Y," and "S" designate the direct beam, the horizon of the sample, the Yoneda peak, and the specularly reflected beam, respectively. The ringlike diffuse scattering is a Debye–Scherrer ring from the polycrystalline arrangement of the Co/CoO particles. Right panel shows a Q_y scan through the Yoneda peak at constant $Q_z = 0.26$ nm^{-1}. Adapted and reproduced with permission from Ref. [219].

spacings. PNR and P-GISANS data evaluation reveals the size of the saturated Co core and the CoO shell of the NPs. In Figure 40 the P-GISANS data are reproduced for a field of 180 mT which is close to saturation, showing the difference intensity $I^+ - I^-$, which is sensitive to the nuclear–magnetic interference cross term, like in SANS. The ringlike intensity is due to the particle form factor. The transmitted beam and specular reflected beam are highlighted.

6.5 TIME DEPENDENT POLARIZED NEUTRON SCATTERING

In the previous sections, we have considered static or quasistatic neutron reflectivity and neutron scattering experiments on time scales of hours to seconds. If higher time resolution is required, neutron scattering has to resort to the frequency domain via inelastic or quasielastic scattering. In backscattering experiments an energy resolution of less than μeV is achieved, corresponding to a few nanoseconds. With spin-echo methods a time resolution of a few microseconds can be reached. These experimental techniques are described in detail by Arai in Chapter 3 of the first volume on Neutron Scattering [14]. Clearly, there is a substantial time or frequency gap between seconds and microseconds or 1 Hz to 10 MHz. This region is highly interesting for nanomagnetism as it covers domain processes such as DW propagation and domain reversal in magnetic nanostructures. Only recently developments have taken place to study with PNR methods the magnetization reversal as a function of switching field frequency covering the range from kilohertz to megahertz [20,220–222].

In Section 6.4.1.4, we have already discussed domain reversal processes by application of quasistatic magnetic fields. Now we consider the case that the in-plane magnetization of a film is reversed in an external alternating magnetic field (AC field) sinusoidally oscillating with the frequency ω applied to the sample as sketched in Figure 41. The period $T = 2\pi/\omega$ is assumed to be short compared to the time integration at each point along a PNR reflectivity curve. Furthermore the AC field amplitude parallel to the neutron polarization axis $H^y_{AC}(t) = H^y_0 \sin(\omega t)$ is taken to be larger than the coercive field H_c of the magnetic film. At this point we have to distinguish between sample magnetization as a function of frequency and what neutrons actually record in time-integrating mode, called AC-PNR, or in time-resolved mode, termed TRAC-PNR.

In the simplest case we assume that the magnetization reversal proceeds by nucleation of opposite domains with 180° DWs propagating at 90° to the applied field in the X-direction. At low frequencies we expect the sample magnetization $M(t)$ to follow the external field reversal. The transient time τ (marked in green at the bottom of Figure 41), during which domain propagation takes place, is short in comparison to the lifetime of the saturation states (marked in red and blue) and to the total period T. We call this regime the adiabatic regime. With increasing frequency, however, the transient state may not be finished before the external field is reversed, i.e., the sample magnetization lags behind the field reversal. This is the nonadiabatic regime, where the sample remains for most of the time in a domain state. At very high

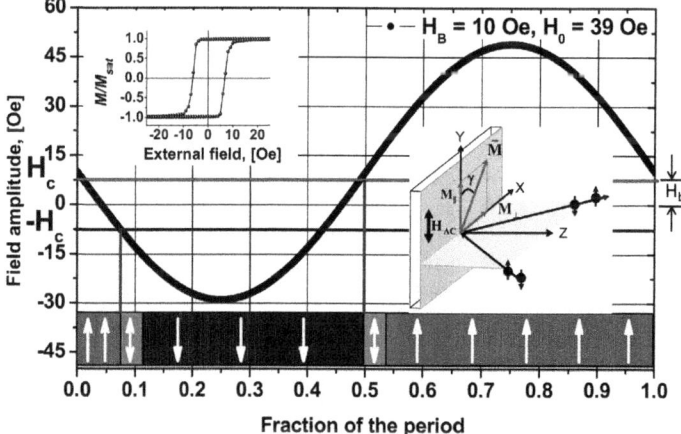

FIGURE 41 Schematics of the AC field applied to a magnetic sample. $H_{AC}(t) = H_0 \cos(\omega t)$ is the high frequency field applied to the sample, where the center line is offset by a bias field H_b. Whenever $H(t) = H_b + H_{AC}(t)$ crosses the coercive field, the sample magnetization starts to reverse, as shown schematically in the lower part. The reversal takes place during some transient time τ. The insets top left and bottom right show the hysteresis along the easy axis and the scattering geometry, respectively. *Reprinted and adapted with permission from Ref. [20].*

frequencies the film magnetization can no longer follow the external field reversal and it will essentially remain in the single domain initial state.

Periodic magnetic field reversals have to be distinguished from pulsed field excitations that switch the magnetization locally during a very short time interval, followed by a free propagation of the DW [223,224]. Those experiments allow to determine the DW velocity as a function of reversal field amplitude. In contrast, AC field experiments aim at the determination and characterization of magnetization reversals as a function of frequency.

Because of lack of sufficiently fast neutron detectors, PNR experiments in AC fields are presently performed in a time-integrating mode, i.e., while the sample is exposed to an alternating magnetic field larger than the coercive field driving the magnetic film from one saturation state to the other, neutron detection takes a time average over a reversal cycle. This method is referred to as AC-PNR.

If over one cycle there are as many up as down domains, the average would cancel and R^{++} would be equal to R^{--}. Therefore the application of a small bias field $H_b > H_c$ is required to prefer—over time—one domain orientation over the other. The sample magnetization $M(t)$ evolves then under the superposition of a DC and AC field $H(t) = H_b + H_{ac}(t)$, schematically sketched in Figure 41. Because of the biased AC field, the magnetic film spends more time in the up magnetization and less time in the down magnetization, i.e., up domains with a weight factor $w_\uparrow$ are more populated than down domains $w_\downarrow = 1 - w_\uparrow$. Due to the bias field imposed asymmetry ($w_\uparrow \leq w_\downarrow$) on the time average, R^{++} and R^{--} reflectivities are split as shown in Figure 42.

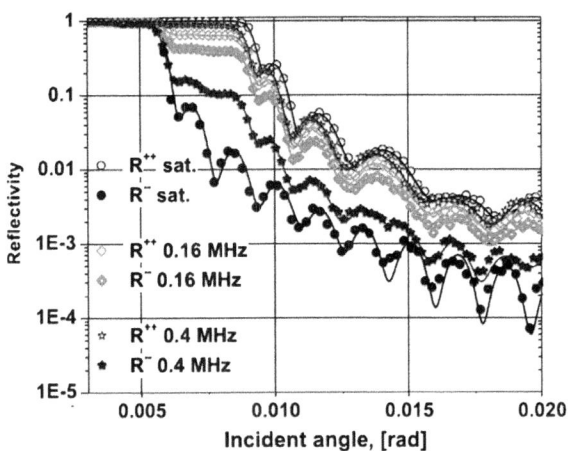

FIGURE 42 Non-spin-flip reflectivities $R^{\pm\pm}$ from a 70-nm-thick Fe(100) film are plotted as a function of incident angles for different AC magnetic field sweep frequencies with a constant amplitude of 39 Oe applied parallel to the Y-axis and parallel to a DC bias field of 10 Oe. The solid lines are least square fits to the data points. *Reprinted with permission from Ref. [20].*

In the experimental setup used by Zhernenkov and coworkers [20,221], the DC and AC fields are provided by two independent Helmholtz coils. The AC coil is part of an LC circuit with a frequency $f = 1/T$ ranging up to 2 MHz. The lower frequency limit is given by the depolarization of the neutron beam and is about 0.1 MHZ.

Figure 42 shows three pairs of reflectivity curves $R^{\pm\pm}$ measured at $f = 0$, 0.16, 0.4 MHz. Above $f = 1$ MHz, the reflectivities can hardly be distinguished from saturation curves and are therefore omitted. The static and high frequency case is characterized by NSF PNR, R^{++} and R^{--}, with well-defined critical angles of total reflection for neutron spins parallel or antiparallel to the mean magnetization. In contrast, for intermediate frequencies $f = 0.16$ MHz and $f = 0.4$ MHz, the total reflection regime shows both critical angles along each of the NSF reflectivities. These NSF curves can be described by the weighted sum $R^{\pm\pm} = [(1 \pm \bar{c})|r^+|^2 + (1 \mp \bar{c})|r^-|^2]/2$ of the reflectivities $|r^\pm|^2$ for neutrons with positive and negative spin projections onto the magnetization direction. The coefficient $\bar{c} = \overline{M(t)}/M_{sat} = \langle \cos \gamma \rangle$ in this equation is determined by the mean magnetization averaged over the AC period T and normalized to the saturation magnetization M_{sat}. For 180° domains $\bar{c} = w_\uparrow - w_\downarrow$.

For low frequencies f the reversal process, including DW nucleation and propagation, is fully completed within one field cycle. This is the adiabatic regime, where the DW propagation follows the external field. Then the magnetization averaged over the period is determined by the AC and DC field amplitudes, and $\bar{c}(f)$ is basically independent of f. This can be seen in Figure 43 for the first two data points. Above a critical frequency $f_c = f_c(H_0, H_b)$ the magnetization reversal is no longer complete during the AC field cycle and the factor $\bar{c}$ starts to depend on ω. In the limit of very high frequencies the magnetization finally stops to react on the AC field remaining in saturation. For a Fe thin film this limit is reached at $f \sim 1$ MHz.

From a more complete analysis presented in Ref. [20], including SF reflectivities it follows that with increasing frequency the reversal is characterized by a coexistence of two types of domains. One of them (80%) is collinear with the Y-axis causing only NSF reflection, while the other one (20%) is oriented orthogonal to the DC and AC fields hence providing small, but not negligible SF reflection (not shown in Figure 42). As a result, the instant magnetization vector makes an angle $\gamma(t)$ so that the mean value $\bar{c}(f) = \langle \cos \gamma \rangle$. At the same time, SF reflectivity is proportional to the mean value $\langle \sin^2 \gamma \rangle = w_\perp$, where $w_\perp = 1 - w_\uparrow - w_\downarrow$ is the statistical weight of 90° domains. These 90° DWs appear to facilitate the propagation of the 180° DWs. Only when crossing the 0.5 MHz barrier, the magnetization reversal is increasingly inhibited and finally stops at about 1 MHz.

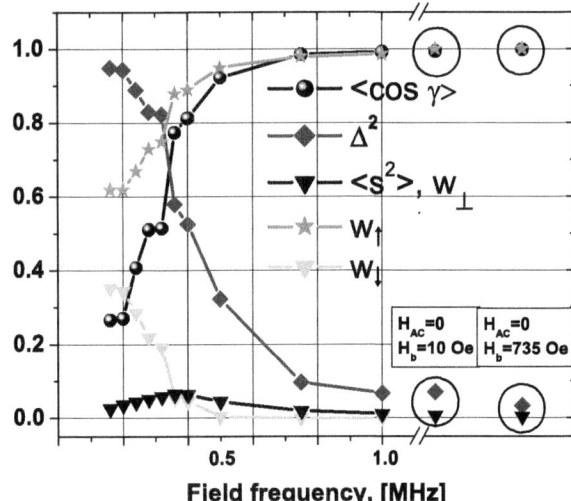

FIGURE 43 Frequency dependencies of fitting parameters: $\bar{c}_\gamma = \langle\cos\gamma\rangle$ and weight factors $w_\uparrow(f)$ and $w_\downarrow(f)$. The two additional parameters, the weight factor $\overline{s^2}_\gamma(f) = w_\perp(f)$ accounting for 90° domains and the dispersion $\Delta^2(f) = \langle\cos^2\gamma\rangle - \langle\cos\gamma\rangle^2$ describing the domain state of the sample, follow from additional spin-flip reflectivities not shown here. The encircled data points represent for comparison of the fit parameters in static fields of 10 Oe and saturation field of 735 Oe. *Reprinted with permission from Ref. [20].*

The present example shows that AC-PNR opens new opportunities for investigating the response of magnetic nano- and heterostructures to high frequencies relevant for spintronic applications. Only one example was shown and discussed. Obviously, there are many more opportunities for varying the basic idea. For instance, the bias field could be replaced by a ferromagnetic/antiferromagnetic heterostructure which exhibits an EB effect. Then the bias field would intrinsically be existent. The AC field could also be applied in the transverse X-direction for studying coherent pendulous-like magnetic oscillations about an easy axis [222]. Furthermore, the experimental technique can be applied to arrays of nanowires, where the aspect ratio and the edge roughness have a large impact on the critical frequency [225], or to NPs, where the dynamical response as a function of temperature would be of interest.

In time averaged AC-PNR experiments, one can only determine the weight factors of the magnetization domains $w_\uparrow$ and $w_\downarrow$. But the time dependence of the magnetization $M(t)$ itself cannot be resolved. This time variation could, in principle, be recorded with time-resolved AC-PNR (TRAC-PNR). However, for frequencies of interest here, TRAC-PNR still faces significant technical difficulties. Those can be overcome by employing a new type of modulation technique termed: modulation of intensity emerging from zero effort (MIEZE), proposed by Gähler and coworkers [226]. Discussing the MIEZE principle would be beyond the scope of this review. The interested reader is referred to the Ref. [227]. A simpler variant of the main idea, which is similar to the heterodyning effect in radio communication, was realized in the thesis work of Zhernenkov [222] and Gorkov [225]. First results turned out to be very promising and will hopefully open a new direction for time-resolved PNR studies.

6.6 SUMMARY, CONCLUSION, AND OUTLOOK

In this book chapter, we describe contributions of neutron scattering to nanomagnetism. Thirty years after introducing the method of PNR by Felcher, this method is widely established for the characterization of magnetic films, multilayers, heterostructures, and lateral magnetic patterns. Often polarized neutron reflectivity and resonant magnetic X-ray scattering are applied to the same system to complement the analysis. Presently, there are at least 20 instruments available worldwide for executing PNR with a perspective of many more instruments still to be built in the coming years. Any neutron source for scientific research features nowadays at least one if not more neutron reflectometers. If a neutron reflectometer is not equipped with polarization option, it can still be retrofitted.

Within the field of PNR, it turns out that specular polarized neutron reflectivity in the saturated state is by far the most popular application. This application does not even require a polarization analysis, if the saturated magnetization vector is parallel to the polarization axis. Then PNR provides a magnetization depth profile in magnetic films and heterostructures, which is increasingly been used for detecting weak magnetic inductions at interfaces between dissimilar material layers.

The most complete layer-by-layer vector magnetometry of thin films and heterostructures is, however, accomplished via quantitative evaluation of all four NSF and SF specular reflectivities. Nowadays programs are available that simultaneously fit a full scope of SF and NSF data appreciably facilitating the data analysis of collinear and noncollinear arrangements of magnetization vectors in a layer sequence. However, one should keep in mind that up to date the neutron spin analysis is applied only in one direction, collinear with the incident polarization vector. This 1D PNR analysis is not complete as it measures only one projection of the magnetization vector and the square of its projection onto the orthogonal axis. Meanwhile, theory and software exists that is able to analyze SF and NSF specular PNR data measured at any orientation between vectors of incident polarization and exit polarization. Experimental realization of 3D PNR would allow further progress in unambiguous determination of noncollinear structures, especially in most complex cases of magnetic domains and DWs. Certainly, magnetic domains can also be imaged by various scanning methods as long as they provide a contrast at the surface. But with PNR noncollinearities and domain states can be depicted still deep inside of heterostructures.

The quantitative analysis of OSS is most challenging as it requires to quantitatively evaluate 2D maps of scattered intensity distributions. Domains larger than the neutron coherence length contribute to specular reflection and therefore can be well distinguished from smaller domains giving rise to off-specular diffuse scattering. Depending on the neutron wavelength and angle of incidence the neutron coherence length can be varied from tenths of millimeters down to hundreds of nanometers, providing sensitivity to various lateral length scales. Present practice is to extract model parameters for the domains by comparing experimental data with results of model simulations in DWBA, but not via a least square routine routinely employed to fit specular PNR. This is, actually, not a severe limitation as most structural parameters for the transverse direction found from the fit of specular PNR are fixed in the model. Only a few free parameters for the lateral domain structure, including mean domain size, mean angle between magnetization vectors of domains, and dispersion of these angles, are sufficient to vary in order to achieve reasonable agreement between 2D sets of experimental data and model calculations.

In future, especially considering noncollinear transverse structures, the DWBA code should be imbedded into that or the other fitting routine. Then, fitting of 2D maps may require appreciable computer resources, which however are available even at the now available PCs.

A new challenge is the investigation of the magnetization reversal processes in the frequency or time domain via specular PNR and off-specular diffuse scattering. This is an emerging field where first results are promising.

With more neutron sources and in particular pulsed sources coming on line in the future, one can expect also substantial improvements for the PNR method. Progress should concern incident beam intensity, background radiation, and resolution. High intensity and low background are the most important improvements. This would allow to measure PNR reflectivities up to larger Q_z values for enhancing the accuracy of fit parameters and the use of smaller samples. Higher intensity would also allow better beam collimation for GISANS application and for increasing the coherence lengths of neutrons. For the future we expect substantial advancements in three emerging fields: (1) 3D vector polarization analysis, (2) extended use of polarized GISANS, and (3) time-resolved studies. The theoretical framework of these fields together with some early experimental realizations is discussed in this review.

ACKNOWLEDGMENTS

Much of the PNR work reported and reviewed here was carried out during funding by BMBF Verbundforschung 05KN7PC1 and before. BPT is grateful for the hospitality at the Institut Laue-Langevin and HZ for the hospitality of the MAINZ Graduate School of Excellence at the Johannes Gutenberg University Mainz while this manuscript was written. We are indebted to many discussions with our colleagues and coworkers, but in particular with our former PhD students Kirill Zhernenkov and Dmitry Gorkov.

REFERENCES

[1] S.D. Bader, Colloquium: opportunities in nanomagnetism, Rev. Mod. Phys. 78 (1) (2006), http://dx.doi.org/10.1103/RevModPhys.78.1.
[2] C.A.F. Vaz, J.A.C. Bland, G. Lauhoff, Magnetism in ultrathin film structures, Rep. Prog. Phys. 71 (2008) 78.
[3] C.L. Dennis, R.P. Borges, L.D. Buda, U. Ebels, J.F. Gregg, M. Hehn, et al., The defining length scales of mesomagnetism: a review, J. Phys. Condens. Matter 14 (2002) R1175.
[4] D.C. Ralph, M.D. Stiles, Spin transfer torques, J. Magn. Magn. Mater. 320 (2008) 1190.
[5] J. Sinova, D. Culcer, Q. Niu, N.A. Sinitsyn, T. Jungwirth, et al., Universal intrinsic spin hall effect, Phys. Rev. Lett. 92 (2004) 126603, http://dx.doi.org/10.1103/PhysRevLett.92.126603.
[6] Y. Tserkovnyak, A. Brataas, G.E.W. Bauer, B.I. Halperin, Nonlocal magnetization dynamics in ferromagnetic heterostructures, Rev. Mod. Phys. 1375 (2005) 77.
[7] G.E.W. Bauer, E. Saitoh, B.J. van Wees, Spin caloritronics, Nat. Mater. 11 (2012) 391.
[8] R.L. Stamps, S. Breitkreutz, J. Åkerman, A.V. Chumak, Y. Otani, G.E.W. Bauer, et al., The 2014 magnetism roadmap, J. Phys. D Appl. Phys. 47 (2014) 333001.
[9] S.A. Wolf, D.D. Awschalom, R.A. Buhrman, J. Daughton, S. von Molnar, M. Roukes, et al., Spintronics: a spin-based electronics vision for the future, Science 294 (2001) 1488.
[10] I. Zutic, J. Fabian, S.D. Sarma, Spintronics: fundamentals and applications, Rev. Mod. Phys. 76 (2004) 323. http://dx.doi.org/10.1103/RevModPhys.76.323.
[11] A. Fert, Nobel lecture: origin, development, and future of spintronics, Rev. Mod. Phys. 80 (2008) 1517. http://dx.doi.org/10.1103/RevModPhys.80.1517.
[12] H. Zabel, Progress in spintronics, Superlattices Microstruct. 46 (2009) 541.
[13] S.D. Bader, S.S.P. Parkin, Spintronics, Annu. Rev. Condens. Matter Phys. 71 (2010). http://dx.doi.org/10.1146/annurev-conmatphys-070909-104123.
[14] F. Fernandez-Alonso, D.L. Price (Eds.), Neutron Scattering, vol. 44, Academic Press, Elsevier, 2013.
[15] G.P. Felcher, R.O. Hillecke, R.K. Crawford, J. Haumann, R. Kleb, G. Ostrowski, Polarized neutron reflectometer: a new instrument to measure magnetic depth profiles, Rev Sci. Instrum. 58 (1987) 609.
[16] http://material.fysik.uu.se/Group_members/adrian/reflect.htm (last accessed 18.03.15).
[17] C. Majkrzak, N. Berk, Exact determination of the phase in neutron reflectometry, Phys. Rev. B 52 (1995) 10827.
[18] C. Majkrzak, N. Berk, J. Dura, S. Satija, A. Karim, J. Pedulla, et al., Phase determination and inversion in specular neutron reflectometry, Phys. B 248 (1998) 338.
[19] H. Leeb, M. Simon, K. Nikolics, J. Kasper, Reconstruction of magnetic profiles from polarized reflectivity data, Phys. B 397 (2007) 50.
[20] K. Zhernenkov, D. Gorkov, B. Toperverg, H. Zabel, Frequency dependence of magnetization reversal in thin Fe(100) films, Phys. Rev. B 88 (2013) 020401(R).
[21] A.B. Shick, S. Khmelevskyi, O.N. Mryasov, J. Wunderlich, T. Jungwirth, Spin-orbit coupling induced anisotropy effects in bimetallic antiferromagnets: a route towards antiferromagnetic spintronics, Phys. Rev. B 81 (2010) 212409.
[22] J. Penfold, M. Tucker, in: F. Fernandez-Alonso, D.L. Price (Eds.), Neutron Scattering, vol. 44, Academic Press, Elsevier, 2013.
[23] J. Fink, E. Schierle, E. Weschke, J. Geck, Resonant elastic soft X-ray scattering, Rep. Prog. Phys. 76 (2013) 056502.

[24] H. Dürr, T. Eimüller, H.J. Elmers, S. Eisebitt, M. Farle, W. Kuch, et al., A closer look into magnetism: opportunities with synchrotron radiation, IEEE Trans. Magn. 45 (2009) 15.
[25] G. Schütz, E. Göring, H. Stoll, Synchrotron Radiation Techniques Based on X-ray Magnetic Circular Dichroism, in: H. Kronmüller, S. Parkin (Eds.), Handbook of Magnetism and Advanced Magnetic Materials, vol. 3, Wiley & Sons, New York, 2007.
[26] V. Lauter-Pasyuk, H.J. Lauter, B.P. Toperverg, L. Romashev, V. Ustinov, Phys. Rev. Lett. 89 (2002) 167203.
[27] S. Couet, K. Schlage, T. Diederich, R. Rüffer, K. Theis-Bröhl, B.P. Toperverg, et al., The magnetic structure of coupled Fe/FeO multilayers revealed by nuclear resonant and neutron scattering methods, New J. Phys. 11 (2009) 013038.
[28] C. Majkrzak, Polarized neutron scattering methods and studies involving artificial superlattices, Phys. B 156–157 (1989) 619.
[29] C.F. Majkrzak, Polarized neutron reflectometry, Phys. B 173 (1991) 75.
[30] G.B. Felcher, Magnetic depth profiling studies by polarized neutron reflection, Phys. B 192 (1993) 137.
[31] H. Zabel, Spin polarized neutron reflectivity of magnetic films and superlattices, Phys. B 198 (1994) 156.
[32] H. Zabel, X-ray and neutron reflectivity analysis of thin films and superlattices, Appl. Phys. A 58 (1994) 159.
[33] C. Majkrzak, Neutron scattering studies of magnetic thin films and multilayers, Phys. B 221 (1996) 342.
[34] J.F. Ankner, G.P. Felcher, Polarized neutron reflectometry, J. Magn. Magn. Mater. 200 (1999) 741.
[35] H. Zabel, R. Siebrecht, A. Schreyer, Neutron reflectometry on magnetic thin films, Phys. B 276–278 (2000) 17.
[36] A. Schreyer, T. Schmitte, R. Siebrecht, P. Bödeker, H. Zabel, S.H. Lee, et al., Neutron scattering on magnetic thin films: pushing the limits, J. Appl. Phys. 87 (2000) 5443.
[37] H. Zabel, K. Theis-Bröhl, Polarized neutron reflectivity and scattering studies of magnetic heterostructures, J. Phys. Condens. Matter 15 (2003) S505.
[38] J.F. Ankner, H. Zabel, Applications of neutron reflectivity to nanoscience: thin films and interfaces, MRS Bull. 28 (2003) 918.
[39] M.R. Fitzsimmons, S.D. Bader, J.A. Borchers, G.P. Telcher, J.K. Furdyna, A. Hoffmann, et al., Neutron scattering studies of nanomagnetism and artificially structured materials, J. Magn. Magn. Mater. 271 (2004) 103.
[40] J.A.C. Bland, A.F. Vaz, Polarized neutron reflection studies on thin magnetic films, in: J.A. C. Bland, B. Heinrich (Eds.), Ultrathin Magnetic Structures III: Fundamentals of Nanomagnetism, Springer Verlag, Berlin, Heidelberg, New York, 2005.
[41] M. Fitzsimmons, C.F. Majkrzak, Modern Techniques for Characterizing Magnetic Materials, 2005.
[42] C. Majkrzak, K. O'Donovan, N. Berk, in: T. Chatterji (Ed.), Neutron Scattering from Magnetic Materials, vol. 9, Elsevier, 2006.
[43] H. Zabel, Neutron reflectivity of spintronic materials, Mater. Today 9 (2006) 42.
[44] H. Zabel, K. Theis-Bröhl, B.P. Toperverg, Polarized neutron reflectivity and scattering from magnetic nanostructures and spintronic materials, in: H. Kronmüller, S. Parkin (Eds.), Handbook of Magnetism and Advanced Magnetic Materials, vol. 3, Wiley & Sons, New York, 2007.
[45] G. Felcher, A. Hoffmann, Domain States Determined by Neutron Refraction and Scattering, in: H. Kronmüller, S. Parkin (Eds.), Handbook of Magnetism and Advanced Magnetic Materials, vol. 3, Wiley & Sons, New York, 2007.

[46] V. Lauter-Pasyuk, Neutron grazing incidence techniques for nano-science, Collection SFN 7 (2007) 221.
[47] H. Zabel, K. Theis-Bröhl, M. Wolff, B.P. Toperverg, Polarized neutron reflectometry for the analysis of nanomagnetic systems, IEEE Trans. Magn. 44 (2008) 1928.
[48] C. Marrows, L. Chapon, S. Langridge, Spintronics and functional materials, Mater. Today (2009) 70. Neutron scattering special issue.
[49] F. Ott, Neutron Scattering on Magnetic Nanostructures, Condensed Matter, Université Paris Sud − Paris X, 2009. I.
[50] T. Saerbeck, Magnetic exchange phenomena probed by neutron scattering, Solid State Phys. 65 (2014) 237.
[51] M.R. Fitzsimmons, I.K. Schuller, Neutron scattering—the key characterization tool for nanostructured magnetic materials, J. Magn. Magn. Mater. 350 (2014) 199.
[52] J.S. Blundell, J.A.C. Bland, Polarized neutron reflectivity as probe of magnetic films and multilayers, Phys. Rev. B 46 (1992) 3391.
[53] F. Radu, V.K. Ignatovich, Generalized matrix method for the transmission of neutrons through multilayer magnetic systems with noncollinear magnetization, Physica B 267−268 (1999) 175.
[54] A. Rühm, B. Toperverg, H. Dosch, Supermatrix approach to polarized neutron reflectivity from arbitrary spin structures, Phys. Rev. B 60 (1999) 16073.
[55] B.P. Toperverg, in: Th. Brückel, W. Schweika (Eds.), Polarized Neutron Scattering, vol. 12, 2002.
[56] B.P. Toperverg, Off-specular polarized neutron scattering from magnetic fluctuations in thin films and multilayers, Appl. Phys. A 74 (2002) S1560.
[57] J. Dawidowski, J.R. Granada, J.R. Santisteban, F. Cantargi, L.A. Rodriguez Palomino, in: F. Fernandez-Alonso, D.L. Price (Eds.), Neutron Scattering, vol. 44, Academic Press, Elsevier, 2013.
[58] L.D. Landau, E.M. Lifshitz, The Classical Theory of Fields, in: Course of Theoretical Physics, vol. II, Butterworth-Heinemann, 1975.
[59] L.D. Landau, E.M. Lifshitz, Quantum Mechanics Non-relativistic Theory, in: Course of Theoretical Physics, vol. III, Butterworth-Heinemann, 1981.
[60] B. Toperverg, O. Schärpf, I. Anderson, Optical theorem for neutron scattering from rough interfaces, Physica B 276−278 (2000) 954.
[61] B. Toperverg, Optical theorem, depolarization and vector tomography, Physica B 335 (2003) 174.
[62] M.T. Rekveldt, J. Phys. Coll. C1 32 (1971) 579.
[63] G.M. Drabkin, A.I. Okorokov, V.V. Runov, JETP Lett. 5 (1972) 324.
[64] A.I. Okorokov, V.V. Runov, Vector analysis of polarization at small-angle neutron scattering, Physica B 297 (2001) 239.
[65] F. Tasset, Neutron beams at the spin revolution, Physica B 297 (2001) 1.
[66] B. Toperverg, O. Nikonov, V. Lauter-Passyuk, H. Lauter, Towards 3d polarization analysis in neutron reflectometry, Physica B 279 (2001) 169.
[67] G. Felcher, S. Adenwalla, V.D. Haan, A.V. Well, Zeeman splitting of surface-scattered neutrons, Nature 377 (1995) 409.
[68] B. Toperverg, A. Rühm, W. Donner, H. Dosch, Polarized neutron grazing angle birefrigent diffraction from magnetic stratified media, Physica B 267−268 (1999) 198.
[69] R. Günther, W. Donner, B. Toperverg, H. Dosch, Birefringent bragg diffraction of evanescent neutron states in magnetic films, Phys. Rev. Lett. 81 (1998) 116.

[70] L.G. Parratt, Surface studies of solids by total reflection of X-rays, Phys. Rev. 95 (1954) 359.
[71] M. Born, E. Wolf, Principles of Optics, Pergamon Press, Oxford, 1975.
[72] B. Toperverg, V. Lauter-Pasyuk, H. Lauter, A. Vorobiev, Grazing incidence neutron diffraction from ferromagnetic films in multi-domain state, Physica B 356 (2005) 51.
[73] S.K. Sinha, E.B. Sirota, S. Garoff, H.B. Stanley, X-ray and neutron scattering from rough surfaces, Phys. Rev. B 38 (1988) 2297.
[74] C. Majkrzak, B.M. Ch Metting, J. Dura, S. Satija, T. Udovic, N. Berk, Determination of the effective transverse coherence of the neutron wave packet as employed in reflectivity investigations of condensed-matter structures. I. Measurements, Phys. A 89 (2014) 033851.
[75] N. Berk, Determination of the effective transverse coherence of the neutron wave packet as employed in reflectivity investigations of condensed-matter structures. II. Analysis of elastic scattering using energy-gated wave packets with an application to neutron reflection from ruled gratings, Physica A 89 (2014) 033852.
[76] R.M. Moon, T. Riste, W.C. Koehler, Polarization analysis of thermal neutron scattering, Phys. Rev. 181 (1996) 920.
[77] F. Mezei, Z. Phys. 255 (1972) 146.
[78] J. Demeter, A. Teichert, K. Kiefer, D. Wallacher, H. Ryl, E. Menéndez, et al., Simultaneous polarized neutron reflectometry and anisotropic magnetoresistance measurements, Rev. Sci. Instrum. 82 (2011) 033902.
[79] A. Schreyer, R. Siebrecht, U. Englisch, U. Pietsch, H. Zabel, ADAM, the new reflectometer at the ILL, Physica B 248 (1998) 349.
[80] http://www.mlz-garching.de/instrumente.html (last accessed 20.03.15).
[81] D. Yamazaki, M. Takeda, I. Tamura, R. Maruyama, A. Moriai, M. Hino, et al., Polarized neutron reflectometer SUIREN at JRR-3, Phys. B 404 (2009) 2557.
[82] S. Basu, S. Singh, A new polarized neutron reflectometer at Dhruva for specular and off-specular neutron reflectivity studies, J. Neutron Res. 44 (2006) 109.
[83] A. Devishvili, K. Zhernenkov, A.J.C. Dennison, B.P. Toperverg, M. Wolff, B. Hjörvarsson, et al., SuperADAM: upgraded polarized neutron reflectometer at the Institut Laue-Langevin, Rev. Sci. Instrum. 84 (2013) 025112.
[84] S. Grigoriev, A. Okorokov, V. Runov, Peculiarities of the construction and application of a broadband adiabatic flipper of cold neutrons, Nucl. Instrum. Methods A 384 (1997) 451.
[85] V. Syromyatnikov, A. Schebetov, D. Lott, A. Bulkin, N. Pleshanov, V. Pusenkov, PNPI wide-aperture fan supermirror analyzer of polarization, Nucl. Instrum. Methods Phys. Res. A 634 (2012) S126.
[86] V.G. Syromyatnikov, V.A. Ulyanov, V. Lauter, V.M. Pusenkov, H. Ambaye, R. Goyette, et al., A new type of wide-angle supermirror analyzer of neutron polarization, J. Phys. Conf. Ser. 528 (1) (2014) 012021. http://stacks.iop.org/1742-6596/528/i=1/a=012021.
[87] A. Petoukhov, K. Andersen, D. Jullien, E. Babcock, J. Chastagnier, R. Chung, et al., Recent advances in polarised ^{3}He spin filters at the ILL, Physica B 385 (2006) 1146.
[88] M. James, A. Nelson, S. Holt, T. Saerbeck, W. Hamilton, F. Klose, The multipurpose time-of-flight neutron reflectometer Platypus at Australia's OPAL reactor, Nucl. Instrum. Methods A 632 (2011) 112.
[89] T. Saerbeck, N. Loh, D. Lott, B.P. Toperverg, A.M. Mulders, M. Ali, et al., Specular and off-specular polarized neutron reflectometry of canted magnetic domains in loose spin coupled CuMn/Co multilayers, Phys. Rev. B 85 (2012) 014411. http://dx.doi.org/10.1103/PhysRevB.85.014411.

[90] F. Ott, C. Fermon, PRISM, the new polarized neutron reflectometer with polarization analysis at the LLB, Neutron News 12 (2001) 27.
[91] C. Mattoni, C. Adams, K. Alvine, J. Doyle, S. Dzhosyuka, R. Golub, et al., A long wavelength neutron monochromator for superthermal production of ultracold neutrons, Phys. B 344 (2004) 343.
[92] P. Courtois, C. Menthonnex, R. Hehn, K. Andersen, V. Nesvizhevsky, O.Z.F. Piegsa, et al., Production and characterization of intercalated graphite crystals for cold neutron monochromators, Nucl. Instrum. Methods Phys. Res. A 634 (2011) S37.
[93] O. Elsenhans, P. Böni, H. Friedli, H. Grimmer, P. Buffat, K. Leifer, et al., Development of Ni/Ti multilayer supermirrors for neutron optics, Thin Solid Films 246 (1994) 110.
[94] T. Veres, L. Cser, Study of the reflectivity of neutron supermirrors influenced by surface oil layers, Rev. Sci. Instrum. 81 (2010) 063303.
[95] http://www.ncnr.nist.gov/resources/n-lengths/elements/he.html (last accessed 15.03.15).
[96] A. Petoukhov, V. Guillard, K. Andersen, E. Bourgeat-Lami, R. Chung, H. Humblot, et al., Compact magnetostatic cavity for polarised He-3 neutron spin filter cells, Nucl. Instrum. Methods A 560 (2006) 480. http://dx.doi.org/10.1016/j.nima.2005.12.247.
[97] M. Wolff, F. Radu, A. Petoukhov, H. Humblot, D. Jullien, K.H. Andersen, et al., Scientific reviews: ^{3}He spin filter at the Institut Laue-Langevin: polarization analysis of diffuse scattering, Neutron News 17 (2) (2006) 26.
[98] S. Mattauch, MARIA − magnetic reflectometer with high incident angle, J. Large Scale Res. Facil. 48 (2014) 6.
[99] http://www.ncnr.nist.gov/programs/reflect/index.html (last accessed on 18.03.15).
[100] B.P. Toperverg, Towards 3d polarization analysis in neutron reflectivity, Phys. B 297 (2001) 169.
[101] X. Zhang, N. Ji, V. Lauter, H. Ambaye, J.P. Wang, Strain effect of multilayer FeN structure on GaAs substrate, J. Appl. Phys. 113 (2013) 17.
[102] X. Zhang, N. Ji, V. Lauter, H. Ambaye, J.P. Wang, Strain induced giant magnetism in epitaxial $Fe_{16}N_2$ thin film, Appl. Phys. Lett 102 (2013) 072411.
[103] V. Lauter, H. Ambaye, R. Goyette, W.T.H. Lee, A. Parizzi, Highlights from the magnetism reflectometer at the SNS, Physica B 404 (2009) 2543.
[104] J. Orna, P.A. Algarabel, L. Morellon, J.A. Pardo, J.M. de Teresa, R.L. Anton, et al., Origin of the giant magnetic moment in epitaxial Fe_3O_4 thin films, Phys. Rev. B 81 (2010) 144420.
[105] S. Singh, M.R. Fitzsimmons, T. Lookman, J.D. Thompson, H. Jeen, A. Biswas, et al., Magnetic nonuniformity and thermal hysteresis of magnetism in a manganite thin film, Phys. Rev. Lett. 108 (2012) 077207.
[106] Y. Ding, D.A. Arena, J. Dvorak, M. Ali, C.J. Kinane, C.H. Marrows, et al., Bulk and near-surface magnetic properties of FeRh thin films, J. Appl. Phys. 103 (2008) 07B515.
[107] R. Fan, C.J. Kinane, T.R. Charlton, R. Dorner, M. Ali, M.A. de Vries, et al., Ferromagnetism at the interfaces of antiferromagnetic FeRh epilayers, Phys. Rev. B 82 (2010) 184418.
[108] C. Baldasseroni, G.K. Pálsson, C. Bordel, S. Valencia, A.A. Unal, F. Kronast, et al., Effect of capping material on interfacial ferromagnetism in FeRh thin films, J. Appl. Phys. 115 (2014) 043919.
[109] M.R. Fitzsimmons, N.W. Hengartner, S. Singh, M. Zhernenkov, F.Y. Bruno, J. Santamaria, et al., Upper limit to magnetism in $LaAlO_3/SrTiO_3$ heterostructures, Phys. Rev. Lett. 107 (2011) 217201. http://dx.doi.org/10.1103/PhysRevLett.107.217201.
[110] P. Korelis, A. Liebig, M. Björck, B. Hjörvarsson, H. Lidbaum, K. Leifer, et al., Highly amorphous $Fe_{90}Zr_{10}$ thin films, and the influence of crystallites on the magnetism, Thin Solid Films 404 (2010) 519. http://dx.doi.org/10.1016/j.tsf.2010.07.084.

[111] A. Bhattacharya, S.J. May, S.G.E. te Velthuis, M.W.X. Zhai, B. Jiang, J.M. Zuo, et al., Metal-insulator transition and its relation to magnetic structure in $(LaMnO_3)/(SrMnO_3)$ superlattices, Phys. Rev. Lett. 100 (2008) 257203. http://dx.doi.org/10.1103/PhysRevLett.100.257203.

[112] M. Ahlberg, M. Marcellini, A. Taroni, G. Andersson, M. Wolff, B. Hjörvarsson, Influence of boundaries on magnetic ordering in Fe/V superlattices, Phys. Rev. B 81 (2010) 214429. http://dx.doi.org/10.1103/PhysRevB.81.214429.

[113] C. He, A.J. Grutter, M. Gu, N.D. Browning, Y. Takamura, B.J. Kirby, et al., Interfacial ferromagnetism and exchange bias in $CaRuO_3/CaMnO_3$ superlattices, Phys. Rev. Lett. 109 (2012) 197202. http://dx.doi.org/10.1103/PhysRevLett.109.197202.

[114] D.K. Satapathy, M.A. Uribe-Laverde, I. Marozau, V.K. Malik, S. Das, T. Wagner, et al., Magnetic proximity effect in $YBa_2Cu_3O_7/La_{2/3}Ca_{1/3}MnO_3$ and $YBa_2Cu_3O_7/LaMnO_{3+\delta}$ superlattices, Phys. Rev. Lett. 108 (2012) 197201. http://dx.doi.org/10.1103/PhysRevLett.108.197201.

[115] M.A. Uribe-Laverde, D.K. Satapathy, I. Marozau, V.K. Malik, S. Das, K. Sen, et al., Depth profile of the ferromagnetic order in a $YBa_2Cu_3O_7/La_{2/3}Ca_{1/3}MnO_3$ superlattice on a LSAT substrate: a polarized neutron reflectometry study, Phys. Rev. B 87 (2013) 115105. http://dx.doi.org/10.1103/PhysRevB.87.115105.

[116] J. Bertinshaw, S. Brück, D. Lott, H. Fritzsche, Y. Khaydukov, O. Soltwedel, et al., Element-specific depth profile of magnetism and stoichiometry at the $La_{0.67}Sr_{0.33}MnO_3/BiFeO_3$ interface, Phys. Rev. B 90 (2014) 041113. http://dx.doi.org/10.1103/PhysRevB.90.041113.

[117] S. Singh, S. Basu, M. Gupta, M. Vedpathakz, R.H. Kodama, Investigation of interface magnetic moment of FeGe multilayer: a neutron reflectivity study, J. Appl. Phys. 101 (3) (2007) 033913. http://dx.doi.org/10.1063/1.2450680.

[118] S. Amir, M. Gupta, A. Gupta, K. A, J. Stahn, Silicide layer formation in evaporated and sputtered Fe/Si multilayers: X-ray and neutron reflectivity study, Appl. Surf. Sci. 277 (2013) 182. http://dx.doi.org/10.1016/j.apsusc.2013.04.021.

[119] S. Singh, S. Basu, C. Prajapat, M. Gupta, A. Poswal, D. Bhattacharya, Depth dependent structure and magnetic properties and their correlation with magnetotransport in Fe/Au multilayers, Thin Solid Films 550 (2014) 326. http://dx.doi.org/10.1016/j.tsf.2013.10.025.

[120] S. Singh, S. Basu, M. Gupta, C.F. Majkrzak, P.A. Kienzle, Growth kinetics of intermetallic alloy phase at the interfaces of a Ni/Al multilayer using polarized neutron and X-ray reflectometry, Phys. Rev. B 81 (2010) 235413. http://dx.doi.org/10.1103/PhysRevB.81.235413.

[121] M. Swain, S. Singh, S. Basu, M. Gupta, Effect of interface morphology on intermetallics formation upon annealing of Al/Ni multilayer, J. Alloys Compd. 576 (2013) 257−261. http://dx.doi.org/10.1016/j.jallcom.2013.04.140.

[122] S. Singh, S. Basu, P. Bhatt, A.K. Poswal, Kinetics of alloy formation at the interfaces in a Ni-Ti multilayer: X-ray and neutron reflectometry study, Phys. Rev. B 79 (2009) 195435. http://dx.doi.org/10.1103/PhysRevB.79.195435.

[123] S. Mayr, W. Kreuzpaintner, B. Wiedemann, J. Ye, A. Schmehl, T. Mairoser, et al., In-situ polarised neutron reflectometry during thin film growth by dc magnetron sputtering, DPG, Verhandlungen (2015). MA 19.64.

[124] A. Paul, S. Mattauch, Chemical and magnetic profile of magnetic semiconductors as probed by polarized neutron reflectivity, J. Phys. D Appl. Phys. 43 (18) (2010) 185002. http://stacks.iop.org/0022-3727/43/i=18/a=185002.

[125] H.J. von Bardeleben, J.L. Cantin, D.M. Zhang, A. Richardella, D.W. Rench, N. Samarth, et al., Ferromagnetism in Bi_2Se_3:Mn epitaxial layers, Phys. Rev. B 88 (2013) 075149. http://dx.doi.org/10.1103/PhysRevB.88.075149.
[126] K. Munbodh, F.A. Perez, C. Keenan, D. Lederman, M. Zhernenkov, M.R. Fitzsimmons, Effects of hydrogen/deuterium absorption on the magnetic properties of Co/Pd multilayers, Phys. Rev. B 83 (2011) 094432. http://dx.doi.org/10.1103/PhysRevB.83.094432.
[127] D. Schumacher, A. Steffen, J. Voigt, J. Schubert, T. Brückel, H. Ambaye, et al., Inducing exchange bias in $La_{0.67}Sr_{0.33}MnO_{3-\delta}$/$SrTiO_3$ thin films by strain and oxygen deficiency, Phys. Rev. B 88 (2013) 144427. http://dx.doi.org/10.1103/PhysRevB.88.144427.
[128] M. Vadalà, A. Nefedov, M. Wolff, K.N. Zhernenkov, K. Westerholt, H. Zabel, Structure and magnetism of Co_2 MnGe−Heusler multilayers with V, Au and AlO_x spacer layers, J. Phys. D: Appl. Phys. 40 (2007) 1289.
[129] B. Toperverg, H. Lauter, V. Lauter-Pasyuk, Larmor pseudo-precession of neutron polarization at reflection, Physica B 356 (2005) 1.
[130] S. te Velthuis, G. Felcher, P. Blomquist, R. Wäppling, Neutron spin rotation in magnetic mirror, J. Phys. Condens. Matter 13 (2001) 5577.
[131] F. Radu, V. Leiner, M. Wolff, V.K. Ignatovich, H. Zabel, Quantum states of neutrons in magnetic thin films, Phys. Rev. B 71 (2005) 214423.
[132] L. Névot, P. Croce, Characterisation of surfaces by grazing X-ray reflection. Application to the study of polishing some silicate glasses, Rev. Phys. Appl. 15 (1980) 761.
[133] W.T. Lee, S. te Velthuis, G. Felcher, F. Klose, T. Gredig, E.D. Dahlberg, Ferromagnetic domain distribution in thin films during magnetization reversal, Phys. Rev. B 65 (2002) 224417.
[134] Y. Yoneda, Phys. Rev. 131 (1963) 2010.
[135] J. Nogués, I.K. Schuller, Exchange bias, J. Magn. Magn. Mater. 192 (1999) 203.
[136] A.E. Berkowitz, K. Takano, Exchange anisotropy − a review, J. Magn. Magn. Mater. 200 (1999) 552.
[137] R.L. Stamps, Mechanisms for exchange bias, J. Phys. D Appl. Phys. 33 (2000) R247.
[138] M. Kiwi, Exchange bias theory, J. Magn. Magn. Mater. 234 (2001) 584.
[139] F. Radu, H. Zabel, Fundamental aspects of exchange bias, in: H. Zabel, S.D. Bader (Eds.), Springer Tracts in Modern Physics, vol. ???, Springer, Berlin, 2007.
[140] S.W. Chen, X. Lu, E. Blackburn, V. Lauter, H. Ambaye, K.T. Chan, et al., Nonswitchable magnetic moments in polycrystalline and (111)-epitaxial permalloy CoO exchange-biased bilayers, Phys. Rev. B 89 (2014) 094419.
[141] J. Demeter, E. Menendez, A. Teichert, R. Steitz, Paramanik, C.V. Haesendonck, et al., Influence of magnetocrystalline anisotropy on the magnetization reversal mechanism in exchange bias Co/CoO bilayers, Solid State Commun. 152 (2012) 292.
[142] F. Radu, M. Etzkorn, R. Siebrecht, T. Schmitte, K. Westerholt, H. Zabel, Interfacial domain formation during magnetization reversal in exchange-biased CoO/Co bilayers, Phys. Rev. B 67 (2003) 134409. http://dx.doi.org/10.1103/PhysRevB.67.134409.
[143] D. Navas, C. Redondo, G.A. Badini Confalonieri, F. Batallan, A. Devishvili, O. Iglesias-Freire, et al., Domain-wall structure in thin films with perpendicular anisotropy: magnetic force microscopy and polarized neutron reflectometry study, Phys. Rev. B 90 (2014) 054425. http://dx.doi.org/10.1103/PhysRevB.90.054425.
[144] O. Hellwig, A. Berger, E.E. Fullerton, Domain walls in antiferromagnetically coupled multilayer films, Phys. Rev. Lett. 91 (2003) 197203. http://dx.doi.org/10.1103/PhysRevLett.91.197203.

[145] J.E. Davies, O. Hellwig, E.E. Fullerton, G. Denbeaux, J.B. Kortright, K. Liu, Magnetization reversal of Co/Pt multilayers: microscopic origin of high-field magnetic irreversibility, Phys. Rev. B 70 (2004) 224434. http://dx.doi.org/10.1103/PhysRevB.70.224434.

[146] P. Miceli, D.A. Neuman, H. Zabel, X-ray refractive index: a tool to determine the average composition in multilayer structures, Appl. Phys. Lett. 48 (1986) 24.

[147] A. Schreyer, J.F. Ankner, T. Zeidler, H. Zabel, M. Schäfer, J.A. Wolf, et al., Noncollinear and collinear magnetic structures in exchange coupled Fe/Cr(001) superlattices, Phys. Rev. B 52 (1995) 16066. http://dx.doi.org/10.1103/PhysRevB.52.16066.

[148] A. Schreyer, J.F. Ankner, T. Zeidler, H. Zabel, C.F. Majkrzak, M. Schäfer, et al., Direct observation of non-collinear spin structures in Fe/Cr(001) superlattices, Europhys. Lett. 32 (1995) 595.

[149] S.G.E. te Velthuis, J.S. Jiang, S.D. Bader, G.P. Felcher, Phys. Rev. Lett. 89 (2002) 1272031.

[150] T. Saerbeck, N. Loh, D. Lott, B.P. Toperverg, A.M. Mulders, A.F. Rodríguez, et al., Spatial fluctuations of loose spin coupling in CuMn/Co multilayers, Phys. Rev. Lett. 107 (2011) 127201. http://dx.doi.org/10.1103/PhysRevLett.107.127201.

[151] F. Brüssing, B. Toperverg, K. Zhernenkov, A. Devishvili, H. Zabel, M. Wolff, et al., Magnetization and magnetization reversal in epitaxial Fe/Cr/Co asymmetric spin-valve systems, Phys. Rev. B 85 (2012) 174409. http://dx.doi.org/10.1103/PhysRevB.85.174409.

[152] A. Bergmann, J. Grabis, B.P. Toperverg, V. Leiner, M. Wolff, H. Zabel, et al., Antiferromagnetic dipolar ordering in $[Co_2MnGe/V]_N$ multilayers, Phys. Rev. B 72 (2005) 214403.

[153] P. Grünberg, Layered magnetic structures: facts, figures, future, J. Phys. Condens. Matter 13 (2001) 7691. http://dx.doi.org/10.1088/0953-8984/13/34/314.

[154] P. Bruno, C. Chappert, Oscillatory coupling between ferromagnetic layers separated by a nonmagnetic metal spacer, Phys. Rev. Lett. 67 (1991) 1602. http://dx.doi.org/10.1103/PhysRevLett.67.1602.

[155] J.C. Slonczewski, Origin of biquadratic exchange in magnetic multilayers (invited), J. Appl. Phys. 73 (10) (1993).

[156] A. Buzdin, Proximity effects in superconductor-ferromagnet heterostructures, Rev. Mod. Phys. 77 (2005) 935.

[157] F.S. Bergeret, A.F. Volkov, K.B. Efetov, Odd triplet superconductivity and related phenomena in superconductor-ferromagnet structures, Rev. Mod. Phys. 77 (2005) 1321.

[158] K.B. Efetov, I.A. Garifullin, A.F. Volkov, K. Westerholt, Spin-polarized electrons in superconductor/ferromagnet hybrid structures, Springer Tracts Mod. Phys. 246 (2013) 85.

[159] J. Hoppler, H. Fritzsche, V.K. Malik, J. Stahn, G. Cristiani, H.U. Habermeier, et al., Polarized neutron reflectometry study of the magnetization reversal process in $YBa_2Cu_3O_7/La_{2/3}Ca_{1/3}MnO_3$ superlattices grown on $SrTiO_3$ substrates, Phys. Rev. B 82 (2010) 174439. http://dx.doi.org/10.1103/PhysRevB.82.174439.

[160] O. Hellwig, L.J. Heyderman, O. Petracic, H. Zabel, Competing interactions in patterned and self-assembled magnetic nanostructures, Springer Tracts Mod. Phys. 246 (2013) 189.

[161] M. Grimsditch, P. Vavassori, The diffracted magneto-optic Kerr effect: what does it tell you? J. Phys. Condens. Matter 16 (2004) R275.

[162] A. Westphalen, M.S. Lee, A. Remhof, H. Zabel, Invited article: Vector and Bragg magneto-optical Kerr effect for the analysis of nanostructured magnetic arrays, Rev. Sci. Instrum. 78 (2007) 121301.

[163] A. Remhof, A. Westphalen, K. Theis-Bröhl, J. Grabis, A. Nefedov, B. Toperverg, et al., Magnetization Reversal Studies of Periodic Magnetic Arrays via Scattering Methods, vol. 94, 2007.

[164] U.B. Arnalds, T.P.A. Hase, E.T. Papaioannou, H. Raanaei, R. Abrudan, T.R. Charlton, et al., X-ray resonant magnetic scattering from patterned multilayers, Phys. Rev. B 86 (2012) 064426.
[165] E. Popova, H. Loosvelt, M. Gierlings, L. Leunissen, R. Jonckheere, C.V. Haesendonck, et al., Magnetization reversal in exchange biased Co/CoO pattern, Eur. Phys. J. B 44 (2005) 491.
[166] B. Toperverg, G. Felcher, V. Metlushko, V. Leiner, R. Siebrecht, O. Nikonov, Grazing incidence neutron diffraction from large scale 2D structures, Physica B 283 (2000) 149.
[167] K. Temst, M.J.V. Bael, H. Fritzsche, In-plane vector magnetometry on rectangular Co dots using polarized neutron reflectivity, Appl. Phys. Lett. 79 (2001) 991.
[168] K. Temst, M.J.V. Bael, J. Swerts, D. Buntinx, C.V. Haesendonck, Y. Bruynseraede, et al., In-plane vector magnetometry on rectangular Co dots using polarized neutron reflectivity, J. Vac. Sci. Technol. B 21 (2003) 2043.
[169] K. Temst, M.V. Bael, J. Swerts, H. Loosvelt, E. Popova, D. Buntinx, et al., Polarized neutron reflectometry on lithographically patterned thin film structures, Superlattices Microstruct. 34 (2003) 87.
[170] K. Temst, E. Girgis, R. Portugal, H. Loosvelt, E. Popova, M.V. Bael, et al., Magnetization and polarized neutron reflectivity experiments on patterned exchange bias structures, Eur. Phys. J. B 45 (2005) 261.
[171] K. Temst, E. Popova, M.J. Van Bael, H. Loosvelt, J. Swerts, D. Buntinx, et al., Magnetization reversal in patterned ferromagnetic and exchange-biased nanostructures studied by neutron reflectivity (invited), J. Appl. Phys. 97 (10) (2005) 10.
[172] J. Mohanty, S. Vandezande, S. Brems, M.V. Bael, T. Charlton, S. Langridge, et al., Magnetization reversal studies of continuous and patterned exchange biased NiFe/FeMn thin films, Appl. Phys. A 109 (2012) 181.
[173] K. Theis-Bröhl, T. Schmitte, V. Leiner, H. Zabel, K. Rott, H. Brückl, et al., CoFe stripes: magnetization reversal study by polarized neutron scattering and magneto-optical Kerr-effect, Phys. Rev. B 67 (2003) 184415. http://dx.doi.org/10.1103/PhysRevB.67.184415.
[174] K. Theis-Bröhl, B.P. Toperverg, V. Leiner, A. Westphalen, H. Zabel, J. McCord, et al., Correlated magnetic reversal in periodic stripe patterns, Phys. Rev. B 71 (2005) 020403. http://dx.doi.org/10.1103/PhysRevB.71.020403.
[175] C. Hamann, J. McCord, L. Schultz, B.P. Toperverg, K. Theis-Bröhl, M. Wolff, et al., Competing magnetic interactions in exchange-bias-modulated films, Phys. Rev. B 81 (2010) 024420. http://dx.doi.org/10.1103/PhysRevB.81.024420.
[176] K. Theis-Bröhl, M. Wolff, A. Westphalen, H. Zabel, J. McCord, V. Höink, et al., Exchange-bias instability in a bilayer with an ion-beam imprinted stripe pattern of ferromagnetic/antiferromagnetic interfaces, Phys. Rev. B 73 (2006) 174408. http://dx.doi.org/10.1103/PhysRevB.73.174408.
[177] K. Theis-Bröhl, A. Westphalen, H. Zabel, U. Rücker, J. McCord, V. Höink, et al., Hyper-domains in exchange bias micro-stripe pattern, New J. Phys. 10 (2008) 093021. http://dx.doi.org/10.1088/1367-2630/10/9/093021.
[178] O. Petracic, Superparamagnetic nanoparticle ensembles, Superlattices Microstruct. 47 (2010) 569.
[179] W.F. Brown Jr., The fundamental theorem of fine-ferromagnetic-particle theory, J. Appl. Phys. 39 (1968) 993.
[180] S.A. Majetich, T. Wen, R.A. Booth, Functional magnetic nanoparticle assemblies: formation, collective behavior, and future directions, ACS Nano 5 (2011) 6081.
[181] A. Dong, X. Ye, J. Chen, C.B. Murray, Two-dimensional binary and ternary nanocrystal superlattices: the case of monolayers and bilayers, Nano Lett. 11 (2011) 1804.

[182] F. Marlow, Muldarisnur, P. Sharifi, R. Brinkmann, C. Mendive, Opals: status and prospects, Angew. Chem. Int. Ed. 48 (2009) 6212.
[183] S. Peng, C. Lei, Y. Ren, R.E. Cook, Y. Sun, Plasmonic/magnetic bifunctional nanoparticles, Angew. Chem. Int. Ed. 50 (2011) 3158.
[184] S. Laurent, D. Forge, M. Port, A. Roch, C. Robic, L.V. Elst, et al., Magnetic iron oxide nanoparticles: synthesis, stabilization, vectorization, physicochemical characterizations, and biological applications, Chem. Rev. 108 (2008) 2064.
[185] A. Ebbing, O. Hellwig, L. Agudo, G. Eggeler, O. Petracic, Tuning the magnetic properties of Co nanoparticles by Pt — capping, Phys. Rev. B 84 (2011) 012405. http://dx.doi.org/10.1103/PhysRevB.84.012405.
[186] N.N. Phuoc, T. Suzuki, R.W. Chantrell, U. Nowak, Spin configuration of ferromagnetic/antiferromagnetic nano-composite particles, Phys. Status Solidi B 244 (12) (2007) 4518. http://dx.doi.org/10.1002/pssb.200777421.
[187] M.J. Benitez, O. Petracic, E.L. Salabas, F. Radu, H. Tüysüz, F. Schüth, et al., Evidence of core-shell behavior in antiferromagnetic Co_3O_4 nanowires, Phys. Rev. Lett. 101 (2008) 097206.
[188] Y. Hu, A. Du, The core-shell separation of ferromagnetic nanoparticles with strong surface anisotropy, J. Nanosci. Nanotechnol. 9 (10) (2009) 5829. http://dx.doi.org/10.1166/jnn.2009.1225.
[189] K.L. Krycka, R.A. Booth, C.R. Hogg, Y. Ijiri, J.A. Borchers, W.C. Chen, et al., Core-shell magnetic morphology of structurally uniform magnetite nanoparticles, Phys. Rev. Lett. 104 (2010) 207203. http://dx.doi.org/10.1103/PhysRevLett.104.207203.
[190] S. Bedanta, W. Kleemann, Supermagnetism, J. Phys. D Appl. Phys. 42 (2009) 013001.
[191] M. Steen Mørup, F. Hansen, C. Frandsen, Magnetic interactions between nanoparticles, Beilstein J. Nanotechnol. 1 (2010) 182.
[192] S. Bedanta, A. B, W. Kleemann, O. Petracic, T. Seki, Magnetic nanoparticles: a subject for both fundamental research and applications, J. Nanomater. 2013 (2013) 952540.
[193] S. Bedanta, O. Petracic, W. Kleemann, Superparamagnetism, in: K.H.J. Buschow (Ed.), Handbook of Magnetic Materials, vol. 23, Elsevier, 2014.
[194] A. Vorobiev, J. Major, H. Dosch, G. Gordeev, D. Orlova, Magnetic field dependent ordering in ferrofluids at SiO_2 interfaces, Phys. Rev. Lett. 93 (2004) 267203. http://dx.doi.org/10.1103/PhysRevLett.93.267203.
[195] I. Torres-Diaz, C. Rinaldi, Recent progress in ferrofluids research: novel applications of magnetically controllable and tunable fluids, Soft Matter 10 (43) (2014) 8584. http://dx.doi.org/10.1039/c4sm01308e.
[196] S.P. Gubin, Y.A. Koksharov, G.B. Khomutov, G.Y. Yurkov, Magnetic nanoparticles: preparation, structure and properties, Russ. Chem. Rev. 74 (2005) 489.
[197] D. Mishra, D. Greving, G.A.B. Confalonieri, J. Perlich, B.P. Toperverg, H. Zabel, et al., Growth modes of nanoparticle superlattice thin films, Nanotechnology 25 (2014) 205602.
[198] Y. Yin, Y. Lu, B. Gates, Y. Xia, Template-assisted self-assembly: a practical route to complex aggregates of monodispersed colloids with well-defined sizes, shapes, and structures, J. Am. Chem. Soc. 123 (2001) 8718.
[199] G.A.B. Confalonieri, V. Vega, A. Ebbing, D. Mishra, P. Szary, V.M. Prida, et al., Template-assisted self-assembly of individual and clusters of magnetic nanoparticles, Nanotechnology 22 (2011) 285608.
[200] S. Bedanta, O. Petracic, E. Kentzinger, W. Kleemann, U. Rücker, A. Paul, et al., Super-ferromagnetic domain state of a discontinuous metal insulator multilayer, Phys. Rev. B 72 (2005) 024419. http://dx.doi.org/10.1103/PhysRevB.72.024419.

[201] J. Gass, P. Poddar, J. Almand, S. Srinath, H. Srikanth, Superparamagnetic polymer nanocomposites with uniform Fe_3O_4 nanoparticle dispersions, Adv. Funct. Mater. 16 (1) (2006) 71−75. http://dx.doi.org/10.1002/adfm.200500335.

[202] V. Barthem, C. Colin, H. Mayaffre, M.H. Julien, D. Givord, Revealing the properties of Mn_2Au for antiferromagnetic spintronics, Nat. Commun. 4 (2013) 2892. http://dx.doi.org/10.1038/ncomms3892.

[203] E. Brok, K. Lefmann, P.P. Deen, B. Lebech, H. Jacobsen, G.J. Nilsen, et al., Polarized neutron powder diffraction studies of antiferromagnetic order in bulk and nanoparticle NiO, Phys. Rev. B 91 (2015) 014431. http://dx.doi.org/10.1103/PhysRevB.91.014431.

[204] A. Wiedemann, in: T. Chatterji (Ed.), Small Angle Neutron Scattering on Magnetic Nanostructures, vol. 10, Elsevier, 2006.

[205] W. Wagner, J. Kohlbrecher, Magnetic small-angle neutron scattering, ETH Zürich, Laboratory Course, 2012.

[206] S. Disch, E. Wetterskog, R.P. Hermann, A. Wiedenmann, U. Vainio, G. Salazar-Alvarez, et al., Quantitative spatial magnetization distribution in iron oxide nanocubes and nanospheres by polarized small-angle neutron scattering, New J. Phys. 14 (1) (2012) 013025. http://stacks.iop.org/1367-2630/14/i=1/a=013025.

[207] S.J. Lister, T. Thomson, J. Kohlbrecher, K. Takano, V. Venkataramana, S.J. Ray, et al., Size-dependent reversal of grains in perpendicular magnetic recording media measured by small-angle polarized neutron scattering, Appl. Phys. Lett. 97 (11) (2010) 112503. http://dx.doi.org/10.1063/1.3486680.

[208] M.P. Wismayer, S.L. Lee, T. Thomson, F.Y. Ogrin, C.D. Dewhurst, S.M. Weekes, et al., Using small-angle neutron scattering to probe the local magnetic structure of perpendicular magnetic recording media, J. Appl. Phys. 99 (8) (2006) 08E707. http://dx.doi.org/10.1063/1.2165798.

[209] K.L. Krycka, J.A. Borchers, R.A. Booth, C.R. Hogg, Y. Ijiri, W.C. Chen, et al., Internal magnetic structure of magnetite nanoparticles at low temperature, J. Appl. Phys. 107 (9) (2010) 09.

[210] D. Mishra, O. Petracic, A. Devishvili, K. Theis-Bröhl, B.P. Toperverg, H. Zabel, Polarized neutron reflectivity from monolayers of self-assembled magnetic nanoparticles, J. Phys. Condens. Matter 27 (2015) 136001.

[211] D. Mishra, M.J. Benitez, O. Petracic, G.A.B. Confalonieri, P. Szary, F. Brüssing, et al., Self-assembled iron oxide nanoparticle multilayer: X-ray and polarized neutron reflectivity, Nanotechnology 23 (2012) 055707.

[212] K. Theis-Bröhl, P. Gutfreund, A. Vorobiev, M. Wolff, B.P. Toperverg, J.A. Dura, et al., Self assembly of magnetic nanoparticles at silicon surfaces, Soft Matter 11 (2015) 4695.

[213] H. Lauter, V. Lauter, B. Toperverg, Reflectivity, off-specular scattering, and GI-SAS: neutrons, Polym. Sci. A Compr. Ref. 2 (2012) 411.

[214] X. Chen, O. Sichelschmidt, W. Kleemann, O. Petracic, C. Binek, J.B. Sousa, et al., Domain wall relaxation, creep, sliding, and switching in superferromagnetic discontinuous $Co_{80}Fe_{20}/Al_2O_3$ multilayers, Phys. Rev. Lett. 89 (2002) 137203. http://dx.doi.org/10.1103/PhysRevLett.89.137203.

[215] S. Bedanta, E. Kentzinger, O. Petracic, W. Kleemann, U. Rücker, T. Brückel, et al., Modulated magnetization depth profile in dipolarly coupled magnetic multilayers, Phys. Rev. B 74 (2006) 054426. http://dx.doi.org/10.1103/PhysRevB.74.054426.

[216] S. Bedanta, O. Petracic, X. Chen, J. Rhensius, S. Bedanta, E. Kentzinger, et al., Single-particle blocking and collective magnetic states in discontinuous $CoFe/Al_2O_3$ multilayers, J. Phys. D Appl. Phys. 43 (2010) 474002.

[217] G. Renaud, R. Lazzari, F. Leroy, Probing surface and interface morphology with grazing incidence small angle X-ray scattering, Surf. Sci. Rep. 64 (8) (2009) 255−380. http://dx.doi.org/10.1016/j.surfrep.2009.07.002.

[218] M. Pannetiera, F. Ott, C. Fermon, Y. Samson, Surface diffraction on magnetic nanostructures in thin films using grazing incidence SANS, Physica B 335 (2003) 54.

[219] K. Theis-Bröhl, M. Wolff, I. Ennen, C.D. Dewhurst, A. Hütten, B.P. Toperverg, Self-ordering of nanoparticles in magneto-organic composite films, Phys. Rev. B 78 (2008) 134426. http://dx.doi.org/10.1103/PhysRevB.78.134426.

[220] K. Zhernenkov, S. Klimko, B.P. Toperverg, H. Zabel, Ac-polarized neutron reflectometry: application to domain dynamics in thin fe film, J. Phys. Conf. Ser. 211 (2010) 012016.

[221] S. Klimko, K. Zhernenkov, B.P. Toperverg, H. Zabel, Development and application of setup for ac magnetic field in neutron scattering experiments, Rev. Sci. Instrum. 81 (2010) 103303.

[222] K. Zhernenkov, Frequency Dependence of Magnetization Reversal in Thin Fe(100) Films (Dissertation), Ruhr-University Bochum, Germany, 2013.

[223] G.S.D. Beach, C. Nistor, D. Knutson, M. Tsoi, J.L. Erskine, Dynamics of field-driven domain-wall propagation in ferromagnetic nanowires, Nat. Mater. 4 (2005) 741.

[224] O. Boulle, G. Malinowski, M. Kläui, Current-induced domain wall motion in nanoscale ferromagnetic elements, Mater. Sci. Eng. Rep. 72 (2011) 159−187.

[225] D. Gorkov, Magnetization Reversal in Magnetic Microstripes (Dissertation), Ruhr-University Bochum, 2015.

[226] R. Gähler, R. Golub, T. Keller, Neutron resonance spin echo—a new tool for high resolution spectroscopy, Phys. B 180−181 (1992) 899.

[227] T. Weber, G. Brandl, R.G.P. Böni, An open-source software package for data treatment in a MIEZE experiment, J. Phys. Conf. Ser. 528 (2014) 012034.

Chapter 7

Nuclear Magnetism and Neutrons

Michael Steiner* and Konrad Siemensmeyer
Helmholtz Zentrum Berlin, Berlin, Germany
Corresponding author: E-mail: steiner@helmholtz-berlin.de

Chapter Outline

7.1 Introduction	436	7.3 Experimental Results from Neutron Diffraction on Cu and Ag 461
7.2 Experimental Background	439	
7.2.1 Nuclear Moments—The Neutron Cross Sections	439	7.3.1 Nuclear Magnetic Ordering of Cu and Ag in Zero and Finite Magnetic Field 461
7.2.2 ULT Experimental Methods and Neutron Techniques	441	
		7.3.1.1 Copper ($I=3/2$) 461
7.2.3 Nuclear Polarization Measurement from Neutron Scattering and Transmission	445	7.3.1.2 Silver ($I=1/2$) 468
		7.3.2 Structures—Cu and Ag 470
		7.3.2.1 Copper 470
7.2.4 Neutron Diffraction Cryostat for ULT Applications	451	7.3.2.2 Silver 471
		7.3.2.3 Copper versus Silver 471
7.2.5 Sample Requirements	453	7.3.3 The Phase Diagrams in a Magnetic Field 472
7.2.6 Spontaneous Nuclear Magnetic Order	453	
		7.3.3.1 Copper ($I=3/2$) 473
7.2.6.1 Nuclear Ordering Observations in Dynamically Polarized Systems	453	7.3.3.2 Silver ($I=1/2$) 474
		7.4 Neutron Diffraction Investigations on Solid ^{3}He 474
		7.5 Applications of Polarized Nuclei in Neutron Diffraction 482
7.2.6.2 Previous Results for Cu and Ag from SQUID and NMR Techniques	455	7.5.1 Neutron Polarization from Polarized ^{3}He Gas Targets 482
		7.5.2 Contrast Variation by Polarized Nuclei in Neutron Scattering 484
7.2.6.3 Theoretical Predictions	457	7.6 Summary 486
		References 486

7.1 INTRODUCTION

Around 60 years ago, two new experimental concepts appeared in the scientific scene, which very quickly demonstrated their exceptional potential for new breakthroughs in many fields of science: the first neutron scattering experiments were performed by Cliff Shull on MnO in 1949 [1] and in 1953, Wu et al. [2] used hyperfine-enhanced nuclear polarization in Co to show symmetry breaking in nuclear decay.

Neutron scattering quickly became the method for investigating magnetic materials on an atomic scale and complemented the already established X-ray techniques for the study of the structure of condensed matter. Shull's first experiment directly proved Néel's prediction of antiferromagnetic (AFM) ordering. Brockhouse's invention of a neutron spectrometer further underlined the importance of this technique [3]; it allowed the measurement of elementary excitations like phonons and magnons. We will see, however, that for studies of nuclear order, Cliff Shull's experiment has to be repeated with a suitable nuclear magnet!

The study of polarized nuclei developed in a very different way: early applications include cooling by nuclear demagnetization around 1956 [4] and NMR became more and more a very important and standard technique to study dynamics in condensed matter. A rather complete description of this field is given by Abragam in 1982 [5]. A highlight referring to our topic is the first observation of nuclear ordering in 1978 [6] in LiH by neutron diffraction. This development would have not been possible without enormous progress in low and ultralow temperature (ULT) physics and techniques [7]. The progress was such that since Kurti's studies microkelvin and nanokelvin temperatures were in reach in the early 1980s. Specialized laboratories like the ULT in Otaniemi (Finland) played a major role in this. One of the driving forces was the discovery of superfluidity in ^{3}He following the progress in ULT techniques. Looking for spontaneous ordering in nuclear magnets in zero field appeared possible. The nuclear moment being only $\sim 10^{-3}$ of a Bohr magneton nuclear order can be expected in the microkelvin temperature regime.

Magnetism benefitted enormously from neutron scattering from the very beginning. Such studies revealed a huge variety of magnetic structures that gave a clear picture of the relation between magnetic structure, macroscopic property, and theory. An important aspect in neutron scattering is that neutrons penetrate through thick materials. This allows the use of complex sample environment without any loss in the quality of the experiments. Thus, neutron scattering could profit from the continuous development of cryostats and magnets. Combining low temperatures, which in the 1970s reached 0.5 K, with strong magnetic fields neutron studies of magnetic phase diagrams were possible opening the door to a new area in the studies of magnetic materials and allowed to test many theoretical concepts for ordering of many-body systems. With the progress in ULT physics at the end of the 1970s, the interest in very low-temperature experiments with neutrons was growing, still predominantly in the field of electronic magnetism. Experiments on Pr [8] showed a very unexpected magnetic phase transition around 0.1 K. It turned out that the hyperfine coupling between the electronic and nuclear spins drives the transition: only the combined system of both can order! This discovery was the beginning of neutron scattering at ULT.

The neutron scattering department at the BER II at the Hahn-Meitner Institute (HMI) in Berlin (now Helmholtz Zentrum Berlin) started in the mid-1970s experiments in the mK range in order to open temperatures from 1 K to 10 mK for neutron experiments. This was achieved by studying systems where ordering occurs in the mK range [9] or where the electronic magnetic moments induced nuclear ordering through hyperfine coupling like $HoVO_4$ in 1985 [10]. In these experiments a ^{3}He- or ^{3}He/^{4}He dilution refrigerator cryostat was used which were commercially available at that time already. Such experiments were complicated in the sense that they always should be ready for cooling when the neutron experiment with its expensive beam time was ready to start. This was more complicated to achieve as with ^{4}He cryostats, the standard for neutron scattering at that time. On the other hand, the above two cooling systems delivered their significant cooling power continuously in the whole T-range. This was considered essential for such experiments. The studies at HMI in Berlin down to 10 mK clearly demonstrated that experiments at such low temperature could become standard in neutron scattering. Unfortunately, however, the base temperature of a dilution refrigerator is around 5 mK and this still is much too high for experiments with nuclear order in zero field.

The next step in neutron scattering studies went down to the microkelvin T-range. Roinel and Goldman [6] could produce nuclear ordering of the protons in LiH and prove this ordering by neutron diffraction. The sample there was cooled by an elegant cooling technique, adiabatic demagnetization in a rotating magnetic field.

While neutron scatterers were working in the mK range most advanced cooling techniques allowed to cool nuclear systems down to nanokelvin temperature or even lower by a double stage nuclear demagnetization system. The ULT laboratory in Otaniemi developed such a system for their search for nuclear ordering in Cu [11,12]. They started this project in the late 1970s and reported in 1986 the observation of nuclear ordering below 60 nK on the basis of NMR and susceptibility data under practically zero field conditions.

The situation around 1985 found those who discovered nuclear order in Cu and those looking for an opportunity to demonstrate the potential of neutron scattering at extreme conditions ready to take on a new challenge, which needed the expertise of both group of scientists. Consequently, a collaboration between the then Hahn-Meitner Institute, the ULT laboratory in Otaniemi, the Riso Nat. Lab. in Roskilde Denmark and the University of Copenhagen was founded to determine the structure of the nuclear order in Cu directly by neutrons. The challenge of this experiment is twofold: firstly, the very complex cooling system needs to operate in a very "unfriendly" environment as compared to an ULT laboratory and secondly, single-shot experiments are only possible. After cooldown, there is a very limited time available for the neutron experiment. A last comment is appropriate here about the visibility of the nuclear moment for the neutrons. While neutrons with their own dipole moment see the electronic magnetic moment in the crystal directly, this is not possible for the nuclear moment, which is 10^{-3} times the electronic magnetic moment in size. In nuclear spin systems the necessary interaction is a term in the neutron-nucleus scattering length b which is spin dependent. Therefore, the cross section for a nuclear magnetic Bragg peak is due to the same physical process as for a structural peak.

The study of the nuclear ordering in Cu is not only an experimental challenge, but it deals with one of the puzzles which are not fully understood since Cliff Shull's pioneering experiment: Shull determined the AFM structure of MnO, which crystallizes in the face-centered cubic (fcc) lattice and found a simple up-down structure. This magnetic structure, however, breaks the fcc symmetry and frustration occurs in this lattice. In fact, it was found in measurements at high resolution that there is a structural distortion occurring at the magnetic phase transition in the fcc lattice of MnO. As Cu has the same fcc lattice structure, the type of AFM nuclear order might give some more inside into the puzzle of frustrated fcc AFM. In this situation, a team of theoreticians did play a very important part in this endeavor. They performed model calculations for the interactions and predicted magnetic structures in Cu and later for Ag as well.

As mentioned above, studies on ^{3}He were doing both, pushing ULT and profiting from ULT progress. ^{3}He is undoubtedly one of the most fascinating quantum many-body systems. In the solid phase, in zero field, a simple AFM nuclear ordered state has been found (for a review see Ref. [13]). The magnetic interaction is produced by particle exchange driven by the enormous zero point motions of the He atoms! It is known from NMR that in a finite external field the nuclear magnetic structure changes, but the structure could not be determined. Several models exist: multiple particle ring-exchange mechanism is proposed, but without a direct determination of the high field structure (by neutron scattering), this problem cannot be solved. There have been attempts, but unsuccessful until now. We will review the state of the research by neutrons on solid ^{3}He.

In the following chapters the relevant neutron cross sections for scattering and absorption, details of the cooling system and the challenges it poses for neutron experiments, the physics of the Cu and the Ag nuclear systems, the experiments and their interpretation are presented. We will then review the neutron scattering experiments on ordering of nuclear spins in insulators like

LiH. An overview will then be given of concepts, techniques, and applications of polarized nuclear spin systems for neutron scattering. An outlook will conclude this review.

7.2 EXPERIMENTAL BACKGROUND

7.2.1 Nuclear Moments—The Neutron Cross Sections

A particularity of the nuclear magnet is that the nuclear spins are pointlike objects in contrast to electronic spins, which are extended in space.

The interaction of neutrons with condensed matter arises from strong nuclear interactions and can be divided into absorption cross sections and scattering cross sections. The dipolar interaction between neutron and nuclear spin is very small compared with the typical scattering cross sections, the reason is of course the small nuclear magnetic moment. Therefore, investigations of nuclear magnetism rely on the spin dependence of the strong neutron—nucleus interaction. Both absorption and scattering cross section depend on the relative orientation between the spin of neutron and nucleus. This opens a path for studies of nuclear magnetic properties using neutron techniques.

The nuclear spin I (which in fact is the total nuclear angular momentum) is determined by nuclear forces for a given isotope. Therefore, in scattering processes, in principal only two different scattering length can appear for the channels $I + 1/2$ and $I - 1/2$ which give rise to scattering length of b^+ and b^-, respectively. The sum of both terms yields the familiar—spin independent—coherent scattering length $\bar{b}$, while the difference, after proper weighting, gives the spin-dependent part of the nuclear scattering length that sometimes also has been converted to the "pseudodipolar" moment b_N [5]. Probably, the main intention of this conversion was to show that the nuclear spin-dependent scattering is comparable in magnitude to conventional dipolar magnetic scattering from electrons, however, the fundamental difference is the nuclear forces as the source of the interaction which is not subject to the dipolar anisotropy. The nuclear spin-dependent scattering length a then is

$$a = \bar{b} + b_N IS \tag{1}$$

where I, S are the nuclear and neutron spin operators, respectively. In case of a single isotope the two constants can be written

and
$$\bar{b} = \left((I+1)b^+ + Ib^-\right)/(2I+1) \tag{2}$$

$$b_N = 2\left(b^+ - b^-\right)/(2I+1) \tag{3}$$

These expressions are very simple and will be used in the following. Numerical values for most isotopes can be found tabulated, along with other relevant scattering length and cross sections, in a number of publications, e.g., see Ref. [14] or Ref. [15]. The actual values of the scattering length of a sample are the weighted average of the isotopic composition. As a consequence, it is almost always better to work with isotopically pure samples, however, the price of enriched materials usually is prohibitively high. Looking more specifically at the nuclear spin scattering lengths of Cu and Ag, one realizes that the intensity gain of enriched samples is large, e.g., more than a factor of 6 (^{65}Cu) and a factor of 40 (^{109}Ag) over the corresponding natural mixtures, which has been the main reason for using isotope material in nuclear ordering experiments. Besides, there is no isotopic incoherent scattering contributing to the background from enriched material.

It should be mentioned that the nuclear spin-dependent cross section b_N is directly related to the spin incoherent cross section which depends on the neutron and nuclear polarization

$$\sigma_{\text{Inc}} = \pi b_N^2 \left(I(I+1) - IP_N P_{\text{Nuc}} - I^2 P_{\text{Nuc}}^2 \right) \quad (4)$$

This is the well-known spin incoherent cross section which appears in addition to the isotopic incoherence.

The intensity for elastic scattering is proportional to

$$I \sim \frac{\lambda^3 |F(Q)|^2}{\sin(2\Theta)} \quad (5)$$

This relation shows that the intensity can be optimized by a proper choice of the wavelength λ, independent on the structure factor $F(Q)$ which contains all the microscopic structure information. For example, it is possible to cover a larger amount of reflections with low wavelength and to reduce absorption, whereas use of long wavelength can increase the measured signal on specific reflections.

We now turn to discuss the structure factor for elastic scattering. It is given by the Fourier transform of the scattering length density, below assumed to be localized at lattice cell sites i. The summation should run over all sites in the (magnetic) unit cell, it contains the nuclear coherent, magnetic, and nuclear magnetic terms.

$$F(Q) = \sum_i a_i \exp(2\pi \vec{r}_i Q) = \sum_i \left(\overline{b}_i \mp p_{\text{Magn}} \vec{e} \, _i f(Q) \mu_i \pm b_{N,i} \vec{P}_{\text{Nuc},i} \right) \\ \times \exp(2\pi \vec{r}_i Q) \quad (6)$$

Blume [16,17] has discussed it for general directions between neutron spin and nuclear moments which leads to a tensorial expression for the cross sections that finds applications in neutron polarimetry. However, this technique so far has not been applied on nuclear spins and for the purpose here the expressions given by Moon et al. [18] are easier to oversee: they assume a fixed neutron spin direction, i.e., the presence of a guide field which allows neutron spin-flip processes to be detected. For a single particle the scattering length for spin flip and non-spin flip reads

$$b_{\text{NSF}} = \overline{b} \mp p_{\text{Magn}} f(Q) \mu_Z^{\perp} \pm b_{\text{N}} P_{\text{Nuc},Z}$$
$$b_{\text{SF}} = \mp p_{\text{Magn}} f(Q) \left(\mu_X^{\perp} + i\mu_Y^{\perp} \right) \pm b_{\text{N}} \left(P_{\text{Nuc},X} + iP_{\text{Nuc},Y} \right) \quad (7)$$

Here, the $\pm$ signs refer to the neutron spin parallel or antiparallel to the z-direction. The direction of the nuclear polarization (P_{Nuc}) and electronic magnetic moments ($\mu^{\perp}$) is given with respect to the neutron spin quantization axis. An inspection of these expressions shows the fundamental difference between the dipolar interaction relevant for electronic moments—only the components perpendicular to the scattering vector q contribute, whereas the spin dependence of the strong interaction is isotropic and all nuclear moments contribute to the scattering length. Further, for the electronic magnetic scattering a form factor $f(Q)$ must be included, while is not the case for the pointlike nuclear moments.

The absorption cross sections are spin dependent in a very similar way. The main difference is that absorption in general is caused by resonances close to the thermal neutron regime and they can have very different character in J. The cross sections depend also on the final product—i.e., situations can arise where several resonances contribute. A particular easy case is proton production only as it takes place, e.g., in ^{3}He neutron capture. Owing to the ^{3}He nuclear spin $I = 1/2$ the absorption can take place in a total $J = 0$ and $J = 1$ channel. The $J = 1$ state needs two parallel neutron spins and violates the Pauli principle, as a result only the Coulomb part of the absorption cross sections remains which is very small. The $J = 0$ channel in contrast is allowed and has a very large absorption cross section of 5300 b. For heavier elements, one may mention ^{149}Sm or ^{157}Gd, the spin dependent, absorption comes from more complex processes that need discussion in terms of the nuclear physics involved—beyond the scope of this review. More details on absorption cross sections and numerical values are given in Refs [19,20].

The nuclear spin-dependent absorption finds broad and very useful application in neutron polarization techniques, as will be discussed below. It can also be used to determine the nuclear polarization from simple transmission experiments. With certain experimental limitations the spin-dependent scattering length has been used successfully also for temperature determination in nuclear spin systems. In our section on temperature determination, we come back to that. The spin-dependent scattering length finds an application for contrast variation in small-angle neutron scattering (SANS) experiments too. And, of course, the spin-dependent scattering length gives us a chance to investigate nuclear ordering with neutron diffraction.

7.2.2 ULT Experimental Methods and Neutron Techniques

Before going into more details of results of the above cross sections, it is worth to give a short overview of the experimental implications of ULT work combined with neutron diffraction. With this chapter the readers should become aware of some important principles of ULT physics and the resulting experimental requirements, e.g., cooling or temperature determination techniques. There are excellent books describing ULT experiments [7,21] and solutions to specific problems so that we can concentrate on issues that are more specific to the combination of neutron and ULT techniques. However, a few basic facts must be mentioned.

From the first law of thermodynamics, it is clear that with approaching $T = 0$ the available cooling power reduces a lot, and the heat load in the experiment needs careful examination. For example, the typical cooling power of a dilution refrigerator at 10 mK is of the order of a few microwatts. At lower temperatures, no continuous cooling techniques exist, and usually nuclear demagnetization techniques are used to reach the sub-microkelvin temperature range. These techniques are "single-shot" techniques, because after precooling the refrigerant in a large magnetic field (typically a metal like Cu, but also other compounds like $PrNi_5$ are possible), it is demagnetized under constant entropy conditions. The precooling temperature achieved in high field determines the entropy reduction that can be used for cooling a sample connected to the demagnetization stage. Figure 1 illustrates the entropy and heat balance

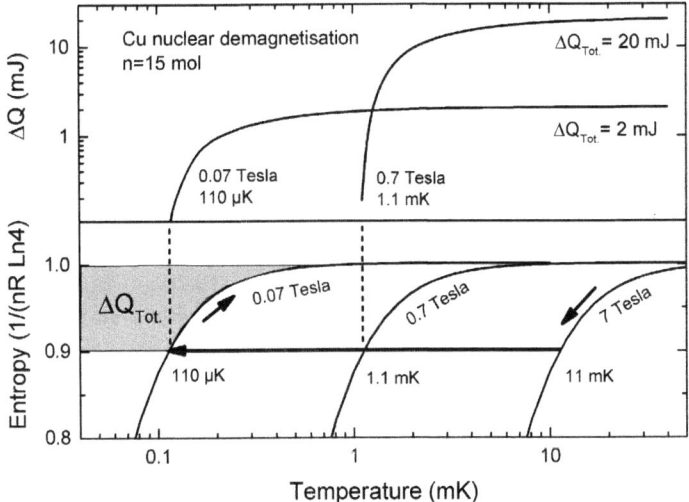

FIGURE 1 The lower section of this figure shows the entropy as a function of temperature for a number of field values. A Cu nuclear stage of 15 mol is assumed. Typical field and temperature values are chosen for precooling (7 T) and the final field (0.07 T) after demagnetization at constant entropy. The final temperature (here 110 µK) allows cooling the sample attached to the nuclear stage provided it stays cold for sufficient time under external heat leaks. The shaded area shows the total amount of heat that can be absorbed, the upper part of the figure shows the head input needed for a certain temperature increase starting from two different field and temperature values. Note that at low field a heat input of 1 mJ doubles the temperature to more than 200 µK, i.e., if 100 µK should be kept for hours, the heat load must be much smaller.

for a realistic example, the nuclear cooling stage that has been used in the Cu nuclear ordering experiments at Risoe.

A practical limit for the time available for experiments at low temperature is set by external heat leak. As seen from Figure 1, if a temperature around 100 µK should be maintained for a couple of hours the total heat input must be very small, not larger than a few nanowatts. Low-temperature laboratories have developed sophisticated methods to achieve such values: main sources for heat input are vibrations of the cryogenic system which lead to eddy currents in a magnetic field or Radio-Frequency (RF) pickup from the surrounding that is dissipated in the system. Electrically shielded rooms and massive vibration damping systems are the standard solutions. In addition, more exotic sources of heat leaks are observed such as the conversion of *ortho-* to *para-*hydrogen which reduces over weeks with the time spend below 4 K. A neutron experiment has specific sources of heating from neutron and γ-ray absorption which can be significant. They will be discussed below for Cu, Ag, and ^{3}He.

A general remark must be made concerning the implementation of ULT experiments in a surrounding for neutron diffraction: the system must be built on a diffractometer where it must be possible to rotate the cryostat and the detector system independently around the same axis. Such requirements in the reactor surrounding are in direct conflict with the ULT needs of a complete elimination of vibrations and electrical RF interference. As a consequence, all successful neutron experiments on nuclear ordering at ULT so far have used dedicated diffractometers and cryostats designed for these experimental needs. A cryogenic system based on the experiments in Otaniemi [11] to reach ULTs was first build at Risoe with modifications necessary to use it on a neutron diffraction instrument.

The instrument mechanics in general must be decoupled from the building by a damping system. The setups for nuclear ordering experiments on Ag and ^{3}He at the HMI in Berlin successfully used advanced active damping systems. Concerning RF pickup, for neutron experiments obviously no shielded rooms can be established. One strategy against electrical interference in a hostile reactor environment was filters to suppress 50 Hz pickup which was observed to be a major contribution to heat leaks. Another strategy was a change of experimental methods, trying to minimize the amount of electrical connections between the ULT parts of the cryostat and the rest of the world, e.g., for the temperature determination. Usually temperatures in the ULT range are measured by NMR techniques, but with neutrons available it was also possible and maybe more efficient to derive temperatures using neutron techniques (see below).

Directly connected to cooling power, considerations are questions of thermal contact and isolation at ULT. This regards on one side the contact between sample and cooling stage that must be made using metals with very high conductivity to avoid large temperature gradients. Even then, the thermal conductivity decreases, at best linearly, with temperature. Impurities that scatter electrons make things worse, they can be avoided with high-purity material and/or made inefficient by a proper heat treatment [22]. Additional aspects come into play for heat transfer between insulators (^{3}He) and metals; the boundary mismatch between phonon (insulator) and electron (metal) dispersion energies leads to the Kapitza boundary impedance that is proportional T^{-4}. To overcome it, very large surfaces are needed, i.e., sintered heat exchangers. We will discuss ^{3}He cells for neutron diffraction in the following sections.

A fundamental aspect of thermal contact and isolation is connected with the fact that the nuclear spin system decouples from the lattice. The concept of a nuclear spin temperature has been developed in detail by Goldmann [23]. One can speak of a nuclear spin and an electronic lattice temperature and they can differ by several orders of magnitude, in particular after cooling the nuclear spins by adiabatic demagnetization. The nuclei are thermally isolated from the lattice with a spin–lattice relaxation time τ_1 which, dependent on material and the electron temperature, can take values up to several hours.

The weak spin–lattice coupling in fact is the mechanism that allows us to demagnetize the nuclei of Cu or Ag into the nano- and picokelvin temperature range, independent on the electron lattice temperature. That is, only the nuclear spin system cools in the demagnetization, while the electron temperature remains high and almost constant. Important is that all heat leaks go to the electrons, at nanokelvin temperature even a 1 nW heat leak to the nuclear spin system would be much too large to ever reach such a temperature by demagnetization.

In a neutron experiment, there are heat loads delivered to the sample by neutron radiation and γ-radiation which deserve special attention because they easily add up to values that are significant on the nanowatt scale. The heat load is deposited in the electron system, therefore, the lattice temperature increases. Apart from the possible direct effect of a larger temperature difference between lattice and nuclear spins, the increase of the lattice temperature also has the indirect effect of faster spin–lattice relaxation rate which leads to a reduction of the insulating effect.

Among sources for heat loads in a neutron experiment, one may mention electromagnetic radiation (i.e., γ's) which directly leads to heat input. It can be well calculated using the Bethe formula, it relates the heat input to the γ-energy and the electron density of the target. Neutron capture is the other event that needs consideration: upon capture prompt γ's emerge, and an excited nucleus is obtained which then decays into a stable isotope under release of energy. For Cu and Ag, one deals with β-decay processes where most of the β-particle energy remains in the sample, whereas the prompt γ's as well as decay γ's contribute less to the heat input. Typical beam heating values between 1 and 10 nW have been found in the case of ^{65}Cu and ^{109}Ag, respectively. The differences are mainly due to the large absorption cross section of ^{109}Ag isotope. The decay time constants are of interest because they delay the release of energy. The ^{65}Cu isotope used in nuclear ordering experiments for scattering length reasons is an unfavorable choice for this parameter with a decay time of $\tau = 5$ min. In case of Ag, the time delays caused by the large Korringa time constant of 10 secK help to cope with the fairly large beam heating load.

Finally, the thermal isolation of the nuclear spin system has the consequence that we are talking about a system where the entropy is constant, in fact, entropy is the thermodynamic variable that is under control in a demagnetization process. Yet it is believed that as long as the spin–spin relaxation time τ_2 is fast enough, an equilibrium temperature can be assigned to the nuclei, although it may be difficult to access the temperature by direct measurements. It is conceivable that critical phenomena at constant entropy need a description different from constant temperature, however, to date we are not aware of any evidence for such differences in the context of nuclear spin ordering.

7.2.3 Nuclear Polarization Measurement from Neutron Scattering and Transmission

After the discussion in the section before the importance of controlling heat flow and relaxation processes in the nuclear spin system is obvious. In the following, we will discuss how to measure it with neutrons. The temperature determination of nuclear spin systems usually is not directly possible, mainly because nuclear spin systems couple only weekly to the electronic and lattice degrees of freedom. In metals like Cu and Ag the spin–lattice relaxation time can be very long, ranging from several minutes to hours. In neutron experiments the temperature determination through the determination of nuclear polarization with neutron diffraction and/or transmission is the obvious way out. These methods provide a good example for the use of spin-dependent absorption and scattering cross sections and will be discussed now, they also demonstrate possible ways to analyze the heat flow in nuclear cooling experiments.

For the simple metals Cu or Ag the intensity at structure reflection is proportional to the unit cell structure factor

$$F^2(hkl) \propto 16(b_{Coh} \pm b_N P_{Nuc})^2 \qquad (8)$$

This dependence is used to determine the nuclear polarization P_{Nuc} which in the paramagnetic state is related to temperature and magnetic field by the well-known Brillouin function. In the paramagnetic state, with known external field therefore the temperature can be derived. This simple method, however, needs calibration because large single crystals usually show extinction effects, i.e., the scattered intensity is no longer proportional to $F^2(hkl)$. Second, it is convenient to use a polarized neutron beam to measure the intensities I^+ and I^- for neutron spins parallel and antiparallel to the field. From these intensities the so-called "flipping ratio" FR can be calculated which depends on the nuclear polarization. For the structure factor (Eqn (8)) with a neutron beam of polarization P_N the dependence of the flipping ratio on nuclear polarization is given by

$$FR = I^+/I^- = |F_+(P_{\text{Nuc}})|^2 / |F_-(P_{\text{Nuc}})|^2 = \frac{b_{\text{Coh}}^2 + 2b_{\text{Coh}}b_N P_N P_{\text{Nuc}} + (b_N P_{\text{Nuc}})^2}{b_{\text{Coh}}^2 - 2b_{\text{Coh}}b_N P_N P_{\text{Nuc}} + (b_N P_{\text{Nuc}})^2}$$

(9)

Compared to the measurement of absolute intensities the flipping ratio is free of many experimental errors, in particular long-term instrumental drift cancels out. However, the need for an extinction correction remains. This type of calibration for thermometry purposes has first been established for experiments on Cu nuclear ordering [24,25]. Different nuclear polarization values have been adjusted and the corresponding flipping ratio has been measured: in high field, the complete polarization range up to saturation was covered and strong extinction corrections were observed. They were attributed to the heat treatment of the sample that leads to large, perfect mosaic grains. A satisfactory model could be established to explain the data using the theory of Becker and Coppens [26,27], as seen from Figure 2.

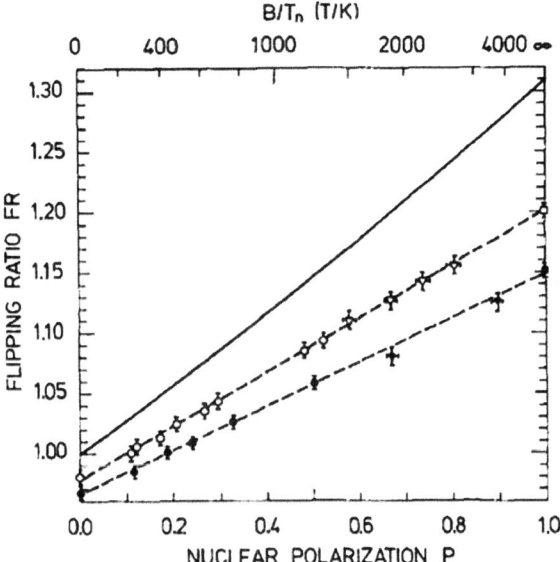

FIGURE 2 Results of calibration measurements on a ^{65}Cu crystal in Risoe [24]. A (200) reflection has been measured. Open symbols are for $\lambda = 2.4$ Å and solid symbols for $\lambda = 3.0$ Å. The solid curve is the extinction free curve with $P_N = 0.95$ and without the correction for the spin-dependent transmission of the μ-metal that appears in the experimental data. The dashed curves are the best linear fits to the data. The top legend shows the conversion of nuclear polarization to the ratio of B/T for paramagnetic ^{65}Cu nuclei. *Reprinted from Ref. [24], http://dx.doi.org/10.1007/BF01312783 with permission of Springer Science + Business Media, ©Springer Verlag 1988.*

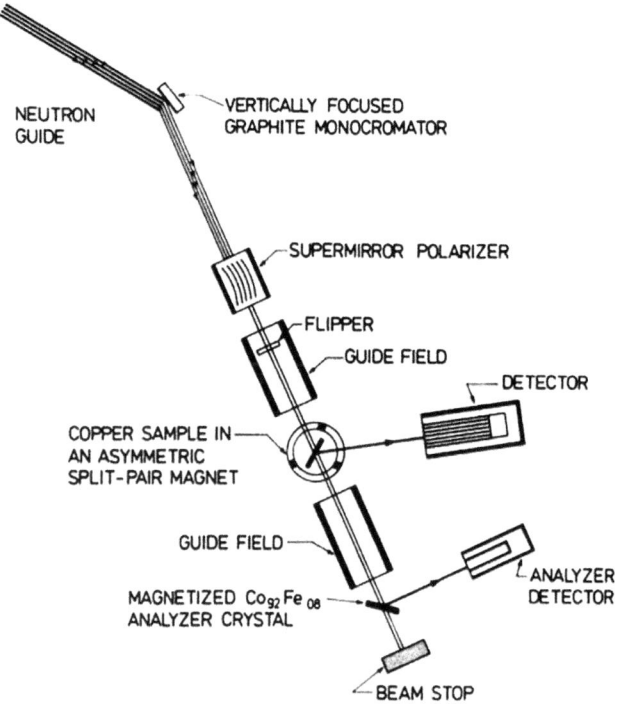

FIGURE 3 The polarized neutron setup at Risoe for nuclear ordering experiments on Cu. Reprinted from Ref. [24], http://dx.doi.org/10.1007/BF01312783 with permission of Springer Science + Business Media, ©Springer Verlag 1988.

Figure 3 illustrates the typical experimental setup for polarized neutrons with polarizer and flipper in front of the sample. Behind the sample an analyzer system is installed to monitor the neutron beam polarization. The sample magnet is an asymmetric split pair Helmholtz coil for the transmission of polarized neutrons. This magnet is also used to cool the Cu sample by adiabatic demagnetization into the ordered state. Nuclear ordering, however, requires the control of magnetic field at the sample in x, y, and z direction on the level of a few microtesla using a small, low field coil system. To shield the low field area from the fairly strong remanent field of the superconducting sample magnet, a cylindrical μ-metal shield surrounds the sample which has impact on the transmission of the polarized beam; the μ-metal must be saturated to allow polarized neutron transmission which limits polarized neutron methods to fields above ~ 0.1 T. Therefore, for Cu nuclear spin ordering, it was not possible to determine the critical entropy and temperature at the phase boundary in low field.

With this setup, however, it has been possible to analyze in detail the heat flow in the two stage demagnetization system and to model the thermal

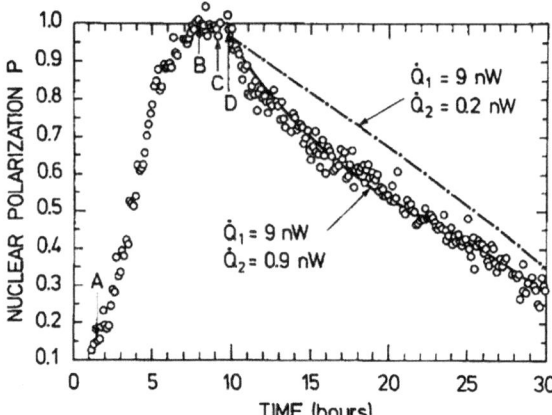

FIGURE 4 The behavior of the nuclear polarization of the copper specimen versus time in an ultralow temperature experiment. The large first nuclear cooling stage is demagnetized from A to C and the sample (second stage) from C to D. The sample heating in the final field of 0.25 T has been modeled using an impedance or $R = 10 \text{ K}^2 \text{ W}^{-1}$ determined earlier from similar data [24] for different heat leaks to the first nuclear stage and the sample. The heat leak to the sample arises mostly from the absorption of neutrons and the following β-decay. *Reprinted from Ref. [24], http://dx.doi.org/10.1007/BF01312783 with permission of Springer Science + Business Media, ©Springer Verlag 1988.*

contacts together with the nuclear spin−lattice relaxation rate. More important, it was possible to estimate the direct deposition of heat into the sample by the absorption of the neutron beam to be of the order of 1 nW [25] (cf. Figure 4). This value was in reasonable agreement with expectations that explain "beam heating" by neutron absorption where the prompt γ's do not contribute significantly, but subsequent β-decay of the neutron capture product causes heating because the electrons released mostly remain in the sample.

Nuclear ordering experiments in Ag in principle offer the same polarized neutron thermometry possibilities as Cu through scattering of polarized neutrons. In addition, Ag has a relatively large absorption cross section for neutrons which is a nuisance on first sight for scattering signal reasons and also the neutron beam heating is increased roughly by a factor of 10 compared to the Cu case. The Ag neutron absorption, however, is spin dependent which opens a second experimental path to determine the Ag nuclear polarization and thus the temperature.

As the scattering length the absorption cross section can be written as a sum of spin-dependent and spin-independent parts:

$$\sigma_{\text{Abs}}(P_{\text{N}}, P_{\text{Nuc}}) = \sigma_0^{\text{Abs}} + \sigma_{\text{N}}^{\text{Abs}} P_{\text{N}} P_{\text{Nuc}} \qquad (10)$$

Important is the magnitude of the spin-dependent absorption effect, which is large enough to be determined without the use of a polarized neutron beam.

Numerically, the values for the ^{109}Ag isotope are $\sigma_0^{Abs} = 91b$ and $\sigma_{Nuc}^{Abs}/\sigma_0^{Abs} = 0.329$ (more data are in Ref. [28]). From this cross section, one can calculate the transmission T of a sample that of course also depends on its effective thickness d and atom density N.

$$T(P_N, P_{Nuc}) = \exp\left(-Nd\sigma_0^{Abs}\right)\left(\cosh\left(Nd\sigma_{Nuc}^{Abs}P_{Nuc}\right) - P_N \sinh\left(Nd\sigma_{Nuc}^{Abs}P_{Nuc}\right)\right)$$
(11)

Spin-dependent transmission measurements are—as spin-dependent scattering—not completely free from a number of correction factors and especially the effective sample thickness needs a calibration measurement to account for a rotation of the sample in the beam [29]. The calibration of spin-dependent neutron absorption in case of Ag was achieved against the spin-dependent scattering effect, i.e., the flipping ratio techniques explained above. Correction factors arise again from spin-dependent transmission of neutrons through magnetic shielding material ("cryoperm" has been used in the Ag experiments instead of µ-metal), eventually for transmission losses when strong Bragg peaks scatter away a part of the beam as well as for spin incoherent scattering. All these corrections, however, can be calculated or measured independently so that altogether a very reliable method is achieved to determine nuclear polarization.

From Eqn (11) for the transmission, it can be seen that spin-dependent absorption will be observed also without using a polarized beam. The absence of the high field limitations of polarized neutron techniques opens new possibilities. First of all, in zero field the (nuclear) paramagnet is unpolarized, and there this important reference can be repeatedly measured just by switching a (small) field on and off as seen in Figure 5. From these data a transmission ratio is determined which similar to a flipping ratio eliminates many experimental sources of error.

With regard to neutron transmission also the AFM state is effectively unpolarized. By switching field between the AFM and paramagnetic phase, the development of temperature with time can be correlated with the scattering signal. This gave the important determination of the entropy-field phase diagram for nuclear spin ordering in Ag, along with the simultaneous measurement of the AFM order parameter using neutron diffraction. A discussion of these results will be postponed to the chapter on nuclear spin ordering experiments of this review.

We would now like to have a closer look on the coupling between nuclear spins and electrons. Neutron thermometry could test theoretical predictions [30] that are difficult to verify, e.g., by NMR methods. The reason is that neutron techniques have no constraints for the applied field, in contrast to NMR.

In metals the electrons obeying Fermi statistics form the lattice heat reservoir—decoupled from the nuclear spins. The coupling to the nuclear spin system usually is described by the Korringa behavior which relates the

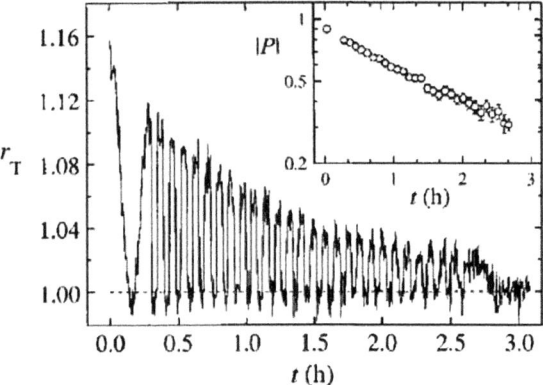

FIGURE 5 Ratio of transmitted intensities r_T for unpolarized neutrons through the polarized ^{109}Ag sample. The first field cycle was made in small steps, whereas the following data were recorded alternately in 300 µT and zero field. The high field signal diminished in the course of time because the spin system warmed up. At the end, a longer period was spent in 300 µT, whereafter a reference level with $M = 0$ was measured in $B = 0$. The deduced nuclear polarization is shown in the inset. *From Ref. [29].*

relaxation time constant τ_1 to the electronic temperature T_e with $\tau_1 = \kappa/T_e$. This relation holds when the Zeeman splitting of the nuclear spin system is smaller than the width of the Fermi edge of the electron gas which given by the electron temperature. Conversely, in high field the relaxation rate is expected to be independent from the electron temperature until T_e reaches values comparable to the nuclear Zeeman spitting. This behavior is seen in Figure 6 where a large

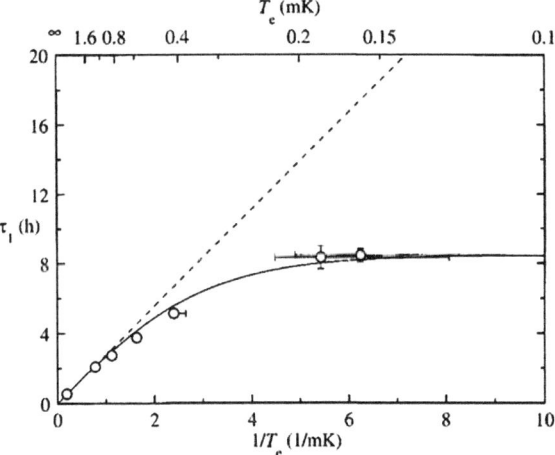

FIGURE 6 Spin–lattice relaxation time τ_1 of ^{109}Ag in a magnetic field of 7 T, as a function of the inverse electronic temperature $1/T_e$. The corresponding Zeeman splitting is 0.34 mK. The dashed line is the conventional Korringa relation $\tau_1 = \varkappa/T_e$, the solid curve represents theoretical expected [30] behavior. A Korringa constant of 10 secK has been used. *From Ref. [31].*

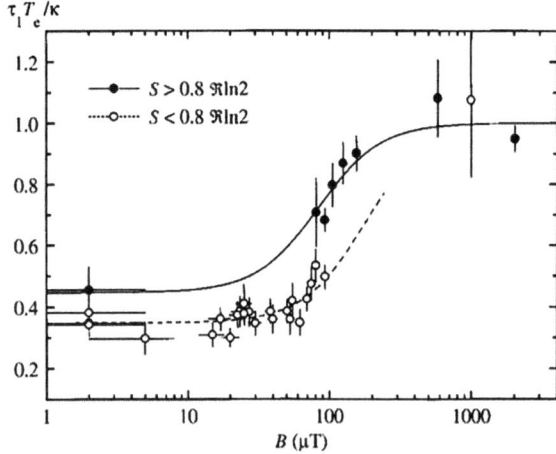

FIGURE 7 Spin–lattice relaxation time τ_1 of ^{109}Ag in low magnetic fields. The data are presented as a product $\tau_1 T_e/\chi$ which is not sensitive to changes in the electronic temperature T_e of the sample. In $B \cong 1$ mT, $\tau_1 = 8$ h was measured which yields $T_e = 0.35$ mK. The curves are fits to a model that accounts for the limited phase space available at high polarization. *From Ref. [31].*

field (7 T) is applied and the nuclear spin–lattice relaxation rate has been determined using neutron transmission.

Similarly, the relaxation depends on the number of relaxation channels available and they depend on the nuclear spin system temperature and field; in applied field the Zeeman energy can be large and dominates over the interaction terms of the Hamiltonian. Then, the relaxation between electron and nuclei must be slower compared to the case where 3D interactions are relevant at low field. This behavior indeed is also seen in Ag where at high polarization the high field case is realized already in a field of 300 µT (Figure 7). The data [31] also show that these effects appear only in strongly polarized systems equivalent to low temperature (or large entropy reduction), they get thermally smeared out at high temperature.

These comments on spin–lattice relaxation of nuclear spin systems may appear as a quite exotic application of neutron thermometry, however, the physics behind is not: it is of interest to control stable (quantum) states that are prepared by some means to be off from the equilibrium. The nuclear spin systems in contact with the electron lattice are probably a first example where this has been achieved.

7.2.4 Neutron Diffraction Cryostat for ULT Applications

Figure 8 is a schematic drawing of the cryogenic system that aims to cool nuclear spins in a metal to nano- and picokelvin temperatures. It consists of a so-called cascade nuclear demagnetization cooling system, which is mounted at the base plate of a powerful dilution refrigerator. It provides a constant

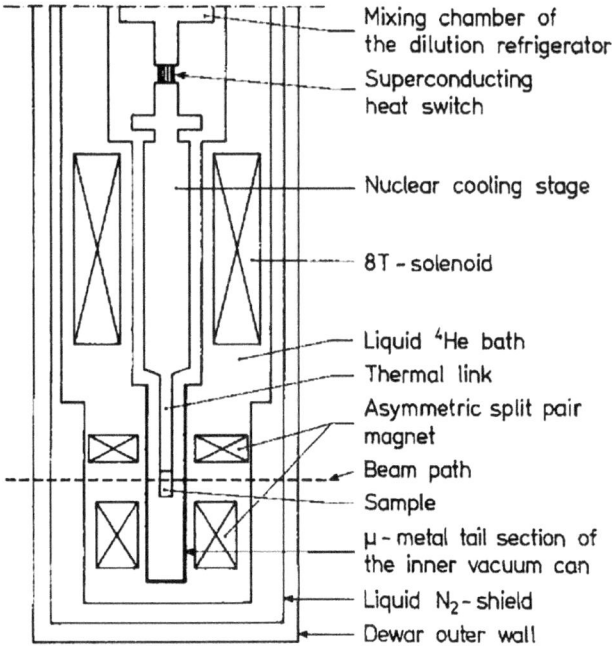

FIGURE 8 Schematic drawing of the lower section of the Risoe cryostat showing the nuclear stages and the magnets around them. *Reprinted from Ref. [24], http://dx.doi.org/10.1007/BF01312783 with permission of Springer Science + Business Media, ©Springer Verlag 1988.*

temperature of 10 mK. There are two demagnetization stages, the first stage consists of 15 mol Cu and the second stage is the sample itself. Both are connected thermally with high conductivity Cu or Ag, the connections are welded and in part annealed. The magnets can be operated independently and they are compensated to keep the thermal mass of the connection sample—first stage low. A superconducting heat switch to the dilution unit ensures good thermal for precooling the system as well as excellent thermal isolation during a demagnetization process. Such a system allows to manipulate the nuclear systems in such a way that at the end the sample nuclear system has "zero" entropy at high field—the starting condition to cool the sample nuclei by demagnetization.

Cooling proceeds in the following way: the complete ULT insert will be cooled to 10 mK by the dilution refrigerator through the superconducting heat switch with both magnets are fully charged. As soon as the system is at 10 mK in thermal equilibrium the heat switch is opened and the cascade demagnetization stages are thermally decoupled from the outer world. At 10 mK and 7 T the entropy at the first nuclear stage is $S \sim 0.9R \ln 4$ (J mol^{-1} K^{-1}) (cf. Figure 1). From there the demagnetization of the first stage takes place, while the magnetic field in the second stage remains constant.

Lowering slowly the first stage field cools the system, ideally in equilibrium to avoid entropy losses. Typically, once the base field of the first stage is reached after several hours, the electron of the complete system is around 100 µK. At that temperature, the sample nuclei entropy is zero, i.e., they are fully polarized. Also, the spin−lattice relaxation rate is slow so that a "fast" demagnetization of the sample cools the nuclei only. The sample demagnetization rate is limited mainly by the need to avoid eddy current heating. It starts at a field of 5 T (Risoe) and is performed down to less than 0.2 mT. If everything goes well then nuclear spins in the sample should be in the ordered state.

7.2.5 Sample Requirements

The samples are part of the cooling system and therefore they must obey stringent requirements: high thermal conductivity needs a low impurity concentration. Impurities would also tend to increase the spin−lattice relaxation rate which cannot be tolerated. Neutron diffraction needs arise from spin-dependent scattering length considerations and consequently isotopically enriched material is used. It turned out that these materials required additional electrochemical purification before a crystal could be grown. Oriented seeds have been used to ensure the right orientation of the samples.

For metals a flat sample geometry is essential to avoid eddy currents, the ^{65}Cu and ^{109}Ag samples had a mass of ~ 2 g and approximate dimensions of $0.5 \times 10 \times 20$ mm^3. The flat geometry has advantages also for beam heating effects because at least in one direction the β-decay products can escape from the sample.

In summary, the experimental setup for neutron diffraction at ULT is demanding because different techniques must be merged. The main challenge is to keep unwanted heat loads on a very low level which needs a careful design of the experiment and also, sometimes, a bit of luck to find the right answer to complex questions.

7.2.6 Spontaneous Nuclear Magnetic Order

7.2.6.1 Nuclear Ordering Observations in Dynamically Polarized Systems

Parallel to brute force cooling techniques using adiabatic nuclear demagnetization mainly at Saclay a technique has been developed to polarize nuclei by dynamic nuclear polarization (DNP) in a high field and at a relatively high temperature [5,32]. The method requires paramagnetic centers (F-type) that can be pumped by microwaves to transfer the spin polarization to the nuclei. The idea behind is that electrons couple to the lattice very well, so they can relax back after transferring their spin to the nuclei. In contrast, the nuclear spin−lattice relaxation is very long, so once a nuclear spin is flipped to a favorable direction with respect to the external field it will remain in that state.

Ideally, it transfers the spin state to neighboring nuclear sites so that spatial equilibrium is achieved. The spin transfer back to the electron system is strongly forbidden because it is saturated.

The method is atom selective, i.e., different nuclei will get polarized at different microwave frequencies. Typical microwave pumping requires a field of several Tesla; temperatures around 0.3 K are sufficient to isolate the nuclear spin system from the surrounding, thus very high nuclear polarization is achieved under conditions that are rather moderate compared with ULT work. Once a high nuclear polarization is reached, the microwave pumping can be stopped. This pumping presents the biggest heat load to the sample, without the system can cool to temperatures of the order of 50 mK which further improves the thermal isolation of the nuclear spins. They can then be demagnetized in the large field by "adiabatic demagnetization in the rotating frame" (ADRF). A circular polarized RF field is applied off resonance and either the magnetic field or the RF frequency is slowly tuned toward resonance. This process is almost equivalent to a brute force nuclear demagnetization.

The ordered nuclear structures after an ADRF in LiH have been the first nuclear spin systems that where investigated by neutron diffraction. LiH has been chosen because the spin-dependent scattering length of both ^{7}Li and H are favorable, in contrast to CaF_2 that also has been investigated after ADRF cooling. With regard to the cooling technique, one must mention that the Li and H nuclei had to be pumped and cooled separately, however, it turned out that both systems after cooling have the same nuclear spin temperature. One special feature of this cooling technique is the fact that both negative and positive temperatures can be established, with significant consequences for the ordered structures. From a simple Curie–Weiss picture, one would expect ferromagnetic (FM) order for a positive Curie–Weiss temperature at positive temperature. Seen from the negative temperature side then antiferromagnetism (AFM) would be expected.

As LiH is an insulator, the nuclei interact only via the dipolar interaction. It crystallizes in a NaCl structure which was amendable for theoretical calculations of the nuclear ordering, also because the interaction is very well known. One aspect of the interaction relevant after an ADRF must be mentioned: as the cooling proceeds in a high field, only the terms of the Hamiltonian that commute with the Zeeman term are relevant which leads to the so-called "truncated" dipolar Hamiltonian. This implies that the direction of the magnetic field with respect to the crystal axis affects the ordering. In other words, as the field sets the direction of the highly polarized nuclei, only structures with moments parallel or antiparallel to the field direction are possible.

In fact, in a series of neutron diffraction experiments on LiH, AFM order has been seen, e.g., for B||[001] ordering of the (110) type was observed, both at positive and negative temperatures [6]. Later, the field was applied along the [011] direction which gave (100) type order at positive temperature and FM at negative temperature [33,34]. The structures observed are in remarkable agreement with calculations [5].

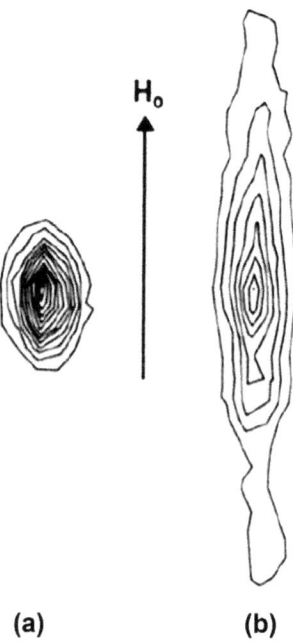

(a)   **(b)**

FIGURE 9 Lines of equal intensity of the neutron beam for the (200) reflection at the center of the rocking curve. (a) Before demagnetization. No dipolar order. The intensity is enhanced by the large nuclear polarization. (b) After demagnetization at $T < 0$. Ferromagnetic structure. *Reprinted from Ref. [33], http://dx.doi.org/10.1051/jphyslet:01980004105012300 with permission of EDP Sciences, ©EDP Sciences.*

The line width of the structural (200) reflection in the FM state shows significant broadening compared to the disordered, sharp line (cf. Figure 1). It has been interpreted to be due to FM domains with a size of 100–150 Å. A comparable domain size was observed for the AF reflections which led to the speculation that their size is determined by the presence of paramagnetic impurities. Their density would justify this picture, however, there is no clear evidence and it is equally possible that slow spin–spin dynamics together with the limited observation time keeps the system from reaching full equilibrium (Figure 9).

Despite their success, the experiments on ADRF cooled nuclei have not been continued. However, the DNP polarization techniques developed by the Saclay group found application in other fields, like the determination of nuclear spin-dependent scattering length [35] or DNP of polarized targets for SANS which is also subject of this review.

7.2.6.2 Previous Results for Cu and Ag from SQUID and NMR Techniques

Between the years 1980 and 1986 mostly Huiku and others [11,12] (Cu) and Hakonen et al. [36,37] (Ag) studied the nuclear spin systems in great detail.

Here we will recall those results only, which will be of most importance for the neutron scattering experiments discussed here. The main experimental tool for their study was SQUID–NMR. This gave detailed results for the behavior of the susceptibility of the system over the full range of interest in field and entropy. The susceptibility was also determined in different crystallographic directions. As a function of temperature the zero field susceptibility indicated AFM both for Cu and Ag. Evidence for further phases in finite field have been found for Cu only.

7.2.6.2.1 Cu

The main results using a polycrystalline (in the earlier experiments) and a single crystalline sample (in the later experiments) are a determination of the critical entropy and temperature which gave $T_N = 58$ nK, along with the measurement of the upper critical field of 0.25 mT. An entropy–magnetic field phase diagram was deduced from the susceptibility results which suggest the existence of at least three magnetic phases as shown in Figure 10. The most convincing result is that for the low B regime, AFM order exists. It disappears for $B > 0.25$ mT.

A jump of the critical entropy has been associated with the occurrence of latent heat and could be taken as an indication for the discontinuous character of the zero field transition. That result would be in line with predictions from renormalization group theory that predicts first-order phase transitions for ordering in the fcc lattice [38].

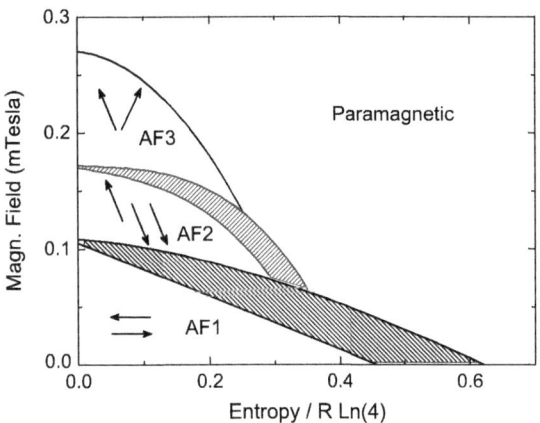

FIGURE 10 The external magnetic field versus entropy phase diagram for the magnetically ordered phases in copper. The three ordered regions are marked by AF1, AF2, and AF3. P stands for the paramagnetic state. The wide bands indicate regions in the first-order phase transitions take place. The transition in zero field occurs at 58 nK according to the temperature measurements on a polycrystalline sample. Some suggested spin arrangements are shown for field along z. *From Ref. [11]*.

All the results known for Cu before the neutron experiments started gave a very firm basis to define the necessary degrees of freedom for a neutron experiment. It also came clear that in neutron experiments connection to the previous experiments had to be made by simultaneous measurement of the susceptibility.

7.2.6.2.2 Ag

Detailed NMR measurements to obtain the susceptibility of the nuclear system have been performed in Otaniemi using a polycrystalline sample with natural isotope distribution. The main result [36,37] is that AF nuclear order occurs in zero field below 560 pK. In contrast to Cu a very simple magnetic S- versus B-phase diagram could be determined with only one phase. The critical field was found to be $B_C = 100$ µT at zero entropy. With a rapid inversion of the external field, it has been possible to invert occupation of Zeeman levels, i.e., negative temperature has been reached. The behavior for $T < 0$ has been studied in detail. It was found that at negative temperatures the nuclear spin system orders ferromagnetically. The critical entropy for FM nuclear order at $T < 0$ was $\sim 0.8R \ln 2$, whereas AF order at $T > 0$ has a critical entropy of $\sim 0.5R \ln 2$. Different to Cu, no hints toward a first-order phase transition have been seen. A very detailed description of these results for Cu and Ag can be found in Ref [39].

7.2.6.3 Theoretical Predictions

A large amount of theoretical work has accompanied the progress in the search for nuclear order and the determination of the ordered states for Cu and Ag by Huiku and colleagues [11,12,36,37,39]. Those theoretical efforts were concentrated on explaining the experimental results like the reduction of critical temperatures against mean field theory expectations and proposing further experiments [40]. The technique used in these experiments to investigate the thermodynamic behavior of Cu and Ag at ULT was SQUID-NMR. This is a particular sensitive technique to measure the homogeneous susceptibility $X(0)$. We shall, however, not review all these papers mentioned above, but rather want to review those papers only, which had most impact on the neutron scattering studies on Cu and Ag, where one is interested in q-dependent properties (Figure 11).

Of central importance is a series of papers that Lindgård et al. [41−43]. have published around the time when the neutron experiments had started running. Lindgård et al. have set up a first - principle (FP) calculation of the interactions in Cu and Ag as a sound basis for studies of the magnetic structures in zero and finite magnetic field taking the specialities of cooling the nuclear spins adiabatically into account as well. Both, Cu and Ag, are simple metals with the high symmetry fcc structure. The "magnetic" interactions in these systems are due the electron−electron mediated Ruderman−Kittel (RK)

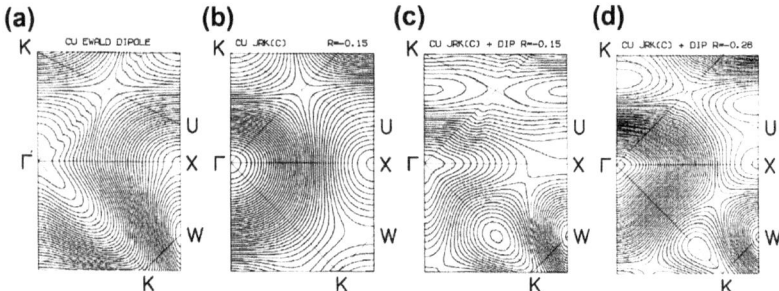

FIGURE 11 Contour plot containing all fcc symmetry points; upper part [1,−1,0] plane, lower part [0,1,0] plane, separation 2 nK. (a) Maximum eigenvalue of dipole interaction maximum (110 nK) near $\Gamma = (0,0,0)$; (b) $J(q)$ for Ruderman−Kittel (RK) interaction. $R = -0.15$ maximum (30 nK) occurs at $X = \pi/a(0,0,2)$. (c) Maximum eigenvalue for dipole + RK interaction for $R = -0.15$, maximum (120 nK) at $Q = \pi/a$ (1,1,0) between Γ and $K = \pi/a(3/2,3/2,0)$; (d) Same for $R = -0.26$, maximum (132 nK) at $Q = \pi/a(\eta,\eta,0)$, $\eta = 1/2$. For $|R| < 0.27$ Q is incommensurate along the Γ-K direction. *Reprinted from Ref. [41], http://dx.doi.org/10.1016/0304-8853(86) 90377-X with permission of Elsevier, ©Elsevier Science Publishers BV 1986.*

interactions and dipole−dipole (DD) interactions. Lindgård et al. [41] performed high-precision FP calculations to obtain the electron bands structure and wave functions. This is necessary to obtain reliable results for the correct range dependence of the RK interaction. The important result is that the RK interaction is dominated by nearest-neighbor contributions, which is in contrast to the expectation from the free electron approximation (FEA), which has been used before (12, 40) and predicts a strong long-range character of the RK interaction. A dimensionless parameter R has been defined to describe the relative strength of the RK and DD interaction. It was determined experimentally [12] to be $R = -0.42 \pm 0.05$. Lindgård et al. find $R = -0.34$ from FP methods. It is found as well that T_N is strongly reduced with respect to T_{MF} due to strong correlation effects. With the band structure calculation it was possible to calculate eigenvalues for each reciprocal lattice vector to obtain stability regions of possible structures, assuming different R-values (see Figure 12). Assuming a second-order PT, one finds for $R = -0.42$ ordering at the X-point (type I AFM on fcc lattice) and $T_N = T_{mf} = 82$ nK, which is too high compared with the measured value of 60 nK.

The calculations show that the obtained structure is stable until $R = -0.28$ with $T_N = 36$ nK. For $R = -0.26$ the ordering vector jumps to $\pi/a(\eta,\eta,0)$ with $\eta = 1.2$. For R toward 0, η decreases and Q moves along the Γ-K ridge. The behavior is qualitatively different to the FEA and that is due to the strong Ruderman-Kittel-Kasuya-Yosida (RKKY) nearest-neighbor contribution. For the calculated R-value of −0.34 the predicted structure is still the type I AFM on the fcc lattice and the Neel temperature $T_N = 0.32T_{MF} = 58$ nK is in good agreement to the measured one. The results of the FP calculations show that the RK and the dipole−dipole interaction are strongly competing in Cu, where R is rather close to the stability limit of the type I AFM on the fcc lattice.

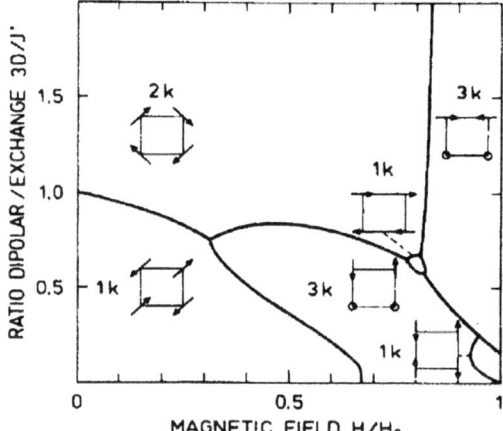

FIGURE 12 The calculated regions of stability at $T=0$ of various multi-k structures as a function of field along (001) and the relative strength of the dipolar and exchange interaction indicated by $3D/J'$. For Cu the latter is between 0.9 and 0.8. For fields along the (101) direction there is a smooth transition between two 1-k structures at $H \sim 0.3J'$. For $3D/J' > 1$ a mean field calculation shows that an incommensurate structure with ordering vector along ΓK is more stable than the type I AFM structures considered here. Quantum fluctuations will make the ΓK phase penetrate to lower D values along the phase separation lines. The various structures are depicted by their projection onto the $x-y$ plane. *From Ref. [42].*

With these results for zero magnetic field a firm basis has been laid down for dealing with the more complicated situation of the behavior of the nuclear spin systems of Cu and Ag in magnetic field. The behavior of Cu, where RK and DD interactions were found to compete strongly, is very interesting: it is tempting to expect that the application of a magnetic field, where the spins will change the angles between them, might destabilize the type I AFM structure and lead to new phases.

In order to further clarify the effect of an external field in Cu, Lindgård [43] has developed a "Theory of Adiabatic Nuclear Magnetic Ordering in Cu." This should lead to a better understanding of the neutron scattering results which had shown three different phases for Cu in a magnetic field. The zero field phase is of the type I AFM structure on a perfect fcc lattice. In this structure, all spins are sitting on triangles, therefore, making the AFM RK interactions fully frustrated. The competing DD interactions only require that the spins are perpendicular to the ordering wave vector.

The adiabatic cooling into the ordered zero field state requires that the spin system passes a phase transition at a critical magnetic field H_c, where it goes from the fully magnetized paramagnetic state into an AFM ordered state. In mean field theory the picture is simple: in an H versus T diagram the isentrops are straight lines through $H = 3\,J'M$ with $J' = 4J + D$, where J and D are RK and dipole strengths, respectively. The slope is given by P_{init}, the initial polarization of the spin system. At the PT at H_c the spin system goes from having the polarization P_{init} to have an AFM order parameter OP > 0. One of the conclusions deduced from the FP calculations was that in Cu RK and DD interactions strongly compete and $T_N < T_{MF}$. Therefore, fluctuations cannot be neglected when deducing a phase diagram for a system which has an infinitely degenerate ground state with respect to linear combinations of possible ordering wave vectors along the cubic directions. Can the application of a magnetic field lift the degeneracy and lead to new stable structures?

Lindgård argues that nonlinear effects will indeed lift the degeneracies and stabilizes new phases in field. He goes through a hierarchy of off-diagonal terms in the Hamiltonian and finds that in field the canting of the spins introduces effective anisotropic interaction terms in both the RK and the DD terms for relatively canted spins and this at the end lifts the degeneracies as expected. The resulting phase diagram at $T = 0$ is shown in Figure 13 as H/H_c versus $3D/J'$. Lindgård predicts for Cu (R = -0.34) with $3D/J' = 0.9$ a transition between the I, II, and III structures for H along (0,0,1). For Cu (R = -0.42 and thus $3D/J' = 0.8$) and for intermediate fields will fluctuate between the 2 k and 3 k structures. For H along $(10-1)$ even an incommensurable structure is possible. For $H = 0$, Lindgård's result agrees with spin wave theory (see, e.g., Ref. [39]). The conclusion of Lindgård's calculation is the prediction that Cu should have in an applied magnetic field three phases. In addition it is suggested that the ordering wave vector is along K direction: $(0,\eta,\eta)$.

In a further paper [43], Lindgård investigates the role and importance of fluctuations. It is emphasized that the FP calculations from Lindgård showed for R = -0.42 (Cu) a ridge along Γ-K with a local maximum at K point which becomes an absolute maximum for R = -0.29. This behavior is due to fluctuations induced by the competition between RK and DD interactions. Applying a magnetic field enhances quantum mechanical fluctuations in the ground state. Analyzing different possible intermediate structures, it was found

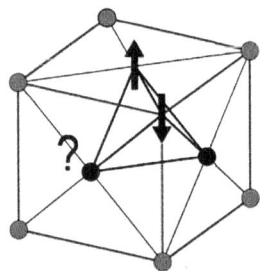

FIGURE 13 The face-centered cubic lattice is formed by edge-sharing tetrahedral, imposing frustration for AFM nearest-neighbor interactions. The simple up-down magnetic ordering is highly degenerate.

that the structure with $\eta = 2/3$ should be stable in Cu at $H = H_c/3$, indeed. This structure is one which is stabilized by fluctuations. This has been found in Monte Carlo simulations (see Frisken and Miller [44]) as well. Thus, two different approaches come to the same conclusion and this gives further confidence that experimental strategies can be built on this.

7.3 EXPERIMENTAL RESULTS FROM NEUTRON DIFFRACTION ON CU AND AG

7.3.1 Nuclear Magnetic Ordering of Cu and Ag in Zero and Finite Magnetic Field

The experiments with neutrons were performed under the same (as much as possible) experimental conditions as they were used in Otaniemi when nuclear ordering was discovered.

As the intensity of the expected Bragg reflection is proportional to the square of the order parameter, one has to do everything possible to reach temperatures as far as possible below the phase transition! An estimate of the expected count rates for the Bragg reflection based on the available neutron flux and the respective scattering lengths shows that the count rates will be quite low and will fall quite quickly when the nuclear system will warm up after cooldown. Making the time for collecting neutrons after the end of the adiabatic demagnetization as long as possible is of key importance for the experiments. Therefore, every parameter in the experiment which gives higher count rates has to be maximized: neutron flux can be gained by using long wavelength neutrons, coarse resolution, the sample must be a good single crystal and as big as tolerable with respect to cooling powder considerations. Under favorable conditions, one should have at least several minutes for a measurement, considering cooling capacity, spin−lattice relaxation, and heat load caused by the neutrons absorbed by the sample. In Ag particularly absorption was found to be real problem based on the experience from the Otaniemi experiments: no more than a few nanowatts heat load on the sample can be tolerated to reach the ordered phase. In order to be able to link the neutron experiments to the experiments performed in Otaniemi, the susceptibility of the nuclear system has been recorded simultaneously by SQUID techniques.

7.3.1.1 Copper (I = 3/2)

Following the results and predictions of Lindgård et al., the search for nuclear magnetic ordering concentrated first on Bragg reflections characteristic for the type I AFM on the fcc lattice. The sample was aligned with one of the (011) directions along the vertical magnetic field, another (100) and one of the (110) directions in the horizontal plane. This orientation gives access to all high-symmetry directions of the fcc lattice. The lowest order AF reflection is then the (1,0,0) reflection. After quite some time of hard work to get all the experimental conditions right, this reflection was seen with the detector fixed at the (1,0,0) position for the first time for about 7 min, see Figure 14. Therefore, $t = 0$ marks the time, when demagnetization was stopped, and the neutron beam was opened. The demagnetization stopped at $B = 0$, which was $B \leq 0.01$ mT. The appearance of this (1,0,0) peak, given the high symmetry of the underlying lattice, is sufficient proof for the magnetic unit cell to be of the AFM fcc type! The characteristic of the observed t-dependence of the peak is an increase of the intensity within the first minutes before the intensity disappeared with time. This peak and its t-dependence were regularly observed for many cooldowns [45].

To obtain more information about the ordered state, full peak profiles were recorded using a linear detector. This allowed to measure the full profile without any movement of any part of the instrument. The so recorded peak profiles at different times after reaching $B = 0$ are shown in Figure 14 (lower). Integrating the profiles yields a t-dependence of the intensity consistent with Figure 14 (upper). If we could measure at ideal resolution the Bragg peak would be a delta function characteristic for a long-range ordered system. With finite resolution the Bragg peak width is given by the resolution width. The peak widths of a series of profiles show that at $t = 0$ a correlation length of about 500 Å which reaches 1000 Å after a couple minutes. This is the resolution limit, the line width does not change further with increasing time. The correlation length is a measure for the size of the ordered regions. It cannot be measured to very high values (>1000 Å) due to the coarse resolution in the experiment. Thus, experiments correspond to ordered regions larger than 1000 Å after a few minutes. The observations can therefore be interpreted as follows: at very early times, ~1 min, the system is not yet fully long-range ordered with a size of the ordered regions being approximately 500 Å. When the intensity is maximum and the size of the ordered regions has grown to 1000 Å, we observe long-range order to the best of our experimental possibilities until the Bragg peak disappears. Thus, we have observed in the Cu nuclear magnetic system type I fcc AFM in zero field and its growing correlation length from 500 Å at early times to long-range order with a correlation length larger than 1000 Å. All symmetry-related (100) type reflections available in the scattering plane have been found.

We know from electronic magnetic systems, that in such AFM systems more detailed information about the Hamiltonian of the system can be obtained by measuring the B-T phase diagram, with if necessary and possible, B being oriented along different symmetry directions. In the present study the necessary magnetic fields are so small ($B < 0.30$ mT) that X-Y-Z coils could be positioned around the sample producing fields in any direction. Only the coils for adiabatic cooling are fixed along the (011) direction. Experiments with this setup have been performed in Otaniemi by NMR and found evidence for a rich phase diagram, which has some similarities to the theoretical predictions by Lindgård and colleagues.

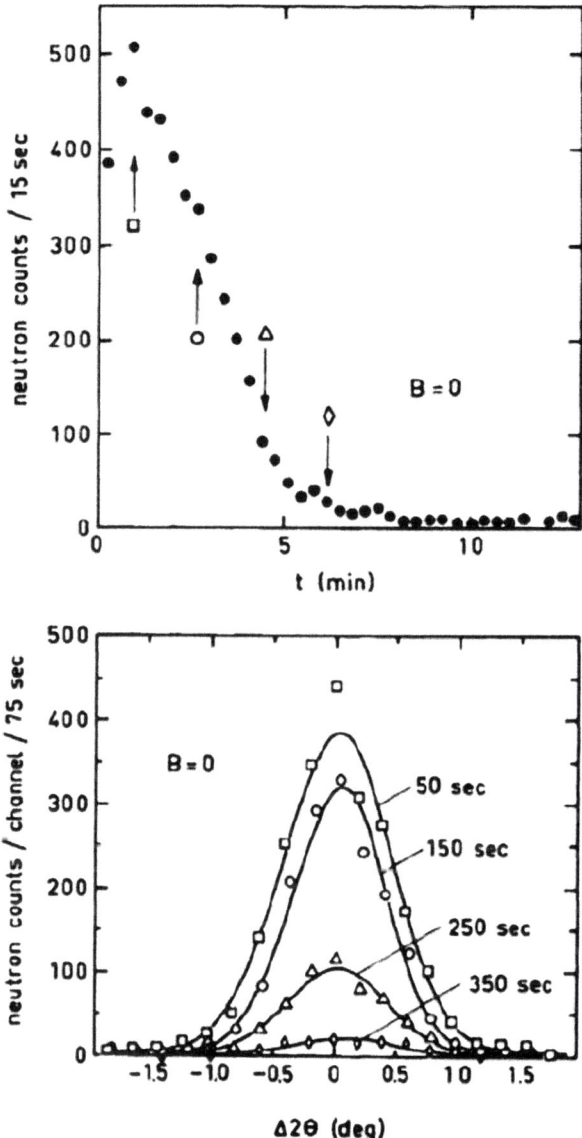

FIGURE 14 Upper: The time dependence of the integrated neutron intensity from the (100) antiferromagnetic Bragg peak after zero field was reached. Lower: The time evolution of the (100) peak, as observed in the linear position-sensitive detector. The four 75 s measuring intervals are centered at the times indicated for each curve by the corresponding symbol in the upper part of the figure. The solid lines are the best Gaussian fits to the experimental points. *Reprinted with permission from Ref. [45], http://dx.doi.org/10.1103/PhysRevLett.60.2418, ©1988 by the American Physical Society.*

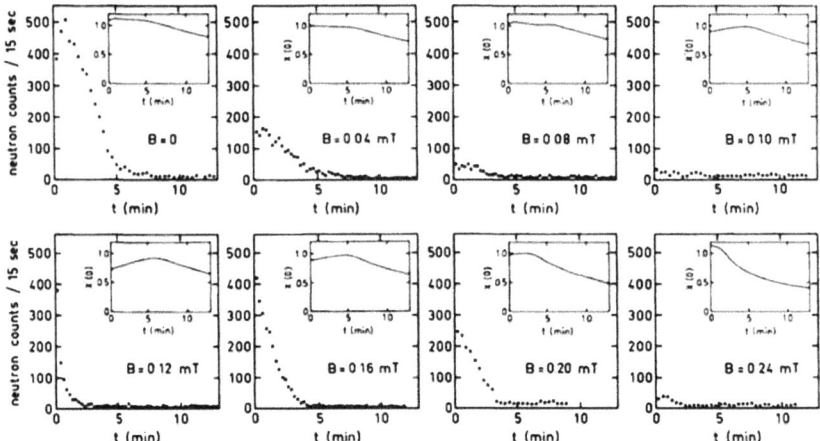

FIGURE 15 The time-dependent (100) intensities, $I(t)$, as measured at different applied magnetic fields. The susceptibility measured simultaneously is shown as well. *Reprinted with permission from Ref. [45], http://dx.doi.org/10.1103/PhysRevLett.60.2418, ©1988 by the American Physical Society.*

Neutron scattering measurements like the one shown in Figure 14 have been performed with very similar parameters as used by NMR in the early experiments: demagnetization was stopped at the chosen field value and the t-dependence of the intensity recorded. In Figure 15, some typical results for fields parallel to (011) between 0 and 0.24 mT are shown together with the simultaneous susceptibility measurements. Surprisingly, the magnetic Bragg peak does not appear at all fields: the observed disappearance of the (1,0,0) reflection at fields beyond 0.24 mT confirms the finding of the upper critical field from NMR data. There is no ordering in this "high" field regime beyond 0.25 mT. But there is another field region, where no (1,0,0) peak could be detected, between 0.08 and 0.11 mT. In addition the maximal intensity and the t-dependence vary significantly with field strength. The time dependence is particularly steep between 0.12 and 0.16 mT. More studies have been done to collect more information on details of field-induced effects. Those included different paths into the ordered state, waiting times at different positions in the phase diagram, and sweep rates during demagnetization. A wealth of different effects on the t-dependence of the (1,0,0) peak has been observed and some seem to indicate strong hysteresis effects indicative for the appearance of a discontinuous character in the phase transitions.

It is tempting to conclude that the phase diagram consists of three areas: the low field regime below 0.08 mT, the high field regime beyond 0.12 mT, and an intermediate field regime between 0.08 and 0.11 mT. In the low and high field regime the nuclear order is of the type I fcc AFM order with different t-dependencies. In the intermediate field regime there is no such order, while NMR suggests convincingly that order must exist, see Figure 10 for $B = 0.01$ mT. Thus, at intermediate fields there must be another type of order! The recorded susceptibility measured simultaneously with the neutron data supports strongly the existence of ordering in the intermediate B range.

This exciting consequence of the studies of the (1,0,0) peak is a very challenging situation for two reasons: it would be a normal situation in structure determination under normal circumstances, where one can measure at constant temperatures and can scan through reciprocal space, wherever one wants to go. Neither the first option constant T nor the second degree of freedom is available, when working at ULT: thinking of rotating the cryostat and the detector bank independently after the stop of demagnetization is frightening. All experience until now said "no nanokelvin under such conditions."

As mentioned above, the particular orientation of the crystal with respect to magnet and horizontal plane was chosen to allow access to all fundamental reflections of the fcc lattice. It is the plane in reciprocal space (see Figure 16), where a phase with a structure different from type I fcc AFM could exist. It was even predicted [42] that scanning along the [1,y,y] direction varying y (see Figure 16, left) a reflection of a new magnetic structure would be likely to occur. Incommensurable structures were not excluded. Scans to verify the prediction required the rotation of the cryostat and the detector simultaneously and have successfully been performed avoiding vibrations due to the mechanical movements during the scan! Figures 17 and 18 show the results: a peak was found at (1,1/3,1/3), which within the resolution is found to be commensurate. Symmetry equivalent reflections like (0,2/3,2/3) and (1,−1/3,−1/3) have been found as well. This new type of a commensurate structure has not been observed before in a fcc AFM in an applied nor in zero magnetic field.

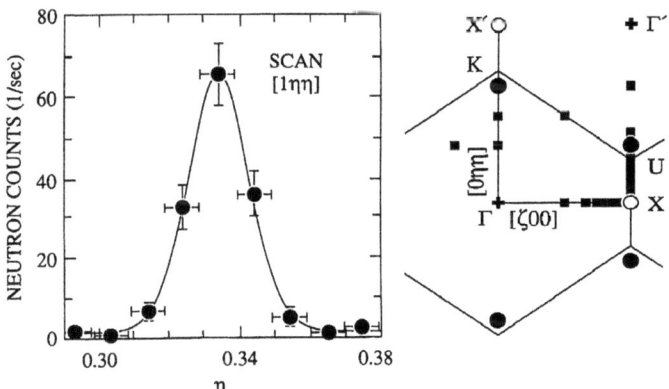

FIGURE 16 Left: Neutron intensity versus position along ΓK [1ηη] showing the discovery of the (1,1/3,1/3) Bragg reflection. Right: The (0,1,−1) scattering plane in the Brillouin zone of a face-centered cubic lattice. The search scans are marked with thick lines. *Reprinted with permission from Ref. [46], http://dx.doi.org/10.1103/PhysRevLett.64.1421, ©1990 by the American Physical Society.*

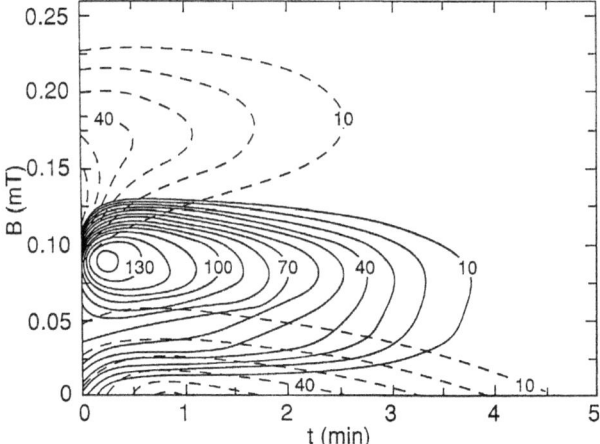

FIGURE 17 The neutron intensity contour diagrams of the (1,1/3,1/3) (solid lines) and the (100) (dotted lines) Bragg peaks as a function of time and external magnetic field. The outermost contours, 10 cts s^{-1}, show approximately when long-range order disappears. *Reprinted with permission from Ref. [46], http://dx.doi.org/10.1103/PhysRevLett.64.1421, ©1990 by the American Physical Society.*

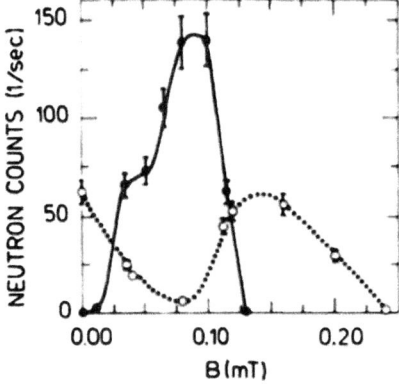

FIGURE 18 Neutron intensities of the (1,1/3,1/3) (full circles) and the (100) (open circles) reflections through the maxima of Figure 17 as a function of the external magnetic field. *Reprinted with permission from Ref. [46], http://dx.doi.org/10.1103/PhysRevLett.64.1421, ©1990 by the American Physical Society.*

Having found and identified the missing order in intermediate external fields, a phase diagram can now be constructed be plotting isointensity contours in the B-t plane for B approximately parallel [0 1 1 0]. This "phase diagram" clearly shows the stability regions of the reflections within their respective field regime. Another observation can be made: in most of the field

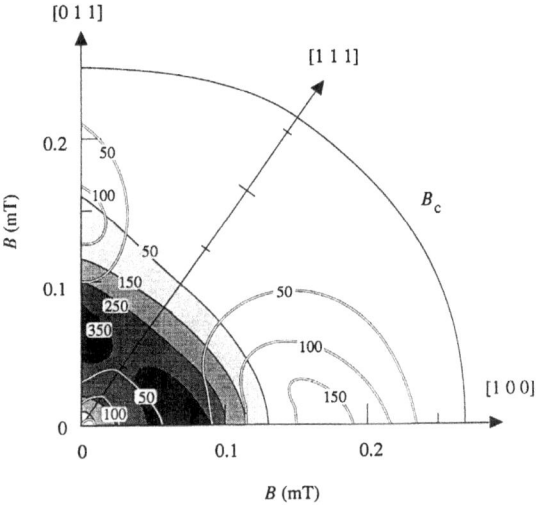

FIGURE 19 Neutron intensity as a function of field. The contours give initial observed intensity after demagnetization in cts per second. Full contours show the (0,2/3,2/3) and dashes lines the (100) reflections. *Figure 19 and caption are reprinted from Ref. [47], http://dx.doi.org/10.1016/ 0921-4526(92)90650-H with permission of Elsevier, ©Elsevier BV 1992.*

regime where order exists, $B < 0.25$ mT, the stability regimes of the two structures do overlap. This is particular evident in the low field regime. Plotting the intensities of the two reflections (1,0,0) and (1,1/3,1/3) versus B (Figure 19) make this strong coexistence of the two types of order even more clear. Only at very low field and in the high field region, $B > 0.12$ mT, the (1,0,0) structure was alone observed.

Putting together all the results obtained, it was possible to obtain a nearly complete diagram, which shows the dependence of the nuclear magnetic order on the field direction with respect to the unit cell axis. The fcc structure of copper possess three high-symmetry directions: the fourfold axis is the [1 0 0] direction, the threefold axis is the [1 1 1] direction, and the twofold axis is the [1 1 0]. As symmetry is a particular important parameter for the stability of magnetic order, we have plotted equal-intensity contours versus B for these three high-symmetry directions of the fcc structure (Figure 20). Surprisingly, the (1,0,0) magnetic Bragg peak is definitely absent at high magnetic fields, $B > 0.12$ mT for B parallel to [1 1 1]. Another observation is important: the critical fields obtained in the neutron scattering experiments agreed reasonably well with the critical fields obtained by susceptibility measurements in Otaniemi. On the other hand with B along [1 1 1], the critical field from susceptibility data do not agree with neutron data (see the line B_c in Figure 20). This discrepancy could not be clarified: either the neutron data missed yet another ordering type, as no useful indication for this was available from

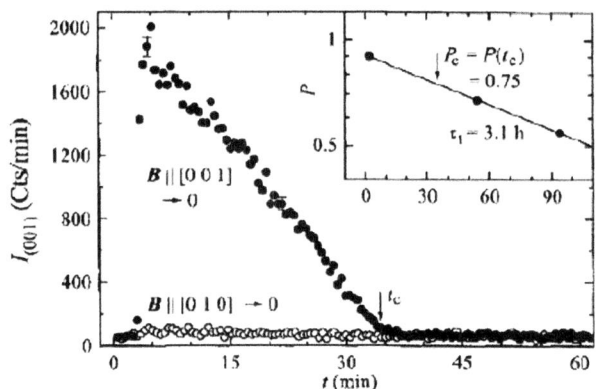

FIGURE 20 Time dependence of neutron intensity, measured by a single counter (30 s point^{-1}) at the (001) position. The initial polarization $P = 0.91 \pm 0.02$ was recorded in a 500 µT field in the paramagnetic phase, whereafter B, in the [001] or [010] direction (full/open circles, respectively), was reduced to zero (± 5 µT) at $t = 3$ min. The (001) reflection appeared immediately below B_c, but only when B||(001). The inset shows the exponential relaxation of nuclear polarization, with $\tau_1 = 3.1$ h. The critical value $P_c = 0.75 \pm 0.02$ was found by interpolation. *Reprinted with permission from Ref. [48], http://dx.doi.org/10.1103/PhysRevLett.75.3744, ©1995 by the American Physical Society.*

existing theories, or the susceptibility data were misleading, since susceptibilities often do not show strong signatures at transitions between two AFM phases.

Having spoken so much about phase transitions in the ordered nuclear spin system in Cu a word of caution is appropriate: we were not able to determine nuclear temperatures in Cu! Nor could we measure the entropy S in the ordered phase. A measurement of T or S was replaced by measurement versus time after the stop of demagnetization. Consequently, we always looked at a system during warm-up. The time for which the nuclear system was in the ordered state was simply too short to perform the complicated measurement procedure to produce a well-defined value for S at a useful spacing during the warm-up. This will be important for further discussions of the data.

7.3.1.2 Silver (I = 1/2)

A few words should be said here about the change of the sample in such experiments: this is a major operation since the sample is a part of the cooling system itself, and thus the major parts of the cascade demagnetization system have to be rebuild, tested, and then reinstalled in the cryostat to test the complete system. Then the system moved to the diffractometer before the neutron experiments could be restarted. As the Ag experiments were planned to be performed at BER II in the HMI in Berlin, the whole team moved to Berlin too. The activities started in 1992 after a break of 2 years for the preparation of the Ag sample and the reinstallation of the instrumentation in Berlin.

The theory, which was so successful in predicting correctly two different types of ordering in Cu, predicted that the type I fcc AFM structure should occur in Ag too as the zero field ordering. The crystal was oriented with the [1 −1 0] axis vertical. The observation of the (1,0,0) peak confirmed the predictions very nicely as shown in Figure 20 [48]. The time the sample remains in the ordered phase was found to be ∼30 min, typically five times longer than in the Cu experiments. The 30 min were long enough to measure nuclear temperatures/entropies under specifically favorable conditions (Section 7.2.3 this paper). The measured Bragg peak was resolution limited, and no specific time dependence as in Cu was observed. By scaling the observed maximal intensity of the (1,0,0) peak to the intensity of the (2,0,0) structural peak, the nuclear polarization in the AFM nuclear-ordered state P_{af} could be determined. It was found to be $P_{af} = 0.48 \pm 0.10$. The search for symmetry equivalent reflections of the (1,0,0) type was unsuccessful! The (1,0,0) type I AFM peak was found with field along [100] and [110] but not along [111]. It was found as well that (1,0,0) was seen for all field directions except a cone with about 50° opening around [111].

The search for other ordered structures predicted by Viertiö and Oja [39,40] in Ag was unsuccessful. In agreement with the result of NMR experiments in silver. The critical field for the (0,0,1) peak was measured to be $B_c = 100$ microtesla in good agreement with analysis of susceptibility data. All the experimental data can now be combined in a simple phase diagram shown in Figure 21. Only one phase boundary from the ordered type I fcc AFM state to the paramagnetic state can be seen. Rotating the external magnetic field toward the [1,0.8,0.8] direction, the critical line in the phase diagram is slightly shifted, see the full and broken lines in Figure 21. The

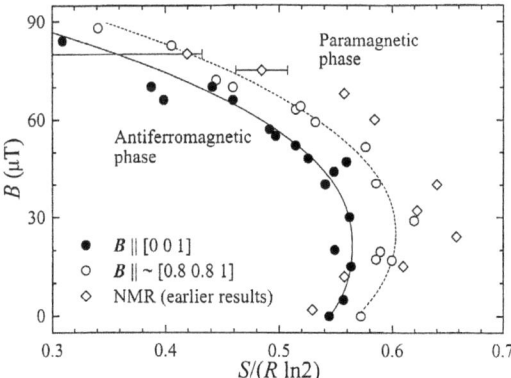

FIGURE 21 Magnetic field versus entropy phase diagram of silver; B was along the [001] or [0.8,0.8,1] direction. Earlier NMR results [36] on a polycrystalline sample are shown for comparison. *Reprinted with permission from Ref. [48], http://dx.doi.org/10.1103/PhysRevLett.75.3744, ©1995 by the American Physical Society.*

observed critical magnetic fields agree reasonably well with NMR results on a polycrystalline Ag sample with natural isotope mixture. It should be emphasized here that the phase diagram so obtained is given in the two thermodynamic variables B and S. S is measured by the nuclear absorption thermometer described in detail earlier in Section 7.2.3.

7.3.2 Structures—Cu and Ag

The observed structure is the well-known type I AFM structure on the fcc structure which both Cu and Ag possess. Both are ideal diamagnets and therefore pure nuclear magnetism can be studied. The type I AFM on the fcc lattice is well known in electronic magnetism, e.g., Shull did his famous very first magnetic structure determination on MnO, which belongs to this class of magnets. On the fcc lattice, next nearest-neighbor AFM interaction does not allow simple up-down alignments. Such a system is fully frustrated and cannot order. This frustration needs to be broken by a lattice distortion, competition between the interactions in the Hamiltonian, etc.

7.3.2.1 Copper

Lindgård et al.'s FP calculation showed that in Cu the ratio of the strengths of RK and DD interaction correspond to $R = -0.42$. This value of R shows that ferro- and antiferromagnetic interactions are competing strongly. The RK interaction is AFM, while the DD interaction is ferromagnetic. The structures found should have domains according to the fcc lattice. This was demonstrated by the search for Bragg reflections of the type (1,0,0), (0,1,0), etc. All such reflections available for the instrument were found. This is an indication that the system is still fcc. The importance of fluctuations induced by the strong competition and by frustration is seen also in the strongly reduced ordering temperature of 60 nK (from NMR measurements) with respect to the value derived within the mean field approximation of $T_{MF} = 3T_N$. Unfortunately, the value of T_N could not be obtained in the neutron experiments, because the time available for experiments did not allow to do so. It was found, however, that the Bragg peak increases in intensity and decreases in width in first few minutes after demagnetization. This shows that the size of the ordered regions grows for at least 2 min until the maximum intensity is reached. The corresponding correlation length grows from 500 Å at zero time to more than 1000 Å after 2 min, when the resolution limit of the experiment is reached. Despite of this the underlying physics of nuclear magnetic ordering in Cu in zero field seems to be nicely explained by available experimental data (susceptibility and neutron scattering) and the theoretical predictions. The influence of the structural frustration is overcome by the competition of short-range RK interaction and long-range DD interaction. The strong fluctuations found in the theory further stabilizes long-range order in the fcc structure of Cu.

The strong hysteresis and coexistence of the two magnetic phases found in Cu can only be stated as an experimental observation. Clarifying experimental test could not be done because of the inherent t-dependence in the system, the short time for experiments in the ordered face, and the lack of measured nuclear thermodynamic, T or S, parameters in Cu.

7.3.2.2 Silver

The situation in Ag is different: the AFM RK interaction dominates. With $R \sim -3$ the DD interaction is only 1/3 of the RK interaction. With this value of R the theory still predicts the type I AFM structure for Ag in zero field. The value for the Neel temperature T_N was determined by susceptibility measurements to be 600 pK. The occurrence of the type I AFM structure on the perfect fcc lattice of Ag is more surprising than in Cu, because the short-range RK interaction dominates strongly (RK = 3DD). The importance of fluctuations for existence of a stable ordering in these frustrated systems was emphasized by Lindgård. The significantly lower-ordering temperature measured in Ag than the mean field value points to strong, probably quantum, fluctuations in Ag. The experimentally determined $P_{af} = 0.48$ at the "lowest" temperatures reached was argued in Ref. [41] to be a consequence of such fluctuations.

It was observed that in Ag the orientation of the demagnetization field with respect to the high-symmetry directions of the cubic crystals [001] and [010] that for the [001] direction only one type (0,0,1) reflection occurred. This means that no symmetry equivalent reflections could be observed, which means that only one of the equivalent domains of the type I AFM fcc exists in zero field. This means that the magnetic domain structure seems to violate cubic symmetry and the type I fcc AFM structure as well.

7.3.2.3 Copper versus Silver

Is this different behavior of Cu and Ag in their domain structure perhaps an indication for a lattice distortion around T_N? As in electronic magnets such small lattice distortions are often observed at T_N in frustrated systems, we can use the concepts from there to estimate the size of such distortions. This requires to compare the energy increase due to a distortion with the energy gain by magnetic ordering. It becomes immediately clear that very small distortions will break frustration: in electronic magnets with an ordering energy of typically 10 K, one finds that the size of a lattice distortion of typically 10^{-3}–10^{-4} Å is enough to break frustration. Scaling this down with the ordering energies of the nuclear system we find that lattice distortions of 10^{-9} Å would be enough to break the cubic symmetry [49]. The detection of such a minute lattice distortion is completely out of reach in neutron experiments of the type considered here. Even in electronic magnets it is sometimes difficult to detect the lattice distortion. In such studies of electronic magnets it was found, however, that very small lattice distortions of 10^{-6} Å could be sufficient to produce nuclear quadrupolar anisotropy. This effect, however, is absent in a perfect cubic lattice.

Can we find an at least qualitative picture of the concepts underlying the zero field structures in Cu and Ag? Considering the theoretical predictions for the type of structures, one can conclude that there is good agreement with experiments in both cases. But there is no concept available to explain the clear difference in the symmetry of the domain distributions between Cu and Ag: in Cu there is no conflict between the experimental findings and the cubic symmetry, although it was impossible to perform very careful and systematic studies of this issue, by carefully measuring correctly all the intensities of all symmetry equivalent reflections due to the short time available for measurements after cooldown. The situation is different for Ag where one domain was found only despite of quite an effort to clarify the situation experimentally within the given experimental boundary conditions.

Several effects could be considered. The (1,0,0) Bragg peak observed is not due to the simple type I fcc AFM but belongs to a more complex domain structure or a structure with coupled order parameters. A test of this would require a detailed measurement of many magnetic reflections throughout the reciprocal lattice and a careful comparison of the intensities; both requirements cannot be met within the limits given by the experimental restrictions; another reason could be found in the dynamics of the growth of a certain domain structure or the change of an already existing domain structure: Ag has a much longer time constant for the nuclear spin–spin relaxation than Cu (^{109}Ag: 9 ms and ^{65}Cu: 0.15 ms) and the 30 min experimental time might not be long enough to obtain an equilibrium domain structure. The system would then stay in the domain, which is build up at the phase transition at B critical; a last difference between Cu and Ag is their spin value 3/2 and 1/2 for Cu and Ag, respectively. Cu is considered a Heisenberg system (close to classical), while Ag should behave like a quantum system. No obvious arguments can be found on the basis of the above concepts to explain the behavior observed in Ag.

There is another observation, which is of interest here in Cu it was observed, that both the intensity and the width of the Bragg peak vary for a couple of minutes after the end of the demagnetization; the intensity raises by 20%, and the peak width is reduced corresponding to strong increase of the correlation length. If we take this as a typical scenario, we must ask why we do not see this in Ag? We do know as said above that the nuclear–nuclear spin relaxation time in Ag is larger than for Cu by at least a factor of 10. Such a slow relaxation could explain the lack of other domains.

7.3.3 The Phase Diagrams in a Magnetic Field

In the studies of magnetic systems the B, T phase diagrams are of fundamental importance because the further information can be obtained about the energy terms in the Hamiltonian; information about anisotropy terms can be identified and studied, because they compete with the Zeeman energy. The competition of the Zeeman term with the other terms may lead to changes in the magnetic structure or reorientations of the spins or both. It is clear that the study of field-induced effects like the (1,1/3,1/3) structure may lead to a deeper understanding of magnetic systems. Therefore, the field-dependent behavior was studied for Cu and Ag as far as possible.

7.3.3.1 Copper (I = 3/2)

As shown in the presentation of the experimental results (see Figures 19 and 21), the application of a magnetic field B produces a broad spectrum of results. The most obvious and important result is the discovery of the theoretically predicted new commensurate reflection (1,1/3,1/3). The tripling of the magnetic unit cell along the b- and c-axis (see Figure 16) fixes the ratio R of exchange and dipole interaction to $R = -0.34$. This important result proves the soundness and correctness of the theoretical approach by Lindgård et al.! The magnetic structures of Cu are fully understood on the basis of this first-principles calculation. When turning to the actual "phase diagrams," we must be careful in the interpretations of experimental results because we do not have the pair of thermodynamic variables B, T or B, S at hand in Cu one normally is dealing with. Therefore, we deal with B, t phase diagrams instead. The B, t phase diagram for B along [011] (Figure 19) clearly shows the stability regions of the two observed structures characterized by the (1,0,0) and the (1,1/3,1/3). It also shows large overlap of existence of the two structures in particular in the low field range. The complexity of the coexistence of the two structures becomes more clear by looking at B scan from $B = 0$ to $B = 25$ mT ($>B$ critical) for the two characteristic reflections (Figure 18); at low field the occurrence of (1,1/3,1/3) quickly suppressing (1,0,0) strongly, yet not totally. Whether the shoulder at the low field side at the B-dependence of the (1,1/3,1/3) is an indication of yet another phase could not be clarified. No clear answer could be found to the question whether the overlap is evidence for coexisting two independent phases or whether they become one phase of a more complex structure with coupled order parameters. The limited q-space accessible did not allow the necessary experimental tests. Another point of interest at phase boundaries is the character of the phase transition. Is it of continuous or discontinuous character (see Figure 10)? This is a quite complex problem here, since the search for hysteresis, which is one possible indicator for the discontinuous character of a phase transition, is practically impossible in a system, which is inherently time dependent due to the given experimental situation. The study, whether the application of a magnetic field along the different high-symmetry directions yields qualitative results again, had an unexpected result (Figure 19): while [011] and [100] are qualitatively identical the behavior along [111] did not show the (1,0,0) structure at all, while susceptibility measurements seem to indicate a B critical along [111] as well!

7.3.3.2 Silver (I = 1/2)

Apparently, the phase diagram of Ag looks much simpler than the one for Cu and it is a phase diagram given in the thermodynamic quantities B and S (see Figure 21). No indication for any other magnetic structure at finite field was found in the given restricted q-space. Lindgård et al. [43] indicated several regions in q-space where structures might be stable, yet regions of stability could not be given. The experimental search was unsuccessful for other theoretical predictions [39,40]. The only hint for some eventual magnetic field effect was a small shift of the phase boundary upon tilting the field toward the [111] direction: the phase boundary is shifted over the whole phase diagram, see Figure 21 and within a cone of about 50° around [100], no (1,0,0) peak could be found. The second result is in good agreement with the behavior of Cu. The close agreement between the phase boundaries obtained by neutrons and NMR, on a polycrystalline sample of natural Ag is very satisfying. The shape of the phase boundary looks very much like the upper part of a spin-flop scenario well known in weak easy axis electronic antiferromagnets. In such systems the B-T phase diagram has two critical lines if B is along the easy axis. With increasing field the first critical field occurs when the Zeeman energy compensates the easy axis energy, and the spins rotate from being parallel to the field by 90° to being perpendicular to the field. This transition is called spin-flop transition and it is of one order! The upper part of the phase diagram including the spin-flop line has a characteristic shape. Comparing this with the phase diagram observed in Ag shows that the shape of the Ag phase diagram looks like this upper part of the spin-flop diagram, it even shows the particular curvature at large S and very low field. It is tempting to argue that in Ag the phase diagram is exhausted by the spin-flop phase, which means the spin system rotates at B critical from being fully polarized parallel to B to the spin-flop state with spins perpendicular to B and with an AFM order parameter of $P_{af} = 48\%$ at the lowest $S = 0.36 R \ln 2$.

7.4 NEUTRON DIFFRACTION INVESTIGATIONS ON SOLID ^{3}HE

The unique crystalline as well as magnetic properties of ^{3}He can mainly be attributed to the low mass of the ^{3}He particles, together with the fact that the ^{3}He nuclei are fermions. Quantum effects lead to large zero point fluctuation. They manifest themselves, e.g., in the simple fact that the liquid even at $T = 0$ solidifies only under pressure of ~ 34 bar. Another quantum effect is the special behavior of the T-P melting curve: it shows a minimum in pressure at 300 mK, this unique feature has been observed (see, e.g., [50]) after the prediction of Pomeranchuk [51]. The low-pressure modification of solid ^{3}He crystallizes in the body-centered cubic (bcc) structure. Further increase of pressure then leads into the hexagonal close-packed phase, even higher pressure to an fcc phase.

The nucleus of the ^{3}He isotope is a fermion with spin $I = 1/2$. Dependent on pressure and temperature a wealth of physical phenomena is observed in condensed ^{3}He, and a very complex phase diagram of liquid and solid results from cooperative phenomena. They comprise two fundamentally different superfluid states (they are usually referred as "A" and "B" phase) [52] in the liquid phases. There, complex vortex states have been observed by NMR techniques [53] and they have been explained in a framework analogue to high-energy physics.

In solid ^{3}He, nuclear magnetic order has been observed at mK temperature both in the bulk solid bcc phase [54,55] as well as in 2D phases formed as thin layers on graphite foils [56,57]. Such ordering temperatures appear very high compared to nano- and picokelvin ordering in the noble metals Cu and Ag. The reason is a unique exchange interaction mechanism: owing to quantum effects connected with the low mass the ^{3}He nuclei are highly delocalized, at $T = 0$ the width of the mean position in the crystal (equivalent to the Debye–Waller factor) amounts to 30% of the lattice constant. The high delocalization makes the exchange of particles on neighboring sites very likely and this gives rise to the magnetic exchange interaction [58] that is observed in solid ^{3}He. This mechanism is unique also because it is the only case known where the direct exchange of particles on adjacent sites leads to magnetic ordering.

These processes are the dominant magnetic interactions and they involve the exchange of two, three, and more ^{3}He atoms. There, exchange of an even number of atoms is ferromagnetic, odd-numbered exchange is AFM [59]. NMR data and theory suggest that the three and four particle exchange processes are dominant, however, higher-order ring exchange with five and six members could not be excluded and may have significant impact on the ordering as well as on the magnetic phase diagram [60]. The topology of the exchange also plays a role: one can imagine fourfold and higher processes to be planar, but they might as well happen in a folded geometry.

The physics of multiple particle exchange was first realized for solid ^{3}He, however, it has been realized now that this could be more common than one would have expected after the overwhelming success of the two-particle Heisenberg exchange interaction model. In particular for cuprates, it has been shown [61] that multiple particle exchange processes are relevant in La_2SrCuO_4 and likely in similar other materials. The question of possible connections of such interaction mechanisms to high T_c superconductivity has been raised, but to date an answer remains open.

This short overview shows that the nuclear magnetic ordering of ^{3}He is a magnetic model system of central importance. In fact, there exists a vast literature where the various properties are investigated by NMR, specific heat, and many other experimental and theoretical methods (for a review see, e.g., Ref. [13]). In contrast, there have been only very few attempts to study ^{3}He nuclear magnetism by neutron diffraction, although there is sufficient justification for such experiments because fundamental questions still remain open.

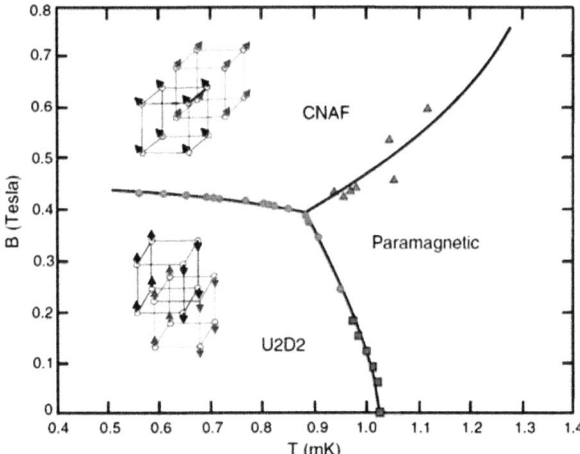

FIGURE 22 The phase diagram for nuclear ordering of solid ^{3}He at a density 24 ccm mol^{-1} close to the melting curve. Note the cell doubling for the U2D2 phase in contrast to the simple canted nuclear antiferromagnet (CNAF) structure in a magnetic field. *Reprinted from Ref. [65], http://dx.doi.org/10.1088/0953-8984/20/10/104246, ©IOP Publishing 2008, reproduced with permission. All rights reserved.*

As far as the nuclear spin ordering in the bcc solid is concerned, the central question is the relevance of the various n-particle ring exchange mechanisms. NMR, magnetization, or specific heat measurements do not provide sufficient information to fit a reliable model up to higher orders of n. Moreover, the U2D2 structure of the zero field phase (cf. Figure 22) inferred from NMR does not exclude a number of other possibilities (e.g., U3D3, etc.). The canted nuclear antiferromagnet (CNAF) structure suggested in high field is of course plausible, but there is lack of experimental evidence for the AF propagation vector. This structure where magnetic field and FM as well as AF interactions compete would be a crucial test for the ratio of odd and even ring exchange (Figure 22).

Neutron diffraction is the only method to solve such structures, however, the experimental barriers are severe. The experimental challenges can be summarized as follows:

1. The ^{3}He absorption is huge which implies samples that must be thin. Connected with absorption is the kinetic decay energy of the capture products, a proton and a triton, that remains in the sample. It typically amounts to several nanowatts, enough to heat up a ^{3}He sample within times too short to see the ordering by neutrons. However, the energy is deposited through successive collisions in ^{3}He over a path length of more than 40 µm and this may help in confined geometries as will be seen below.

2. The sample must be cooled well below 1 mK, for metals not too difficult, but for an insulator the Kapitza boundary resistance is a severe limit for

heat transfer. The ^{3}He solid therefore can only be cooled in a micro- or nano-grained metal sinter with a large surface. The sinter is also an efficient tool to carry away the beam heating: the Bethe formula tells that owing to the much higher electron density in the sinter, the kinetic energy of the charged capture products from absorption mainly goes to the sinter, not to the ^{3}He with a low electron density ratio $\sim 1/30$ to $1/40$. In other words, without a sinter material neutron scattering from ^{3}He can never be observed, the free ^{3}He solid would warm up immediately.

3. A single crystal must be grown and oriented in situ at pressures around 30—40 bar at temperatures around 1 K inside the sinter material that limits the coherence of the growth process. One could also ask: Is the ^{3}He solid inside a sinter equivalent to bulk ^{3}He? This question, fortunately, has been solved positive [62].

4. The high pressure poses strict requirements on the mechanical strength of the sample cell. In particular, under varying pressure any mechanical deformation must be so small that no bulk ^{3}He can build up. Free ^{3}He, not confined in a sinter as mentioned above, would not cool and if, it would heat up immediately in the neutron beam. In addition, it adds up to the unwanted absorption. Such requirements together with limitations in the material choice dictated by neutron transmission and background reasons limit the possible size of ^{3}He neutron pressure cells and therefore the signal.

5. Altogether, the diffraction signal in general will be weak because of absorption, geometry restrictions, and due to the weak localization that results in an unfavorable Debye—Waller factor. In addition, the measurement time below T_N is expected to be short, not more than a few minutes after precooling and exposing the sample to the neutron beam.

Here the to date solutions achieved on these issues will be discussed: so far, there have been three serious attempts to investigate ^{3}He nuclear ordering with neutron diffraction. In the 1980s, two groups set up experiments, one at the spallation source in Argonne and another at the CENG in Grenoble. Later, another setup has been constructed at the former HMI in Berlin.

To deal with the absorption, all setups share fairly thin sample cells, with sample thickness between 0.1 mm (CENG) and 60 μm (HMI). These numbers correspond to a transmission of approximately 30% for the respective cells and wavelength used. Values of the Argonne cell are not known to us. Figure 23 shows the scheme of the HMI cell.

For thermal transport reasons, the cell body must be made from high conductivity Cu that contains the volume for the sample and the sinter material to cool the sample. The very soft Cu needs stabilizing bars to support it under the high pressure needed for solidification of the ^{3}He. Of course, also the support structure is subject of mechanical deformation but this can well be calculated from simple mechanics. It is also possible to determine cell deformations

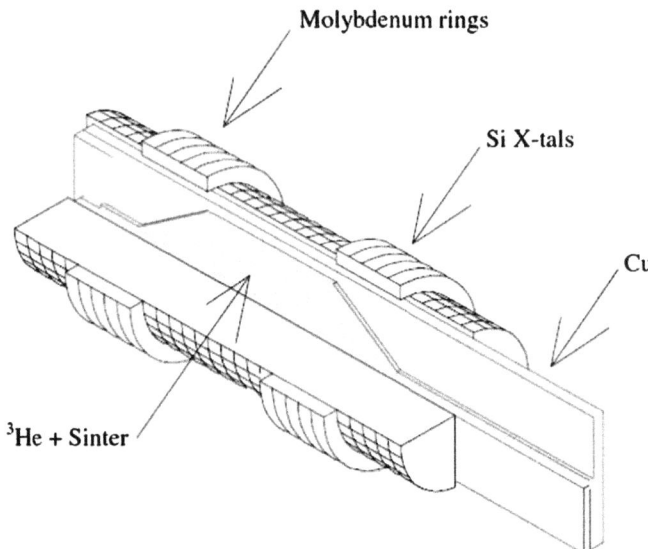

FIGURE 23 The HMI ^{3}He sample pressure cell assembly consists of a Cu boat that is supported by single-crystalline Si rods. The assembly is kept together by molybdenum rings with a very tight fit. They are mounted hot to achieve a prestrain on the assembly. The Cu parts of the cell are soft soldered along the edges. They consist of an inner plate with the He cell contours, in the beam area they are filled with Ag sinter. Below and above the sinter there are unrestricted volumes with seed areas for crystal growth. The effective ^{3}He thickness in the beam is measured by neutron absorption to be ~ 60 μm, the sample surface here is 20×13 mm^2. The inner plate is covered with high-conductivity Cu foils for good thermal contact. The Ag sinter has been bonded to one side of the cell. The free Cu on top of the cell is used for the thermal contact to the cooling stage. Reprinted from Ref. [65], http://dx.doi.org/10.1088/0953-8984/20/10/104246, ©IOP Publishing 2008, reproduced with permission. All rights reserved.

experimentally while loading the cell at known temperature. The absorption as a function of applied pressure should go with the ^{3}He density which gives a sensitive test for cell deformations. At HMI, such tests were done at $T \sim 1$ K with less than 10% change of the effective ^{3}He cell thickness up to 60 bar pressure.

Such values of free volume appear acceptable and may even help in forming single crystals in a sinter material with micrometer porosity, because a small free volume is available over the entire length of the sinter inside the cell. The growth of single crystals indeed has not been a too big problem at CENG and HMI. Both successful cells at CENG and HMI used submicron size Ag powder to form a sinter that typically reaches a surface of 20–50 m^2 cm^{-3}. At Argonne, "black Pt nanopowder" has been pioneered which potentially gives much larger surface [63]. This type of sinter material gives very good thermal contact, however, attempts at HMI to grow crystals in this type of sinter have not been successful (unpublished) which could be attributed to the

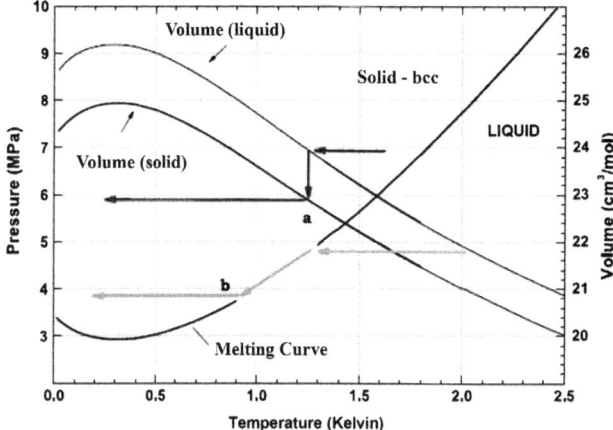

FIGURE 24 Solidification of ^{3}He; the curves show the melting curve and the molar volume on the melting curve for the liquid and solid phase. Above the melting curve minimum, there are two principal methods as indicated in this figure. The constant volume methods would need special heating systems along the filling capillary which otherwise gets blocked and does not allow to replace the volume change. Usually crystals are made with a "blocked" capillary which leads to a reduction in pressure indicated here by the points a and b. bcc, body-centered cubic. *Reprinted from Ref. [65], http://dx.doi.org/10.1088/0953-8984/20/10/104246, ©IOP Publishing 2008, reproduced with permission. All rights reserved.*

smaller pore size. It is known that the ^{3}He melting curve depends on geometry restrictions [64], therefore, one could imagine that solid formed in larger capillaries of the sinter traps liquid ^{3}He in nanopores which solidifies under very different conditions. This includes the possibility that it never solidifies there. The principal scheme of solidification at constant volume is shown in Figure 24. In reality, due to temperature gradients, the solidification is accompanied with density changes that can well be monitored with neutron transmission. Figure 25 shows a typical result.

For orientation of ^{3}He crystals and the observation of the U2D2 structure one looks after the (110) and (1/2 00) reflections. A measurement of these reflections needs different measurement strategies. For orientation, at least two different (110) type reflections are needed which requires coverage in q-space. The time available is not limited because orientation happens at relatively high temperatures around 1 K. Therefore, intensity losses due to short wavelength are acceptable to achieve larger coverage in q-space (cf. Eqn (5)). In contrast, for the weak nuclear magnetic reflection (1/2 00) with very limited time of observation it is an advantage to change to larger wavelength. This strategy has been used at the CENG using a main wavelength of 1.7 Å and the corresponding $\lambda/2$ wavelength. They were separated by filters. At HMI it was possible to switch between two monochromators, a Ge monochromator set to 1.4 Å and a pyrolytic graphite (PG) monochromator at 2.4 Å wavelength. At a

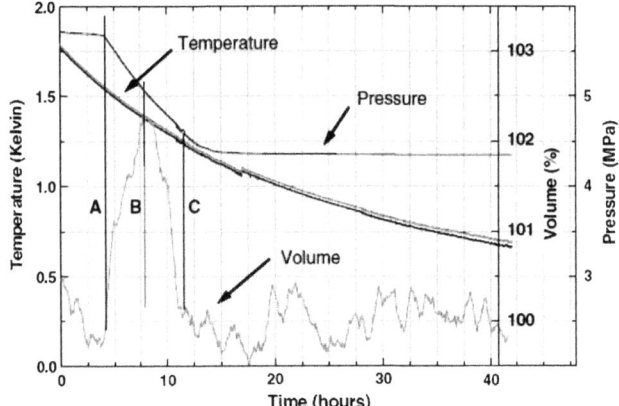

FIGURE 25 Pressure and specific volume during crystal growth as a function of temperature; At time "A" crystal growth sets in, usually somewhere in the filling capillary of the pressure cell. The ^{3}He molar volume is measured in the center of the sinter using neutron transmission and shows an increase of volume due to the solidification outside the cell. At time "B" crystal growth has reached the neutron beam area and the ^{3}He volume change reverses direction. The initial and final ^{3}He molar volumes are similar which points to an annealing of the crystal that equalizes all pressure differences with time. *Reprinted from Ref. [65], http://dx.doi.org/10.1088/0953-8984/20/10/104246, ©IOP Publishing 2008, reproduced with permission. All rights reserved.*

pulsed source the situation is different because in principle, without filtering, all wavelengths are present in a single pulse. This could be an advantage for orientation purposes, however, we are not aware of published diffraction results from the Argonne experiment.

A typical search result from the HMI experiment [65] is shown in Figure 26. For a cell position approximately perpendicular to the incoming beam there is little absorption and it is not too difficult to find the first (110) reflection because it has low absorption. With this reflection given, an axis is defined for further search because other (110) reflections must lie on cones around that axis. The cell position for their observation is of course not favorable because the absorption will be much larger. However, it turned out that crystals can be systematically oriented with such methods both at CENG and HMI.

These efforts of course all aimed for the observation of nuclear ordering, first of all in the U2D2 phase. In fact, at the CENG a very weak signal has been observed [66] for a few minutes after cooldown and attributed to the (1/2 00) reflection that would be consistent with the expected U2D2 structure. It has been correlated with the transmission signal to deduce the nuclear temperature and the transition temperature. However, this experiment has not been reproduced.

One must say that fundamental questions on this unique model system remain open until today. In particular, the CNAF structure which is a prediction

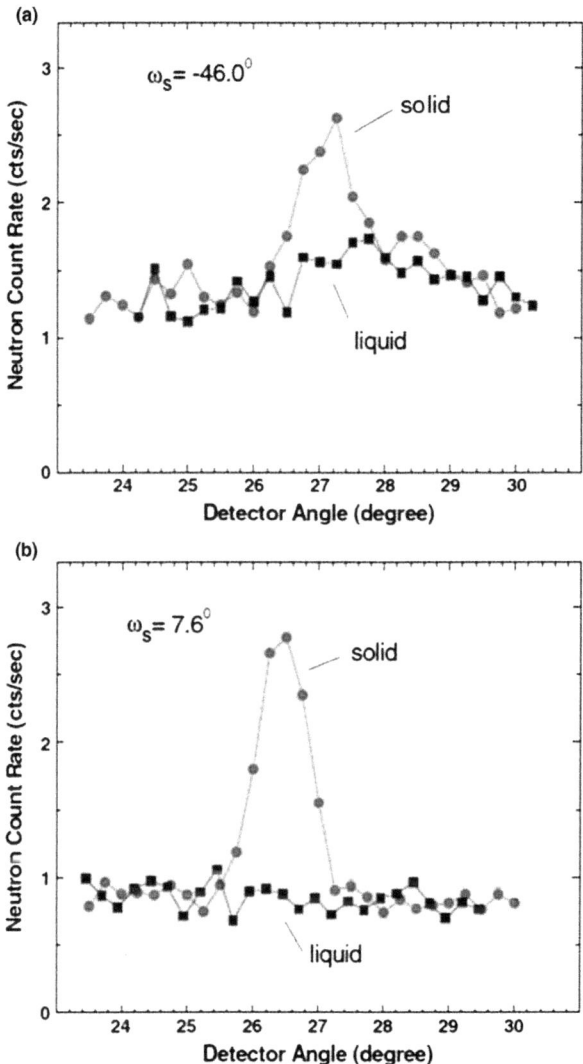

FIGURE 26 Two independent reflections of the (110) type from a solid ^{3}He crystal. The scans are shown as a function of detector angle, taken at different cryostat rotation angle ω that have been located in search scans. The wavelength here was 1.4 Å, with this orientation the wavelength was changed to 2.4 Å. This yields a larger signal and less background, the signal/background improvement is $\sim a$ factor 3. *Reprinted from Ref. [65], http://dx.doi.org/10.1088/0953-8984/20/10/104246, ©IOP Publishing 2008, reproduced with permission. All rights reserved.*

of the multiple particle exchange model has not been determined. Also, given only one observation of the U2D2 phase with very low intensity from a low flux instrument (CENG workers report an incoming flux of 10^4 n s^{-1} cm^{-2}), it would be highly desirable to confirm this result. The work at HMI, in particular for the cell design and growth geometry, could be a solid basis for further investigations on which other investigators can build to save valuable development time in such experiments.

7.5 APPLICATIONS OF POLARIZED NUCLEI IN NEUTRON DIFFRACTION

7.5.1 Neutron Polarization from Polarized ^{3}He Gas Targets

The most important application from polarized nuclear spin to date is the use of ^{3}He spin polarized gas as neutron polarizer and analyzer. It is pumped by optical methods. Initially, these techniques have been developed for use in nuclear physics experiments, however, it was soon realized that polarized neutron techniques can benefit compared to conventional single-crystal polarizers or supermirror benders. The main advantage of ^{3}He polarizers is the fact that the gas cells can be tailored to the experimental needs (e.g., wide-angle coverage) and act as insertion devices that have no impact on the instrumental resolution. Second, they are not limited to one single wavelength which opens a path for neutron polarization experiments using white beams in time-of-flight experiments.

The absorption cross sections usually are written for the neutron spin directions "up" and "down" as $\sigma^+_{^3He} \approx 5$ barn and $\sigma^-_{^3He} \approx 5300$ barn, the small Coulomb absorption term "up" is usually neglected. The numbers here are for neutrons of 1800 m s^{-1}, and in this context one must mention that the cross section is wavelength dependent with $1/\lambda$. This dependence is relevant for white beam applications but the correction does not cause problems. Then, in total one obtains the following expressions for transmission T of an unpolarized beam and its polarization P_N:

$$T = \cosh(N\sigma_{\uparrow+}lP_{^3He})$$
$$P_N = \tanh(N\sigma_{\uparrow+}lP_{^3He}) \quad (12)$$

When the technique started in the end of the 1990s [67], the possible ^{3}He polarization values were $\sim 40\%$ calling for optimization of transmission versus polarization. A "quality" factor $Q = P_N^2 T_N$ has been defined and is used to optimize the cell thickness and ^{3}He pressure for a given wavelength. Typical numbers are 3 bar pressure and cell thickness of ~ 5 cm. This optimization was dictated also by the fact that the ^{3}He polarization is not constant over time. Collisions of ^{3}He atoms with the wall are one reason for the loss of polarization, leading to an intensive search for absolutely nonmagnetic cell materials and to the use of coatings, e.g., by cesium. The gas atoms also interact with each other which include spin exchange processes. This leads to ^{3}He polarization loss, also an external field homogeneity below 10^{-4} is required. To date these technical problems are solved, the relaxation times of typical cells reach 200 h and more, long enough for most neutron experiments (Figure 27).

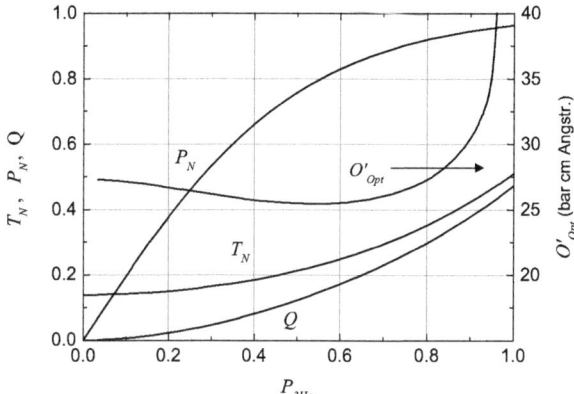

FIGURE 27 Characteristics of a ^{3}He neutron spin filter as a function of ^{3}He polarization $P_{^3He}$. The left ordinate applies to the polarization (P_N) and transmission (T_N) of the neutron beam as well as to the quality factor $Q = P_N^2 T_N$, all calculated for constant opacity $O' = 27$ barÅcm, The right ordinate applies to the optimized opacity O'_m. *Reprinted from Ref. [67], http://dx.doi.org/10.1016/S0168-9002(96)00923-0 with permission of Elsevier, ©Elsevier BV 1997.*

To end this short discussion of spin-dependent absorption in ^{3}He, the methods for polarizing the gas must be mentioned. In contrast to other nuclear polarization experiments that require very low temperature, the ^{3}He nuclei are polarized at room temperature or above by optical pumping; two methods are used, the so-called metastable exchange optical pumping (MEOP) [68] and the spin exchange optical pumping (SEOP) [69] processes. For the user, the main difference is that the MEOP spin pumping process needs a gas discharge at ~ 1 mB pressure which is too low for direct applications. Thus, this process needs a compressor stage and usually a central laser pumping station with a compressor that can serve several neutron experiments with polarized ^{3}He, but the cells need to be transported between pumping station experiment. In the SEOP process the ^{3}He gas contains a small amount of Rb atoms which are used in the pumping process. The spin-polarized Rb transfers the polarization to the ^{3}He. This method works at high pressure and is usually installed at the experimental sites, often with cells that are especially adapted for wide-angle spin analysis. The total performance of both methods to date is very similar, in effect ^{3}He polarizations up to $\sim 80\%$ are obtained [70], and the emerging neutron polarization is a matter of cell pressure and geometry, it can reach 95%.

Last, one may say that the success of ^{3}He polarizers as wide-angle insertion devices has superseded ideas to use polarized nuclear spin targets as neutrons polarizer. One idea was to use the large incoherent cross section of $^1_1$H as a spin filter. The protons should be polarized by microwave techniques at low temperature (~ 0.3 K) and in a high magnetic field (~ 3 T). This leads to a spin filter that scatters away the unwanted spin polarization direction. Another idea was the use of spin-polarized Sm nuclei as a spin filter, in this case the spin-dependent transmission would have been used. In ferromagnetic $SmCo_5$ no large external field is needed because the Sm nuclei get polarized by the strong hyperfine field. However, the temperatures still must be in the mK range. For wide-angle neutron polarization, these methods suffer from the enormous experimental effort connected with the need for millikelvin temperatures and, especially for the case of $SmCo_5$, from unwanted side effects like the Co absorption. Microwave pumping techniques on protons, however, have been successfully used for contrast variation in small-angle scattering experiments which we discuss next.

7.5.2 Contrast Variation by Polarized Nuclei in Neutron Scattering

SANS is one of the most requested methods for studies of soft matter like polymers and biological samples. These materials are characterized by structures, which consist of large substructures in an even larger global structure. Contrast variation has been of great importance for the improvement of the sensitivity in SANS. As H is the dominant species in such materials and as the two isotopes 1H and 2H (D) have very different scattering lengths H-exchange became the standard procedure for contrast variation in SANS. They have the following scattering parameters:

for ^{1}H: $b_c = -3.47\ b$ and for ^{2}D: $b_c = 6.67\ b$.

Two properties are of importance: the ^{1}H has a negative scattering length and a very large incoherent cross section due to the large spin-dependent part of b and ^{2}D has a positive scattering length. The incoherent cross section is a nuisance, because it corresponds to a large background produced by any sample containing ^{1}H. By mixing the two isotopes of hydrogen in the structure and using appropriate H_2O-D_2O mixtures, it is possible to vary the contrast systematically. Naturally, this procedure is standard in the studies of soft matter, polymers, and biological samples, where the structures of interest are very complex and have huge unit cells. While the preparation of H_2O/D_2O mixtures is easy, specific deuteration of building blocks of the global structure is often nontrivial and may require special chemistry to achieve a proper deuteration of the particular building block. Therefore for each step of a series of contrasts, a new sample has to be made, which cost time, rescaling of the different results in a series may introduce systematic errors, and sometimes the making of deuterated samples is a research project by itself. On the other hand, performing the neutron experiments follows standard procedures!

Very early, in 1974, Hayter et al. [71] suggested to use the nuclear polarization of the proton to vary the contrast in proton-containing samples. These first attempts were not satisfying, because the polarization induced in their sample by DNP was not homogenous. The concept was followed and improved by collaborations between ILL in Grenoble, Orphee in Saclay, and PSI in CH and GKSS in Geesthacht. One of the key goals was the design and building of a new sample and to optimize the cryostat. These requirements are a major hurdle for a widespread use of this technique: it needs significant and expensive equipment for the experiments, low temperature, and NMR expertise, which is not usually available at SANS instruments. In the early 1990s, first encouraging results were achieved by a group lead by Stuhrmann at GKSS [72]. An important contribution to the success came from polarized target specialists of the high-energy physics labs resulting in a nuclear polarization of $P_n = 95\%$! On the side of the concept itself it is important to note that they now used polarized neutron scattering from a sample which was partially deuterated and contained protons polarized up to 95%. The above paper also describes the steps in the development in the 5 years before the publication. The results show proof that the technique used can produce high-quality data which can reliably analyzed leading to state-of-the-art results.

In the last 20 years the field has seen strong progress in the understanding of the physics of the processes relevant for experiments, like relaxation processes and in the mathematical description of the scattering functions to deduce reliable results. It has been observed that polarization is not homogeneous around the magnetic impurities, see, e.g., Ref. [35] and ref. therein: they studied the time evolution of the buildup and the relaxation of nuclear polarization in DNP by time-dependent SANS neutron scattering of a frozen solution of EHBA complexes, a standard magnetic ion for use in DNP, at low concentration in a highly deuterated matrix. SANS and NMR were measured simultaneously to record the behavior of the protons around the magnetic impurity and the protons far away in the bulk, respectively. As the relevant time scales are of the order of seconds, one has to use stroboscopic techniques to record the desired time dependencies. They found that a cloud of 20 protons around an impurity will be highly polarized within seconds. This polarization gradient between the "close" and the "far" protons in the bulk remains for a short time before spin diffusion brings the polarization into equilibrium. This opens up the possibility to a produce a very strong contrast enhancement locally if a paramagnetic impurity can be placed at important positions in the structure of interest.

This enhanced polarization cloud around the paramagnetic impurities has become the focus of work performed using contrast variation by proton polarization, by Stuhrmann [73].

All this can be summed up by saying: while the H-D exchange remains the standard technique for contrast variation in soft matter, polymers, and biological samples, it has become clear that contrast variation by nuclear polarization is an excellent amplifier of an existing contrast. And for such very focused experiments, a designated instrument might be a very useful extension of the present instrument suite in the future.

7.6 SUMMARY

A general reminder should be made here about the important restrictions, which will limit the interpretation of the experimental data in comparison to experiments on electronic magnets:

- In the neutron scattering experiments described here, the signal is due to nuclear–nuclear cross section and not to nuclear magnetic–electronic magnetic moment cross section. This prohibits the determination of the spin direction with respect to the crystal lattice.
- To reach nuclear ordering temperature, one needs very specific and complex cooling techniques, which may influence the obtained results significantly; in Cu and Ag, one uses single-shot cooling, which means that the system will warm up continuously after demagnetization. This also means that thermodynamic parameters like T or S are difficult to obtain or even cannot be determined at all. For LiH the cooling principle with rotating field imposes that the full Hamiltonian is not operating but the truncated dipolar part only. On the other hand, experiments on insulator materials are possible and the requirements for the cooling system are not as dramatic as they are for Cu and Ag or other metals.

The main conclusions from the experiments are:

- The nuclear ordering has been determined for LiH using cooling in a rotating field and has been found to be as predicted. At negative temperatures AFM order turns into fm order.
- For both Cu and Ag, neutron scattering has proven nuclear magnetic long-range ordering at zero field. The magnetic structure is of the type I AFM on the fcc lattice for both. For Cu in finite field there exists a new, commensurate AFM structure in addition.
- The magnetic phase diagrams have been determined for both systems. In Cu the phase diagram is complex and shows hysteresis and kinetics of the ordering.
- The efforts to observe nuclear ordering in ^{3}He have failed so far.
- Finally, for new topics at low temperature coming up to be investigated on new, high-flux neutron sources, a word of care may be allowed that emerges from the experience gained in ULT work with neutrons discussed here; beam heating problems go with flux. The large increase of flux at spallation sources of the Megawatt (MW) class can lead to thermalization problems already at millikelvin temperatures and above, especially with nonmetals that have poor conductivity and are subject to the Kapitza resistance.

REFERENCES

[1] C.G. Shull, E. Wollan, G.A. Morton, W.L. Davidson, Phys. Rev. 73 (1949) 842.
[2] C.S. Wu, E. Ambler, R.W. Hayward, D.D. Hoppes, R.P. Hudson, Phys. Rev. 105 (1957) 1413.

[3] see e.g. B. Brookhouse, Nobel Lecture (1994).
[4] N. Kurti, F.N.H. Robinson, S. Francis, D.A. Spohr, Nature 178 (1956) 450.
[5] A. Abragam, M. Goldmann, Nuclear Magnetism: Order and Disorder, Oxford University Press, 1982.
[6] Y. Roinel, V. Bouffard, G.L. Bachella, M. Pinot, P. Mériel, P. Roubeau, O. Avenel, M. Goldmann, A. Abragam, Phys. Rev. Lett. 41 (1978) 1572.
[7] O.V. Lounasmaa, Experimental Principles and Methods below 1K, Academic Press London, 1974.
[8] K.A. McEwen, W.G. Stirling, C. Vettier, Phys. Rev. Lett. 41 (1978) 343.
[9] M. Steiner, L. Bevaart, Y. Ajiro, A.J. Millhouse, K. Ohlhoff, G. Rahn, H. Dachs, U. Scheer, B. Wanklyn, J. Phys. C Solid State Phys. 14 (1981) L597.
[10] M. Steiner, in: IAEA Proceedings "Neutron Scattering in the Nineties", vol. 185, 1985.
[11] M.T. Huiku, M.T. Loponen, Phys. Rev. Lett. 49 (1982) 1288.
[12] M.T. Huiku, T.A. Jyrkkio, J. Kyynarainen, M.T. Loponen, O.V. Lounasmaa, A.S. Oja, J. Low Temp. Phys. 62 (1986) 433.
[13] E.D. Adams, J. Low Temp. Phys. 135 (2004) 695.
[14] http://www.ncnr.nist.gov/resources/n-lengths/ (last accessed 07.08.15).
[15] L. Koester, H. Rauch, L. Seymann, At. Data Nucl. Data Tables 49 (1991) 65.
[16] M. Blume, Phys. Rev. 130 (1963) 1670.
[17] M. Blume, R.I. Schermer, Phys. Rev. 166 (1968) 554.
[18] R.M. Moon, T. Riste, W.C. Koehler, Phys. Rev. 181 (1969) 920.
[19] S.F. Mughabghab, Atlas of neutron resonances, Elsevier Sci. (April 17, 2006).
[20] https://www-nds.iaea.org/relnsd/NdsEnsdf/neutroncs.html (last accessed 07.08.15).
[21] F. Pobell, Matter and Methods at Low Temperature, Springer, 2007.
[22] M.T. Huiku, M.T. Loponen, T.A. Jyrkkiö, et al., in: U. Eckern, et al. (Eds.), Proc. 17th Int. Conf. Low Temp. Phys, North-Holland, Amsterdam, 1984, p. 133.
[23] M. Goldmann, Spin Temperature and NMR in Solids, Oxford Univ. Press, 1970.
[24] T.A. Jyrkkiö, M.T. Huiku, K.N. Clausen, K. Siemensmeyer, K. Kakurai, M. Steiner, Z. Phys. B 71 (1988) 139.
[25] T.A. Jyrkkiö, M.T. Huiku, K. Siemensmeyer, K.N. Clausen, J. Low Temp. Phys. 74 (1989) 435.
[26] P.J. Becker, P. Coppens, Acta Crystallogr. A 30 (1974) 129.
[27] P.J. Becker, P. Coppens, Acta Crystallogr. A 31 (1975) 417.
[28] K. Lefmann, Nuclear Magnetic Ordering in Silver (Dissertation), Risoe Report 850, 1995.
[29] K. Lefmann, J.T. Tuoriniemi, K.K. Nummila, A. Metz, Z. Phys. B 102 (1997).
[30] P. Jauho, P.V. Pirilä, Phys. Rev. B 1 (1970) 21.
[31] J.T. Tuoriniemi, K.K. Nummila, K. Lefmann, R.T. Vuorinen, A. Metz, Z. Phys. B 102 (1997) 433.
[32] A. Abragam, M. Goldmann, Principles of dynamic nuclear polarization, Rep. Prog. Phys. 41 (1978) 395.
[33] Y. Roinel, G.L. Bachella, O. Avenel, V. Bouffard, M. Pinot, P. Roubeau, P. Mériel, M. Goldmann, J. Phys. Lett. 41 (1980) L123.
[34] Y. Roinel, V. Bouffard, C. Fermon, M. Pinot, F. Vigeron, G. Fournier, M. Goldmann, J. Phys. 48 (1987) 837.
[35] B. van den Brand, H. Glättli, I. Grillo, P. Hautle, H. Jouve, J. Kohlbrecher, J.A. Konter, E. Leymarie, S. Mango, R.P. May, H.B. Stuhrmann, O. Zimmer, Europhys. Lett. 59 (2002) 62.
[36] P.J. Hakonen, K.K. Nummila, R.T. Vuorinen, O.V. Lounasmaa, Phys. Rev. Lett. 68 (1992) 365.
[37] P.J. Hakonen, S. Yin, K.K. Nummila, Europhysics Lett. 15 (1991) 677.

[38] P. Bak, S. Krinsky, D. Mukamel, Phys. Rev. Lett. 36 (1976) 52.
[39] A.S. Oja, O.V. Lounasmaa, Rev. Mod. Phys. 69 (1997) 1.
[40] A.S. Oja, Phys. Scr. 36 (1987) 462.
[41] P.-A. Lindgård, X.-W. Wang, B.N. Harmon, JMMM 54–57 (1986) 1052.
[42] P.-A. Lindgård, PRL 61 (1988) 629.
[43] P.-A. Lindgård, JMMM 90&91 (1990) 138.
[44] S.J. Frisken, D.J. Miller, PRL 61 (1988) 1017.
[45] T.A. Jyrkkiö, M.T. Huiku, O.V. Lounasmaa, K. Siemensmeyer, K. Kakurai, M. Steiner, K. N. Clausen, J.K. Kjems, Phys. Rev. Lett. 60 (1988) 2418.
[46] A.J. Annila, K.N. Clausen, P.A. Lindgård, O.V. Lounasmaa, A.S. Oja, K. Siemensmeyer, M. Steiner, J.T. Tuoriniemi, H. Weinfurter, Phys. Rev. Lett. 64 (1990) 1421.
[47] K. Siemensmeyer, M. Steiner, H. Weinfurther, K.N. Clausen, P.A. Lindgård, A.J. Annila, O. V. Lounasmaa, A.S. Oja, J.T. Tuoriniemi, Phys. B 180 & 181 (1992) 29.
[48] J.T. Tuoriniemi, K.K. Nummila, R.T. Vuorinen, O.V. Lounasmaa, A. Metz, K. Siemensmeyer, M. Steiner, K. Lefmann, K.N. Clausen, F.B. Rasmussen, Phys. Rev. Lett. 75 (1995) 3744.
[49] K. Siemensmeyer, M. Steiner, Z. Phys. B 89 (1992) 305.
[50] D.S. Greywall, P.A. Busch, Phys. Rev. B 36 (1987) 6853.
[51] I. Pomeranchuk, Zh. Eksp. Teor. Fiz 20 (1950) 919.
[52] D.D. Osheroff, W.J. Gully, R.C. Richardson, D.M. Lee, Phys. Rev. Lett. 29 (1972) 920.
[53] M.M. Salomaa, G.E. Volovik, Rev. Mod. Phys. 59 (1987) 533.
[54] D.D. Osheroff, M.C. Cross, D.S. Fisher, Phys. Rev. Lett. 44 (1980) 792.
[55] D.D. Osheroff, J. Low Temp. Phys. 87 (1992) 297.
[56] H. Fukuyama, J. Phys. Soc. Jpn. 77 (2008) 111013.
[57] C. Bäuerle, Y.M. Bunkov, A.S. Chen, S.N. Fisher, H. Godfrin, Low Temp. Phys. 110 (1998) 333.
[58] D.J. Thouless, Proc. Phys. Soc. 86 (1965) 893.
[59] M. Roger, J.H. Hetherington, J.M. Delrieu, Rev. Mod. Phys. 55 (1983) 1.
[60] M. Roger, C. Bäuerle, Y.M. Bunkov, A.S. Chen, H. Godfrin, Phys. Rev. Lett. 80 (1998) 1308.
[61] A.M. Toader, J.P. Goff, M. Roger, N. Shannon, J.R. Stewart, M. Enderle, Phys. Rev. Lett. 94 (2005) 197202.
[62] E.A. Schuberth, M. Kath, L. Tassini, C. Millan-Chacartegui, Eur. Phys. J. B 46 (2005) 349.
[63] P. Roach, Y. Takano, R.O. Hilleke, M.L. Vrtis, D. Jin, Cryogenics 26 (1986) 319.
[64] D.N. Bittner, E.D. Adams, J. Low Temp. Phys. 97 (1994) 519.
[65] V. Boiko, S. Matas, K. Siemensmeyer, J. Phys. Cond. Matter 20 (2008) 104246.
[66] A. Benoit, J. Bossy, J. Flouquet, J. Schweitzer, J. Phys. Lett. 46 (1985) L923.
[67] R. Surkau, J. Becker, M. Ebert, T. Großmann, W. Heil, D. Hofmann, H. Humblot, M. Leduc, E.W. Otten, D. Rohe, K. Siemensmeyer, M. Steiner, F. Tasset, N. Trautmann, Nuc. Instr. Meth. A 384 (1997) 444.
[68] F.D. Colegrove, L.D. Sghearer, G.K. Walters, Phys. Rev. 132 (1963) 2561.
[69] M.A. Bouchiat, T.R. Carver, C.M. Varnum, Phys. Rev. Lett. 5 (1960) 373.
[70] E. Babcock, B. Chann, T.G. Walker, W.C. Chen, T.R. Gentile, Phys. Rev. Lett. 96 (2006) 083003.
[71] J.B. Hayter, G.T. Jenkin, J.W. White, Phys. Rev. Lett. 33 (1974) 696.
[72] W. Knop, M. Hirai, H.-J. Schink, H.B. Stuhrmann, R. Wagner, J. Zhao, O. Schärpf, R.R. Crichton, M. Krumpolc, K.H. Nierhaus, A. Rijllart, T.O. Niinikoski, J. Appl. Cryst 25 (1992) 155.
[73] H.B. Stuhrmann, Journ. Phys. Conf. Ser. 351 (2012) 012002.

Index

Note: Page numbers followed by "f" indicate figures, and "t" indicate tables.

A

$A_2B_2O_7$, 87
$A_2BB'O_6$, 248–249, 248f
A_2BX_4, 251–252
Absorption cross section, 439
ACr_2O_4, 242–243, 245–246
Adiabatic demagnetization in the rotating frame (ADRF), 454
Adiabatic limit, 15
$AFe_3(OH)_6(SO_4)_2$, magnetic order in kagomé lattice of, 263, 263f
AFM-coupled superlattices, 400–401
Ag_2CrO_2, 259–260
Ag_2MnO_2, 259–260
Ag_2MO_2, 259–260
Ag_2NiO_2, 259–260, 259f
$AgCrO_2$, 258
$AgFeO_2$, 258
Alkali metal iron chalcogenide, 146–147, 161–162
Alloys, and diffuse scattering, 275–276, 276f
Alternating magnetic field (AC field), 417–420, 417f
Amperian current loops, 4–5, 5f
Analyzer, neutron, 34–35, 447
Angle dispersive reflectometer, 366, 370
Antiferromagnetic coupling, 397–401, 398f
Antiferromagnetic order, 146, 151–152, 157, 187–189, 449
  metallic parent compounds, 162–168
  and superconductivity, 146–147, 160–162, 168–169, 179–182
Antiferromagnetic structures, 208f, 242–243, 244f, 248–250, 256–257, 268–270, 273–274
Antiferromagnetism (AFM), 208–209, 208f, 455–456
  Type I, 457–458
  $TlCuCl_3$, 80–81
Antisymmetry, 223–224, 237–238
Aurivillius phases, 251–252
Axial vector function, 222–223
Axial vector representation matrix, 228–229

B

Ba_2CoGeO_7, 314t
Ba_2LaRuO_6, 248–249
Ba_2LnSbO_6, 248–249
Ba_2YRuO_6, 248–249
$Ba_3NbFe_3Si_2O_{14}$, 263–264
$Ba_3NiNb_2O_9$, 260
$BaCo_2(AsO_4)_2$, 262, 265f
$BaFe_2As_2$, 253–254, 254f
$BaM_2(XO_4)_2$, 262
$BaNd_2O_4$, 271f, 272
$BaNi_2(AsO_4)_2$, 262, 265f
$BaNi_2(PO_4)_2$, 262, 265f
$BaNi_2(VO_4)_2$, 262, 265f
BasIreps (computer program), 231–233
Basis vectors, 229–230
  multiple magnetic sites, 230
  pyrochlores, 237–240, 239f, 239t, 240f
  spinel compounds, 244, 245t
$BaTiO_3$, 314t
Be filter, 366–368, 370–371, 372f
Beam heating, 444–445
Beam optics, 12–13
Beam trajectories, 10
Belov–Neronova–Smirnova (BNS) notation, 226
$BiFeO_3$, 297–300, 314t
  contour map of neutron diffraction, 319f
  domains, 315
  lattice, 298f
  neutron intensity mapping, in reciprocal space, 299f
  normalized time-of-flight powder diffraction pattern, 298f
  polarized neutron scans, 300f
  spin dynamics, 323
  spin-density wave (SDW), 301f
  thin films and multilayers, 316–319
Bilbao Crystallographic Server, 226, 231–233, 235–237
Biquadratic coupling, 362–363, 398–399
Black-white groups, 224–225
Bloch domain walls, 394–395, 395f

489

490   Index

Blocking temperature, 407–408
Bohr magneton, 204–205
Boltzmann distribution, 52–53
Bootstrap configuration, 36–37
Born approximation (BA), 13, 344, 348–352, 355, 357, 409
 applicability range of, 344
Born scattering amplitude, 349–351
Bose factor, 56
Bose–Einstein condensation (BEC), 74, 76f
 of triplet excitations, 77
Bragg reflection, 362–363, 396–397, 397f, 403–407
 estimation of, 461
Bragg sheets, 396–397, 398f
Bragg's law, 57, 215
Bravais lattices, 215–221, 234–235
 symmetry points of Brillouin zones of, 220t
Brillouin function, 445–446
Brillouin zone, 215–216, 218–219
 of Bravais lattices, symmetry points, 220t

C

$Ca_{0.4}K_{0.6}(N_{0.3})_{1.4}$, 30f
Ca_2MnO_4, 251–252, 252f
Ca_2WMnO_6, 248–249
$Ca_3Co_{1-x}Mn_xO_6$, powder neutron diffraction patterns, 310f
Ca_3CoMnO_6, 307–313, 314t
 crystal structure, 310f
 powder neutron diffraction patterns, 311f
$Ca_3Mn_2O_7$, 252–253, 252f
Canted nuclear antiferromagnet (CNAF), 476, 476f, 480–482
Cascade demagnetization system, 451–452
$CdCr_2O_4$, 242–243
CdV_2O_4, 242–243
Ce-based heavy fermions, 101–102
 $CeCoIn_5$, 111–112
 $CeIrIn_5$, 112
 $CeRhIn_5$, 109–110
 $CeT(In_1L_xM_x)_5$, 107–108
$CeCoIn_5$, 107–108, 111–112
$CeCu_6L_xAu_x$, 102–107
$CeIrIn_5$, 107–108, 112
$CeMIn_5$, 107–108
Centrosymmetric groups, 294t
$CeRhIn_5$, 107–110, 112
$CeT(In_1L_xM_x)_5$, 107–108
Chalcogenides, square lattices of metal atoms coordinated by, 253–256, 254f–255f

Chemical pressure (doping)
 $CeCu_6$, 102–105
 URu_2Si_2, 122–125
Chemisorption, 62–63
Chromate-based spinels, 242–243, 245–246
Circular envelope, 221–222
Classical phase transition (CPT), 44–46
Closed cycle refrigerator (CCR) cryostats, 59, 60f
Cloud of close-by protons around the magnetic impurities, 485
Co_2MnGe/Al_2O_3, polarized neutron reflectometry from, 382f
$CoAl_2O_4$, magnetic structure of, 246–247
Cobalt hydroxy-sebacate, 267–268
$CoCr_2O_4$, 314t
 magnetic structure of, 245–246, 246f
Coexistence of phases, 471
CoFeB, 299–300
Coherence ellipsoid, 365, 387
Coherence length, 365–366, 388, 390, 396–398, 412–413, 422
 longitudinal, 385–386, 388
 transverse, 386
Coherence volume, 386–388, 387f, 391, 394–395, 404
Coherent magnetization rotation, 390f, 391
Coherent neutron scattering, 54
Colorless groups, 224–225
Colossal magnetoresistance (CMR) manganites, 208–209
Commensurate magnetic structures, 218–222, 225–228, 232
Competing interaction, importance of, 471
$CoNb_2O_6$, 128, 130–138
 energy scans, 136f
 QCP, 134, 136
 zero-field spin excitations, 135f
Constant entropy, 442–443
Continuous (or steady) high magnetic fields, 64
Continuous quantum phase transitions, 46–49
 critical exponents in, 48
Contrast variation
 for small angle neutron scattering), 484
 using polarized protons, 484–485
  major difficulties, 484–485
  prospects, 485
  recent developments, 484–485
Cooling power, 442–443
Cooling solid ^3He, 476–477

Copper (Cu)
  high symmetry directions of fcc lattice, 467–468
  magnetic structure, 457–458, 473
  nuclear magnetic ordering in zero and finite magnetic field, 461–468, 463f–468f
  phase diagram, 456, 456f, 473
CORELLI spectrometer, 274–275
Corner-sharing triangles, 262–264, 268
Correction factors, 449
Correlation effects, 457–458
Correlation length, 46–48
  along the imaginary time dimension, 49
Crednerite, 260–262, 261f
Critical edges for total reflection, 358–359, 362–363, 384–385
Critical exponents, 48t
Critical phenomena, 445
Critical wave number, 355–356, 360
Cross section, differential scattering, 211–212, 212f, 215
Cryogen-free systems, 60–62, 62f
Cryomagnets, 65
Cryostats
  closed cycle refrigerator cryostats, 59, 60f
  liquid helium bath cryostat, 59–60, 61f
  neutron diffraction cryostat for ULT applications, 451–453, 452f
Crystal anisotropy, 340
Crystal growth in confinement, 477
Crystal structure
  Ca_3CoMnO_6, 310f
  $CeTIn_5$, 108f
  $CoNb_2O_6$, 131f
  $MnWO_4$, 304f
  URu_2Si_2, 121f
Crystallographic phase problem, 234–235
Cs_2CuCl_4, 260–262
$Cu_3Co_2SbO_6$, 262, 266f
$Cu_3Ni_2SbO_6$, 262
Cubic rare earth titanate pyrochlores, 87–88
$CuCrO_2$, 258, 259f
$CuFeO_2$, 314t
  magnetic structure of, 258–259
$CuMnO_2$, 260–262, 261f
CuO, 314t
Cuprate, 99, 145–147, 164–166, 187, 190–192, 192f
  compared with iron-based superconductors, 160–161, 164

  hole-doped, 147–160, 176–177
  in-plane nearest-neighbor antiferromagnetic order in, 254f
  Curie–Weiss analysis, 454
  $Yb_2Ti_2O_7$, 93–94
Cycloid magnetic structures, propagation vectors of, 221–222

D

de Broglie model, 2–3, 50
Debye–Waller factor, 214
Delafossites, 258
Delta chain. See Sawtooth chain
Demagnetization
  adiabatic demagnetization in the rotating frame (ADRF), 454
  cascade demagnetization system, 451–452
  nuclear demagnetization, 442–443
demagnetized state, 384, 391
Density functional theory, 333
Detwinned samples, 181–182, 191
Diamond anvil cells, 70, 71f
Diamond (computer program), 236–237
Diamond lattices, in spinels, 242–247
Differential scattering cross section, 349–350
Diffuse magnetic scattering, 274–275
  from disordered alloys, 275–276, 276f
  in frustrated magnets, 278–281, 279f–280f
  from systems with reduced dimensionality, 276–278, 277f
Dilution refrigerator, 442–443
Dion–Jacobson phases, 251–252
Dipolar interaction, 441, 454
Dipole, forces acting on magnetic moment in, 5, 5f, 15f
Dipole approximation, magnetic form factor in, 212–213
Disordered magnetic structures, 274–281
Dispersion, 341, 384, 385f, 420f, 422
DISPIRAL (computer program), 232
Distorted Wave Born Approximation (DWBA), 373–374, 390, 398f, 409, 422
  off-specular neutron scattering in, 363–366
Domain fluctuations, 388–390
Domain formation, 384, 390–391, 390f
Domain nucleation, 392–394

Domain structure in Ag, 471
Domain walls (DWs), 340, 394−395
  of film with perpendicular anisotropy, 395f
  propagation, 390f, 416, 419
Double differential cross section, 52−54
Double perovskites, edge-sharing tetrahedra in, 248−250, 248f, 250f
Drabkin flippers, 368
$Dy_2Ti_2O_7$, 87−88, 93−94, 97
Dynamic critical exponent, 48−49
Dynamic nuclear polarization (DNP), 453−455, 455f
Dynamic structure factor, 55
Dzyaloshinskii−Moriya interactions, 116−117, 262

E

Edge-sharing equilateral triangles, planes of, 258−260, 259f−260f
Edge-sharing nonequilateral triangles, planes of, 260−262, 261f
Electrically shielded rooms, 443
Electromagnons, 325−328
Electron correlations, 166, 191
Electron doping, 161−163, 176−177, 187−189
Electronic magnetic moment, 441
Electronic nematicity, 153−154, 191
Elliptical envlope, 221−222
Energy conservation, neutron scattering, 52−53
Energy scan, 77, 136f
  in $CeCu_{6-x}Au_x$, 104f
  in $CoNb_2O_6$, 134, 135f
  in $TbMnO_3$, 329f
Energy spectrum
  of $LiHoF_4$, 130f
  of spin dimer system, 76f
Energy transfer, 52
Enriched samples, 439−440
Entropy, 442−443, 442f
Entropy (S) − Field (B) phase diagram, 449, 456
Equal layer thicknesses, enhancing interface sensitivity by, 401−403
Equilateral triangles, planes of edge-sharing, 258−260, 259f−260f
$Er_2Ti_2O_7$, 87−90, 87f
  elastic scattering, 89f
  energy *versus* direction in reciprocal space slice for, 90, 91f
  magnetic structure of, 239−240, 240f
  microscopic spin Hamiltonian for, 90−93, 92f
  QPT in, 88−90
  spin excitations in, 90−93
  spin-wave spectrum, 90, 91f
Evolving order with time, 485
Exceed, neutron time-of-flight diffractometer, 70−71
Exchange bias (EB), 391−394, 404−407, 420
Exchange interaction, 475
Expectation values of the three spin components, 9−10
Extinction, 445−446
Extinction length, 359

F

Fcc AFM reflection (1,0,0), 461−462
  correlation length, 462
  critical fields, 464
  field dependency of (1,0,0), 462
  long range order, 462
  measuring time, 462
  peak profiles, 462
  reciprocal space scan, 465
Fe, spin dynamics in, 32f
Fe thin film, 358−359, 359f, 419
  domain walls, 394
  non-spin-flip curve, 381f, 418f
  saturated state, 377−378, 377f
  on silicon substrate, 375f
  spin-flip reflectivity curve, 381f
$Fe_{1+y}Te$, 253−254, 254f−255f
Fe_2SiO_4, 268−270
Fe-Cr alloys, ferromagnetic-disorder scattering, 276, 276f
Fermi pseudopotentials, 6, 345
Fermi statistics, 449−451
Fermi's golden rule, 52−53
Ferroelectric (FE) polarization, 291−292, 315, 316f, 317−319, 318f, 323
  along *a*-axis of $MnWO_4$, 305f
Ferromagnet (FM)
  and neutron beam depolarization, 31−32, 34−35
  domains, 455
$FeSc_2S_4$, 246−247
FeV_2O_4, 243−244, 244f
Field cooling (FC)
  effect of exchange bias after, 404−405
  magnetization reversal of Co/CoO bilayer after, 394f
  zero field cooling, 407−408, 408f

Index **493**

Field-induced magnetic order, in one-
  dimensional quantum systems,
  272—274
Field-induced quantum phase transitions, in
  TlCuCl$_3$, 73—79
First principles (FP) calculations, 457—458
First-order (or discontinuous) phase
  transition, 45, 456
  in KMnF$_3$, 46f
Flexible protein molecule, 28, 28f
Flipping ratio, 371—372, 381—382, 398—399,
  445—446
Focusing neutron guide, 368—370
Form factor, 441
  magnetic, 210, 212—214, 276f
Forte coil, 35—36, 35f
Fourier series components
  amplitude and direction of, 233—234
  propagation vector in, 215—218
  helical magnetic structures, 221—222
  magnetic structures, 218—219
  spin density wave/sine wave structures,
    219—221
  representation analysis, 229—230
  symmetry analysis of couplings between,
    232—233
Fourier transforms, 8—9, 440
FPStudio (computer program), 236—237
Free electron approximation (FEA), 457—458
Free-energy density, homogenous form of,
  47—49
Frequency, of neutron, 3
Fresnel amplitudes, 356—357
Fresnel corrections coils, 18—19, 19f
Frustrated magnets, diffuse magnetic
  scattering in, 278—281, 279f—280f
Fulde—Ferrell—Larkine—Vchinnikov state,
  110
Full frustration, 459
FULLPROF (computer program), 233—234

G

Gaussian approximation, radial integrals,
  213—214, 214f
Gd$_2$Sn$_2$O$_7$, 239—240, 240f
Geometrical optics, in neutron scattering
  experiments, 10—12
Gray groups, 224
Grazing incidence Bragg diffraction,
  352—353, 363—364
Grazing incidence kinematics, 352—359,
  353f, 358f—359f

Grazing incidence small angle neutron
  scattering (GISANS), 363—364, 407,
  413—415, 415f, 422
  polarized-GISANS (P-GISANS),
    413—416, 416f
Ground state
  magnon density-of-states, 325f
  of spin system, 80—83
GSAS (computer program), 233—234

H

Haldane chains, 266—267, 272—274
Half order peaks, 362—363, 396—399
Hamiltonian, 322—323
  for slow neutrons, 3—4, 6
  propagation through space and matter, 6
  spin Hamiltonian, 90—93, 92f, 235—236,
    238
  truncated dipolar Hamiltonian, 454
  unperturbed Hamiltonian, eigenstates, 7—8
  Zeeman energy type Hamiltonian, 5—6
Heat balance, 442—443, 442f
Heat flow, 445
Heat leak, 443
Heat load, 442—443
Heavy fermions, 99—102
  CeCoIn$_5$, 111—112
  CeCu$_{6-x}$Au$_x$, 102—107
  CeIrIn$_5$, 112
  CeRhIn$_5$, 109—110
  CeT(In$_1$L$_x$M$_x$)$_5$, 107—108
  QCPs, 101—102
    local QC, 100—101
    spin-density-wave QC at, 100
    rare earth intermetallic, 100—101
Heesh groups. *See* Black-white groups
Height—height correlation length,
  363—364
Heisenberg model, 8—9, 164—168, 167f
Helical magnetic structures, propagation
  vectors of, 221—222
Helium closed cycle refrigerator, 59
Helium-3
  crystals, orientation of, 479—480
  gas polarizers, 482
    advantages, 482
  neutron spin filter, 483f
  and neutrons, 475—476
  nuclear magnet, 475
  sample pressure cell assembly, 478f
  solidification of, 479f—481f
  sorption system, 62—63, 63f

494 Index

Helium-3 (*Continued*)
  transmission spin filter, 368, 372, 372f, 409, 411−412
  unique many body quantum system, 474
Helium-3/helium-4 dilution refrigerators, 63−64, 64f−65f
Helmholtz coil for polarized neutrons, 447
Hertz−Millis−Moriya (HMM) spin fluctuation theory, 100−102, 106f
Heterostructures, interfacial magnetism in, 379
Heusler multilayer
  antiferromagnetically coupled, 397−398, 398f
  polarized neutron reflectometry from, 382f
Hexaferrite, 314t
$HgCr_2O_4$, 242−243
h-$HoMnO_3$, 314t
High magnetic field, neutron scattering, 64−67
High-energy magnetic excitations, iron-based superconductors, 176−179
Highly polarized ^{3}He, 484−485
High-pressure cells
  hydrostatic cells, 69, 70f
  large-volume (clamped) cells, 69, 71f
  neutron scattering, 68−71
  opposed anvil cells, 70−71
High-transition-temperature superconductor
  hole-doped cuprate superconductors. *See* Hole doped cuprate
  iron-based superconductors. *See* Iron-based superconductors
$Ho_2Ru_2O_7$, 240−241, 241f
$Ho_2Ti_2O_7$, 31, 32f, 87−88, 93−94, 97
Hole doped cuprate, 147−160, 176−177
  magnetic excitations evolution upon doping, 151−156
  magnetic order and spin waves in the parent compounds, 147−151
  neutron scattering *versus* RIXS, 156−158
  resonance mode in superconducting state, 158−160
$HoMnO_3$, 294−296
  crystal structure, 295f
  domains, 315−316
  spin dynamics, 322−323
  temperature dependence of normalized total counts, 315−316, 317f
  temperature *versus* magnetic field phase diagram for, 297f

Honeycomb lattices, 264−265, 265f−266f
Horizontal cryomagnetic system, 65−66
Huygens' principle, 10
Hybrid high magnetic fields, 64
Hydrostatic cells (piston−cylinder devices), 69, 70f
Hyperscaling relations, 48
  for QPT, 48−49

I

Impurity concentration, 453
In-beam correction elements, 25−26
Incident energy, 52
Incoherent averaging, 385−386
Incoherent neutron scattering, 54
Incommensurable structure, 460
Incommensurate magnetic order, 151−156, 152f, 163−164, 167−168, 169f
Incommensurate magnetic structures, 218, 227−228, 232
Inelastic neutron scattering, 52
  $CeCoIn_5$, 111f
  $CoNb_2O_6$, 131−134, 135f, 137, 137f
  $Er_2Ti_2O_7$, 91f
  spin dynamics. *See* Spin dynamics
  $TlCuCl_3$, 75f, 82f, 86, 86f
  $Yb_2Ti_2O_7$, 96f, 98f−99f
Inelastic polarized neutron scattering, 326f, 329f
Infinite coherence length, 11−12
Inorganic Crystal Structure Database, 211f
Instrumental resolution, 373−374, 387−388
Intensity map
  from magnetic nanoparticles, 410, 410f
  multilayers, 396−397, 397f
  Heusler multilayer, 397−399, 398f
  off-specular scattering, 399f
  for non-spin-flip and spin-flip scattering, 388, 389f
  2D small-angle neutron scattering, 411f
  stripe arrays, 406f
Intensity modulated neutron spin echo (IMNSE), 31−32
Intensity of elastic scattering, 440
Interaction parameters, 6−7
Interaction potential, 344, 346−349, 364−365
Interface sensitivity, 341−342
  enhancing by equal layer thicknesses, 401−403

Interfacial magnetism, in heterostructures, 379
Interference pattern, 11–12
Interlayer exchange coupling (IEC), 398–399
Intermediate scattering function, 55
Inverse proximity effect, 401
Iron chalcogenide, 161–162
Iron pnictide, 145–147, 160–161, 189–191, 192f, 253
   and cuprates, 155–156, 160–161, 164
   magnetic exchange energy, 158–159
   magnetic order, 182
   magnetism in, 163f, 164, 166, 168, 187–189
   parent compounds, 161–164, 173–176, 182–186
   resonance in, 179–182
Iron-based superconductors
   antiferromagnetically ordered metallic parent compounds, 162–168
   compared with cuprates, 160–161, 164
   compounds, 160–162
   high-energy magnetic excitations, 176–179
   introduction, 160–162
   low-energy magnetic excitations, polarization dependence of, 186–190
   magnetic order and magnetic excitations evolution with doping, 168–176
   magnetism in $A_xFe_{2-y}Se_2$ compounds, 182–186
   phase diagrams, 160–162
   resonance mode in iron-based superconductors, 179–182
Irreducible representations (IRs), 222–223, 228–231, 237–238
ISODISTORT (computer program), 228
Isopropyl ammonium (IPA)-$CuCl_3$, field-induced magnetic order in, 272–273
Isotope, 439
ISOTROPY Software Suite, 226, 231–233
Isovalent doping, 170–171
Itinerant magnets
   MnSi, 115–120
   URu_2Si_2, 120–126
   weak itinerant ferromagnets, 112–114

J

J_1–J_2 model, 270
Jahn–Teller distortion, 260–262
JANA (computer program), 233–234
JANA2006 (computer program), 228, 232–233
Jarosites, 262–263
Jordan–Wigner transformation, 127–128
Josephson's identity, 48

K

K_2CoF_2, 250–251
K_2MnF_2, 250–251
K_2NiF_4, 250–252, 252f
Kagomé lattices, 262–264, 262f–263f, 268
Kapitza boundary impedance, 443–444
$KCr_3(SO_4)_2(OD)_6$, 263
$KCuCl_3$, progressive Zeeman splitting of the triplet modes, 77f
Kiessig fringes, 374–376, 388, 412–413
Kondo screening, 101–102, 104–105
Korringa behavior, 449–451
Kovalev tables, 230–231
Kramers–Kronig relation, 104f
K-SEARCH (computer program), 232

L

$LaAlO_3/SrTiO_3$ superlattice, 379, 379f
$La_{0.75}Pr_{0.25}Co_2P_2$, 256
La_2NaOsO_6, 248–249
La_2NaRuO_6, 248–249
LaFeAsO, 253–254, 254f
Landau theory, 312–313, 323, 398
Langasites, 263–264
Large moment antiferromagnetic (LMAF) phase, 125
Large-volume (clamped) cells, 69, 71f
Larmor precession, 9–10, 13, 24–25
Latent heat, 456
Lateral domain size, 384–385, 398–399
Laterally patterned magnetic nanostructures, 403–416
Lateral magnetic structures, 398, 403–404, 421
Layered structures, 250–265
LCMO, 401–403, 402f
$LiCrSi_2O_6$, 266–267
Lifting of degeneracy, 459–460
$LiHoF_4$, 128–130
Liquid helium bath cryostat, 59–60, 61f
$Ln_3Ga_5SiO_{14}$, 263–264
Longitudinal coherence length, 385–386, 388

Low-energy magnetic excitations, iron-based
    superconductors, 186–190
LuMnO$_3$, 320
    antiferromagnetic peak in, 321f
o-LuMnO$_3$, temperature dependence of
    dielectric constant for, 321f

M

M$_3$(VO$_4$)$_2$, 263–264
Magic box, 372
Magnetic Bragg reflections, 76–77,
    81–83, 83f
Magnetic chains, triangular-based, 268–270,
    269f
Magnetic crystallographic information files
    (mCIF), 235, 267–268
Magnetic dipoles, 4–5
Magnetic domains, 390, 392–395, 421
    in AFM-coupled multilayers,
        397–401
    off-specular scattering (OSS), effect of,
        363–364
    state, 384–390, 394
Magnetic domain state, 384, 388–390,
    389f
Magnetic field(s)
    inside magnetic matter, 4
    reversal, 391, 418
Magnetic films, 342, 374–376, 418, 421
    homogeneous, 403f
    with perpendicular anisotropy, 394, 395f
Magnetic form factor, 346, 353
Magnetic frustration
    breaking, 470
    diffuse magnetic scattering in frustrated
        magnets, 278–281, 279f–280f
    full frustration, 459
    and one-dimensional magnetic lattices,
        270–272
Magnetic heterostructures, 342
    exchange-biased, 391–392
Magnetic induction, 4–5, 345–346,
    350–352, 354, 357, 379,
    382–383, 421
Magnetic inelastic neutron scattering,
    TlCuCl$_3$, 82f. See also Spin dynamics
Magnetic interaction vector, time-
    independent, 211–212, 212f
Magnetic moment, of neutron, 1–2, 55
Magnetic moment density, 4–5
Magnetic multilayers, 396–403
Magnetic nanoparticles, 407–413
    model containing ferrimagnetic core and
        magnetic shell, 411f
    self-assembled
        Co/CoO, 416f
        PNR of, 412f
    small-angle neutron scattering intensity
        patterns from, 410f
Magnetic neutron interaction, 346, 382
Magnetic neutron scattering, 341–344, 394,
    407
Magnetic order, 291–292
Magnetic point groups, 224, 226
    admissible, 225, 225t
Magnetic representation matrix,
    228–229
Magnetic roughness, 363–364, 383–384
Magnetic scattering differential cross
    section, 56
Magnetic scattering length, 343, 346–347,
    377f–378f
Magnetic scattering length density (mSLD),
    343, 352–354, 367–368, 374–377,
    378f, 382–384, 401–403,
    410–411
Magnetic structures
    complex, limitations of neutron scattering
        for, 234–235
    of CoNb$_2$O$_6$, 131
    determination
        steps, 231–234
        using neutron diffraction, 206–211
    diffuse magnetic scattering
        from disordered alloys, 275–276
        in frustrated magnets, 278–281
        from systems with reduced
            dimensionality, 276–278
    disordered, 274–281
    layered structures, 250–265
        combined square-planar metal-oxygen/
            metal-pnictogen lattices, 256–257
        layers of honeycomb lattices,
            264–265
        planar kagomé lattices, 262–264
        planes of edge-sharing equilateral
            triangles, 258–260
        planes of edge-sharing nonequilateral
            triangles, 260–262
        square lattices of metal atoms
            coordinated by pnictogen/chalcogen
            atoms, 253–256
        square-planar metal-oxygen/halogen
            layers, 251–253

Index   497

magnetic scattering
  magnetic propagation vector formalism, 211–215
  scattering amplitude from magnetic order, 211–231
  quasi-one-dimensional lattices, 266–274
  field-induced magnetic order in one-dimensional quantum systems, 272–274
  frustrated interchain interactions, 270–272
  long-range magnetic order in simple chain-based structures, 266–268
  triangular-based magnetic chains, 268–270
  standard description for, 235
  symmetry considerations, 222–231
  magnetic superspace groups, 227–228
  representation analysis, 228–231
  Shubnikov space groups, 223–226
  three-dimensional networks, 237–250
  corner-sharing tetrahedral lattices in pyrochlore oxides, 237–242
  diamond lattices in spinel systems, 246–247
  edge-sharing tetrahedra in double perovskites, 248–250
  interpenetrated tetrahedra and diamond lattices in spinel compounds, 242–246
  types of, 219f
Magnetic superspace groups, 227–228
Magnetic superstructure, Bragg peaks, 362–363
Magnetic susceptibility, 56
Magnetic symmetry approach, 222–223
Magnetic unit cell structure factor, 214
Magnetite, ferrimagnetic structure of, 208–209
Magnetization depth profile, 421
Magnetization reversal, 390f, 391–394
Magnetization rotation, 380–384, 413
  coherent, 390f, 391
Magnetization tilt angle, 380, 384–385
MAGNEXT (computer program), 233
Magnon condensation, 272
Majorana limit, 15
Maximizing measurement time, 461
Maxwell equations, 4–5
Maxwell's theory, 4
mCIF. See Magnetic crystallographic information files (mCIF)

Metal-organic frameworks (MOFs), 267–268
Mezei spin flipper, 366–368, 372, 372f, 411f
MgV_2O_4, 242–243
MIEZE. See Modulation of intensity emerging from zero effort (MIEZE)
Mineral hubnerite. See $MnWO_4$
Mixing coefficients, 229–230
β-Mn, 278–280
  structure, 279f
β-$Mn_{0.8}Co_{0.2}$, magnetic diffuse scattering patterns for, 278–280, 279f
Mn_2GeO_4, 268–270
Mn_2SiO_4, 268–270
MnO, neutron powder diffraction patterns for, 208–209, 208f
$MnSc_2S_4$, 246–247
MnSi, 115–120
  applied pressure, 118–120
  measured with polarized neutron inelastic scattering, 117f
  primitive unit cell, 115f
  QCPs, 120
MnV_2O_4, 243–244, 245f
$MnWO_4$, 302–307, 314t
  crystal structure, 304f
  electric polarization and spin structure, 306f
  magnetic structure of, 330f
  spin dynamics, 328–329
  spin-wave dispersion, 331f
Mode coupling theory, 30
Modulation of intensity emerging from zero effort (MIEZE), 22
MODY (computer program), 231–233
Momentum conservation, neutron scattering, 52
Momentum transfer, 13–14
Monochalcogenides, 253
Monte Carlo simulations, 460–461
Mosaic grains, 446
Mott insulator, 146, 190–191, 192f
Multiferroic properties of compounds, 314t
Multiferroics, 291–293
  domains
    $BiFeO_3$, 315
    $HoMnO_3$, 315–316
  future directions, 332–333
  spin dynamics
    $BiFeO_3$, 323
    electromagnons, 325–328

Multiferroics (*Continued*)
  $HoMnO_3$, 322−323
  $MnWO_4$, 328−329
  $RbFe(MoO_4)_2$, 330
  $(Sr-Ba)MnO_3$, 323−325
  YMn_2O_5, 329
  symmetry considerations, 293−294
  thin films and multilayers
    $BiFeO_3$, 316−319
    $TbMnO_3$, 319−320
  type-I (proper) MFs, 294−301
    $BiFeO_3$, 297−300
    $HoMnO_3$, 294−296
    $(Sr-Ba)MnO_3$, 300−301
  type-II (improper) MFs, 301−313
    Ca_3CoMnO_6, 307−313
    $MnWO_4$, 302−307
    $Ni_3V_2O_8$, 305
    $RbFe(MoO_4)_2$, 302−307
    $TbMnO_3$, 302−307
    YMn_2O_5, 307−313
Multi-k magnetic structures, 216−217, 217f
Multilayers, 396−403
  Co/Cu, 399f
  high resolution, 383f
  polarized neutron reflectometry measurements, 411−412, 414f
Muon spin spectroscopy, 234−235

N

$Na_3Co_2SbO_6$, 262
$NaCrGe_2O_6$, 266−267
$NaFeGe_2O_6$, 267
α-$NaFeO_2$, 258−259
$NaMnGe_2O_6$, 267
$NaMnO_2$, 260−262
$NaNiO_2$, 260−262
Nanoparticles (NPs)
  core, 407−412
  shell, 407−412, 415−416
Natural mixtures, 439−440
Nd_2CuO_4, 322−323
Nd_2NaOsO_6, 249, 250f
Nd_2NaRuO_6, 249, 250f
$Nd_3Ga_5SiO_{14}$, 263−264
$NdCo_2As_2$, 255−256, 255f
$NdCo_2P_2$, 255−256, 255f
$NdCoAsO$, 254−255, 255f
Nearest neighbor contributions, 457−458
Néel domain walls, 394−395, 395f
Negative temperatures, 454, 457

Neutron beam polarizers and analyzers, 34−35
Neutron beam propagation, high precision rules for, 12−13
Neutron diffraction, 146, 164, 170−171, 182, 191
  magnetic structure determination using, 206−211
Neutron diffractometer, 443
Neutron energy transfer, 13−14
Neutron grazing incidence diffraction, conventional, 33f
Neutron magnetic moment, 1−2
Neutron momentum transfer, 23
Neutron optics, 10−12, 368−373
Neutron polarimetry, 440−441
Neutron polarization techniques, 441
Neutron powder diffraction, 232, 234
  for alloys, 276
  patterns for MnO, 208−209, 208f
Neutron reflectivity, 341−342.
  See also Polarized neutron reflectivity (PNR)
Neutron scattering, 51−58, 145−147, 158−160, 164−166, 172f, 181f, 191−192
  beam propagation in, 12−13
  on $CeCoIn_5$, 111−112
  on $CoNb_2O_6$, 131
  cross sections, 52−54
  cryogenics, 59−64
  general principles, 51−52
  geometry, 53f
  high magnetic field, 64−67
  inelastic, 173−176, 181−182
  limitations for complex magentic structures, 234−235
  magnetic cross section, 55−56
  multiferroics, 293
  polarized, 147, 161−162, 187−189, 347−352
  versus RIXS, 156−158, 179
  space−time correlation, 54−55
  spin labeling, 13−24
  time-of-flight spectrometers, 58, 58f
  on $TlCuCl_3$, 75
  triple-axis spectrometer, 57−58, 57f
  unpolarized, 187−189
  wave *versus* geometrical optics in, 10−12
  See also Small-angle neutron scattering (SANS)

Index   499

Neutron single-crystal diffraction, 205–206, 232, 234
Neutron spectroscopy techniques, 130
Neutron spin echo (NSE), 24–32
  for elastic scattering at small angels, 33–40
    neutron beam polarizers and analyzers, 34–35
    polarized neutron beams transportation and spin-injection devices, 35–36
    precession region and magnetic shielding, 36–37
    spin-echo reflectometry, 38–40
    spin-echo small-angle scattering, 34f, 38
  in magnetism, 30–32
  for nuclear scattering, 26–30
Neutron spin flipper coil operated by DC current, 15f
Neutron transparency, 68t
New phases, 459
$Ni(C_5H_{14}N_2)_2N_3(ClO_4)$ (NDMAZ), field-induced magnetic order in, 273–274
$Ni(C_5D_{14}N_2)_2N_3(PF_6)$ (NDMAP), field-induced magnetic order in, 273–274
Ni_2SiO_4, 268–270, 269f
$Ni_3V_2O_8$, 263–264
$NiCr_2O_4$, 245–246, 246f
Niggli–Indenbom theorem, 231
NMR methods, 449
  SQUID and NMR, 455–457
Non-centrosymmetric groups, 294t
Noncollinear magnetization, 396, 398–399, 421
Nonequilateral triangles, planes of edge-sharing, 260–262, 261f
Non-Fermi liquid (NFL) behavior, 99
Nonlinear effects, 460
Non-spin flip (NSF) scattering, 209–210, 352, 421
Nuclear demagnetization, 442–443
Nuclear magnet, 439
Nuclear polarization by neutrons, 441, 445–451, 446f–448f, 450f–451f
  dynamic nuclear polarization (DNP), 453–455, 455f
Nuclear pseudomagnetism, 6

Nuclear scattering length, 345–347, 403–404
  densities for Fe film on Si substrate, 375f
Nuclear scattering length density (nSLD), 358, 367–368, 374–377, 382–384, 401–403
Nuclear spin temperature, 444

O
Off-specular neutron scattering, 342
  in DWBA, 363–366
Olivines, 268–270
One-dimensional (1D) PNR analysis, 358, 384, 421
One-dimensional (1D) potential, specular polarized neutron reflectivity for, 359–363
One-dimensional quantum systems, field-induced magnetic order in, 272–274
Opechowski–Guccione (OG) notation, 226
Opposed anvil cells, 70–71
Optical diffraction grating, 38
Optimal doping, 159–160, 164, 172f, 178–179, 187–189, 189f–190f
Orange cryostats, 59–60, 61f
Order parameter, 46, 48–49
Order-by-disorder mechanism, 246
Ordered (nuclear) system, 454
Overdoped, 151–152, 157, 159–160, 171–176, 179–182, 181f, 187–189, 191
Oxide superlattices, 379, 401–403

P
Parallelogram-shaped effective precession fields, 22, 23f
Paramagnetic centers, 453–454
Paramagnetic groups, 224, 230–231
Paramagnetic state, 445–446
Parent compound, 102, 151–152, 154f, 156f, 168, 189–190
  antiferomagnetic order, 157, 160–168, 190–191
  of cuprate, 145–146, 160–161
  of iron pnictide, 145–147, 161–164, 173–176, 182–186, 191
  magnetic order and spin wave in, 147–160
Paris–Edinburgh presses, 70
Parratt recursion algorithm, 360–361
Particle form factor, 409, 415–416

Particle scattering processes, probability of, 6–7
Pauli spin, 6
Permutation representation matrix, 228–229
Perovskite manganites, 208–209, 209f
Perovskites, 237
 double, edge-sharing tetrahedra in, 248–250, 248f, 250f
Phase boundary
 shape of, 474
 stability of, 474
Phase diagram, 50–51, 474–475
 AFM spin dimer system, 82f
 $Ba_{1-x}Na_xFe_2As_2$, 170, 170f
 $Ba(Fe_{1-x}Cr_x)_2As_2$, 170–171, 171f
 $Ba(Fe_{1-x}Mn_x)_2As_2$, 170–171
 $Ba(Fe_{1-x}Ru_x)_2As_2$, 171f
 $BaFe_2As_2$, 191
 $CeIrIn_5$, 114f
 $CeRhIn_5$, 109f–110f
 $Er_2Ti_2O_7$, 89f, 93, 94f
 iron pnictides and cuprates, 192f
 iron-based superconductors, 161–162, 162f, 168
 $La_{2-x}Ba_xCuO_4$, 155
 $La_{2-x}Sr_xCuO_4$, 151f, 153–154
 $LiHoF_4$, 129f
 MnSi, 118f
 $MnWO_4$, 328–329
 $RbFe(MoO_4)_2$, 308f
 spin Ising model system, 127f–128f
 $Sr_{1-x}Ba_xMnO_3$, 302f
 URu_2Si_2, 126f
 $Yb_2Ti_2O_7$, 95f
Phase separation, 152, 155–156, 161–162, 182
Phase transitions, categories, 45
Physisorption, 62–63
Planck's relation between energy and frequency, 2–3
Pnictides
 combined square-planar metal-oxygen/metal-pnictogen lattices, 256–257
 square lattices of metal atoms coordinated by, 253–256, 254f
Pnictogen. See Pnictides
Point-like classical particle, 9–11
Polarization analysis, 234, 349–351, 358f, 383, 403–404, 411–412
 1D, 362–363
 3D, 362
 SANS, 409, 411–412, 411f

Polarization axis, 358–359, 362–363, 374–376, 375f, 379–380, 384–385, 388–391, 389f, 414f, 417, 421
Polarization vector, 349–350, 352, 357, 382–383, 421
 3D, 352
Polarized GISANS (P-GISANS), 413–416, 416f
Polarized neutron beams transportation and spin-injection devices, 35–36
Polarized neutron reflectivity (PNR), 319–320, 341f, 374, 409, 421
 for 1D potential, 359–363
 from 100-nm-thick Fe film, 377f
 of FeN on GaAs substrate, 378f
Polarized neutron reflectometers, 366–368, 401–403
 instrumental considerations
  data detection and acquisition, 373
  data modeling and fitting, 373–374
  neutron optics, 370–373
  neutron reflectometers design, 366–370
Polarized neutron reflectometry (PNR), 205–206, 342–344, 352, 358f, 390f, 399f
 1D, 358, 421
 AC-PNR, 417–418, 420–421
 Co/Cr/Fe/Cr superlattices, 400f
 domain state, 384–385
 domain walls, 394–395, 395f
 Heusler multilayer Co_2MnGe/Al_2O_3, 382–383, 382f
 $LaAlO_3/SrTiO_3$ supelattice, 379f
 laterally patterned magnetic nanostructures, 403–407
 magnetic NPs, 407, 413–415
 self-assembled NP multilayers, 412–413, 412f
 magnetization reversal, 390–391, 390f
 modeling and fitting of data, 373–374
 multilayers, 401–403, 414f
 non-spin-flip, 359f
 spin-flip, 359f
 $TbMnO_3$, 320f
 thin films, saturated state, 376–380
 time-resolved AC-PNR (TRAC-PNR), 417, 420–421
 using monochromatic incident beam, 366–367

Polarized neutron scattering, 147, 161–162,
  187–189, 299–300, 300f–301f, 305,
  315–316, 317f, 319f
  cross section of, 347–352
Polarized neutrons, 209–210, 233–234,
  274–278, 277f
Polarized nuclei, applications in nuclear
  diffraction, 482–485
Polarized protons
  contrast variation using, 484–485
  major difficulties, 484–485
  prospects, 485
  recent developments, 484–485
Polarized version of SANS (SANSPOL),
  343, 407, 409–411, 413–415
  with polarization analysis, 411f
Polarizer, flipper, 447
Poly(3-hexylthiophene-2,5-diyl) (P3HT);
  [6,6]-phenyl-C61-butyric acid methyl
  ester (PCBM), 39–40
Polystyrene spheres, and NSE spin
  labeling, 38
Position sensitive detectors (PSDs), 232,
  372f, 373, 390, 411–412
Powder averaging, 234
Precession region and magnetic shielding,
  36–37
Pressure-induced quantum phase transitions,
  in $TlCuCl_3$, 79–86
Probabilistic scattering and absorption by
  nuclear reactions, 12–13
Projection operator technique,
  229–230
Propagation vectors, 211–215,
  216f–217f
  helical magnetic structures, 221–222
  identification of, 232
  magnetic structures, 218–219
  representation analysis, 228–231
  spin density wave/sine wave structures,
    219–221
Proximity effect, 340, 376,
  401–403
Pseudodipolar moment, 439
Pulse tube refrigerator (PTR), 61–62, 62f
Pulsed high magnetic fields, 64
Pulsed resistive magnets, 67
Pyrochlore magnets, 87–88
Pyrochlores, corner-sharing tetrahedral
  lattices in, 237–242, 239f, 239t,
  240f–242f
Pyroxenes, 266–267

Q

Quantization axis, 347, 352, 355, 360–361,
  374–376
Quantum critical point (QCP), 49–50
Quantum critical region, 50–51
Quantum critical scaling, 48–49
Quantum criticality (QC), 51, 84–86
  in $CeCu_{6-x}Au_x$, 102, 106–107
  heavy fermions, 99
  local, 100–101
  spin-wave-density, 100
Quantum effects, 474
Quantum fluctuations, 44–45
Quantum magnetic systems, one-dimensional
  quantum systems, 272–274
Quantum mechanics, 2–3
Quantum phase transitions (QPTs)
  defined, 44–45
  in heavy fermions, 99–112
  in itinerant magnets, 112–126
  in $TlCuCl_3$, 73
  in transverse field Ising systems, 126–138
  in XY pyrochlore magnets, 87–88
Quantum probabilistic scattering, 13
Quantum spin liquid, 278–280
Quasi-Laue diffractometers, 232
Quasi-one-dimensional lattices, 266–274

R

$R_2B_2O_7$, 87
R^{3+} ions, 87
Radial integrals, of magnetic ions, 213–214,
  214f
Radiation frequency, 3
Radiation wavelength, 3
Rare earth intermetallic heavy fermions,
  100–101
Rb_2CoF_2, 250–251
$Rb_2Fe_2O(AsO_4)_2$, 270
$RbAg_2Fe[VO_4]_2$, 260
$RbFe(MoO_4)$, 314t
$RbFe(MoO_4)_2$, 302–307
  dispersion measurements, 332f
  electric field, 309f
  low-temperature $P3$ structure of, 307f
  magnetic structure of, 260
  spin dynamics, 330
Reciprocal lattice vector, 215
Refl1D program, 373–374
Reflection amplitude, 356, 382
  matrix, 357, 361–362

REMn$_2$O$_5$, 309
Representation analysis, 222−223, 228−231, 237−238
Representation/corepresentation analysis, 231, 305−307, 309−313
Resonance mode, 441
  in A$_x$Fe$_{2-y}$Se$_2$, 182−186, 186f
  in iron-based superconductors, 179−182, 180f−181f, 183f
  in near-optimal doped Ba(Fe$_{0.94}$Co$_{0.06}$)$_2$As$_2$, 187−189
  in spin space, 187
  in superconducting state, 158−160, 160f, 178−179, 190−191
Resonance neutron spin echo (RNSE), 19−21, 24f, 36−37, 37f
Resonant inelastic X-ray scattering (RIXS), 147, 157f, 179, 179f
  neutron scattering *versus*, 156−158
Returning vector, 224, 228−229
Reverse Monte Carlo (RMC) refinement, 232, 278−280
RF-spin flipper, 373f
Ridge of specular reflection, 388
RIETAN-2000 (computer program), 233−234
Rietveld refinement, 232
Ruddlesden−Popper phases, 251−252
  magnetic structure of, 252f
Ruderman-Kittel (RK) interactions, 458
Ruderman−Kittel−Kasuya−Yosida (RKKY) exchange, 101, 398−399
Russell−Saunders approximation, 213

S

SANSPOL. *See* Polarized version of SANS (SANSPOL)
Sapphire anvil cells, 70
SARAh (computer program), 231−233, 237−238
Saturated state, 374−380
Saturation, 446
Sawtooth chain, 268−270, 269f
Scattering amplitude, 344−353, 355, 357, 364, 386−387
  Born scattering amplitude, 349−351
  from magnetic order, 211−231
Scattering cross section, 439
  for polarized neutrons, 347−352
Scattering function, 55−56
  intermediate, 55

Scattering length density (SLD), 343, 352−353, 359, 374−376, 410−411
  magnetic (mSLD), 343, 352−354, 367−368, 374−377, 378f, 382−384, 401−403, 410−411
  nuclear (nSLD), 358, 367−368, 374−377, 382−384, 401−403
Scattering matrix, 347−349, 364−365
Scattering potential, 344−346, 348−349, 353, 355−356, 359, 362−364
Scattering vector, 353−354, 370−371, 374−377, 380, 382, 385−386, 394, 409, 411−414
  magnitude of, 52
Schrödinger equation, 6−8, 13
Second-order (or continuous) phase transitions, 45−46
SERGIS (spin-echo labeling technique for SANS), 33
SESAME (spin-echo labeling technique for SANS), 33
Series-connected hybrid magnet, 65−66, 67f
Shape anisotropy, 340, 403−405, 407−408
Shape function, 354
Shubnikov space groups, 223−226, 233
Silver (Ag)
  critical fields, 457
  domain structure in, 471
  magnetic structure, 457−458
  maximal measurement time, 469
  nuclear magnetic ordering in zero and finite magnetic field, 468−470, 469f
  phase diagram, 474
Simple chain-based structures, long-range magnetic order in, 266−268
Simulated annealing (SAnn) method, 233−234
Sine wave magnetic structures, propagation vector of, 219−222
Single domain state, 371−372, 374−376, 379−380, 384−385, 407
Single color groups, 224
Slow neutrons, particle properties and interaction of, 2−7
Small-angle neutron scattering (SANS), 33, 38, 39f, 205−206, 293, 343, 371−372, 409−410
  contrast variation for, 484
  GISANS, 363−364, 407, 413−415, 415f, 422
    polarized-GISANS (P-GISANS), 413−416, 416f

intensity patterns from magnetic nanoparticles, 410f
on nanoparticles, 411–414
polarization analysis, 409, 411–412, 411f
2D, 411f. *See also* Neutron scattering
Solenoid-type pulsed magnets, 67
Specular ridge, 387–388, 392–394, 396–398
specular polarized neutron reflectivity for 1D potential, 359–363
Spherical neutron polarimetry, 239–240, 307, 309–312
Spin density matrix, 349–350
Spin dependent scattering, 439
Spin dimer systems, 71–73
  magnetic properties, 74
  TlCuCl$_3$, 73
Spin dynamics
  BiFeO$_3$, 323
  electromagnons, 325–328
  HoMnO$_3$, 322–323
  MnWO$_4$, 328–329
  RbFe(MoO$_4$)$_2$, 330
  (Sr-Ba)MnO$_3$, 323–325
  YMn$_2$O$_5$, 329
Spin flip (SF) maps, 388–390
Spin flip reflectivity, 352, 377
Spin flippers, 366, 368–370
  Mezei, 366–368, 372, 372f, 411f
  RF, 373f
Spin flop scenario, 474
Spin glass, 151–152, 151f, 168
Spin Hamiltonian, 235–236, 238
Spin incoherent scattering, 440
Spin labeling, 13–24
  defined, 2
  neutron parameters choices, 22–24
  practical spin labeling, 18–22
Spin ladder, 272–273
Spin matrix, 345
Spin system, ground state, 80–83
Spin wave, 146, 164–166, 179
  A$_x$Fe$_{2-y}$Se$_2$, 185f
  BaFe$_2$As$_2$, 164–166, 165f, 176–177, 187–189, 188f
  Brillouin zone, 164–166
  in FeTe, 167–168, 167f
  high-energy, measurement of, 164
  itinerant model, 166
  K$_x$Fe$_{2-y}$Se$_2$, 185f
  La$_2$CuO$_4$, 149f, 150, 157
  NaFeAs, 188f

in parent compounds, 146–151, 149f–151f, 159–160
Rb$_{0.8}$Fe$_{1.6}$Se$_2$, 182–186, 185f
in YBa$_2$Cu$_3$O$_{6+x}$, 150f, 159
Spin-density wave (SDW), 151f, 168
  BiFeO$_3$, 299–300, 301f
  propagation vector of, 219–221
  QC, 100
  TbMnO$_3$, 328f
Spin-echo reflectometry, 38–40
Spin-echo small-angle scattering, 38
Spinels
  diamond lattices in, 242–246, 247f
  interpenetrated tetrahedra and diamond lattices in, 242–246, 244f–245f, 245t, 246f
  magnetic diffuse scattering in, 280–281, 280f
Spin-flip (SF) scattering, 209–210
Spin-flop phase, 362–363, 363f
Spin-lattice relaxation time, 444
Spintronics, 340–341, 420
  antiferromagnetic, 342–343
Spin-wave dispersion
  BiFeO$_3$, 320
  HoMnO$_3$, 322f
Spin-wave theory, 90–93
Spiral magnetic structures, propagation vectors of, 221–222
Split-type pulsed magnets, 67
Spontaneous nuclear magnetic order, 453–461
  fcc, 460f
  contour plot of symmetry points, 458f
  dynamically polarized systems, 453–455, 455f
  SQUID and NMR, 455–457
  stability regions, 459f
Square lattices, of metal atoms coordinated by pnictogen/chalcogen atoms, 253–256
Square-planar metal-oxygen/halogen layers, 251–253
Square-planar metal-oxygen/metal-pnictogen lattices, 256–257
SQUID and NMR, 455–457
Sr$_2$CoOsO$_6$, 249, 250f
Sr$_2$CrOsO$_6$, 249
Sr$_2$FeOsO$_6$, 249
Sr$_2$LnSbO$_6$, 248–249
Sr$_2$LuRuO$_6$, 248–249
Sr$_2$Mn$_3$As$_2$O$_2$, 256–257, 257f
Sr$_2$Mn$_3$CuAs$_2$O$_2$, 257f
Sr$_2$Mn$_3$Pn$_2$O$_2$, 256–257, 257f

$Sr_2Mn_3Sb_2O_2$, 257f
Sr_2MoMnO_6, 248−249
Sr_2WMnO_6, 248−249
(Sr-Ba)MnO_3, 300−301, 314t
  spin dynamics, 323−325
$SrDy_2O_4$, 271−272
$SrEr_2O_4$, 271−272
$SrHo_2O_4$
  diffuse scattering from, 276−278, 277f
  magnetic structure of, 271−272, 271f
$SrTb_2O_4$, 271−272
$SrTiO_3$ (STO), 316−317, 319−320, 379, 379f, 401−403
$SrYb_2O_4$, 271−272
Stability regions, 457−458
Stable (quantum) states, 451
Stars of propagation vector, 216−217, 217f
Stereograms, computer-generated, 232
Structure factor, 409, 440
Sul-Cu_2Cl_4, field-induced magnetic order in, 272−273, 273f
Superconducting heat switch, 451−452
Superferromagnetic state, 407−408, 413
Superfit (program), 373−374
Super-iterative algorithm, 373−376
Supermatrix (SM), 361−362, 373−374
Supermirrors, 366−368, 371−372, 411−412, 411f
Superparamagnetic state, 407−408, 413
Superspace formalism, 222−223
Superspace groups, 227−228
Symmetry analysis, 205
  of couplings between Fourier components, 232−233
Symmetry of magnetic structures, 222−231
  magnetic superspace groups, 227−228
  models, 233
  representation analysis, 228−231
  Shubnikov space groups, 223−226
Symmetry related intensities, 467−468

T

$Tb_2Ti_2O_7$, 241−242, 242f
$TbMnO_3$, 302−307, 314t, 325−326
  energy scan, 329f
  field dependence of the magnetic Bragg scattering from, 304f
  fitted polarized neutron reflectivity, 320f
  magnetic excitation spectrum, 326f
  polarization, 327f
  thin films and multilayers, 319−320

Temperature
  in nuclear spin systems, 441
  size of ordering temperature, 470
  versus tuning parameter, 50−51, 50f
Thermal fluctuations, 44−45
Thermal neutron regime, 441
Thermodynamics, 442−443
$\theta - 2\theta$ scans, 78f, 81−83, 83f
Thickness oscillations, 358−359, 374−377, 388
Thin films, 374−395
  domain walls, 394−395
  magnetic domain state, 384−390
  magnetization reversal, 390−394
  magnetization rotation, 380−384
  saturated state, 374−380
Three-dimensional (3D) antiferromagnetic (AFM) metals, 100
Three-dimensional (3D) polarization vector analysis (3DPVA), 352, 358, 362
Tilted field approach for inelastic scattering, 24
Time-dependent experiments, 416−421
  DW propagation, 390f, 416, 419
Time-of-flight spectrometers, 58, 58f, 90
Time-resolved PNR (TRAC-PNR), 417, 420−421
$TlCuCl_3$, 73
  crystal structure, 74f
  energy dispersion of the magnetic excitation modes, 73−74, 76f
  field-induced QPT in, 73−79
  phase boundary, 79f
  pressure-induced QPT, 79−86
  progressive Zeeman splitting of the triplet modes, 77f
  spin dynamics, 77
  spin structure in the magnetic field-induced ordered phase, 76−77, 78f
Top-loading pulse tube refrigerator, 61−62, 62f
Total reflection, 355−356, 358−360, 362−363, 371, 374−376, 382, 384−385, 387−388, 409, 419
Transition-metal intermetallic weak ferromagnets, 100−101
Transition-metal oxides, 100−101
Translation groups, 225−226
Transmission, 447
  amplitude, 356, 361−362
  experiments, 441
Transmittivity, 355

Transverse coherence length, 386
Transverse field quantum Ising model, 126–130
Trial and error approach, 205, 233, 235
Triangular lattices
 planes of edge-sharing equilateral triangles, 258–260
 planes of edge-sharing nonequilateral triangles, 260–262
Triangular-based magnetic chains, 268–270, 269f
Triple-axis spectrometer, 57–58, 57f
Truncated dipolar Hamiltonian, 454
Tuning parameter, 50–51
Two-dimensional (2D) antiferromagnet materials, 100
Type I antiferromagnetism, 457–458
Type I spin ordering, 248–249
Type-I (proper) MFs, 294–301
 $BiFeO_3$, 297–300
 $HoMnO_3$, 294–296
 $(Sr-Ba)MnO_3$, 300–301
Type-II (improper) MFs, 301–313
 Ca_3CoMnO_6, 307–313
 $MnWO_4$, 302–307
 $RbFe(MoO_4)_2$, 302–307
 $TbMnO_3$, 302–307
 YMn_2O_5, 307–313

U

Ultralow temperature (ULT)
 high pressure at, 436–437
 impurities at, 443–444
 needs, 443
 thermal contact at, 443–444
 thermal isolation at, 443–444
Unconventional superconductor, 99, 146–147, 159–160
Underdoped, 151–152, 155, 159–160, 168–169, 179–182, 181f, 187–189
Unperturbed Hamiltonian, eigenstates, 7–8
Untwinned samples, 151f, 153–154, 153f
$URu_{1.9}Re_{0.1}Si_2$, incommensurate fluctuations, 124f
URu_2Si_2, 120–126
 applied pressure, 125–126
 chemical pressure, 122–125
 contour plot of scattering, 122f
 excitation spectrum of, 121f

V

Vanadates, 263–264
 based spinels, 242–244
Vector magnetometry, 421
Vertical cryomagnetic system, 65, 66f
VESTA (computer program), 236–237
Vibration damping, 443

W

Walliston prism-type magnetic DC coils, 23
Warren function, 256–257
Wave functions, 7–8
Wave optics, in neutron scattering experiments, 10–12
Wave packets, 7–9, 11
Wave vector. See Propagation vectors
Wave vector transfer, 349, 352–353, 363–365, 385–386
Wavelength dispersive reflectometer, 366
Wavelength resolution, 370–371
Weak itinerant ferromagnets, 112–114
Wigner corepresentations, 230–231

X

x–T phase diagram of $TlCuCl_3$, 85f
$XoNb_2O_6$, continuum spectrum, 133f
X-ray resonant magnetic scattering (XRMS), 343, 404, 421
X-rays
 scattering, 68
 synchrotron, 10–11
XY pyrochlore magnets, 55

Y

$Yb_2Ti_2O_7$, 87–88, 87f, 93–95
 Curie-Weiss analysis, 93–94
 microscopic spin Hamiltonian in, 97–99
 spin excitations in, 95–99, 96f
 and temperatures, 94
YBCO, 401–403, 402f
YMn_2O_5, 307–313, 314t
 constant-Q scan on single crystals, 331f
 crystal and magnetic structures in the LTI phase, 311f
 electrical polarization, 312f
 low-temperature incommensurate magnetic structure, 313f
 spin dynamics, 329
$YMnO_3$, 322–323

Yoneda enhancement, 388, 416f
Yoneda wings, 387–388

Z

Zeeman energy, 13
    type Hamiltonian, 5–6
Zeeman interaction, 13, 55–56
Zeeman splitting, 449–451
Zero field cooling (ZFC), 407–408, 408f

Zero field neutron spin echo (ZFNSE), 19–20
Zigzag chains, 268–270
    quasi-one-dimensional magnetic systems containing, 271f
$ZnCr_2O_4$, 242–243
$ZnFe_2O_4$, magnetic diffuse scattering in, 280–281, 280f
$ZnLn_2S_4$, 268–270
ZnV_2O_4, 242–243

Edwards Brothers Malloy
Ann Arbor MI. USA
December 8, 2015